Multidimensional Approaches to Reservoir Fisheries Management

The symposium
Multidimensional Approaches to Reservoir Fisheries Management
was organized by the

American Fisheries Society
Southern Division, Reservoir Committee

Support for the symposium was provided by

U.S. Department of the Interior
Fish and Wildlife Service, Division of Federal Aid

Tennessee Valley Authority

U.S. Army Corps of Engineers

Support for the publication of this proceedings was provided by

U.S. Department of the Interior
Fish and Wildlife Service, Division of Federal Aid

Multidimensional Approaches to Reservoir Fisheries Management

Edited by

Leandro E. Miranda
National Biological Service

Dennis R. DeVries
Auburn University

American Fisheries Society Symposium 16

Proceedings of the Third National Reservoir Fisheries Symposium
Multidimensional Approaches to Reservoir Fisheries Management

Held at Chattanooga, Tennessee, USA
12–14 June 1995

American Fisheries Society
Bethesda, Maryland
1996

The American Fisheries Society Symposium series is a registered serial. Suggested citation formats follow.

Entire book

Miranda, L. E., and D. R. DeVries, editors. 1996. Multidimensional approaches to reservoir fisheries management. American Fisheries Society Symposium 16, Bethesda, Maryland.

Article within the book

Ney, J. J. 1996. Oligotrophication and its discontents: effects of reduced nutrient loading on reservoir fisheries. Pages 285–295 in L. E. Miranda and D. R. DeVries, editors. Multidimensional approaches to reservoir fisheries management. American Fisheries Society Symposium 16.

Cover illustration by Delana Bettoli

Library of Congress Catalog Card Number: 96-078695

ISBN-0-913235-92-X

ISSN 0892-2284

Printed in the United States of America on recycled, acid-free paper

American Fisheries Society
5410 Grosvenor Lane, Suite 110
Bethesda, Maryland 20814-2199, USA

Contents

SOCIOECONOMICS AND OPTIMUM YIELD

HABITAT MANAGEMENT

AQUATIC MACROPHYTES AS HABITAT FOR RESERVOIR FISH

Preface

Reservoirs are complex environmental and biological systems built by humans for a variety of societal needs. Within reservoirs there is a great diversity of interactive components ranging longitudinally from riverine to lacustrine, and latitudinally from limnetic to littoral. Also, among reservoirs there is substantial diversity, ranging widely in morphometrical, operational, and biological characteristics. The use of reservoirs for fishing must compete with other often conflicting uses such as flood control, hydropower generation, navigation, water supply, and aesthetics. Moreover, reservoir use and operation priorities are continuously coevolving with societal demands. Given these complexities, reservoir fisheries managers must work at the interface of numerous diverse disciplines, such as hydrology, limnology, fish biology, and sociology.

To be effective, reservoir fisheries managers must have working knowledge of multiple disciplines. The goal of this symposium was to expand the knowledge base and the ability of fisheries managers to administer reservoir fisheries by focusing on the multidimensionality of reservoir management. Thus, this symposium included a diversity of subjects across a variety of topics essential to management of reservoir fisheries, but mostly outside the realm of traditional fisheries management activities. The range of topics covered was wide, but not exhaustive, and by design there was an emphasis on human dimension issues.

The symposium was convened in Chattanooga, Tennessee, on 12-14 June 1995. This was the third national symposium organized by the Reservoir Committee of the Southern Division of the American Fisheries Society. Members of the Reservoir Committee participated in all aspects of organizing the symposium, from identification of the session themes to local arrangements. The symposium and these proceedings were arranged into an introductory session and eight technical sessions. Each technical session addressed a selected theme, with most of the papers invited by the Program Committee to address that theme. Each member of the Program Committee was responsible for organizing a session; they diligently identified relevant issues and authors, guided authors through manuscript development, administered peer review, and chaired the session. The American Fisheries Society's Book Department coordinated the editing and production of this book.

We hope that these proceedings will prompt reservoir fisheries managers to look beyond traditional boundaries of our discipline. Often, the solution to reservoir fisheries problems may lay beyond these boundaries. As such, we encourage reservoir fisheries managers to have a working knowledge of multiple disciplines, and be prepared to interface with other disciplines in seeking solutions to reservoir fisheries problems.

L. E. Miranda
D. R. DeVries
Editors

Symposium Committees

Steering

Leandro E. Miranda, *National Biological Service*

Program Committee and Session Chairs

Michael C. Alexander, *U.S. Army Corps of Engineers*
Douglas J. Austen, *Illinois Department of Natural Resources*
Phillip W. Bettoli, *Tennessee Technological University*
James R. Jackson, *North Carolina State University*
Steve Quinn, *In-Fisherman, Incorporated*
Richard W. Luebke, *Texas Parks and Wildlife Department*
Stephen P. Malvestuto, *Fishery Information Management Systems*
Harold L. Schramm, *National Biological Service*
Michael J. Van Den Avyle, *National Biological Service*

Posters

Scott L. Van Horn, *North Carolina Wildlife Resources Commission*
Donald J. Degan, *Duke Power Company*

Local Arrangements

James J. McHugh, *Alabama Department of Conservation and Natural Resources*
Donny R. Lowery, *Tennessee Valley Authority*
Wayne Poppe, *Tennessee Valley Authority*
Richard W. Christie, *South Carolina Wildlife and Marine Resources Department*
Douglas C. Peterson, *Tennessee Wildlife Resources Agency*
Franklin E. Massie, *Virginia Power*
Bert E. Pierce, *West Virginia Department of Natural Resources*
Carl T. Swor, *U.S. Army Corps of Engineers*

Finances

Michael L. Armstrong, *Arkansas Game and Fish Commission*
Eugene R. Gilliland, *Oklahoma Department of Wildlife Conservation*
Gene R. Wilde, *Texas Parks and Wildlife Department*

Registration, Publicity, and Printing

Scott R. Hale, *Kentucky Department of Fish and Wildlife Resources*
Michael C. Duval, *Virginia Commission of Game and Inland Fisheries*
Mark A. Webb, *Texas Parks and Wildlife Department*
Kenneth F. Kurzawski, *Texas Parks and Wildlife Department*

Reviewer Acknowledgments

The Editors and Program Committee thank the following people who contributed technical reviews of manuscripts.

S. Marshall Adams
Leslie M. Ager
Lothian Ager
Michael S. Allen
Richard O. Anderson
Mark B. Bain
Peter Bayley
T. Doug Beard
Eric Bergersen
Carol Bisogni
Drew Bodaly
Raphael Brock
Pete Bromley
Al Brown
Tommy L. Brown
Gerard L. Buynak
Colleen Caldwell
Richard L. Cailteaux
Robert F. Carline
Timothy N. Churchill
Howard A. Clonts
Richard A. Cole
Michael Colvin
Mike Conlin
Charles C. Coutant
Johnie Crance
Bill Davies
Donald Degan
Ron Dent
Eric Dibble
Phil Doerr
Todd Driscoll
Donald Dycus
Ellen S. Ebert
Anthony J. Fedler
Carlos Fetterolf
Ralf Fiedel
James Fletcher
Leska Fore
Allen A. Forshage
Mary Freeman
Joe N. Fries
Donald Gabelhouse
Jerry Germann
Larry M. Gigliotti
Eugene R. Gilliland
Timothy J. Goeman
Bobby Grinstead
William C. Guest
Steve Gutreuter
Cindy Halbert
Jeffrey Hahn
John Hains
Scott Hardin
Fred A. Harris
Don Hayne
Roy Heidinger
Gary Hickman
Robert Howells
Michael D. Hudgins
Elise R. Irwin
Barry Johnson
David L. Johnson
Peter Johnson
Jeanne C. Jones
Roger I. Jones
James Karr
Robert Kennedy
James P. Kirk
Barbara Knuth
Thomas J. Kwak
Wilson Laney
Charles Madenjian
Randall Manning
Joe Margraf
Frank T. McBride
Tom McDonough
Earl Meredith
Gary G. Mittelbach
Alvin D. Mills
Eric Morgan
Ian A. Munn
Val Nash
Richard L. Noble
Larry Olmsted
Donald Orth
Teofilo Ozuna
Ron Pasch
Donald L. Pereira
John Pitlo
Steven Plakas
Gene R. Ploskey
William C. Provine
Lawrence L. Pugh
Frank Rahel
Jeffrey Reed
Mike Reed
Bill Reeves
Kevin Rogers
Roger Rulifson
Ross Self
Susan Shipman
Bruce D. Shupp
Scott M. Smith
Thomas M. Steeger
Norman P. Stucky
Gregory L. Summers
Cindy Swanson
Les Sztramko
Danny Thompson
Scott Van Horn
Mark Webb
A. Steven Weithman
Gene R. Wilde
Cindy A. Williams
Robert W. Wiley
Kirk Winemiller
Katie Young
Gene Zuerlein
Alexander V. Zale

Symbols and Abbreviations

The following symbols and abbreviations may be found in this book without definition. Also undefined are standard mathematical and statistical symbols given in most dictionaries.

A	ampere
AC	alternating current
Bq	becquerel
C	coulomb
°C	degrees Celsius
cal	calorie
cd	candela
cm	centimeter
Co.	Company
Corp.	Corporation
cov	covariance
DC	direct current District of Columbia
D	dextro (as a prefix)
d	day
d	dextrorotatory
df	degrees of freedom
dL	deciliter
E	east
E	expected value
e	base of natural logarithm (2.71828 . . .)
e.g.	(exempli gratia) for example
eq	equivalent
et al.	(et alii) and others
etc.	et cetera
eV	electron volt
F	filial generation Farad
°F	degrees Fahrenheit
fc	footcandle (0.0929 lx)
ft	foot (30.5 cm)
ft^3/s	cubic feet per second ($0.0283\ m^3/s$)
g	gram
G	giga (10^9, as a prefix)
gal	gallon (3.79 L)
Gy	gray
h	hour
ha	hectare (2.47 acres)
hp	horsepower (746 W)
Hz	hertz
in	inch (2.54 cm)
Inc.	Incorporated
i.e.	(id est) that is
IU	international unit
J	joule
K	Kelvin (degrees above absolute zero)
k	kilo (10^3, as a prefix)
kg	kilogram
km	kilometer
l	levorotatory
L	levo (as a prefix)
L	liter (0.264 gal, 1.06 qt)
lb	pound (0.454 kg, 454 g)
lm	lumen
log	logarithm
Ltd.	Limited
M	mega (10^6, as a prefix) molar (as a suffix or by itself)
m	meter (as a suffix or by itself) milli (10^{-3}, as a prefix)
mi	mile (1.61 km)
min	minute
mol	mole
N	normal (for chemistry) north (for geography) newton
N	sample size
NS	not significant
n	ploidy nanno (10^{-9}, as a prefix)
o	ortho (as a chemical prefix)
oz	ounce (28.4 g)
P	probability
p	para (as a chemical prefix)
p	pico (10^{-12}, as a prefix)
Pa	pascal
pH	negative log of hydrogen ion activity
ppm	parts per million
qt	quart (0.946 L)
R	multiple correlation or regression coefficient
r	simple correlation or regression coefficient
rad	radian
S	siemens (for electrical conductance) south (for geography)
SD	standard deviation
SE	standard error
s	second
T	tesla
tris	tris(hydroxymethyl)-aminomethane (a buffer)
UK	United Kingdom
U.S.	United States (adjective)

USA	United States of America (noun)
V	volt
V,	Var variance (population)
var	variance (sample)
W	watt (for power)
	west (for geography)
Wb	weber
yd	yard (0.914 m, 91.4 cm)
α	probability of type I error (false rejection of null hypothesis)
β	probability of type II error (false acceptance of null hypothesis)
Ω	ohm
μ	micro (10^{-6}, as a prefix)
′	minute (angular)
″	second (angular)
°	degree (temperature as a prefix, angular as a suffix)
%	per cent (per hundred)
‰	per mille (per thousand)

Introduction

American Fisheries Society Symposium 16:3–11, 1996

Development of Reservoir Fisheries Management Paradigms in the Twentieth Century

L. E. MIRANDA

National Biological Service, Mississippi Cooperative Fish and Wildlife Research Unit[1]
Post Office Drawer BX, Mississippi State, Mississippi 39762, USA

Abstract.—I describe the development and evolution of reservoir fisheries management paradigms in the Twentieth Century. The "biological deserts" paradigm evolved rapidly early in this century concurrently with an increase in reservoir construction. This paradigm imparted the belief that reservoirs were unproductive, and called for stringent harvest restrictions, construction of hatcheries to maintain fish populations through stocking, and research into methods for improving fish production. Research during the late 1930s and 1940s led to the belief that reservoirs were in fact productive, but that fish communities needed adjustment to achieve maximum yield. During the 1950s this new "biomanipulation" paradigm influenced development of a network of conceptual, instrumental, and methodological commitments that guided reservoir fisheries management and research during most of the second half of the Twentieth Century. Concurrently, increases in angling effort, decreases in reservoir construction, and changes in angler attitudes led to reformulation of the "biomanipulation" paradigm to accommodate social issues. I suggest that a reconceptualization of fisheries management precipitated by the need to respond to new social concerns has substantially altered the "biomanipulation" paradigm, and that we may be witnessing its replacement by a new paradigm.

The word "paradigm" according to most dictionaries signifies an accepted model or pattern. Kuhn (1970) adapted the word to describe a conceptual system that serves as a model or pattern that directs scientific endeavor within a discipline. Kuhn characterized a paradigm as a network of conceptual, instrumental, and methodological commitments that guides the direction and methodology of scientific research. According to Kuhn, the advent of a paradigm is often signaled by the formation of specialized groups of scientists, publication of specialty journals and books, and the claim for a special place in the academic curricula. The scientific community that shares a common paradigm has often been "instructed" into that paradigm by reading similar textbooks and journals, and hence may also share the same rules and standards for scientific practice, a common view of problems to be solved, and how they should be approached. Thus, the paradigm defines what should be studied, what questions should be asked, what tools should be used, and what rules should be followed in interpreting the results.

Periodically, the adequacy of a paradigm may be challenged by observations that contradict the existing concepts, leading to intensified research to try to make the anomalous observations conform to the existing paradigm. If the anomalies cannot be explained within the context of the existing paradigm, a new paradigm may develop that leads to a new set of commitments and a new basis for practice within a scientific discipline. A shift in paradigm is seldom just an incremental addition to existing knowledge; often, it involves the reconstruction of earlier concepts and the reevaluation of prior apparent facts, and imposes entirely new questions to address.

In this paper, I describe the development of reservoir fisheries management paradigms during the Twentieth Century. The original paradigm evolved rapidly and shifted to the current paradigm. The current paradigm was reformulated, and we may now be in the midst of another paradigm shift. I track these shifts, and document how they affected concepts, methods, techniques, and applications in reservoir fisheries management, using a sampling of papers on reservoir fisheries management published since the early 1900s. While discussing these papers I note the important issues and try to identify the paradigm that might have raised those issues. Although Jenkins (1961, 1970) has reviewed the history of reservoir fisheries management, my account differs in that his was descriptive whereas mine is interpretive.

Reservoir Construction—The Catalyst for a Paradigm

In 1900 there were about 100 reservoirs with surface areas greater than 500 acres in North America (Jenkins 1970). During the early part of the Twentieth Century, advances in engineering and

[1]The Unit is sponsored by the Mississippi Department of Wildlife, Fisheries and Parks, Mississippi State University, and the National Biological Service.

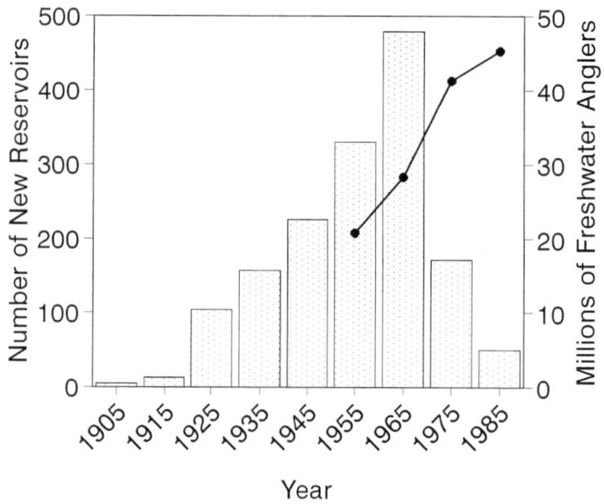

FIGURE 1.—Number of new reservoirs (bars) and angler effort (line) during 1900–1990 in the United States. Angler effort represents all freshwater anglers (excluding Great Lakes) as reported by USFWS (1991).

earth-moving equipment facilitated a rapid rate of reservoir construction that peaked in the 1960s and decreased greatly by the late 1980s (Figure 1). Today, over 1,700 reservoirs have surface areas greater than 500 acres in North America, and their aggregate area exceeds 11 million acres, but few new reservoirs are being constructed.

Early Signs of an Emerging Paradigm

Reservoir fisheries management is a young science, emerging early this century in response to reservoir proliferation. Although early fisheries biologists had at their disposal basic facts about fish biology and limnology applicable to natural lakes, reservoirs were different. Ellis (1937:64) was one of the first authors to articulate this difference in his comparison of natural lakes with reservoirs:

> The student of impounded waters on the other hand is impressed at the outset however that conditions in impounded waters often vary abruptly and over wide ranges; that no biological adjustment on the part of fishes and other aquatic species is possible to some of these changes; and, that the environment supplied by impounded waters is new and unstable as compared with that of natural lakes. In fact in many impounded waters changes of catastrophic magnitude as far as the aquatic organisms are concerned occur at irregular and unannounced intervals, and often with a frequency disastrous to the aquatic faunae.
>
> Therefore, before an intelligent program for the utilization of any body of impounded water for fisheries projects can be established it is essential that the limitations and hazards of such waters in the locality where they are created, be defined, for generalities from natural lakes are not readily applicable to many impounded waters.

Because they were different, these constructed environments required new management and research approaches and prompted investigations that generated unique knowledge and ultimately a new set of concepts and beliefs (i.e., the paradigm).

The "Biological Deserts" Paradigm

As reservoir construction increased in the early part of the Twentieth Century, inventories of limnological characteristics and fish species in reservoirs multiplied (e.g., Henshall 1920; Galtsoff 1924; Scott 1927; Richardson 1928; Coker 1930; Coleman 1930; Moore 1931, 1933; Wickliff 1933; Wickliff and Roach 1937). Henshall (1920:146) warned about low productivity in western U.S. reservoirs:

> It has been said that the immense storage reservoirs to be constructed in connection with the reclamation projects will furnish a home and a haven for millions of fish—but not on your life. These reservoirs will be built in narrow mountain gorges, where the water will be too deep for fish life to exist, and the rocky bottom and sides will forever preclude the existence of fish food.

Cahn (1937:399), announcing the formation of the Tennessee Valley Authority's (TVA) fisheries program, indicated that below Norris and Wheeler reservoirs, fish hatcheries were constructed "where fish (particularly the largemouth bass) will be propagated for the sole purpose of stocking." Cahn implied that these hatcheries were built in anticipation of the need to stock these reservoirs to maintain suitable fish production. Cahn (1937:398) further stated:

> It is a familiar fact, which has been emphasized by Mr. Jay N. Darling, Dr. M. M. Ellis, and others, that impounded lakes become, after the first few years of their existence, decidedly lacking in fish productivity. Mr. Darling has referred to the results of dam construction as 'biological deserts'.

The "biological deserts" paradigm influenced attempts to manage reservoir fisheries during the 1920s and 1930s. The belief that after an initial period of high productivity, reservoirs became largely unproductive elicited an emphasis on harvest restrictions, such as closed seasons and gear restrictions, and on stocking. The perceived need for stocking motivated construction of numerous hatcheries. Many fisheries managers believed that natural reproduction was inadequate to sustain reservoir fish populations, and that fishing could be improved by releasing hatchery-reared fish. Titcomb (1918:11), commenting on proposals to liberalize fishing laws during World War I, indicated:

> proper regulations for the protection of fish are as essential to the stocking of our waters as is fish culture and all efforts relevant thereto, and that if the fishery laws of the various states were fundamentally right before the war they are just as essential today as they were then.

Two years later Henshall (1920:146–147) wrote in reference to stocking in western U.S. reservoirs:

> Close seasons for all game- and food-fishes during the breeding periods should be established, and severe penalties should be imposed for the violation of such laws. . . . By the continual stocking of waters, intelligently, with fish artificially propagated, a fair amount of fish life may still be maintained in inland waters, even in the older states.

Origins of Reservoir Fisheries Management

Management programs specifically designed for reservoir fisheries appear to have originated in the 1930s in response to the need for increasing fish production in the "biological deserts." In the 1930s and early 1940s, the TVA was beginning a navigation and flood control program that required construction of a series of dams throughout the Tennessee River and some of its tributaries. To mitigate the "biological deserts" that would be induced by reservoirs, the TVA announced the formation of the Biological Readjustment Unit. Cahn (1937:398) wrote:

> There are innumerable examples of aquatic deserts, and to list them would be to name practically every large dam so far constructed. In an effort to counteract the known effects of dam construction on the biological resources of the areas involved, the Tennessee Valley Authority created the Biological Readjustment Unit within the Forestry Division of its organization, and charged it with the specific functions which its name indicates: the readjustment of the biological factors of the territory to the newly created physical conditions. The duties of this unit involve both game and fisheries problems.

Apparently formation of the unit emerged out of the TVA's concern for the thousands of people whose livelihood depended on the fisheries of the Tennessee River, and the TVA's belief that reservoirs would reduce fish production. Cahn (1937: 398–399) further wrote:

> In days past, the Tennessee River has afforded an abundant supply of fish to serve as food for the people of the valley. A survey of the Tennessee River in the immediate vicinity of Decatur, Alabama, for example, showed that on that portion of the river which will eventually become a part of Wheeler Lake there are at the present time approximately 7,550 fishermen dependent annually upon the river for either their food or their livelihood. . . . If the history of other impounded waters repeats itself—and there seems to be no reason why it should not—the 'biological desert' looms in the offing. With the large population of the area dependent upon the water for food, the Tennessee Valley Authority recognizes its moral responsibility to make every effort to keep up and to build up the fisheries industry in order that this specter of an aquatic desert may not become a reality.

From Cahn's passages it is clear that the TVA's intentions were to investigate methods to enhance fish production. Concurrently, fisheries investigations were ongoing elsewhere in the United States, including Michigan (Hubbs 1938), Texas (Harris and Silvey 1940), Wyoming (Simon 1938), New Mexico (Greenbank 1937), California (Felin 1940), and New York (Moore 1933; Burdick 1940), and in various states by Ellis (1942). From investigations conducted from 1936 to 1939 in 33 reservoirs Ellis (1942:80) provided several management recommendations to mitigate "biological deserts":

> The limitations of impounded waters for fish productivity can in part be offset by: (1) Controlling the draw-off . . . so that fairly constant water level will be maintained during the nesting season of the most desirable species of fish. (2) By judicious restocking . . . from hatchery or hatchery ponds with vigorous specimens of the desired species. (3) By the construction where feasible of lateral areas of shallow water of constant level and tributary to the main impoundment, in which adult fish can nest, young fish can find refuge and fish food can be produced. (4) By the use of various devices such as floating nests, floating shelters, the planting of exposed portions of the reservoir during low water. (5) By the artificial enrichment of soil and water with specific substances such as phosphates, seed meal and organic waste, in such limited areas as seem to justify the expense of these procedures.

Although most of these approaches are in use today, later I will argue that they are now applied with a different rationale.

Disenchantment during this period with perceived nonproductivity of reservoirs and devastating effect on anadromous fishes was epitomized by a resolution passed at the 1941 meeting of the American Fisheries Society (AFS 1942:52):

> Whereas, these immense projects [high dams], present and planned for the future wholly ignore the biology of water, existing aquatic biological resources, and the public's right to enjoy them; and there are present threats to blanket the country with high dams;
>
> Therefore be it resolved, that the American Fisheries Society . . . oppose the authorization or construction of any more high dams regardless of their stated purpose

> until such time as biologists have an adequate role in the preliminary planning of all such projects.

Thus, under the auspices of the "biological deserts" paradigm, the 1930s and early 1940s witnessed the origins of reservoir fisheries management in North America.

The "Biological Deserts" Paradigm is Questioned in the 1940s

Results from investigations initiated in the 1930s and early 1940s began to identify anomalies in the "biological deserts" paradigm. Eschmeyer and Tarzwell (1941:15) reporting on their work in various TVA reservoirs indicated:

> the catch in the reservoirs is now undoubtedly far greater than it was in the river before the dams were built. There is little reason to believe that these impoundments will tend to become 'biological deserts' after a few seasons.

Moffett (1943:184) in a preliminary report of his work in Lake Mead, Nevada, concluded that:

> under present conditions, 1. The introduction of forage fish into Lake Mead is believed unnecessary.... 2. There is no closed season ... [and] none should be applied.... 3. No further stocking of Lake Mead is advisable.... 4. Introduction of aquatic plants should not be attempted.... 6. Regulate the water level ... from April through June so that rapid falls in levels do not occur.

Tarzwell (1944:139,155,156), also reporting on work conducted in the TVA reservoirs, wrote:

> Contrary to preconceived notions, the investigations conducted to date indicate that fish populations in these reservoirs are large ... legislation permitting netting should be liberal and legalize all types of gear ... to provide food for the war effort, to protect the sport fishery, and to utilize the present market for the coarse species which has been created by the war, it is desirable that the commercial fishery in TVA main-stream reservoirs be expanded immediately.

In 1944–1945 Tennessee opened its reservoirs to year-round fishing, followed by Nebraska and Ohio, and subsequently other states. By the late 1940s, studies in TVA reservoirs confirmed that reservoir fish populations had not been diminished by fishing (Miller 1950), and that population changes were caused primarily by natural biological phenomena (Stroud 1948). In the Midwest, Bennett (1947) showed that the chemical treatment or draining of small impoundments to remove undesirable populations entirely eliminated "aquatic desert" conditions, and that once these impoundments were restocked with desirable fish they did become very productive. Research in the 1930s and 1940s initiated the collapse of the "biological deserts" paradigm, and eventually led to development of a new paradigm with a new set of research questions and management approaches.

The "Biomanipulation" Paradigm

A new "biomanipulation" paradigm emerged from results of investigations conducted in the 1940s. Although it became evident that fish abundance did not decline drastically after impoundment, biologists learned that community composition in the reservoir was different from that in the original stream, that fish abundance and composition varied depending on reservoir characteristics, and that less desirable fish species became relatively more abundant as the reservoir aged (Eschmeyer and Tarzwell 1941; Tarzwell 1944). Thus, the new paradigm focused on manipulation of reservoir ecosystems to achieve maximum fisheries benefits. With the "biomanipulation" paradigm, a new set of concepts, tools, approaches, and questions developed rapidly.

During the 1950s the idea that reservoirs were underutilized renewable resources in need of biological adjustment generated various methodological approaches directed at increasing harvest and manipulating fish community composition. The perspective in the 1950s was represented by Parsons' (1958:342) statements:

> Fish populations should be managed to give greater productivity, improve species composition, and increase availability. In many reservoirs commercial fishing may prove an effective and profitable tool for population control. Sport fishermen should be educated into more effective methods of fishing and into the acceptance of a wider range of species. The intensity of angling should be increased through the provision of more and better facilities.

Consequently, research and management of reservoir fisheries focused largely on control of species composition, availability of fish, and promotion of fishing. Several strategies were developed whose rationale could not have been conceived under the auspices of the "biological deserts" paradigm. Drastic water level fluctuation was promoted as a powerful tool for restructuring community composition through influences on spawning, predator–prey dynamics, aquatic macrophytes, and primary production (Wood 1951; Parsons 1958; Hulsey 1959); under the "biological deserts" paradigm, water level fluctuation was seen merely as something that needed stabilization during the spawning season (see quote from Ellis' 1942 above). Stocking began to be seen as a remedial tool rather than as an ecumenical method for increasing fish abundance. Introductions of prey and predator fish species not originally present in the reservoir basin were advocated as a means to fill unoccupied niches, control overabundant prey species, and provide additional species to the angler's creel (Fetterolf 1957; Kimsey 1958). Selective kills of overabundant populations of gizzard shad *Dorosoma cepedianum* became a common practice for adjusting size structure and biomass of fish populations, although the method was controversial and results short lived (Bowers 1955; Mathis and Hulsey 1959; Wyatt and Zeller 1965; reviewed by DeVries and Stein 1990). Commercial fishing for rough fish, including commercial harvest of lightly fished game fish populations, was urged to reduce competition for food and space with game fish; however, a continual conflict with recreational fishing interests was perceived to be a major drawback to this approach (Irwin 1957; Parsons 1958; Riggs 1958; Shields 1958; Hulsey 1959). Brush shelters, seen earlier as devices for increasing spawning success and reducing mortality, were now promoted to attract sportfish and increase harvest, and incomplete timber clearing was promoted to increase habitat diversity and reduce the need for brush shelters (Irwin 1957; Parsons 1958; Hulsey 1959). Increased sportfish harvest was encouraged through increased emphasis on developing fishing facilities (e.g., boat ramps and parks, docks and harbors including heated docks for winter fishing), publishing accounts of fishing successes and depth distribution diagrams in newspapers to educate anglers as to where to catch the most fish, elimination of no-fishing (sanctuary) areas, control of pleasure boating, and liberal harvest regulations (Irwin 1957; Parsons 1958). Many of the approaches advanced by the "biomanipulation" paradigm, particularly in the southeastern United States, were likely influenced by results from H. S. Swingle's farm pond research program of the 1940s and 1950s (e.g., Hulsey 1959).

Research in the Last Half of the Century is Guided by the "Biomanipulation" Paradigm

The "biomanipulation" paradigm provided reservoir fisheries managers the foundation for building successful management programs. Nevertheless, considerable work remained to solve problems identified by the paradigm. Thus, researchers and managers concentrated on investigating a range of narrowly focused problems that the biomanipulation paradigm identified as relevant. These investigations generated observations about reservoir fish, habitat, and fisheries published in technical journals and symposia proceedings in the last half century. I identify three foci of these investigations.

Research that Reveals Facts Germane to the "Biomanipulation" Paradigm

The first focus involves research that is particularly revealing of the nature of reservoir fisheries. This includes data on fish abundance, catch rate, community composition, and life history. Attempts to describe these characteristics, and to increase the accuracy and precision with which estimates are obtained, occupy a substantial portion of our literature. We have designed increasingly sophisticated methodology and instrumentation to enhance the quality of our estimates. Hydroacoustic and geographical information system technologies are just two recent examples of the lengths to which research workers will go to obtain information relevant to the paradigm. Some fisheries scientists have acquired their reputation not from any novelties of their discoveries, but from the precision, reliability, and scope of the methods they have developed for the redetermination of a previously known fact revealed by the paradigm.

Research that Generates Confirming Observations

A second focus has been to generate observations that can be compared directly with predictions made by the paradigm. Examples of this endeavor include research on how fisheries are affected by manipulating water level or by introducing new species. The "biomanipulation" paradigm predicts that both of these manipulations will influence a reservoir fishery. Hence, both have been researched often, and not surprisingly, have been demonstrated to affect fisheries. Thus, the paradigm has guided the problem to be investigated and often predicted the outcome.

Articulation of Facts

The third focus of reservoir fisheries research has consisted of empirical work to support the paradigm principles. For instance, because the paradigm suggests that reservoir fish communities can be manipulated to obtain maximum benefits, the paradigm implies that there are community composition optima. Thus, the paradigm has sent researchers in quest of constants or laws that identify these optima. Examples include Jenkins and Morais' (1978) available prey:predator optima, Jenkins' (1975) fish community structure pyramid, Davies et al.'s (1982) species composition values, and Anderson's (Anderson and Gutreuter 1983) proportional stock density targets.

Reformulation of the "Biomanipulation" Paradigm

The "biomanipulation" paradigm that originated in the 1950s has been reformulated, but still guides present management approaches and investigations. Perhaps the forces that have most strongly directed change have been a progressive increase in angling effort (Figure 1), a substantial decrease in the rate of reservoir construction, and changes in angler attitudes. These changes prompted re-examination of some of the original concepts.

The increase in effort and associated increase in exploitation prompted the return of harvest restrictions (Fox 1975; Redmond 1986). However, this time the rationale for regulations was different. Under the "biological deserts" paradigm, regulations were considered necessary to prevent overexploitation of limited stocks, and to allow adults to spawn at least once. Around the middle of the Twentieth Century the new "biomanipulation" paradigm suggested harvest was low relative to population size and regulations were unnecessary. However, as fishing effort increased during the second half of the Twentieth Century, the "biomanipulation" paradigm was reformulated and regulations reemerged due to the perceived need to control overharvest of abundant stocks, adjust size composition of stocks, and maintain enough predators to control prey abundance in the system. The latter two concepts, developed in the late 1960s and early 1970s, would have been unthinkable during the early years of the "biomanipulation" paradigm, because fishing effort was perceived to be low.

In the 1950s commercial fishing was considered an effective technique for removing rough fish from reservoirs, although the conflict with recreational fishing was recognized. Parsons (1958:346–347) wrote:

> Commercial fishing appears to be the best means by which a higher and more balanced yield of fish can be taken from reservoirs. A continual conflict has existed for many years between sport and commercial fishing interests. . . . Because of this controversy and problems of marketing food fishes, no strong demand has been made for commercial fishing rights. . . . Commercial fisheries, properly regulated, can greatly increase the take of fish from some reservoirs. . . . Since sport fish constitute only 1.2 percent of the commercial gear catches from Alabama waters of TVA reservoirs. . . . there is little probability that expanded commercial fishing will result in excessive mortality of sport fish.

By 1967, biologists had begun to recognize that commercial fishing as a management tool had fallen short of expectations (Greenwood and Boussu 1967). Yet, commercial fishing was still seen optimistically as a potentially useful management tool in need of further development (Greenwood and Boussu 1967; Schueler and Sullivan 1967). By 1983, the perceived value of commercial fishing in reservoir fisheries management was fading. Fritz and Wight (1986:201) wrote:

> At this time, it appears we are at the crossroads of either rejecting or further exploring the roles of commercial fishing in reservoir fishery management.

By 1993, commercial fishing was thought incapable of removing enough undesirable species to benefit fisheries, and description of its use as a management technique was relegated to a single paragraph in a recent inland fisheries management textbook (Kohler and Hubert 1993). The inability to make commercial fishing economically feasible combined with conflicts with the increasing number of recreational anglers contributed largely to the abandonment of commercial fishing as a management tool.

The rationale for fish stocking and introduction has also evolved. In the 1950s introduction of piscivorous fishes was recommended to fill vacant niches, convert excessive prey biomass into sport fish, and provide additional prey species. Stocking was recommended in renovated waters or those depleted by pollution. Although these recommendations are still considered valid, other rationales were developed in the 1970s and 1980s to restructure the "biomanipulation" paradigm. For introductions, additional rationales included increasing community stability through enhanced prey and predator diversity and redistribution of fishing effort among a broader range of target species (Keith 1986; Stroud 1986). For stocking, additional rationales include altering the population gene pools to, for instance, increase abundance of trophy fish (e.g., stocking of Florida strain largemouth bass *Micropterus salmoides floridanus*). More recently, as a result of evolving environmental standards, stockings and introductions are being scrutinized closely. Emphasis on natural systems with native biota and optimal biodiversity more frequently shapes current fish culture and stocking policies (Schramm and Piper 1995). These rationales would not have evolved had angling effort and public attitudes remained stable since the 1950s.

Selective fish removals were common in the 1950s, but use of this biomanipulation tool has decreased (Wiley and Wydoski 1993). Abandonment of this technique has been influenced by an accumulation of evidence that suggest that the success of selective fish removals are generally short-lived, and by increased negative public opinion towards killing fish (by an increasingly environmentally aware public). Currently, selective removals of undesirable fish are being replaced by holistic control approaches that consider the interaction of each species with its ecosystem (including other species).

Are We in the Midst of a Paradigm Change?

Although the value of fishing as a recreational experience was recognized in the 1950s, the goal of reservoir fisheries management, as dictated by the "biomanipulation" paradigm, was chiefly to maximize yield. Parsons (1958:343) wrote:

> The question of inefficient utilization of reservoir fishery resources has but one answer—that of increasing fish yield. Attempts to achieve increased yield should not, however, seriously reduce the pleasures of sport fishing.

The goal of maximizing yield was dictated by users, who were interested in catching fish primarily for food and fun. To achieve this goal, fisheries biologists had to focus principally on attaining maximum yield of popular target species, while protecting stocks and their environment to ensure future yields. However, since the 1950s the number of users has increased, and their needs appear to have changed. Although the constituency is dominated by the same kind of users, anglers have become more informed, more eager to participate in the management process, and more diverse. Articulate components of the angling community are now concerned with catch-and-release fishing, competitive fishing, and maintenance of environmental quality. In fact, segments of this constituency are now demanding participation in the reservoir fisheries management process. To expand public involvement in resource management planning and policymaking, fisheries managers have had to reformulate and stretch the "biomanipulation" paradigm to consider a vast array of social concerns along with traditional habitat- and fish-oriented management obligations.

The reformulated paradigm has forced a change in the responsibilities, research needs, and expertise of reservoir managers. This reformulation, which I suggest has been driven primarily by the need to accommodate social concerns, may or may not be sufficiently comprehensive to qualify as a paradigm shift. At one extreme, it could be argued that the reconceptualization of fisheries management brought on by the need to respond to new concerns has substantially displaced the conceptual framework through which reservoir fisheries biologists have viewed aquatic resources. For instance, for many fisheries management agencies, it would now be inconceivable to develop fisheries management plans without first assessing angler attitudes. For some managers, comanagement is becoming a common term. At the other extreme, it could be argued that the "biomanipulation" paradigm as conceived midcentury is still guiding most aspects of reservoir fisheries management, and that the recent emphasis on social aspects is merely a necessary expansion, or reformulation, of this paradigm. Kuhn (1970) indicated that during the transition period between two paradigms there is a large overlap between the applications of the old and new paradigms. Thus, we may be embracing two paradigms, and witnessing the replacement of the "biomanipulation" paradigm—time will tell.

Multidimensional Approaches to Reservoir Fisheries Management

These new emphases in reservoir fisheries management prompted the Reservoir Committee of the Southern Division of the American Fisheries Society to organize this Third Reservoir Symposium. The first symposium, held in 1967, focused on fish ecology and habitat characteristics, and only one of the 38 articles published in the proceedings (Reservoir Committee 1967) considered social issues. The second symposium, held in 1983, also centered largely on fish and their environment, but its proceedings (Hall and Van Den Avyle 1986) included sections on regulating angler harvest and on economic valuation. The proceedings of the third symposium looks very different. It includes eight sections, four which focus directly on social aspects of reservoir fisheries management, and three that include at least one article on this subject.

Acknowledgments

Discussions with Randy Jackson helped elucidate many of the views presented in this paper. Carlos Fetterolf, Don Jackson, Hal Schramm, and Mike Van Den Avyle provided helpful reviews.

References

AFS (American Fisheries Society). 1942. Reports of committees. Transactions of the American Fisheries Society 71:51–57.

Anderson, R. O., and S. J. Gutreuter. 1983. Length, weight, and associated structural indices. Pages 283–300 *in* L. A. Nielsen and D. L. Johnson, editors.

Fisheries techniques. American Fisheries Society, Bethesda, Maryland.

Bennett, G. W. 1947. Fish management—a substitute for natural predation. Transactions of the North American Wildlife Conference 12:276–285.

Bowers, C. C. 1955. Selective poisoning of gizzard shad with rotenone. Progressive Fish-Culturist 17:134–135.

Burdick, G. E. 1940. Studies on the invertebrate fish food in certain lakes, bays, streams and ponds of the Lake Ontario watershed. New York Conservation Department, Supplementary 29th Annual Report of 1939, 21:147–166, Albany.

Cahn, A. R. 1937. The fisheries program in the Tennessee Valley. Transactions of the American Fisheries Society 66:398–402.

Coker, R. E. 1930. Studies of the common fishes of the Mississippi River at Keokuk. Bulletin of the U.S. Bureau of Fisheries 45:141–225.

Coleman, G. A. 1930. Biological survey of the lakes, reservoirs and streams of San Diego County. California Fish and Game 16:111–113.

Davies, W. D., W. L. Shelton, and S. P. Malvestuto. 1982. Prey-dependent recruitment of largemouth bass: a conceptual model. Fisheries 7(6):12–15.

DeVries, D. R., and R. A. Stein. 1990. Manipulating shad to enhance sport fisheries in North America: an assessment. North American Journal of Fisheries Management 10:209–223.

Ellis, M. M. 1937. Some fishery problems in impounded waters. Transactions of the American Fisheries Society 66:63–75.

Ellis, M. M. 1942. Fresh-water impoundments. Transactions of the American Fisheries Society 71:80–93.

Eschmeyer, R. W., and C. M. Tarzwell. 1941. An analysis of fishing on the TVA impoundments during 1939. Journal of Wildlife Management 5:15–41.

Felin, F. 1940. The seasonal fluctuation of benthic macrofauna and limnetic plankton in Searsville Lake. A contribution to the biology of fluctuating reservoirs. Doctoral dissertation. Stanford University, Stanford, California.

Fetterolf, C. M., Jr. 1957. Stocking as a management tool in Tennessee reservoirs. Proceedings of the Annual Conference Southeastern Association of Game and Fish Commissioners 10(1956):275–284.

Fox, A. C. 1975. Effects of traditional harvest regulations on bass populations and fishing. Pages 392–398 *in* H. Clepper, editor. Black bass biology and management. Sport Fishing Institute, Washington, DC.

Fritz, A. W., and H. L. Wight. 1986. Commercial fishing as a reservoir management tool. Pages 196–202 *in* Hall and Van Den Avyle (1986).

Galtsoff, P. S. 1924. Limnological observations in the upper Mississippi, 1921. Bulletin of the U.S. Bureau of Fisheries 39:347–438.

Greenbank, J. 1937. A chemical and biological study of the waters of Elephant Butte Reservoir as related to fish culture. Master's thesis. University of New Mexico, Albuquerque.

Greenwood, M. R., and M. F. Boussu. 1967. Efficient fishing systems—potential for increasing effectiveness of commercial fisheries in reservoir management and utilization. Pages 467–478 *in* Reservoir Committee, editors. Reservoir fisheries resources symposium. American Fisheries Society, Southern Division, Reservoir Committee, Bethesda, Maryland.

Hall, G. E., and M. J. Van Den Avyle, editors. 1986. Reservoir fisheries management: strategies for the 80's. American Fisheries Society, Southern Division, Reservoir Committee, Bethesda, Maryland.

Harris, B. B., and J. K. G. Silvey. 1940. Limnological investigations of Texas reservoir lakes. Ecological Monographs 10:111–143.

Henshall, J. A. 1920. Concerning the protection of fish, fish food and inland waters. Transactions of the American Fisheries Society 49:141–147.

Hubbs, C. L. 1938. Can annual fish crops be maintained? Transactions of the North American Wildlife Conference 3:131–138.

Hulsey, A. H. 1959. A proposal for the management of reservoirs for fisheries. Proceedings of the Annual Conference Southeastern Association of Fish and Game Commissioners 12(1958):132–143.

Irwin, W. G. 1957. The management of large impoundments for fish production. Proceedings of the Annual Conference Southeastern Association of Game and Fish Commissioners 10(1956):271–275.

Jenkins, R. M. 1961. Reservoir fish management—progress and challenge. Sport Fishing Institute Special Publication, Washington, DC.

Jenkins, R. M. 1970. Reservoir fish management. Pages 173–182 *in* N. G. Benson, editor. A century of fisheries in North America. American Fisheries Society, Special Publication 7.

Jenkins, R. M. 1975. Black bass crops and species associations in reservoirs. Pages 114–124 *in* H. Clepper, editor. Black bass biology and management. Sport Fishing Institute, Washington, DC.

Jenkins, R. M., and D. I. Morais. 1978. Prey–predator relations in the predator-stocking-evaluation reservoirs. Proceedings of the Annual Conference Southeastern Association of Fish and Wildlife Agencies 30(1976):141–157.

Keith, W. E. 1986. A review of introduction and maintenance stocking in reservoir fisheries management. Pages 144–148 *in* Hall and Van Den Avyle (1986).

Kimsey, J. B. 1958. Fisheries problems in impounded waters in California and the lower Colorado River. Transactions of the American Fisheries Society 87: 319–332.

Kohler, C. C., and W. A. Hubert, editors. 1993. Inland fisheries management in North America. American Fisheries Society, Bethesda, Maryland.

Kuhn, T. 1970. The structure of scientific revolutions, second edition. University of Chicago Press, Chicago.

Mathis, W., and A. Hulsey. 1959. Rough fish removal from Lake Catherine, Arkansas. Proceedings of the Annual Conference Southeastern Association of Game and Fish Commissioners 13(1959):197–203.

Miller, L. F. 1950. Fish harvesting on two TVA mainstream reservoirs. Transactions of the American Fisheries Society 80:2–10.

Moffett, J. W. 1943. A preliminary report on the fishery

of Lake Mead. Transactions of the North American Wildlife Conference 8:179–186.

Moore, E. 1931. A follow-up program on the impounded waters of the Sacandaga Reservoir. Transactions of the American Fisheries Society 60:139–142.

Moore, E. 1933. Certain minnows showing adaptability to conditions in impounded waters. Transactions of the American Fisheries Society 62:290–291.

Parsons, J. W. 1958. Fishery management problems and possibilities on large southeastern reservoirs. Transactions of the American Fisheries Society 87:333–355.

Redmond, L. C. 1986. The history of development of warmwater fish harvest regulations. Pages 186–195 *in* Hall and Van Den Avyle (1986).

Reservoir Committee. 1967. Reservoir fisheries resources symposium. American Fisheries Society, Southern Division, Reservoir Committee, Bethesda, Maryland.

Richardson, R. E. 1928. The bottom fauna of the middle Illinois River, 1913–1925. Illinois Natural History Survey Bulletin 17:387–475.

Riggs, C. D. 1958. Unexploited potentials of our freshwater commercial fisheries. Transactions of the American Fisheries Society 87:299–308.

Schramm, H. L., Jr., and R. G. Piper, editors. 1995. Uses and effects of cultured fishes in aquatic ecosystems. American Fisheries Society Symposium 15.

Schueler, R. L., and J. K. Sullivan. 1967. Quantifying current and potential commercial fishery habitat and production on the inland waters of midwest U.S. exclusive of the Great Lakes. Pages 489–498 *in* Reservoir Committee, editors. Reservoir fisheries resources symposium. American Fisheries Society, Southern Division, Reservoir Committee, Bethesda, Maryland.

Scott, F. M. 1927. Introduction to the limnology of Searsville Lake. Stanford University Publications, University Series in Biological Science 5:1–84.

Shields, J. T. 1958. Fish management problems of large impoundments on the Missouri River. Transactions of the American Fisheries Society 87:356–364.

Simon, J. R. 1938. The management of Wyoming irrigation reservoirs. Transactions of the North American Wildlife Conference 3:433–439.

Stroud, R. H. 1948. Growth of the basses and black crappie in Norris Reservoir, Tennessee. Journal of the Tennessee Academy of Science 23:31–99.

Stroud, R. H., editor. 1986. Fish culture in fisheries management. American Fisheries Society, Fish Culture Section and Fisheries Management Section, Bethesda, Maryland.

Tarzwell, C. M. 1944. The possibilities of a commercial fishery in the TVA impoundments and its value in solving the sport and rough fish problems. Transactions of the American Fisheries Society 73:137–157.

Titcomb, J. W. 1918. Importance of a permanent policy in stocking inland waters. Transactions of the American Fisheries Society 47:11–21.

USFWS (U.S. Fish and Wildlife Service). 1991. 1991 national survey of fishing, hunting, and wildlife-associated recreation. U.S. Government Printing Office, Washington, DC.

Wickliff, E. L. 1933. Are newly-impounded waters in Ohio suitable for fish life? Transactions of the American Fisheries Society 62:275–277.

Wickliff, E. L., and L. S. Roach. 1937. Some studies of impounded waters in Ohio. Transactions of the American Fisheries Society 66:78–86.

Wiley, R. W., and R. S. Wydoski. 1993. Management of undesirable fish species. Pages 335–354 *in* C. C. Kohler and W. A. Hubert, editors. Inland fisheries management in North America. American Fisheries Society, Bethesda, Maryland.

Wood, R. 1951. The significance of managed water levels in developing the fisheries of large impoundments. Journal of the Tennessee Academy of Science 26: 214–235.

Wyatt, H. N., and H. D. Zeller. 1965. Fish populations dynamics following a selective shad kill. Proceedings of the Annual Conference Southeastern Association of Game and Fish Commissioners 16(1962):411–417.

American Fisheries Society Symposium 16:12–22, 1996

Understanding Reservoir Systems with Experimental Tests of Ecological Theory: A Prescription for Management

ROY A. STEIN, MARY T. BREMIGAN, AND JOHN M. DETTMERS
Ohio State University, Department of Zoology, Aquatic Ecology Laboratory
1314 Kinnear Road, Columbus, Ohio 43212–1194, USA

Abstract.—Enhancement of angler opportunities in U.S. reservoirs necessitates creative management approaches. One such approach, which builds on a successful manager–researcher partnership, derives from understanding reservoir food webs through experimental tests of ecological theory (in this case, the trophic cascade hypothesis). An explicit example involves predator control of gizzard shad *Dorosoma cepedianum*, a species that dominates Ohio reservoirs, suppresses zooplankton, outcompetes larval life stages of sport fishes, and regulates community composition. In pond experiments, hybrid striped bass *Morone saxatilis* × *M. chrysops* can control age-0 gizzard shad, increasing zooplankton density and size, thus improving conditions for sport-fish recruitment. Combining pond results with small-scale laboratory experiments, field values of crustacean zooplankton size and productivity, and gizzard shad density, we generated an ecologically based conceptual model of recruitment environments across Ohio reservoirs smaller than 500 ha. With this model, managers can apportion their limited resources according to prioritized management goals. To attain these goals, additional features associated with the community, such as anglers (as top predators with unique characteristics), and watersheds (as determinants of reservoir water quality) must be included. Whole-system manipulations, in which management actions permit hypothesis testing (i.e., adaptive management), combine with small-scale experiments to yield an understanding that provides a quantitative foundation for effective reservoir management.

Ecological theory can guide experiments across multiple spatiotemporal scales, providing a mechanistic understanding of reservoir community structure and function. Testing theory, however, often involves only the discovery and quantification of a field pattern that happens, by chance, to match an explanatory theory (Wiens 1989, 1992). Too frequently, Wiens (1992:166) submitted, investigators fail to pursue those underlying mechanistic processes that actually explain field patterns, thus ignoring "the problem of multiple causation—[for] the same [field] pattern may be produced by several processes," any two of which may call for diametrically opposed management actions. To reduce this possibility, we describe a research approach with a central tenet of understanding those ecological processes that drive ecosystem structure and function (sensu Parrish et al. 1995), an understanding that can provide a basis for reservoir management.

Our research approach uses small-scale experiments, whole-system manipulations, and documented field patterns to understand system structure and function. The work itself has been guided by ecological theory, facilitated by manager–researcher partnerships, and tested by an adaptive management process. In our reservoir ecosystems, we consider anglers to be integral to the biotic community and watersheds to be major determinants of water quality. Through an understanding of these systems, we develop specific management recommendations, well-grounded in ecological principles, that should improve recreational fishing opportunities.

Theory and Management

First, we explore how basic ecological theory and applied management interact to strengthen our approach. Typically, basic research requires considerable time before its results, if ever, are applied (Comroe and Dripps 1976; Lederman 1984). In contrast, applied research deals with extant, human-oriented problems and ideally generates a viable answer that can be applied in the short term. In general, basic research provides the raw material for solving applied problems (Bloch 1986; Gavin 1989; Babbitt 1995). In this context, basic and applied science are not discrete, distinct entities, but rather lie at opposite ends of a continuum that encompasses all of science (Figure 1). Research projects along this continuum differ in their relative degree of application to existing human problems. Experimental exploration, guided by basic theory, can be productively employed whether it be for the basic understanding of food webs or for the improvement of a sport fishery through biomanipulation (Figure 1).

Through experiments, applied research draws explicitly upon basic theory, providing (1) a frame-

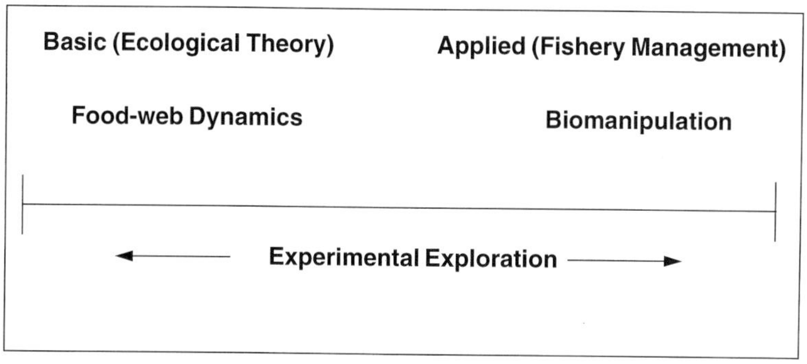

FIGURE 1.—Our perception of the relation between basic and applied science. Ecological theory emanates from basic science, whereas fishery management derives from applied science. Experimental exploration provides ecological insight into both basic (food web structure) and applied (biomanipulation) science.

work for answering questions and (2) guideposts for problem resolution. Below, we describe the theoretical framework that has guided our search for answers to applied management questions.

Conceptual Framework

The trophic cascade hypothesis serves as our conceptual framework. Originally developed for the marine rocky intertidal zone (Paine 1966), this hypothesis has been extended to north-temperate lakes (Carpenter et al. 1985; Carpenter and Kitchell 1993). It has stimulated basic research into the structure and function of lake food webs and has guided efforts to manage these communities (Carpenter and Kitchell 1993).

The trophic cascade hypothesis proposes that effects of top predators cascade from the top down because interactions are transmitted through the food web from piscivores to planktivores, herbivores, and primary producers (Figure 2A). Via strong food web linkages in north-temperate lakes, greater piscivore consumption of planktivores reduces consumption of zooplankton, and causes shifts in composition and abundance of crustacean zooplankton, higher grazing rates on phytoplankton, and increased water clarity (Oksanen et al. 1981; Carpenter et al. 1985, 1987; McQueen et al. 1989; Jassby et al. 1990; DeMelo et al. 1992; Persson et al. 1992; Carpenter and Kitchell 1993). In this context, the trophic cascade hypothesis has been used to manage lake communities via biomanipulation for which the primary benefits include enhanced predator populations and improved water clarity (Shapiro and Wright 1984; Benndorf 1990; Gophen 1990; Kitchell 1992).

To explore the generality of the trophic cascade hypothesis, we sought to determine if reservoir food webs respond to biomanipulation attempts in a manner similar to food webs in north-temperate lakes. In general, both abiotic and biotic conditions in reservoirs differ from conditions in north-temperate lakes. Most north-temperate lakes are relatively deep and clear, are typically classified as oligo- to mesotrophic, and are most abundant above 42°N latitude. In contrast, reservoirs, which typically have larger watershed-to-surface-area ratios than north-temperate lakes, are most abundant between 30° and 40°N latitude. Size and trophic conditions vary widely across reservoirs, often limiting the generality of conclusions drawn from research on particular subsets of reservoirs. Our research has focused on reservoirs within Ohio (an artificial ecological distinction, but a pragmatic one from a management viewpoint) that are smaller than 500 ha, shallow (2–7 m maximum depth), turbid (Secchi depths < 0.75 m), and eutrophic to hypereutrophic. When reservoirs are discussed below, we refer specifically to Ohio reservoirs with these abiotic characteristics.

In contrast to north-temperate lakes, reservoir food webs appear resistant to top-down control due in part to the predominance of gizzard shad (Figure 2B). Gizzard shad dominate reservoirs because they escape top-down control by resident predators through high fecundity (Parrish and Vondracek 1989), rapid growth, and generally low resident piscivore biomass (Johnson et al. 1988; Stein et al. 1995). Despite suppressing crustacean zooplankton, gizzard shad are not resource limited because they can consume detritus beyond their larval stage (Stein et al. 1995).

Reservoir food webs appear regulated from the "middle out" by abundant, omnivorous gizzard

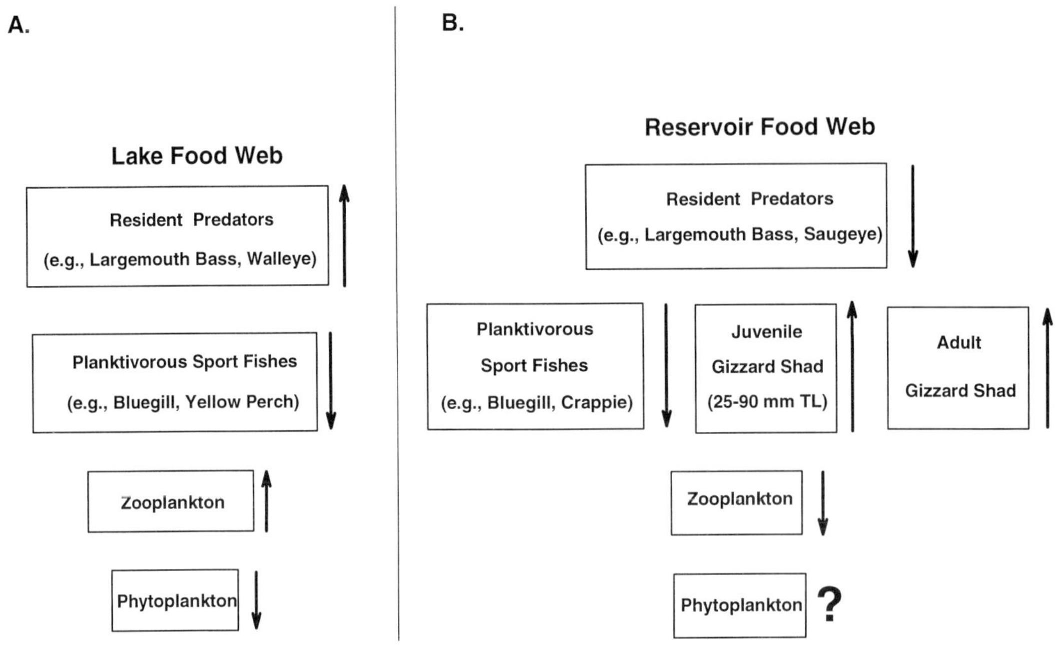

FIGURE 2.—Food webs of (**A**) north-temperate lakes and (**B**) small Ohio reservoirs (<500 ha) with trophic levels characterized as black boxes with representative species or sizes. These black boxes do not contain all taxa associated with the trophic levels portrayed. Arrows reflect the response (up arrow, increase in production; down arrow, decrease in production) of a specific trophic level to interactions with other trophic levels. In north-temperate lakes, these responses derive from piscivores, whereas in Ohio reservoirs, they derive from juvenile gizzard shad. Species are largemouth bass *Micropterus salmoides*, walleye *Stizostedion vitreum*, bluegill *Lepomis macrochirus*, yellow perch *Perca flavescens*, saugeye *Stizostedion canadense* × *S. vitreum*, crappie *Pomoxis* spp., and gizzard shad *Dorosoma cepedianum*.

shad. When age-0 gizzard shad suppress crustacean zooplankton (Dettmers and Stein 1992, 1996), they reduce growth and survival of other age-0 zooplanktivorous fishes through exploitative competition for zooplankton in early summer (Guest et al. 1990; DeVries et al. 1991). Sampling in four Ohio reservoirs over 5 years demonstrated that density of age-0 bluegills was inversely related to density of age-0 gizzard shad; only at larval gizzard shad densities of less than $10 \cdot m^{-3}$ did bluegill successfully recruit (J. E. Garvey, Ohio State University, unpublished data). Ongoing manipulative experiments have revealed that exploitative competition by gizzard shad for limited zooplankton reduces larval bluegill growth and survival (Garvey, unpublished data). If bluegill recruit poorly, then growth of age-0 largemouth bass, which feed on age-0 bluegills, will also decline (DeVries et al. 1991; N. S. Donovan, Ohio State University, unpublished data). Though abundant, age-0 gizzard shad typically grow too quickly to be consumed by age-0 largemouth bass (Adams and DeAngelis 1987; Johnson et al. 1988). Without bluegill or gizzard shad prey, age-0 largemouth bass grow slowly and are likely to have poor survival. Hence, reservoir food webs, driven by middle-out processes, do not typically support abundant predator populations.

In north-temperate lakes, biomanipulation not only results in abundant predators, but this management strategy also can improve water clarity via the trophic cascade. By contrast, improved water clarity in reservoirs appears unlikely because (1) turbidity in reservoirs more frequently derives from inorganic sources; (2) most reservoirs are dominated by small zooplankton and lack large *Daphnia* (the principal grazer responsible for algal declines in lakes, McQueen et al. 1986; Vanni et al. 1990; Elser and Goldman 1991; Carpenter et al. 1993); and (3) abundant gizzard shad act as nutrient pumps by consuming sediment and resuspending nutrients that otherwise would be unavailable to phytoplankton (Vanni 1995).

Given these differences in food web structure and abiotic characteristics between lakes and reservoirs, we evaluate the applicability of the trophic cascade hypothesis to Ohio reservoirs and explore the options available for managing these systems. We then illustrate how experimental insights, coupled with reservoir classification and whole-system manipulations via adaptive management, can provide ecologically sound recommendations for improving recreational fishing opportunities.

Experimental Exploration

Below, we summarize experiments across scales from 10-L containers to 0.4-ha ponds. These experiments generated quantitative insight into whether (1) introduced piscivores can generate top-down responses through to zooplankton and phytoplankton, (2) zooplankton can persist as a function of zooplankton productivity and age-0 gizzard shad abundance, and (3) zooplankton size can differentially influence foraging and survival of large- and small-gaped larval fishes.

To assess the potential for top-down effects in reservoir communities dominated by gizzard shad, we used eight 0.4-ha ponds at the Hebron State Fish Hatchery, Ohio. We first added gizzard shad in spawning condition to all eight ponds and then hybrid striped bass *Morone saxatilis* × *M. chrysops* (180–400 mm total length [TL]) to four randomly chosen ponds. To quantify the responses of lower trophic levels, we sampled age-0 gizzard shad, crustacean zooplankton, and phytoplankton.

Hybrid striped bass reduced age-0 gizzard shad (20–30 mm TL) within 9 d (Dettmers et al. 1996). Crustacean zooplankton abundance declined and average size increased in the presence of piscivores; phytoplankton biovolume did not respond to our manipulation (Dettmers et al. 1996). Thus, hybrid striped bass can induce top-down responses affecting zooplankton abundance and composition in hatchery ponds characterized by high zooplankton production.

We used in-reservoir enclosures (1 m^3) to quantify the conditions under which zooplankton might persist and thus enhance recruitment of those fishes whose larvae depend on limited zooplankton. Enclosures spanned the range of age-0 gizzard shad abundance (0–61 $\cdot$ m^{-3}) and crustacean zooplankton production (< 1–1,089 $mg \cdot m^{-3} \cdot d^{-1}$, dry weight) observed in Ohio reservoirs. At each level of productivity, zooplankton persistence was challenged by adding each of four densities of age-0 gizzard shad. We incorporated zooplankton abundance, zooplankton production, and age-0 gizzard shad density from each of 36 enclosures into a multiple regression model of zooplankton persistence (Dettmers and Stein 1996).

Our regression explained over 94% of the variation in zooplankton persistence among enclosures; values from the pond experiment independently verified the following quantitative conclusion (Dettmers et al. 1996). For zooplankton to persist at more than 100 individuals/L, the minimum density required for recruitment by larvae of some sport fishes (Werner and Blaxter 1980; Eldridge et al. 1981; Li and Mathias 1982), we predicted that abundance of age-0 gizzard shad (25–50 mm TL) must be less than 10 $\cdot$ m^{-3} and zooplankton production must exceed 220 $mg \cdot m^{-3} \cdot d^{-1}$, dry weight (Dettmers and Stein 1996).

To determine if reservoirs can be managed to constrain gizzard shad recruitment, we must quantify those ecological factors that drive gizzard shad recruitment. In addition to zooplankton abundance, zooplankton size likely affects recruitment of gizzard shad as it does for other larval fishes (Frank and Leggett 1986; Crowder et al. 1987; Miller et al. 1988; Mills et al. 1989a, 1989b). Notably, zooplankton size varies predictably between north temperate lakes and turbid Ohio reservoirs (Figure 3). In north-temperate lakes, zooplankton assemblages are often characterized by large taxa, such as *Daphnia* and large copepods (Brooks and Dodson 1965; Mills et al. 1987; Hall and Ehlinger 1989), whereas assemblages in turbid reservoirs commonly are dominated by small taxa, such as *Bosmina*, small copepods, and rotifers (Bremigan et al. 1991; DeVries and Stein 1992; Figure 3).

These size and taxon dissimilarities should differentially affect recruitment of large- versus small-gaped larval fishes because gape size constrains size of zooplankton that can be consumed (Wong and Ward 1972; Schael et al. 1991). In particular, larval gizzard shad have a smaller gape than larval bluegills and should feed on smaller zooplankton. To evaluate these differential effects, we conducted laboratory experiments in which larval bluegills and gizzard shad were fed crustacean zooplankton assemblages that varied in size. Bluegills selected larger zooplankton (0.6–0.8 mm, sizes characteristic of north-temperate lakes) than did gizzard shad (0.2–0.4 mm, sizes characteristic of reservoirs; Bremigan and Stein 1994). To test if zooplankton size affected survival, we added larval gizzard shad (15 mm TL) to 1-m^3 enclosures dominated by small or large crustacean zooplankton. After 2 weeks, gizzard shad survival was 300% higher when combined with small zooplankton than with large zooplankton, confirming that availability of preferred zooplankton size promotes survival of larval gizzard

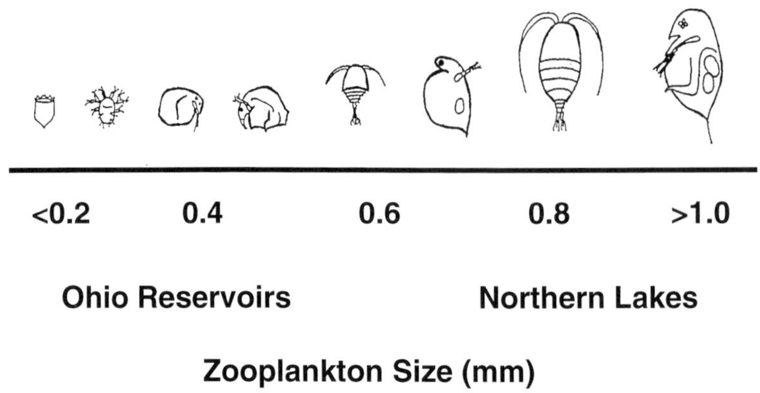

FIGURE 3.—The gradient of dominant zooplankton sizes in freshwater bodies. Small, turbid Ohio reservoirs (<500 ha) are dominated by rotifers and small crustacean zooplankton including *Bosmina*, copepod nauplii, and small copepod species. North-temperate lakes are dominated by large copepods and *Daphnia*.

shad (M. T. Bremigan and R. A. Stein, Ohio State University, unpublished data).

Given these results, larval recruitment of gizzard shad and other species may vary predictably across zooplankton assemblages. Reservoirs dominated by small zooplankton may represent ideal foraging environments for gizzard shad, whereas those dominated by larger zooplankton may favor larval bluegills or crappies (Schael et al. 1991; D. R. DeVries, Auburn University, unpublished data). Large-gaped larval sport fishes may be placed in "double jeopardy" in turbid reservoirs, where they are forced to compete with abundant larval gizzard shad for less-than-desirable zooplankton prey.

Management Within the Community

Our small-scale experiments, motivated by ecological theory, provide an understanding of reservoir fish communities that permit an exploration of management options. In this section, we use these experimental results to categorize reservoir recruitment environments explicitly with respect to their potential to favor specific reservoir fishes differentially. We then recommend specific management actions for each of these recruitment environments and follow the fate of one such management option (i.e., stocking hybrid striped bass) in mechanistic detail. Next, we explore how anglers should be incorporated into this approach, continuing with an overview of whole-system manipulation that demonstrates the real-world validity of our approach.

Classification of Ohio Reservoirs

Given that we have empirically identified the potential role of zooplankton productivity and size in successful fish recruitment, we now can characterize, by extrapolation, the relative suitability of Ohio reservoirs as recruitment environments for gizzard shad and resident sport fishes (Table 1). Whereas this framework organizes our thoughts with regard to biotic contributions to recruitment variability, our designations (e.g., high versus low, large versus small, etc.) have some basis in reality (Schael et al. 1991; DeVries et al. 1992; Bremigan and Stein 1994), and are qualitative benchmarks.

In general, recruitment declines as zooplankton productivity declines; gizzard shad should recruit more successfully with small zooplankton, but less successfully with large zooplankton (Table 1, rows A and B versus rows C and D). We base our predictions for bluegill, and hence largemouth bass, recruitment on the resultant effects of gizzard shad abundance. Specifically, high zooplankton productivity and small zooplankton size combine to yield an excellent recruitment environment for gizzard shad and a poor one for large-gaped sport fishes (e.g., bluegill and ultimately largemouth bass, Table 1, row A). Opposite predictions occur regardless of productivity at large zooplankton size (Table 1, rows C and D). In reservoirs characterized by low productivity and small zooplankton (Table 1, row B), conditions that compromise recruitment of resident sport fishes, we recommend stocking piscivores at a life stage at which they depend on fish prey rather than zooplankton. In our case, we support continued stocking of saugeye by the Ohio Division of Wildlife (ODW) at a size (~35 mm TL) at which larval gizzard shad prey dictate their growth and success (Stahl and Stein 1994; Stahl et al. 1996).

TABLE 1.—Our conceptual understanding of zooplankton characteristics, predicted recruitment of gizzard shad and resident sport fishes, and derived management implications for Ohio reservoirs. Rows are labelled A–D to correspond with table references in the text.

	Zooplankton characteristic		Predicted recruitment of reservoir fishes		Management implications	
Scenario	Size	Productivity	Gizzard shad	Sport fishes	Action	Goals
A	Small	High	High	Low	Stock hybrid striped bass	Hybrid striped bass fishery; increased sport fish recruitment
B	Small	Low	Medium	Low	Stock saugeye	Saugeye fishery
C	Large	High	Low	High	None	Resident sport fishery
D	Large	Low	Low	Medium	None	Resident sport fishery

Once classified, Ohio reservoirs less than 500 ha can be managed categorically to improve sport fish recruitment (Table 1). To illustrate, we consider the application of just one commonly used management tool, stocking, for reservoirs with strong gizzard shad recruitment.

When gizzard shad enjoy an excellent recruitment environment (Table 1, row A), then stocking hybrid striped bass to reduce gizzard shad abundance may be warranted. Upon piscivore addition, we speculate that a chain of complex ecological interactions, manifested over varying time intervals, is set into motion, ultimately leading to a reservoir with a recruitment environment more conducive to sport fishes (Table 1, rows C and D) than to gizzard shad (Table 1, rows A and B). While speculative, the relatively long-term scenario below is based on an extensive data set that includes experimental results and field patterns documented across numerous lakes and years.

Upon hybrid striped bass stocking, age-0 gizzard shad abundance would be reduced, easing typical midsummer declines in zooplankton abundance and size (Figure 4). Thus, even in the first biomanipulation season, moderate improvement in sport fish recruitment conditions could be realized. Across several years of continued stocking of hybrid striped bass, more complex interactions among nutrients, turbidity, zooplankton size, and gizzard shad may further drive recruitment conditions from those favoring gizzard shad (Table 1, rows A and B; Figure 4) to those favoring sport fishes (Table 1, rows C and D). As gizzard shad populations decline, reductions in the nutrient recycling they cause should follow, reducing phytoplankton and biotically derived turbidity. Because turbidity can negatively affect large zooplankton (Arruda et al. 1983), reduced turbidity should increase dominant zooplankton size and constrain gizzard shad recruitment even further. This positive feedback loop with negative consequences for gizzard shad potentially reduces gizzard shad more rapidly than might be expected with hybrid striped bass predation alone. If large zooplankton can be substantially enhanced (directly, via reduced zooplanktivory by age-0 gizzard shad longer than 25 mm, and indirectly, via reduced turbidity), then enhanced recruitment of sport fishes may occur (Stein et al. 1995).

In developing this suite of cause and effect interactions, we speculate that the invasion potential of gizzard shad initially would be limited if dominant zooplankton are large and turbidity and water column phosphorus are low. In these reservoirs, invasion would be quite gradual until gizzard shad became sufficiently abundant to influence zooplankton size, phosphorus, and turbidity. Upon reaching this level of influence, the gizzard shad population could then quickly come to dominance through middle-out processes (Stein et al. 1995). The positive feedback loop favoring gizzard shad operates as follows: as gizzard shad increase, all system-wide effects, including reduced zooplankton size, poor recruitment of resident predators such as largemouth bass, and increasing phosphorus and turbidity, interact to create an environment increasingly favorable to gizzard shad (Stein et al. 1995; Bremigan and Stein, unpublished data; Figure 4).

When Ohio reservoirs support large zooplankton, regardless of productivity, we speculate that these environments would support poor gizzard shad recruitment and good sport fish recruitment (Table 1, rows C and D); under these scenarios, no stocking would be required. Simple reliance on resident sport fish production should satisfy the angling public. Clearly then, managers should seek to alter the dominant zooplankton size from small to large in order to promote sport fisheries without stocking (Table 1).

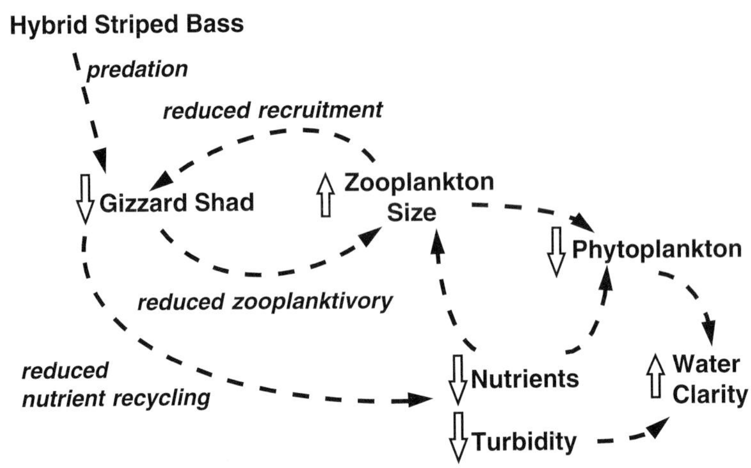

FIGURE 4.—Schematic representation of the probable set of interactions within a reservoir food web after hybrid striped bass have been stocked. Black arrows show the directions of biological interactions. White arrows indicate increases (up) or decreases (down) in abundance or size as a result of those interactions.

Consideration of Anglers

As reservoir conditions improve to support additional sport fishes (be they stocked or resident), they become more attractive to anglers, requiring a quantitative assessment of angler impact on the community (Magnuson 1991; Kitchell and Carpenter 1993). Historically, this assessment has quantified angler catch rates and effort using creel surveys (Walters 1986; Matlock 1991; Hilborn and Walters 1992). However, creel surveys frequently are not conducted over sufficiently long time periods or large enough spatial scales to reflect complex angler dynamics, including angler responses to publicity about catch rates on nearby waters (Johnson and Carpenter 1994). It may be more appropriate to consider anglers as another trophic level in the biotic community (Johnson and Staggs 1992). As are predatory fishes, anglers are both size- and species-selective, but they likely differ from fishes in functional and numerical response curves (Johnson and Carpenter 1994). Only by considering anglers as dynamic, flexible predators will we continue to improve our understanding of how anglers operate in reservoir communities.

Statistically rigorous creel surveys, coupled with detailed diet data for other trophic levels, provide insight into system-specific functional links between anglers and the fish community that permit managers to determine whether their management actions will be successful. For example, in Ohio reservoirs we expect one of two stable states to develop (Stein et al. 1995). Either gizzard shad dominate with resultant poor recruitment of sport fishes, or gizzard shad are rare with the potential for good sport fish recruitment. If a reservoir community is dominated by gizzard shad, only intensive, multiyear stocking of hybrid striped bass will likely effect gizzard shad control. In this case, reservoir management becomes an all-or-nothing proposition. Stocking a few piscivores will not provide the predator density required for sustained gizzard shad control. As a result, managers must be prepared to defend what might be termed an extraordinary stocking policy: large size at stocking to ensure survival (Wahl et al. 1995), high annual rates of stocking, and severe constraints on angler harvest. Only with these exceptional measures, which may be politically unpopular, can the gizzard shad stable state common to Ohio reservoirs be shifted to one that supports sport-fish recruitment.

Whole Reservoirs: the Relevant Scale for Managers

Our research approach to understanding reservoirs demonstrates how ecological theory drives small-scale experimentation that provides insight for management. Yet, the most appropriate scale for testing these insights is the whole system, the typical purview of the resource manager. Coupling the manager's wisdom and administrative mandate with the researcher's experimental expertise generates a synergy sufficiently dynamic to successfully address large-scale resource management issues. The manager–researcher partnership would best be served by embracing adaptive management strategies (Walters 1986; Lee 1993), especially in systems characterized by extreme variability among years and sites. Specific hypotheses (derived from specific management actions) coupled with replicates and control or reference systems at the whole-system scale, characterize adaptive management and distinguish it from traditional trial-and-error methods.

Selection of replicates for rigorous evaluation of management actions requires careful consideration. As with any field experiment, the pool from which replicates are chosen depends upon the systems to which one will extend experimental inferences. Extremely variable systems may require many replicates, perhaps too many to be feasible, yet choosing highly similar systems may limit the extension of the results to an unacceptably small subset of reservoirs.

As an example, consider how fish should be stocked to maximize survival. For simplicity, assume two methods exist for stocking, point stocking (all at one site) or scatter stocking (distribute fish around the reservoir, see Wahl et al. 1995). To test which method increased survival for saugeye, we matched eight pairs of reservoirs, based on historical survival rates documented by the ODW in collaboration with our laboratory. We then randomly chose one reservoir within each pair to scatter stock, and the other was point-stocked (N. S. Donovan, Ohio State University, unpublished data). By classifying reservoirs, then pairing those with similar characteristics, we could reduce variability within each tested pair, which in turn provided us with the statistical power to detect a treatment effect. Saugeye survival did not differ between reservoirs within each pair; hence, stocking technique did not influence saugeye survival. Our experimental approach in this case generated a sufficiently general conclusion that the ODW could adopt point stocking as policy, yielding savings in time and money.

The esocid stocking program in Ohio provides another example that demonstrates the value of adaptive management to determine the most appropriate size, time, and esocid to stock. To judge the relative success of muskellunge *Esox masquinongy*, northern pike *E. lucius*, and their hybrid, tiger muskellunge *E. masquinongy* × *E. lucius*, we simultaneously stocked different sizes of each taxon into several reservoirs (Wahl and Stein 1989, 1991, 1993) and quantified sources of mortality (Wahl et al. 1995). Muskellunge performed quite well, especially if stocked at 250 mm TL in early autumn (Wahl and Stein 1989, 1991, 1993). These experimental results have led to modifications in stocking practices for Ohio esocids.

Still a third example, illustrated by supplemental forage stocking, illustrates the need to determine if the experimental manipulations actually caused measurable system responses. In collaboration with the ODW, we established threadfin shad *Dorosoma petenense* in two Ohio reservoirs to determine if this species increased growth of white crappies *Pomoxis annularis* and largemouth bass. We collected premanipulation data and estimated postmanipulation threadfin shad abundance and predator growth rates and diets in both manipulated and reference systems. Though abundant, threadfin shad did not appear in predator diets nor did its presence increase predator growth (Bremigan et al. 1991). The ODW no longer considers threadfin shad addition as a viable management tool.

Carefully designed, whole-system experiments that include both food web manipulations and ecosystem monitoring (including the angler) are rare. However, the insights gained from an experiment on Lake Mendota (Kitchell 1992), where managers and both agency and academic researchers collaborated, have increased our understanding of complex ecosystems. In Ohio, we have been fortunate to participate in whole-system experiments with the ODW that have led to better management. Partnerships between managers and researchers, with each bringing complementary skills to bear on management problems, can serve the resource well.

Management Beyond the Community

We now move beyond the community by briefly delving into the importance of watersheds in influencing reservoir water quality, the biotic community, and ultimately the potential for successful management. In so doing, we also explore how variability in both abiotic and biotic characteristics across reservoirs influences our ability to generalize our conclusions. For example, not all Ohio reservoirs support dense populations of gizzard shad; peak larval densities range less than 1 to $100 \cdot m^{-3}$. In a statewide survey, we are contrasting reservoirs that support high densities of gizzard shad with those that do not to explore factors underlying gizzard shad dominance. From this survey we have documented that larval gizzard shad abundance is correlated positively with water column phosphorus and turbidity (Bremigan, unpublished data). In preliminary analyses, these systems tended to have zooplankton assemblages dominated by smaller taxa than those systems with lower nutrients and turbidity. With these relations, we have documented a field pattern only without the explanatory mechanisms necessary for system understanding (Wiens 1989, 1992). Even so, we can use these relationships in the short term to guide policy as we design and complete experiments whose results will shed light on underlying ecological mechanisms. We believe that watershed management, coupled with biotic manipulations, could reduce the suitability of the environment for gizzard shad recruitment by increasing zooplankton size (see Figure 4; Table 1). We endorse the perspective that enhancement of fish communities and water quality must derive, in part, from watershed management through improved land use practices (Addis and Les 1996, this volume).

In applying our conclusions to reservoirs across the United States, we must be aware of the very broad range of abiotic and biotic variables represented. For example, in work on Virginia and Alabama reservoirs, concerns regarding reductions in nutrient inputs have been raised in light of a positive relation between total fish biomass and phosphorus (Ney 1996; Maceina et al. 1996; both this volume). However, if phosphorus exceeds 40 μg/L, non-sport, detritivorous fishes would be reduced first during any oligotrophication (Maceina et al. 1996). Therefore, in Ohio reservoirs (50–300 μg/L phosphorus), reduced phosphorus loading should reduce gizzard shad biomass without adversely affecting sport-fish production.

Conclusions

Through explicit examples derived from our reservoir research in collaboration with the Ohio Division of Wildlife, we have illustrated how a mechanistic understanding of reservoir communities can provide the foundation for their effective management. This understanding can best be attained by coupling experimental tests at a variety of scales with the adaptive management process at the whole-reservoir scale. More explicitly, this approach works well when insights gained from small-scale experiments form the basis for hypotheses to be tested at the whole-reservoir scale, the scale most relevant to managers. To assure success, features beyond the reservoir community itself should be incorporated. In particular, both a quantitative appreciation for the impact of anglers as top predators and a recognition of the critical contribution of watersheds to water quality will improve the probability of success of any management scheme. Coupling a landscape perspective that includes interdisciplinary expertise with a research approach will lead to uncommon management insight into reservoir–watershed ecosystems.

Acknowledgments

In a quick read of a preliminary draft, Libby Marschall, Deanna Stouder, Rusty Wright, Brett Johnson, Dave Wahl, Mike Vanni, Nick Donovan, Randy Miller, and Jim Kitchell provided advice, insight, and wise counsel. We thank the members of the Aquatic Ecology Laboratory for contributing their thoughts and ideas during the past few years. Finally, a productive agency–university partnership evolves through time from interactions with individuals; in that spirit, we extend our appreciation to D. Barry Apgear, Gary Isbell, Ken Paxton, and Randy Miller, the ODW biologist–administrators, who taught us much of management and the real world. This work was funded by Federal Aid in Sport Fish Restoration Project F-69-P, administered jointly by the U.S. Fish and Wildlife Service and the Ohio Division of Wildlife, by National Science Foundation grants DEB-9107173 and DEB-9407859, and by Electric Power Research Institute grant 91-07 to R.A.S.

References

Adams, S. M., and D. L. DeAngelis. 1987. Indirect effects of early bass–shad interactions on predator population structure and food web dynamics. Pages 103–117 *in* W. C. Kerfoot and A. Sih, editors. Predation: direct and indirect impacts on aquatic communities. University Press of New England, Hanover, New Hampshire.

Addis, J. T., and B. L. Les. 1996. Expanding perspectives on gaining public support for management. Pages 42–46 *in* L. E. Miranda and D. R. DeVries, editors. Multidimensional approaches to reservoir fisheries management. American Fisheries Society Symposium 16.

Arruda, J. A., G. R. Marzolf, and R. T. Faulk. 1983. The role of suspended sediments in the nutrition of zooplankton in turbid reservoirs. Ecology 64:1225–1235.

Babbitt, B. 1995. Science: opening the next chapter of conservation history. Science 267:1954–1955.

Benndorf, J. 1990. Conditions for effective biomanipulation: conclusions derived from whole-lake experiments in Europe. Hydrobiologia 200/201:187–203.

Bloch, E. 1986. Basic research and economic health: the coming challenge. Science 232:595–599.

Bremigan, M. T., E. M. Lewis, M. B. Jones, R. A. Stein, and D. R. DeVries. 1991. Evaluating effects of stocking threadfin shad on young-of-year crappie, bluegill, and largemouth bass in Ohio lakes. Ohio Department of Natural Resources, Federal Aid in Sport Fish Restoration, Final Performance Report, Project F-57-R, Columbus.

Bremigan, M. T., and R. A. Stein. 1994. Predator gape

and zooplankton size: exploring a size-dependent recruitment mechanism in fishes. Canadian Journal of Fisheries and Aquatic Sciences 51:913–922.

Brooks, J. L., and S. I. Dodson. 1965. Predation, body size, and composition of plankton. Science 150:28–35.

Carpenter, S. R., and J. F. Kitchell. 1993. The trophic cascade in lakes. Cambridge University Press, New York.

Carpenter, S. R., J. F. Kitchell, and J. R. Hodgson. 1985. Cascading trophic interactions and lake productivity. BioScience 35:634–639.

Carpenter, S. R., J. A. Morrice, J. J. Elser, A. St. Amand, and N. A. Mackay. 1993. Phytoplankton community dynamics. Pages 189–209 *in* S. R. Carpenter and J. F. Kitchell, editors. The trophic cascade in lakes. Cambridge University Press, New York.

Carpenter, S. R., and nine coauthors. 1987. Regulation of lake primary productivity by food web structure. Ecology 68:1863–1876.

Comroe, J. H., Jr., and R. D. Dripps. 1976. Scientific basis for the support of biomedical science. Science 192:105–111.

Crowder, L. B., M. E. McDonald, and J. A. Rice. 1987. Understanding recruitment of Lake Michigan fishes: the importance of size-based interactions between fish and zooplankton. Canadian Journal of Fisheries and Aquatic Sciences 44(Supplement 2):141–147.

DeMelo, R., R. France, and D. J. McQueen. 1992. Biomanipulation: hit or myth? Limnology and Oceanography 37:192–207.

Dettmers, J. M., D. R. DeVries, and R. A. Stein. 1996. Quantifying responses to hybrid striped bass predation across multiple trophic levels: implications for reservoir biomanipulation. Transactions of the American Fisheries Society 125:491–504.

Dettmers, J. M., and R. A. Stein. 1992. Food consumption by larval gizzard shad: zooplankton effects and its implications for reservoir communities. Transactions of the American Fisheries Society 121:494–507.

Dettmers, J. M., and R. A. Stein. 1996. Quantifying linkages among gizzard shad, zooplankton, and phytoplankton in reservoirs. Transactions of the American Fisheries Society 125:27–41.

DeVries, D. R., and R. A. Stein. 1992. Complex interactions between fish and zooplankton: quantifying the role of an open-water planktivore. Canadian Journal of Fisheries and Aquatic Sciences 49:1216–1227.

DeVries, D. R., R. A. Stein, J. G. Miner, and G. G. Mittelbach. 1991. Stocking threadfin shad: consequences for young-of-year fishes. Transactions of the American Fisheries Society 120:368–381.

Eldridge, M. B., J. A. Whipple, D. Eng, M. J. Bowers, and B. M. Jarvis. 1981. Effects of food and feeding factors on laboratory reared striped bass larvae. Transactions of the American Fisheries Society 110:111–120.

Elser, J. J., and C. R. Goldman. 1991. Zooplankton effects on phytoplankton in lakes of contrasting trophic states. Limnology and Oceanography 36:64–90.

Frank, K. T., and W. C. Leggett. 1986. Effect of prey abundance and size on the growth and survival of larval fish: an experimental study employing large volume enclosures. Marine Ecology Progress Series 34:11–22.

Gavin, T. A. 1989. What's wrong with the questions we ask in wildlife research? Wildlife Society Bulletin 17:345–350.

Gophen, M. 1990. Biomanipulation: retrospective and future development. Hydrobiologia 200/201:1–11.

Guest, W. C., R. W. Drenner, S. T. Threlkeld, F. D. Martin, and J. D. Smith. 1990. Effects of gizzard shad and threadfin shad on zooplankton and young-of-year white crappie production. Transactions of the American Fisheries Society 119:529–536.

Hall, D. J., and T. J. Ehlinger. 1989. Perturbation, planktivory, and pelagic community structure: the consequence of winterkill in a small lake. Canadian Journal of Fisheries and Aquatic Sciences 46:2203–2209.

Hilborn, R., and C. J. Walters. 1992. Quantitative fisheries stock assessment. Chapman and Hall, New York.

Jassby, A. D., T. M. Powell, and C. R. Goldman. 1990. Interannual fluctuations in primary production-direct physical effects and the trophic cascade in Castle Lake, California. Limnology and Oceanography 35: 1021–1038.

Johnson, B. M., and S. R. Carpenter. 1994. Functional and numerical responses: a framework for fish–angler interactions. Ecological Applications 4:808–821.

Johnson, B. M., and M. D. Staggs. 1992. The fishery. Pages 350–375 *in* J. F. Kitchell, editor. Food web management: a case study of Lake Mendota. Springer-Verlag, New York.

Johnson, B. M., R. A. Stein, and R. F. Carline. 1988. Use of a quadrat rotenone technique and bioenergetics modeling to evaluate prey availability to stocked piscivores. Transactions of the American Fisheries Society 117:127–141.

Kitchell, J. F. 1992. Food web management: a case study of Lake Mendota. Springer-Verlag, New York.

Kitchell, J. F., and S. R. Carpenter. 1993. Variability in lake ecosystems: complex responses by the apical predator. Pages 111–124 *in* M. J. McDonnell and S. T. A. Pickett, editors. Humans as components of ecosystems. Springer-Verlag, New York.

Lederman, L. M. 1984. The value of fundamental science. Scientific American 251(5):40–48.

Lee, K. N. 1993. Compass and gyroscope: integrating science and politics for the environment. Island Press, Washington, DC.

Li, S., and J. A. Mathias. 1982. Causes of high mortality among cultured larval walleyes. Transactions of the American Fisheries Society 111:710–721.

Maceina, M. J., and five coauthors. 1996. Compatibility between water clarity and quality black bass and crappie fisheries in Alabama. Pages 296–305 *in* L. E. Miranda and D. R. DeVries, editors. Multidimensional approaches to reservoir fisheries management. American Fisheries Society Symposium 16.

Magnuson, J. J. 1991. Fish and fisheries ecology. Ecological Applications 1:13–26.

Matlock, G. C. 1991. Use of surveys in decision making. Pages 1–4 *in* D. Guthrie and seven coeditors. Creel and angler surveys in fisheries management. American Fisheries Society Symposium 12.

McQueen, D. J., M. R. Johannes, J. R. Post, T. J. Stewart, and D. R. S. Lean. 1989. Bottom-up and top-down impacts on freshwater pelagic community structure. Ecological Monographs 59:289–309.

McQueen, D. J., J. R. Post, and E. L. Mills. 1986. Trophic relationships in freshwater pelagic ecosystems. Canadian Journal of Fisheries and Aquatic Sciences 43:1571–1581.

Miller, T. J., L. B. Crowder, J. A. Rice, and E. A. Marschall. 1988. Larval size and recruitment mechanisms in fishes: toward a conceptual framework. Canadian Journal of Fisheries and Aquatic Sciences 45:1657–1670.

Mills, E. L., D. M. Green, and A. Schiavone, Jr. 1987. Use of zooplankton size to assess the community structure of fish populations in freshwater lakes. North American Journal of Fisheries Management 7:369–378.

Mills, E. L., M. V. Pol, R. E. Sherman, and T. B. Culver. 1989a. Interrelationships between prey body size and growth of age-0 yellow perch. Transactions of the American Fisheries Society 118:1–10.

Mills, E. L., R. Sherman, and D. S. Robson. 1989b. Effect of zooplankton and body size on growth of age-0 yellow perch (*Perca flavescens*) in Oneida Lake, New York. Canadian Journal of Fisheries and Aquatic Sciences 46:880–886.

Ney, J. J. 1996. Oligotrophication and its discontents: effects of reduced nutrient loading on reservoir fisheries. Pages 285–295 *in* L. E. Miranda and D. R. DeVries, editors. Multidimensional approaches to reservoir fisheries management. American Fisheries Society Symposium 16.

Oksanen, L., S. D. Fretwell, J. Arruda, and P. Niemela. 1981. Exploitation ecosystems in gradients of primary productivity. American Naturalist 118:240–261.

Paine, R. T. 1966. Food web complexity and species diversity. American Naturalist 100:65–75.

Parrish, D. L., M. E. Mather, and R. A. Stein. 1995. Problem-solving research for management: a perspective. Fisheries 20(10):6–12.

Parrish, D. L., and B. Vondracek. 1989. Population dynamics and ecology of Lake Erie gizzard shad. Ohio Department of Natural Resources, Federal Aid in Sport Fish Restoration, Annual Performance Report, Project F-61-R, Columbus.

Persson, L., S. Diehl, L. Johansson, G. Andersson, and S. F. Hamrin. 1992. Trophic interactions in temperate lake ecosystems: a test of food chain theory. American Naturalist 140:59–84.

Schael, D. M., L. G. Rudstam, and J. R. Post. 1991. Gape limitation and prey selection in larval yellow perch (*Perca flavescens*), freshwater drum (*Aplodinotus grunniens*), and black crappie (*Pomoxis nigromaculatus*). Canadian Journal of Fisheries and Aquatic Sciences 48:1919–1925.

Shapiro, J., and D. I. Wright. 1984. Lake restoration by biomanipulation. Round Lake, Minnesota-the first two years. Freshwater Biology 14:371–383.

Stahl, T. P., and R. A. Stein. 1994. Influence of larval gizzard shad (*Dorosoma cepedianum*) density on piscivory and growth of young-of-year saugeye (*Stizostedion vitreum vitreum* × *S. canadense*). Canadian Journal of Fisheries and Aquatic Sciences 51:1993–2002.

Stahl, T. P., and five coauthors. 1996. Factors affecting survival of age-0 saugeye *Stizostedion vitreum* × *S. canadense* stocked in Ohio reservoirs. North American Journal of Fisheries Management 16:378–387.

Stein, R. A., D. R. DeVries, and J. M. Dettmers. 1995. Food-web regulation by a planktivore: exploring the generality of the trophic cascade hypothesis. Canadian Journal of Fisheries and Aquatic Sciences 52: 2518–2526.

Vanni, M. J. 1995. Nutrient transport and recycling by consumers in lake food webs: implications for algal communities. Pages 81–95 *in* G. A. Polis and K. O. Winemiller, editors. Food webs: integration of patterns and dynamics. Chapman and Hall, New York.

Vanni, M. J., C. Luecke, J. F. Kitchell, Y. Allen, and J. Temte. 1990. Effects on lower trophic levels of massive fish mortality. Nature 344:333–335.

Wahl, D. H., and R. A. Stein. 1989. Comparative vulnerability of three esocids to largemouth bass (*Micropterus salmoides*) predation. Canadian Journal of Fisheries and Aquatic Sciences 46:2095–2103.

Wahl, D. H., and R. A. Stein. 1991. Field testing an esocid bioenergetics model. Transactions of the American Fisheries Society 120:230–246.

Wahl, D. H., and R. A. Stein. 1993. Comparative life history attributes of muskellunge, northern pike, and their hybrid. Canadian Journal of Fisheries and Aquatic Sciences 50:1961–1968.

Wahl, D. H., R. A. Stein, and D. R. DeVries. 1995. An ecological framework for evaluating the success and effects of stocked fishes. Pages 176–189 *in* H. L. Schramm, Jr. and R. G. Piper, editors. Uses and effects of cultured fishes in aquatic ecosystems. American Fisheries Society Symposium 15.

Walters, C. J. 1986. Adaptive management of renewable resources. Macmillan, New York.

Werner, R. G., and J. H. S. Blaxter. 1980. Growth and survival of larval herring, *Clupea harengus*, in relation to fry density. Canadian Journal of Fisheries and Aquatic Sciences 37:1063–1069.

Wiens, J. A. 1989. The ecology of bird communities, volume 2. Cambridge University Press, Cambridge, UK.

Wiens, J. A. 1992. Ecology 2000: an essay on future directions in ecology. Bulletin of the Ecological Society of America 73:165–170.

Wong, B., and F. J. Ward. 1972. Size selection of *Daphnia pulicaria* by yellow perch (*Perca flavescens*) fry in West Blue Lake, Manitoba. Journal of the Fisheries Research Board of Canada 29:1761–1764.

American Fisheries Society Symposium 16:23–27, 1996

Resolving Conflicts in Reservoir Operations: Some Lessons Learned at the Tennessee Valley Authority

CHRISTOPHER D. UNGATE

Tennessee Valley Authority, Reservoir Operation and Planning Review
400 West Summit Hill Drive, Knoxville, Tennessee 37902, USA

Abstract.—The public's dependence on water resources for survival increasingly conflicts with its demand for clean bodies of water with scenic views to use for recreation and leisure. Many recent attempts to resolve such conflicts have been marked by frustration and failure. In contrast, the Lake Improvement Plan developed by the long-embattled Tennessee Valley Authority (TVA) has succeeded at relatively low cost, in a comparatively short time. This success can be attributed to three factors: an excellent team of leaders, specialists, and scientists; good timing; and the TVA's broad mandate. This paper describes aspects of the TVA experience that contributed to the success of its plan, emphasizing the need for institutions with a broad, multipurpose mandate, leaders with the authority to make necessary decisions, and a public that has been included in the decision-making process.

Water resources are inherently limited in supply and used simultaneously and sequentially for multiple purposes, such as for drinking water, wastewater disposal, fishing, and recreation. These uses can and do conflict with each other. Conflict in use and management of water resources is increasing because the human population is growing and societal needs are shifting from "survival" issues, such as good water quality for public health and industrial growth, to quality-of-life issues, such as living and engaging in recreation on a clean lake with scenic views.

What increases the potential for conflict is that most major water resource developments—like the Tennessee Valley Authority (TVA) reservoir system—were built more for survival purposes, such as navigation, flood control, hydropower, irrigation, and water supply. The constituencies that support these projects are enfranchised by law. Many people take these benefits for granted or decry their environmental impact and increasingly value other benefits, such as fisheries, water quality, recreation, wildlife, and endangered species. In many cases, these benefits have an economic value simply because they exist, a value that may be even greater than the value of their use.

Reinforcing this conflict further are the interests of entrepreneurs and big businesses, who see an opportunity to make money in water-related recreation and lifestyles, such as recreational fishing, boating, jet skis, resorts, and second homes. More people have the time and money to find a refuge in the great outdoors from the stress and complexity of modern life.

For more than 50 years, the TVA managed a system of 30 major dams on the Tennessee River and its tributaries primarily for navigation, flood control, and power production. Today's world is much different from that of the 1930s. Few people remember the severe flooding that once occurred at Chattanooga, Tennessee, or recall being able to walk across the Tennessee River during drought periods. Today, the benefits of reservoir operation are often taken for granted. Although people still value the low-cost hydropower benefits afforded by the system, many are less willing to accept the associated degradation of water quality and other recreation benefits. Instead, they are demanding improved water quality and additional opportunities for high-quality recreation experiences on TVA lakes.

Recognizing these changes, in 1987 the TVA Board of Directors authorized a study of the priorities for operating the Tennessee River System. In February 1991, the board adopted a Lake Improvement Plan calling for three major changes in system operation: minimum flows below all TVA dams; aeration of releases from 16 dams in the eastern half of the Tennessee Valley to improve dissolved oxygen in tailwaters; and improved summer lake levels on 10 tributary storage reservoirs (TVA 1991). The board also decided to pay the costs of water quality improvements (about US$44 million for aeration equipment, plus about $4 million per year for operating and maintenance costs) with power revenues and to pay the costs of lake level improvements (about $2 million per year for lost energy value) with Congressional appropriations. I discuss the process used by the TVA to formulate and select alternatives for reservoir operation outlined in the Lake Improvement Plan.

How Are Agencies Handling Conflicts Over Water Use?

Federal and state agencies have not handled water use conflicts well in some cases. The Environmental Protection Agency gave the U.S. Army Corps of Engineers (Corps) the lowest possible rating for its environmental impact statement on the revision of the Missouri River Master Manual. States along the Missouri River have sued the Corps and each other over the effects of too much or too little flow from upstream storage projects and their effects on recreational reservoir levels, commercial navigation, and habitat for fish and wildlife.

The upper Mississippi River basin states are pondering recommendations on how to manage that river after the devastating 1993 flood. Environmental and navigation interests are polarized as the Corps begins a study costing more than $40 million on how to rehabilitate the 29 navigation locks above Cairo, Illinois.

The $11+ million Comprehensive Study of the Alabama–Coosa–Tallapoosa and Apalachicola–Chattahoochee–Flint river systems is due to end this fall after two schedule extensions. Only then can the Corps begin drafting an environmental impact statement on alternative operating plans for its reservoirs on these rivers. The study was begun after Alabama sued Georgia and the Corps in 1990 over a proposed reallocation of storage on Lanier, Allatoona, and Carter reservoirs for Atlanta's water supply. There is still no resolution in sight 5 years after the litigation began.

The process for deciding how to operate the Bureau of Reclamation and Corps of Engineers dams on the Columbia River is now dictated by the courts. The Bonneville Power Administration, the Bureau of Reclamation, and the Corps began a System Operation Review in 1990, but they have been unable to decide how to complete it. This is an interesting case of the Endangered Species Act being in conflict with itself; alternative operations that might improve habitat for salmon threaten reservoir-dependent species upstream that also are in a protected status.

The Tennessee Valley Authority's Success with the Lake Improvement Plan

The TVA's Lake Improvement Plan successfully reached a conclusion, took only 41 months, and cost only $4 million—why did it succeed where other efforts have failed? How could TVA have accomplished this? After all, this is the same agency that tried to build Columbia Dam; got an exemption from the Endangered Species Act to close Tellico Dam; was a major destroyer of wetlands in its fight with malaria; and impounded two-thirds of large river habitat in the most diverse river fishery on the North American continent and in the area of the most diverse mussel fauna on earth. I believe the success of the Lake Improvement Plan process can be attributed to people, timing, and the institution.

First, we had a visionary, John Waters, then a director of TVA, who loved the river, rafted it from Sevierville, Tennessee, to its mouth at the Ohio River and on to New Orleans, and wanted to make his mark on it. We had good scientists, engineers, planners, foresters, process experts, economists, and analysts. We had a few good leaders who wanted to improve reservoir operations and a team of about 150 men and women at TVA who gave their best to the development of the Lake Improvement Plan.

Second, we had the fortune of good timing. Senators Robert Stafford of Vermont and Jim Sasser of Tennessee wanted TVA to take responsibility for water quality below TVA dams. Senator Sasser made a tour of the valley in June 1987, pointing out problems and demanding that TVA solve them. This prompted the TVA Board to begin development of the Lake Improvement Plan in October 1987. Problems with constructing and managing nuclear power plants brought Marvin Runyon to TVA as Chairman of the Board in 1988. His negotiation skills were essential to making the final decision to adopt the Lake Improvement Plan.

Third, we benefited from TVA's broad mandate for management of the Tennessee River for "navigation, flood control, reforestation, proper use of marginal lands, agricultural and industrial development, national defense, and other purposes," as listed in the preamble to the Tennessee Valley Authority Act of 1933. We also had an operational TVA myth (and here I borrow from Hargrove 1994: 64) reinvented for the 1990s:

> The TVA myth was based on four values, each of which was part of a dichotomy: professional expertise supported by statutory missions versus grass roots democracy; and a multipurpose authority versus the power company. Lilienthal thought it very important to keep all four values in balance, and the importance of doing so was the legacy he left his successors. [David Lilienthal was one of the original TVA Board members.]

Thus, we sought to incorporate public desires into the plan and enhance reservoir operations to fulfill multiple statutory purposes, while assuring that hydropower benefits were not reduced.

How did we bring these three elements together to make a decision? John Waters, the visionary, saw the opportunity created by good timing and invoked the institution's mandate, myth, and expertise to begin a process of reevaluating the operative priorities of TVA's multipurpose dams. And in another stroke of good timing, Marvin Runyon forged a coalition of distributors, large industrial customers, and elected officials to implement the recommendations arising from the process. Thus, we were lucky to have good engineering, good biology, and good political science.

Institutional Framework for Revising Operating Guidelines

I have discussed the development of the Lake Improvement Plan with groups in a variety of river basins across the country, including many of the basins I listed previously (Ungate 1992). In most of these basins, interests are polarized and litigation is the primary means of addressing conflict. Only one organization adopted the particular approach that TVA used to develop the Lake Improvement Plan, and it has failed as yet to reach a decision.

I could tell you the war stories of the Lake Improvement Plan and the details of how we formulated and selected alternatives for reservoir operations. But I won't, for three reasons.

First, readers of this volume are among the many fully qualified and well-intentioned people employed by federal agencies, state water quality agencies, state fish and game agencies, and other levels of government who could devise their own processes and analyses to deal with conflicts in the management of reservoirs charged to their care. I am confident that readers, like those of us at the TVA who worked on the Lake Improvement Plan, could successfully develop a process to determine what to do. In other words, people are not the problem.

Second, the forces of change are upon us. Demands for good quality water and aquatic life, for recreational opportunities, and for good environmental quality are creating many opportunities to reexamine how we manage water resources. The human population continues to increase, and development around reservoirs increases proportionately, begging the question of why reservoirs are managed the way they are. In other words, timing is not the problem. There are many opportunities.

And third, in my view, good professionals respond to the incentives that are placed in front of them. They have an inherent interest in doing good work. The need for their expertise is evident. However, institutions do not empower them to do the work that is needed, despite the opportunities that present themselves from time to time. Perhaps our institutions are not up to the task.

Need for a Decision Maker

What features of TVA's institutional framework allowed the people of TVA to take advantage of the timing to change how we operate our reservoirs? As strange as this may seem, one very important feature was that a decision maker existed and was empowered to decide. Under the Tennessee Valley Authority Act, the TVA Board is fully empowered to decide how to operate TVA reservoirs and is accountable for its decisions. On the Columbia River, the Bureau of Reclamation, the Corps, and the Bonneville Power Administration must jointly decide how to operate their dams and sell hydropower through Bonneville Power Administration contracts. Because these agencies were unable to reach a decision, the courts are now orchestrating the decision-making process. On the upper Mississippi River, responsibility for operating 29 lock-and-dam facilities is spread across three Corps districts and two divisions. The Mobile District's authority to reallocate storage in Lanier, Carter, and Allatoona reservoirs for water supply is limited and was one of the factors that led to litigation over this issue in 1990. It is rare to find a single organization whose leadership is empowered to decide how to manage a system of reservoirs.

Importance of a Broad Mandate

A second important feature of TVA's institutional framework that helped assure the success of the Lake Improvement Plan is the agency's multipurpose authority for the Tennessee River basin. The TVA cannot easily walk away from an issue related to natural resource management and claim that it does not have a stake in the outcome. Other agencies and levels of government may have primacy on a particular natural resource issue, but TVA is involved no matter what the situation. The TVA mandate is not narrow, as it is for the Corps or the Bureau of Reclamation. On the upper Mississippi and Missouri rivers, the Corps's mandate hinges on navigation, and it is unable or at least unwilling to address the broad range of concerns, which has resulted in polarization, litigation, and inefficiency.

As Hargrove (1994) pointed out, TVA often has let the concerns of its power company override its multipurpose mandate in the Tennessee watershed. This did not occur with the Lake Improvement Plan, because of good leadership and the work of the staff who prepared the basis for decisions. Nevertheless, I am sure that improvements are possible in the framework that governs how TVA fulfills its multipurpose mandate.

Involvement of Key Stakeholders

The third important feature of TVA's institutional framework that explains the success of the Lake Improvement Plan is its "reinvented" tradition of balancing the judgment of its professional experts with its practice of building support for its programs among grass roots organizations and community leaders. As Hargrove (1994) correctly observed, this "tradition" was really a myth in TVA's early days. The original operating guidelines for TVA reservoirs were developed entirely by TVA engineers. There was no balance, except to the extent that members of Congress, as representatives of the electorate, set out the operating priorities in the Tennessee Valley Authority Act. In my view, TVA began learning to balance good science and engineering in response to the desires of river system users and the general public, as it reconciled the growth of the environmental movement in the 1960s and 1970s with its roots in the conservation movement of the first few decades of the Twentieth Century.

The TVA was the conservation movement's great experiment in how a system of large dams in a river basin could control floods, provide water for dependable navigation, and produce electric power to pay for the dams. The agency would reforest the watershed and reduce erosion from agriculture to assure that the dams would not fill with sediment. It would destroy wetlands and manage the reservoir margin to control mosquitoes and malaria. It would promote forestry to grow trees and create jobs. It would assure that the reservoirs would not become biological deserts. The TVA was a product of the idea that natural resources could be managed in a sustainable way so that future generations could benefit from them. These benefits were primarily seen as economic in nature—for the benefit of human uses.

It took two decades for TVA river managers to recognize that the conservation movement had evolved into the environmental movement. They now have added ecological health and biological diversity to the mix of benefits they attempt to improve. Whereas they once resisted the public involvement provisions of the National Environmental Policy Act, they now embrace them as a means of building grass roots support for river management decisions and programs. Publications such as *RiverPulse* are designed to transfer the knowledge gained by TVA professionals to the general public as a way of building consensus on required actions to change land or reservoir management or to clean up the river.

All key interest groups were actively involved in the development of the Lake Improvement Plan. Public involvement in developing the plan involved more than just conducting public information meetings. A level playing field was created where all interested parties have an equal chance to make their case known to the decision maker, and data and analyses are appraised in an open, reviewable process by anyone who has a stake in the outcome. The TVA was given the first Federal Environmental Quality Award by the President's Council on Environmental Quality for making "the public a full partner in its decision making."

In other river basins, not integrating public involvement with the study process undermines the capability of agencies to build support for their final decisions. Concerns about the effectiveness of public involvement have been expressed in the Alabama–Coosa–Tallapoosa and Apalachicola–Chattahoochee–Flint Comprehensive Study, the ongoing navigation study on the upper Mississippi River, and the Master Manual revision process on the Missouri River.

Relevance to Other River Basins and Agencies

To establish or revise operating guidelines for reservoirs (or make any other river management decision), we need river basin management institutions whose leaders have the responsibility and commensurate authority to make decisions. Such institutions should also have a broad, multipurpose authority for river basin management and should derive their accountability from the customers they serve through effective involvement of all stakeholders in decision-making processes. Most river basins do not have a management institution with these basic characteristics. I think this is why there is a high degree of conflict today in the management of water resources.

If it were not for the current effort to reinvent government, my remarks would be only of academic interest. I believe that we as professionals and as members of professional organizations have an opportunity to help shape how water resource developments will be managed in the future. We already have seen two bills in the 104th Congress that would privatize TVA river management or give the TVA's power program the responsibility for running river management programs. We are about to see other bills to sell the U.S. Department of Energy agencies that market the power from the U.S. Army Corps of Engineers and Bureau of Reclamation dams. This could affect how these dams are managed for hydropower. It is conceivable that these dams, or the hydropower facilities in them, could be sold along with the power-marketing agencies.

The most influential member of Congress who advocated the establishment of TVA in 1933 was Senator George Norris of Nebraska. He advocated public ownership and multipurpose development of river resources and local control by nonpolitical, professional experts. When the debate about restructuring government addresses the issue of managing the federal investment in water resource developments, such as reservoirs, I hope these same points will be made to the new generation of legislators. I also hope that the insights learned from the process of developing TVA's Lake Improvement Plan, which complement Senator Norris's arguments, become part of that debate.

References

Hargrove, E. C. 1994. Prisoners of myth: the leadership of the Tennessee Valley Authority, 1933–1990. Princeton University Press, Princeton, New Jersey.

TVA (Tennessee Valley Authority). 1991. Tennessee River and reservoir system operation and planning review. Tennessee Valley Authority (TVA/RDG/EQS-91/1), Knoxville.

Ungate, C. D. 1992. "Equal consideration" at TVA: changing system operations to meet societal needs. Hydro Review 11(4):28–37.

Administrating and Marketing Reservoir Fisheries

American Fisheries Society Symposium 16:31–37, 1996

Reservoir Fisheries and Agency Communication

TOMMY L. BROWN

Cornell University, Human Dimensions Research Unit
Fernow Hall, Ithaca, New York 14853–3001, USA

Abstract.—Communication is a critical aspect of reservoir fisheries management. Fisheries are managed for the benefits derived by the public; therefore, "know your audience" is an important motto for biologists to heed. For audiences to become better known to reservoir managers, a variety of communication opportunities must be developed and tapped by fisheries staff. These include news releases, brochures, and conservation magazine articles that communicate from the agency to the public, as well as meetings, angler surveys, and other mechanisms that allow feedback from the public. Education, public relations, public involvement, and marketing—all important aspects of fisheries management—have close interrelationships with communication and are discussed in this paper. The Strawberry Reservoir in Utah is used to illustrate a good communication effort in conjunction with reservoir management.

Communication is unquestionably one of the more important components of fisheries management; however, agency fisheries staff often have little formal preparation in communication beyond college courses in public speaking and writing. Such courses are important in developing these specific communication skills, but additional training is needed to examine the interface between communication and other agency functions.

I provide a general overview of the role of communication in fisheries management, with application to reservoir fisheries. I also review some important principles of effective communication and link these principles to successful fisheries programs. I then show interrelationships between communication and the other constituency-oriented management functions of education, public relations, public involvement, and marketing evaluation. Finally, I present a brief case study in which many of these principles were applied in a reservoir management situation.

Natural resource management agencies across North America vary considerably as to their organizational structure. As a result, it is impractical to recommend a particular organizational chart that would show the specific location and function of a communication unit and the linkages between that unit and other management functions. Rather, in generic fashion, some of the important internal and external communication challenges facing fisheries agencies will be developed in this paper. Determining how to develop an improved fisheries communication program from an organizational perspective will be left to fisheries staff.

Role of Communication in Fisheries Management

Fisheries management is a multidisciplinary activity. Nielsen (1993) depicts fisheries management as three overlapping circles representing organisms, habitats, and people. Krueger and Decker (1993) characterize the management environment as having ecological, political, sociocultural, and economic components. The process of fisheries management includes five broad steps: (1) choice of goals, (2) selection of objectives, (3) identification of problems, (4) implementation of actions, and (5) evaluation of actions. Communication, in broad terms, is a continuing two-way process between managers and the public that facilitates management. Good communication, both within an agency (internal communication) and with the media, anglers, and other stakeholder groups outside an agency (external communication) is critical to successfully progressing through the five management steps.

A good communication program uses multiple methods to reach a variety of audiences. These audiences include the public at large and individual stakeholder groups who have particular interests (or stakes) in fisheries management decisions. Black bass *Micropterus* spp. anglers, shoreline anglers, riparian landowners, and the local business community are examples of stakeholder groups. Communication also must be directed toward political groups such as the governor's office, the state legislature, and particularly in the case of many reservoir fisheries, units of local government.

Effective internal communication is critical for the efficient functioning of agency staff, and for the coordination and accurate presentation of the agency's policies and programs. Maintaining effective internal communication is no small task. A central fisheries agency office often must communicate with field offices and with other units or bureaus such as education, public relations, environmental protection, law enforcement, wildlife, and parks. Effective intercommunications among these units are critical to the effective implementation of agency programs.

External communication programs, while organized in a number of ways within fisheries agencies, generally serve the following four functions: education (usually informal), public relations, public involvement, and marketing. After presenting some fundamental principles of communication, I examine the interface between communication as a broad agency program and each of the above four functions.

Fundamentals of Communication

Any textbook definition would characterize communication as an interaction or exchange of information (the message) between two parties. This definition seems so straightforward as to be simplistic, yet it is easily forgotten. Fishing regulations can be specified in a guidebook that accompanies each license sold, but until each angler reads and understands the message, full communication has not occurred. Regulations can be enacted that make it the responsibility of the angler to read the guidebook, understand it, and be responsible for heeding all fisheries regulations. Such actions clarify where the responsibility lies, but are not necessarily effective in communicating regulations (or in creating a positive agency image). Thus, the mechanism and channel of communicating regulations need to be examined carefully, and in some cases multiple approaches may be needed.

Getting the message out is important, but that is only one part of the communication process. Getting feedback from anglers, organizations, the business community, and any other important stakeholder group is also important. Thus, means need to be developed for two-way communication between agency staff and stakeholders. The public involvement process, which usually occurs formally when new regulations or policies are proposed, is one part of this feedback process. However, informal means of obtaining feedback also need to be developed.

One means of obtaining informal feedback is through telephone contact with members of the public who call for information. In addition to answering a specific question, agency personnel should take advantage of the phone call opportunity in two ways. First, try to understand the basis for the call. Was it simply to inquire about a regulation, or did the caller also attempt to communicate that the regulation was not covered or was vague in the regulations guide, or that some aspect of the regulation was unfair? Second, if the call is related to a particular reservoir, the caller can be asked a question or two about the fishery as a means of informal information gathering. In doing so, staff must remember that these callers are not representative of everyone who fishes the reservoir, and policy changes should rarely if ever be based on such information alone. However, this informal feedback can help agency staff determine whether a study of a representative sample of users is needed, or whether some broader public involvement process should be initiated on the topic in question. Seeking angler feedback also demonstrates the interest of the agency in anglers' perspectives and ideas and helps enhance anglers' image of the agency.

The communication process has five basic components: communicator, message, audience, channel, and feedback. Each of these must be appropriate for the task at hand for effective communication to occur. For further information on basic communication refer to Fazio and Gilbert (1986) and to Decker and Krueger (1993).

Know Your Audience

The preceding material on informal feedback mechanisms leads to the broader topic of knowing your audience. By "knowing," we refer to a particular audience's identity (e.g., place of residence, age, sex, and other demographic characteristics), fishing preferences, attitudes, values, and behaviors. This information should be obtained by periodic assessment and involve professionally conducted creel and angler surveys. See Pollock et al. (1994) for how to conduct such surveys, and Hunt and Ditton (1996, this volume).

A number of studies have shown that managers' perceptions of particular stakeholder groups are often at some variance from the true beliefs, attitudes, or behaviors of those groups. Moreover, the actual values of managers may differ from those of constituents. Brunson (1992) cites an example in which forest outputs were valued differently by professional foresters and the public and itemized several reasons why these differences develop. First, as people spend more time in natural resource agencies, they tend to be "captured" gradually by the most vocal or dominant interest groups they deal with. Foresters who work in areas where clearcut-ting is a common harvesting method gradually accept clearcuts from an aesthetic perspective. If a particular reservoir's primary organized constituents are highly skilled black bass anglers, biologists over time may discount the needs of general anglers which may conflict with regulations developed for the specialist group.

Brunson (1992) cites four reasons for communications failure, or why managers may not be in tune with constituents. First, managers over time develop "a cult of expertise." Technical and scientific knowledge is a basis for fisheries knowledge, but in adopting a technical and scientific management paradigm, managers sometimes ignore the values of the public, or may appear arrogant to the public; thus the managers' communications are not regarded seriously.

Second, and a corollary to the above, managers, like other professional groups, adopt and use their own language or jargon, which sometimes provides a barrier to effective communication with the public. Jargon provides a parsimonious means for managers to discuss complex concepts with each other, but jargon is not readily understood by the public; the public may even assign unfavorable and untrue motives to why managers use such language in their communications.

Third, managers may send mixed messages either in dual communications or in their communications versus their actions. If individual managers or biologists publicly disagree about a problem or a solution, the public may get mixed messages on the topic. Or, if a manager announces a new policy direction before everyone in the agency agrees to back the decision, a retraction or modification may become necessary. Such actions give the appearance of an agency that is unorganized, incompetent, or both.

The fourth barrier to communication cited by Brunson (1992) is the assumption that all publics share a value system. Hunters and members of the animal rights movement obviously have different values. However, distinctions are not always obvious without research data. Carl (1982) illustrates this by citing a rotenone treatment to the Huron River in southern Michigan to reduce populations of rough fish so that seven sport fish species could be introduced. Fishing effort before and after these actions remained relatively constant. However, a group of largely African American anglers who enjoyed fishing for the rough fish were replaced by a group of largely white anglers who preferred the introduced species. Even less obvious than this example might be a situation in which regulations favoring the taking of fewer large fish are implemented in place of those which allowed the taking of more smaller fish. Managers and biologists might personally favor such a regulations change. But does the public who fishes the reservoir? Know your audience!

Communication and Education

Six primary functions of fisheries agency communications with an educational focus are listed here:

1. Educate anglers and the public about fisheries and related agency programs (this function also has a marketing and a communications focus). These communications should be timely, proactive, and often should use multiple methods. Press releases to the mass media, brochures at kiosks or businesses adjacent to specific fisheries and at sport shows, and speakers at fishing forums are examples of a multi-method approach. It may be helpful to develop standard protocols of how various types of new information will be distributed. However, such a practice should be in the form of guidelines rather than unalterable policy. Any given news item may have several appropriate methods of communication, or a normally appropriate method might be inefficient or undesirable at a specific point in time.
2. Communicate regulations to the public. The communication of fishing regulations has left much to be desired in many fisheries agencies. This important task has not always been seen in the context of a communications function, and it is very difficult to effectively communicate regulations to every resident and nonresident angler who uses a particular fishery. Higher quality communications and a heightened awareness about the communications process may be necessary to more effectively communicate regulations. Some anglers have complained that it is difficult to find, decipher, and interpret the many fishing regulations found in state fishing manuals today. In addition to striving to keep the regulations as simple as possible, agencies need to put more effort into ways of helping anglers find and interpret those regulations. Charts that depict regulations by species and by waterway would be one means of simplifying the communication of regulations. A better indexing system for the more lengthy regulations guides would facilitate finding a specific regulation. It should be remembered that a regulations manual is not a book that one picks up and reads cover to cover. Rather, it is primarily a reference that an angler

consults when fishing for a particular species on a particular reservoir. Thus, the guide should be organized accordingly.

Fisheries law enforcement has sometimes fallen into the trap of declaring that it is the responsibility of the public to know and heed the law, thereby implying that law enforcement staff have little if any responsibility for communicating the laws (regulations) to the public. Fortunately, this is changing and enforcement of the law is now being coupled with a responsibility for public service that includes teaching and communication of laws and regulations. The ultimate objective is to have the public understand the need for the regulations and to heed them as a result. This is best accomplished by having conservation officers communicate regulations through presentations at public meetings. In addition, conservation officers should be given flexibility to issue warnings to first time offenders where regulations violations are judged to be minor.

3. Communicate advisories or warnings to anglers. Unfortunately, some reservoirs contain toxic substances to the extent that certain fish are either unsafe to eat at all, or should be eaten only in limited quantities. Generally, this is less of a problem in reservoirs than in other types of waters. Warnings may also be posted regarding swimming restrictions or shallow water. If a condition is constant throughout a given season, the warning should be included in the regulations guide. In addition, on-site warnings should be posted in places where they will be seen by anglers.
4. Teach fisheries and aquatic resource principles and concepts.
5. Teach fisheries conservation and stewardship principles.
6. Teach fishing skills.

Within a fisheries agency, items 4–6 would usually occur in informal situations outside the classroom, such as at camps, nature centers, or sporting shows. Other occasional opportunities may exist such as writing a column for the local newspaper, appearing on local radio or television, or giving a talk in a classroom setting. In each situation, education can occur only through good communication. For good communications to occur, it is important that you know your audience, their age and level of knowledge and experience, and if possible, their interests. Each of these factors can help you design communications, whatever the medium, that will more successfully get your message to your audience.

Communication and Public Relations

Many elements of public relations are simply extensions of courtesy and good manners. In fact, Fazio and Gilbert (1986) define public relations as planned effort to influence public opinion through good character and responsible performance. Character and performance need to be backed up by an effective external communications program.

Fazio and Gilbert's (1986) first principle of public relations is that every agency action makes an impression on its publics. An agency "action" is not limited to a policy or regulation, or enforcement of these, but should be interpreted in the broadest sense to include even informal communications. Fisheries biologists represent their agency in the eyes of the public in their every action. Biologists communicate nonverbally as well as verbally, in the things they do and the attitudes they display. A publicly employed biologist, regardless of one's job description, is a public servant. This means that biologists owe the public courtesy, respect, and a good-faith effort to listen to their questions, complaints, and suggestions and respond with information or referrals when appropriate.

An important principle in management and also in public relations is that proactive is better than reactive. Applied to communications, this means that when public input is needed, it is sought, and when decisions are made, they are communicated along with the rationale for the decisions. The public actually consists of many groups, and thus a comprehensive communications strategy is needed to reach each group in a timely manner.

Good public relations often include the opportunity to provide informal education to the public, and agency staff in turn can often learn a great deal from the public. Good public relations also help instill public confidence in the fisheries agency. This confidence is particularly valuable in times when license fees need to be increased or when catch limits need to be made more restrictive.

Communication and Public Involvement

It is difficult to design an effective public involvement process or to evaluate the process until a clear objective is set for any particular public involvement scenario. According to Iacofano (1990), the four main goals of public involvement are to: (1) exchange information, (2) obtain viewpoints of all relevant interest groups, (3) mediate public interests in an attempt to reflect the desires of competing interest groups (at least to some degree), and (4) permit agency responsiveness to public concerns. The purpose of a public involvement process must first be clarified by agency staff, then communicated effectively to the various publics who will be involved.

Public involvement requires a format that facilitates both effective communication and interaction. Formats for public involvement are dealt with by McMullin (1996, this volume). It is critical that public involvement occur during the planning process or when a policy or regulation is being evaluated. If anglers or other citizens sense that public involvement is occurring solely because it is required or solely to enhance the agency's image, they will become mistrustful of the agency.

Good communication must go hand in hand with the public involvement process if it is to succeed. All communicating parties need to understand the issue or topic (including appropriate historical background), the public involvement process being used, and in general, how the public involvement process fits into the decision-making process.

Communication and Marketing

Kotler (1975) notes that to survive and succeed, any organization must (1) attract sufficient resources, (2) convert these resources into products, services, and ideas, and (3) distribute these outputs to various consuming publics. This is equally true of a clothing manufacturer, an airline, a university, and a fisheries management agency. Because the organization and the customer or clientele work with each other on a voluntary basis (in which no force is exerted), the organization relies primarily upon exchange mechanisms to function. The agency provides fishing opportunities, information, and conservation services. The public provides the agency fishing license dollars, tax dollars, and political support.

A professional marketer, according to Kotler, is someone who is very good at understanding, planning, and managing exchanges. The marketer knows how to research and understand the needs of existing and potential clientele, how to design a valued offering (product or service) to meet those needs, how to communicate the offer effectively, and how to present the offering in a convenient place and time.

From the above, it should be clear why fisheries agencies not only should, but must be involved in marketing. Although biologists may not have been trained to think of their profession in this way, fisheries management is ultimately a social or public service. If fisheries organizations don't stay attuned to the public market for fisheries experiences, they risk being perceived as indifferent, perhaps stubborn, and ultimately, unnecessary (Witter and Adams 1994).

Communication at many levels is necessary for effective marketing. Many angler surveys are in large part market research surveys. They are designed to be an exchange medium, to provide information from a representative sample of anglers on topics such as where they fish, what they prefer to fish for, what kind of factors in combination provide for a satisfying fishing experience, and which options from those that managers can responsibly provide (while maintaining the integrity of particular fisheries and habitats) would be most preferable to anglers.

Where do the questions used in angler surveys come from? In part, they may come from professional resources such as colleagues in other state and federal agencies, and academics. But many of the ideas for the questions about angler preferences and management options should be generated as a result of exchanges with anglers. Telephone information services, fishing forums, public involvement obtained during planning, and informal conversations with anglers all provide some communication to fisheries staff about angler preferences. From all of this input, the most promising suggestions can be extracted and incorporated into an angler survey to determine the breadth and strength of angler interests.

Witter and Adams (1994) use MARKET as an acronym to describe six marketing practices that are useful in the public sector, including conservation agencies.

Maintain traditional sponsors
Analyze products and services
Respond to the underserved and unserved
Know the competition
Educate outside and inside the agency
Talk to the customer.

Communication in one form or another is integral to accomplishing each of these practices, including "analyze products and services." In addition to requiring internal communication within the agency to array the products and services offered, analyzing them requires some input as to how well they meet public needs and whether or not gaps exist within the current offerings. Obtaining this input requires communication with various constituencies or stakeholder groups.

A Case Study: Strawberry Reservoir in Utah

Strawberry Reservoir, approximately 4,900 ha in area, dates from 1920 and is located in central Utah within a 1-h drive of one million people. Due to a variety of factors, including siltation of tributary beds from overgrazing, the reservoir's cutthroat trout *Oncorhynchus clarki* population declined during the 1970s and 1980s and was replaced largely by the illegally introduced Utah sucker *Catostomus ardens* and Utah chub *Gila atraria*. Because freshwater lakes and reservoirs are quite scarce in Utah, angling organizations began in the 1980s to demand higher quality fisheries in Strawberry Reservoir.

Local fisheries biologists determined that chemical rehabilitation would be needed before stocking of sport-caught species would be successful. Between 2 and 3 years before treatment, an interagency team from the U.S. Forest Service, U.S. Bureau of Reclamation, Central Utah Water Conservancy District, Utah State Health Department, and sportfishing organizations was formed and devised a plan for rehabilitation.

Treatment of the reservoir occurred in August 1990. Because the treatment occurred on federal lands and Federal Aid funding was used, the National Environmental Policies Act (42 U.S.C. §§ 4321–4361) process was required. At implementation, approximately 4,100 kg of rotenone powder were used in the lake over a period of 6 d and 16,000 L of liquid rotenone were used in the tributaries over a 60-d period. Over 250 people, including many volunteers, and the Utah National Guard were involved. The total cost of the project was approximately US$3.5 million. Following treatment, the reservoir was stocked with Bonneville cutthroat trout *O. c. utah*, kokanee salmon *O. nerka*, and sterilized rainbow trout *O. mykiss*. The reservoir was reopened to fishing in January 1991. Today the reservoir has excellent fishing; anglers spent over 1.25 angler-hours fishing in 1994 and had a mean catch rate of 0.6 fish per hour. The primary emphasis of reservoir users is on fish and wildlife related recreation.

What distinguishes Strawberry Reservoir from some less successful attempts to rehabilitate reservoir fisheries is (1) the lengthy planning of the interagency planning committee and lining up the assistance needed for a project of this size to be successful, and (2) the involvement of the media in a positive way from the planning stages through treatment and beyond. By the time public hearings were held, considerable publicity about plans for the reservoir had been given to the media and was transmitted to residents. Some concerns did arise in the public hearings over possible impacts on beaver, raptors, and endangered species. These concerns were taken into account as completely as possible.

Communication with anglers was important initially, especially to maintain the reservoir's population of cutthroat trout. Although an eight-fish limit was maintained, anglers were encouraged to release their catch of cutthroat. Signs and illustrations were used at the reservoir to help anglers identify the cutthroat trout, and creel agents also provided assistance. Recent angler surveys have shown that 97% of anglers can identify the cutthroat trout, and about 50% of those caught are released. A one-fish cutthroat trout limit was imposed in 1993. (Unpublished data, R. Spateholts, Utah Division of Wildlife.)

Summary

Communication is a necessary and critical aspect of reservoir fisheries management. Because the publics that fisheries programs serve are so varied, communication programs must use multiple means to effectively reach agency constituencies. Communication necessarily involve two parties—in this case the agency and some component of the public. Just as effective mechanisms are needed to communicate fisheries programs and regulations to the public, feedback from the public is also important, and the agency must establish adequate means to assure effective feedback.

Strong linkages exist between communication and education, public relations, the public involvement process, and marketing. An agency's communication section should establish and maintain a strong linkage with any of these functions that are located elsewhere in the agency.

References

Brunson, M. 1992. Professional bias, public perspectives, and communication pitfalls for natural resource managers. Rangelands 14:292–295.

Carl, L. M. 1982. Social impacts of a stream reclamation project on urban anglers. North American Journal of Fisheries Management 2:164–170.

Decker, D. J., and C. C. Krueger. 1993. Communication: catalyst for effective fisheries management. Pages 55–74 *in* C. C. Kohler and W. A. Hubert, editors. Inland fisheries management in North America. American Fisheries Society, Bethesda, Maryland.

Fazio, J. R., and D. L. Gilbert. 1986. Public relation and

communication for natural resource managers, 2nd edition. Kendall/Hunt, Dubuque, Iowa.

Hunt, K. M., and R. B. Ditton. 1996. Using survey research in support of fisheries management: the 1994 Texas statewide angler survey. Pages 236–244 *in* L. E. Miranda and D. DeVries, editors. Multidimensional approaches to reservoir fisheries management. American Fisheries Society Symposium 16.

Iacofano, D. I. 1990. Public involvement in environmental planning: a proactive theory for environmental planning program design and management. Garland Publishing Inc., New York.

Kotler, P. 1975. Marketing for nonprofit organizations. Prentice-Hall, Englewood Cliffs, New Jersey.

Krueger, C. C., and D. J. Decker. 1993. The process of fisheries management. Pages 33–53 *in* C. C. Kohler and W. A. Hubert, editors. Inland fisheries management in North America. American Fisheries Society, Bethesda, Maryland.

McMullin, S. L. 1996. Natural resource management and leadership in public arena decision making: a prescriptive framework. Pages 54–63 *in* L. E. Miranda and D. DeVries, editors. Multidimensional approaches to reservoir fisheries management. American Fisheries Society Symposium 16.

Nielsen, L. A. 1993. History of inland fisheries management in North America. Pages 3–31 *in* C. C. Kohler and W. A. Hubert, editors. Inland fisheries management in North America. American Fisheries Society, Bethesda, Maryland.

Pollock, K. H., C. M. Jones, and T. L. Brown. 1994. Angler survey methods and their applications in fisheries management. American Fisheries Society Special Publication 25, Bethesda, Maryland.

Witter, D. J., and C. E. Adams. 1994. Fish, wildlife, and public service: why resource professionals must market. Fisheries 19(9):20–21.

American Fisheries Society Symposium 16:38–41, 1996

A Partnership Approach to Reservoir Fisheries Management

J. LARRY WILSON

University of Tennessee
Post Office Box 1071, Knoxville, Tennessee 37901–1071, USA

DAVID A. TOMLJANOVICH

Tennessee Valley Authority, Aquatic Biology Laboratory, Norris, Tennessee 37828, USA

DOUGLAS C. PETERSON

Tennessee Wildlife Resources Agency
6032 W. Andrew Johnson Highway, Talbott, Tennessee 37877, USA

CHRISTOPHER O'BARA

Tennessee Technological University
Box 5033, Cookeville, Tennessee 38505, USA

Abstract.—We describe a program wherein state and federal fishery professionals, research staff and students from two universities, and volunteer anglers joined together in a partnership approach to fisheries management in Norris Reservoir, Tennessee. Major issues and concerns perceived by anglers were identified and solutions to selected problems were proposed and implemented. Two major projects were undertaken by the group: (1) a comprehensive baseline reservoir sampling effort involving various agencies and volunteers, and (2) a habitat enhancement program that used many volunteers. Volunteers donated over 6,500 h to these projects while state, federal, and university personnel provided supervision, equipment, and supplies, and coordinated the sampling and work efforts. Working together in these cooperative efforts enabled biologists to gain a better understanding of angler views and concerns and anglers to gain insights into the complexity of reservoir fisheries management.

The Norris Reservoir Task Force (NRTF) is an interagency group formed in January 1992 to address fish management issues on Norris Reservoir. The task force was formed by the Tennessee Wildlife Resources Agency (TWRA) in response to increasing complaints and dissatisfaction by local anglers over the fishing quality on Norris Reservoir. Angler complaints centered around the perception by some anglers that striped bass *Morone saxatilis*, through predation and indirect effects, had an undesirable influence on native game fish stocks. Other angler concerns included lack of cover, lack of closed seasons to promote successful spawning, and dramatic water level fluctuations.

The task force was initially composed of nine members representing the TWRA (3 members), the Tennessee Valley Authority (TVA, 2 members), the University of Tennessee–Knoxville (UTK, 1 member), Tennessee Technological University (TTU, 1 member), and two anglers from the public sector. Two additional anglers were added to the NRTF in 1993 to improve angler representation and involvement. These additions improved credibility of the task force with the angling community and provided greater representation from the principal angler groups (i.e., those seeking black basses *Micropterus* spp., crappies *Pomoxis* spp., walleye *Stizostedion vitreum*, and striped bass).

Since inception of the task force, meetings have been held regularly, usually monthly. The focus of the meetings has been to examine existing data, to solicit input from both the public and biologists experienced with the Norris Reservoir fisheries, to develop recommendations and workplans, and to present results of studies recommended by the group. The usual meeting format has been a round-table discussion at which various members reported on assigned projects and identified subjects for discussion by the group. At times, meetings have been opened to the public (non–task-force members) for input on specific issues, such as revisions of creel limits and proposed stockings of crappies, striped bass, and walleye. These open forums have been run as panel discussions to keep the group focused on specific topics; this format helped task force members gather input from local anglers and also tended to prevent the meeting from being dominated by complaints.

Task Force Goal and Objectives

The original goal of the task force was to identify factors limiting fishery production and to develop optimal fisheries management strategies for Norris Reservoir. The initial objectives were to (1) identify the key issues and concerns as perceived by the fishing public, (2) compile and disseminate accurate fisheries information about the Norris Reservoir fisheries, and (3) develop and implement workplans and directions for future research and management efforts, including habitat enhancement, individual species management plans, and follow-up evaluation of fisheries management actions.

During its initial meetings the NRTF agreed that before any solutions could be identified, a well-documented evaluation of the existing Norris Reservoir fishery was needed. In 1992, a week-long sampling effort involving various agencies and anglers was implemented to establish baseline data for evaluation of future management strategies; similar sampling efforts have taken place annually since then. This massive cooperative effort was the benchmark for the partnership approach to improving the reservoir fishery. It also led to a succession of cooperative efforts, one of which has been the development of a habitat enhancement program. These cooperative ventures, the annual data collection activity and the habitat enhancement program, will be discussed in detail in the following sections.

Data Collection Program

Discussions by the NRTF indicated that information on the basic biology of several important game fish species, as well as on the abundance of the forage base, was lacking. To aid in the development of management plans, the NRTF decided to implement a data collection program which would integrate university and agency researchers, as well as interested anglers, in sampling activities. Work plans were developed that outlined the data collection and analysis protocols, and field collections were initiated in August of 1992.

Norris Reservoir has two major tributaries, the Clinch and Powell Rivers. It extends 116 km up the Clinch River and 90 km up the Powell River. It has 1,288 km of shoreline with a maximum width of 1.9 km and a surface area of 13,840 ha (TVA 1991). Throughout the year, annual water level fluctuations average 9.2 m. The reservoir reduces to winter pool by January 1 and returns to maximum pool by the end of May; during maximum pool it maintains an elevation of 306 m (Tennessee Valley Authority, unpublished data).

Field collections were divided into two main parts. First, a week-long, intensive sampling was undertaken annually in mid-August. Sampling techniques included cove rotenone applications, electrofishing, gill netting, seining, hydroacoustics, trawling, angling, and determination of primary productivity. Second, the limnetic component of the reservoir fish community was evaluated twice per month. Agency and university personnel and volunteers have conducted evaluations of predator and prey abundances and condition of the fish community in selected limnetic areas of the reservoir since May 1993. Sampling activities included gill netting, hydroacoustics, and trawling.

Funding and personnel for sampling were provided by TWRA, TVA, TTU, and UTK. Over 40 volunteers (anglers and concerned citizens) were involved in daily field tasks, contributing nearly 2,000 h during a 3-year period. These volunteers were active in all data collection activities including operating boats, running gill nets, dipping fish in rotenone treatments, electrofishing, and data entry and validation. Volunteer efforts enabled the NRTF to evaluate many aspects of the aquatic ecosystem that would have been impossible if only agency and university resources had been available.

Evaluations of water quality and primary productivity within selected areas of the reservoir included undergraduate and graduate students and volunteers. Seasonally, students and volunteers obtained water quality and limnological samples twice monthly, including oxygen and temperature profiles and zooplankton tows, at 11 locations throughout the reservoir. Immediately after sampling, a student graphed the oxygen and temperature profiles and furnished them to TVA Printing Services, where the graphs were reproduced and mailed within 3 d to over 300 anglers. This partnership was beneficial for anglers, TVA, and the students. The anglers were able to adjust their fishing efforts with information about oxygen and temperature patterns, TVA was applauded for providing timely information, and the students received academic credit and practical experience for their participation. Several graduate students (TTU, UTK) participated in this effort and used the data collected for Master's theses. Volunteers also were active in collection and entry of water quality and primary productivity data and provided over 700 h of on-site labor.

Habitat Enhancement Program

Discussions by the NRTF suggested the lack of physical habitat might limit spawning success and survival of age-0 fish of some game species, notably black basses and crappies. Also, Thomas (1983) found that anglers on Norris Reservoir who fished around physical structures placed to attract fish caught significantly more fish than those who did not fish attractors. As a result, a fish habitat enhancement program was begun in January 1993. This program was implemented in winter to take advantage of the annual drawdown.

Altogether, 8,329 habitat structures were installed over three winters, including 736 spawning benches, 4,277 brush attractors, 1,051 stake beds (31,530 stakes), 115 tire–brush attractors, 60 sunken wave breakers, and 2,000 planted baldcypress trees *Taxodium distichum*. The spawning structures were constructed and distributed throughout 13 coves in the reservoir. These structures, modified from those described by Hoff (1991), resembled a simple bench approximately 2.4 m long, constructed of two cedar support posts and a 5 × 25–36-cm cedar plank. They were placed in two, three, or four rows, which corresponded to water depths of 3.5–6 m during the spring spawning period. Cover for age-0 fish was cedar and various hardwood trees or branches scattered throughout the coves where the spawning structures were placed. Both the benches and the trees were anchored securely to prevent them from being dislodged during inundation.

Habitat enhancement efforts involved volunteers, TVA, and TWRA. One hundred six volunteers donated over 8,200 h of their time to the project. Habitat enhancement work was performed on 29 weekdays and 3 Saturdays, with an average of 6 volunteers on weekdays and 15 on Saturdays. Volunteers ranged in age from 17 to 70 years and were mostly avid anglers. However, a few individuals who rarely fished contributed as much time as those who fished quite often. Each volunteer was reimbursed for mileage at US$0.21/mile and food at $3.00/d.

The TWRA contributed 491 h of on-site labor, 285 h of cutting and hauling cedar logs, and 12,000 board feet of cedar valued at $3,100. The TVA contributed 388 h of on-site labor, provided project coordination, and preparation. Both agencies provided work boats, tools, and equipment. The total cost for TVA and TWRA salaries, materials, and equipment amounted to US$38,430. Free boat slips valued at $575 were donated by two dock owners.

The response of the volunteers more than met expectations. Much of the work was done in adverse weather, but this did not seem to affect volunteer enthusiasm or participation. This partnership effort served not only to enhance fish habitat and collect needed data, but to build better relationships between anglers and the agencies involved in reservoir operations and fisheries management. Both the anglers and the biologists developed a greater level of respect for the insight that each had into the fishery. Also, friendships grew out of the experience, a clear demonstration of the success of the partnership. And, although there may never be complete agreement on how to best manage the fishery, anglers and biologists were willing to listen to each other rather than simply dismiss the other's points of view.

Another partnership between TVA and UTK was formed to help evaluate fish habitat. The TVA provided vehicles, boats, scuba tanks, and sampling equipment and the UTK students conducted the sampling. Direct observations of smallmouth bass *Micropterus dolomieu* nests under spawning benches and of fish that used brush and stake beds as cover were made by five students who received credit for special project coursework. These students spent 294 h in the field and made 1,086 observations.

Summary

The cooperative approach has provided several valuable products otherwise not possible if only a single agency or university had been involved. First, much needed information was collected that has been essential in the development of the fishery management plans. Second, on-site informal communication took place in which the public and the professional biologists freely interacted and openly discussed the issues at hand. This provided the public a link to the information, and more importantly, provided biologists with insights into the concerns of the public. Third, in addition to academic credit, undergraduate students received valuable experience in real-life applications of fisheries management. Fourth, graduate students were able to collect data for thesis projects, observe fisheries management activities in action, and witness the impact of their efforts.

Volunteers have played a very important role in the successes achieved to date by the NRTF. Their enthusiasm was contagious and their genuine interest in improving the fishery are indicators that future anglers will benefit from their hard work.

We encourage the partnership approach to fisheries management and offer several important keys to a successful volunteer program:

1. Design projects that address real problems that have a high probability of being solved successfully. They should include activities that volunteers believe are worth their efforts.
2. Make sure that a project leader is present at all activities to be an example and to ensure safety.
3. Identify and support a leader among the volun-

teers who will take responsibility for recruiting and notifying other volunteers to fulfill work schedules.

4. Have all equipment and materials present at the site on time and in good working order. Start on time. The volunteers are there to work and will not want to waste time waiting for anything.
5. Give clear instructions as to what will be done, how long one will work, the importance of safety, and agency policies.
6. Give the volunteers as much responsibility as allowable (e.g., driving boats, backing trailers, operating posthole diggers).
7. Make it fun, but have goals that require hard work. At the end of the work day, everyone should feel that a significant amount of work was accomplished.
8. Be sensitive to when it is time to quit for the day. Avoid the tendency to work volunteers a little longer than was agreed.
9. Be sensitive to varying abilities and health of the volunteers and their tendency to work beyond their limits.
10. Let volunteers know they are appreciated. Keep an accurate account of time they work and distance they drive. Throughout the project, express gratitude for their efforts. After the project is completed, reward them with appropriate recognition and awards (e.g., banquet, T-shirts, reimbursement for food and gas expenses, personal letters).
11. Invite the media to feature volunteers at work. Encourage the media to emphasize the role of volunteers in the project.
12. Keep in touch with volunteers between activities. Let them know the results of other activities undertaken for the project.
13. Always speak positively about projects, agencies, and volunteers. Do not let projects become a forum for negativism towards user groups and agencies. Rather, let them be opportunities for building understanding and respect among the varied user and management groups.

References

Hoff, M. H. 1991. Effects of increased nesting cover on nesting and reproduction of smallmouth bass in northern Wisconsin lakes. Pages 39–43 *in* D. C. Jackson, editor. The first international smallmouth bass symposium. Mississippi Agriculture and Forestry Experiment Station, Mississippi State.

Thomas, R. G. 1983. Effects of fish attractors on sport fishing success on Norris Reservoir, Tennessee. Master's thesis. University of Tennessee, Knoxville.

American Fisheries Society Symposium 16:42–46, 1996

Expanding Perspectives on Gaining Public Support for Management

James T. Addis and Betty L. Les

Wisconsin Department of Natural Resources, Division of Resource Management, Box 7921, Madison, Wisconsin 53707, USA

Abstract.—Many societal changes are occurring throughout the world, including decentralization of information through computer networks, shifts from representative democracy to participatory democracy, and reduced reliance on and trust of expert opinions. At the same time, concerns have arisen over biological diversity, resulting in new expectations for the role of resource managers in its conservation. These changes require resource management professionals to expand their perspective on their work, on who their publics are, and on how they can go about gaining constituent support. We describe an ecosystem management decision model developed by the Wisconsin Department of Natural Resources for use in planning management projects and stakeholder involvement. The model requires managers to examine the ecological, socioeconomic, and institutional context of proposed projects as well as their own everyday work. It also requires an analysis of scale from each of the three perspectives. The model is biocentric, viewing humans as part of the ecosystem, and it ties human sustainability to ecosystem sustainability. The proposed model incorporates ideas from total quality management and adaptive environmental assessment. It is a beginning point from which more defined approaches to future ecosystem management can be developed.

Gaining and retaining public support is a key indicator of success for fisheries agencies (McMullin et al. 1991). The traditional fisheries management decision-making paradigm involved bringing representatives from various stakeholder groups together to view a proposed program or project. Frequently, these representatives were organizational leaders, political leaders, or other surrogates trusted by various stakeholder groups to represent their interests. The representatives met with resource managers to examine the expert definition of the problem and the recommended solution. Once an approach was formulated and agreed upon, the stakeholder representatives returned to their constituency and explained (sold) the project to them.

Much has been written over the past 15 years to indicate that this approach may no longer work, beginning with trends analysis by social scientists. Naisbitt (1982) predicted 10 major shifts that would transform our society, predictions that were reaffirmed by Naisbitt and Aburdene (1990). Especially relevant to natural resource managers are the predicted shifts from representative democracy to participatory democracy, from "either–or" to multiple options, and from hierarchies to networking.

In the shift to a highly participatory democracy, individual stakeholders will prefer self-representation, rather than representation by leaders or other surrogates. According to Naisbitt (1982), the guiding principle of this new paradigm is that people whose lives are affected by a decision must be included in the process of arriving at that decision. This is a formidable challenge for fisheries managers, given the number of stakeholders potentially affected by a proposed management action. Several polling efforts and special public involvement activities may be required to make sure all affected parties have been consulted. Having multiple options is often a key to success in this process.

Naisbitt's (1982) predicted shift from hierarchies to networking is rooted in the failure of hierarchies to solve society's problems. Observing that networking begins by circumventing or working around bureaucracy, he predicted that networking would result in the elimination of bureaucracies. Toffler (1980) and Yankelovich (1981) also predicted this trend. Yankelovich called the trend a movement toward an "anarchy of institutions." Fisheries agencies are not the only organizations affected by these trends. The U.S. government, most state governments, and many private companies are being reorganized to flatten their hierarchies and to downsize. Osborne and Gaebler (1992) described ways in which public sector agencies are responding to many of the forces described by Naisbitt.

Naisbitt also predicted the demise of the expert. Today's better educated and better informed participative stakeholder expects to be involved with defining the goals and objectives and with identifying alternative means to achieve them. Resource managers who portray themselves as experts possessing "the answer" face the most opposition from the stakeholders they affect. Escalating product liability, malpractice suits, and other challenges by lay people reflect a reduced reliance on the word of experts.

The social trend toward more inclusive stakeholder involvement is only one of the forces affecting efforts to gain public support for resource management decisions. An equally strong trend has been emerging on another front: biological diversity and the role of resource managers in its conservation. This trend was sparked by public concern for loss of species and awareness of the interconnectedness of life (Moyle and Leidy 1992; Moyle and Williams 1990; Sebens 1994; P. B. Moyle, University of California-Davis, unpublished). Like most professional resource management societies, the American Fisheries Society (AFS) has encouraged introspection and analysis of these issues (Blockstein 1992; Cairns and Lackey 1992; Titus 1992; Upton 1992; Williams and Rinne 1992). These efforts have culminated in a draft policy statement by AFS on the conservation of biological diversity (Winter and Hughes 1995).

A Transitional Model for Decision-Making

As natural resource professionals, our challenge is to deal with all of these trends at the same time. Like many resource management agencies, Wisconsin's Division of Resource Management within the Department of Natural Resources (WDNR) seeks to integrate ecological (functional), socioeconomic (human), and organizational (institutional–jurisdictional) contexts into our management planning. We initially adopted the term "integrated resource management" to describe the way we blend human needs and values with ecosystem capability and sustainability (Peterson et al. 1993). However, over the past 2 years, this approach converged with efforts to develop an agency strategy for the conservation of biological diversity (Addis et al. 1995). Today, we simply call our approach "ecosystem management."

In this approach, we attempt to account for broad ecosystem, human, and institutional connections, even when the project may be focused on a particular component of the ecosystem such as a single lake or reservoir. It is based on acknowledgment of uncertainty and the need to act, and it uses a science-based approach that allows us to continually increase our understanding of ecosystems and adapt our practices as we learn more (WDNR 1995). Thus, our ecosystem management approach is grounded in the principles of adaptive management (Holling 1978; Walters 1986), and the Deming model of continuous quality improvement (Walton 1986), both of which embody systems thinking.

The heart of WDNR's ecosystem management approach is its ecosystem management decision model (Figure 1). It is designed to be used daily by natural resource managers, but it is also useful for specific projects and long-term planning. The model follows the "plan, do, check, act" management cycle. By examining several alternatives and their consequences in ecological, socioeconomic, and institutional contexts in the "plan" phase, the model provides a framework for thinking about projects from their inception. Whatever alternative we decide to implement ("do") is then monitored ("checked") for actual results in all three contexts, and we revise our management ("act") according to those results. The process is continued as is or adapted by making needed changes.

Success lies in carefully considering all three contexts—ecological, socioeconomic, and institutional—and finding an alternative that is acceptable in all three. We cannot consider the ecological context alone in making management decisions. The "correct" ecological decision may be impossible or unworkable given the realities of our society, the economy, and the institutions in which we work. Similarly, the "best" economic decision may be ecologically disastrous. Our hope is to provide an open public process in which social, economic, and ecological perspectives are evaluated and weighed early on, before positions become hard and fast. We search for acceptable alternatives that help preserve long-term options and improve the quality of our everyday lives and the lives of the generations to come.

We do not recommend a cookbook formula for conducting this search for alternatives. Instead, we propose a thought process guided by a series of questions related to each of the three decision-making perspectives: ecological, socioeconomic, and institutional (Table 1). The answers to the questions posed in Table 1 will depend on the specific circumstances or the context in which they are asked. For example, a decision to restore a damaged community is appropriate in some situations but inappropriate (or impossible) in others. Similarly, a decision to stock fish may fulfill a user demand at the expense of native species, or it may replace a missing predator and restore balance to an ecosystem. Few management practices are inherently desirable or undesirable. The appropriateness of any management practice or environmental decision depends on its context.

Consideration of stakeholders is imbedded in the socioeconomic perspective (Table 1). However, before managers can design an effective program for

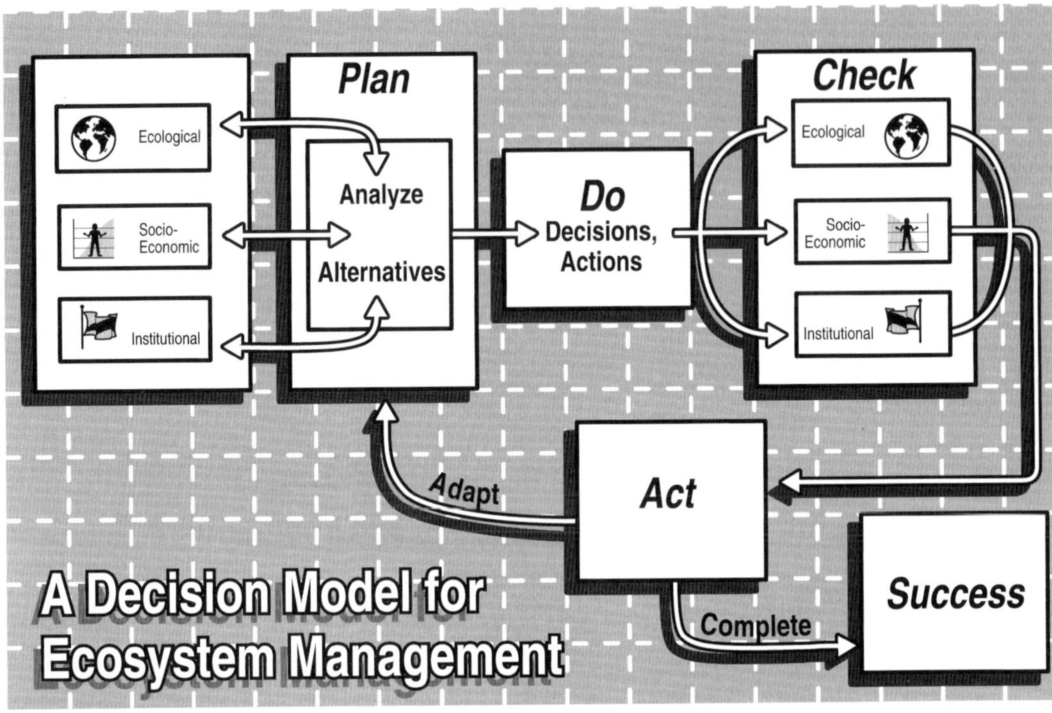

FIGURE 1.—An ecosystem management decision model requiring natural resource managers to integrate three perspectives: ecological, socioeconomic, and institutional (from Addis et al. 1995).

stakeholder involvement, they must examine the scale of the proposed project. This examination of scale should occur from all three perspectives represented in the model. From the ecological perspective, scale examination will help managers determine the appropriate geographic area for analysis based upon biotic and abiotic factors. In many cases, the analysis area will be the watershed, but in some cases it may be more extensive. From a socioeconomic perspective, scale examination will help define the relationship of the proposed project to human endeavors. Managers likely have experienced the surprise resistance to a management action from people who live and work far from the site of the action. Scale must be defined broadly enough to include the ultimate level of impact of our actions. From the institutional perspective, examination of scale may define needs for interagency cooperation and private partnerships.

Thus, the model we are proposing goes beyond the traditional view of an ecosystem, which usually views the human component as a perturbator of the system rather than as an integral part of it. Our proposed model includes humans as just another part of the system. Managers must look beyond the scale of the lake, reservoir, or stream that is the site of the proposed management action, to the scale at which humans operate. By putting humans inside the system, our model also advocates a biocentric approach to ecosystem management, viewing human sustainability as inextricably linked to ecosystem sustainability. In this view, human uses of the ecosystem are not limitless, but rather are constrained by the need to maintain ecological integrity (Stanley 1995).

The ecosystem management decision model complicates efforts to manage public support for agency projects. Moving from a site-specific scale to a larger point of view increases the number and variety of stakeholders who may want to influence the outcome of a given project. Also, a larger scale may put some decision-makers outside the organizational or jurisdictional authority of their governmental unit and require that the project be interjurisdictional. However, designing and scoping a project without comprehensive stakeholder involvement is likely to prolong if not prevent development and implementation of major projects, as dispossessed stakeholders exercise their veto powers (IPP 1978).

TABLE 1.—Implementation framework for the ecosystem management decision model (adapted from Addis et al. 1995).[a]

Context	Relevant questions	Considerations in developing answers
Ecological scale	At what spatial and temporal scales should the decision be defined?	Size of the affected system and its parts Current inventory of what is there, including measures of biological diversity Species with critical needs or special status Presence of ecological gradients and corridors Patterns of community distribution Existing patterns of disturbance or change Current management practices, land use, and land ownership
Capability	What is the system's past, current, and potential future ecological capability? Where is the landscape headed?	Natural and potential capability (e.g., presettlement vegetation as one indicator) History of successional stages Potential for recovery, enhancement, or expansion Potential to be self-sustaining
Function	How can we improve or protect the system's function? How well is it working now?	Interrelationships between abiotic and biotic components; missing components (composition, structure) Connectivity Fragmentation Resilience Gene flow Energy and nutrient flow Food webs
Socioeconomic scale	At what spatial and temporal scale should the decision be defined? What are the magnitudes of social and economic effects?	Number of people affected, directly and indirectly Cost and funding for the project Time period for completion Time period of social or economic impacts Scope of social or economic impacts (local, state, regional, national, international)
Land use	What are the past, present, and potential future land uses?	Current and previous land uses Projected changes in land use Alternative future land uses Indirect effects on adjacent or regional land uses Land ownership patterns
Economic impacts	What are the opportunities and threats for business and employment?	Direct and indirect impacts on local and state businesses Potential new businesses created Impacts on current and projected property values Impact on employment Sources of raw materials or other resources for business needs Relationships to transportation and utility networks Relationship to national and international economy
Stakeholder interests	Who are the stakeholders, and how can they best be involved? What are their various perspectives, needs, and values?	Major and minor stakeholders (public and private) Role of elected officials Public involvement and information strategies Relationship of stakeholders to DNR and to each other Opportunities for partnerships with stakeholders
Institutional legal framework	What institutional support or constraints come from outside DNR? What is DNR's legal relationship to local, state, and federal governments?	Legal authority (statutes and administrative rules) Legal constraints Processes for obtaining decisions from appropriate authorities (local government, legislature, federal agencies) Need for new legal authorities or changes in existing laws State budget development, federal grants, etc.
Internal agency policies	What internal policies or procedures support or hinder a proposed action?	Strategic plans Budget and staffing Manual codes and handbooks
Interagency cooperation and private partnerships	What kinds of relationships with other public agencies or with private interests are needed?	Existing and potential partners and cooperators Institutional constraints of partners and cooperators, such as their ability to enter into long-term agreements Innovative approaches to cooperation and partnership

[a]DNR refers to the Wisconsin Department of Natural Resources, in which this framework was developed. The name of any natural resource management agency could be substituted.

By broadening the scope of the management paradigm we hope to provide more inclusive processes and strategies. The role of stakeholders has begun to change from that of constituents to that of partners—who may actually carry out the project as their own, or who may work as full decision-making partners to implement the project. In the words of Osborne and Gaebler (1992) we must move from rowing to steering wherever we can. This will mean a transformation in how managers view their role, a transformation from resource managers to resource management leaders. Effectively engaging, leading, and partnering with stakeholders will be key determinants of success.

Conclusions

What does all this mean to fisheries managers? First, we must view our efforts to implement ecosystem management as a journey rather than a destination, and we must recognize that we are only at the beginning of the journey. Secondly, we must recognize that most of the skills and tools we will need to manage ecosystems successfully are not in our present toolbox. Our training and experience are at different scales, and historically we have operated within a limited professional perspective. Fortunately, the fisheries and resource management literature is replete with authors seeking to understand these emerging ideas and to apply them to their work. Symposia like this one are important to the process of transforming our profession. Day-to-day management during this transition will be ambiguous and messy. Like any period of organizational change, we must blend comfortable old processes with experimental new ones.

References

Addis, J., and thirteen coauthors. 1995. Wisconsin's biodiversity as a management issue: a report to Department of Natural Resources managers. Wisconsin Department of Natural Resources, Madison.

Blockstein, D. E. 1992. An aquatic perspective on U.S. biodiversity policy. Fisheries 17(3):26–30.

Cairns, M. A., and R. T. Lackey. 1992. Biodiversity and management of natural resources: the issues. Fisheries 17(3):6–10.

Holling, C. S., editor. 1978. Adaptive environmental assessment and management. Wiley, New York.

IPP (Institute for Participatory Planning). 1978. Citizen participation handbook for public officials and other professionals serving the public. Institute for Participatory Planning, Laramie, Wyoming.

McMullin, S. L., S. R. Amend, and L. A. Nielsen. 1991. Managing information about how we are managing: multiple perspectives on the factors that determine agency effectiveness. Transactions of the North American Wildlife and Natural Resources Conference 56:162–68.

Moyle, P. B., and R. A. Leidy. 1992. Loss of biodiversity in aquatic ecosystems: evidence from fish faunas. Pages 127–169 *in* P. L. Fielder and S. K. Jain, editors. Conservation biology: the theory and practice of nature conservation, preservation and management. Chapman and Hall, New York.

Moyle, P. B., and J. E. Williams. 1990. Biodiversity loss in the temperate zone: decline of the native fish fauna of California. Conservation Biology 4(3):275–284.

Naisbitt, J. 1982. Megatrends, ten new directions transforming our lives. Warner, New York.

Naisbitt, J., and P. Aburdene. 1990. Megatrends 2000, the next 10 years . . . major changes in your life and world. Pan Books by Sidgwick & Jackson LTD, London.

Osborne, D., and T. Gaebler. 1992. Reinventing government, how the entrepreneurial spirit is transforming the public sector, from schoolhouse to statehouse, city hall to the Pentagon. Addison Wesley, Reading, Massachusetts.

Peterson, T. L., L. E. Claggett, and J. T. Addis. 1993. Managing watersheds for fisheries and wildlife: an integrated approach to natural resource management. Transactions of the North American Wildlife and Natural Resources Conference 58:296–310.

Sebens, K. P. 1994. Biodiversity of coral reefs: what are we losing and why? American Zoologist 34:115–133.

Stanley, T. R. 1995. Ecosystem management and the arrogance of humanism. Conservation Biology 9:255–262.

Titus, T. R. 1992. Biodiversity: the need for a national policy. Fisheries 17(3):31–34.

Toffler, A. 1980. The third wave. William Morrow, New York.

Upton, H. F. 1992. Biodiversity and conservation of the marine environment. Fisheries 17(3):20–25.

Walters, C. 1986. Adaptive management of renewable resources. Macmillan, New York.

Walton, M. 1986. The Deming management method. Putman, New York.

Williams, J. E., and J. N. Rinne. 1992. Essay: biodiversity management on multiple-use federal lands: an opportunity whose time has come. Fisheries 17(3):4.

Winter, B., and R. Hughes. 1995. AFS draft position statement—biodiversity. Fisheries 20(4):20–26.

WDNR (Wisconsin Department of Natural Resources). 1995. Science-based resource management in the Wisconsin Department of Natural Resources. Wisconsin Department of Natural Resources, Madison.

Yankelovich, D. 1981. New rules, searching for self-fulfillment in a world turned upside down. Random House, New York.

American Fisheries Society Symposium 16:47–53, 1996

Multiple Resource Management on an Interjurisdictional River

KENT D. KEENLYNE
U.S. Fish and Wildlife Service, 420 South Garfield Avenue Pierre, South Dakota 57501–5408, USA

Abstract.—A historical review of attempts to develop interjurisdictional resource management on the 3,768-km-long Missouri River is presented. Resource concerns are multijurisdictional and involve a broad spectrum of issues, including such things as water rights, habitat needs for rare and endangered species, and management of sport fisheries in a changing environment. These concerns are superimposed on a mixture of state and federal laws that presuppose removal of water from the river under rules of western water law. This paper describes formation of an interjurisdictional committee to consider fish and wildlife resource issues, including discussion of the committee's operation, results to date, and reflection on the significance of such a group in future deliberations on the river.

The Missouri River has a long history of conflict. The river has been described as a river of change, noted for its tendency to erode banks and seek new beds much as would a "traveling man" (Jackson and Stucky 1988). The river has been a source of human conflict, first in the struggle to make the river more domesticated and then to control its vagaries during years of flood or drought, and now in battles over rights to use its waters. The continual conflict between upper- and lower-basin interests and emerging conflicts over tribal rights to water and river resources have led to the river being characterized as both the seam that ties people together and the tear that keeps them apart (Keenlyne 1988).

The Missouri River is one of the most highly federalized rivers in the United States. The river is considered navigable from the headwaters in southwestern Montana (Figure 1) to the mouth near Saint Louis, Missouri, so development activities fall under U.S. Army Corps of Engineers (Corps) permit requirements for most of its length. The Environmental Protection Agency (EPA) also scrutinizes the river under the Clean Water Act (33 U.S.C. §§ 1251–1376). A third of the river is impounded by seven federal dams and a third is channelized as a federal navigation project; flows in the remaining reaches are regulated by releases from the federal dams by the Corps (Hesse et al. 1989b). There are also three designated Wild and Scenic River sections on the Missouri River: a 240-km section above Fort Peck Reservoir, a 63-km section below the Fort Randall Dam, and a 93-km section below the Gavins Point Dam. The uppermost section is managed by the Bureau of Land Management, whereas the lower two units are managed by the National Park Service. All three units qualified for inclusion under the Wild and Scenic Rivers Act because of their outstanding recreational and wildlife values.

There are also 19 Native American tribes that claim rights to water from the Missouri River (Guhin 1985). Although tribal allocation of water must come through the state appropriation process under western water law, the treaty relationship of the federal government with the individual tribes assures strong federal agency trust responsibility roles in what is perceived as the eventual (and inevitable) dividing of the waters on the Missouri River (Guhin 1985).

Dieffenbach (1948) explained that the development of the Missouri basin for power generation, irrigation, navigation, and flood control were all commendable goals, but he warned that development would be at the expense of fish and wildlife. Lord et al. (1975) indicated that water allocation would eventually be the most critical issue facing fish and wildlife on the Missouri River. Keenlyne (1993) observed, however, that there is not now, and never has been, enough water in the system to provide for all the benefits supposedly promised in the Pick–Sloan development plans of 1944, which serve as the basis for basin development. Challenges to the wisdom of depleting the Missouri River for water development or the substance for any such mandate to do so have been made only recently (Keenlyne 1988).

Lord et al. (1975) explained that full consideration of the implications of water development or allocation depend on the availability of information and the nature of the decision-making process itself. The purpose of this paper is to describe efforts to merge governmental interests in obtaining coordinated information on Missouri River resources and the means being taken to assure that biologists will participate in future decision making.

FIGURE 1.—The Missouri River basin, showing federal dams and head of navigation at Sioux City, Iowa. The abbreviation RK is river kilometer from the mouth.

Study Area

The 3,768-km-long Missouri River is described as an integration of eastward-flowing streams that were diverted by advancing lobes of glaciers about 115,000 years ago (Hesse 1987). At one time, the upper basin emptied into the Hudson Bay, but its present course generally outlines the southern extent of the Pleistocene ice sheet (Hesse et al. 1989b). Western feeder streams cross vast expanses of highly erodible, unconsolidated sedimentary materials and its eastern streams flow through erodible glacial till providing the substance that gives the Missouri River its nickname of "Big Muddy" (Keenlyne 1994).

The river arises in the mountains of southwestern Montana in an area of high winter precipitation, flows across the semiarid plains, and then through a zone of high summer precipitation. The river has two periods of high water: the March rise accompanies the usual snowmelt from the prairies, and the June rise consolidates mountain snowpack melt with spring rains in the lower basin (Hesse et al. 1989b). The Missouri River valley ranges from 2.4 km in width in the mountainous headwaters to more than 27.4 km in the downstream portion of the basin (Hesse et al. 1989b).

This diverse series of geomorphic and hydrologic features, which Vannote et al. (1980) described as a river continuum, leads to a river with unusual habitat diversity and species richness. At least 156 native fish species have been found in the Missouri River (Hesse et al. 1989a), along with 12 exotic fish species (Hesse et al. 1989b). Plants, birds, insects, and other forms of life that use the river are similarly diverse in relation to the wide array of habitats found (Keenlyne 1994).

This river's development has a long history. Snagging and clearing of trees began in 1838, the first dam was built on the river in 1909, and channelization of the lower river started in 1912 (Hesse and Sheets 1993). Hesse and Sheets (1993) termed most of the earlier works largely "ephemeral" in relation to effects on wildlife resources because the river was quick to rework them and continue with its business of transporting water and sediments to the Mississippi River. The Missouri River became forever changed after passage of the Flood Control Act of 1944, which authorized the construction of huge federal dams on the upper river and its tributaries according to the Pick–Sloan plan, and a similar omnibus act for water development in 1945, which authorized a 2.8-m navigation channel and levee system from Sioux City, Iowa, to the mouth (Hesse 1987).

Construction of the dams under the Pick–Sloan authorization began in 1946 and the system first filled to operating levels in 1967 (Sveum 1988). The upper three Corps reservoirs (Fort Peck, Garrison, and Oahe) rank respectively as the third, fourth, and fifth largest storage reservoirs in the United States falling behind only the Hoover and Glen Canyon dams on the Colorado River system (Sveum 1988). The commercial navigation project on the lower Missouri River extended to Omaha by 1955, to Sioux City by 1964, and was not considered complete until 1980 (Sveum 1988).

Cumulatively, the various construction projects have created tremendous changes in the river. River silt loads have been reduced by 81%, whereas sand fractions have increased by 260% (Hesse et al. 1989b). Thermal stratification, which now occurs in the Oahe, Garrison, and Fort Peck reservoirs, alters water temperatures below these dams because water is released from the middle of the water columns (Hesse et al. 1989b). In addition, water is captured in spring and released in summer to support navigation, a practice that eliminates the normal spring rises over most of the river and affects not only the trigger for fish spawning but also the cycles of insect life, plant growth, and plant reproduction (Hesse et al. 1989a). Morphometrically, the river has lost nearly all sandbars, sloughs, chutes, and backwaters through either damming, the channelization works, or ongoing degradation of the channel below the dams (Hesse 1987). We are just beginning to recognize the long-term implications of structural changes to the future ecology of the river (Hesse 1987).

Loss of productivity has also occurred as a result of development. The loss of floodplain plant materials through inundation by the reservoirs and disconnection of the river from its remaining floodplain by eliminating most overbank flows together have reduced organic matter to the river by 65% (Hesse et al. 1989b). In the lower river, benthic habitat has been reduced by 67% and the standing crop of drift by 88% (Morris et al. 1968). More than 21% of the native fish species are listed as rare, endangered, threatened, or of special concern by the seven states through which the Missouri River flows (Hesse et al. 1989a); more than 50 species of plants and animals that utilize the Missouri River are either listed under the federal Endangered Species Act (16 U.S.C. §§ 1531–1543) or are in one of the federal review categories, and the list is expected to continue to grow if the downward trend in habitat condition is not reversed (Keenlyne 1994).

Structure and Function of the Interjurisdictional Organization

Dieffenbach (1948) predicted that dams on the Missouri River could seriously reduce the range of large river fishes, such as sturgeons *Scaphirhynchus* spp. and paddlefish *Polyodon spathula*, that backwaters could dry up because of the dams and levees, and that disturbing the water cycle could have serious effects on the life processes of a variety of native species. He also predicted that, after an initial boom in production, sport fish would decline in the reservoirs and that extreme fluctuations would impose severe limitations on many species. Unfortunately, his predictions came true.

Politically, little was done (or could have been) to curb the resource problems predicted by Dieffenbach, because the Missouri River development program was seen as a nationally critical effort to put this country back to work after World War II (Keenlyne 1993). Even though fish and wildlife were considered authorized purposes for such projects, these resources were ignored for reasons of expedience during construction of the huge public works (Hesse 1987). Under the circumstances, the best biologists could do was to document changes and wait for a better time to implement biological considerations.

Biologists in the basin heeded Dieffenbach's predictions, however, and began to document the impacts that development was having on fish and wildlife. Most of the responsibility for such assessments fell on state fish and wildlife agencies even though the river is highly federalized; exceptions were a Bureau of Commercial Fisheries station in Mobridge, South Dakota, which operated in the 1960s, and the now-defunct North Central Reservoirs Investigation (NCRI) teams that were formed in 1962 to study the four reservoirs constructed in South Dakota (Nelson and Walburg 1977). The NCRI stations were disbanded in the late 1970s, and the task of monitoring has become a responsibility of the states again.

The first formal attempt to coordinate fish and

wildlife management concerns occurred under the auspices of the American Fisheries Society (AFS) through formation of an Upper Missouri River AFS chapter in 1963. This chapter included state, federal, and university fishery personnel from Montana, North Dakota, South Dakota, and Nebraska. The chapter formed an ad hoc committee in 1969 specifically to monitor and coordinate work on the river and to provide fishery recommendations to the Corps Reservoir Control Center, which operated the system (C. Berry, National Biological Service, personal communication). The group met with the Corps each spring to provide fishery recommendations on pool elevations and project releases. Coordinated recommendations ceased once Nebraska and Montana formed their own AFS chapters.

The formation of the next coordinating group, the Missouri River Natural Resources Committee (MRNRC), is explained in its first annual report (MRNRC 1987). In 1985 a U.S. Fish and Wildlife Service (USFWS) representative recommended to the Corps at a public meeting on operations that multiyear operation plans (rather than annual plans) would better serve the river system's natural resources. The Corps requested the USFWS to help organize a meeting with state resource agencies in hopes of developing a process for obtaining coordinated resource recommendations for operations. The Corps observed that the existing process through the AFS frequently produced recommendations that conflicted with official agency positions and that what was needed was a formal process involving official agency delegates.

The USFWS and Corps personnel met in 1985 with official resource delegates of the seven states along the Missouri River main stem (Montana, North Dakota, South Dakota, Nebraska, Iowa, Kansas, and Missouri) to explore organizational options. After several presentations on different types of coordinating organizations, state agencies selected an organization similar to the Upper Mississippi River Conservation Committee. Because a strong federal presence existed on the river, the states asked the federal agencies to provide funding and staff for a nonvoting coordinator position. The Corps agreed to provide funds to the USFWS which would then provide the coordinator. The coordinator serves as staff to the committee under the discretion of the chairperson through a memorandum of understanding signed by the MRNRC and the USFWS.

The MRNRC is run by an Executive Board made up of one officially designated delegate from each state fish and wildlife resource agency. Delegates represent the Director and continue their appointments until replaced, in writing, by their respective agency head. Chairmanship is rotated annually among state delegates. The Chairman makes all statements and signs all letters of position on behalf of the MRNRC. All MRNRC actions are performed under a Constitution and Bylaws signed by the seven state agency heads or their designees.

The MRNRC also has nonvoting ex-officio and cooperating-entity delegates. Agencies must request to join and the official delegates then decide whether the agency will be an ex-officio member (an agency that contributes significant resources to the management of fish and wildlife resources) or a cooperating agency. The Corps has two ex-officio delegates (one from the Reservoir Control Center and one from the divisional environmental staff) and the USFWS also has two ex-officio delegates (one from each of its two regions along the river). The Western Area Power Administration joined the group as a cooperating agency in 1990 and became an ex-officio member in 1994. The Bureau of Reclamation became a cooperating agency in 1992 and the National Park Service in 1993. Although all 26 of the basin tribes have been invited to attend annual meetings, none have asked to become members.

The MRNRC can also form temporary or standing committees to address specific issues. In 1990 a Rare and Endangered Fish Subcommittee was formed. It became a standing Fish Technical Section in 1993. A Least Tern and Piping Plover Management Team was formed in 1992 and it became a standing committee in 1993. Members of technical committees are also officially designated by their respective agencies. Technical committee chairs can be either from voting or from nonvoting entities and are appointed by the MRNRC chairperson.

Multijurisdictional Accomplishments

The Pick–Sloan plan for the Missouri River was passed by Congress in 1944. In the 50 years since, have we overcome the problem identified by Dieffenbach (1948) and his contemporaries of not having the opportunity to express concerns about the impact of river development on fish and wildlife resources? Have we overcome the problems outlined by Lord et al. (1975) of being able to adequately express the full impact of river development and water allocation on fish and wildlife and to make effective inroads into the decision-making process? Has the formation of the MRNRC solved all our interjurisdictional resource problems and provided us with an avenue to assure that fish and wildlife will receive equal consideration in water management and water allocation decisions on the Missouri River in the future? The answers are "no," but we are better off than we used to be.

We have made progress. State biologists have observed that federal environmental laws can be useful tools with which to influence the decision-making process of federal agencies that control management of the Missouri River (Hesse 1987; Hesse et al. 1989a). More importantly, federal laws can be useful for meeting management goals rather than simply serving as a club to force submission. For example, having a federally listed species not only authorizes federal agencies to expend federal dollars to further the cause of the species, but it allows construction and permit-approving agency biologists to incorporate biological recommendations to further the needs of the listed species. In the case of the Missouri River, the endangered-species involvement also opened the door to another federal process: preparation of an environmental impact statement on the master manual for operating the system. This document will provide the first opportunity for the public to see how the system is actually operated and for what purposes; it also provides the first real opportunity for biologists to provide scientific information on how system operations affect fish and wildlife.

What remains to be done to assure environmentally sensitive river management? Federal policy and law already mandate that fish and wildlife should be on an equal par with other management considerations (Hesse 1987; Sheer 1988), but we still do not have a systems approach to protecting the health of this multijurisdictional river and all its resources (Hesse et al. 1989a). Has the federal government—or the Corps, which actually decides how water is released and regulated on the river—done a bad job? Not necessarily so. But, as Sheer (unpublished) states, it is an improper role for the Corps or any other executive agency to allocate the benefits of multipurpose projects; these decisions should be left to policymakers and the public. The need for navigation still drives the decision-making process for operating the system (D. Sveum, U.S. Army Corps of Engineers, unpublished). Until the public (through its political representatives) wrests the decision from the Corps on how multipurpose benefits are to be allocated, the job of the fishery and wildlife biologist remains unfinished.

Lessons Learned

What are the lessons we have learned in our efforts to manage a multiresource, interjurisdictional river? Forming broad-based groups such as the MRNRC is a critical first step for success; only through such organizations can biologists look at the river as a whole and have a formal process for discussing issues and making decisions about the long-term needs of the river. A uniformly supported recommendation has a better chance of being adopted than does a group of mixed messages. In 1995 the recommendations of this group were accepted by the Corps in developing its annual operating plan for the main-stem reservoir system. The recommendations included additional spring flows above navigation targets to emulate the natural hydrograph and means to institute multiyear reservoir operations. This feature to benefit fish and wildlife will become incorporated in the new master manual for operations.

We have also learned some lessons about what is needed to make such a group function effectively. Official status is necessary so that the representative can speak for the respective agency. Similarly, designation should be in writing by the agency director and any change in delegation authority should also be in writing to avoid confusion and inability to reach a quorum if a vote is needed. A charter (constitution and bylaws), signed by each of the voting members, is an important early step to provide purpose (or direction) and to place sideboards on the general mode of operation. Having a single agency representative forces coordination within an agency and helps avoid situations where individuals attempt to protect only their primary program interests. Rotation of leadership is advised because it helps raise the level of understanding of more people and spreads the burden of responsibility.

Maintaining options to create subcommittees (technical committees) is also important because executive board members seldom have time (or may lack expertise) to delve into all issues that come before the group. The subcommittees examine issues in greater detail and make reasoned recommendations to the executive board, which is in a position to make commitments on the part of the agencies. Subcommittees also serve as a means of helping more individuals learn about complex multijurisdictional resources and serve as a training ground for future leadership.

The function of the coordinator needs to be well understood when the position is created. The coordinator plays a key role in maintaining continuity and usually serves as the "organizational memory" for procedural processes and past positions taken by the group. Consequently, to remain effective, the coordinator should not represent or function as a spokesperson for any agency. If the organizational chairperson rather than the coordinator signs all correspondence and makes presentations on positions of the group, much confusion can be avoided. This is particularly true if the coordinator happens to be an agency employee. One lesson the MRNRC learned the hard way is that the commitment of time for a federally supplied coordinator should be spelled out in a memorandum of agreement signed by both the employee's sponsoring agency and any cooperating agency supplying funds. The coordinator becomes the principal staff of the organization and needs sufficient time to perform committee functions.

Peterman (1988), an original charter delegate to the MRNRC, noted that the ultimate success of a biologist's efforts to obtain resource protection was not in development of technical information but in social acceptance of management goals. Unfortunately, gaining public acceptance for positive resource actions is much more complex than simply providing factual information on impacts. On the Missouri River, we face a long list of highly respected untruths and volumes of officially certified, politically motivated, nonfacts. An interjurisdictional organization such as the MRNRC can do a great service to the public by preparing coordinated informational materials on present conditions of the resources, explaining perspectives on how resources change and what yearly fluctuations do to resource abundance, and providing educated judgments on future trends in the resources that reflect such ongoing processes as sedimentation and eutrophication. Coordinated efforts in preparing media releases, informational articles in state natural resource publications, and tying into the federal communication networks can help the public understand and prepare for the future of their resources.

The question about who votes as official members depends on the purpose of the group. In the MRNRC, each state game and fish agency has one vote. Federal agencies should be involved in information exchange to smooth the way for developing positions that may be feasible, but federal employees who have nonvoting status may be able to discuss resource options that do not represent the "official" position of their agency. Federal agencies often follow directives that conflict with state positions or those of other federal agencies. Because state game and fish agencies share interest, they are more likely to reach consensus. Tribes have not chosen to join the MRNRC as yet because many do not have sufficient staff or resources to participate fully whereas others are actively challenging water rights issues (which can include fishery resources) and are legally advised not to compromise any jurisdictional rights by participating in the MRNRC.

A sanctioned institutional arrangement that includes state, tribal, federal, and public entities that share the river is the critical feature still lacking in solving interjurisdictional problems on the Missouri River (Keenlyne 1988). This river may well remain a river of conflict for years to come. The future of natural resources looks brighter, however. Biologists are now organized and the public is beginning to understand that everything in the ecosystem is interrelated. We now understand that rivers are the canaries of society and what happens in the hinterlands will eventually show up in our rivers. As a result of studies initiated after the Missouri and Mississippi rivers "Flood of '93," proposals on how we manage our nation's floodplains are under way, including possible establishment of a more comprehensive Water Resource Council. Such a revised council may provide the vehicle for developing the institutional arrangement necessary to address the interjurisdictional issues on large rivers such as the Missouri and serve as a forum that eventually develops instructions on how multipurpose benefits from this vast resource are to be allocated. It will be critical for organizations like the MRNRC to be included in resource deliberations of such basin councils and for them to share with the public what has happened to their rivers and the fish and wildlife that depend on them and what the future of the resources could be.

References

Benson, N. G., editor. 1988. The Missouri River, the resources, their uses and values. American Fisheries Society, North Central Division, Special Publication 8, Bethesda, Maryland.

Dieffenbach, R. 1948. River development programs and their relationship to fish and wildlife resources. Journal of Wildlife Management 12:96–104.

Guhin, J. P. 1985. The law of the Missouri. South Dakota Law Review 30:346–487.

Hesse, L. W. 1987. Taming the wild Missouri River: what has it cost? Fisheries 12(2):2–9.

Hesse, L. W., G. R. Chaffin, and J. Brabander. 1989a. Missouri River mitigation: a system approach. Fisheries 14(1):11–15.

Hesse, L. W., and six coauthors. 1989b. Missouri River fishery resources in relation to past, present, and future stresses. Canadian Special Publication of Fisheries and Aquatic Sciences 106:352–371.

Hesse, L. W., and W. Sheets. 1993. The Missouri River hydrosystem. Fisheries 18(5):5–14.

Jackson, J. P., and N. P. Stucky. 1988. History relived on the mighty Missouri. Pages 3–12 *in* Benson (1988).

Keenlyne, K. D. 1988. Economic development from Lewis and Clark to Pick–Sloan. Pages 33–37 *in* Benson (1988).

Keenlyne, K. D. 1993. Historical overview of the upper Missouri River and its tributaries. Pages 1–6 *in* K. F. Higgins and M. R. Brashier, editors. Proceedings, the Missouri River and its tributaries: piping plover and least tern symposium. South Dakota State University, Brookings.

Keenlyne, K. D. 1994. Conserving a river ecosystem—a Missouri River partnership. Pages 655–664 *in* Proceedings watershed '93—a national conference on watershed management. U.S. Environmental Protection Agency (840-R-94-002), Washington, DC.

Lord, W. B., S. K. Tubbesing, and C. Althen. 1975. Fish and wildlife implication of upper Missouri basin water allocation. University of Colorado, Institute of Behavioral Science, Program on Technology, Environment, and Man. Monograph 22, Boulder.

Morris, L. A., R. N. Langemeier, T. R. Russell, and A. Witt, Jr. 1968. Effects of main stem impoundments and channelization upon the limnology of the Missouri River, Nebraska. Transactions of the American Fisheries Society 97:380–388.

MRNRC (Missouri River Natural Resources Committee). 1987. First annual proceedings of the Missouri River Natural Resources Committee. MRNRC, Pierre, South Dakota.

Nelson, W. R., and C. H. Walburg. 1977. Population dynamics of yellow perch (*Perca flavescens*), sauger (*Stizostedion canadense*), and walleye (*S. vitreum vitreum*) in four main stem Missouri River reservoirs. Journal of the Fisheries Research Board of Canada 34:1748–1763.

Peterman, L. 1988. Montana's program for stream preservation in the Missouri River drainage. Pages 99–105 *in* Benson (1988).

Sheer, D. 1988. Customer involvement in reservoir operations. Pages 5–9 *in* A review of the Missouri River main stem system operation. Missouri Basin States Association, Topeka, Kansas.

Sveum, D. J. 1988. Regulation of Missouri River main stem reservoirs. Pages 51–66 *in* Benson (1988).

Vannote, R. L., G. W. Minnshall, K. W. Cummins, J. R. Sedell, and C. E. Cushing. 1980. The river continuum concept. Canadian Journal of Fisheries and Aquatic Sciences 37:130–137.

American Fisheries Society Symposium 16:54–63, 1996

Natural Resource Management and Leadership in Public Arena Decision Making: A Prescriptive Framework

STEVE L. MCMULLIN

Virginia Polytechnic Institute and State University, Department of Fisheries and Wildlife Sciences
Blacksburg, Virginia 24061-0321, USA

Abstract.—Finding the proper balance between biology and public involvement and then effectively involving the public in making natural resource management decisions are among the most difficult tasks facing resource managers. The training of resource managers and the culture of their profession result in strongly held values and beliefs that biological and ecological principles, rather than the human dimension, should govern management of resources; however, constituents expect to participate in making management decisions. Adaptation of the Vroom–Yetton model of leader decision-making styles, a contingency model of leadership, offers a normative framework that may be useful for resource managers in making management decisions about natural resources. To use the model, managers answer "yes" or "no" to seven questions, thus identifying various types of problems and the most appropriate public involvement strategies for each. I discuss application of the model to resource management decisions commonly faced by reservoir managers, including choice of sites for reservoir access, stocking density of sport fishes, preservation of shorelines and river corridors in desirable states, and development of a reservoir management plan when one or more interest groups strongly oppose the manager. The Vroom–Yetton model is intuitively appealing because it encourages managers to recognize the different contexts in which resource management problems must be considered and to use a variety of public involvement techniques rather than only one or two techniques for all situations. Although empirical studies support the Vroom–Yetton model in organizational contexts, studies of its application to public involvement in natural resource management are needed.

Determining the appropriate role of the public in making natural resource management decisions and finding the most effective means of public involvement have always been among the most challenging problems facing resource managers. Successfully resolving these problems to achieve a balance between biological and human dimensions is critical to achieving management effectiveness. Two of the eight common characteristics of effective fish and wildlife agencies described by McMullin (1993a) are (1) having open, responsive relationships between managers and their publics and (2) performing effectively in the political arena while maintaining an image of nonpolitical, scientific professionalism. In this paper, I describe the roots of the resource manager's struggle with effective public involvement and present a normative framework for assisting managers in making public-arena resource allocation decisions.

Most agencies responsible for management of fish and wildlife resources in the United States were established on the principles of scientific resource management. Scientific resource management evolved in the United States during the late nineteenth century as a combination of the sustained-yield silviculture developed in Germany and the organizational efficiency espoused by Frederick Winslow Taylor (Flader 1976). Gifford Pinchot was one of the best-known proponents of scientific resource management; the U.S. Forest Service, which he founded, served as a model for many scientific resource management agencies formed later (McGeary 1960). Although scientific resource management does not preclude incorporation of human dimension considerations into decision making, it traditionally has emphasized biological factors more than social, economic, and political factors.

In marked contrast to contemporary views of government, the prevailing view during the Progressive Era (late nineteenth and early twentieth centuries) was that government experts could best manage resources in the public interest. Scientific resource management prescribed the placement of decision-making authority in the hands of trained professionals. As fish and wildlife resources were becoming depleted in the early part of this century, the public began to abandon the economic and politically driven resource management that was common at the time and to trust fish and wildlife professionals to bring populations back to healthy levels. Recovery of white-tailed deer *Odocoileus virginianus* and wild turkey *Meleagris gallopavo* are often cited as evidence of the success of scientific resource management (Downing 1987; Lewis 1987).

At least as early as the 1950s, natural resource managers began showing concern about public acceptance of management programs (Huber 1953; Lively 1953). The proposed solution was nearly always the same—educate the public about scientific principles of fish and wildlife management to generate acceptance. However, the relationship between professionals and their publics began to change dramatically in the 1960s, when President Johnson's Great Society initiatives mandated public involvement in formulating programs (Rich and Rosenbaum 1981). Environmental legislation of the 1970s mandated significant public involvement in setting goals and choosing options for natural resource management. At the same time, rapid growth of single-issue interest groups was important in changing the relationship between agencies and their publics (Langenau 1982).

Evolution of public involvement in fisheries management paralleled that in other natural resource fields. Nielsen (1976) traced the evolution of fisheries management from its historical beginnings, through management for maximum sustained yield (traditional scientific resource management), to more recent efforts to achieve optimum sustained yield. The latter, by definition, incorporates social and economic concerns and, in practice, also encompasses political concerns. The optimum-yield concept was formally acknowledged in the language of the Magnuson Fishery Conservation and Management Act (16 U.S.C. §§ 1801–1882), which directed regional marine fishery management councils to consider not only biology, but also a variety of other factors, in making management decisions.

Evolution from scientific resource management to more meaningful public involvement is forcing many natural resource managers to face difficult philosophical adjustments. The training of most resource managers and the culture associated with natural resource management professions reinforce the scientific resource management philosophy (Kennedy 1985). Human dimensions and natural resource policy are being incorporated into more university resource management curricula, but most graduates quickly discover that they need more training in these areas (Adelman et al. 1994). Although employees of effective state fish and wildlife agencies believed constituents should have a major role in setting management goals, they were less inclined to believe constituents should have a major role in developing or selecting management alternatives (McMullin 1993b). Magill (1988) described resource managers as "reluctant public servants" because their values, based in scientific resource management, often prevented them from meaningfully involving the public in making decisions.

Checkoway (1981) documented the shortcomings of public meetings as a method of involvement, but he also noted that such meetings are required at all levels of government, are increasing in number and use, and are often the only participation method employed. Many natural resource managers, tired of the "management in a fishbowl" approach (Tipple and Wellman 1989), have turned to alternative forms of public involvement (Gasson and White 1993); however, lack of knowledge or training in the techniques of public involvement often compounds resource managers' problems. When an alternative technique results in some success, it is often applied to a wide variety of situations. When that technique is not as successful in other situations as it was in its original use, frustration with public involvement increases. The net result of growing public expectation of meaningful involvement in making difficult resource allocation decisions, combined with limited knowledge of public involvement techniques, has left many resource managers wondering if they should be leaders in the decision-making process or if they should be responding to public opinion.

A potential solution to the resource manager's public involvement quandary may be found in the social science literature on leadership. Leadership and public involvement have much in common. Both resource managers and organizational leaders have authority and responsibility to make decisions. Both resource managers and organizational leaders have learned that participation in making decisions can increase commitment (by the public or by organization members) to those decisions (Angle and Perry 1981; Zimmerman and Rappaport 1988). Contingency theories of leadership, which prescribe adaptation of leader behavior to the contingencies of different situations, should prove useful to resource managers. In the remainder of this paper, I discuss adaptation of one such contingency theory, the Vroom–Yetton model of leader decision-making styles (Vroom and Yetton 1973), to public involvement in making resource management decisions.

The Vroom–Yetton Model and its Application to Public Involvement

The Vroom–Yetton model focuses on one aspect of leadership, involving subordinates in making decisions. Vroom and Yetton (1973) identified five decision-making styles and developed a decision tree that leads the user to a prescription for one or more of the styles for each of 13 different problem types. Problem types are defined by answering "yes" or "no" to a series of seven questions (the decision tree). Subsequent revision of the model added questions and expanded the range of responses, generating a more complex decision tree that requires use of a computer to reach a recommended course of action (Vroom and Jago 1988). The simpler version of the model is adapted here.

I have renamed and adapted the five decision-making styles identified by Vroom and Yetton (1973) to reflect aspects of dealing with the public in resource management decision making. Thomas (1990, 1993) also adapted the Vroom–Yetton model to decision making in the public arena. His adaptation simplified the original model substantially and reduced the prescription to a single choice for each problem type. The decision-making styles and corresponding public involvement techniques, according to my adaptation of the model, are summarized in Table 1. The first decision-making style is the autocratic style (A1), in which the manager makes a unilateral decision from currently available information. The autocratic style does not use public involvement in making decisions.

The second style is the advisory style (A2), in which the manager obtains needed information before making an autocratic decision. Information may be obtained through informal personal contact between the manager and knowledgeable individuals, solicited written comments, surveys, or newspaper inserts that include a mechanism for public response (Table 1).

The third style is the consultative style (C1), in which the manager clearly retains decision-making authority but discusses a problem with various interest groups individually to get their input; the manager makes no attempt to bring different interest groups together. Consequently, little or no interaction among the interest groups occurs. The most commonly used public involvement technique in this category is meeting with sporting clubs or relatively homogenous community groups. Other C1-style public involvement techniques include open house meetings, computer simulations, and focus groups (Table 1).

The fourth style is the consultative-group style (C2), in which the manager discusses the problem with various interest groups simultaneously but clearly retains decision-making authority. An important aspect of this decision-making style is that interest groups with differing views are brought together to hear what others have to say about the issue. Public involvement techniques that fit into this category include the standard public meeting, workshop meetings (where meeting attendees form small groups to attempt to resolve differences), and advisory committees. Among C2 public involvement methods used less often are fishbowl planning and interactive television or radio programs (Table 1). The policy Delphi panel is a unique C2 technique because it allows exchange of information without face-to-face interaction (USDOT 1976).

The fifth decision-making style, shared decision making (S1), elevates the role of constituents in the decision-making process. In an S1 process, the manager discusses a problem with a group that is representative of the various concerned publics and together they attempt to reach a consensus in solving the problem. The manager's role may be that of chair of the group, facilitator of the process (a manager who has a deep personal interest in the problem's solution should defer to a neutral person to fill the role of facilitator), or simply that of a resource person. Decision-making authority is either shared with the group or expressly delegated to the group. The manager must be willing to accept and implement any decision arrived at by the group. Citizen task forces and charrettes (Table 1) rarely have been used by natural resource managers, except in the case of some comanagement arrangements in marine fisheries. (I distinguish between advisory committees and citizen task forces, the latter receiving shared or delegated decision-making authority, but the terms have been used interchangeably in the literature of public involvement.)

The Vroom–Yetton decision tree, adapted to public involvement in natural resource management decision making, is diagramed in Figure 1. The first three questions in the decision tree (A, B, and C in Figure 1) concern attributes of the problem and the manager's knowledge of the problem. The first question asks if it is necessary to make a high-quality decision, defined as one that "solves" a problem consistent with the resource agency's goals. (I assume the agency has secured public input in developing clearly stated goals. If that basic assumption is not met, it is unlikely that any prescriptive framework for designing a decision-making process will enhance management effectiveness.) Not all problems require high-quality decisions. A manager need not be overly concerned about the quality of routine decisions that have little effect on agency goals. The second question asks if the resource manager has sufficient information to make a high-quality decision. In cases where the manager lacks information, particularly regarding human dimension aspects of the problem, autocratic decision making is inappropriate and more participatory methods are prescribed. The third question asks if

TABLE 1.—Natural resource management decision-making styles (adapted from Vroom and Jago 1988) and associated public involvement techniques.

Decision-making style	Public involvement techniques
Autocratic (A1)	None
Advisory (A2)	1. Individual personal contact: informal contacts between manager and constituents to discuss issues and receive input. Objectives are to receive information and have interactions 2. Solicited written comments: manager requests specific written comments on a proposed course of action, such as a draft management plan. Objective is to receive information 3. Surveys: mail, telephone, or personal interview surveys conducted to gather quantitative information on public opinion regarding proposed actions or alternative management strategies. Objective is to receive information 4. Newspaper inserts: information about the proposed action is published in a newspaper or magazine and includes a brief survey to facilitate response by readers. Objectives are to give and receive information
Consultative (C1)	5. Computer simulation or games: computer simulations designed to allow constituents to test alternative strategies and learn about the system in question. Objectives are to give and receive information 6. Open house meeting: agency personnel are available at a specified time and place to answer questions and receive input from constituents, who drop in at their convenience. Agency may have information available for constituents in the form of brochures, posters, displays, etc., but make no formal presentations. Objectives are to give and receive information and have interactions 7. Interest group or club meetings: manager presents information to a somewhat homogeneous group and receives input. Objectives are to give and receive information and have interactions 8. Community meetings (homogeneous audience): manager presents information to a relatively homogenous group from one community or region and receives input. Objectives are to give and receive information and have interactions 9. Focus group: marketing technique consisting of a facilitated discussion among a relatively homogeneous group of 8–10 people chosen to represent particular characteristics or points of view, usually videotaped in a special facility. Objective is to receive information
Consultative-group (C2)	10. Formal public hearing: public meeting chaired by a hearing officer (often an agency official), run according to formal rules. Minutes of meeting are kept and input from constituents is usually limited to formal testimony. Agency personnel may make informational presentations, but primary purpose of hearing is to receive testimony from the public. Objectives are to give and receive information and have interactions 11. Open public meeting (heterogeneous audience): public meeting with less formal procedures than a hearing, usually consisting of informational presentations by agency personnel, followed by questions and comments from the audience. Objectives are to give and receive information and have interactions 12. Workshop meeting: audience at public meeting is broken up into smaller groups to discuss problems and attempt to resolve them. Agency personnel may present information before breaking into work groups, and each group normally is facilitated by an agency person or a trained volunteer. Objectives are to give and receive information and solve problems 13. Call-in radio or television show: listeners (viewers) call in to ask questions or offer comments in response to information presented by agency officials. Objectives are to give and receive information and have interactions 14. Advisory committee: committee of key individuals representing major interests relative to a specific issue or action. Agency personnel usually answer questions and provide information to committee members, but committee is purely advisory with no decision-making authority. Objectives are to give and receive information, have interactions, and solve problems 15. Fishbowl planning: iterative process of open public meetings, workshop meetings, and written comments designed to identify alternative strategies, pros and cons of those strategies, and positions of various interest groups. Alternative strategies are described in a brochure that is distributed before each public meeting, and meeting attendees comment on alternatives and suggest revisions. Smaller workshop meetings are held to rewrite alternatives and develop new ones. The iterative process is designed to eventually produce an alternative that is widely supported. Objectives are to give and receive information, have interaction, and solve problems 16. Policy Delphi panel: planning technique involving representatives of interests and "experts" who respond to a questionnaire. Responses are summarized by agency personnel and sent back to the panel with a revised questionnaire. Process continues until panel opinions converge. Panel members have mediated rather than direct interaction. Objectives are to receive information and solve problems
Shared decision making (S1)	17. Citizen task force: similar to advisory committee in that key representatives of various interests are selected to review information and advise managers, except decision-making authority is expressly shared or delegated. Objectives are to give and receive information, have interactions, and solve problems 18. Charrette: multiday workshop designed to develop solutions to complex problems. Repeated meetings of a group with constant membership may last weeks or be condensed to a few days of intensive activity (with members staying at a 24-h facility such as a resort or hotel). Members define a problem, analyze it, clarify issues, develop proposals (with assistance from agency staff), refine proposals, and present them to the general public. Objectives are to give and receive information, have interactions, and solve problems

Decision Points

A. Is it necessary to make a high quality decision?
B. Does the resource manager have sufficient information to make a high quality decision?
C. Is the problem structured?
D. Is acceptance of the decision by the public important for effective implementation?
E. If the manager were to make a unilateral decision, is it reasonably certain it would be accepted by the public?
F. Does the public share the agency's goals to be attained in solving the problem?
G. Is conflict among interest groups over preferred solutions likely?

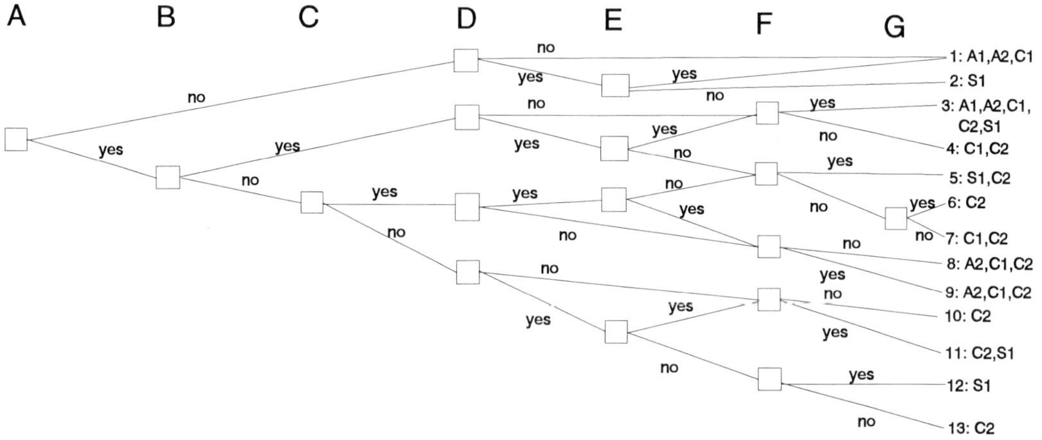

FIGURE 1.—The Vroom–Yetton decision tree, modified and adapted for public involvement in making resource management decisions. At the right are numbered types of problems (1–13) and the applicable decision-making styles. Decision-making styles are abbreviated as follows: autocratic (A1), advisory (A2), consultative (C1), consultative-group (C2), and shared decision making (S1). (Adapted and modified from Vroom and Jago 1988.)

the problem is structured, that is if it has known potential alternative solutions and criteria against which those alternatives can be evaluated. Unstructured problems, with ill-defined alternatives and ramifications, are more likely to be successfully resolved by more participatory methods (Vroom and Jago 1988).

The fourth and fifth questions (D and E in Figure 1) concern public acceptance of the decision. The fourth question asks if acceptance of the decision by the public is important for effective implementation. If acceptance is important, the public usually should be actively involved in making the decision. The fifth question asks if a unilateral decision by the manager is likely to be accepted by the public. Whenever decision acceptance is doubtful, autocratic (A1) and advisory (A2) decision-making styles should be avoided. Participation tends to produce greater acceptance if properly managed; however, participation may result in increased conflict if inappropriate techniques are used. Autocratic decisions may be accepted if they are seen as clearly within the manager's base of power. The most likely of French and Raven's (1959) bases of power to apply to resource managers are legitimate power (the power derived from the public's belief that the manager has the duty and responsibility to make the decision) and expert power (the power derived from the public's belief that the manager has the superior expertise required to make the decision).

The final two questions (F and G in Figure 1) concern congruence of public and agency goals and the likelihood of conflict among interest groups. The sixth question asks if the public shares the agency's goals that are to be attained in solving the problem. Even if the agency has secured public input in developing clearly stated goals, it is common for some interest groups to disagree with the goals. Each time a problem is considered, managers should identify the potentially affected interests and attempt to determine their positions. Groups that do not share agency goals are more likely to oppose or even sabotage problem-solving efforts. Their participation is critical to success but must be carefully managed. The seventh and final question asks if conflict among interest groups over preferred solutions is likely. Conflict among interest groups is common in resource management and results in an increase in the time needed to make management decisions. Conflict may result in polarization and continued divisiveness, but it is also a sign that interest groups should interact more, rather than less, to resolve their differences.

The decision tree may assist resource managers in determining the public involvement techniques that should be most effective in resolving various types of problems. The key to successful use of the decision tree (and, it is hoped, successful problem resolution) is recognition of the need to use appropriate public involvement techniques and avoid use of those techniques less likely to lead to successful problem resolution. The conceptual basis for the decision tree is a set of rules that eliminates decision-making styles that are likely to jeopardize either decision quality or decision acceptance. Those rules (adapted from Vroom and Jago 1988) are as follows.

1. Information rule—If decision quality is important and the manager does not possess enough information to make the decision alone, eliminate autocratic (A1) style.
2. Goal congruence rule—If decision quality is important and interested publics do not share agency goals in solving the problem, eliminate shared decision making (S1).
3. Unstructured problem rule—If decision quality is important, the manager lacks information needed to solve the problem alone, and the problem is unstructured, the method for solving the problem should provide for interaction among the various publics who have relevant information. Therefore, eliminate autocratic (A1), advisory (A2), and consultative (C1) styles, all of which lack an interaction component.
4. Acceptance rule—If acceptance by the public is important for effective implementation and it is not reasonably certain that an autocratic decision would be accepted, eliminate autocratic (A1) and advisory (A2) styles.
5. Conflict rule—If decision acceptance by the public is important for effective implementation, acceptance of a unilateral decision is not reasonably certain, and conflict among interest groups over possible solutions is likely, the method for solving the problem should enable those in disagreement to resolve their differences with full knowledge of the differences. Therefore, eliminate autocratic (A1), advisory (A2), and consultative (C1) styles because they do not permit interaction among interests.
6. Fairness rule—If decision quality is not important but acceptance is, and if acceptance is uncertain, the decision process should be designed to generate acceptance; interaction and negotiation among interest groups with full or shared responsibility for determining what is fair and equitable is appropriate. Therefore, eliminate all styles except shared decision making (S1).
7. Acceptance priority rule—If decision acceptance is important but not reasonably certain and the publics involved share the agency's goals, then methods that provide equal partnership generate greater acceptance without risking decision quality. Therefore, eliminate all styles except shared decision making (S1).

Application of the Model

The following examples illustrate use of the model to identify the most appropriate public involvement techniques. In the first (hypothetical) case, suppose the agency plans to build a boat ramp at one of two sites, both of which are located in the same area of a reservoir. The entire reservoir shoreline is in public ownership. Either site meets the agency goal for reservoir access (decision quality is not important). The manager knows from previous contacts with reservoir users that more access to this area of the reservoir is desired (decision acceptance probably is not needed or at least is reasonably certain). This is a type 1 problem (Figure 1). The model prescribes autocratic (A1), advisory (A2), or consultative (C1) decision- making styles. Because the decision does not have a quality requirement and any decision is likely to be accepted, there is no need to slow down the decision-making process by involving the public. The cautious manager may wish to announce a tentative decision and gauge public reaction before making the final decision (advisory or consultative styles). If the manager is not certain about public acceptance, it may be prudent to treat this as a type 2 problem, for which shared decision making (S1) is the recommended style. The reader probably has already noticed that for many types of problems a range of decision-making styles may be appropriate. In these cases, the manager must consider trade-offs between the effort required to secure public input and the likelihood of decision acceptance. As decision acceptance becomes increasingly more certain, less-participatory methods become more efficient and effective; however, as decision acceptance becomes more questionable, opting for the short-term efficiency of less participatory methods may result in long-term ineffectiveness and even greater effort to implement the decision.

The second example is a problem facing the Virginia Department of Game and Inland Fisheries in managing Smith Mountain Lake, a large reservoir. The problem, determining the number of fingerling striped bass *Morone saxatilis* to stock annually, seems to be essentially a biological decision (within the economic and facility-related constraints of what the agency's hatchery system can produce). Previous studies have shown that a stocking rate of 200,000–250,000 fish/year is optimum (Moore et al. 1991). However, one user group (the Smith Mountain Lake Striper Club) has been especially critical of the agency's stocking policy and vocal in its desire for greater stocking rates. Other groups are interested in management of the reservoir but have been quiet on the issue of stocking rate. This problem clearly has a quality requirement because stocking striped bass at higher rates could have substantial effects on the agency budget and its ability to stock fish in other waters. The manager knows how members of the striper club feel but lacks information on the position of other groups. Options available to the manager are relatively clear (it is a structured problem) and the decision is clearly within the manager's domain of power (decision acceptance probably is not needed for effective implementation that is, although striper club members are vocal, they are not likely to legally challenge the manager's decision). This is a type 8 or type 9 problem, depending on congruence of agency and public goals. In either case, the model prescribes either the advisory (A2), consultative (C1), or consultative-group (C2) decision-making styles. The manager's choice of decision-making styles should be governed by the probability that the decision will be accepted without a major challenge and by the amount of public interaction deemed necessary to make the decision. At a minimum, the manager should meet with striper club members to increase their awareness of why a particular option is chosen and to decrease the likelihood of the group formally challenging the decision. Consultative-group (C2) techniques that bring interest groups face-to-face would be appropriate if other interest groups felt strongly that the stocking rate should not be increased.

The third example involves management of a 148-km reach of the lower Wisconsin River corridor in southwest Wisconsin. Increasing rates of recreational use of the river and associated bottomlands, coupled with increasing development of river corridor lands for home construction, has caused serious concern about management of the area (a high-quality decision was needed). Resource managers clearly lacked information needed to resolve unstructured problems. The Wisconsin Department of Natural Resources owned or leased approximately 12,000 ha of land in the river corridor. Subdivision of large agricultural parcels into smaller lots threatened to reduce availability of recreational opportunities. Recreationists were concerned about crowding, litter, maintaining the river's scenic quality, and other problems associated with high use levels. Area residents generally agreed that disjointed local zoning ordinances were not adequate to preserve the rural character of the river corridor. All parties wanted a better system for preserving the character of the lower Wisconsin riverway but were leery of too much government intervention (acceptance of the decision was important but not reasonably certain). Because all parties were in agreement that a better management system was needed, the involved public shared the agency's goals. This is a type 12 problem, the ideal situation for a citizen task force (a shared decision-making style, S1). When there are many potential solutions to a problem, and public acceptance of the final solution is more important than the management details, the most constructive role for the resource manager is that of facilitator or information source. The manager can serve as staff to a citizen task force composed of a representative subset of the concerned publics. The Wisconsin legislature followed the recommendation of the Department of Natural Resources in creating a Lower Wisconsin Riverway Board composed of representatives of each affected county and three members representing various recreational groups (all of whom were appointed by the governor). The riverway board developed a management plan that resolved many difficult issues with minimal controversy. Agency managers acknowledged that many elements of the plan were accepted because the board had greater credibility with the public than did the agency.

Finally, consider unstructured problems in which solutions demand acceptance from fragmented publics that are not likely to agree with the agency nor share the agency's goals. These are what social scientists refer to as "wicked problems," problems that have no solutions, only temporary and imperfect resolutions (Rittel and Webber 1973). Because the interested publics do not share the agency's goals, the manager must retain decision-making authority; however, the high probability of intergroup conflict requires the groups to work out their differences together. These are type 13 problems, best suited to the consultative-group decision-making style (C2). A primary goal of the resource manager in dealing with wicked problems is to generate informed consent—that is, a solution that opponents will not challenge, even if they do not agree with the manager (IPP 1981). It is in making decisions related to wicked problems that resource managers demonstrate their proficiency in dealing with the public.

For example, in developing a fisheries management plan for Canyon Ferry Reservoir, managers for the Montana Department of Fish, Wildlife and Parks had to contend with a vocal group of anglers who were adamant in their demands for introduction of walleye *Stizostedion vitreum* in the reservoir. Agency managers were opposed to any new introductions because the reservoir already supported heavily used fisheries for rainbow trout *Oncorhynchus mykiss* and yellow perch *Perca flavescens*. Introduction of predaceous walleye would please at least one group of anglers; however, a previous study (Colby and Hunter 1989) indicated that introduced walleye had the potential to affect adversely the existing fisheries and therefore other reservoir users. Obviously, the situation demanded a high-quality decision. Agency managers still felt they lacked information on walleye–rainbow trout interactions. The range of management options was undetermined (unstructured problem), and it was clear that acceptance of the decision was critical and uncertain. At least one group (the walleye anglers) did not share the agency's goals.

Development of a management plan for Canyon Ferry Reservoir had all the earmarks of a wicked problem. User groups were strongly polarized and every management alternative left at least one group strongly opposed to the management plan. Agency managers pursued several C2 strategies to deal with the problem. They involved constituents in organizing a trout–walleye interactions workshop, in which several invited experts discussed with the public and fishery managers potential effects of introducing walleye in the reservoir. An advisory committee was established to assist the agency in developing the management plan. Numerous public meetings and open houses were held. In addition to the C2 strategies, the agency also conducted surveys to assess public opinion regarding reservoir management. In the final plan, agency managers decided not to introduce new species to the reservoir (MDFWP 1992) and the decision was not contested. However, in a perfect illustration of the frustration of dealing with wicked problems, walleyes were illegally introduced and are now well established in the reservoir.

Discussion

The modified Vroom–Yetton model can assist resource managers in classifying problems quickly and in making resource managers aware that different problems require different public involvement approaches. When answers to the seven questions result in a prescription for autocratic (A1) or advisory (A2) decision-making styles, the manager may rapidly and confidently make a decision. Incremental use of the model may also be beneficial. If, for example, the manager lacks sufficient information to answer the seven questions, he or she may use advisory (A2) or consultative (C1) methods to learn more about the problem so as to answer the questions and then employ appropriate public involvement techniques to resolve the problem. Although the model does not specify the use of more than one public involvement technique, managers should understand that one technique alone is rarely adequate to deal with complex problems.

A common concern among resource managers is the investment of effort and money required to implement effective public involvement programs. Greater concern with short-term efficiency in decision making than with long-term effectiveness can lead to use of inappropriate public involvement techniques, poor quality decisions, or lack of decision acceptance by constituents. When constituents believe they have been dealt with unfairly, they are more likely to challenge a decision through administrative appeal or in the courts. Defending an appealed decision is usually more costly than properly incorporating public involvement into decision making at the outset.

Larkin (1988) described the dilemma facing managers of many coastal fisheries as the impossible task of trying to simultaneously optimize management for recreational, commercial, and artisanal fisheries. Reservoir fishery managers may face similar problems in trying to balance demands of recreational anglers, tournament anglers, guides, resort owners, commercial fishers, and, perhaps, Native Americans. More and more frequently, resource managers dealing with these especially complex and contentious issues are calling for a return to traditional scientific resource management. Although it is clear that the scales have been tipped too far toward politics and economics in the management of many commercial fisheries (Larkin 1991), I believe it would be a mistake to attempt to make decisions about these resources using less participatory styles. More science is needed in making these decisions, but biological science alone cannot resolve problems that are primarily allocative in nature. Hilborn and Luedke (1987) and Pringle (1985) demonstrated that participatory decision making improved acceptance among commercial fishers without compromising the high-quality of the decision.

Empirical tests of the Vroom–Yetton model in organizational contexts have generally supported its validity (Vroom and Jago 1978; Zimmer 1978; Tjosvold et al. 1986). In six studies of the model's validity, leader behavior conforming to the model's prescriptions resulted in solutions that were deemed successful 62% of the time. When leaders used decision-making styles other than those prescribed by the model, only 37% of the solutions were deemed successful (Vroom and Jago 1988). Miner (1984) rated validity and usefulness of the Vroom–Yetton model as high compared with other theories of leadership.

Adaptation of the Vroom–Yetton model to public involvement in natural resource decision making has intuitive appeal. Its prescriptions for flexible decision-making styles by resource managers correspond to constituents' desires for managers who are strong leaders but who also are open to public input in equitable decision-making processes (McMullin 1993a). Policymakers ranked factors that contributed to public support and satisfaction with fish and wildlife agencies (including openness to public input) as most important in determining effectiveness of agencies (McMullin et al. 1991).

Intuitive appeal alone is not sufficient for determining the model's utility in natural resource decision making. Thomas (1993) analyzed 42 public decisions described in published accounts and concluded that the effectiveness of decisions (defined as how smoothly the process worked and how well the eventual decision worked) was significantly enhanced when public managers followed the model's prescription. Manipulative empirical tests of the model raise ethical concerns because public agencies cannot afford to jeopardize their credibility with constituents by purposely employing techniques of public involvement that are considered unlikely to succeed. However, it is possible to test hypotheses concerning decision effectiveness by reviewing cases after the fact, as Thomas (1993) has done. Additional documentation of relationships between success and failure when resource managers follow or fail to follow the model's prescriptions in resolving natural resource issues should increase managers' confidence in using the model as a decision-making aid.

Acknowledgments

I thank Eric Hallerman, Donna Minnis, James Parkhurst, and R. Ben Peyton for their review and helpful comments on drafts of the manuscript. Dick Luebke, L. E. Miranda, Susan Shipman, and one anonymous reviewer provided comments that improved the paper.

References

Adelman, I. R., D. J. Schmidly, and Y. Cohen. 1994. Educational needs of fisheries professionals: results of a survey. Fisheries 19(11):17–25.

Angle, H. L., and J. L. Perry. 1981. An empirical assessment of organizational commitment and organizational effectiveness. Administrative Science Quarterly 26:1–13.

Checkoway, B. 1981. The politics of public hearings. Journal of Applied Behavioral Science 17:566–582.

Colby, P., and C. Hunter. 1989. Environmental assessment of the introduction of walleye beyond their current range in Montana. Report to the Montana Department of Fish, Wildlife and Parks. OEA Research, Helena, Montana.

Downing, R. L. 1987. Success story: white-tailed deer. Pages 45–57 *in* H. Kallman, editor. Restoring America's wildlife. U.S. Fish and Wildlife Service, Washington, DC.

Flader, S. L. 1976. Scientific resource management: an historical perspective. Transactions of the North American Wildlife and Natural Resources Conference 41:17–30.

French, J. R. P., Jr., and B. Raven. 1959. The bases of social power. Pages 150–167 *in* D. Cartwright, editor. Studies in social power. University of Michigan Institute for Social Research, Ann Arbor.

Gasson, W., and J. White. 1993. Comprehensive management through teamwork in the Wyoming Game and Fish Department. Transactions of the North American Wildlife and Natural Resources Conference 58: 221–228.

Hilborn, R., and W. Luedke. 1987. Rationalizing the irrational: a case study in user group participation in Pacific salmon management. Canadian Journal of Fisheries and Aquatic Sciences 44:1796–1805.

Huber, W. W. 1953. Conservation education—the key to resource management. Transactions of the North American Wildlife Conference 18:631–638.

IPP (Institute for Participatory Planning). 1981. Citizen participation handbook for public officials and other professionals serving the public. Institute for Participatory Planning, Laramie, Wyoming.

Kennedy, J. J. 1985. Viewing wildlife managers as a unique professional culture. Wildlife Society Bulletin 13:571–579.

Langenau, E. E. 1982. Bureaucracy and wildlife: a historical overview. International Journal for the Study of Animal Problems 3:140–157.

Larkin, P. A. 1988. The future of fisheries management: managing the fisherman. Fisheries 13(1):3–9.

Larkin, P. A. 1991. The wet revolution: fisheries production to the year 2025. Pages A22–A35 *in* J. P. Harville, editor. Proceedings of the North American fisheries leadership workshop. American Fisheries Society, Bethesda, Maryland.

Lewis, J. B. 1987. Success story: wild turkey. Pages 31–43 *in* H. Kallman, editor. Restoring America's wildlife. U.S. Fish and Wildlife Service, Washington, DC.

Lively, C. E. 1953. The social side of conservation—some reflections on the conservation movement. Transactions of the 18th North American Wildlife Conference 18:36–43.

Magill, A. W. 1988. Natural resource professionals: the reluctant public servants. Environmental Professional 10:295–303.

McGeary, M. N. 1960. Gifford Pinchot: forester politician. Princeton University Press, Princeton, New Jersey.

McMullin, S. L. 1993a. Characteristics and strategies of effective state fish and wildlife agencies. Transactions of the North American Wildlife and Natural Resources Conference 58:206–210.

McMullin, S. L. 1993b. Approaches to management effectiveness in state fish and wildlife agencies. Doctoral dissertation. Virginia Polytechnic Institute and State University, Blacksburg.

McMullin, S. L., S. R. Amend, and L. A. Nielsen. 1991. Managing information about how we are managing: multiple perspectives on the factors that determine agency effectiveness. Transactions of the North American Wildlife and Natural Resources Conference 56:162–168.

MDFWP (Montana Department of Fish, Wildlife and Parks). 1992. Canyon Ferry Reservoir/Missouri River fisheries management plan 1993–1998. Montana Department of Fish, Wildlife and Parks, Helena.

Miner, J. B. 1984. The validity and usefulness of theories in an emerging organizational science. Academy of Management Review 9:296–306.

Moore, C. M., R. J. Neves, and J. J. Ney. 1991. Survival and abundance of stocked striped bass in Smith Mountain Lake, Virginia. North American Journal of Fisheries Management 11:393–399.

Nielsen, L. A. 1976. The evolution of fisheries management philosophy. U.S. National Marine Fisheries Service Marine Fisheries Review 38(12):15–23.

Pringle, J. D. 1985. The human factor in fishery resource management. Canadian Journal of Fisheries and Aquatic Sciences 42:389–392.

Rich, R. C., and W. A. Rosenbaum. 1981. Citizen participation in public policy. Journal of Applied Behavioral Science 17:439–445.

Rittel, H. W. J., and M. Webber. 1973. Dilemmas in a general theory of planning. Policy Sciences 4:155–169.

Thomas, J. C. 1990. Public involvement in public management: adapting and testing a borrowed theory. Public Administration Review 50:435–445.

Thomas, J. C. 1993. Public involvement and governmental effectiveness: a decision-making model for public managers. Administration and Society 24:444–469.

Tipple, T. J., and J. D. Wellman. 1989. Life in the fishbowl: public participation rewrites public foresters' job descriptions. Journal of Forestry 87(3):24–27,30.

Tjosvold, D., W. C. Wedley, and R. H. G. Field. 1986. Constructive controversy, the Vroom–Yetton model, and managerial decision making. Journal of Occupational Behaviour 7:125–138.

USDOT (U.S. Department of Transportation). 1976. Effective citizen participation in transportation planning. Volume II. A catalog of techniques. U.S. Department of Transportation, Washington, DC.

Vroom, V. H., and A. G. Jago. 1978. On the validity of the Vroom–Yetton model. Journal of Applied Psychology 63:151–162.

Vroom, V. H., and A. G. Jago. 1988. The new leadership: managing participation in organizations. Prentice Hall, Englewood Cliffs, New Jersey.

Vroom, V. H., and P. W. Yetton. 1973. Leadership and decision-making. University of Pittsburgh Press, Pittsburgh, Pennsylvania.

Zimmer, R. J. 1978. Validating the Vroom–Yetton normative model of leader behavior in field sales force management and measuring the training effects of TELOS on the leader behavior of district managers. Doctoral dissertation. Virginia Polytechnic Institute and State University, Blacksburg.

Zimmerman, M. A., and J. Rappaport. 1988. Citizen participation, perceived control and psychological empowerment. American Journal of Community Psychology 16:725–750.

Environmental Assessment

American Fisheries Society Symposium 16:67–72, 1996

Algal Off-Flavors in Reservoirs

PETER W. PERSCHBACHER

University of Arkansas at Pine Bluff, Agriculture Department
Box 4005, Pine Bluff, Arkansas 71601, USA

DWIGHT MILLER AND ERIC D. CONTE

U.S. Food and Drug Administration, National Center for Toxicological Research
Jefferson, Arkansas 72709, USA

Abstract.—Historically, the proliferation of certain naturally occurring algae and the musty and earthy odors and flavors they produce have greatly reduced the quality of drinking water in reservoirs around the world. These algae can similarly affect fish flavor because off-flavor compounds in water become concentrated in lipid-rich fish tissue. The only successful treatment for off-flavor water in reservoirs has been the application of copper sulphate crystals to algal beds. Biocontrol with filter-feeding fish is currently being examined in pond research and may hold promise. Off-flavor compounds in drinking water and fish flesh are detected by taste and odor testing or by gas chromatography. We have developed a low-cost test for musty and earthy off-flavors based on colorimetric procedures and designed for use by fisheries and waterworks researchers and managers. The major compounds producing musty and earthy off-flavor, 2-methylisoborneol (MIB) and geosmin, are first extracted from a 1-L water sample by solid-phase adsorbent. The solvent is combined with the color-producing compound in concentrated sulfuric acid. Color development is completed in 30 min for MIB and approximately 20 h for geosmin and read with color comparators or with a colorimeter. Concentrations in fish are similarly determined but require distillation in a microwave oven. The sensitivity level is 1 part per billion for a 1-L sample.

Nuisance algae have caused problems with water taste and odor at reservoirs in the United States and around the world. The most common sensory deterioration in drinking water (AWWARF 1987) and in farmed fish (Tucker and Martin 1991) in the United States occurs with the proliferation of cyanobacteria (still commonly known as blue-green algae), which produce two metabolic compounds that impart musty and earthy off-flavors and odors. Thirty percent of South Africa's reservoirs have serious problems with earthy, muddy off-flavor water (Wnorowski and Scott 1982), and in Japan about 40% of 400 reservoir water samples tested positive for musty odor (Matsumoto and Tsuchiya 1988).

Considerable research in the United States has been conducted by the Metropolitan Water District of Southern California and by aquaculturists to characterize and control episodes of musty- and earthy-tasting water. Reports of flavor problems of wild-caught fish in reservoirs, rivers, and farm ponds have been scant, although in one incident, commercial harvest of fish in a Canadian reservoir was interrupted by muddy off-flavors that made the catch unmarketable (Tabachek and Yurkowski 1976; Yurkowski and Tabachek 1980).

One difficulty in researching and acting on musty and earthy flavors is that objectionable (off-flavor) taste and odor occur at low concentrations (parts per trillion in drinking water, parts per billion in fish flesh). Gas chromatography–mass spectrometry (GC–MS) is the only available quantitative test, but each determination costs more than US$100. An inexpensive test for detecting and measuring the major off-flavor compounds, 2-methylisoborneol (MIB) and geosmin, in reservoir water and fish would enable frequent monitoring so that preventative or corrective action could be taken to maintain or restore water quality. In this paper we review algal off-flavors and cite drinking water and fish culture case histories. We describe an inexpensive colorimetric test for off-flavor in water and fish flesh, comment on cyanobacterial control, and suggest research to assess off-flavor potential and remediation in reservoirs.

Waterborne Off-Flavors

Ideal potable water has been characterized as odorless, colorless, and tasteless (AWWARF 1987). The presence of objectionable flavors and odors depends on whether the water inflows are of surface or groundwater origin, on water residence times and mixing regimes, on seasonal patterns in natural productivity and water use, and on the nature of industrial, agricultural, or urban effluents (AWWARF 1987). Algae are the major sources of drinking water off-flavors in the United States, according to 82% of water utility personnel (AWWARF 1987). More than 50 algal genera can produce odor and taste in

TABLE 1.—Common algal off-flavors from drinking water reservoirs (after Palmer 1977).

Off-flavor	Descriptors	Cause(s)
Musty or earthy	Potato bin, moldy[a]	Cyanobacteria and the diatom *Synedra*
Aromatic	Vegetable, skunk, garlic	Flagellates and diatoms
	Fishy	Larger numbers of flagellates and diatoms
Grassy		Large numbers of green algae
Septic	Pigpen, foul, putrefactive	Dying and decaying algal masses and scums

[a]Weedy, swampy, marshy, peat, strawlike, and woody are thought to be related.

drinking-water reservoirs (Palmer 1977). Alga-related odors have been described as musty or earthy, aromatic, grassy, and septic (Table 1). The most common of these are musty or earthy (AWWARF 1987), resulting from the metabolic byproducts (MIB and geosmin) of certain cyanobacteria. Water with MIB often has a musty odor, described as camphorous at high intensities. Geosmin imparts a fresh earth odor.

Musty and Earthy Off-Flavors in Reservoir Water and Fish

The cyanobacteria responsible for musty and earthy off-flavors are planktonic- or benthic-growing, filamentous forms in the genera *Aphanizomenon*, *Lyngbya*, *Phormidium* or *Symploca*, and (notably) *Anabaena* and *Oscillatoria* (AWWARF 1987). Cyanobacterial development and off-flavors in reservoir water often occur in periods of low rainfall. For example, complaints of bad-tasting water increased by 65% in France during drought (AWWARF 1987), and when the Yodo River in Japan was temporarily dammed during drought, it developed an earthy flavor from planktonic *Anabaena* (Hishida et al. 1988). Musty taste and odor are recurring problems in late summer when rainfall decreases and water clears in a large reservoir serving Norman, Oklahoma (N. Campbell, Norman Water Treatment Plant, personal communication). Decomposition of floating blue-green algal scums may stimulate growth of bacteria, including actinomycetes that also produce MIB and geosmin. Algal beds in rivers leading into reservoirs may be a source of off-flavor. In fact, the first mention of off-flavor in fish in the scientific literature involved Scottish salmon in a river with beds of algae (Thayson 1936).

Fish absorb MIB and geosmin from the water through their gills (Lovell and Sackey 1973). The compounds bioconcentrate in the fatty tissue, specifically in peritoneal fat and subepidermal adipose tissue (Martin et al. 1990). Musty and earthy flavors will persist for several days to months after MIB and geosmin can no longer be detected in water. Purging rates depend on water temperature, MIB and geosmin levels in the flesh, and the fat or lipid level in the flesh (Johnsen and Lloyd 1992). Thus, fatty and moderately fatty fishes, (e.g., cyprinids, clupeids, ictalurids, and salmonids) may retain the undesirable taste longer than lean fishes (e.g., centrarchids, percids, and serranids). Lipid-rich tissue also purge more slowly than lean tissue. Higher levels of off-flavors are found under the skin and in the reddish flesh under the lateral line, commonly known as the mud line. This flattened wedge of dark muscle between the dorsal and ventral muscle masses contains high fat levels (Lagler et al. 1962). By removing the mud line and the skin, considerable amounts of off-flavors can be eliminated. Benthic or midwater fishes (e.g., ictalurids, acipenserids, and cyprinids) may absorb higher levels of off-flavors than surface fishes because of their potential for close association with algal beds.

Reservoir Case Histories

Lake Matthews and Lake Phoenix, California

Lake Matthews, in the Southern California Water District, receives water from the Colorado River (Izaguirre et al. 1983). Musty off-flavor was noted in the Colorado River before its occurrence in the reservoir. The average depth of the reservoir is 56 m and the temperature range is 11–27°C. *Oscillatoria curviceps* was identified as the causative cyanobacterium, forming benthic mats on rocks and in shallow water near shore. Mats were found at depths of 2–11 m in warm weather and were not visible from shore. The highest incidence of flavor problems was at water temperatures of 24–25°C (September–October) with a Secchi depth of 2.3–3.3 m. The lowest water temperature at which beds developed was 20°C. Treatment, which was started when levels of MIB 0.3 m above the mats exceeded 5 parts per trillion (ppt), consisted of adding copper sulphate to the beds; however, copper accumulation in the sediments led to suspension of this practice.

Lake Phoenix is a hilly, wooded reservoir (mean depth of 15 m) in the Marin Water District of northern California (Izaguirre et al. 1983). Problems similar to those in Lake Matthews occurred at water temperatures between 9.5 and 23.3°C. The alga responsible was presumed to be a benthic *Oscillatoria tenuis*.

Cedar Lake, Manitoba, and Saline Lakes, Saskatchewan

Cedar Lake, Manitoba, is a eutrophic, shallow reservoir with a mean depth of 3.5 m (Yurkowski and Tabachek 1980). In August 1977 commercial catches of walleye *Stizostedion vitreum* were the source of complaints of bad, muddy flavor. Subsequently, northern pike *Esox lucius*, lake herring *Coregonus artedi*, and lake whitefish *Coregonus clupeaformis* were found to be contaminated and the season was closed. Minced walleye flesh boiled with distilled water and analyzed by GC–MS contained both MIB and geosmin.

Rainbow trout *Oncorhynchus mykiss* grown in eutrophic, saline lakes in Saskatchewan were similarly affected from April to October (Tabachek and Yurkowski 1976). Eleven species of blue-green algae (seven species of *Oscillatoria*, three of *Lyngbya*, and one of *Symploca*) suspected of producing muddy off-flavor were isolated and cultured; GC–MS confirmed MIB and geosmin production.

Fish Culture Ponds

Fish culture ponds receive high organic loading and are conducive to off-flavor development (Lovell 1979). Ponds containing common carp *Cyprinus carpio* in Israel (Aschner et al. 1969) and ponds with channel catfish *Ictalurus punctatus* in the United States (Martin et al. 1988) seasonally exhibit high levels of musty and earthy off-flavors. High water temperature and feed application rates (Brown and Boyd 1982; Armstrong et al. 1986) were correlated with levels of earthy and musty off-flavors in Alabama. Dominance by the cyanobacteria *Oscillatoria* spp. was correlated with levels of MIB in fish ponds in Oklahoma (Lorio et al. 1992). In this case, when MIB levels were low, the delay in uptake to objectionable levels in fish flesh might allow early corrective action to avoid a deterioration in fish quality.

In catfish ponds in soft waters, such as occur in eastern Alabama, geosmin-caused off-flavor predominated and was produced by several straight-chain planktonic cyanobacteria of the genus *Anabaena* (van der Ploeg and Boyd 1991). Spring and early summer were the periods of greatest concern. A planktonic cyanobacterium similar to *O. chalybea* was responsible for MIB (Martin et al. 1991). Off-flavor caused by MIB predominated in fish ponds in hard waters, such as in western Mississippi, and was of higher incidence in late summer and early fall (Tucker and Martin 1991; van der Ploeg et al. 1992).

These examples of off-flavor occurrences in reservoirs and ponds have in common the development and, in many cases, dominance of benthic or planktonic blue-green algae. Warm and enriched water promoted their growth. Reservoirs with these algal patterns may be subject to musty and earthy off-flavor problems in water and fish.

Detection and Monitoring of Off-Flavor Compounds

Detection of MIB and geosmin relied solely on taste and odor (organoleptic) testing until the advent of GC–MS procedures allowed measurement of these compounds at ppt levels; however, GC–MS determinations require expensive equipment and materials and are time consuming. Organoleptic testing is the only practical method available to quality control managers, but it remains subject to the ability of an individual to detect geosmin and MIB and it involves variable rejection criteria.

Methods for MIB and geosmin determination by GC–MS in water and seafood samples have not been standardized (Karahadian and Lindsay 1994). Typically, these tertiary alcohols are extracted by an organic solvent such as methylene chloride. The solvent is sent to a laboratory for analysis; however, not many laboratories are capable of performing this procedure. Some larger water districts maintain GC–MS capability for MIB and geosmin. Because accuracy of detection depends on conditions of extraction and analysis, continued research to improve efficiency is needed. For example, the use of functionalized solid-phase absorbents has resulted in very high extraction efficiencies (D. Miller, U.S. Food and Drug Administration, unpublished data).

Enzyme-linked immunosorbent assays (ELISAs), in which antibodies to MIB or geosmin are linked to colorimetric development, have been developed for water samples (Chung et al. 1990, 1991). Although ELISAs provide rapid field tests, sensitivity is 1 part per million, which is substantially greater than the low parts-per-trillion and parts-per-billion levels at which off-flavor occurs in water and fish, respectively. Therefore, water samples must be concentrated (Chung 1992). A major supplier of ELISAs has expressed interest in examining concentration techniques for MIB and geosmin detection.

Recently, a colorimetric method to quantify MIB in water and in fish flesh was developed collaboratively by the University of Arkansas at Pine Bluff and the U.S. Food and Drug Administration's Na-

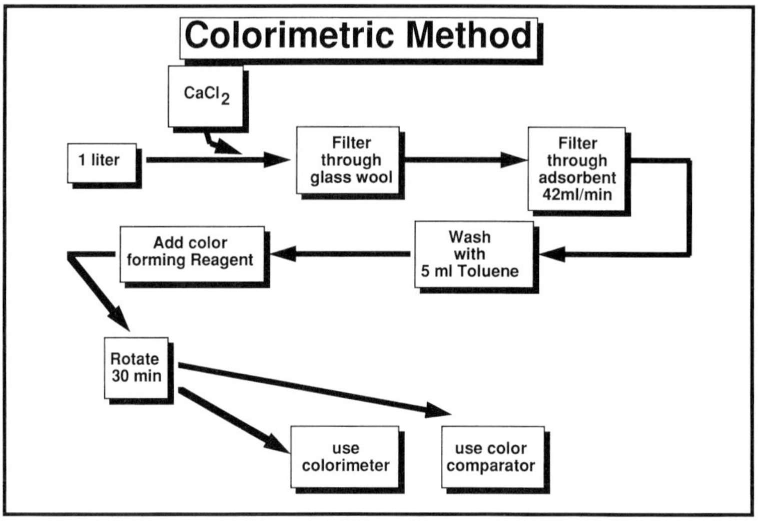

FIGURE 1.—Flow diagram of a colorimetric method for detection and measurement of 2-methylisoborneol and geosmin concentrations in water.

tional Center for Toxicological Research. The same technique was also found to be applicable to geosmin determination. The test has the advantages of being simple, inexpensive, and relatively fast. The most toxic compound used is concentrated sulfuric acid, and all chemicals used are available commercially. At a rate of 12 samples per hour, the cost is approximately US$4.00 per sample. To test a water sample, 1 L of filtered water is pumped through a solid-phase extraction device. The MIB or geosmin is captured on the absorbent and is eluted in place with 1–5 mL of toluene solvent. Liquid–liquid extraction (LLE) can also be used. The toluene is then combined with a 1% vanillin and concentrated sulfuric acid solution and agitated for 30 min to produce the color change (Figure 1). The standard solution is yellow and, at increasing concentrations of MIB or geosmin, changes from light orange to blood red. Color development of geosmin requires approximately 20 h, which provides a possible means of discrimination between the two compounds based on rate constants. In addition, maximal absorption peaks differ by 20 nm.

Extraction from 1 L of water results in a sensitivity of 1 part per billion (ppb). Extraction from greater volumes will be required to obtain sensitivities in parts per trillion. Closed-loop stripping technology (McGuire et al. 1981) and LLE techniques for geosmin and MIB from volumes larger than 1 L and the resulting colorimetric sensitivities need to be investigated. Interference from other alcohols in water samples might occur, although interferences have not been a problem thus far (D. Miller, unpublished data).

Measurement of MIB and geosmin levels in fish flesh involves purging at 30% power for 3 min in a microwave oven modified to contain an extraction flask and trapping on a solid-phase adsorbent in a chilled collection chamber outside the microwave oven. After collection, colorimetric development and measurement are the same as for water samples.

Any colorimetrically measurable level of MIB or geosmin in water from a 1-L sample would result in taste problems in fish flesh. At tissue levels of approximately 1 ppb, MIB and geosmin may be detectable depending on the background flavor of the fish species (Tucker and Martin 1991).

Control of Algal Composition

Approaches to algal management in reservoirs include phosphorus inactivation, modification of water circulation, algicides, nutrient diversion, water flushing, and food web manipulation (Cooke et al. 1993). Phosphorus inactivation is costly and temporary, and efforts to manage water circulation have produced inconsistent results in 15 reservoirs (Cooke et al. 1993). Copper sulphate algicide is the only treatment used successfully for off-flavor abatement in drinking water reservoirs (Izaguirre et al. 1983; AWWARF 1987). Results are short term (up to several weeks), and repeated additions are required. Chemical control by copper is therefore costly and leads to accumulation in sediments. Furthermore, all algae are affected by copper sulphate addition (Tucker and Martin 1991). Because adequate levels of algae are needed to sustain oxygen levels, absorb polluting nutrients, and provide food to zooplankton and other aquatic organisms, elimination or impairment of all algal populations can reduce water quality.

Long-term control of undesirable algal diversity and abundance will require nutrient diversion, water flushing, or food web manipulation (Cooke et al. 1993). Nutrient diversion could reduce the overall productivity of the reservoir and may involve regulation of numerous nonpoint pollution sources. Removal of 10% of algal biomass daily by flushing or through filter-feeding organisms has been proposed as a way to favor small, fast-growing noncyanobacteria, which do not produce off-flavors (van der Ploeg 1992). Flushing of 10% daily, even if feasible, merely moves the problem downstream and cannot be considered a sustainable solution. In experimental channel catfish production ponds, planktonic cyanobacteria known to produce off-flavors and other water quality problems appear to be controllable by the addition of filter-feeding herbivorous fishes, such as Nile tilapia *Oreochromis niloticus* and blue tilapia *O. aureus* (Torrans and Lowell 1987; Perschbacher and Lorio 1993). In Israeli reservoirs, control of off-flavor from benthic *Oscillatoria* by *O. aureus* and common carp has been reported (Leventer 1981). Where water temperatures fall below 9°C, tilapia will die and will need to be replaced each spring. Filter-feeding candidates with greater cold tolerance include silver carp *Hypophthalmichthys molitrix*, threadfin shad *Dorosoma petenense*, and clams. In areas where these fishes are exotic, prevention of reproduction (e.g., sterile triploid or monosex populations) could be used.

Future Approaches and Needs

Surveys of flavor tainting of reservoir fish and water supplies are needed to assess the presence and extent of off-flavor problems. The control of flavor- and odor-producing algae in multiuse reservoirs may be an area for cooperation between water supply and fishery agencies. In another matter of reservoir water quality (provision of standing timber), interaction between these groups (Layher 1984) led to the establishment of guidelines in Kansas and other states. Research programs on morphoedaphic features, water characteristics, and seasonality relating to off-flavor occurrences are desirable. Food web manipulations have the most promise for controlling cyanobacteria that produce musty and earthy off-flavors without decreasing productivity and should be investigated in afflicted reservoirs. Improved and standardized monitoring and detection methodologies are essential components of successful management.

References

Armstrong, M. S., C. E. Boyd, and R. T. Lovell. 1986. Environmental factors affecting flavor of channel catfish from production ponds. Progressive Fish-Culturist 48:113–119.

Aschner, M., C. Laventer, and I. Chorin-Kirsch. 1969. Off flavor in carp from fish ponds in the coastal plain and the Galil. Bamidgeh 19:23–25.

AWWARF (American Water Works Association Research Foundation). 1987. Identification and treatment of tastes and odors in drinking water. American Water Works Association, Denver, Colorado.

Brown, S. W., and C. E. Boyd. 1982. Off-flavor in channel catfish from commercial ponds. Transactions of the American Fisheries Society 111:379–383.

Chung, S-Y. 1992. Attempts to improve the sensitivity of an enzyme-linked immunosorbent assay for 2-methylisoborneol. Water Science and Technology 25:89–95.

Chung, S-Y., P. B. Johnsen, and P. H. Klesius. 1990. Development of an ELISA using polyclonal antibodies specific for 2-methylisoborneol. Journal of Agricultural and Food Chemistry 38:410–415.

Chung, S-Y, J. R. Vercelloti, P. B. Johnsen, and P. H. Klesius. 1991. Development of an enzyme-linked immunosorbent assay for geosmin. Journal of Agricultural and Food Chemistry 39:764–769.

Cooke, G. D., E. B. Welch, S. A. Peterson, and P. R. Neurath. 1993. Restoration and management of lakes and reservoirs. CRC Press, Lewis Publishers, Boca Raton, Florida.

Hishida, Y., K. Ashitani, and K. Fujiwara. 1988. Occurrence of musty odor in the Yodo River. Water Science and Technology 29:193–196.

Izaguirre, G., C. J. Hwang, S. W. Krasner, and M. J. McGuire. 1983. Production of 2-methylisoborneol by two benthic Cyanophyta. Water Science and Technology 15:211–220.

Johnsen, P. B., and S. W. Lloyd. 1992. Influence of fat content on uptake and depuration of the off-flavor 2-methylisoborneol by channel catfish (*Ictalurus punctatus*). Canadian Journal of Fisheries and Aquatic Sciences 49:2406–2411.

Karahadian, C., and R. C. Lindsay. 1994. Chemical isolation of odoriferous components. Pages 275–283 *in* J. W. Kiceniuk and S. Ray, editors. Analysis of contaminants in edible aquatic resources. VCH Publishing Inc., New York.

Lagler, K. F., J. E. Bardach, and R. R. Miller. 1962. Ichthyology. Wiley, New York.

Layher, W. G. 1984. Compatibility of multiple uses: potable water supplies and fisheries. Fisheries 9(6):2–11.

Leventer, H. 1981. Biological control of reservoirs by fish. Bamidgeh 33:3–33.

Lorio, W. J., P. W. Perschbacher, and P. B. Johnsen. 1992. Relationship between water quality, phytoplankton community and off-flavors in channel catfish (*Ictalurus punctatus*) production ponds. Aquaculture 106: 285–292.

Lovell, R. T. 1979. Flavor problems in fish culture. Pages 186–190 *in* T. V. R. Pillay and W. A. Dill, editors. Advances in aquaculture. Fishing News (Books) Ltd., London.

Lovell, R. T., and L. A. Sackey. 1973. Absorption by channel catfish of earthy-musty flavor compounds synthesized by cultures of blue-green algae. Transactions of the American Fisheries Society 102:774–777.

Martin, J. F., G. Izaguirre, and P. Waterstrat. 1991. A planktonic *Oscillatoria* species from Mississippi catfish ponds that produces the off-flavor compound 2-methylisoborneol. Water Research 25:1447–1451.

Martin, J. F., C. P. McCoy, C. S. Tucker, and L. W. Bennett. 1988. 2-methylisoborneol implicated as a cause of off-flavor in channel catfish *Ictalurus punctatus* (Rafinesque) from commercial culture ponds in Mississippi. Aquaculture and Fisheries Management 19:151–157.

Martin, J. F., S. M. Plakas, J. H. Holley, and J. V. Kitzman. 1990. Pharmacokinetics and tissue disposition of the off-flavor compound 2-methylisoborneol in the channel catfish (*Ictalurus punctatus*). Canadian Journal of Fisheries and Aquatic Sciences 47:544–547.

Matsumoto, A., and Y. Tsuchiya. 1988. Earthy-musty odor-producing cyanophytes isolated from five water areas in Tokyo. Water Science and Technology 20: 179–183.

McGuire, M. J., S. W. Krasner, C. J. Hwang, and G. Izaguirre. 1981. Closed-loop stripping analysis as a tool for solving taste and odor problems. American Water Works Association Journal 73:530–537.

Palmer, C. M. 1977. Algae and water pollution. Municipal Environmental Research Laboratory, U.S. Environmental Protection Agency (EPA-600/9-77-036), Cincinnati, Ohio.

Perschbacher, P. W., and W. J. Lorio. 1993. Filtration rates of catfish pond phytoplankton by Nile tilapia *Oreochromis niloticus*. Journal of the World Aquaculture Society 24:424–427.

Tabachek, J-A., and M. Yurkowski. 1976. Isolation and identification of blue-green algae producing muddy odor metabolites, geosmin and 2-methylisoborneol, in saline lakes in Manitoba. Journal of the Fisheries Research Board of Canada 33:25–35.

Thayson, A. C. 1936. The origin of earthy or muddy taint in fish. I. The nature and isolation of the taint. Annals of Applied Biology 23:99–104.

Torrans, L., and F. Lowell. 1987. Effects of blue tilapia/ channel catfish polyculture on production, food conversion, water quality and channel catfish off-flavor. Proceedings of the Arkansas Academy of Science 41:82–86.

Tucker, C. S., and J. F. Martin. 1991. Environment-related off-flavors in fish. Pages 133–179 *in* D. E. Brune and J. R. Tomasso, editors. Aquaculture and water quality. The World Aquaculture Society, Baton Rouge, Louisiana.

van der Ploeg, M. 1992. Dynamics of off-flavors in preharvest channel catfish. Arkansas Aquafarming 10(2): 1–4.

van der Ploeg, M., and C. E. Boyd. 1991. Geosmin production by cyanobacteria (blue-green algae) in fish ponds at Auburn, Alabama. Journal of the World Aquaculture Society 22:207–216.

van der Ploeg, M., C. S. Tucker, and C. E. Boyd. 1992. Geosmin and 2-methylisoborneol production by cyanobacteria in fish ponds in the southeastern United States. Water Science and Technology 25:283–289.

Wnorowski, A. U., and W. E. Scott. 1982. Incidence of off-flavors in South African surface waters. Water Science and Technology 25:225–232.

Yurkowski, M., and J-A. Tabachek. 1980. Geosmin and 2-methylisoborneol implicated as a cause of muddy odor and flavor in commercial fish from Cedar Lake, Manitoba. Canadian Journal of Fisheries and Aquatic Sciences 37:1449–1450.

American Fisheries Society Symposium 16:73–84, 1996

Application and Modification of the Fish Health Assessment Index Used for Largemouth Bass in the Catawba River, North Carolina–South Carolina

David J. Coughlan, B. Kim Baker, Donald G. Cloutman,[1] and W. Mark Rash

Duke Power Co., Environmental Division
13339 Hagers Ferry Road, Huntersville, North Carolina 28078, USA

Abstract.—Health of largemouth bass *Micropterus salmoides* was assessed by the fish health assessment index (FHAI) procedure at various sites along the Catawba River, North Carolina–South Carolina, in 1993 and 1994. Tissues, organs, and blood from 15 fish per location were evaluated and scored for deviations from normal appearances or values to derive an FHAI score for each location. Minimally affected locations and locations near thermal discharges had the lowest scores, indicating relatively good fish health. Higher FHAI scores were noted at locations influenced by urbanization, industrialization, hydroelectric facility tailraces, and various combinations of these activities. We propose modifications of some variables and scoring methods, including use of fish between 250 and 450 mm; incorporation of a relative weight metric; deletion of parasite scoring; development of graded scales (as opposed to "all or none" scales) to score liver, eyes, and gills; and establishment of metrics to score gross abnormalities and gill raker deformities. The modified FHAI procedure is easy to apply and appears to produce scores that correspond to ecological health.

The fish health–condition assessment procedure (gross systematic evaluation of various organs, structures, and blood) was initially developed in the 1970s for evaluation of hatchery produced trout (R. Goede, Utah Division of Wildlife Resources, personal communication). This procedure was based on assumptions that observable changes in certain tissues and organs occur when fish are sufficiently stressed (Goede and Barton 1990). Fish health–condition assessments for fish populations are reported as three indices (normality, severity, and feeding; Goede 1993) and require interpretation by a trained biologist. Positive correlation of fish health indices and the success of various stocking programs in Utah demonstrated that "normal" fish survive in the wild better than "nonnormal" fish. The utility of this procedure for assessing fish health was further demonstrated by applications to free-ranging trout populations (Goede and Barton 1990) and ultimately to other fish species.

Adams et al. (1993) modified the above procedure to numerically score deviations from the "normal" appearance or the normal value of a prescribed list of metrics; numerical scores are summed to produce one number describing the health of a population. Fish health can be quantitatively assessed with the fish health assessment index (FHAI) procedure; higher scores indicate relatively poorer health. The FHAI is not designed to be a diagnostic procedure; its value lies in the ability to indicate areas where further study should be directed and to compare fish health scores within sites over time or among sites evaluated at the same time. The FHAI has been used extensively to study health of largemouth bass *Micropterus salmoides* in Tennessee Valley Authority reservoirs (Brown and Hickman 1990). This species has several characteristics that make it well suited for the FHAI procedure; a top predator in the aquatic food chain, it is easily captured, fairly nonmigratory, widely distributed, and highly valued by the fishing public.

The FHAI procedure appeared to provide an appropriate tool for a systematic investigation of largemouth bass health throughout the Catawba River, North Carolina–South Carolina. The river has 11 hydroelectric impoundments, and the watershed is characterized by limited agriculture, silviculture, expanding residential development, and regions of urbanization and industrialization. All of these activities affect water quality through surface runoff, erosion, sedimentation, water withdrawal, recreation, sewage treatment, and pollution. Fish integrate the effects of water quality variables through both direct contact and dietary pathways; water quality may also act in combination with biotic and other abiotic factors to influence fish health (Adams et al. 1993). In this paper we document the health of largemouth bass sampled from diverse

[1]Present address: Mississippi State University, Department of Fisheries and Wildlife, Box 9690, Mississippi State, Mississippi 39762, USA.

habitats along the Catawba River in 1993 and 1994, assess temporal and spatial trends in FHAI data, and propose modifications to the FHAI procedure.

Methods

The Catawba River originates on the Atlantic slope of the Appalachian Mountains and travels 362 km through the Piedmont region of North Carolina and South Carolina before becoming the Wateree River and ultimately a part of the Santee–Cooper basin in South Carolina (Figure 1). Fish health assessments were conducted during September and October at 26 (1993) and 27 (1994) Catawba River sites representing main-stem reservoir and riverine habitats with varying levels of anthropogenic influence. We subjectively classified habitats as (A) minimal impact locations lacking obvious upstream National Pollutant Discharge Elimination System discharges and characterized by limited land use activities, (B) areas influenced by thermal discharges, (C) hydroelectric facility tailraces, (D) areas near industrial or sewage effluents, (E) areas remotely influenced by industrial or sewage effluents, and (F) locations influenced by two or more classifications B–E.

Fifteen largemouth bass were collected by daytime electrofishing at each location. The minimum size of fish used was 250 mm total length (TL). The upper size was unlimited in 1993 but capped at 450 mm TL in 1994. Fish were anesthetized (tricaine methanesulfonate), the visceral and pericardial cavities were opened, livers were examined immediately for coloration, and blood samples were taken. Fish were tagged, set aside, and evaluated for the FHAI variables (Table 1) described by Adams et al. (1993) upon completion of the blood sampling. Otoliths were collected from all fish and sectioned for age determination (Jearld 1983). Parasites in viscera, gills, and eyes from eight largemouth bass from each 1993 location were relaxed and preserved in a heated formalin solution (Cable 1977) immediately after completion of the FHAI. Parasites were identified and enumerated in the laboratory (Hoffman 1967; Roberts 1970; Beverley-Burton 1984). We calculated FHAI scores both by the criteria of Adams et al. (1993) and by modified criteria developed during this study (Table 1).

Results

In 1993, FHAI scores ranged from 18 to 94 and averaged 42 (Figure 2; Table 2). Minimal impact locations (A) and thermal discharges (B) exhibited the lowest scores, while locations influenced by combinations of activities (F), hydroelectric facility tailraces (C), and regions of urbanization and industrialization (D) had the highest scores (Table 3). Locations warranting further study during 1993 (i.e., those with the highest scores) were, in decreasing order, 21, 20, and 6 and 27 (tied).

In 1994, FHAI scores ranged from 21 to 62 and averaged 37 (Figure 2; Table 2). The high score at location 6 in 1993 was not repeated in 1994. Location 4 (33 km upstream from location 6) exhibited the highest score in 1994. The 1993 peaks at locations 20, 21, and 27 were repeated in 1994, though their departures from the other scores were less pronounced. Average FHAI scores of all categories of anthropogenically influenced locations in 1994 were the same as or lower than 1993 scores; ranks remained the same for all but the two lowest scoring categories (Table 3). Locations warranting further study in 1994 were, in decreasing order, 4, 21, and 20.

Of all the metrics evaluated during this study, liver discoloration was the most frequently observed abnormality. The percentage of fish with anomalous livers was slightly higher in 1994 than in 1993 (Table 4). At each minimal impact location, at least one fish had abnormal liver coloration; at the highest scoring site in 1993, all 15 fish had discolored livers. Liver scores exceeded or equaled those for trunk kidney or spleen at 48 of 53 surveyed locations over the two years. Trunk kidney scores (typically due to swelling) were highest at three locations (none were areas with unusually high FHAI scores) and spleen scores were highest at two minimal impact locations. The only spleen anomalies we saw were nodules and one fish with a satellite spleen. A lower percentage of fish had anomalous trunk kidneys and spleens in 1994 than in 1993 (Table 4).

Of 795 largemouth bass examined during 1993 and 1994, we did not observe a single anomalous thymus, pseudobranch, or hindgut. Opercles of four fish exhibited slight shortening, and fin erosion was of minor importance. Hemorrhagic skin lesions were typically mild and infrequent except at locations with elevated scores, where a majority of the fish displayed mild lesions. Blood variables were minor contributors to FHAI scores and did not vary by location classification. A slight decrease in hematocrit and increase in leucocrit in 1994 compared to 1993 may have resulted from improved fish-holding facilities in our electrofishing boat (Table 4).

Locations with the highest FHAI scores in 1993 also had the highest average fish weights. A significant positive linear relation between average

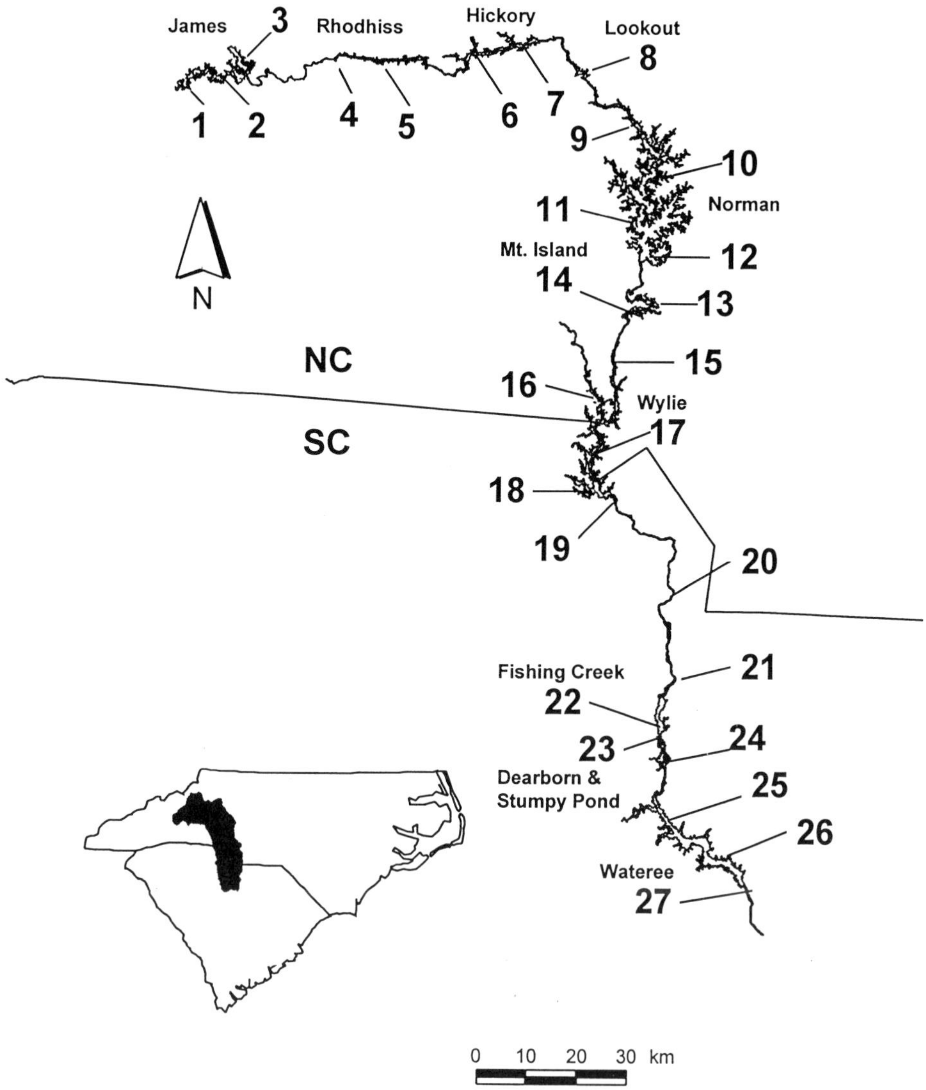

FIGURE 1.—Sampling locations on the Catawba River, North Carolina–South Carolina.

weight and FHAI score was noted (Figure 3). Larger fish deviated more from normal (i.e., they scored higher) than smaller fish. Similarly, average fish age and FHAI scores exhibited a significant positive linear relationship that was best described as

$$\text{FHAI} = -3.54 + 10.13\ (\text{age})$$

($N = 26, F = 6.66, P = 0.016, r^2 = 0.22$); fish age in 1993 ranged from 2 to 13 years (mean, 4.5 years).

The exclusion of fish larger than 450 mm in 1994 may have contributed to the lack of a relationship ($N = 27, F = 3.73, P = 0.07$) between average weight and FHAI score (Figure 3). However, average fish age and FHAI scores continued to exhibit a significant positive linear relationship that was best described as

$$\text{FHAI} = 1.06 + 10.53\ (\text{age})$$

($N = 27, F = 11.05, P = 0.003, r^2 = 0.31$); fish age in 1994 ranged from 2 to 10 years (mean, 3.4 years).

Parasites in fish were identified and enumerated from five locations associated with the lowest, highest, and three other unique FHAI scores. Seven

TABLE 1.—Tissues, organs, and blood variables assessed during the fish health assessment index (FHAI) procedure, the conditions associated with them, designations used in the field, and substituted values used to calculate FHAI scores for Catawba River largemouth bass health (modified from Goede and Barton 1990 and Adams et al. 1993).

Tissue, organ, or variable	Condition	Designation	Substituted value
Liver	Normal; solid red or light red color	A	0
	"Fatty" liver; light tan color as "coffee with cream"		
	Color moderate[a]	C1	15
	Color severe	C2	30
	Cysts or nodules	D	30
	Focal discoloration; change of color in local areas or foci of liver	E	30
	General discoloration of whole liver		
	Color moderate[a]	F1	15
	Color severe	F2	30
	Other; any observation that does not fit above categories	OT	30
Gills	Normal with no apparent aberrations	N	0
	Frayed; erosion of tips of lamellae resulting in "ragged" appearing gills	F	30
	Clubbed; swelling of gill lamellae tips	C	30
	Marginate; light gill margin, discolored lamellar tips	M	30
	Pale; whole gills light, discolored	P	30
	Other; any observation that does not fit above categories		
	Mild[a]	OT1	10
	Moderate[a]	OT2	20
	Severe	OT3	30
Gill rakers[a]	Normal		0
	Slightly deformed or missing (<5 rakers)		10
	Moderately deformed or missing (5–10 rakers)		20
	Severely deformed or missing (>10 rakers)		30
Pseudobranch	Normal; flat with no aberrations	N	0
	Swollen; convex in appearance	S	30
	Lithic; mineral deposits, amorphous white spots	L	30
	Swollen and lithic	X	30
	Inflamed	I	30
	Other; any observation that does not fit above categories	OT	30
Thymus	Normal appearance; no hemorrhage		0
	Mild hemorrhage		10
	Moderate hemorrhage		20
	Severe hemorrhage		30
Mesenteric fat	No fat between pyloric ceca	0	
	Less than 50% of ceca covered with fat	1	
	50% of ceca covered with fat	2	
	More than 50% of ceca covered with fat	3	
	Ceca totally covered with fat	4	
Bile	Straw color, bladder empty	0	
	Straw color, bladder full	1	
	Grass green color, bladder full	2	
	Dark green, bladder full	3	
Spleen	Normal; black, very dark red, or red	B	0
	Normal; granular appearance	G	0
	Nodular; nodules or fistulas of various sizes	N	30
	Enlarged	E	30
	Other; any observation that does not fit above categories	OT	30
Hindgut	Normal; no inflammation or reddening		0
	Slight inflammation or reddening		10
	Moderate inflammation or reddening		20
	Severe inflammation or reddening		30
Sex	Male	M	
	Female	F	
Trunk kidney	Normal; firm, lying relatively flat dorsally along the ventral surface of the vertebral column	N	0
	Swollen; enlarged or swollen, wholly or in part	S	30
	Mottled; gray discoloration	M	30
	Granular; granular appearance or texture	G	30
	Urolithic; white or cream-colored mineral deposits in kidney tubules, nephrocalcinosis	U	30
	Other; any observation that does not fit above categories	OT	30

TABLE 1.—Continued.

Tissue, organ, or variable	Condition	Designation	Substituted value
Eyes	Normal clear eyes (lens); no aberrations	N	0
	Lenticular opacity (blind)		
	One eye	B1	15[a]
	Both eyes	B2	30
	Exopthalmia; swollen or protruding eye		
	One eye	E1	15[a]
	Both eyes	E2	30
	Hemorrhagic or bleeding		
	One eye	H1	15[a]
	Both eyes	H2	30
	Missing		
	One eye	M1	15[a]
	Both eyes	M2	30
	Other; any observation that does not fit above categories		
	One eye	OT1	15[a]
	Both eyes	OT2	30
Opercles	Normal; no shortening, gills completely covered		0
	Slight shortening, a very small portion of the gills exposed		10
	Moderate shortening, a small portion of the gills exposed		20
	Severe shortening, a considerable portion of the gills exposed		30
Skin	Normal; no hemorrhagic areas		0
	Mild hemorrhagia on skin surface (<10%)		10
	Moderate hemorrhagia on skin surface (10–60%)		20
	Severe hemorrhagia on skin surface (>60%)		30
Fins	Normal; no active erosion		0
	Light active erosion		10
	Moderate active erosion with some hemorrhaging		20
	Severe active erosion with hemorrhaging		30
Parasites	No observed parasites		0
	Few observed parasites, parasites in just one organ		10
	Moderate parasite infestation, parasites observed in several organs		20
	Numerous observed parasites, extensive infestation in several organs		30
Gross abnormalities[a,b]	No visible gross abnormalities	N	0
	Tumors visible on external surfaces	E	30
	Tumors visible on internal surfaces	I	30
	Lordosis of vertebral column	L	30
	Scoliosis of vertebral column	S	30
	Skeletal deformities or broken bones of head and jaws	D	30
	Skeletal deformities or broken bones of remaining bony structures	B	30
	Other; any observation that does not fit above categories	OT	30
Hematocrit (%)[c]	Normal range (30–45)		0
	Above normal range (>45)		10
	Below normal range (19–<30)		20
	Well below normal range (<19)		30
Leucocrit (%)[c]	Normal range (0–<4)		0
	Above normal range (≥4)		30
Plasma protein (g/dL)[d]	Normal range (3–7)		0
	Above normal range (>7)		10
	Below normal range (<3)		30
Relative weight (%)[a]	≥85.00		0
	≥70.00 and <85.00		15
	<70.00		30

[a]Variables used to calculate modified scores only.
[b]All gross abnormalities were noted if more than one occurred, score only (in the observer's opinion) the most severe.
[c]Measured after centrifugation for 5 min in a microhematocrit centrifuge.
[d]Measured with a protein refractometer.

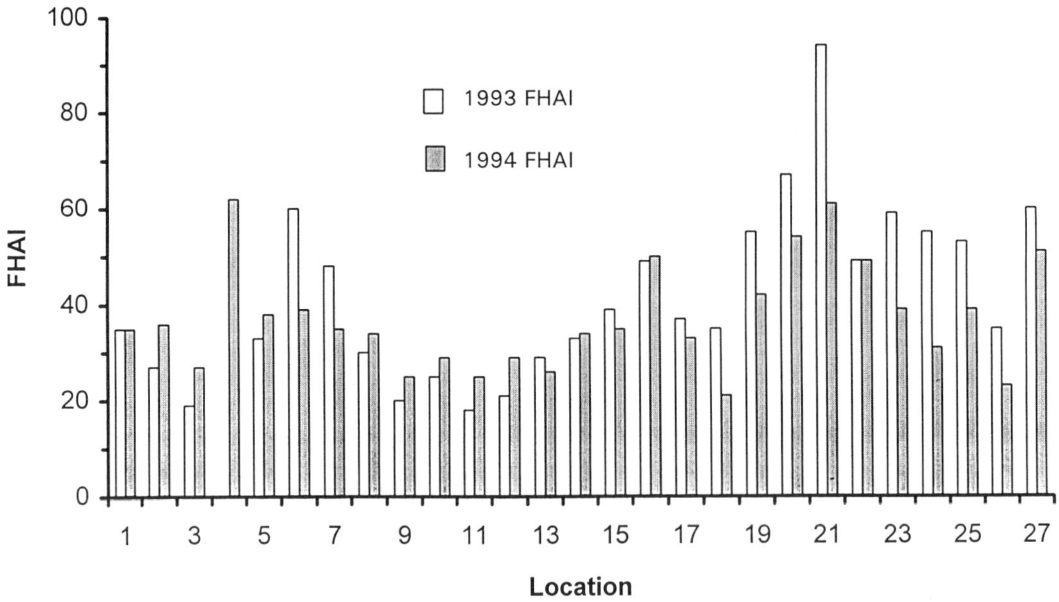

FIGURE 2.—Fish health assessment index (FHAI) scores for largemouth bass collected from 27 locations in the Catawba River.

core species (defined as a parasite species found in more than 50% of the fish examined at a location) were identified; 25 additional species were found in very low numbers and considered of doubtful significance. Four of the core species were not scored by the FHAI procedure. Three gill monogeneans (*Onchocleidus principales*, *Haplocleidus furcatus*, and *Syncleithrium fusiforme*) were visible only microscopically and would not normally be scored. The acanthocephalan *Neoechinorhynchus cylindratum* was found in the intestine and would not have been scored because the intestine was not surveyed. Only three core species, the copepod *Ergasilus centrarchidarum*, metacercariae of the digenetic trematode *Posthodiplostomum minimum*, and pleurocercoids of the tapeworm *Proteocephalus ambloplitis*, were visible grossly and accounted for the majority of our FHAI parasite scores (Figure 4). The locations with the lowest and highest FHAI scores in 1993 had fish infested with 18 and 16 parasite species, respectively.

Discussion

In 1993 and 1994, FHAI scores identified similar locations with elevated scores, though the 1994 data did not show the prominent peaks seen in 1993. One of our a priori concerns with the recently developed FHAI procedure was the potential problem of repeatability over time. This study indicated that the FHAI procedure consistently indicated sites where largemouth bass have reduced health.

We have viewed the FHAI information as a tool to prioritize areas warranting further study and avoided rigorous statistics, though they can be applied. Predetermined "concern" levels (average FHAI scores of 90 and greater; Brown 1993) have been used to indicate locations warranting further study. This value would equal scores of approximately 75 and greater based on the differential blood scoring procedures described in the FHAI procedure of Brown and Hickman (1990) and Adams et al. (1993). Based on a "concern" value of 75, only location 21 evaluated in 1993 would warrant further study, which suggests limited impacts of human activities on Catawba River largemouth bass.

The positive relationship between FHAI scores and both weight and age indicated FHAI scores could be biased high if collections contained many large, old individuals within a location. This concern is especially relevant when locations with numerous large fish are compared to nutrient-limited locations that experience heavy sportfishing pressure and are characterized by small individuals. Sampling by electrofishing, which is more effective in capturing larger than smaller individuals of a spe-

TABLE 2.—Mean fish health assessment index (FHAI) scores ± SD (Adams et al. 1993) and modified FHAI scores for samples of 15 largemouth bass collected from 27 locations in the Catawba River during 1993 and 1994.

Location	Location description[a]	FHAI 1993	FHAI 1994	Modified FHAI 1993[b]	Modified FHAI 1994
1	D	35 ± 20	35 ± 28	20 ± 20	27 ± 29
2	A	27 ± 27	36 ± 26	14 ± 14	30 ± 24
3	A	19 ± 14	27 ± 23	10 ± 15	18 ± 23
4	D		62 ± 22		42 ± 24
5	E	33 ± 26	38 ± 18	19 ± 18	36 ± 18
6	D	60 ± 29	39 ± 25	50 ± 34	29 ± 26
7	E	48 ± 40	35 ± 21	20 ± 25	16 ± 15
8	A	30 ± 24	34 ± 20	21 ± 25	24 ± 18
9	A	20 ± 18	25 ± 22	20 ± 21	27 ± 21
10	B	25 ± 27	29 ± 22	36 ± 26	27 ± 23
11	A	18 ± 15	25 ± 17	17 ± 17	25 ± 15
12	B	21 ± 17	29 ± 18	24 ± 15	31 ± 16
13	A	29 ± 19	26 ± 21	24 ± 21	26 ± 18
14	B	33 ± 20	34 ± 19	27 ± 22	34 ± 16
15	D	39 ± 28	35 ± 19	22 ± 18	26 ± 17
16	F	49 ± 24	50 ± 36	37 ± 24	33 ± 23
17	A	37 ± 30	33 ± 23	21 ± 25	17 ± 15
18	B	35 ± 25	21 ± 14	21 ± 20	9 ± 15
19	C	55 ± 39	42 ± 28	37 ± 36	27 ± 18
20	D	67 ± 21	54 ± 25	55 ± 17	39 ± 23
21	F	94 ± 29	61 ± 27	69 ± 26	41 ± 27
22	D	49 ± 29	49 ± 28	32 ± 26	35 ± 25
23	F	59 ± 40	39 ± 21	43 ± 41	27 ± 23
24	E	55 ± 38	31 ± 31	33 ± 34	20 ± 24
25	E	53 ± 34	39 ± 28	34 ± 24	25 ± 22
26	A	35 ± 23	23 ± 15	23 ± 22	15 ± 12
27	C	60 ± 28	51 ± 27	40 ± 31	30 ± 23
Average		42 ± 18	37 ± 11	30 ± 14	27 ± 8

[a]Locations are classified as: (A) minimal impact; (B) thermal discharge; (C) hydroelectric facility tailrace; (D) industrial or sewage effluent; (E) remote influence of industrial or sewage effluent; (F) combinations of B through E.
[b]Means based on variable numbers of fish.

cies (Reynolds 1983), may further bias FHAI scores upward. Large individuals might receive higher FHAI scores because of degenerative changes associated with old age or relatively longer duration of exposure to environmental perturbations. Clearly, old age or high weight can confound the interpretation of FHAI results.

Studies involving bioenergetics and physiological health use fish of uniform size, and it seems reasonable that this criterion should apply to the FHAI procedure. Goede and Barton (1990:96) stated that "comparisons of stressed fish against normal or control conditions should be based on fish of the same age, strain, sex, and season." We propose restricting the FHAI procedure to fish of a specified size range (≥250 mm and ≤450 mm for largemouth bass) to provide samples that are homogeneous across locations. The 450-mm size limit was a compromise between a smaller upper size limit that would restrict the age range of fish to a few years and the

TABLE 3.—Means, SDs, and (in parentheses) ranks (1–6; low–high) for fish health assessment index (FHAI; Adams et al. 1993) and modified FHAI scores by location type on the Catawba River during 1993 and 1994. Means followed by the same letter within a column are not significantly different as determined by Duncan's multiple-range test (α = 0.05).

Code	Description	FHAI 1993	FHAI 1994	Modified FHAI 1993	Modified FHAI 1994
A	Minimal impact	26.9 ± 7.3 (1) z	28.6 ± 4.9 (2) z	18.8 ± 4.8 (1) z	22.8 ± 5.4 (1) z
B	Thermal discharge	28.5 ± 6.6 (2) z	28.3 ± 5.4 (1) z	27.0 ± 6.5 (3) zy	25.3 ± 11.2 (3) z
C	Hydroelectric facility tailrace	57.5 ± 3.5 (5) yx	46.5 ± 6.4 (5) yx	38.5 ± 2.1 (5) yx	28.5 ± 2.1 (4) z
D	Industrial or sewage effluent	50.0 ± 13.6 (4) yx	45.7 ± 11.1 (4) yx	35.8 ± 16.0 (4) zy	33.0 ± 6.7 (5) z
E	Remote industrial or sewage effluent	47.3 ± 9.9 (3) y	35.8 ± 3.6 (3) zy	26.5 ± 8.1 (2) zy	24.3 ± 8.7 (2) z
F	Combinations of B through E	67.3 ± 23.6 (6) x	50.0 ± 11.0 (6) x	49.7 ± 17.0 (6) x	33.7 ± 7.0 (6) z

TABLE 4.—Overall occurrence of atypical major organs and mean hematological variables (SD) of Catawba River largemouth bass by year.

Year	Organ			Hematological variable		
	Liver (%)	Trunk kidney (%)	Spleen (%)	Hematocrit (%)	Leucocrit (%)	Plasma protein (g/dL)
1993	41.2	21.5	10.3	39.4 (4.80)	0.6 (0.35)	6.2 (0.87)
1994	44.4	7.9	6.2	36.4 (4.31)	0.9 (0.54)	6.2 (0.77)

need to minimize fish collection and holding times without undue imposition of stress. Additionally, the 250–450-mm size range would eliminate the sacrifice of large "trophy" bass for FHAI procedures. The inclusion of an upper size limit in our 1994 FHAI work allowed comparison of similar sized fish and may have contributed to the lack of significance between weight and FHAI score. The continued significant positive relationship between age and FHAI score may indicate the need for an even smaller size range or an age-adjusted FHAI score.

Largemouth bass collected from one site (location 10) in 1993 were emaciated, but otherwise appeared healthy (FHAI = 25). Our impression of these fish was one of poorer health than the FHAI implied. Lengths (TL, mm) and weights (g) enter the fish health–condition assessment procedures (as condition factor) of Goede and Barton (1990), but not the FHAI score of Adams et al. (1993). We

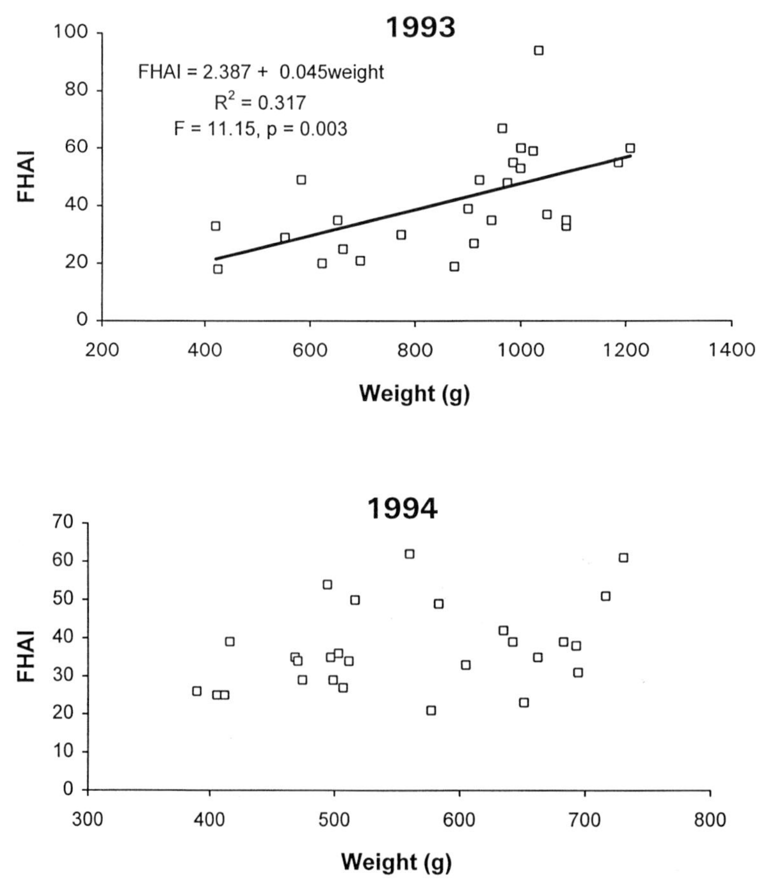

FIGURE 3.—Fish health assessment index (FHAI) score as a function of average weight for largemouth bass collected from the Catawba River during 1993 and 1994. A significant relationship was found in 1993 but not in 1994.

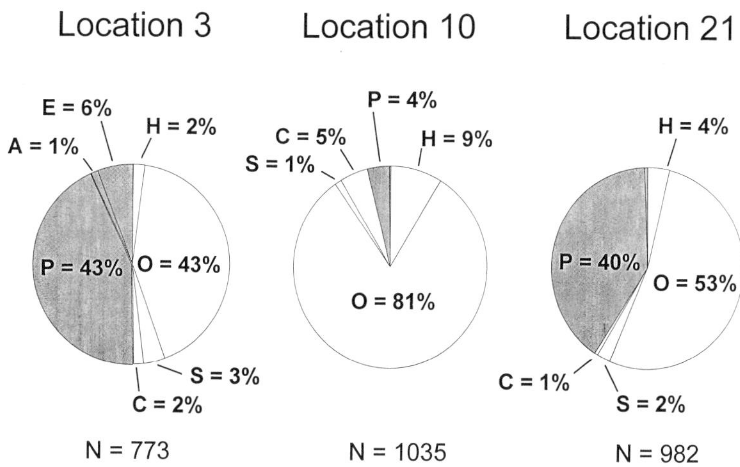

FIGURE 4.—Average composition of core parasites in the viscera, gills, and eyes of eight largemouth bass from each of three locations in the Catawba River in 1993. Location 3 was a minimal impact site with one of the lowest fish health assessment index (FHAI) scores, location 10 represented a site with low FHAI score but with emaciated fish, and location 21 had the highest FHAI score. Shaded areas indicate grossly visible parasites: (A) *P. ambloplitis*, (C) *N. cylindratum*, (E) *E. centrarchidarum*, (H) *H. furcatus*, (O) *O. principales*, (P) *P. minimum*, (S) *S. fusiforme*; *N* is the average number of parasites found in eight largemouth bass.

suggest the inclusion of a metric that considers emaciated fish as unhealthy and vulnerable to additional stressors.

Relative weight is a comparison of measured lengths and weights against a standard weight–length regression for a species based on populations from across its geographic distribution (Murphy et al. 1991). The largemouth bass standard weight equation of Henson (1991, cited by Murphy et al. 1991) is

$$\log_{10} \text{weight} = -5.528 + 3.273 \log_{10} \text{TL}.$$

We propose that a relative weight metric be incorporated into the FHAI by calculating and scoring relative weight for each fish (Table 1).

Parasites are a normal component of fish populations; however, their utility as a fish health monitoring tool is unsupported for two reasons. First, an evaluation of the relative detriment of parasite burdens based on grossly visible parasites, without some consideration of unobserved parasites, is tenuous. Unobserved parasites, both microscopic and internal, were generally more numerous than parasites visible on external surfaces (Figure 4). It could be speculated that the tremendous numbers of "unobserved" *O. principales* on the gills of fish at location 10 contributed to the emaciated condition of these fish in 1993. Indeed, bacterial pathogens probably pose a greater fish health concern than parasites (Overstreet 1993), but they receive no attention in fish health evaluation procedures. We understand the need to keep FHAI procedures simple and do not endorse detailed examinations of fish for parasitic or bacterial infections. We suggest that chronic stress resulting from harmful parasites, bacteria, or contaminants, acting alone or in combination, will be readily detected by changes in relative weight. Second, toxicant-induced immunosuppression and subsequent impaired ability to resist parasites is the basis for relating heavy parasite burdens to degraded environments (Poulin 1992). However, parasites and their intermediate hosts, when applicable, exhibit a wide range of sensitivities to innumerable toxicants in various combinations and concentrations (Poulin 1992; Overstreet 1993). Generalizations about the interactions between fish parasites, the susceptibility of fish hosts, and pollutants are not possible (Overstreet 1993). Within a single fish species in a contaminated environment, infection by one parasite species may decrease while that by another may increase (Khan 1991; Khan et al. 1992). We suggest continued monitoring of fish parasites as a means of collecting historical data; however, fish parasites should not be incorporated into the FHAI.

The narrower size range of largemouth bass used in 1994 probably contributed to the lower scores obtained for that year; however, meteorological conditions may also have been a contributing factor. The summer of 1993 was hotter and drier than that of 1994. Total rainfall and mean temperature from June through October 1993 were 39.3 cm and 22.9°C compared to 64.0 cm and 21.7°C during the same period in 1994. Flood conditions existed throughout the Catawba River in August 1994.

Recommendations

Based on the results of this 2-year FHAI study, we gained valuable insights into the methodology and can propose several refinements to the procedure.

1. Establish an upper length limit to reduce the size range of fish used in the FHAI procedure; appropriate size ranges will vary by species and may vary by geographic region. Studies evaluating FHAI scores by size-class of fish in pristine and degraded environments would help elucidate the relation of size to score.
2. Incorporate a relative weight metric to account for fish with abnormal weights.
3. Remove the parasite metric from the FHAI scoring procedure.
4. Develop a graded scale (at a minimum 0, 15, and 30) to replace the "all-or-nothing" metrics currently available for assessing liver and eyes (Table 1). We have examined numerous livers exhibiting only slight discoloration that would appropriately be classified as 15, but which receive the same score as one exhibiting gross discoloration under the current procedure. Similarly, a fish with only one missing eye or lenticular opacity receives the same score (30) as a fish with two problematic eyes. We think a score of 15 would better represent intermediate conditions.
5. Develop a graded scale (10, 20, and 30) to supplement the "all-or-nothing" metrics currently available for frayed, clubbed, marginate, or pale gills (Table 1). This additional designation could be used for otherwise normal gills displaying anomalies (e.g., missing sections of filaments) that do not correspond to the current designations. We also suggest the application of partial scores based on the relative degree of damage.
6. Establish a metric that accounts for gross anomalies. Previous experiences and the observation of a fish with lordosis in 1993 indicated the need to include a gross abnormality parameter. This metric would be of benefit in instances where skeletal deformities, tumors, etc. are observed. Though rare in our work, these anomalies should be scored when encountered.
7. Evaluate a metric that accounts for gill raker anomalies. The structure of gill rakers was thought to be sensitive to contaminants (Goede, personal communication) and was evaluated during the 1994 survey. Our data did not indicate any relation between gill raker deformities and location FHAI score, but we think further study in degraded locations is warranted. The ease of observation and scoring add minimal effort to the FHAI procedure.
8. Promote consistent application of the FHAI procedure. We suggest that experienced practitioners use photographic documentation of the various anomalies and conditions as aids in training others.

Effects of the Recommendations on Our Data

We recalculated our 1993 FHAI scores to examine the effects of incorporating these recommendations (hereafter referred to as the modified FHAI): use of fish between 250 and 450 mm TL, incorporation of relative weight, use of a graded score for eyes, elimination of the parasite metric, and inclusion of gross abnormalities. No information allowing the use of graded scales was available for livers or gills in 1993. The elimination of fish larger than 450 mm resulted in unequal sample sizes among sites for the 1993 modified FHAI scores. The 1993 modified FHAI scores were lower at 23 of 26 sites, while displaying the same pattern as the unmodified scores (Table 2; Figure 5). The site with emaciated fish (location 10) yielded a below-average FHAI, but an above-average modified FHAI, the latter indicating relatively poorer health. As with the original index, locations 21, 20, and 6 would warrant further study based on modified FHAI scores. The ranking of anthropogenic influence categories for modified FHAI scores remained essentially the same as for FHAI scores in 1993 (Table 3). No relationship existed ($F = 0.852$, $P = 0.365$) between average fish weight and 1993 modified FHAI score.

Scores calculated for the 1994 modified FHAI included 15 fish per site and recommendations 1–7 (Figure 5). Modified 1994 FHAI scores were the same as or lower than FHAI scores for 25 of 27 sites. Locations 4, 21, and 20 scored the highest with the modified method, although they did not stand out as prominently from other locations as in 1993. Ranks of 1994 modified FHAI scores varied from those of the original index, and no statistical difference was found between categories of anthropo-

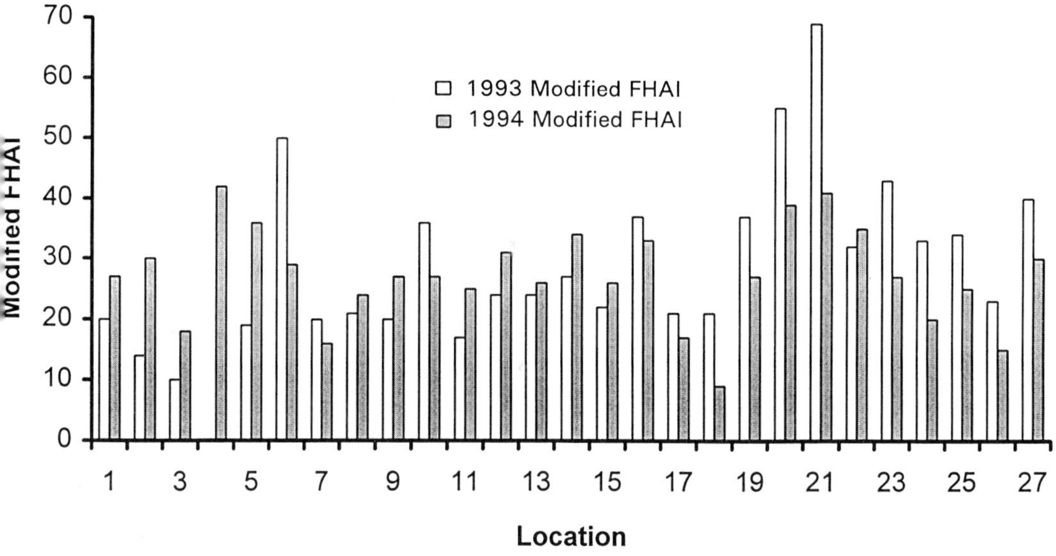

FIGURE 5.—Modified fish health assessment index (FHAI) scores for largemouth bass collected in 27 locations in the Catawba River.

genic influence (Table 3). No relationship (F = 0.057, P = 0.81) was found between average weight and modified FHAI score in 1994.

The size range imposed on largemouth bass samples helped minimize the bias towards higher scores caused by larger fish and made comparisons among locations more meaningful. We think the more homogeneous size range, the use of relative weight to score emaciated fish, the lack of a parasite metric, the inclusion of graded scores for eyes, gills, and liver, and the use of a gross anomaly metric resulted in modified FHAI scores that are an improvement over the FHAI procedure of Adams et al. (1993) and that better reflect the health of largemouth bass in the Catawba River.

Conclusion

The FHAI procedure measures changes in the appearance or value of a prescribed set of metrics. These changes are assumed to result from the integrated response of fish to various stressful physical, chemical, and biological conditions. Our surveys of the Catawba River indicated that largemouth bass FHAI scores may reflect the ecological health of or water use activities at a location, though the impacts of anthropogenic influence on the FHAI scores were not dramatic. Reservoir managers can use FHAI scores to prioritize locations in need of further study, to evaluate one location over time, and to compare several locations sampled at the same time. These scores can be readily communicated to nontechnical personnel, who often allocate the resources required for aquatic studies. We suggest that a long-term fish health monitoring program is most appropriate, because annual variation may mask trends. The FHAI procedure is a valuable tool that is cost-effective and easy to use; however, we suggest certain modifications that would make the FHAI more representative of largemouth bass health.

References

Adams, S. M., A. M. Brown, and R. W. Goede. 1993. A quantitative health assessment index for rapid evaluation of fish condition in the field. Transactions of the American Fisheries Society 122:63–73.

Beverley-Burton, M. 1984. Monogenea and turbellaria. Pages 5–209 *in* L. Margolis and Z. Kabata, editors. Guide to the parasites of fishes of Canada, part I. Canadian Special Publication of Fisheries and Aquatic Sciences 74.

Brown, A. M. 1993. Fish community health in four western North Carolina reservoirs utilizing the fish health assessment index. Report of Tennessee Valley Authority to Western Piedmont Council of Governments, Hickory, North Carolina.

Brown, A. M., and G. D. Hickman. 1990. Fish health program 1989. Tennessee Valley Authority, Knoxville.

Cable, R. M. 1977. An illustrated laboratory manual of parasitology, 5th edition. Burgess, Minneapolis, Minnesota.

Goede, R. W. 1993. Fish health/condition assessment

procedures. Utah Division of Wildlife Resources, Logan.

Goede, R. W., and B. A. Barton. 1990. Organismic indices and an autopsy-based assessment as indicators of health and condition of fish. Pages 93–108 *in* S. M. Adams, editor. Biological indicators of stress in fish. American Fisheries Society Symposium 8.

Henson, J. C. 1991. Quantitative description and development of a species-specific standard growth form for largemouth bass, with application to the relative weight index. Master's thesis. Texas A&M University, College Station.

Hoffman, G. L. 1967. Parasites of North American freshwater fishes. University of California Press, Berkeley and Los Angeles.

Jearld, A., Jr. 1983. Age determination. Pages 301–324 *in* L. A. Nielsen and D. L. Johnson, editors. Fisheries techniques. American Fisheries Society, Bethesda, Maryland.

Khan, R. A. 1991. Effect of oil-contaminated sediment on the longhorn sculpin (*Myoxocephalus octodecemspinosus*) following chronic exposure. Bulletin of Environmental Contamination and Toxicology 47:63–69.

Khan, R. A., D. Barker, R. Hooper, and E. M. Lee. 1992 Effect of pulp and paper effluent on a marine fish *Pseudopleuronectes americanus*. Bulletin of Environmental Contamination and Toxicology 48:449–456.

Murphy, B. R., D. W. Willis, and T. A. Springer. 1991 The relative weight index in fisheries management status and needs. Fisheries 16(2):30–38.

Overstreet, R. M. 1993. Parasitic diseases of fishes an their relationship with toxicants and other environmental factors. Pages 111–156 *in* J. O. Couche an J. W. Fournie, editors. Pathobiology of marine an estuarine organisms. CRC Press, Boca Raton, Florida.

Poulin, R. 1992. Toxic pollution and parasitism in freshwater fish. Parasitology Today 8(2):58–61.

Reynolds, J. B. 1983. Electrofishing. Pages 147–163 *in* L. A. Nielsen and D. L. Johnson, editors. Fisherie techniques. American Fisheries Society, Bethesda Maryland.

Roberts, L. S. 1970. *Ergasilus* (Copepoda: Cyclopoida) revision and key to species in North America. Transactions of the American Microscopical Society 89 134–161.

American Fisheries Society Symposium 16:85–97, 1996

Assessing the Reservoir Fish Assemblage Index: A Potential Measure of Reservoir Quality

GARY D. HICKMAN AND THOMAS A. MCDONOUGH

Tennessee Valley Authority, Aquatic Biology Laboratory, Norris, Tennessee 37828-2017, USA

Abstract.—A reservoir fish assemblage index (RFAI), based on various aspects of resident fish communities, was developed to help assess the condition of water resources in Tennessee River valley reservoirs. Validity of the index depends on obtaining samples representative of existing fish assemblages, selecting biological attributes (metrics) that reliably reflect human disturbance over the possible range of degradation, and specifying reference conditions against which study sites can be evaluated. We used boat electrofishing to collect in littoral areas and experimental gill netting to collect limnetic bottom-dwelling fish species. Reservoirs were initially separated by operational characteristics (run-of-river mainstream and tributary storage). Tributary reservoirs were further categorized by physicochemical features generally associated with ecoregion geology and by hydrological characteristics. Reference conditions for each reservoir class were established from 1990 to 1994 standard sampling of Tennessee Valley Authority reservoirs, and inferences of biologists with knowledge of the reservoir system, the distribution of resident biota, and anthropogenic effects. Reference conditions were also developed for each reservoir zone (inflow, transition, and forebay). Our current sampling effort (15 shoreline electrofishing transects 300 m in length and a minimum of 10 successful overnight experimental gill-net sets) met representative sample requirements. Combining results from the two gear types increased RFAI precision. The RFAI detected significant differences among reservoirs and yielded consistent rankings of reservoirs across years. Use of the index in other river systems is necessary to test its performance under a wider range of conditions than is available along the Tennessee River. Correlation with known human impacts remains a critical need in determining if the RFAI provides a reliable method of assessing reservoir environmental quality.

Reauthorization of the Clean Water Act, along with increased public awareness of the importance of water quality, has contributed to increased interest in biomonitoring. Karr et al. (1986) and the Ohio Environmental Protection Agency (OEPA 1987) noted that biomonitoring often detects environmental quality problems when chemical and sediment toxicity tests indicate acceptable conditions. The Tennessee Valley Authority (TVA) recognized the need to assess aquatic environmental quality in Tennessee Valley reservoirs and initiated a "vital signs" monitoring program in 1990 (Dycus and Meinert 1991). This program included two biomonitoring components (fish and benthic macroinvertebrates) along with sediment toxicity, chlorophyll, and physicochemical measurements to obtain a holistic view of reservoir environmental quality.

Fish are often included in aquatic monitoring programs because of their importance in the aquatic food web and because they have long life cycles that allow them to integrate conditions over time. Gammon's (1976) approach to environmental quality was to measure various fish community attributes with his "index of well-being." The OEPA (1987) later modified this index to evaluate responses of riverine fish communities to environmental stressors. Karr (1981) developed an index (index of biotic integrity or IBI), which demonstrated that human impacts on midwestern streams could be identified by measuring various attributes of resident fish communities. Karr and Dudley (1981) defined biotic integrity as "a balanced, integrated, adaptive community of organisms having species composition, diversity, and functional organization comparable to that of natural habitat of the region." The IBI is a multimetric index based on comparisons of existing community characteristics to those present at relatively undisturbed reference sites. These reference sites provide a benchmark against which regional expectations for a healthy community are derived. The index, which is now routinely used to determine stream quality, has been adapted to Tennessee Valley streams for fish (Saylor and Scott 1987) and invertebrates (Kerans and Karr 1994). Karr and Dionne (1991) discussed biological integrity assessment needs for lakes and reservoirs and indicated that development of a fish index was critical to determining reservoir environmental quality.

Reservoirs provide a more difficult challenge than streams, rivers, or lakes in determining biotic integrity because they are artificial systems that lack natural reference sites. Jennings et al. (1995) discussed the various conditions making reservoirs unnatural systems and noted that their relatively temporary nature on an evolutionary time scale precludes expectation of adaptive communities. These concerns led Jennings et al. (1995) to conclude that the term "biotic integrity" is inappropriate in reservoir applications; they recommended instead the term "reservoir fish assemblage index" (RFAI).

In support of the TVA's aquatic-resource monitoring program, Dionne and Karr (1992) first analyzed long-term cove rotenone data from Tennessee Valley reservoirs. They concluded that this sampling technique provided biased data unsuitable for indexing reservoir fish populations because samples were restricted to shallow water at the back of coves and coves were purposefully not sampled if they were associated with human influence such as residential development. Jennings et al. (1995) then examined TVA's annual fall electrofishing sampling program. The conceptual framework of the stream IBI was maintained, but an alternative method was required to derive reference conditions. Although a preliminary index was developed with a potential to provide a reliable assessment of fish assemblage status, index scores exhibited high annual variation. Examination of variability in supplemental intensive sampling studies revealed that sampling error was the primary cause of high variability of RFAI scores. Analysis of 1989–1992 data led Jennings et al. (1995) to suggest that increased sampling effort and continued refinement of metrics probably would lead to a more reliable index. These recommendations had already been incorporated into our 1993 and 1994 monitoring program. This paper reassesses RFAI variability after these sampling protocol modifications and a slightly different method of establishing reference conditions were implemented. Specific objectives of this paper are to define adequate sample size for TVA lakes, test precision of metrics used in the RFAI, identify habitat effects, determine an appropriate sample allocation method (i.e., random, stratified random, or fixed station), and evaluate results from combining electrofishing and experimental gill-net sampling data.

Methods

Field Sampling

Soballe et al. (1992) noted that general patterns of reservoir structure and function can be determined by representative sampling of resident fish communities. A representative sample includes the majority of resident species with individual species in approximate proportion to their true existenc (Karr et al. 1986). All methods of sampling fis communities are selective. Because reservoir habi tats are diverse horizontally, longitudinally, and ver tically, a cost-efficient representative sample of th entire community is not possible with a single gea Meador et al. (1993) and Weaver et al. (1993 recommended a multigear approach to take advar tage of differences in gear selectivity and efficienc to achieve a more accurate representation of th fish community structure. We selected shorelin electrofishing for the shallow littoral zone and ex perimental gill netting for the limnetic bottom zon

Major zones (inflow, transition, and forebay a described by Kimmel and Groeger 1984) of Ter nessee River mainstream reservoirs and foreba zones of tributary reservoirs were sampled annuall with shoreline electrofishing and experimental gi netting at fixed locations between September an mid-November, 1990–1994 (Hickman et al. 199 Scott et al. 1992; Brown et al. 1993; Hickman et a 1994). Only two areas were sampled on mainstrear reservoirs where zones were indistinguishable. Ex cessive current limited sampling in mainstream ir flow areas to electrofishing. On tributary reservoi larger than 2,000 ha, one transition zone (the nur ber depending on the number of major arms of th reservoir) was also sampled. Stream IBIs were use to monitor inflows of tributary reservoirs.

During sampling in 1990 and 1991, 10 electrofish ing transects of 10 min each were run along with 1 overnight experimental gill-net sets (five 6.1-m par els with mesh sizes of 2.5, 3.8, 5.1, 6.4, and 7.6 cn respectively) in each appropriate reservoir zone. I 1992, electrofishing runs were changed to a dis tance-based approach; with each standard samp included 300 m of shoreline. Based on a prelimina analysis of electrofishing results through 1992, ir cluding an intensive sampling study at four inflo zones (Jennings et al. 1995), sampling effort in 199 was increased to 15 standard electrofishing runs. I addition, gill netting increased to 12 nets per site t assure 10 successful net sets. A global positionin system was used to measure distance during ele trofishing runs and identify standard sampling loc tions. Electrofishing and gill-net sampling was don in all habitats in approximate proportion to the occurrence within the zone. Primary and seconda habitats (in terms of area) were noted for eac sample.

All sport-fish species and channel catfish *Ictalur punctatus* were measured (total length, mm) an weighed (g). The remaining species captured wer counted and released. During electrofishing, fis observed but not captured were included if positive identification could be made; estimated numbers were used when high densities of identifiable fish were encountered. Age-0 fish were counted separately and, as in stream IBI calculations (Karr 1981), were excluded from proportional and abundance metrics because of sampling inefficiencies for this age-group. Only fish that had been measured were inspected externally for signs of disease, parasites, and anomalies (including natural hybridization). Other species groups often included several individuals that were observed but not captured; thus, disease and other proportions were not obtainable for these groups. Karr (1981) used occurrence of natural hybrids (i.e., those known not to be part of a fisheries management program) as a metric in stream IBI. We treated natural hybrids as an anomaly because most field crews were unable to distinguish natural hybrids in their true proportions, and thus we could not use this category as a full metric.

Identification of appropriate sampling protocol is a major hurdle to developing a meaningful RFAI. The basic premise is that the entire fish fauna of a particular area has been sampled in its true relative abundance without taxon-specific bias. Jennings et al. (1995) found sampling error to be high during early development of this index when samples were limited to 10 electrofishing runs. To further resolve sample adequacy and help determine sources of variation in sample data, intensive samples (45 standard electrofishing runs and 24 overnight experimental gill-net sets) were collected from two reservoir forebay zones (Chickamauga and Watts Bar reservoirs) once per week during October 1993. Dominant habitat features in each electrofishing run and at each gill-net site were recorded to determine habitat influences on RFAI metric results. A 50-m shoreline section between each electrofishing run was not sampled to mitigate the effects of one sample on the next.

TABLE 1.—Metrics used in the Reservoir Fish Assemblage Index and expected response to degradation.

Metric	Expected response to degradation
Taxon richness and composition	
Total number of species	Decrease
Number of *Lepomis* sunfish species	Decrease
Number of sucker species	Decrease
Number of intolerant species	Decrease
Percent individuals as tolerant species	Increase
Percent dominance (numerical percentage of most common species)	Increase
Trophic composition	
Number of piscivorous species	Decrease
Percent individuals as omnivores	Increase
Percent individuals as invertivores	Decrease
Reproductive composition	
Lithophilic spawning species	Decrease
Abundance	
Total number of individuals	Decrease
Fish health	
Percent with diseases, parasites, or anomalies[a]	Increase

[a]Anomalies include natural hybrids.

Reservoir Fish Assemblage Index

The current RFAI incorporates 12 fish community metrics from five general categories (Table 1). Nine of the metrics were used by Jennings et al. (1995). The other three are either variations of metrics suggested by Jennings et al. or new metrics: number of piscivorous species, percentage numerical dominance of the community by one species, and percentage of fish with diseases, parasites, or anomalies. Tolerance classifications and designations of trophic and reproductive guilds follow those identified by Dycus and Meinert (1994).

Separation of reservoirs into similar classes or groups is a critical initial step in developing reference conditions. Care must be taken to compare only those reservoirs for which similar communities would be expected. Independent scoring criteria were developed for reservoirs based on operational characteristics (run-of-river mainstream, tributary storage). Tributary reservoirs were further categorized to reflect variations in physicochemical features generally related to geological characteristics of different ecoregions and varying hydrological characteristics. Independent scoring criteria were also developed for reservoir longitudinal sections (i.e., inflow, transition, and forebay), each gear type, and combined gears.

Because reservoirs are artificial, no standard reference sites exist, and expected conditions must be obtained from other sources. Alternative approaches include using historical or preimpoundment conditions, predictive models, best observed conditions, or professional judgment. Preimpoundment conditions are inappropriate because impoundments eliminate prior habitats. The state of understanding of reservoir fish communities precludes effective predictive modeling of species richness, composition, and abundance, leaving only observed best conditions and professional judgment as alternative ways of establishing reference conditions for reservoirs.

Jennings et al. (1995) used the best observed

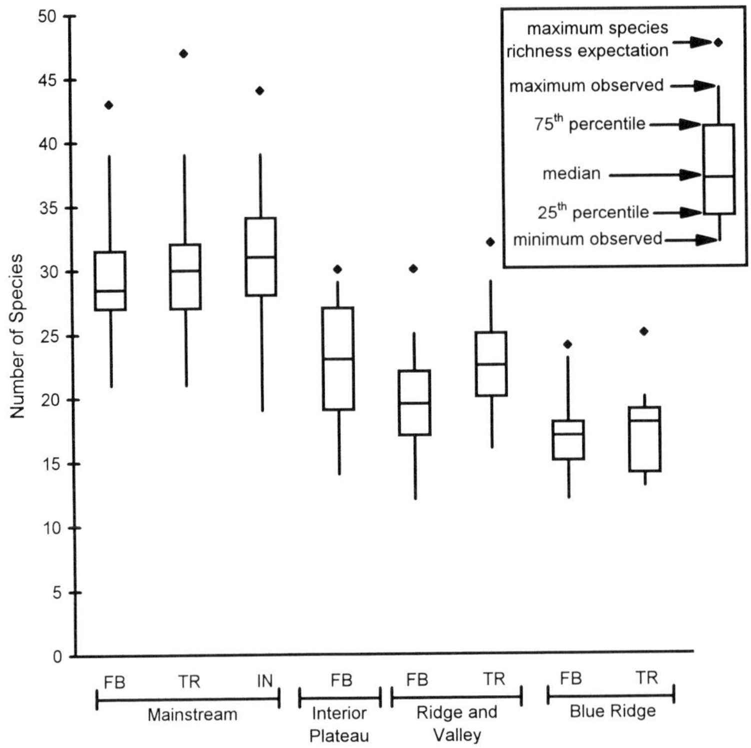

FIGURE 1.—Box-and-whisker plots of the species richness metric showing observed range and 25th and 75th percentiles for each class of reservoir site (1990–1994) compared with expected maximum species richness expectations in Tennessee River valley reservoirs. (FB = forebay; TR = transition; IN = inflow.)

conditions for similar reservoirs to establish reference conditions. This method is based on the assumption that the data set includes a wide range of conditions from good to poor for each community characteristic or metric. This assumption was not met in our reservoir system because the number of reservoirs within each group was limited. Therefore, we used 1990–1994 sampling results as the basis for reference conditions; for species richness metrics, we used a composite of all species collected throughout the 1990–1994 period from similar sites instead of using best observed conditions. Species richness expectations were adjusted according to inferences of experienced biologists knowledgeable about the reservoir system, resident fish species, susceptibility of individual species to electrofishing and gill netting, and human impacts on these species. Figure 1 compares the observed species richness range in each class of reservoir sites with maximum species richness expectations obtained with our approach. Our expectations are higher than would be obtained from maximum observed species richness based on individual samples. We feel this conservative scoring criterion is warranted because we lack reference sites. After maximum species richness was determined for each gear type, reservoir type, and longitudinal zone, the range of individual metric scores (0–95% of the expected maximum score) was trisected, as suggested by Fausch et al. (1984).

Scoring criteria for abundance and proportional metrics were determined by trisecting observed ranges after outliers (extreme values we could not trust) were omitted. We adjusted values for some metrics because the narrow range of the conditions measured did not include human-induced conditions observed in other reservoir groups. Scoring criteria for the fish health metric are those described by Karr et al. (1986).

For each metric, scoring criteria were used to separate sites into one of three categories we believed to represent relative degrees of degradation. Sample results are compared to reference conditions and assigned a corresponding value: least degraded - 5, intermediate - 3, and most degraded - 1 (Jennings et al. 1995). These categories are based on "expected" fish community characteristics in the absence of human impacts other than impoundment. Individual metric scores for a site are summed to obtain the RFAI score.

Statistical Analysis

Species accumulation curves and a bootstrap algorithm were used to determine the level of sampling effort necessary to collect representative samples. With species accumulation curves, a representative sample is assumed to be attained when the number of species collected approaches the total number of species expected. Fore et al. (1994) and Jennings et al. (1995) found the bootstrap method originally reported by Efron (1981) to be a reliable method to evaluate the precision of IBI-type indices in terms of variability among metric and index scores at a site. The bootstrap algorithm was used on the forebay special studies data to obtain an assessment of adequate sample size.

Sampling schemes (random, stratified random, and fixed) were analyzed by applying the bootstrap method on the special forebay-sampling results. Fixed-station sampling is sometimes used when variance among sites is greater than variation within specific sites; when this is not the case, simple random or stratified random sampling schemes are used. Stratified-random sampling is usually more efficient than simple-random sampling because it ensures that each stratum (habitat) is adequately sampled, thereby increasing estimate accuracy (Elliott 1977). Precision is usually increased because variance is computed from within-stratum differences, with variation among strata removed from sampling error. Variability of 100 random samples selected from the forebay special studies according to a simple-random design was compared to variability of samples selected from a stratified-random design. To determine if gear had any impact on appropriate sampling strategy, variabilities with both gears separately and combined under the two sampling strategies were compared. Comparisons were also made between results obtained from forebay special studies at fixed stations at Watts Bar and Chickamauga reservoirs.

Combined-gear RFAI was computed from pooled electrofishing and experimental gill-net sampling results for specific metrics (taxon richness, reproductive composition, and fish health). Abundance and proportional metrics used jointly (electrofishing and gill-net results) were worth one-half the combined weight for each metric value. Scoring criteria for combined metrics were developed with techniques described above.

Results of 1993 and 1994 standardized sampling for TVA's vital signs monitoring program were compared to the scoring criteria, and each of the 12 metrics was rated as discussed previously. Kendall's coefficient of concordance (Zar 1984) was used to compare reservoir rankings between years. We assumed that without new impacts or changes in impacts, rankings would be relatively similar.

Multivariate analysis of variance (MANOVA) was used on both forebay special-study data sets (all four weekly sample periods pooled) to assess the extent to which habitat differences among collection sites affect RFAI variability. Each of the 45 electrofishing and 24 gill-net sites in the forebay special studies were categorized by dominant or primary habitat features. For this analysis, we calculated the 12 metrics for individual electrofishing and gill-net samples. We used MANOVA to test for differences in community structure and function among habitats. When MANOVA indicated significant differences among habitats, univariate ANOVA (in which P-values were adjusted by the Bonferroni procedure) and graphical analyses (limited to the three metrics having the highest F-statistic) were used to determine which metrics and habitats accounted for these differences.

Results and Discussion

Variability of RFAI scores in bootstrap resampling analysis of forebay special studies at Watts Bar and Chickamauga reservoirs decreased with increasing sample size for both collection methods (Figure 2). As sample size increased, the rate of variability decrease lessened. Increasing the sampling effort beyond 15–20 electrofishing runs or 10–12 gill-net samples provided little improvement in precision. Confidence interval widths (± 2 SDs) included about 96% of samples. We wanted these error bands to be no wider than the range of individual RFAI categories (10), as described by Hickman et al. (1994). This level of precision was not consistently achieved for either collection method with our current sampling effort of 15 electrofishing transects and 10 overnight experimental gill-net sets. Our desired level of precision was achieved in all four sampling periods in Chickamauga Reservoir but in none of the four sample periods in Watts Bar with 15 electrofishing samples. Ten gill-net samples yielded this level of precision in three periods in Chickamauga Reservoir but in only one sample period in Watts Bar. Even by doubling current ef-

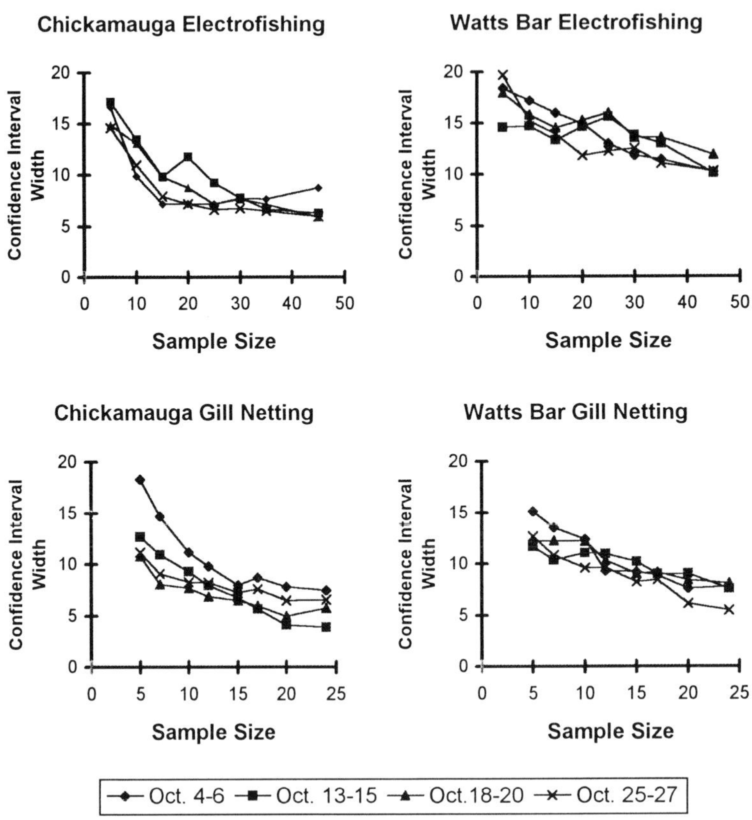

FIGURE 2.—Confidence interval widths (±2 SDs) of reservoir fish assemblage index scores from bootstrap samples of 100 randomly drawn sets of varying numbers of electrofishing runs and gill-net samples taken in Chickamauga and Watts Bar reservoir forebays, 1993.

fort, our desired level of precision was never reached with electrofishing in Watts Bar Reservoir.

Variability in RFAI from the bootstrap resampling analysis based on our current sampling effort (15 electrofishing runs and 10 gill-net sets) was lower for combined gears than for individual gears (Figure 3). Our desired level of precision was achieved in both reservoirs in all four sample periods for the combined-gear index. Between-sample variability was also lower for the combined-gear index than for individual gears (Figure 3). The standard deviation of combined-gear RFAI (for pooled data from all four sample periods) was 42% lower than the electrofishing RFAI and 28% lower than the gill-net RFAI in Chickamauga Reservoir. In Watts Bar Reservoir the SD of the combined index was 44% lower than the SD of the electrofishing index and 30% lower than that of the gill-net index. Within-sample period variability of individual-gear RFAI could be reduced by considerably increasing effort; however, for our purposes, this is prohibitively costly. Combining sample results of the two gears provided a more cost-effective approach to reducing RFAI variability both within and between periods.

Jennings et al. (1995) reported that a minimum of 15 electrofishing runs was necessary at mainstream inflow zones to approach the asymptote of species accumulation curves derived from 20 runs per sampling effort. In our study, intensive electrofishing sampling on forebay zones of two mainstream reservoirs (Chickamauga and Watts Bar) revealed that species accumulation curves (based on 45 electrofishing runs per sampling effort) began to flatten at fewer than 10 runs (Figure 4). After 15 runs, 78–84% of the total species collected during the 45 runs for that sample period had been collected. Gill-net samples (24 overnight sets per sampling period) from each of these forebay areas revealed a leveling off of species accumulation curves beginning at six

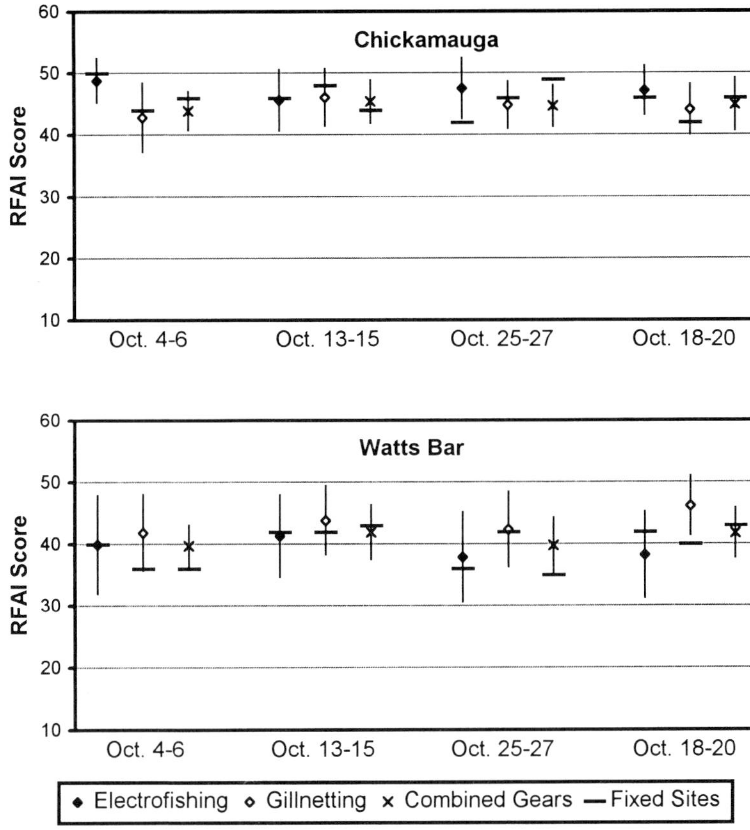

FIGURE 3.—Means and confidence intervals (±2 SDs) of reservoir fish assemblage index (RFAI) scores from bootstrap samples of 100 randomly selected sets of 15 electrofishing runs, 10 gill-net samples, and combined electrofishing and gill-net samples taken in Chickamauga and Watts Bar reservoir forebays, 1993. Fixed-station RFAI scores are included for comparison.

sets. Ten overnight net sets resulted in collection of 78–93% of the total species collected in 24 sets at either reservoir forebay zone. Combining the results of electrofishing and gill-net sampling from a site revealed that 15 runs and 10 overnight sets resulted in collection of 84–95% of the species obtained from all 45 runs and 24 overnight sets. Plafkin et al. (1989) noted that the length of a reference station should be considered sufficient if the sample includes 80–90% of the total species expected from the site with a 50% increase in sampling distance. If this principle is applied to sampling effort instead of to distance, 15 electrofishing runs and 10 overnight gill-net sets easily provide a sample representative of the population reasonably expected with a 50% increase in sampling effort.

In both reservoirs, combining the results of electrofishing and gill-net samples provided a more efficient means of sampling fish species than was obtained from a comparable level of effort with single gears. In Chickamauga Reservoir, doubling our current level of effort for each gear would provide 25–28 electrofished species (30 samples) and 18–21 gillnetted species (20 samples). Combining collection methods at the current level of effort would result in 30–32 species. In Watts Bar Reservoir, doubling the current sampling effort would provide 24–27 species by electrofishing and 20–23 species by gill netting. Combining collection methods at the current level of effort increased species numbers to between 32 and 37.

A reliable index should consist of metrics that are able to distinguish significant differences among reservoirs and, to a lesser extent, among locations within reservoirs. Except during considerable shifts in yearly environmental conditions, the RFAI should not reveal significant annual variation within individual sites. Jennings et al. (1995) found sub-

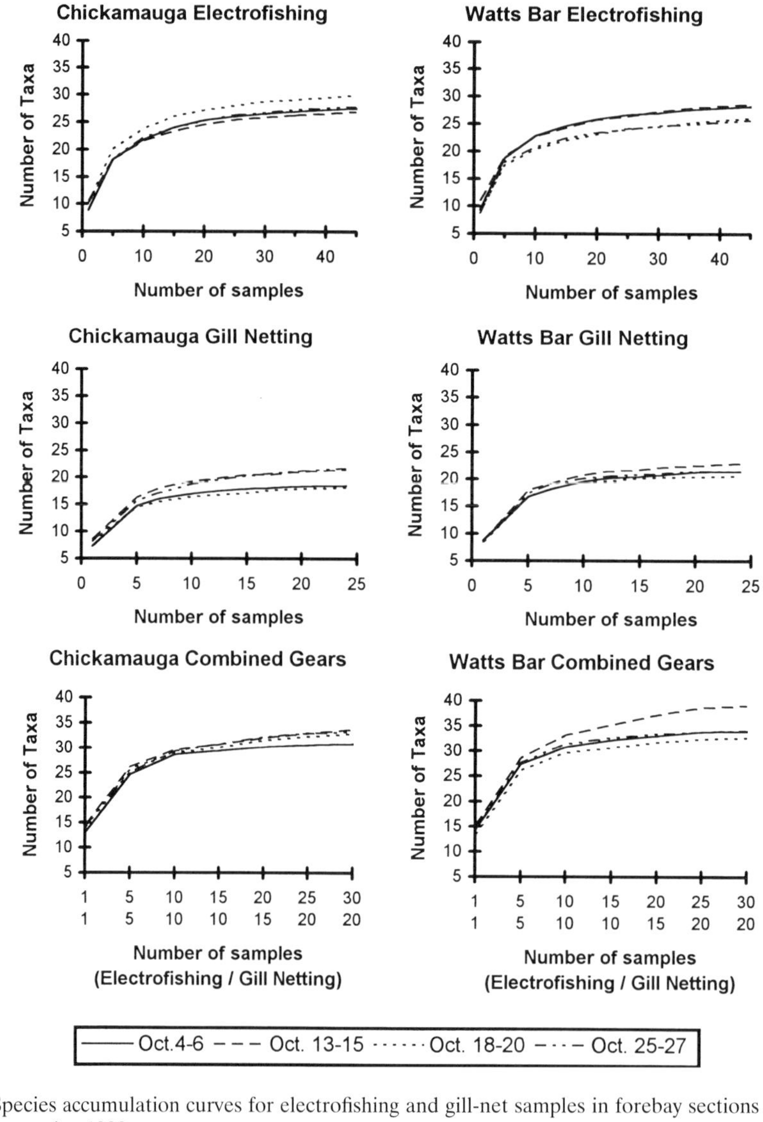

FIGURE 4.—Species accumulation curves for electrofishing and gill-net samples in forebay sections of Watts Bar and Chickamauga reservoirs, 1993.

stantial annual fluctuation in mainstream reservoir rankings based on RFAI scores generated from 10 electrofishing runs. Our current RFAI, which is based on increased sampling effort and combined gear results (15 electrofishing runs and 10 gill-net sets), reflected differences among reservoir locations that were generally repeatable between years (Kendall's W, $P < 0.001$). For individual metrics, ANOVA indicated significant differences between reservoirs in at least one class of reservoir sites (Table 2). For all metrics combined, significant differences were detected among reservoirs 40% of the time. Annual variations in RFAI classification were significant only 9% of the time.

Habitat influences on the 12 RFAI metrics were estimated from intensive studies in Watts Bar and Chickamauga forebay. Each metric was computed separately for individual electrofishing run and gill-net set. Significant differences were revealed by MANOVA among habitats for both gear types in both reservoirs (Table 3). Although MANOVA indicated significant differences in community structure and function in Chickamauga Reservoir gill netting, univariate ANOVAs did not detect signifi-

TABLE 2.—Summary of analysis of variance for 12 reservoir fish assemblage index metrics tested for significant differences between sites and years. Values represent the number of times significant differences ($P < 0.05$) were detected between sites and years within classes of reservoir sites.

Metric	Gear	Number of comparisons	Number of significant differences: Sites	Years
Tolerant species	Combined	8	4	0
Lithophilic species	Combined	8	6	2
Piscivorous species	Combined	8	1	0
Anomalies (%)	Combined	8	3	1
Total species	Combined	8	4	0
Sunfish species	Combined	8	5	1
Sucker species	Combined	8	3	0
Abundance	Electrofishing	8	2	1
	Gill netting	7	1	2
% Dominance	Electrofishing	8	4	2
	Gill netting	7	3	1
% Invertivores	Electrofishing	8	3	0
	Gill netting	7	1	0
% Omnivores	Electrofishing	8	4	0
	Gill netting	7	4	0
% Tolerant	Electrofishing	8	3	1
	Gill netting	7	2	1
All metrics		131	53 (40%)	12 (9%)

cant differences for any of the metrics examined separately (Table 3). In Watts Bar Reservoir gill netting, significant habitat differences were detected for seven metrics. Samples from gill nets set near bluffs generally differed significantly from samples from nets set over mud substrate in Watts Bar Reservoir. This tendency is evident in plots of three metrics having the highest F-statistics (two of these metrics are shown in Figure 5). Differences between one specific habitat and the others accounted for much of the differences within each reservoir in electrofishing samples (Table 3; Figure 5). Chickamauga Reservoir electrofishing sites with abundant brush were most often significantly different from other habitat types. Riprap sites differed most frequently from other habitats in Watts Bar Reservoir with respect to electrofishing samples. Habitats accounting for most of the variability in electrofishing data reflect features of shoreline and riparian zone, often modified by human activities.

In our intensive forebay studies, mean RFAIs based on simple-random samples were very similar to those of stratified random samples within the same sample period for both reservoirs and gears (Figure 6). We only present findings from 15 electrofishing runs and 10 gill-net sets because results are similar. Variability was only moderately lower for stratified-random samples than for simple-random samples. The slight reductions in variability are the result of more consistent allocation of sample effort among strata. Because RFAI is computed from pooled electrofishing runs, gill-net sets, or both, among-stratum variability can not be separated from total variability. Variability of RFAI within reservoir zones probably could be reduced further by using stratified-random sampling; however, habitat-specific scoring criteria would be required. This approach would result in index scores that are adjusted for habitat differences. Because we feel the index should reflect physical habitat as determined by land use patterns and riparian zone characteristics, we suggest that expectations not be adjusted for differences in habitat.

Fixed-sample RFAIs derived from the intensive sampling data set were also comparable to those generated from random sampling. Variability of RFAI was not considerably reduced when fixed-station sampling was used. Fixed-station index scores were neither consistently higher nor lower than means in random sampling and fell outside ±2 SDs only once for each collection method (Figure 3). The combined-gear index for fixed-station sampling more closely approximated the mean for random sampling and had lower between-sample period variability.

Conclusions and Recommendations

The most difficult aspect of indexing environmental conditions based on resident fish communities is obtaining a representative sample. Our current standard samples were sufficient to capture a ma-

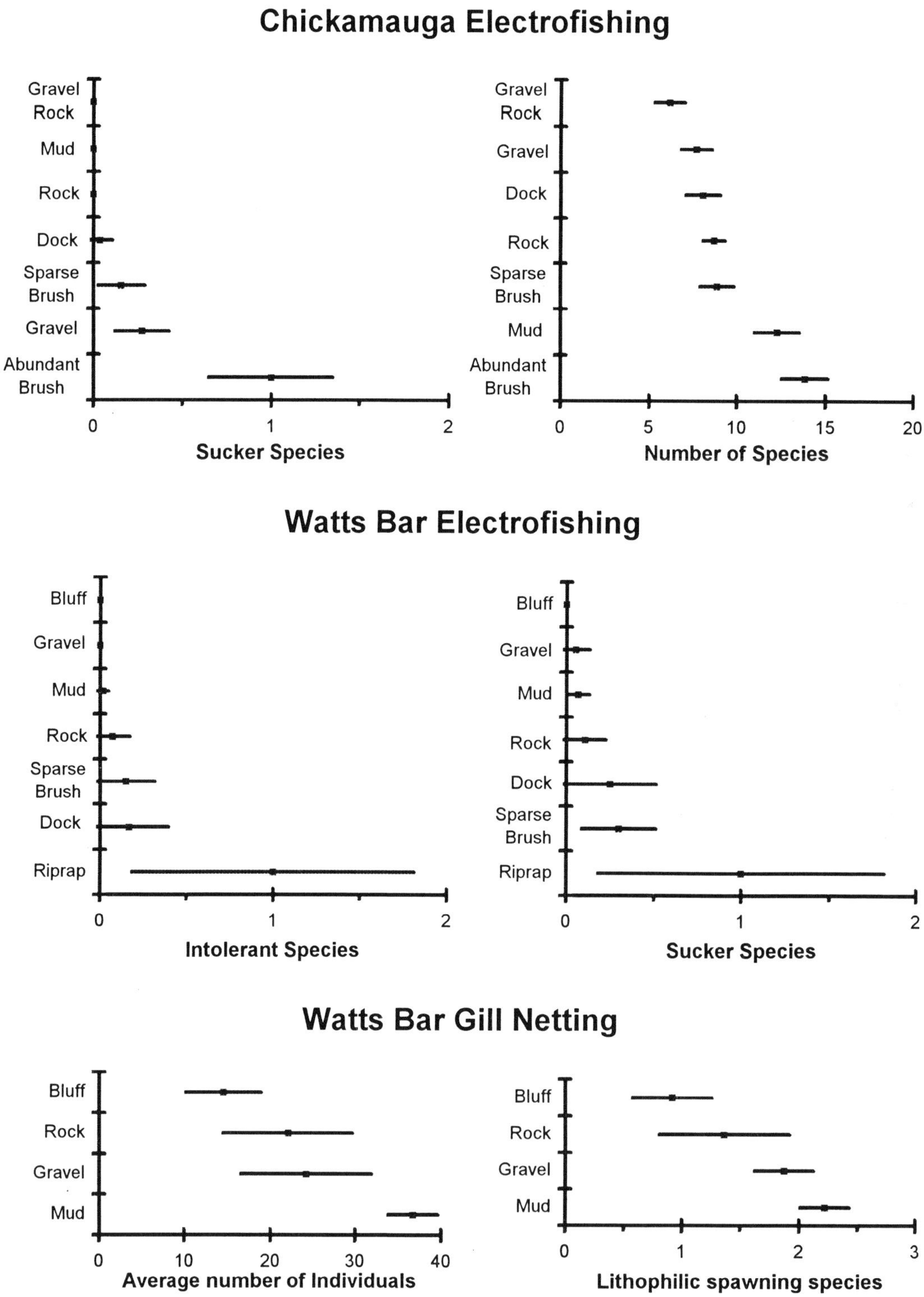

FIGURE 5.—Comparisons of means (±2 SEs) for Chickamauga (electrofishing) and Watts Bar (electrofishing and gill netting) reservoirs by habitat category for three metrics having the highest *F*-statistics in forebay intensive studies, 1993. Plots are not presented for Chickamauga Reservoir gill netting because univariate analyses of variance did not detect significant differences.

TABLE 3.—Results of multivariate analysis of variance (MANOVA) and analysis of variance (ANOVA) testing for differences in reservoir fish assemblage index metrics among habitat categories in Chickamauga and Watts Bar reservoir forebay intensive studies. Values are F-statistics; asterisks denote significance at $P = 0.05$*, $P = 0.01$**, or $P = 0.001$***. Probability levels of MANOVA were based on Wilk's lambda. Probability levels of ANOVAs were adjusted by using the Bonferroni procedure.

Metric	Electrofishing		Gill netting	
	Chickamauga	Watts Bar	Chickamauga	Watts Bar
All metrics (MANOVA)	4.34***	3.07***	1.68*	2.90***
Number of individuals	3.85**	2.92	0.17	25.76***
Dominance	1.65	0.68	2.77	2.36
% Invertivores	3.94**	1.60	1.18	2.14
Intolerant species	7.41***	13.19***	0.68	13.18***
Lithophilic spawning species	6.41***	3.26	2.03	16.87***
% Omnivores	7.46***	1.11	0.37	4.87*
% Anomalies	2.45	0.73	0.55	1.73
Piscivorous species	14.30***	1.62	3.86	10.52***
Number of species	15.57***	2.85	2.43	13.68***
Sucker species	16.06***	7.85***	0.00	6.15**
Sunfish species	3.44*	3.74*	0.50	3.70
% Tolerant species	7.41***	2.77	0.26	2.65

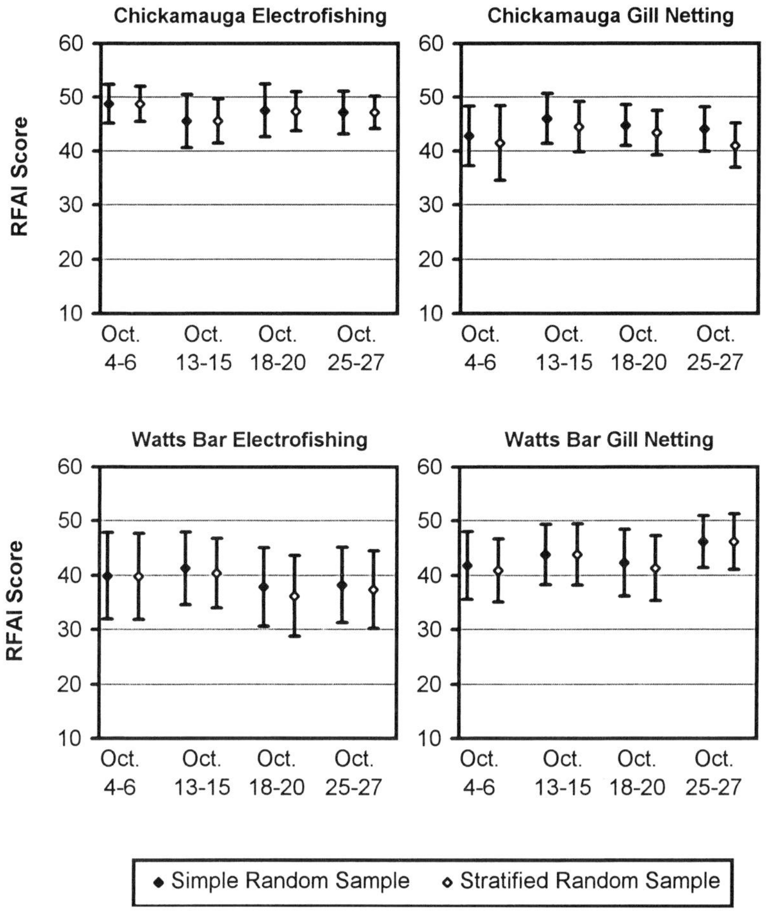

FIGURE 6.—Comparisons of means (±2 SDs) of reservoir fish assemblage index (RFAI) scores in simple-random versus stratified-random sampling designs. Comparisons were based on 100 randomly selected sets of 15 electrofishing runs and 10 gill-net samples in Chickamauga and Watts Bar forebays, 1993.

jority of the species present in the respective zones (littoral and benthic limnetic). Species accumulation curves revealed that a higher percentage of resident fish species could be sampled if samples obtained with the two gears were combined. Combining results of the two collection methods for specific metrics (taxon richness, reproductive composition, and fish health) and using joint abundance and proportional metrics (electrofishing and gill-net results worth one-half weight for metric value) reduced the variability in RFAI scores.

Potential improvements in RFAI include using a simple-random sampling design rather than our fixed-station design to enhance statistical validity with little increase in variability. Use of the index in other river systems is necessary to test its performance under a wider range of conditions than is available along the Tennessee River. Correlation with known human impacts remains a critical need in determining if the RFAI provides a reliable method of assessing reservoir environmental quality.

Acknowledgments

We appreciate the work of Al Brown, Gary Jenkins, George Peck, and Anita Masters, who collected the majority of field data; Linda Torbett, who processed and managed our data; Steve Bloom, who provided statistical analysis and graphics; and Charlie Saylor and Ed Scott, Jr., who provided considerable biological inferences from their extensive knowledge of the reservoir system and resident fish distribution. Without the insights and expertise of these individuals, many of the improvements in sampling technology and scoring criteria designations would not have been possible. Special thanks are also due to James Karr and key researchers under his guidance, including Michele Dionne, Martin Jennings, and Leska Fore; all provided useful input and ideas and laid the foundation of the RFAI. The editorial suggestions of James Karr, Leska Fore, and Donald Dycus significantly improved the manuscript.

References

Brown, A. M., G. D. Jenkins, and G. D. Hickman. 1993. Reservoir monitoring, 1992 summary of fish community results. Tennessee Valley Authority, Water Resources, Norris.

Dionne, M., and J. R. Karr. 1992. Ecological monitoring of fish assemblages in Tennessee River reservoirs. Pages 259–281 *in* D. H. McKenzie, D. E. Hyatt, and V. J. McDonald, editors. Ecological indicators, volume I. Elsevier Applied Science, London.

Dycus, D. L., and D. L. Meinert. 1991. Reservoir monitoring, 1990 summary of vital signs and use impair ment monitoring on Tennessee Valley reservoirs Tennessee Valley Authority, Water Resources, Chat tanooga.

Dycus, D. L., and D. L. Meinert. 1994. Tennessee Valle reservoir and stream quality, 1993 summary of vita signs and use suitability monitoring—volume 1. Ten nessee Valley Authority, Water Management, Chat tanooga.

Efron, B. 1981. Nonparametric estimates of standard er ror: the jackknife, the bootstrap and other methods Biometrika 68:589–599.

Elliott, J. M. 1977. Some methods for the statistical anal ysis of samples of benthic invertebrates. Freshwate Biological Association, Scientific Publication 25 Ambleside, Louisiana.

Fausch, K. D., J. R. Karr, and P. R. Yant. 1984. Regiona application of an index of biotic integrity based o stream fish communities. Transactions of the Ameri can Fisheries Society 113:39–55.

Fore, L. S., J. R. Karr, and L. L. Conquest. 1994. Statis tical properties of an index of biological integrity use to evaluate water resources. Canadian Journal o Fisheries and Aquatic Sciences 51:1077–1087.

Gammon, J. R. 1976. The fish populations of the middl 340 km of the Wabash River. Purdue University Water Resources Research General Technical Re port 86, West Lafayette, Indiana.

Hickman, G. D., A. M. Brown, and G. Peck. 1994. Ten nessee Valley reservoir and stream quality, 1993 sum mary of reservoir fish assemblage results. Tennesse Valley Authority, Norris.

Hickman, G. D., E. M. Scott, and A. M. Brown. 1991 Reservoir vital signs monitoring, 1990 fish communit results. Tennessee Valley Authority, Water Re sources, Norris.

Jennings, M. J., L. S. Fore, and J. R. Karr. 1995. Biolog ical monitoring of fish assemblages in Tennessee Val ley reservoirs. Regulated Rivers 11:263–274.

Karr, J. R. 1981. Assessment of biotic integrity using fis communities. Fisheries 6(6):21–27.

Karr, J. R., and M. Dionne. 1991. Designing surveys t assess biological integrity in lakes and reservoir Pages 62–72 *in* Biological criteria— research and reg ulation. U.S. Environmental Protection Agency (EPA 44015–91-005), Office of Water, Washington, DC.

Karr, J. R., and D. R. Dudley. 1981. Ecological perspec tive on water quality goals. Environmental Manage ment 5(1):55–68.

Karr, J. R., K. D. Fausch, P. L. Angermeier, P. R. Yan and I. J. Schlosser. 1986. Assessing biological integ rity in running waters: a method and its rational Illinois Natural History Survey Special Publication 5

Kerans, B. L., and J. R. Karr. 1994. Development an testing of a benthic index of biotic integrity (B-IBI for rivers of the Tennessee Valley. Ecological Appli cations 4:768–785.

Kimmel, B. L., and A. W. Groeger. 1984. Factors con trolling primary production in lakes and reservoirs: perspective. Pages 277–281 *in* Lake and reservoi management. U.S. Environmental Protection Agenc (EPA 440/5/84–001), Washington, DC.

Meador, M. R., T. F. Cuffney, and M. E. Gurtz. 1993. Methods for sampling fish communities as part of the national water-quality assessment program. U.S. Geological Survey (OFR03-04), Raleigh, North Carolina.

OEPA (Ohio Environmental Protection Agency). 1987. Biological criteria for the protection of aquatic life, volume II: users manual for biological field assessment of Ohio surface waters. Ohio Environmental Protection Agency, Division of Water Quality Monitoring and Assessment, Columbus.

Plafkin, J. L., M. T. Barbour, K. D. Porter, S. K. Cross, and R. M. Hughes. 1989. Rapid bioassessment protocols for use in streams and rivers: benthic macroinvertebrates and fish. U.S. Environmental Protection Agency (EPA/444/4–89-001), Washington, DC.

Saylor, C., and E. M. Scott, Jr. 1987. Application of the index of biotic integrity to existing TVA data. Tennessee Valley Authority, Water Resources, Chattanooga.

Scott, E. M., G. D. Hickman, and A. M. Brown. 1992. Reservoir vital signs monitoring, 1991 fish community results. Tennessee Valley Authority, Water Resources, Norris.

Soballe, D. M., B. L. Kimmel, R. H. Kennedy, and R. F. Gaugush. 1992. Reservoirs. Pages 421–474 *in* C. T. Hackney, S. M. Adams, and W. H. Martin, editors. Biodiversity of the southeastern United States aquatic communities. Wiley, New York.

Weaver, M .J., J. J. Magnuson, and M. K. Clayton. 1993. Analyses for differentiating littoral fish assemblages with catch data from multiple sampling gears. Transactions of the American Fisheries Society 122:1111–1119.

Zar, J. H. 1984. Biostatistical analysis. Prentice Hall, Englewood Cliffs, New Jersey.

American Fisheries Society Symposium 16:98–106, 1996

A Review of the Basic Principles and Assumptions Used to Issue Fish Consumption Advisories

Robert E. Reinert
University of Georgia, D.B. Warnell School of Forest Resources, Athens, Georgia 30602, USA

Barbara A. Knuth
*Cornell University, Department of Natural Resources, Fernow Hall
Ithaca, New York 14853-3001, USA*

Michael A. Kamrin
*Michigan State University, Institute of Environmental Toxicology
C-231 Holden Hall, East Lansing, Michigan 48824-1206, USA*

Quentin J. Stober
U.S. Environmental Protection Agency, College Station Road Athens, Georgia 30605, USA

Abstract.—Contaminants become concentrated in fish directly from water and through the aquatic food chain. Because fish live in an environment that is a sink for pollutants, they accumulate more contaminants than do terrestrial animals. In the past several years, fish consumption advisories issued by states have grown in number and type; this is a measure of the increased concern about the possible public health effects of contaminants in sport fish. These advisories have caused a great deal of concern among the general public and especially among anglers. One of the most controversial aspects is the accuracy of the relatively new advisories that warn of possible carcinogenic effects of fish consumption, because our ability to detect trace amounts of contaminants in the environment far surpasses our ability to predict the health effects they cause. Because epidemiological data are lacking, the human cancer risk of low concentrations of contaminants is most often estimated with mathematical models. Many assumptions are built into these models, however, and they produce at best crude and conservative estimates of the actual health risk. To do a better job of informing the public, fisheries professionals need to understand how these advisories are generated; otherwise, anglers may act inappropriately—from totally disregarding the advisories to being afraid to eat fish from any body of water. Proper risk communication can put the health risks associated with eating contaminated fish in perspective by comparing them with other types of risks, inform anglers of how to decrease health risks associated with eating contaminated fish such as by changing fishing habits and using proper techniques for cleaning and cooking fish, and increase anglers' concern for water quality. Anglers can learn to protect their health and still enjoy fishing.

Federal and state agencies and the public have become concerned about the potential human health risks from consuming chemically contaminated fish and shellfish. Over the past several years, this concern has been reflected in an increasing number of fish consumption advisories. At least 43 states issued fish consumption health advisories in 1994 (USEPA 1995). Historically, the earliest advisories occurred in the 1970s (Connelly and Knuth 1993). Advisories have been issued in response to a variety of chemicals, including PCBs, chlordane, dioxin, mercury, selenium, lead, mirex, dieldrin, and toxaphene (USEPA 1995). These advisories are management tools to inform the public about toxic substances in fish while still allowing benefits of the fishery to accrue. The increasing number of these advisories and, in particular, the transition from fish consumption advisories based on U.S. Food and Drug Administration (FDA) action levels to advisories based on U.S. Environmental Protectio Agency (EPA) risk assessment procedures hav caused confusion among fishery professionals, an glers, and the public. Why do advisories alway seem to be issued for fish and not for terrestria animals? What are risk assessment and risk man agement? What are the differences between th previous and recent advisories? How good are ou analytical techniques? How are fish consumptio advisories established, and how accurately do the actually describe the health risk associated wit eating fish? Anglers and the public deserve answer to these questions so they can better understand process that has economic, social, and huma health implications for sport fisheries. Because fish eries professionals are often only peripherally in volved with the process of generating and issuin fish consumption advisories, they have tended to le others try to answer these questions. Fisheries pro fessionals, however, often constitute the first line of communication with sport anglers and therefore need to be able to answer the questions anglers have about fish consumption advisories (Knuth 1991).

Fish as the Perfect Biological Filter

In designing the perfect system for filtering contaminants out of water, one would probably come up with something that looked like a fish or at least a fish gill. To enhance oxygen uptake from water, the surface area of a fish gill is large, many times greater than the surface area of the rest of the fish. The respiratory tissue of the gill is extremely thin—1–10% the diameter of a human hair. This extremely thin tissue is all that separates contaminants in the water from the fish's blood. Virtually all water pumped across the gill contacts this tissue, and most contaminants in water pass rapidly across the gill membrane into the blood; the blood carries them to various body tissues, where they may be deposited in concentrations exceeding those in the external water (bioconcentration). Fish, like terrestrial animals, also may acquire contaminants from their food (bioaccumulation), and fishes at the top of an aquatic foodchain accumulate higher concentrations of contaminants than fishes and other organisms that are lower in the chain (biomagnification).

Not only are fish highly efficient concentrators, but their environment is also the ideal repository for contaminants. Runoff from land, direct input from point sources, and input from airborne pollutants add to the contaminant load in an aquatic system. Also, the physicochemical nature of water tends to keep at least some contaminants in solution and suspension in the water column, where they are available for direct uptake by fish.

In contrast, terrestrial animals live in an environment where concentrations of contaminants are generally lower than in aquatic environments. Unlike fish, which are constantly exposed to contaminants in water as well as in their food, terrestrial animals accumulate the majority of their contaminants from their food and have relatively little exposure to airborne contaminants.

Risk Assessment and Risk Management

Fish consumption advisories are based on concepts of risk assessment and risk management. Risk assessment is a scientific procedure used to estimate the probability of adverse health effects from a specified source under particular exposure conditions. There are several methods for conducting risk assessments, each based on different underlying assumptions. In 1983 the National Academy of Sciences provided a framework for conducting risk assessments in *Risk Assessment in the Federal Government: Managing the Process*, often referred to as the "Red Book." Risk assessment includes four steps: hazard identification, dose–response assessment, exposure assessment, and risk characterization. Risk management is the process of integrating risk assessment data with social, economic, and political information to decide whether and how to reduce or eliminate the potential risks identified.

A simple analogy that demonstrates the use of risk assessment and risk management is the setting of highway speed limits. Gathering and analyzing data about the relation between increased speed and associated human fatalities is risk assessment. Because risk of fatal automobile accidents increases with speed, the risk is decreased if the speed is decreased. However, people traveling on highways do not want to increase the time required to reach their destinations. Speed limits are thus the result of integrating data on increased speed and fatal accident rates with the social and economic practicality of traveling between points in a reasonable amount of time. Expressway speed limits in many states were recently increased by some 20% despite risk assessment data showing that increased speed results in more traffic fatalities. The public and legislatures are willing to balance the increased risk of an accident with the convenience of a shorter travel time.

Differences between Food and Drug Administration- and Environmental Protection Agency-Based Fish Consumption Advisories

Although FDA action levels and EPA risk assessment procedures both use principles of risk assessment and risk management, they are designed to protect different segments of the population. The purpose of FDA action levels established under the authority of the Food, Drug, and Cosmetics Act (21 U.S.C. § 301 et seq.) is to protect the general public from contaminants in fish shipped in interstate commerce, while balancing these risks against the potential economic and other benefits associated with fish consumption (USEPA 1989). Action levels are developed in response to national needs and are based on national patterns of consumption that are often different from those of local sport or subsistence anglers (USEPA 1989). In contrast, the purpose of the EPA risk assessment procedure is to provide states with a means for informing sport and subsistence anglers about health risks associated with contaminated fish they catch from local waters (USEPA 1989). These subpopulations of anglers are potentially at greater risk than the general population because they tend to eat larger quantities of fish and because they often fish the same locations repeatedly.

For several reasons, the newer EPA risk-based assessment approach generally gives a higher estimate of health risk for a given level of contaminant than that used to develop FDA tolerance guidelines. The two agencies use different risk assessment methodologies based on different assumptions (USEPA 1994), and fish consumption rates vary in scope from national (FDA) to local (EPA). Also, FDA action levels are based not only on risk assessment but also on risk management considerations, such as the economic impact on the commercial fishing industry (USEPA 1989). For example, the FDA clearly indicates that its rationale for the current 2-mg/kg action level for PCBs was a balance between public health protection and the economics involved in the loss of commercial fish to the consumer (USFDA 1984). In contrast, the EPA approach for fish consumption advisories gives full priority to protection of public health.

Although the FDA procedure was not designed to be used for sport-fish consumption advisories, it was used in this way by the states from the early 1970s to the late 1980s because other methods did not exist. In the past several years, much of the confusion about fish consumption advisories has resulted because of the replacement of the FDA-based advisories with the more conservative EPA-based advisories. That some states use different combinations of the FDA and EPA procedures to formulate their advisories further adds to the disparities in consumption advisories among states.

The EPA has published a series of four guidance documents to assist states in developing and issuing health advisories by the agency's recommended procedures. The EPA emphasizes that states may modify the methods and procedures to reflect local conditions, such as cultural and dietary characteristics, as well as social and economic impacts, and such health issues as the benefits of fish consumption (USEPA 1994). Thus, although states may be moving toward adoption of a new risk assessment protocol for health advisories, variation may still exist among states in how data are interpreted and advisories are issued. The EPA and state agencies involved in issuing health advisories recognize that a consistent, risk-based approach demands standard procedures, assumptions, and decisions (RTI 1990), but these agencies also recognize the need for local flexibility.

How Low Can We Go?

During the mid-1960s to mid-1990s, analytical instruments have become more sensitive and sample cleanup techniques have improved so that the chemist's ability to detect contaminants in fish and water have increased about a millionfold. In the early 1960s, the limit of detection of many substances was at the level of parts per million, which is equivalent to about 74 mL of a substance in enough water to fill a 75,700-L railroad tank car. Now chemists can detect trace amounts of most contaminants at the level of parts per trillion, which is equivalent to about 74 mL of material in enough water to fill 1 million railroad tank cars. This many tank cars would make a train long enough to stretch from the east coast of the United States to the west coast more than three times. Chemists can now detect some heavy metals in water at concentrations at the level of parts per quadrillion (0.001 part per trillion).

With this increased analytical ability, chemists can find trace amounts of contaminants in most bodies of water and in most fish, and increasing attention is being paid to what effect, if any, trace amounts of contaminants have on fish and on animals, including humans, that eat the fishes. Our ability to detect low environmental concentrations of contaminants has far surpassed our ability to accurately assess their effects, however.

Risk Assessment Models and Calculation of Risk Assessment Values

For the purpose of risk assessment, contaminants are often placed in one of two categories—those that are known to cause or are suspected of causing cancer in humans and those that have other toxic effects. The methods used to calculate health risks associated with these two classes of compounds are different.

Risk Assessment for Carcinogens

Because many fish consumption advisories are based on contaminants that are either known or suspected carcinogens, it is important to understand some of the basic characteristics of cancer. Development of a cancer appears to proceed in a multistage process during which a single cell progresses through a number of changes (SCHWA 1985). Some of the changes may be repaired, or damaged cells may be destroyed by the body's defense systems. The time required for a cancer to complete all stages of development is often the better part of a human lifetime, and for much of this time the transformed cells are undetectable (SCHWA 1985). Because of the long latency period and natural repair, it is difficult to use epidemiology to determine if low environmental concentrations of a material induce cancer in humans. Following segments of the population for the required length of time is difficult. Even when it can be done, individual differences in lifestyle and exposure to other materials make it extremely difficult to determine the specific cause of a cancer (SCHWA 1985).

Because reliable human epidemiological cancer data involving environmental exposures are lacking, animal bioassays provide most of the information used to predict carcinogenic effects on humans. Scientists use mathematical models to extrapolate from effects of high doses administered to experimental animals to effects of low doses corresponding to levels found in the environment on humans. Depending on the model chosen, the estimated increase in cancer incidence can differ by several orders of magnitude (Brown 1982; SCHWA 1985). The model used by EPA is a version of the linearized, multistage, no-threshold model developed by Crump (USEPA 1980). This model leads to estimates of cancer risk that are conservative (i.e., that yield the highest risk values) (USEPA 1980, 1989). In addition to its conservative nature in extrapolating from high to low doses, the EPA model is also conservative in extrapolating from rodents to humans, and its method of compensating for the size difference between humans and rodents differs from the FDA model. Finally, because good information about threshold levels for carcinogens does not exist, the EPA model assumes there is no threshold below which there are no effects.

Figure 1 shows three curves that represent possible extrapolations for data from high-dose rodent exposures to effects of low doses on humans. Depending on the model used to generate the curve, any of the three extrapolations could be derived from the same data. For the curve representing the threshold or no observable adverse effect level (NOAEL), any dose below the threshold level would have no observable adverse effect on humans. In contrast, the other two extrapolations predict an effect for any dose above zero. Of those two, the linear low-dose extrapolation, which is the one recommended by the EPA, is the most conservative (i.e., it predicts a greater adverse effect at low-dose exposures than does the nonlinear low-dose extrapolation).

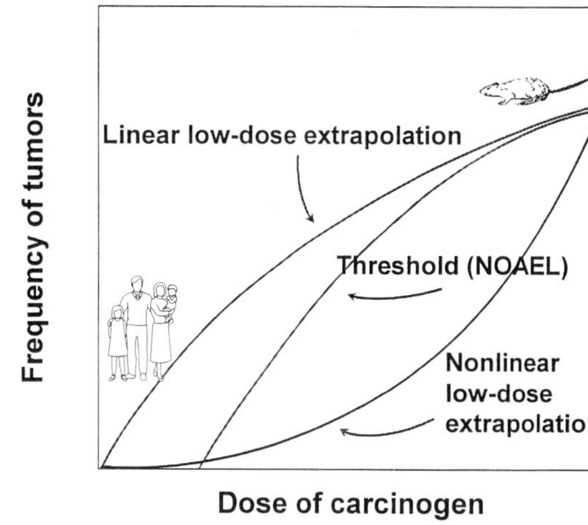

FIGURE 1.—Three curves that represent possible extrapolations for data from high-dose rodent exposures to effects of low doses on humans. The linear low-dose and nonlinear low-dose curves predict an adverse effect for any dose above zero. The threshold curve predicts a level (L) below which there are no observable adverse effects (NOAEL).

In any dose–response curve there is a degree of uncertainty. Scientists calculate confidence limits from the quantity and extent of the data; such limits are the upper and lower values within which the estimate of mean risk (or "best estimate") should fall. The EPA model considers increased cancer risk to be the 95% upper-bound estimate of the dose–response curve (USEPA 1980). This procedure generally leads to the highest (most conservative) estimate of the risk. If the best estimate or the lower-bound estimate were used, the risk value would be much lower and could even be zero or close to zero. Thus, the numbers reported as an estimate of increased cancer risk include margins of safety and are conservative estimates of risk to human health. For example, in Figure 1 the linear low-dose curve represents the upper 95% estimate of the increased cancer risk not the best estimate of the risk.

The 95% upper bound is expressed mathematically as $Q1^*$, the carcinogenic potency factor or slope factor (USEPA 1989). The formula $P = (X)(Q1^*)$ represents the increased lifetime cancer risk (P) caused by exposure to a daily dose (X) with a potency factor ($Q1^*$) for 70 years (the projected human life span). Dividing the risk (P) by the slope factor ($Q1^*$) gives the daily dose (X) for that level of risk. Thus, one can calculate a dose for any level of risk, and the dose increases with the risk (e.g., the daily dose with an increased cancer risk of 1 in 10,000 would be 10 times greater than the dose for the same material with an increased cancer risk of 1 in 100,000). The daily dose is expressed as milligrams of contaminant per kilogram of human body weight per day. With the daily dose, an average size for the human body, a time factor in weeks or months, the size of each meal of fish, and the concentration of contaminant in the meal, one can calculate the number of meals that can safely be consumed over a given time at a given increased cancer risk. For example, assume that at an increased cancer risk of 1 in 100,000, the daily dose for an animal carcinogen is 0.5 μg/kg human body weight. Multiplying this by an average body weight for a human (e.g., 70 kg) and a time (e.g., 1 month or 365 d/12) gives the dose for an average-sized human for 1 month. In this example it would be about 1,065 μg (or 1.065 mg) for a 70-kg human for 1 month. If the size of the meal of fish is 0.227 kg and the concentration of contaminant in the fish is 2,348 μg/kg, the dose in the meal would be (0.227 kg)(2,348 mg/kg) = 533 μg (or 0.533 mg). The allowable contaminant dose for the month divided by the dose in the fish meal gives the number of meals that the angler can eat per month. In this example, the angler could eat two meals per month at an increased cancer risk above that of the rest of the population that is not greater than 1 in 100,000.

Risk Assessment for Noncarcinogens

For contaminants that produce health effects other than cancer (e.g., nerve damage or birth defects), it is assumed that, unlike for cancer-causing compounds, there is a threshold below which there are no adverse effects. With noncarcinogens, risks are not expressed as a probability of occurrence of a health effect but rather as levels of exposure estimated to be without any harm. The daily dose for such contaminants is expressed as a reference dose (RfD). The RfD is an estimate of daily exposure to the human population (including sensitive subgroups) that is likely to be without appreciable risk of deleterious effects during a lifetime (USEPA 1993). It is calculated by dividing the no observable adverse effect (or the lowest observable adverse effect) level in experimental animals by an uncertainty factor. The magnitude of the uncertainty factor depends on the quality of the toxicity data. If there are good data on effects of the contaminant on humans, the uncertainty factor would be lower than if the majority of the data were from animal exposures. For example, if the lowest observable adverse effect level for a toxicant is a daily dose of 0.5 μg/kg and the uncertainty factor is 10, the RfD daily dose for that toxicant would be 0.5/10 or 0.05 μg/kg. Risk managers use the RfD with the same elements (i.e., human body weight, time, meal size, and concentration of contaminant in the meal) used to develop fish consumption advisories for carcinogens.

Concerns About the Risk Assessment Process

Animal bioassays that use high doses of chemicals to assess carcinogenic effects of low environmental doses are coming under increasing criticism, primarily because many chemicals that cause cancer at high doses may not cause cancer at the low doses that are more comparable to human exposures (Cohen and Ellwein 1990). Extrapolation from rodents to humans has also been questioned because of differences in life span, metabolic rate, and biochemical and pharmacokinetic differences (SCHWA 1985; Ames et al. 1987). Also, the actual shape of the lower end of the carcinogen dose–response curve has been hotly debated. A noncarcinogen has a threshold dose below which there is no observable detrimental effect on the animal. Conversely, a cancer may in theory develop from a single transformed cell. Therefore, cancer could develop from a nonthreshold effect initiated by very small doses of a carcinogen reaching the right cell at the right time (SCHWA 1985).

Even if there is no threshold for a carcinogen, marked alterations in metabolic pathways that occur at high experimental doses but not at low environmental doses could result in nonlinearity of the dose–response curve for some animal carcinogens at low doses (Gehring and Blau 1977; Hart and Fishbein 1986). These alterations could result in disproportionately high incidences of cancer at high doses. If a carcinogen does have a threshold or if the dose–response curve is not linear at low environmental concentrations, the cancer risk could be much lower than that predicted by the model (NRC 1994).

The only assumption on which all factions involved in the risk assessment controversy agree is that a decreased dose leads to decreased risk. The present conservative approach of EPA is based on the assumption that any detectable amount of a carcinogen has the potential for inducing cancer (i.e., there is no threshold). The EPA has adopted this position because cancer researchers cannot denine with any degree of certainty the minimum ls at which substances cause cancer (USEPA 4). Another argument suggested in favor of such onservative approach is that exposure to low centrations of a variety of substances could have itive or synergistic effects (SCHWA 1985). wed in this manner, EPA assumes that at low ironmental levels the dose–response curve is lin- and therefore no level of exposure is free from .

nother conservative assumption in the EPA risk ssment process for carcinogens is that humans sume contaminated fish for 70 years at a con- t dose (USEPA 1989). However, many com- nds listed as animal carcinogens (e.g., chlori- d hydrocarbon insecticides and PCBs) have e into common use only in the last 40–50 years. , concentrations of many of these contaminants quatic systems have declined because of regula- actions taken over the last 30 years. For exam- average PCB levels in fillets of coho salmon *orhynchus kisutch* from Lake Michigan declined n 1.93 mg/kg in 1980 to 0.39 mg/kg in 1984 (De lt et al. 1988). The EPA linear model translates decline to about a fivefold decrease in the mated cancer risk. Average DDT levels in bloat- *Coregonus hoyi* in Lake Michigan declined from mg/kg in 1969 to 0.67 mg/kg in 1986 (Hessel- et al. 1990). This decline would result in about -fold decrease in the estimated cancer risk.

iven the high degree of uncertainty associated the risk assessment process, there is a tempta- to delay issuing fish consumption advisories l more reliable information is available. How- , agencies charged with protecting human th cannot afford to wait. Chemicals that cause cer in experimental animals and noncarcinogens cause a variety of deleterious effects are ent in the environment and are continuously umulated by fish. Because of the lack of knowl- about the low-dose effects of these chemicals humans, EPA has adopted a conservative ap- ach toward estimating increased cancer risks other health risks. Despite the many shortcom- of interspecies extrapolation models or RfDs, are the primary tools available for predicting cts of environmental and dietary levels of con- inants on humans.

ecause use of the EPA risk assessment process tate fish consumption advisories is relatively , and because of the many associated uncertain- the process is under continual review by EPA state agencies. In the future, more states prob- will replace older FDA-based fish health advisories with advisories based on the newer EPA risk assessment process. Also, as new techniques for predicting cancer risk and other health risks (such as developmental and reproductive effects) are refined, states probably will incorporate them into their fish consumption advisories. Consequently, even where concentrations of contaminants in fish are unchanged, health risks suggested by the advisories may change.

Public Perceptions and Health Advisories

State fish consumption advisories can lead the public, particularly anglers, to the perception that fish are the only food source that contains contaminants. The sensitive instrumentation now available makes it possible to detect trace amounts of contaminants in many other foods. Also, dietary risks from natural carcinogens may be much more important than risks from synthetic pesticide residues or contaminants in food (Ames et al. 1987; Schuplein 1990). The same risk assessment techniques discussed earlier can be applied to any food. Based on these calculations, the lifetime cancer risk associated with drinking 1 pint of milk per day for 70 years is estimated to be 1 in 7,143 (Bro et al. 1987). One contaminant in milk is aflatoxin, produced by a mold that grows on corn and peanuts that may be used in feed grains. Similarly, eating 4 tablespoons of peanut butter per day for 70 years results in an estimated increased lifetime cancer risk of 1 in 1,666 (Bro et al. 1987). Peanut butter also can contain trace amounts of aflatoxin. Risk comparisons in fish consumption advisories should include dietary risks from other foods, specifically alternative protein sources. These comparisons would be helpful to persons who need to replace sport fish in their diet if they follow fish consumption advisories (Wendt 1986; Clark et al. 1987).

Many anglers also perceive the only risk involved with fishing is the health risk from eating contaminated fish. Indeed, a risk factor can be associated with everything we do, including driving to the lake and going fishing. Estimates of such risks are derived from actuarial tables. Thus, they are a different class of risk from the health risks associated with eating contaminated fish, which are based on laboratory and epidemiological studies. However, such risks do offer anglers a way to place the health risks of eating contaminated fish in perspective with a variety of risks encountered in daily life. The lifetime risk of death from accidents is 1 of every 59 deaths in the population. Similarly, 1 of every 400 deaths will be the result of a boating accident (Clark et al. 1987). Thus, driving to the lake and being out on the water also involve risk.

Anglers' perceptions of risks and benefits from sportfishing influence their fishing behavior (Velicer and Knuth 1994). As risk communicators, fishery professionals should ensure that these perceptions correspond to our best understanding of risks so that anglers can make sound decisions that will protect their health and still encourage continued enjoyment of the sport. Risk communicators can provide comparisons and describe an array of alternative behaviors that provide incremental reductions in risk (Knuth 1990). Anglers who are properly informed about advisories may decide that fishing in other areas, fishing for other species, selecting different size-classes of fish, and reducing the frequency of fish consumption are more attractive alternatives than to stop fishing altogether. In areas where advisories suggest not eating any fish, we can encourage anglers to practice catch and release.

A simple way that anglers can decrease contaminant-related health risks associated with eating fish is to trim and cook fish properly (Reinert et al. 1972; Skea et al. 1979; Foran et al. 1989; Gall and Voiland 1990;). Because many organic contaminants are stored primarily in fish fat, removing fatty areas can greatly reduce the amounts of these contaminants ingested and consequently reduce the health risk. Skinning fish removes the fatty layer between the skin and the flesh. Filleting removes fatty areas around the fins. Other fatty areas that can be removed are those along the top of the backbone, the lateral line, and the belly. Baking, broiling, or grilling fish on a rack allows fats containing organic contaminants to drain off. Puncturing the skin also helps fats to drain off. Although these methods may also result in some reduction of heavy metals, the reduction will not be as great as it is for organic contaminants because heavy metals are more generally concentrated in muscle tissue.

Risk Management Applications in Fishery Management

Fishery professionals manage risk by communicating with anglers, choosing which fish species to manage, and working with other agencies to decrease levels of toxic chemicals in aquatic systems. Communication with anglers about contaminants in water and fish and about fish consumption advisories is one aspect of risk management. Understanding what anglers know about contaminants, their attitudes toward risk, and their fishing behavior can help fishery professionals become more effec risk communicators (Velicer and Knuth 1994).

Fishery management objectives and decisions be important components of risk management. F species differ physiologically in their capacity accumulate chemical contaminants. Water bo differ in their contaminant levels and susceptib to future pollution. Anglers differ in their fis behavior and use of caught fish. Uses include s sistence consumption, occasional consumption, phies, and catch and release. Fishery managers come risk managers through the decisions t make regarding which species to stock (e.g., fat longer-lived fish are more likely to accumulate c taminants), numbers of fish to stock (e.g.. hig availability and high catch rates encourage m frequent consumption), which locations to man (e.g., more or less polluted), and what kinds fisheries to promote (e.g., catch and release). though many concerns dictate which fishery m agement strategies will be implemented, issues affect the contaminant level in fish should defini be considered in fisheries affected by chemical c taminants.

Conclusions

Fish consumption advisories can cause u formed anglers to polarize into one of two gro At one extreme are anglers who refuse to eat fish from any body of water, and at the other treme are anglers who pay no attention to the visories and continue to eat unlimited amount fish. There are several strategies that can be use increase anglers' understanding and adherence fish consumption advisories. First, states should velop a more uniform approach to the formula of fish consumption advisories. This need for c sistency in approach has been recognized by st (RTI 1990) and by anglers. For example, a majo of licensed anglers answering a mail survey ab health advisories in the Great Lakes expressed c cern that too many contradictions exist between government agencies writing advisories. The Co cil of Great Lakes Governors recognized the n for consistency in approach and directed an ei state multiagency task force to develop a unif advisory protocol. As of 1995, that protocol had yet been adopted throughout the Great Lakes gion but rather was the focus of an indepen scientific review. This process illustrates the mense complexity of coordinating programs as tentially controversial as health advisories ac jurisdictional boundaries. Second, anglers shoul le aware of the assumptions used in the devel- nent of advisories. For example, the EPA risk essment procedure for carcinogens assumes a year lifetime consumption of fish. With this in- nation, anglers might choose to adjust their con- nption of fish according to their lifetime con- nption history. Also, a single number as the mate of health risk for carcinogens (e.g., an eased cancer risk of 1 in 100,000) implies a ree of certainty that, in fact, does not exist (Fes- den-Raden et al. 1987). Risk assessments that tain the full range (i.e., upper-bound, best, and er-bound estimates) of risk estimates produced nterspecies extrapolation models would provide managers with a more complete view of the . However, risk managers must find effective ns of communicating this complex array of in- nation to the angler who only wants to know if fish are safe to eat. States should publish guides explain their advisories in a manner that is easy anglers to understand, but these guides also uld contain information that anglers need to ce an informed decision about fish consumption. evision, radio, newspapers, and magazines are important sources of information for anglers ble and Udd 1990). Risk communicators need lo a better job of using these media to inform lers about fish consumption advisories.

pecial consideration should to be given to mi- ties, especially non-English-speaking minori- and certain groups of subsistence anglers who hard to reach with standard advisories (Velicer Knuth 1994). These groups often consume ter amounts of fish than does the general pop- ion and their fish often come from contami- d urban waters (Reinert et al. 1991). Also, e high-volume fish consumers are more likely n other anglers to eat whole or untrimmed fish have significantly higher levels of contaminants inert et al. 1991).

ommunication problems associated with ex- ning fish consumption advisories involve us in a sic "bad news–good news" situation. The bad s is that we live in a world contaminated with nical compounds. Aquatic systems are sinks for e compounds, and fish have a remarkable ca- ty to concentrate them. Relatively few of the dreds of chemicals that have been identified in atic systems are monitored routinely. Finally, we little information about the chronic effects of y of these compounds on fish and humans and n less information about their additive or syner- c effects.

he good news is that aquatic systems are remarkably resilient. If a contaminant is prevented from entering these systems, its concentration declines in water, sediments, and fish. Dramatic declines in DDT and PCB concentrations in the Great Lakes from the mid-1970s to the 1990s provide good evidence of this process (De Vault et al. 1986, 1988; Hesselburg et al. 1990). Also, fish consumption advisories increase public interest in and concern for water quality (Connelly and Knuth 1993). Proper application of risk communication can increase anglers' understanding of fish consumption advisories, keep them interested in fishing, and help channel their legitimate concern into actions that will result in stricter water quality regulations. The end result of such actions will be improved water quality, which will benefit both the health of the fish and the people who eat them.

References

Ames, B. N., R. Magan, and L. S. Gold. 1987. Ranking possible carcinogenic hazards. Science 236:271–280.

Bro, K. M., W. C. Sonzagni, and M. E. Hanson. 1987. Relative risks of environmental contaminants in the Great Lakes. Environmental Management 11:495–505.

Brown, C. C. 1982. High to low-dose extrapolation in animals. Pages 57–79 *in* J. V. Rodricks and R. G. Tardiff, editors. Assessment and management of chemical risks. American Chemical Society, Washington, DC.

Cable, T. T., and E. Udd. 1990. Effective communication of toxic chemical warnings to anglers. North American Journal of Fisheries Management 10:382–387.

Clark, M. J., L. Fink, and D. De Vault. 1987. A new approach for the establishment of fish consumption advisories. Journal of Great Lakes Research 13:367–374.

Cohen, S. M., and L. B. Ellwein. 1990. Cell proliferation in carcinogenesis. Science 249:1007–1011.

Connelly, N. A., and B. A. Knuth. 1993. Great Lakes fish consumption health advisories: angler response to advisories and evaluation of communication techniques. Cornell University, Department of Natural Resources, Human Dimensions Research Unit Series 93-3, Ithaca, New York.

De Vault, D. S., M. J. Clark, G. Lahvis, and J. Warren. 1988. Contaminants and trends in fall run coho salmon. Journal of Great Lakes Research 14:23–33.

De Vault, D. S., and six coauthors. 1986. Contaminant trends in lake trout, *Salvelinus namaycush*, from the upper Great Lakes. Archives of Environmental Contamination and Toxicology 15:349–356.

Fessenden-Raden, J., J. M. Fitchen, and J. S. Heath. 1987. Providing risk information incommunities: factors influencing what is heard and what is accepted. Science, Technology, and Human Values 12:94–101

Foran, J. A., M. Cox, and D. Croxton. 1989. Sport fish consumption advisories and projected cancer risks in the Great Lakes basin. American Journal of Public Health 79:322–325.

Gall, K. L., and M. Voiland. 1990. Contaminants in sportfish: managing risks. Cornell Cooperative and Sea Grant Extension Fact Sheet, Ithaca, New York.

Gehring, P. J., and G. E. Blau. 1977. Mechanisms of carcinogenesis: dose response. Journal of Environmental Pathology and Toxicology 1:163–179

Hart, R. W., and L. Fishbein. 1986. Interspecies extrapolation of drug and genetic toxicity data. Pages 3–35 *in* D. B. Clayson, D. Krewski, and I. Munro, editors. Toxicological risk assessment, volume I—biological and statistical criteria. CRC Press, Boca Raton, Florida.

Hesselberg, R. J., J. P. Hickey, D. A. Nortrup, and W. A. Willford. 1990. Contaminant residues in the bloater, *Coregonus hoyi*, of Lake Michigan, 1969–1986. Journal of Great Lakes Research 16:121–129.

Knuth, B. A. 1990. Risk communication: a new dimension in sport-fisheries management. North American Journal of Fisheries Management 10:374–381.

Knuth, B. A. 1991. Risky (but necessary) business: involving fishery professionals in the health advisory process. Fisheries 16(1):2–3.

NRC (National Research Council). 1994. Science and judgement in risk assessment. National Academy Press, Washington, DC.

Reinert, R. E., B. A. Knuth, M. A. Kamrin, and O. J. Stober. 1991. Risk assessment, risk management, and fish consumption advisories in the United States. Fisheries 16(6):5–12.

Reinert, R. E., D. Stewart, and H. L. Seagram. 1972. Effects of dressing and cooking on DDT concentrations in certain fish from Lake Michigan. Journal of the Fisheries Research Board of Canada 29:525–529.

RTI (Research Triangle Institute). 1990. Current state fish and shellfish consumption advisories and bans. RTI, Research Triangle Park, North Carolina.

Schuplein, R. J. 1990. Perspectives on toxicological risk—an example: food-borne carcinogenic risk. Pages 351–372 *in* D. B. Clayton, I. C. Munro, P. Shubik, and J. A. Swenberg, editors. Progress in predictive toxicology. Elsevier Science Publishers, Amsterdam, the Netherlands.

SCHWA (State of California Health and Welfare A cy). 1985. A policy for chemical carcinogens: gu lines for chemical carcinogen risk assessment their scientific rationale. Department of Health vices, Berkeley, California.

Skea, J. C., and six coauthors. 1979. Reducing leve Mirex, Aroclor 1254, and DDE by trimming cooking Lake Ontario brown trout and smallm bass. Journal of Great Lakes Research 5:153–15

USEPA (U.S. Environmental Protection Agency). 1 Water quality criteria documents. Federal Reg 45:231:79318–79378, Washington, DC.

USEPA (U.S. Environmental Protection Agency). 1 Assessing human health risks from chemically taminated fish and shellfish: a guidance ma USEPA Office of Marine and Estuarine Prote and Office of Water Regulation and Stand USEPA-503/8-89-002, Washington, DC.

USEPA (U.S. Environmental Protection Agency). 1 ABCs of risk assessment. USEPA Journal 19(1) 15, Office of Communications, Education, and P Affairs, 175-N-93-014, Washington, DC.

USEPA (U.S. Environmental Protection Agency). 1 Guidance for assessing chemical contaminant for use in fish advisories, volume II: risk assess and fish consumption limits. USEPA, EPA 823-B 004, Washington, DC.

USEPA (U.S. Environmental Protection Agency). 1 Guidance for assessing chemical contaminant for use in fish advisories, volume IV: risk comm cation. USEPA, EPA 823-R-95-001, Washing DC.

USFDA (U.S. Food and Drug Administration). 1 Polychlorinated biphenyls (PCBs) in fish and s fish; reduction of tolerances; final decision. Fe Register 49:10:21514–21520, Washington, DC.

Velicer, C. M., and B. A. Knuth. 1994. Community taminant risks for sport-caught fish: the import of target audience assessment. Risk Analysis 14: 841.

Wendt, M. E. 1986. Low income families' consum of freshwater fish caught from New York state w Master's thesis. Cornell University, Ithaca, York.

ican Fisheries Society Symposium 16:107–118, 1996

Methylmercury Accumulation by Reservoir Fish: Bioenergetic and Trophic Effects

David W. Rodgers

Ontario Hydro Technologies, Natural Sciences Unit
800 Kipling Avenue, Toronto, Ontario Canada M8Z 5S4

Abstract.—Bioenergetics-based equations of contaminant accumulation were coupled to an existing fish bioenergetics model to model mercury (Hg) accumulation in reservoir fish. The simulations demonstrate that the increase in weight-specific Hg concentrations observed in reservoir fish after impoundment probably results from an increase in ambient concentrations of methylmercury (MeHg), rather than from changes in growth rate. An increase in growth substantially reduced weight-specific Hg concentration in the fish but had much less effect on age-specific Hg concentration. These results support earlier field observations that Hg concentrations in fish often correlate better with fish age than with fish weight and suggest that at least some of the variability observed in weight-specific Hg concentrations in fish is caused by individual differences in growth. Bioenergetics-based models provide a mechanism for quickly extrapolating the effects of changes in dietary MeHg or other variables to Hg concentrations in predatory fish and also provide a check of the assumptions of the bioenergetics models. Furthermore, the modeling approach provides a mechanism for determining the effects of fisheries management programs on fish contaminant accumulation. Management or experimental procedures such as "biomanipulation," which attempt to maintain desirable species or water quality conditions within impoundments by regulating trophic interactions, would also be expected to affect the concentrations of contaminants such as Hg within the fish.

'he increase in mercury (Hg) concentrations in after reservoir flooding has emerged as one of most perplexing and unexpected consequences mpoundment (Bodaly et al. 1984). Elevated Hg ·ls have been observed in fish from reservoirs ıout known natural or anthropogenic sources of that are located over a wide geographic range, uding North America, Europe, and Asia (Aberıy and Crumbie 1977; Meister et al. 1979; Loius et al. 1983; Bodaly et al. 1984; Jones et al. 6; Yingcharoen and Bodaly 1993). Jackson 38) suggested that increased fish Hg concentras after flooding may be a general, if not univer- effect of impoundment. Mercury accumulation ish is of particular concern because consumption ontaminated fish is the major route for transfer Ig from the aquatic environment to fish-eating s and mammals, including humans.

lercury accumulation in fish results from the ıplex interactions of a series of environmental ıponents, including Hg supply, Hg methylation, ›hic interactions, and fish bioenergetics (Figure The increase in Hg within reservoir fish appears esult in part from increased Hg supply caused by sion of flooded soils and inundation of vegeta-, with the suspension of the humic horizon of ·ded soils being particularly critical (Louchouarn l. 1993). Because the large majority of Hg accuated by fish is in the form of methylmercury (MeHg[1]) (Lockhart et al. 1972; Westoo 1973; Bloom 1992), changes in MeHg dynamics within reservoirs are also critical to increased MeHg in reservoir fish after impoundment. In particular, decomposition of newly inundated vegetation and organic material in soils after flooding results in elevated concentrations of dissolved and particulate organic material, which stimulate microbial activity in general, including microbial production of MeHg (Winfrey and Rudd 1990).

Once formed, MeHg is efficiently accumulated and transferred by biota, including fish, because of the efficient uptake of MeHg from water and diet (Matida et al. 1971; deFreitas et al. 1975; Phillips and Buhler 1978; Rodgers and Beamish 1981, 1982), combined with slow clearance of assimilated MeHg (Jarvenpaa et al. 1970; deFreitas et al. 1975; Ruohulta et al. 1975; Sharpe et al. 1977; Rodgers and Beamish 1982). Consequently, Hg accumulation by fish may be substantially modified by factors affecting either fish diet or growth rates (Norstrom et al. 1976; Rodgers 1994) or by the structure of food chains that transfer MeHg to fish (Cabana and Rasmussen 1994).

Synthesis of the cascade of effects resulting from

[1]Although the large majority (>90%) of Hg in fish is MeHg, only Hg is measured in most routine determinations for fish. Thus, fish concentrations are reported as Hg but are assumed to be MeHg.

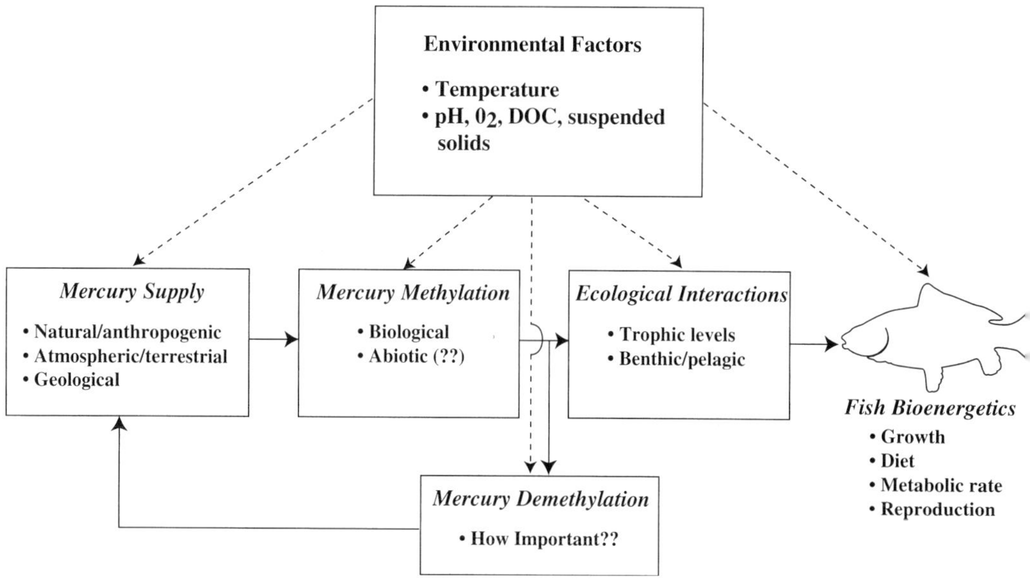

FIGURE 1.—Critical pathways of mercury accumulation by fish (DOC is dissolved organic carbon).

a specific environmental disturbance may help explain, although perhaps not predict, the large variation in Hg accumulation observed in fish from apparently similar bodies of water. Fisheries management practices may affect some of the critical linkages. Accordingly, this paper focuses on the last two compartments of Figure 1—fish bioenergetics and trophic interactions—as they may affect Hg accumulation of reservoir fish and as they be influenced by specific fisheries management practices.

Fish Bioenergetics and MeHg Accumulation

Consistent with the central role of fish in the dynamics of environmental Hg, various models have been applied to describe MeHg accumulation by fish. Bioenergetics-based models are particularly useful in describing MeHg accumulation by fish because common units of energy are used to describe uptake of MeHg from food and water and elimination of assimilated MeHg (Fagerström et al. 1974; Norstrom et al. 1976). These robust models can use any available information about fish biology (water temperature, diet, and growth rates) and ambient MeHg concentrations (Norstrom et al. 1976; Rodgers 1994) to characterize MeHg accumulation in fish.

I used the University of Wisconsin's PC-based "fish bioenergetics model 2" (Hewett and Johnson 1992) to estimate fish bioenergetic parameters. The model (or its components) has been applied to a variety of fisheries concerns with a range of spe of North American freshwater fish including no ern pike *Esox lucius*, walleye *Stizostedion vitre* and lake trout *Salvelinus namaycush* (Kitchell et 1977; Stewart et al. 1983; Wahl and Stein 1991) my applications, the proportion of maximum rat consumed by each year-class of the fish spe modeled was estimated using a proportional c sumption model (Hewett and Johnson 1994), ba on field temperature regimes, annual growth ra and diets. This was followed by a bioenergetics in which fish food consumption R (g/week) metabolism Q (kcal/week) were estimated un these conditions. The calculated values from bioenergetics run provided the basis for subsequ calculations of MeHg accumulation. Table 1 su marizes the specific variables used in my simu tions of fish bioenergetics and MeHg accumulati

The change in Hg within the fish (dP/dt) defined as the difference between uptake (from and water) and clearance of assimilated MeHg, terms being described in common units of ene (Norstrom et al. 1976) as follows:

$$dP/dt =$$

$$\text{uptake (diet)} + \text{uptake (water)} - \text{clearance.}$$

Uptake of MeHg from diet (g/week) was cal lated as the product of the efficiency of uptake MeHg from the diet (e_{pf}), the concentration

ABLE 1.—Values of variables used in simulations of fish bioenergetics and methylmercury (MeHg) accumulation.

Variable	Lake Simcoe walleye	Lake Simcoe northern pike	Tadenac Lake lake trout
(years)	0–10	0–10	2–8
ıl weight (g)	0.01	0.01	45
weight (g)	3,688	6,228	2,500
perature	Nonstratified[a], 2–22°C	Nonstratified, 2–22°C	Stratified[b], 4–10°C
respiration	Default FB2[c], walleye	Modified FB2[d], pike	Default FB2, lake trout
at first spawning (years)	3[a]	2[a]	5[b]
ortion of mass lost at spawning	Default FB2, 0.107	Default FB2, 0.13	Default FB2, 0.068
caloric density (cal/g wet weight)	Default FB2, walleye, 1,000 (constant)	Default FB2, pike, 860 (constant)	Default FB2, lake trout, varies with size[e]
caloric density (age range: /g wet weight)	0–2: 1,000; 3–10:1,200	0–1: 1,000; 2–10: 1,200	2–4: 1,000; 4–6: transition[f]; 6–8: 1,500
MeHg concentration (age range: /g wet weight)	0–2: 0.047[a]; 3–10: 0.174	0–1: 0.037[a]; 2–10: 0.101	2–4: 0.033; 4–6: transition[f]; 6–8: 0.22[b]

lathers (1984).
lacCrimmon et al. (1983).
B2 = fish bioenergetics model 2 of Hewett and Johnson (1992).
'he default value of the metabolic rate in fish bioenergetics model was halved.
'or fish up to 1,472 g, caloric density = 1,362 + 0.736 (weight); for larger fish, caloric density = 2,172 + 0.186 (weight).
t age-4, diet shifts progressively from 100% diet 1 (1,000 cal and 0.033 μg Hg/g wet weight) on day 0 to 50% diet 1 and 50% diet 500 cal and 0.22 μg Hg/g wet weight) on day 365; at age-5, diet shifts from 50% diet 1 and 50% diet 2 on day 0 to 100% diet 2 on 365. Diet 1 consists of invertebrates, diet 2 of fish.

Hg in the diet (C_{pf}, g/g), and the total food sumption (R) as calculated by the bioenergetics gram:

$$\text{uptake (diet)} = e_{pf} \cdot C_{pf} \cdot R. \quad (2)$$

Jptake of MeHg from water was calculated from relationship between MeHg uptake, oxygen sumption (O_2con, mg O_2/h), and MeHg concenion in the water (C_{pw}, g/g) observed for rainbow ıt *Oncorhynchus mykiss* (Rodgers and Beamish 1); O_2con was calculated by dividing Q by (168 · , where 168 is the number of hours in a week q_{ox} was the caloric equivalent of oxygen (3.42 mg):

$$\text{ake (water)} = 10^{1.445} \cdot (O_2\text{con} \cdot C_{pw} \cdot 10^9)^{1.134} \cdot (168/10^9). \quad (3)$$

intercept ($10^{1.445}$) and exponent 1.134 were d parameters reported by Rodgers and Beamish 31). The factors of 10^9 and $168/10^9$ converted the s of uptake (ng MeHg/h) used by Rodgers and mish (1981) to the units (g/week) used in the ulations.

n my simulations, I used a reference concentra of waterborne MeHg (C_{pw} = 0.4 ng/L) that was 10% of that employed by Norstrom et al. 76); however, the concentration of 4.0 ng/L used Norstrom et al. (1976) was based on the very ted environmental measurements available at time. With continuing improvements in the ection and analysis of both Hg and MeHg in the environment, reported concentrations in natural water have decreased by two to three orders of magnitude, and if anything, a value of 0.4 ng/L is at the high end of the range of concentrations observed in oxygenated natural waters (Watras et al. 1994). At this concentration, waterborne MeHg is estimated to provide less than 10% of the total MeHg uptake (Harris and Snodgrass 1993; Rodgers 1994).

Clearance of incorporated MeHg was calculated as the product of W^{ζ}, the weight of the fish (g) raised to the power ζ (reference value = 0.58: Norstrom et al. 1976), the body burden of pollutant in the fish (P - grams), and the clearance coefficient $k_{cl}(T)$ (Rodgers 1994):

$$\text{clearance} = W^{\zeta} \cdot P \cdot k_{cl}(T). \quad (4)$$

The clearance coefficient $k_{cl}(T)$ was the product of k_{cl} (reference value = $0.202 \cdot g^{-\zeta}$/week: Norstrom et al. 1976) and a temperature-dependent factor with a Q_{10} of 2 and a base temperature (T_b) of 10°C (Rodgers 1994):

$$k_{cl}(T) = k_{cl} \cdot 2^{(T - T_b)/10}. \quad (5)$$

A temperature-dependent clearance factor was introduced because the temperature-independent clearance function of Norstrom et al. (1976) produced extreme annual oscillations in fish Hg concentration resulting from high overwinter clearance (Rodgers 1994).

In the simulations, fish spawned at the age reported by the particular field study and lost the proportion of mass given in the default value of the fish bioenergetics model 2 program for that species (Hewett and Johnson 1992). The loss of Hg from the fish at spawning was equated to the product of the mass lost at spawning and the Hg concentration of the fish at the time.

Postimpoundment Changes in Body Burdens

The base case for northern pike and walleye simulations is a detailed study of bioenergetics and Hg accumulation in these species in Lake Simcoe, Ontario (Mathers 1984; Mathers and Johansen 1985). The diet of both species changed as fish grew. From age 0 through age 2, walleyes ate invertebrates with an estimated average Hg concentration of 0.047 μg/g and an assigned caloric content of 1,000 cal/g (Mathers 1984[2]); from age 3 on, walleyes ate fish, principally rainbow smelt *Osmerus mordax* (69% of diet by weight) having an estimated average Hg concentration of 0.174 μg/g and an assigned caloric content of 1,200 cal/g (Mathers 1984). From age 0 through age 1, northern pike consumed a mixed diet of invertebrates and fish with an estimated average Hg concentration of 0.037 μg/g and an assigned caloric content of 1,000 cal/g (Mathers 1984). From age 2, northern pike consumed a mixed fish diet (35% white sucker *Catostomus commersoni*, 28% yellow perch *Perca flavescens*, and 20% lake herring *Coregonus artedi*) with an estimated average Hg concentration of 0.101 μg/g and an assigned caloric content of 1,200 cal/g (Mathers 1984).

Coupling the output from the fish bioenergetics model 2 program for walleyes to the equations describing MeHg accumulation with the above values produced a reasonable simulation of Hg for Lake Simcoe walleyes (Figure 2A). With northern pike, however, the unmodified fish bioenergetics model 2 program for pike produced unreasonably high concentrations of Hg ("1"s in Figure 2B). When the metabolic rate of northern pike was reduced by halving "RA" (one of the terms describing active respiration; Hewett and Johnson 1994), as suggested by Wahl and Stein (1991), estimated metabolism and food consumption were reduced and yielded a more reasonable simulation of pike Hg concentrations ("2"s in Figure 2B). These res are thus consistent with the conclusions of W and Stein (1991) that the unmodified fish bioe getics program overestimates food consumption northern pike and that a substantial reductio metabolic rate is required to substantially imp model predictions.

Because reservoir impoundment can affect growth rates and ambient MeHg concentration determined the response of Hg concentration northern pike and walleyes from age 0 through 10 under the following conditions.

1. Growth and environmental MeHg were th found in Lake Simcoe (base).
2. Growth was increased by 50% by multiplying observed annual weight increment by 1.5 each age-class.
3. MeHg concentrations in food and water w increased by 50%.
4. Both growth and MeHg concentrations in f and water were increased by 50%.

A 50% increase in ambient MeHg may be a c servative estimate of the potential changes in Me after flooding. Rudd (1995) observed at leas 10-fold increase in the concentration of waterbo MeHg within the first month of flooding of L 979 in the Experimental Lakes Area of northw ern Ontario. Concentrations of both Hg and Me in plankton and benthic invertebrates from re voirs in the La Grande hydroelectric complex northern Quebec were more than 10 times gre than concentrations in a nearby natural l (Plourde et al., in press; Tremblay and Lucotte press). Similarly, a 50% increase in fish growth a flooding may be conservative, because the weigh an age-10 northern pike roughly doubled fr 2,500 to 5,000 g after flooding of the LG-2 Re voir of the La Grande hydroelectric complex (V don et al. 1991; Morrison and Thrien 1995).

For northern pike and walleyes under both b and enhanced growth rates, a 50% increase in concentrations of MeHg in water and diet resu in an equivalent increase of approximately 50% Hg concentrations of the fish (Figures 3, 4, comp the "1" curves with the "3" curves and the curves with the "4" curves). The response to creased growth was more complex. For both no ern pike and walleyes, faster growth shifted the concentration versus weight curve to the right; th at an equivalent weight, the concentration of Hg a fish growing at the faster rate was appreciably than that of the slower-growing fish. For a 3,00 walleye, the Hg concentration of the faster-grow

[2]For both walleyes and northern pike, I have delayed the age of feeding transition by 1 year from that reported by Mathers (1984) on the basis of very limited data for age-0 through age-2 fish. In my simulations, predicted values matched the observed residues much better when the dietary transition was delayed by 1 year.

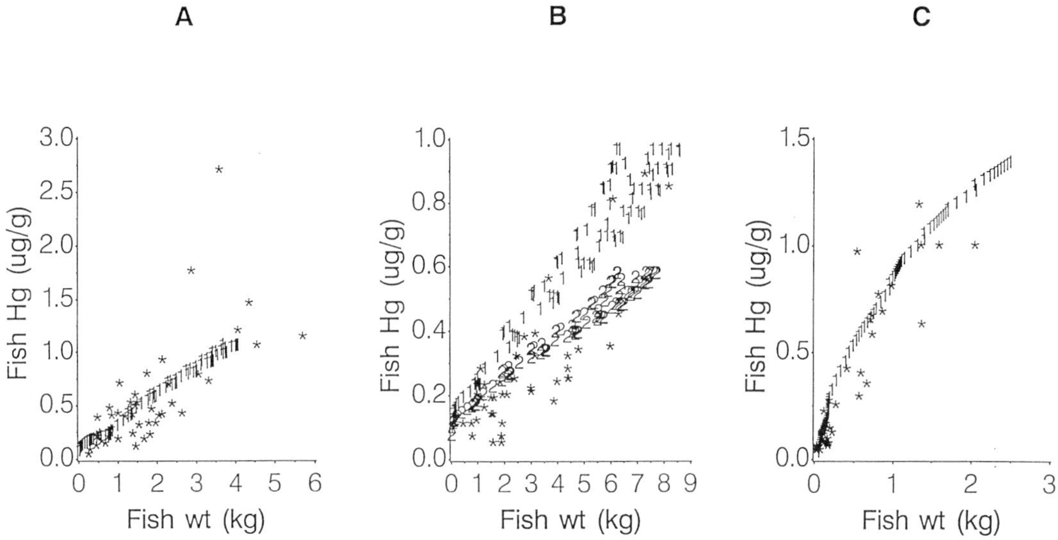

FIGURE 2.—Simulated ("1"s and "2"s) and observed (*) Hg concentrations (μg/g) related to fish weight (wt) for (**A**) Lake Simcoe walleyes, (**B**) Lake Simcoe northern pike, and (**C**) Tadenac Lake lake trout. For pike, "1" represents simulations with the default fish bioenergetic model 2 program, and "2" represents simulations in which metabolic rate was reduced by half.

fish was approximately 70% of that of the slower-growing fish at equivalent environmental Hg concentrations (Figure 3A). Similarly, for a 5,000-g northern pike, the Hg concentration of the faster-growing fish was approximately 83% of that of the slower-growing fish at equivalent environmental MeHg concentrations (Figure 4A). The reduction in Hg concentrations at a given weight in faster-growing fish is consistent with the "growth dilution effect" elucidated by Norstrom et al. (1976); it results from a reduction in the total metabolic cost required for faster-growing fish to reach a given weight. In effect, net food conversion efficiency is higher in faster-growing fish; consequently, less food is required to produce a unit weight of fish tissue and thus less Hg is consumed and accumulated. The greater relative difference between fast- and slow-growing walleyes, compared with northern pike, is consistent with the higher relative annual metabolic rate (cal/g fish) of walleyes (Mathers and Johansen 1985). Growth dilution of contaminants appears to be robust with respect to model structure because generally similar results are obtained with very different models, such as the individual-based model of PCB accumulation in Lake Michigan coho salmon *Oncorhynchus kisutch* and chinook salmon *O. tshawtscha* of Stow and Carpenter (1994).

In contrast, growth rate had much less effect on Hg concentration of the fish at a given age than at a given weight (Figures 3B, 4B). At a given level of environmental MeHg, the predicted concentration of Hg in an age-10 walleye or northern pike differed by less than 10% between fast- and slow-growing fish (Figures 3, 4), with the higher Hg concentration being observed in faster-growing fish. Again, this general result appears to be robust with respect to model structure because Stow and Carpenter (1994) also reported that fish grown under different growth scenarios attained similar PCB concentrations at the same age in their simulations. These results thus support much earlier field observations that Hg concentrations in fish often correlate better with fish age than with fish weight (Bache et al. 1971; Scott 1974) and suggest that at least some of the variability observed in weight-specific Hg concentrations of fish is caused by individual differences in growth.

My results clearly demonstrate that increased fish growth rates after impoundment would reduce weight-specific Hg concentrations in fish. Increased weight-specific Hg concentrations observed in fish within reservoirs after impoundment thus appears to result from increased ambient MeHg concentrations. The elevated concentrations of both Hg and MeHg observed in plankton and benthic invertebrates from reservoirs in the La Grande hydroelectric complex in northern Quebec (Plourde et al., in press; Tremblay and Lucotte, in press) support the

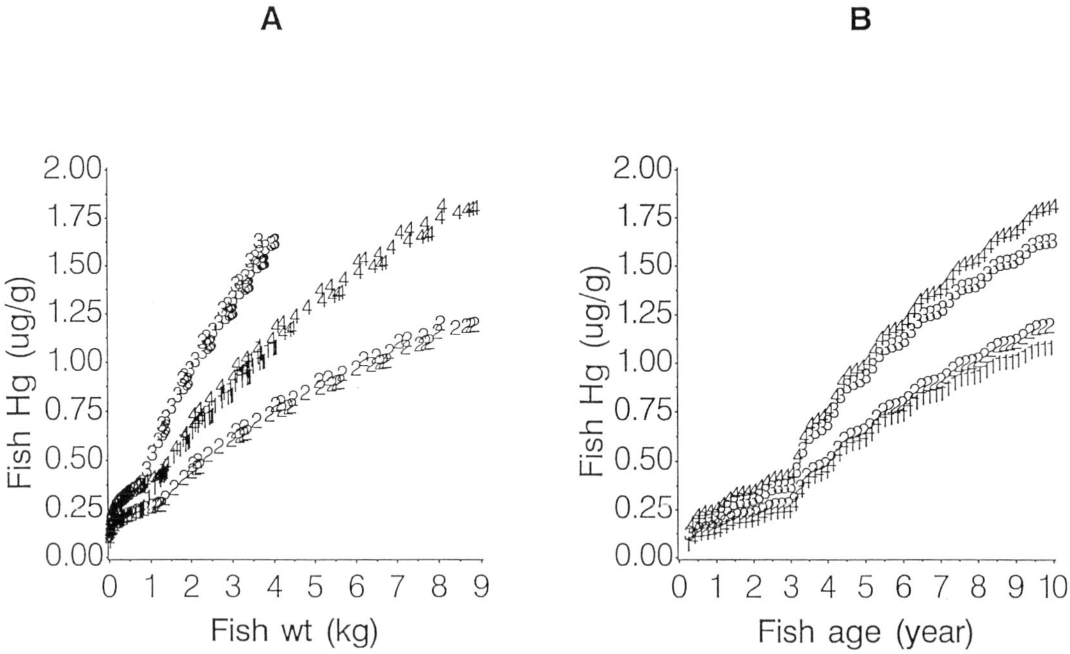

FIGURE 3.—Simulated Hg concentrations (μg/g) in Lake Simcoe walleyes with respect to weight (wt) (**A**) and age in years (y) (**B**) under base conditions ("1"s) and after a 50% increase in growth ("2"s), a 50% increase in environmental MeHg ("3"s) or a 50% increase in growth and environmental MeHg ("4"s).

hypothesis of a general increase in environmental MeHg throughout the food chain after reservoir flooding.

Growth rate differences also have a pronounced effect on the total quantity of Hg contained in fish. In my simulations, the faster-growing fish contained 2–3 times more total Hg ($C_{pf} \cdot W$) at age 10 than did slower-growing fish. Furthermore, if the increases in fish Hg concentrations and growth rates are accompanied by increases in population size, the total quantity of Hg contained within the fish population could increase manyfold after impoundment.

Effects of Weight at Stocking on MeHg Accumulation in Lake Trout

My simulations were based on a study of Hg accumulation by lake trout in Tadenac Lake, a Precambrian Shield lake in central Ontario, in which fish were stocked at age 2 (approximately 45 g and 0.04 μg/g Hg) and then recaptured between 2 and 6 years after stocking (MacCrimmon et al. 1983). Age-2 to age-4 lake trout consumed an invertebrate-based diet (estimated average Hg concentration, 0.033 μg/g; assigned caloric content, 1,000 cal/g: Norstrom et al. 1976). Age-4 and age-5 lake trout shifted progressively from an invertebrate-based to a fish-based diet (Table 1), and trout older than age-5 (>30 cm total length and >200 g wet weight) consumed only fish, largely rainbow smelt (estimated average Hg concentration, 0.22 μg/g: MacCrimmon et al. 1983; assigned caloric content, 1,500 cal/g: Foltz and Norden 1977). Coincident with the change in diet, both growth and Hg accumulation of lake trout increased dramatically, such that at age 8, lake trout weighed roughly 2,500 g with an Hg concentration of approximately 1.0 μg/g (MacCrimmon et al. 1983). The other values used in the simulations are summarized in Table 1.

Use of these values produced a reasonable fit between observed and predicted Hg concentrations of lake trout in Tadenac Lake (Figure 2C). To assess the effects of differences in stocking weight, I estimated the Hg concentrations in the lake trout stocked at two and four times the original stocking weight (45, 90, and 180 g, respectively). I assumed that the Hg concentration of stocked lake trout was 0.04 μg/g for all stocking weights and that the shift from an invertebrate- to fish-based diet was size related, such that all lake trout larger than 30 cm or 200 g consumed only fish. Each doubling in stocking

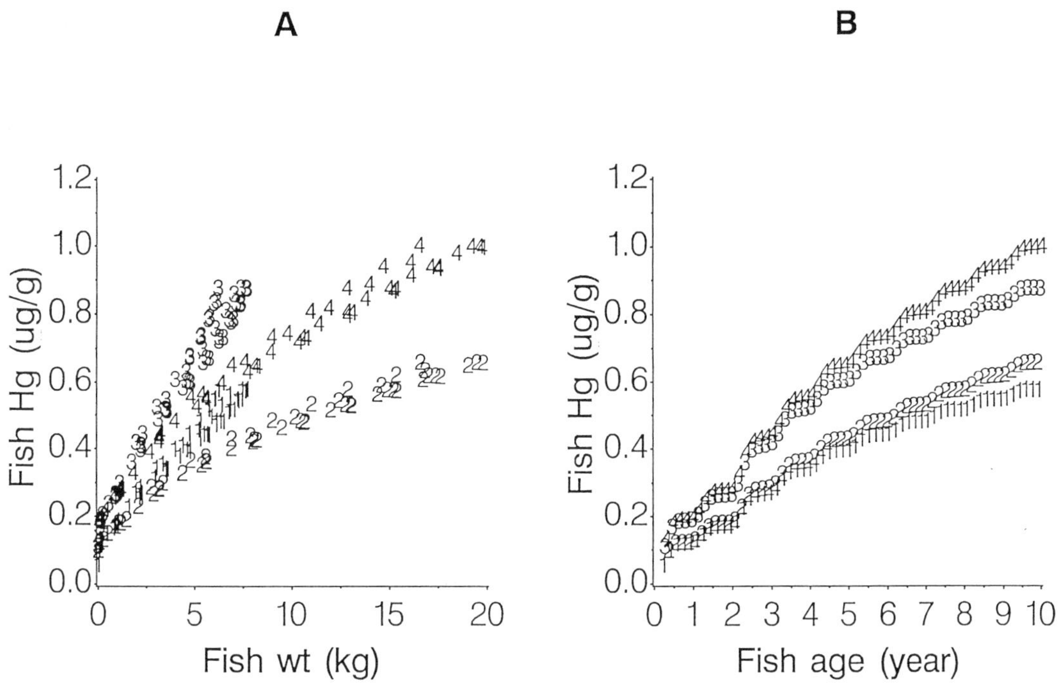

FIGURE 4.—Simulated Hg concentrations (μg/g) in Lake Simcoe northern pike with respect to weight (**A**) and age in years (**B**). Abbreviations and number symbols (in curves) are the same as in Figure 3.

weight thus reduced the time for the lake trout to reach 2,500 g (age 8 for an age 2 lake trout stocked at 45 g; fish stocked at 45 g reach approximately 90 g at age 3 and 180 g at age 5. Under these conditions, the effects of the first doubling in stocking weight (45 to 90 g) would be marginal (Figure 5A), because in both cases the fish would consume an invertebrate diet for a period of years after stocking and reach similar Hg concentrations before and during the rapid increase in Hg concentration with the shift to a fish diet. In contrast, lake trout stocked at 180 g would have an appreciably lower Hg concentration at the switch to a fish diet and this difference would be maintained throughout the simulation (Figure 5A). This difference, although relatively small, may be relevant to fish consumption advisories and restrictions (such as the Canadian limit of 0.5 μg/g Hg for the sale of commercial fish), because a 500-g lake trout would be expected to have a Hg concentration of approximately 0.4 μg/g if it were stocked at 180 g and approximately 0.6 μg/g if stocked at 45 or 90 g. A different perspective is obtained, however, when Hg concentration of lake trout under different stocking scenarios is plotted against time in the lake (Figure 5B). From this perspective, the Hg concentration of the 180-g lake trout would initially rise much more quickly than that of fish stocked at smaller sizes, because fish stocked at 180 g would switch very quickly to the high-Hg fish diet. These results are complementary to those of Madenjian and Carpenter's (1993) individual-based model of PCB accumulation in lake trout, wherein an increase in stocking size significantly increased PCB concentrations of the cohort at age 5, at least in part because larger stocked fish consumed adult alewifes *Alosa pseudoharengus* longer. The effects of stocking size on contaminant accumulation in fish may thus vary with the magnitude of the change relative to the life history of the fish, particularly with respect to food habits and growth patterns.

Trophic Interactions

Concentrations of MeHg tend to increase or magnify with passage through the food web as a consequence of the efficient uptake and slow elimination of MeHg by biota; the highest concentrations are observed in the older and larger individuals of piscivorous species (see Thomann 1981 and Fagerstrom 1991 for a general discussion of the modeling and biomagnification of persistent chemicals). Although numerous studies have demon-

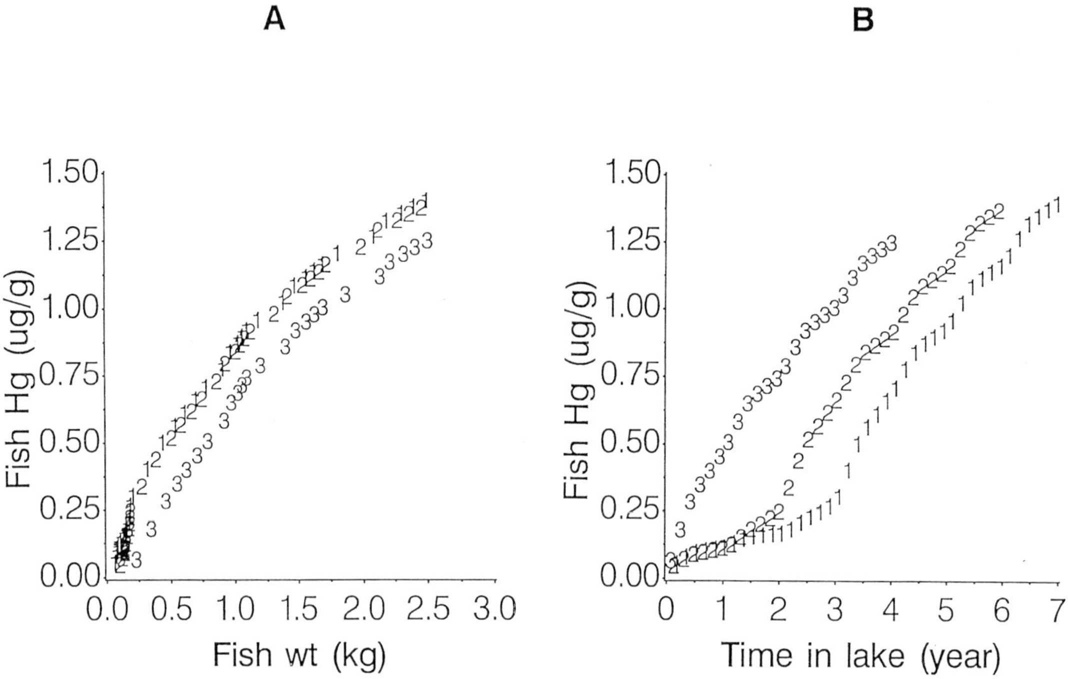

FIGURE 5.—Simulated Hg concentrations (μg/g) with respect to weight (wt) (**A**) and number of years (y) in lake (**B**) for Tadenac Lake lake trout stocked at 45 g ("1"s), 90 g ("2"s), or 180 g ("3"s).

strated higher Hg concentrations in predatory fish than in their prey (Bache et al. 1971; Scott 1974; Philips et al. 1980), the effects of food web structure on Hg concentrations in fish are best illustrated by a recent investigation by Cabana et al. (1994). In that study, Hg concentrations of lake trout from lakes in southern Ontario were significantly affected by both fish weight and estimated food chain length; Hg concentrations of fish from lakes with food webs that included both pelagic fish and the crustacean *Mysis relicta* were approximately 3.5 times greater than Hg concentrations of lake trout from lakes where these two trophic components were absent. In the same group of lakes, pelagic food chain length did not significantly affect Hg levels in smallmouth bass *Micropterus dolomieu*, which exploit the littoral rather than the pelagic food web. Although there were substantial differences in the median weight of lake trout from the lakes with different prey classes and weight was also a significant variable in regressions predicting fish Hg, Cabana and Rasmussen (1994) reported a strong correlation between Hg concentrations in lake trout and enrichment of the stable isotope of nitrogen, ^{15}N, which generally corresponds with trophic position. Similarly, Kidd et al. (1995) reported a consistent relationship between ^{15}N enrichment and fish Hg for a series of lakes in northern Ontario and the Yukon.

Changes in food web structure may also be responsible, at least in part, for the postimpoundment increase in Hg of reservoir fish. After impoundment of the La Grande hydroelectric complex in northern Quebec, the significance of benthos to energy flux of reservoir fish increased substantially (Morrison and Thrien 1995). The combination of increased energy transfer through the benthos and elevated MeHg concentrations observed in benthic invertebrates in these new reservoirs (Tremblay and Lucotte, in press) may thus account for some of the dramatic postimpoundment increases in fish Hg concentrations in these reservoirs.

As a top predator, humans also significantly affect the fish communities in many aquatic systems and thereby may change MeHg dynamics. In particular, intensive fishing pressure may lead to a considerable reduction in Hg concentrations of individual fish. This decrease in fish Hg concentration appears to result from both removal of MeHg from the aquatic system (fish may contain more than 50% of the MeHg in the lake: Harris 1991; Watras et al. 1994) and increased growth of the remaining fish. In Lake Hacksjön, Sweden, Göthberg (1983) reported that the removal of fish, mostly roach *Rutilus rutilus*, Eurasian perch *Perca fluviatilis*, and pike (52 kg/ha over 2 years), resulted in a 45–65% decrease in fish Hg concentrations. The changes in fish Hg persisted for at least 5 years, even though increased growth rates in pike (and potential growth dilution) were only apparent in the first 2 years after intensive fish removal. Verta (1990) also reported decreased fish Hg concentrations after intensive fish removal (approximately 30 kg/ha over 2 years) from Lake Hakojärvi, Finland. Decreased Hg in pike was largely attributed to changes in growth rate or growth dilution, but the decrease in burbot *Lota lota* and roach was apparently caused by lower concentrations of MeHg in diet or water. Although intensive fishing may be feasible as a means of reducing fish Hg concentrations in small lakes, it is expensive and may not be technically feasible for lakes larger than 1 km^2 (Lindqvist 1991). Furthermore, fishing pressure of this intensity may be incompatible with recreational fisheries management, because Göthberg (1983) noted that few pike larger than 1 kg were collected from Lake Hacksjön after intensive fishing.

Other human activities may also significantly affect Hg in fish by affecting fish community structure. Björklund et al. (1984) noted that the roach was one of the most sensitive species to regional acidification of waters in Sweden and was absent from water with a summer pH less than 6.0. In lakes without roach, pike consumed other fish, particularly perch and smaller pike, that had higher Hg concentrations than roach. Consequently, Hg concentrations of pike from acid lakes, where roach were absent, were appreciably higher than the Hg concentrations of pike in lakes with a "normal" species composition.

Fisheries Management Implications

Although increased fish Hg concentrations after reservoir flooding may be a general effect of impoundment, the extent and duration of elevated Hg concentrations in fish depend on site-specific factors. Bioenergetics-based models provide the means to extrapolate quickly the effects of observed changes in ambient concentrations of MeHg to Hg concentrations in predatory fish; my simulations indicate that the postimpoundment increase in fish Hg concentrations results from changes in MeHg concentrations rather than from changes in growth rates. Furthermore, when combined with preimpoundment data on fish growth rates and Hg concentrations, these models may also be useful for estimating the relative magnitude of the potential postimpoundment increase in fish Hg.

This modeling approach also provides a mechanism for estimating the effects of fisheries management practices on fish contaminant accumulation. In particular, management practices that maximize individual fish growth will result in lower weight-specific fish Hg concentrations as a consequence of growth dilution, if dietary MeHg concentrations are equivalent (see also Stow and Carpenter 1994). The high sensitivity of the simulations to the specific fish life history characteristics indicates that as much site-specific information as possible should be incorporated into the model. In my simulations, the effects of changes in stocking size on lake trout Hg concentrations were contingent on the magnitude of the change relative to the life history of the fish. Stocking may also affect fish Hg concentrations indirectly through its effects on growth rates and diets of all fish in the reservoir.

Conversely, management or experimental procedures such as "biomanipulation," in which attempts are made to maintain desirable species or water quality conditions within impoundments by regulating trophic interactions (Stein and Bremigan 1996, this volume), may also be expected to affect the concentrations of contaminants such as Hg within the fish. These effects would result from changes in growth rate or trophic interactions. Consequently, contaminant analysis of the fish before, during, and after biomanipulation, coupled with bioenergetics-based modeling of contaminant accumulation, may provide insight into the effects of the biomanipulation on both contaminant accumulation by fish and trophic interactions.

Although intensive fishing may provide a means of reducing fish Hg concentrations, the resulting fishery may be unattractive to recreational anglers. Nonetheless, several management options could be applied to minimize anglers' exposure to contaminants. Harvest limits that encourage consumption of fish from lower trophic levels and "slot sizes" that encourage consumption of smaller individuals of contaminated species should reduce anglers' exposure to contaminants. Similarly, practices encouraging catch and release of the largest fish will reduce consumption of these fish, which are also likely to contain the highest Hg concentrations. Interestingly, some of these management strategies are similar to those recommended to preserve "quality" angling in intensively fished waters.

For species such as lake trout that normally depend upon the pelagic food web, longer food webs generally result in higher Hg concentrations (Cabana et al. 1994; Cabana and Rasmussen 1994). Introduction of food organisms such as *Mysis relicta* or smelt is thus inconsistent with the objective of minimizing contaminant accumulation in pelagic fish. The consequences of introducing littoral species are less clear, probably because of the greater complexity of littoral food webs. In some lakes that contain lake trout, the presence of yellow perch affects lake trout Hg concentrations by reducing the importance of the pelagic food web to lake trout (J. B. Rasmussen, McGill University, personal communication).

Bioenergetics-based models of contaminant accumulation offer an exciting tool for exploring and understanding the integrated effects of many factors on contaminant accumulation in fish and may also prove useful as a check of the assumptions of the bioenergetics models. In my simulations, the results of the unmodified fish bioenergetics model 2 program produced unreasonably high concentrations of Hg in the northern pike, and a significant reduction in metabolic rate, as suggested by Wahl and Stein (1991), provided more realistic estimates of Hg in the pike.

A bioenergetics-based approach is also central to recent whole-lake Hg-cycling models (Harris 1991; Hudson et al. 1994), although at present, whole-lake models do not include many of the complex factors commonly required to describe reservoir limnology. In combination, whole-lake cycling and detailed bioenergetics-based fish accumulation models may provide a useful framework for investigating the potential effects of changes in management of existing reservoirs. For example, because both net Hg methylation rate and fish Hg concentrations are positively correlated with epilimnetic temperature of Canadian Shield lakes (Ramal et al. 1993; Bodaly et al. 1993), a modeling approach could be applied to examine the potential effects on fish Hg concentrations of changing the thermal structure of the reservoir by switching from hypolimnetic to epilimnetic drawdown.

Acknowledgments

I am indebted to Alistair Mathers and Chris Wren for freely sharing their data on fish from Lake Simcoe and Tadenac Lake. I also thank Drew Bodaly, Steve Griffiths, Reed Harris, Barry Johnson, John Huckabee, Alistair Mathers, and an anonymous referee for review and discussion of various drafts of this paper.

References

Abernathy, A. R., and P. M. Crumbie. 1977. Mercury accumulation by largemouth bass (*Micropterus salmoides*) in recently impounded reservoirs. Bulletin of Environmental Contamination and Toxicology 17:595–602.

Bache, C. A., W. H. Gutenmann, and D. L. Lisk. 1971. Residue of total mercury and methylmercury salts in lake trout as a function of age. Science 172:951–952.

Björklund, I., H. Borg, and K. Johansson. 1984. Mercury in Swedish lakes: its regional distribution and causes. Ambio 13:118–121.

Bloom, N. S. 1992. On the chemical form of mercury in edible fish and marine invertebrate tissue. Canadian Journal of Fisheries and Aquatic Sciences 49:1010–1017.

Bodaly, R. A., R. E. Hecky, and R. J. P. Fudge. 1984. Increases in fish mercury levels in lakes flooded by the Churchill River diversion, Northern Manitoba. Canadian Journal of Fisheries and Aquatic Sciences 41: 682–691.

Bodaly, R. A., J. W. M. Rudd, R. J. P. Fudge, and C. A. Kelly. 1993. Mercury concentrations in fish related to size of remote Canadian Shield lakes. Canadian Journal of Fisheries and Aquatic Sciences 50:980–987.

Cabana, G., and J. B. Rasmussen. 1994. Modelling food chain structure and contaminant bioaccumulation using stable nitrogen isotopes. Nature 372:255–257.

Cabana, G., A. Tremblay, J. Kalff, and J. B. Rasmussen. 1994. Pelagic food chain structure in Ontario lakes: a determinant of mercury levels in lake trout (*Salvelinus namaycush*). Canadian Journal of Fisheries and Aquatic Sciences 51:381–389.

deFreitas, A. S. W., M. J. Gidney, A. E., McKinnon, and R. E. Norstrom. 1975. Factors affecting whole body retention of methylmercury in fish. ERDA (Energy Research and Development Administration) Symposium Series 15:441–451.

Fagerström, T., B. Åsell, and A. Jernelöv. 1974. Model for accumulation of methyl mercury in northern pike, *Esox lucius*. Oikos 25:14–20.

Fagerström, T. 1991. Biomagnification in food chains and related concepts. Oikos 62:257–260.

Foltz, F. F., and C. R. Norden. 1977. Seasonal changes in food composition and energy content of smelt (*Osmerus mordax*) in Lake Michigan. Transactions of the American Fisheries Society 106:230–234.

Göthberg, A. 1983. Intensive fishing—a way to reduce the mercury level in fish. Ambio 12:259–261.

Harris, R. C. 1991. A mechanistic model to examine mercury in aquatic systems. Master's thesis. McMaster University, Hamilton, Ontario, Canada.

Harris, R. C., and W. J. Snodgrass. 1993. Bioenergetic simulations of mercury uptake and retention in walleye (*Stizostedion vitreum*) and yellow perch (*Perca flavescens*). Water Pollution Research Journal of Canada 28:217–236.

Hewett, S. H., and B. L. Johnson. 1992. Fish bioenergetics model 2—an upgrade of a generalized bioenergetic model of fish growth for microcomputers. University of Wisconsin-Madison Sea Grant Institute WIS -SG- 92- 250, Madison.

Hudson, R. J. M., S. J. Gherini, C. J. Watras, and D. B. Porcella. 1994. Modelling the biogeochemical cycle of mercury in lakes: the mercury cycling model (MCM) and its application to the MTL study lakes. Pages 473–523 *in* C. J. Watras and J. W. Huckabee, editors. Mercury pollution: integration and synthesis. Lewis Publications, Boca Raton, Florida.

Jackson, T. A. 1988. The mercury problem in recently formed reservoirs of northern Manitoba (Canada): effects of impoundment and other factors on the production of methyl mercury by microorganisms in sediments. Canadian Journal of Fisheries and Aquatic Sciences 45:97–121.

Jarvenpaa, T., M. Tillander, and J. K. Miettinen. 1970. Methylmercury: half-time of elimination in flounder, pike and eel. Suomen Kemistilehti B43:439–442.

Jones, M. J., and five coauthors. 1986. Mercury release in hydroelectric reservoirs. Canadian Electrical Association (Report 185 G 399), Montreal.

Kidd, K. A., R. H. Hesslein, R. J. P. Fudge, and K. A. Hullard. 1995. The influence of trophic level as measured by $\delta^{15}N$ on mercury concentrations in freshwater organisms from freshwater food webs. Water, Air and Soil Pollution 80:1011–1015.

Kitchell, J. F., D. J. Stewart, and D. Weininger, 1977. Applications of a bioenergetics model to yellow perch (*Perca flavescens*) and walleye (*Stizostedion vitreum vitreum*). Journal of the Fisheries Research Board of Canada 34:1922–1935.

Lindqvist, O., editor. 1991. Mercury in the Swedish environment: recent research on causes, consequences and corrective methods. Water, Air, and Soil Pollution 55.

Lockhart, W. L., J. F. Uthe, A. R. Kenney, and P. H. Mehrle. 1972. Methylmercury in northern pike (*Esox lucius*): distribution, elimination and some biochemical characteristics of contaminated fish. Journal of the Fisheries Research Board of Canada 29:1519–1523.

Lodenius, M., A. Seppänen, and M. Herranen. 1983. Accumulation of mercury in fish and man from reservoirs in northern Finland. Water, Air, and Soil Pollution 19:237–246.

Louchouarn, P., M. Lucotte, A. Mucci, and P. Pichet. 1993. Geochemistry of mercury in two hydroelectric reservoirs in Quebec, Canada. Canadian Journal of Fisheries and Aquatic Sciences 50:269–281.

MacCrimmon, H. R., C. D. Wren, and B. L. Gots. 1983. Mercury uptake by lake trout, *Salvelinus namaycush*, relative to age, growth, and diet in Tadenac Lake with comparative data from other precambrian shield lakes. Canadian Journal of Fisheries and Aquatic Sciences 40:114–120.

Madenjian, C. P., and S. R. Carpenter. 1993. Simulation of the effects of time and size of stocking on PCB accumulation in lake trout. Transactions of the American Fisheries Society 122:492–499.

Matida, Y., and six coauthors. 1971. Toxicity of mercury to aquatic organisms and accumulation of the compounds by the organisms. Bulletin of Freshwater Fisheries Research Laboratory (Tokyo) 21:197–225.

Mathers, R. A. S. 1984. The effects of feeding ecology on mercury accumulation in walleye (*Stizostedion vitreum*) and pike (*Esox lucius*) in Lake Simcoe, Ontario. Master's thesis. Queen's University, Kingston, Ontario.

Mathers, R. A., and P. H. Johansen. 1985. The effects of feeding ecology on mercury accumulation in walleye (*Stizostedion vitreum*) and pike (*Esox lucius*) in Lake Simcoe. Canadian Journal of Zoology 63:2006–2012.

Meister, J. F., J. DiNunzio, and J. A. Cox. 1979. Source and level of mercury in a new impoundment. Journal of the American Water Works Association 71:574–576.

Morrison, K., and N. Thérien. 1995. Changes in mercury levels in lake whitefish (*Coregonus clupeaformis*) and northern pike (*Esox lucius*) in the LG-2 reservoir since flooding. Water, Air and Soil Pollution 80:819–828.

Norstrom, R. J., A. E. McKinnon, and A. S. W. deFreitas. 1976. A bioenergetics-based model for pollutant accumulation by fish. Simulation of PCB and methylmercury residue levels in Ottawa River yellow perch (*Perca flavescens*). Journal of the Fisheries Research Board of Canada 33:248–267.

Phillips, G. R., and D. R. Buhler. 1978. The relative contributions of methylmercury from food or water to rainbow trout (*Salmo gairdneri*) in a controlled laboratory environment. Transactions of the American Fisheries Society 107:853–861.

Phillips, G. R., T. E. Lanhart, and R. W. Gregory. 1980. Relation between trophic position and mercury accumulation from the Tonge River Reservoir, Montana. Environmental Research 22:73–80.

Plourde, Y., M. Lucotte, and P. Pichet. In press. Contribution of suspended particulate matter and zooplankton to MeHg contamination in mid-northern Québec (Canada) reservoirs. Canadian Journal of Fisheries and Aquatic Sciences.

Ramal, P. S., C. A. Kelly, J. W. M. Rudd, and A. Furutani. 1993. Sites of methyl mercury production in remote Canadian Shield lakes. Canadian Journal of Fisheries and Aquatic Sciences 50:972–979.

Rodgers D. W. 1994. You are what you eat and a little bit more: bioenergetic-based models methylmercury accumulation in fish revisited. Pages 427–439 *in* C. J. Watras and J. W. Huckabee, editors. Mercury pollution: integration and synthesis. Lewis Publications, Boca Raton, Florida.

Rodgers, D. W., and F. W. H. Beamish. 1981. Uptake of waterborne methylmercury by rainbow trout (*Salmo gairdneri*) in relation to oxygen consumption and methylmercury concentration. Canadian Journal of Fisheries and Aquatic Sciences 38:1309–1315.

Rodgers, D. W., and F. W. H. Beamish. 1982. Dynamics of dietary methylmercury in rainbow trout, *Salmo gairdneri*. Aquatic Toxicology 2:271–290.

Rudd, J. W. M. 1995. Sources of methyl mercury to freshwater ecosystem: a review. Water, Air and Soil Pollution 80:697–713.

Ruohulta, M., and J. K. Miettinen. 1975. Retention and

excretion of ^{203}Hg-labeled methylmercury in rainbow trout. Oikos 26:385–390.

Scott, D. P. 1974. Mercury concentration of white muscle in relation to age, growth and condition of four species of fishes from Clay Lake, Ontario. Journal of the Fisheries Research Board of Canada 31:1723–1729.

Sharpe, M. A., A. S. W. deFreitas, and A. E. McKinnon. 1977. The effect of body size on methylmercury clearance by goldfish (*Carassius auratus*). Environmental Biology of Fishes 2:177–183.

Stein, R. A., M. T. Bremigan, and J. M. Dettmers. 1996. Understanding reservoir systems with experimental tests of ecological theory: a prescription for management. Pages 12–22 *in* L. E. Miranda and D. R. DeVries, editors. Multidimensional approaches to reservoir fisheries management. American Fisheries Society Symposium 16.

Stewart, D. J., D. Weininger, D. V. Rottiers, and T. A. Edsall. 1983. An energetics model for lake trout, *Salvelinus namaycush*. Application to the Lake Michigan population. Canadian Journal of Fisheries and Aquatic Sciences 40:681–698.

Stow, C. A., and S. R. Carpenter. 1994. PCB accumulation in Lake Michigan coho and chinook salmon: individual-based models using allometric relationships. Environmental Science and Technology 28:1543–1549.

Thomann, R. V. 1981. Equilibrium model of fate of microcontaminants in diverse aquatic food chains. Canadian Journal of Fisheries and Aquatic Sciences 38:280–296.

Tremblay, A., and M. Lucotte. In press. Accumulation of mercury and methylmercury in insect larvae of hydroelectric reservoirs. Canadian Journal of Fisheries and Aquatic Sciences.

Verdon, R., and five coauthors. 1991. Mercury evolution (1978–1988) in fishes of the La Grande Hydroelectric Complex, Québec, Canada. Water, Air, and Soil Pollution 56:405–417.

Verta, M. 1990. Changes in fish mercury concentrations in an intensively fished lake. Canadian Journal of Fisheries and Aquatic Sciences 47:1888–1897.

Wahl, D. H., and R. A. Stein. 1991. Food consumption and growth of three esocids: field tests of a bioenergetic model. Transactions of the American Fisheries Society 120:230–246.

Watras, C. J., and twenty-one coauthors. 1994. Sources and fate of mercury and methylmercury in Wisconsin lakes. Pages 153–177 *in* C. J. Watras and J. W. Huckabee, editors. Mercury pollution: integration and synthesis. Lewis Publications, Boca Raton, Florida.

Westoo, G. 1973. Methylmercury as percentage of total mercury in fish and viscera of salmon and sea trout of various ages. Science 181:567–568.

Winfrey, M., and J. W. M. Rudd. 1990. Environmental factors affecting the formation of methylmercury in low pH lakes: a review. Environmental Toxicology and Chemistry 9:853–869.

Yingcharoen, D., and R. A. Bodaly. 1993. Elevated mercury levels in fish resulting from reservoir flooding in Thailand. Asian Fisheries Science 6:73–80.

American Fisheries Society Symposium 16:119–132, 1996

Assessing Contaminant Distribution and Effects in a Reservoir Fishery

MARK S. BEVELHIMER AND S. MARSHALL ADAMS

*Oak Ridge National Laboratory[1] Environmental Sciences Division
Oak Ridge, Tennessee 37831–6036, USA*

Abstract.—A federally mandated remedial investigation was begun in 1989 to investigate the transport, fate, and distribution of waterborne contaminants released from the U.S. Department of Energy (DOE) Oak Ridge facilities and to assess potential risks to human and ecological health associated with these contaminants. The investigation took place on Watts Bar Reservoir, a 15,621-ha mainstem reservoir on the Tennessee River, which receives releases from three DOE facilities via two small streams and the Clinch River. A large component of the study was fisheries related, and included contaminant analysis of several fish species collected throughout the reservoir, fish community surveys, bioindicator analyses on two species, and reproductive success and toxicity exposure tests (the latter are not discussed here). The results of these studies were used in both human health and ecological risk assessments to determine the need for remedial action. This paper summarizes the fisheries aspects of the project and demonstrates the importance of a multitiered approach to evaluation of the contamination effects on a fishery. The contaminant analyses showed significant annual variation in polychlorinated biphenyl (PCB) concentrations in fish, though trends were not consistent among sites. The PCB concentrations in catfish *Ictalurus* spp. were normally 2–3 times higher than those in largemouth bass *Micropterus salmoides* and shad *Dorosoma* spp. As expected, mercury levels were highest at those sampling sites closest to the DOE facilities that historically had released mercury. Differences among sampling sites identified by the community survey, health assessment, and bioindicator studies appeared related to proximity to runoff from DOE facilities. Precise relationships with individual contaminants could not be determined. A human health risk assessment indicated that the consumption of mercury- and PCB-contaminated fish from Watts Bar Reservoir presents a potential health risk. Angler creel surveys indicated that the response of anglers to consumption advisories posted for Watts Bar Reservoir is mixed. While angling effort for some species has declined, harvest rates have not declined as expected.

Contaminated aquatic systems are a major problem throughout the world. Watts Bar Reservoir, Tennessee, which borders the U.S. Department of Energy's (DOE) Oak Ridge Reservation (ORR), has been the site of an extensive investigation since the ORR was added to the National Priority List as a Superfund site in December 1989. Most sites identified for investigation and possible remediation are on the ORR and inaccessible to the public. However, given public use of the off-site surface waters, specifically Watts Bar Reservoir, off-site contamination is a major concern. Under Section 117 of the Comprehensive Environmental, Response, Compensation, and Liability Act (CERCLA; 42 U.S.C. §§ 9601 to 9675) and the Resource Conservation and Recovery Act (RCRA; 42 U.S.C. §§ 6901 to 6991i), the DOE was required to perform a remedial investigation to: (1) determine the nature and extent of contamination in off-site surface waters and sediments downstream of the ORR; (2) quantify any risk to human health and the environment due to existing contamination; and (3) identify, evaluate, and implement potential corrective measures and remedial alternatives. This was accomplished in the form of a Remedial Investigation (RI) by staff at Oak Ridge National Laboratory (ORNL; ESD-ORNL 1995).

The Clinch River RI consisted of a variety of studies including: (1) evaluation of contaminants in water, sediment, fish, and waterfowl; (2) fish and benthic invertebrate community surveys; (3) evaluation of contaminant effects on fish and piscivorous wildlife with bioindicator assessments, toxicity tests, and measures of fish reproductive success; and (4) ecological and human health risk assessments. Phase I of sampling and analysis occurred during 1989–1990, after which preliminary data analysis and screening-level risk assessments were performed (Hoffman et al. 1991; Suter 1991). Based on Phase I results (Cook et al. 1992), a second phase of sampling was performed during 1991–1994, focusing on a reduced list of contaminants. Depending on the final risk assessment results, remedial actions will be implemented which could include removal

[1]Managed by Lockheed Martin Energy Research Corp. under contract DE-AC05-96OR22464 for the U.S. Department of Energy.

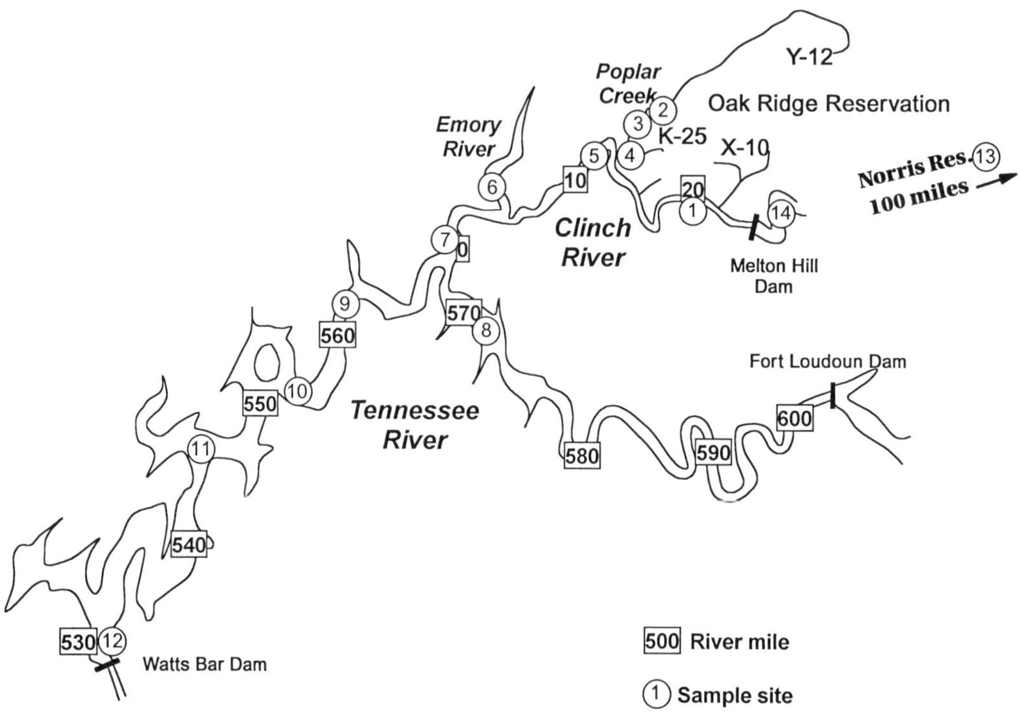

FIGURE 1.—Map of Watts Bar Reservoir with sampling locations and river miles indicated. Sample site numbers refer to numbers in Table 1.

or treatment of contaminated sediments, treatment of water, or restrictions on public use of the resource, such as fish and wildlife consumption advisories or restrictions on contact with water. The purpose of this paper is to demonstrate the importance of performing an integrated assessment that includes a variety of analyses to address a complex fishery problem. In this paper we will summarize: (1) the fisheries studies performed during the RI of the Clinch River and Watts Bar Reservoir; (2) the assessment of effects or potential effects of contamination on the fish community, and ultimately on human health; and (3) the responses of anglers to fish consumption advisories. More specific details of these studies and analyses can be found in the RI report (ESD-ORNL 1995).

Site Description

Watts Bar Reservoir

Watts Bar Reservoir is a 15,621-ha mainstem reservoir on the Tennessee River in eastern Tennessee and is used for municipal water supply, boating, fishing, swimming, and residential development. This system has a drainage area of 44,833 km^2 and 1,127 km of shoreline. The reservoir extends 116 km from Watts Bar Dam at Tennessee River mile (TRM) 530 to Fort Loudoun Dam at TRM 602, and includes the Clinch River from its mouth at TRM 568 (Clinch River mile, CRM, 0) to Melton Hill Dam at CRM 23 (Figure 1). The reservoir was impounded in 1942, just prior to the initial operations and subsequent contaminant releases by three DOE facilities on the ORR.

Contamination of the Watts Bar Reservoir fishery was first identified in the mid 1980s (TVA 1985; Dycus and Hickman 1988; Dycus 1989). In 1986, a "do not consume" advisory was issued for catfish *Ictalurus* spp. in the upper Tennessee River arm of the reservoir (the tailwaters of Fort Loudoun Dam) due to high polychlorinated biphenyl (PCB) concentrations. The next year, this advisory was expanded to include striped bass *Morone saxatilis*. As more species and more locations were sampled, the advisory was expanded further. In 1989, the advisory was extended to include the entire upper Tennessee River arm (above TRM 568), and a precautionary advisory was issued for largemouth bass *Micropterus salmoides*, carp *Cyprinus carpio*, sauger *Stizostedion canadense*, white bass *Morone chrysops*, and smallmouth buffalo *Ictiobus bubalus* for this part of the reservoir and for catfish in the Clinch River arm. The precautionary advisory warned that children, pregnant women, and nursing mothers should not consume these species, while other persons should limit consumption to 0.544 kg per month. At present, the advisory issued for the upper Tennessee River arm in 1989 includes the lower Watts Bar Reservoir as well. The advisory for the Clinch River arm now includes a "do not consume" warning for striped bass and a precautionary advisory for sauger and catfish. Advisories were generally issued when PCB concentrations in fish fillets approached the U.S. Food and Drug Administration's (FDA) action level of 2.0 mg/kg.

The Oak Ridge Reservation

The ORR includes three DOE facilities: ORNL, an energy-related research and development lab; K-25, a waste management facility that was formerly the Oak Ridge Gaseous Diffusion Plant; and Y-12, a weapons production plant. All of these facilities became operational in the mid-1940s. The ORR is located on the Clinch River arm of the reservoir beginning at CRM 11 and extending upstream of Melton Hill Dam to CRM 43. Effluent discharges and runoff from all three facilities eventually entered the Clinch River arm of Watts Bar Reservoir via either White Oak Creek at CRM 21 or Poplar Creek and its tributaries at CRM 12.

Contaminants released from the three DOE facilities have included a variety of radionuclides, metals, and organic compounds that originated from research, industrial, and waste management activities (DOE 1988). The majority of the contaminant load in the reservoir today is a result of releases prior to 1980, primarily in the 1950s and 1960s (Turner et al. 1984). In recent years, source reduction activities on the ORR have curtailed the release of contaminants to the off-site environment, and all surface water discharges are now under National Pollutant Discharge Elimination System (NPDES) permits. The distribution of contaminants in the sediments confirms their history of release from the Oak Ridge DOE facilities. Peak concentrations of mercury and cesium-137 in the sediment, corresponding to peak releases during the 1950s and 1960s, are buried beneath several centimeters of less-contaminated sediment (Turner et al. 1984).

In addition to contaminants of ORR origin, the Clinch River also receives waste from municipal water treatment facilities and urban runoff from the city of Oak Ridge, Tennessee. The Tennessee River arm of Watts Bar Reservoir receives wastes from industrial and municipal water treatment facilities and runoff from urban and agricultural areas. Major sources of contaminants to Fort Loudoun Reservoir just upstream of Watts Bar Reservoir include runoff from the city of Knoxville and surrounding industry. The Little River drainage is known to contribute a large quantity of PCBs to Fort Loudoun Reservoir, though the precise source has not been identified and remediated.

TABLE 1.—Fish collection sites for Clinch River Remedial Investigation, 1989–1994. CRM = Clinch River mile, PCM = Poplar Creek mile, ERM = Emory River mile, and TRM = Tennessee River mile.

Site	Location	Name
		Watts Bar Reservoir sites
1	CRM 20	Jones Island, just downstream of mouth of White Oak Creek
2	PCM 5.3	Poplar Creek at mouth of East Fork Poplar Creek which receives runoff from Y-12
3	PCM 4.6	Poplar Creek at mouth of Mitchell Branch which receives runoff from K-25
4	PCM 1	Poplar Creek embayment
5	CRM 10	Brashears Island
6	ERM 1	Emory River embayment
7	CRM 1	Kingston City Park
8	TRM 570	Long Island
9	TRM 561	Mid Watts Bar Reservoir
10	TRM 556	Thief Neck Island
11	TRM 545	Half Moon Island
12	TRM 530	Lower Watts Bar Reservoir
		Reference sites
13	CRM 125	Norris Reservoir (Clinch River arm)
14	CRM 24	Melton Hill Reservoir

Methods

Contaminant Distribution

To determine the contaminant concentrations in fish, we collected over 1,200 fish during 1989–1994 at 12 sites throughout Watts Bar Reservoir as well as from upstream "reference" reservoirs for analysis (Table 1, Figure 1). All regions of the reservoir were represented in sampling, particularly those close to potential sources of contamination. Poplar Creek embayment was sampled at three locations because the K-25 facility is located there and the Y-12 facility is located on one of its major tributaries, East Fork Poplar Creek. The CRM 20 site is also of special interest because it is immediately downstream of the mouth of White Oak Creek, which drains the watershed surrounding ORNL and is heavily contaminated.

Fish were collected by electroshocking, gillnetting, and trapping (slat baskets). Species were selected based on angling popularity, desirability for consumption, contaminant assimilation properties, trophic status, and size. Catfish (both channel catfish *Ictalurus punctatus*, and blue catfish *I. furcatus*) are a popular food fish particularly for subsistence anglers and are omnivorous, bottom-dwelling fish with relatively high lipid content. Fish with higher lipid contents generally accumulate higher concentrations of hydrophobic contaminants such as PCBs. Largemouth bass are an extremely popular sport fish and are mid-sized piscivores as adults with a somewhat limited home range. Striped bass (including striped bass × white bass hybrids), popular as a sport fish, are large piscivores known to range throughout the reservoir on a seasonal basis (Cheek et al. 1983). Bluegill *Lepomis macrochirus*, also a popular food fish, are small invertivores with a relatively small home range. The contaminant body burden of fish with small home ranges should be more representative of contamination at the capture location compared to fish with larger home ranges. Gizzard shad *Dorosoma cepedianum* and threadfin shad *D. petenense* are important food items for striped bass and largemouth bass in Watts Bar Reservoir, but are not consumed by humans. Composite samples (i.e., several fish of the same species homogenized and analyzed as one sample) of carp and white crappie *Pomoxis annularis* were also collected from a single site. The lengths of fish collected were usually within the range of the typical harvestable size for each species; an occasional larger-than-average fish was also analyzed. Due to the cost of chemical analysis, not all fish could be analyzed for all contaminants. With a few exceptions, catfish, largemouth bass, shad, and striped bass were analyzed for PCBs and pesticides, and bluegill and largemouth bass were analyzed for several heavy metals, particularly mercury. Composite samples of five fish each for bluegill, white crappie, and carp from CRM 1 were analyzed for PCBs, and 25 catfish samples (15 from CRM 1 and 10 from CRM 10) were analyzed for metals. Bluegill and largemouth bass were collected in the spring, while other species were, with a few exceptions, collected in the fall.

Fish were processed for analysis as they typically would be prior to cooking and eating. Fillets were removed from all species except shad, which were analyzed whole. Catfish were skinned, while other species were scaled and analyzed with the skin intact. Samples were homogenized and analyzed by standard U.S. Environmental Protection Agency (EPA) methods (Cook et al. 1993). Samples collected during the first phase of the project in 1989 were analyzed for PCB and pesticides, semivolatile organics, metals, and radionuclides (Cook et al. 1992). Samples collected from 1991 to 1994 were analyzed for a reduced list of priority contaminants (i.e., PCBs, chlordanes, mercury, arsenic, selenium, and beryllium) based on preliminary risk assessment results (Cook et al. 1993); PCBs were reported as Aroclors (1016, 1221, 1232, 1242, 1248, 1254, and 1260), the common manufactured mixtures. An unbalanced analysis of variance (ANOVA) was performed to evaluate the effects of sample location, species, and year on the total PCB concentration in the fish samples. Similar data (primarily PCB and mercury concentrations) collected by the ORNL Biological Monitoring and Abatement Program (BMAP) (Loar et al. 1992) and by the Tennessee Valley Authority (TVA) during the past 10 years (D.L. Dycus, TVA, unpublished data) were also considered in our analysis.

Ecological Effects

The assessment of effects on aquatic organisms can be extremely complex because response to a contaminant can be affected by and sometimes masked by responses to natural environmental factors such as temperature and fluctuations in food resources. Not only are potential environmental stressors of multiple origins (both natural and anthropogenic), but they often vary temporally and spatially as well. Although the effect of contaminants may be death following acute exposure, effects are more often long-term and chronic, making their detection and assessment more difficult. Here we used several methods to assess possible chronic effects.

The Health Assessment Index (HAI, Adams et al. 1993) methodology was performed on largemouth bass and bluegill collected at 11 sites during 1993–1994. The HAI involves capture and sacrifice of fish, followed by a field necropsy that provides a health profile of individual fish based on the occurrence of anomalies observed in the tissues and organs of the individuals sampled. The assessment includes internal and external visual examination for tumors, discoloration, hemorrhages, and other aberrations as well as some simple blood measurements (e.g., hematocrit, leukocrit, and plasma protein). A score is given for each assessment variable; a higher score is indicative of a relatively more damaged or stressed condition.

A bioindicator evaluation was also performed on those fish used for the HAI analysis. The response of the fish community to contamination may be manifested in many ways and occur at several levels of biological organization. The bioindicator (or biomarker) approach attempts to integrate an array of stress responses spanning multiple levels of biological organization (Adams 1990; McCarthy and Shugart 1990; Peakall 1994), and typically includes a select set of bioindicators appropriate for the situation at hand. In this case, we used representatives of five functional response groups (Adams et al. 1992): detoxification activity (e.g., by 7-ethoxyresorufin-*O*-deethylase [EROD] and microsomal protein); organ dysfunction (e.g., serum protein, creatinine, and transferase enzymes); histopathology (e.g., liver parasites and tissue anomalies); nutritional status (e.g., stomach fullness and serum triglycerides); and overall fish health (e.g., condition factor, visceral somatic index, and liver somatic index). Our analysis included a comparison among sites for individual bioindicators. A multivariate approach (canonical discriminant analysis) was used with representative bioindicator variables from each of the functional response groups plus the HAI score for an integrated site analysis (Adams et al. 1994). The Mahalonobis (least squares) distance between group means was used to compare the relative statistical distance among sites. Some sites were excluded from this analysis because of low sample size.

We performed electrofishing and gill-net surveys at eight Watts Bar Reservoir sites and one reference site during 1993 and 1994 to investigate possible effects of contaminants on fish community structure. We surveyed three sites in the Clinch River arm, three sites in the Poplar Creek embayment, and one site each in lower Watts Bar Reservoir, the Tennessee River arm, and Norris Reservoir, the latter being the reference site. Sampling at each site consisted of five 10-min electrofishing runs and five net-nights of gillnetting. A variety of habitat types were sampled including inshore and offshore areas. Species richness (number of species) was calculated for each site using 1993 and 1994 data combined.

An ecological risk assessment was performed as part of the RI to estimate ecological risks due to contaminants in Watts Bar Reservoir. One component of the assessment was specifically designed to estimate risks to fish. Screening benchmarks were developed based on Suter and Mabrey (1994). Information used in the assessment included contaminant concentrations in water, fish body burdens, water toxicity results using both invertebrates and fish, fish community survey data, and bioindicator measurements. A level of effect of more than 20% reduction in any of the endpoint parameters (i.e., survival, growth, reproduction, species richness) is considered potentially important (Suter 1993).

Human Effects and Resource Use

A baseline human health risk assessment was performed to quantify the carcinogenic risk and noncarcinogenic hazard associated with human exposure to contaminants in fish and other media from Watts Bar Reservoir assuming no remediation or institutional controls are applied. Risk was calculated for a maximum exposure scenario, that is, for a person who consumes fish on a regular basis (380 g/wk) for an extended period of time (30 yr). Risk was calculated for several contaminants at each site and for each species separately to account for those anglers that consumed primarily a single species. Carcinogenic contaminants whose associated risk exceeded an increased lifetime cancer risk of 1×10^{-4} (or 1 in 10,000), which is the maximum acceptable risk according to the U.S. Environmental Protection Agency (EPA; EPA 1989), and noncarcinogenic contaminants whose associated hazard index exceeded 1.0, were identified for further evaluation and possible remediation.

Since 1977, the Tennessee Wildlife Resource Agency has performed creel surveys on Watts Bar Reservoir to obtain angling effort and harvest data (O'Bara 1994; C. J. O'Bara, Tennessee Technological University, personal communication). We used these data to evaluate angler response to the posting of consumption advisories for various species. For comparison, we also evaluated survey data from nearby Chickamauga Reservoir which has no advisories. Chickamauga Reservoir is immediately downstream of Watts Bar Reservoir and is similar in size, limnology, overall fishing pressure, and species composition.

Results

Contaminant Distribution

Both inorganic (arsenic, beryllium, mercury, and selenium) and organic compounds (PCBs and chlordanes) were detected in a majority of the samples. We limit our discussion here to mercury and PCBs, which, based on prior sampling and analysis (Cook et al. 1992), were of greatest concern from both ecological and human health perspectives.

Mercury was detected in 53% of the bluegill (N = 159), 80% of the largemouth bass (N = 135), and 96% of the catfish (N = 25) samples collected from Watts Bar Reservoir (reference sites not included) during Phase II (1991–1994) of the RI sampling. Seven largemouth bass, all from Poplar Creek, had mercury concentrations that exceeded the FDA recommended action level of 1.0 mg/kg. This represents nearly a quarter (7 of 29) of the largemouth bass analyzed from Poplar Creek from 1991 to 1994. Mean mercury concentrations in largemouth bass at the three sites in Poplar Creek ranged from 0.6 to 1.3 mg/kg, while those throughout the rest of Watts Bar Reservoir ranged from 0.2 to 0.4 mg/kg. The highest concentrations were found at Poplar Creek mile (PCM) 5.3, near the mouth of East Fork Poplar Creek which drains the primary source of mercury to the system (the Y-12 facility). Mercury levels in bluegill showed a similar spatial pattern, although concentrations were approximately 50% of those found in largemouth bass. Mercury concentrations found in catfish were intermediate to those found in largemouth bass and bluegill.

We tested for several PCB Aroclors, but detected only 1254 and 1260. Excluding samples collected from Norris Reservoir (CRM 125) where PCB levels were low and the incidence of nondetection was high, Aroclors 1254 and 1260 were detected in 98% of the samples. The 1254 and 1260 values were summed to estimate total PCB concentrations. In the few samples where either Aroclor was not detected above the detection limit (i.e., <0.01 mg/kg), we assigned a value of one half the detection limit for that Aroclor for the purpose of calculating the total PCB concentration.

The FDA action level for PCBs of 2.0 mg/kg was exceeded by the mean total concentration for only one species at only one site (catfish at CRM 1 in 1991), but was exceeded in 35 individual fish taken from the Clinch (16 catfish), upper Tennessee (6 catfish), and Emory (1 catfish) Rivers, and lower Watts Bar Reservoir (11 catfish and 1 striped bass). Mean total PCBs exceeded 1.0 mg/kg for catfish at nearly every (eight of ten) site sampled in Watts Bar Reservoir, for striped bass at two of three sites, and for largemouth bass and shad at one of nine sites each.

The ANOVA results indicated that there were significant species, site, and year effects, but significant interaction effects cast some uncertainty on this conclusion (Table 2). Duncan's multiple-range test found PCB levels in catfish greater than those found in striped bass, which were greater than those found in largemouth bass and shad, which were similar (Figure 2). The among-species differences identified by ANOVA are confirmed by the site-by-year means for each species (i.e., samples of each species collected at the same time and place). In 13 instances where largemouth bass and catfish were collected from the same site simultaneously, largemouth bass mean PCB concentration ranged from 19 to 84% (mean = 37%) of those in catfish. In three instances where both striped bass and catfish were sampled together, striped bass mean PCB concentrations ranged from 54 to 105% (mean = 78%) of those in catfish. On average, shad PCB concentrations were roughly 25% of those in catfish and 70% of those in largemouth bass. The carp and bluegill composite samples from CRM 1 in 1992 had PCB concentrations about one half of those in catfish, and the white crappie composite sample had PCB concentrations about one fourth of those in catfish.

Differences among sites determined by the ANOVA and Duncan's tests did not identify any consistent upstream-to-downstream pattern in PCB concentrations. Likewise, there is no indication of a consistent increase or decrease in concentrations over time.

Data on PCB concentrations in catfish from the Clinch River RI were combined with results from the TVA and ORNL BMAP sampling to further evaluate spatial and temporal differences. Because the same collection sites were not maintained through time and across sampling programs, sites were combined into reaches of 8–32 km in length. Rigorous statistical analysis was not possible on this collection of data because some results were available only as site means and not individual values. However, these data suggest that on average catfish from Watts Bar, Melton Hill, and Fort Loudoun Reservoirs all have PCB levels that are higher than in Norris Reservoir, the reference site (Figure 3). In addition, the data suggest that catfish from Melton Hill Reservoir and the Clinch River arm of Watts Bar Reservoir have slightly lower PCB concentrations than those from Fort Loudoun Reservoir and the Tennessee River arm of Watts Bar Reservoir. However, the trend is not consistent from year to year (Figure 4).

Ecological Effects

Except for one site in Poplar Creek (PCM 4.6), bluegill HAI scores did not differ from those at the reference site (CRM 125); the score at PCM 4.6 was nearly twice that of the reference site. Greater among-site variation was observed for largemouth bass (Figure 5). Two sites in Poplar Creek (PCM 1 and PCM 4.6) and one in the Clinch River just below the mouth of White Oak Creek (CRM 20) had HAI scores for largemouth bass that were sig-

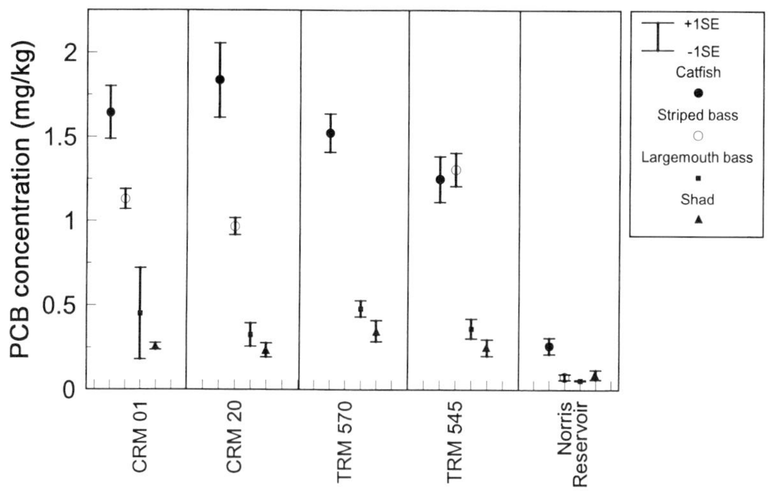

FIGURE 2.—Mean (±1 SE) total PCB concentration (mg/kg) in fish tissue of four species at four sites in Watts Bar Reservoir and one reference site in Norris Reservoir.

nificantly higher than at the reference site (ANOVA, $\alpha = 0.05$).

Among-site differences were apparent for several bioindicator variables. Those that were most important in discriminating among sites for both bluegill and largemouth bass were EROD and microsomal protein (both indicators of contaminant exposure; Jimenez and Stegeman 1990) and the visceral-somatic index (an indicator of lipid storage and general condition). Largemouth bass from two Poplar Creek sites had significantly higher EROD values than those from the reference site (Norris Reservoir), indicating exposure to higher levels of contaminants (Figure 5).

For both bluegill and largemouth bass, the integrated health response (i.e., canonical discriminant analysis of key bioindicators and HAI) revealed significant differences among sites. The integrated health response of male largemouth bass at the Poplar Creek site (PCM 1) is the most different site from the reference site (CRM 125) with the other

TABLE 2.—Analysis of variance results for total PCB in sport fish collected during Phase II sampling for the Clinch River Remedial Investigation (1991–1994).

Source	df	*F*	*P*
Site	12	17.19	0.0001
Year	3	37.11	0.0001
Site * year	10	11.53	0.0001
Species	3	170.61	0.0001
Year * species	5	0.00	1.0000
Site * species	14	2.04	0.0139

sites (CRM 1, 10, 20, and TRM 530) being intermediate (Figure 6). Not enough male largemouth bass were collected at the two upstream Polar Creek sites (PCM 4.6 and 5.3) for inclusion in the analysis. For bluegill, the Poplar Creek sites (PCM 1 and 4.6) also were significantly different from the reference site.

The community survey results suggest that there may be a relationship between species richness and proximity to Poplar Creek, which receives runoff from both the K-25 and Y-12 facilities. Significantly fewer species (26–28) were found at the four sites nearest these facilities (i.e., the three Poplar Creek sites and the Clinch River site just downstream of Poplar Creek, CRM 10) than at the other five sites (30–37; Figure 5). The most species were found at CRM 1, CRM 20, and TRM 570. However, the possibility that these differences are attributable to biological and physical differences unrelated to contaminant exposure, such as differences in habitat and hydrology among sites, cannot be discounted.

Significant ecological risks were identified in Poplar Creek, but not in the Clinch River arm of the reservoir or in lower Watts Bar Reservoir. The fish community of Poplar Creek was found to be at risk of experiencing a 20% or more reduction in species richness and total abundance based on toxicity testing results, ambient water concentrations of dissolved metals, community survey data, and bioindicator results. While no single line of evidence was adequate as a sole indicator of toxic effects in Poplar Creek, several lines of evidence were consistent

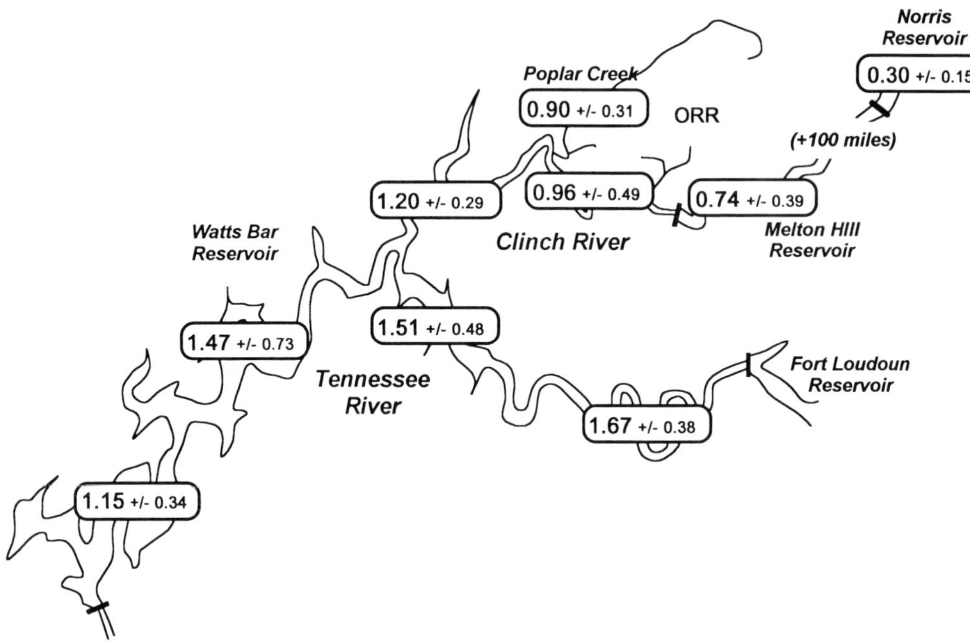

FIGURE 3.—A summary of mean PCB concentrations (mean ± 1 SE mg/kg) in catfish from Watts Bar Reservoir and upstream reservoirs based on data collected during 1987–1993.

with toxic effects, and a high probability of risk to the fish community was found.

Human Effects and Resource Use

Of the various pathways of exposure considered during the risk assessment, the biggest risk to human health was posed by the consumption of contaminated fish. The human health risk assessment predicted that PCB concentrations in largemouth bass, catfish, and striped bass throughout the reservoir would result in an increased cancer risk that exceeds the EPA acceptable level of 1×10^{-4} (EPA

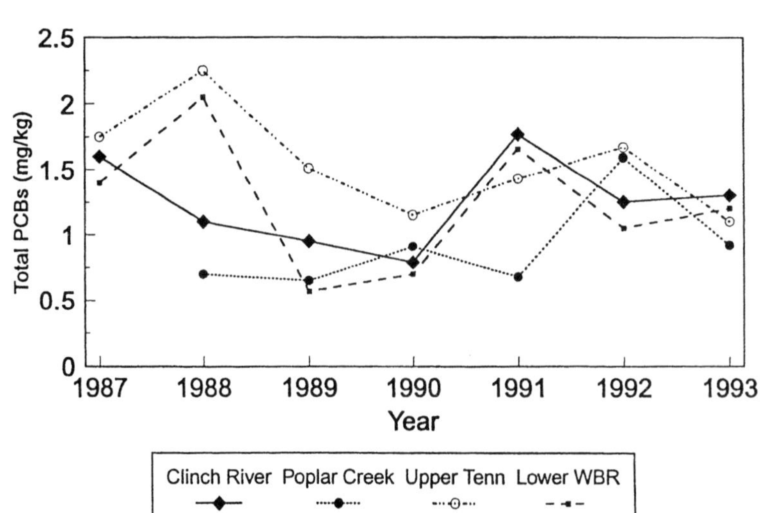

FIGURE 4.—Yearly mean PCB concentrations (mg/kg) in catfish in four areas of Watts Bar Reservoir (WBR).

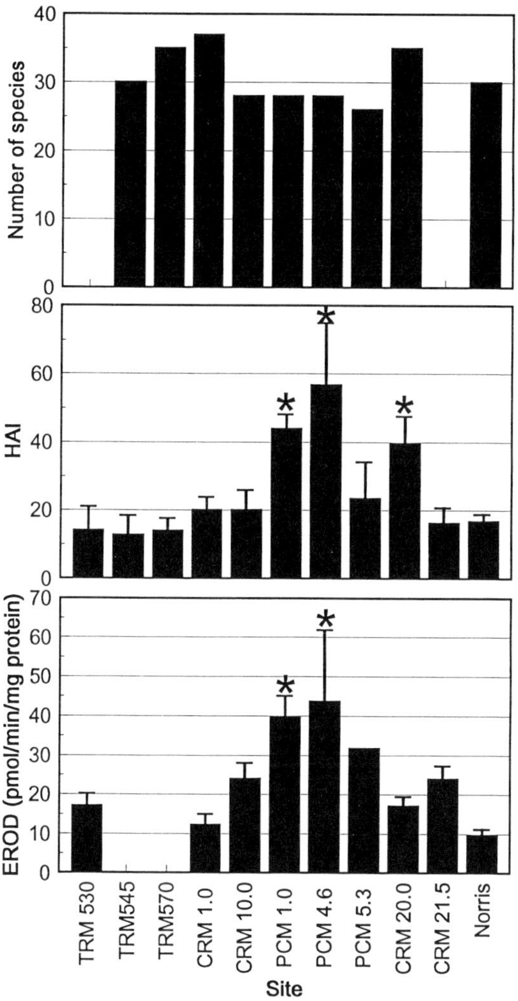

FIGURE 5.—Selected results from community survey (species richness), health assessment (HAI), and bioindicator analyses (EROD) for largemouth bass performed during the remedial investigation on Watts Bar Reservoir. Asterisks indicate a significant difference from the reference site (Norris Reservoir).

1989) for individuals consuming a steady diet of these fish for many years. The risk associated with catfish and striped bass actually exceeded 10^{-3}, which means that even if consumption rates were reduced by ten times, acceptable risk would still be exceeded. The assessment also indicated that high mercury concentrations in largemouth bass in Poplar Creek and the lower Clinch River could be a noncarcinogenic hazard, particularly for children. Chlordane, arsenic, and selenium were also deemed contaminants of concern based on high noncarcinogenic hazard values.

The creel survey summary includes one species group (black bass—largemouth bass and smallmouth bass *Micropterus dolomieu*) with a precautionary advisory, two groups (catfish—*Ictalurus* spp., and striped bass—striped bass and striped bass × white bass hybrids) with "do not consume" advisories, and two groups (crappie—*Pomoxis annularis* and *P. nigromaculatus*, and bluegill—*Lepomis* spp.) with no advisories. Angling effort for black bass in Watts Bar Reservoir has gradually increased since 1977 while harvest seems to have fluctuated around 30,000–40,000 fish for the last 10 years (Figure 7). These data do not indicate any obvious decrease in black bass harvest since the precautionary advisory was issued in 1989. A decrease in harvest of 41% between 1992 and 1993 might reflect the actual posting of the advisories at public access points in September 1992, but a similar decrease was found at Chickamauga suggesting that natural fluctuations in year-class strength or other factors could be the primary reason for the decline in harvest. Some of the decline in harvest might also be explained by the increasing practice of catch-and-release fishing which is growing in popularity throughout the country, particularly among bass fishing clubs and bass anglers.

The survey data suggest a decline in effort by those anglers seeking catfish (Figure 7). The estimated angler effort (in terms of hours fished) for catfish during the last 6 years (1988–1993) in Watts Bar Reservoir is only 58% of what it was during the 6 years prior to the first consumption advisories (1980–1985). The first advisories for catfish were in 1986 and 1987 for a small part of upper Watts Bar Reservoir. For the same time periods, effort in Chickamauga increased by 36%. Although effort in Watts Bar Reservoir decreased markedly, actual harvest over the same periods decreased by only 9%. Unlike black bass, which are sought as much for sport as for food, catfish are most likely sought for food only. Based on these data, the advisories do not appear to be having the desired effect on consumption of catfish, which is the most contaminated of the species or groups tested. A decline in catfish harvest of nearly 80% from 1992 to 1993 following the posting of the advisory suggests that greater public awareness may result in greater compliance. It should be noted, however, that there was also a corresponding decline in catfish harvest in Chickamauga Reservoir.

Striped bass, like black bass, are desired more for sport than for food, which could partially explain the steady increase in effort during the last 17 years despite a "do not consume" advisory (Figure 7).

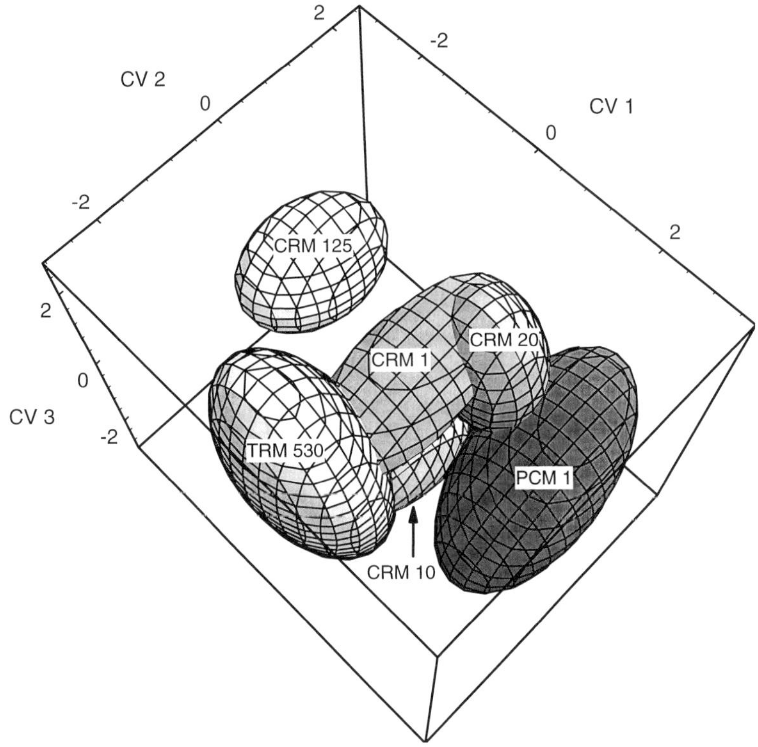

FIGURE 6.—Segregation of integrated health responses for male largemouth bass collected from five sites in Watts Bar Reservoir (CRM 20, CRM 10, CRM 1, PCM 1, and TRM 530) and one reference site (Norris Reservoir - CRM 125) based on canonical discriminant analysis. Ellipsoids represent the mean integrated responses of bass within a site for 1993–1994.

Unlike black bass and catfish, however, the trend in harvest estimates appears to be increasing. Comparisons with Chickamauga Reservoir for this species are not appropriate because of large differences in population size between the two reservoirs, largely due to differences in management objectives for this nonnative species.

Neither bluegill nor crappie are included in the consumption advisories, and, as expected, angling effort and harvest estimates show no obvious trends related to the issuance of advisories for other species (Figure 7). Harvest estimates of crappie during the last 5 years have been consistently lower than for most previous years, but the reasons for this can not easily be assigned to the public's perception of health threat because corresponding harvest rates in Chickamauga have also been very low.

Discussion

The distribution of contaminants within a reservoir depends on several factors, including reservoir retention time, sedimentation rate, sediment remobilization, food web interactions, and physical and chemical properties of the contaminant (Soballe et al. 1992). Because retention time is related to flow, we expect the distribution mechanisms of contaminants in reservoirs to be intermediate between those of lakes and rivers. In a river, most contaminants are likely to be transported away from the

FIGURE 7.—Angler effort (1,000 hrs) for and estimated harvest (1,000s of fish) of five groups of fish at Watts Bar and Chickamauga reservoirs, 1977–1993. During this time, different levels of consumption advisories were in effect for these groups of fish at Watts Bar Reservoir, while no advisories existed at Chickamauga Reservoir. "A"s along the x-axis indicate when advisories were issued, revised, or expanded. These data are based on creel surveys conducted by the Tennessee Wildlife Resources Agency (TWRA).

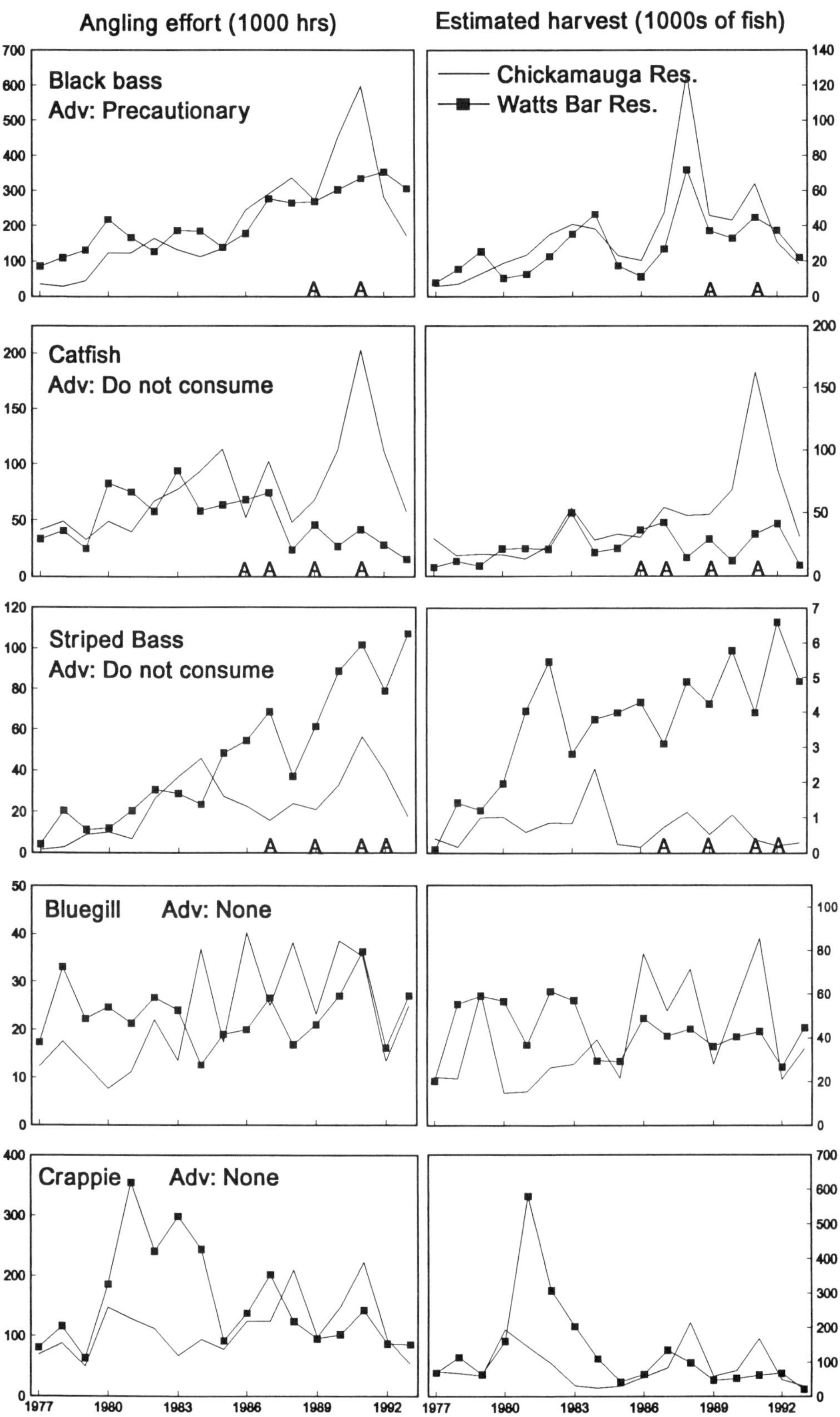

Angling effort (1000 hrs)
Estimated harvest (1000s of fish)
Chickamauga Res.
Watts Bar Res.
Black bass
Adv: Precautionary
Catfish
Adv: Do not consume
Striped Bass
Adv: Do not consume
Bluegill
Adv: None
Crappie
Adv: None

source rather quickly, while in a lake the distribution of contaminants is likely to be a function of the source location. This is particularly true of soluble contaminants such as arsenic and selenium. However, sediment-associated (or particle-reactive) contaminants such as PCBs, mercury, and cesium-137 have different distributional mechanisms. The high sedimentation rates typical of many reservoirs will likely result in a deposition of sediment-associated contaminants (e.g., PCBs, mercury, and cesium-137) such that after several years of curtailed releases, the majority of these contaminants will be buried sufficiently deep in the sediments so as to be unavailable biologically. Reservoir flow conditions can also result in depositional areas (e.g., original river channels, backwater areas, wide basins) where contaminants accumulate away from the original source.

The spatial distribution of mercury in fish in Watts Bar Reservoir was directly related to the distance from the primary source, DOE's Y-12 facility. The release of mercury from Y-12 is much reduced from in the past (DOE 1995), and levels in Poplar Creek are not likely to increase unless proposed clean-up activities in the Poplar Creek's East Fork floodplain result in the resuspension and transport of mercury to Poplar Creek.

The PCBs in Watts Bar Reservoir have come from several sources, and the spatial distribution of PCBs shows widespread contamination consistent with multiple sources. Levels in the upper Tennessee River arm are equal to or slightly higher than the Clinch River arm, suggesting that the ORR is not the primary source of PCBs to the reservoir even though a few small impoundments on the reservation are known to have greater concentrations of PCBs than the reservoir (Goddard et al. 1991). That the ORR has been a contributor of PCBs to the reservoir is apparent from the differences between levels in the Clinch River arm and those in upstream Melton Hill Reservoir.

Differences in contaminant concentration among fish species within a system is common (Verta 1990; Hebert and Haffner 1991). Such differences are primarily a function of the rate of contaminant ingestion or absorption (which are directly related to fish size, age, and diet) and the assimilation efficiency (which is related to lipid content and various biochemical and physiological interactions). For example, the highest PCB concentrations are usually found in long-lived species with a large terminal size, a high lipid content, and which are either bottom feeders or top carnivores. Among-species differences can be exaggerated or masked due to size selectivity in collection. In Watts Bar Reservoir, the species with the highest PCB concentrations were catfish, a bottom-feeding group with high lipid content, and striped bass, a large piscivore whose primary prey are high-lipid shad. These among-species differences can have several management implications. Consumption advisories need to be tailored to individual species or species groups and sometimes even to size ranges within species. Public acceptance of and compliance with consumption advisories will likely depend on the relative popularity of the species and whether a species is sought primarily for sport purposes or as a source of food.

When contaminants are identified as harmful to ecological or human health, their production, use, and release is usually reduced or discontinued, resulting in a gradual decline in concentrations in the environment due to biodegradation, biological unavailability, or removal from the system. For example, human health risks associated with exposure to PCBs were recognized in the 1970s, and the manufacture of PCBs in the United States was banned in 1976. Environmental contamination was expected to decline as a result. An apparent decline in PCB concentrations in Watts Bar Reservoir catfish in 1989 and 1990 raised hopes that the consumption advisories might soon be lifted, but subsequent data indicated that levels remained high and were not decreasing after the ban as quickly as anticipated. Factors that contribute to annual variation in PCB concentration probably include the amount of precipitation, incidence of flood events, and biological productivity. High rainfall may dilute contaminant concentration resulting in reduced bioaccumulation or may even flush contaminants from the reservoir, but could also have the opposite effect (i.e., cause an increase in contaminants) as a result of high runoff of both urban and rural land or resuspension of previously deposited contaminants. Depending on the relationship between contaminant assimilation by fish and food web dynamics, annual variation in productivity could have a significant effect on the availability of contaminants in the food web.

The observed ecological effects and the risk assessment results point to Poplar Creek as the most likely place for significant impacts to the fish community. Although concentrations of mercury and a few other contaminants in fish and sediments from Poplar Creek are elevated compared to reference sites (Ashwood et al. 1986), none are high enough to suggest by themselves a direct link to observed effects. Given that a wide variety of contaminants have been released into Poplar Creek from two of the DOE facilities, we suspect that the observed effects may be a result of multiple contaminants in the ecosystem.

There is not yet any indication that human health problems have occurred as a result of contamination in Watts Bar Reservoir, although epidemiology studies are underway (Bruce et al. 1993). However, risk calculations suggest that consumption of PCB-contaminated fish could result in a higher lifetime cancer risk if contaminated fish are regularly consumed over many years. A CERCLA record of decision which outlines remediation activities for the contaminants in Watts Bar Reservoir has not yet been issued, but is likely to include a continuation of the fish consumption advisories as the best remedial action to reduce human health risks related to contaminants in fish. The effectiveness of these advisories to actually reduce the consumption of contaminated fish has not been evaluated. However, estimates of angling effort and fish harvest, as determined by creel surveys during the last 17 years, suggest that compliance with the advisories is limited. Because actual posting of signs describing the advisory are recent (1992), it may take additional time for a more dramatic angler response to become evident. Alternatively, it is quite possible that many anglers are ignoring the advisories.

Fish consumption advisories are common throughout the United States for a variety of aquatic systems and species as a result of a variety of contaminants, primarily mercury, chlordane, PCBs, and dioxin (Cunningham et al. 1990). From a public resource perspective, the ramifications of environmental contamination include reduced recreational opportunities, reduced angling success due to an unhealthy fish community, and the loss of a food source. The identification of precise impacts and causes can be difficult, costly, and time consuming. Remediation to correct the problem can be even more difficult and costly.

Acknowledgements

We thank numerous members of the Clinch River Environmental Restoration Program that contributed to data collection (both field and laboratory) and analysis, analytical support, and risk assessment. We also thank reviewers for providing thoughtful comments that improved the final manuscript.

References

Adams, S. M., editor. 1990. Biological indicators of stress in fish. American Fisheries Society Symposium 8.

Adams, S. M., A. M. Brown, and R. W. Goede. 1993. A quantitative health assessment index for rapid evaluation of fish condition in the field. Transactions of the American Fisheries Society 122:63–73.

Adams, S. M., W. D. Crumby, M. S. Greeley, M. G. Ryon, and E. M. Schilling. 1992. Relationships between physiological and fish population responses in a contaminated stream. Environmental Toxicology and Chemistry 11:1549–1557.

Adams, S. M., K. D. Ham, and J. J. Beauchamp. 1994. Application of canonical variate analysis in the evaluation and presentation of multivariate response data. Environmental Toxicology and Chemistry 13: 1673–1683.

Ashwood, T. L., C. R. Olsen, I. L. Larsen, and D. P. Lowry. 1986. Sediment contamination in streams surrounding the Oak Ridge gaseous diffusion plant. Oak Ridge National Laboratory (ORNL/TM-9791), Oak Ridge, Tennessee.

Bruce, J. M., J. E. Buddenbaum, J. K. Lamb, and T. E. Widner. 1993. Dose reconstruction feasibility study, volume II—part A. Tennessee Department of Health, Division of Environmental Epidemiology, Nashville.

Cheek, T. E., M. J. Van Den Avyle, and C. C. Coutant. 1983. Distribution and habitat selection of adult striped bass, *Morone saxatilis* (Walbaum), in Watts Bar Reservoir, Tennessee. Oak Ridge National Laboratory Environmental Sciences Division (ORNL/TM-8447), Oak Ridge, Tennessee.

Cook, R. B., and sixteen coauthors. 1992. Phase 1 data summary report for the Clinch River remedial investigation: health risk and ecological risk screening assessment. Oak Ridge National Laboratory Environmental Sciences Division (ORNL/ER-155), Oak Ridge, Tennessee.

Cook, R. B., and twenty-two coauthors. 1993. Phase 2 sampling and analysis plan, quality assurance project plan, and health and safety plan for the Clinch River remedial investigation: an addendum to the Clinch River RCRA facility investigation plan. Oak Ridge National Laboratory Environmental Sciences Division (DOE/OR/01-1111), Oak Ridge,Tennessee.

Cunningham, P. A., J. M. McCarthy, and D. Zeitlin. 1990. Results of the 1989 census of state fish/shellfish consumption advisory programs. Research Triangle Institute, Research Triangle Park, North Carolina.

DOE (U.S. Department of Energy). 1988. Historical radionuclide releases from current DOE Oak Ridge Operations Office facilities. Oak Ridge Operations Office (OR-890), Oak Ridge, Tennessee.

DOE (U.S. Department of Energy). 1995. Mercury abatement report on the U.S. Department of Energy Oak Ridge Y-12 plant for fiscal year 1995. Oak Ridge Y-12 Plant (Y/ER-251). Oak Ridge, Tennessee.

Dycus, D. L. 1989. PCB studies on fish from Watts Bar, Ft. Loudoun, Tellico, and Chilhowee Reservoirs, 1987. Tennessee Valley Authority, Water Resources (TVA/AWR-89/10), Chattanooga.

Dycus, D. L., and G. D. Hickman. 1988. PCB levels in fish from Fort Loudoun Reservoir, Fort Loudoun tailrace, Tellico Reservoir, and Chilhowee Reservoirs, autumn 1986 to winter 1987. Tennessee Valley

Authority, Water Resources (TVA/ONRED/AWR-88/19), Knoxville.

EPA (U.S. Environmental Protection Agency). 1989. Risk assessment guidance for Superfund volume 1: human health evaluation manual (part A) interim final. U.S. Environmental Protection Agency (EPA/540/1–89/002), Office of Health and Environmental Assessment, Washington, DC.

ESD-ORNL (Environmental Sciences Division—Oak Ridge National Laboratory and Jacobs Engineering Group, Inc.). 1995. Remedial investigation/feasibility study for the Clinch River/Poplar Creek operable unit. ORNL Environmental Restoration Division (ORNL/ER-315/V1&D1), Oak Ridge, Tennessee.

Goddard, P. L., A. J. Legeay, D. S. Pesce, and A. M. Stanley. 1991. Site descriptions of environmental restoration units at the Oak Ridge K-25 site, Oak Ridge, Tennessee. Oak Ridge National Laboratory Environmental Restoration Division (K/ER-47), Oak Ridge, Tennessee.

Hebert, C. E., and G. D. Haffner. 1991. Habitat partitioning and contaminant exposure in cyprinids. Canadian Journal of Fisheries and Aquatic Sciences 48:261–266.

Hoffman, F. O., and five coauthors. 1991. Preliminary screening of contaminants in the off-site surface water environment downstream of the U.S. Department of Energy Oak Ridge Reservation. Oak Ridge National Laboratory Environmental Sciences Division (ORNL/ER-9), Oak Ridge, Tennessee.

Jimenez, B. D., and J. J. Stegeman. 1990. Detoxification enzymes as indicators of environmental stress on fish. Pages 67–79 *in* S. M. Adams, editor. Biological indicators of stress in fish. American Fisheries Society Symposium 8.

Loar, J. M., and twenty coauthors. 1992. First annual report on the biological monitoring and abatement program at Oak Ridge National Laboratory. Oak Ridge National Laboratory, Environmental Sciences Division (ORNL/TM-10399), Oak Ridge, Tennessee.

McCarthy, J. F., and L. R. Shugart, editors. 1990. Biomarkers of environmental contamination. Lewis Publishers, Boca Raton, Florida.

O'Bara, C. J. 1994. TWRA creel survey, 1993. Prepared for Tennessee Wildlife Resources Division, Fish Management Division, Nashville, Tennessee. Center for the Management, Utilization, and Protection of Water Resources, Tennessee Technological University, Cookeville.

Peakall, D. 1994. The role of biomarkers in environmental assessment: introduction. Ecotoxicology 3:157–160.

Soballe, D. M., B. L. Kimmel, R. H. Kennedy, and R. F. Gaugush. 1992. Reservoirs. Pages 421–474 *in* C. T. Hackney, S. M. Adams, and W. H. Martin editors. Biodiversity of the southeastern United States—aquatic communities. Wiley, New York.

Suter, G. W., II. 1991. Screening level risk assessment for off-site ecological effects in surface waters downstream of the U.S. Department of Energy Oak Ridge Reservation. Oak Ridge National Laboratory, Environmental Sciences Division (ORNL/ER 8), Oak Ridge, Tennessee.

Suter, G. W., II. 1993. Ecological risk assessment. Lewis Publishers, Ann Arbor, Michigan.

Suter, G. W., II., and J. B. Mabrey. 1994. Toxicological benchmarks for screening of potential contaminants of concern for effects on aquatic biota: 1994 revision. Oak Ridge National Laboratory, Environmental Sciences Division (ES/ER/TM-96/R1), Oak Ridge, Tennessee.

Turner, R. R., C. R. Olsen, and W. J. Wilcox. 1984. Environmental fate of mercury and ^{137}Cs discharged from Oak Ridge facilities. Pages 329–338 *in* D. D. Hemphill, editor. Trace substances in environmental health-XVIII, 1984: a symposium. University of Missouri, Columbia.

TVA (Tennessee Valley Authority). 1985. Instream contaminant study—task 4, fish sampling and analysis. Office of Natural Resources and Economic Development, Knoxville, Tennessee.

Verta, M. 1990. Changes in fish mercury concentrations in an intensively fished lake. Canadian Journal of Fisheries and Aquatic Sciences 47:1888–1897.

Competitive Fishing and Angling Perspective

American Fisheries Society Symposium 16:135–144, 1996

Understanding the Diversity among Largemouth Bass Anglers

ROBERT B. DITTON

Texas A&M University, Department of Wildlife and Fisheries Sciences
College Station, Texas 77843–2258, USA

Abstract.—A marketing perspective is needed for targeting recreational fisheries programs and services. Whereas many state fisheries agencies survey their angler populations on a regular basis, the results provide only an aggregate or "average angler" profile of their clientele. There is a need for market segmentation whereby anglers with similar characteristics and behavioral tendencies are partitioned into groups. I discuss four approaches for segmenting largemouth bass *Micropterus salmoides* anglers (i.e., by fishing frequency, residence location, gender, and tournament participation) in support of agency programs. As demographics change over the next 25 years, it may become necessary for managers to learn more about African Americans, Hispanics, and other minority groups in the angler population as well as their fishing habits and preferences in order to develop new program initiatives for recruiting members of these groups to recreational fishing. Many angler surveys are insufficient from a marketing perspective because sample size is too small to allow for segmentation of the angler population and subsequent investigations of within-group differences. In addition to the approach used by marketing specialists, the emerging concept of recreation specialization provides a means for understanding the diversity among various groups of largemouth bass anglers, for example. Largemouth bass fishing is viewed as a social world made up of various subworlds. I discuss various practical applications of market segmentation as a part of fisheries management.

Fishing for black bass *Micropterus* spp. is a popular leisure activity. Of the 30.2 million anglers in the United States who fished freshwater (other than the Great Lakes) in 1991, 12.9 million spent 158 million days fishing for black bass (USFWS 1993a). Of the 2.5 million resident and nonresident anglers who fished in Texas freshwater in 1991, 1.1 million spent 13.5 million days fishing for black bass (USFWS 1993b).

In neither of these U.S. Fish and Wildlife Service (USFWS) reports, was there further analysis of these black bass anglers or other species-specific information that would be useful to fisheries management. The purpose of the National Survey of Fishing, Hunting, and Wildlife-Associated Recreation, conducted by the USFWS since 1955, is "to gather information on the number of anglers . . . as well as how often they participate and how much they spend on these activities." For a further understanding of black bass fishing and its participants, fishery managers must sponsor supplemental state- or regional-level studies for the various categories of human dimensions inquiry they require.

In this paper I discuss and demonstrate approaches for segmenting the largemouth bass *Micropterus salmoides* angler market in support of fishery management efforts. In addition to approaches currently used by marketing specialists, I describe the emerging concept of recreation specialization as a theoretical means for understanding the diversity among various sub-groups of largemouth bass anglers. This paper explores largemouth bass fishing as a social world made up of various subworlds, and presents practical applications for managers throughout the paper.

Aggregate results presented in the USFWS surveys (1993a, 1993b) are often used to calculate means or other measures of central tendency for angler groups. For example, freshwater anglers averaged 12.3 d/year and 12.4 d/year of black bass fishing in the United States and Texas, respectively (USFWS 1993a, 1993b). Babbie (1989) uses the concept of ecological fallacy to suggest we mislead when we analyze aggregated data in order to understand individual behavior. Also, means are misleading as they portray an average black bass angler instead of focusing on the various segments or types of anglers (i.e., novices, club anglers, tournament anglers, etc.) we would expect to find within this angler group. Means mask differences found within the population (Shafer 1969). Imagine McDonalds trying to meet the dining needs of its average patron rather than focusing on the various clientele groups that eat at their establishments (i.e., before-work breakfast clientele, after-work clientele, weekend clientele, etc.). By tradition, most successful private sector establishments would appear to have a good understanding of their markets. In contrast, public sector marketing efforts are in their infancy.

Several writers have recognized the need for more attention to a marketing perspective among natural resources managers (Salwasser et al. 1989;

Witter and Adams 1994). This perspective has not been adopted widely because many managers confuse marketing with promotion. Because many fish stocks are fully used or even overexploited, marketing of recreational fishing has little appeal to some (Kutkuhn 1994). Aside from its private sector roots, a marketing perspective is similar to a human dimensions perspective on recreational fisheries, which has traditionally included development of angler profiles in support of management efforts.

Natural resources agencies are increasingly encouraging attention to marketing (or human dimensions) efforts for several reasons: (1) a general lack of public recognition of fish and wildlife agencies and the services and benefits agencies provide, (2) the implications of demographic change on who pays and is able to pay for agency services in the future, (3) a greater concern for equity or fairness among constituencies instead of serving only those who shout the loudest, (4) a greater concern for customer satisfaction in evaluating agency efforts, and (5) a way to identify and resolve angler group conflicts.

Whereas public sector marketing is recent, many state fisheries agencies have a history of surveying their angler population on a regular basis. However, most of these angler surveys are generally insufficient from a marketing standpoint, in that sample size is too small to allow for segmentation of angler groups. Most statewide surveys seek to know about their angler clientele from an aggregate perspective.

A marketing perspective needs to be designed into a statewide survey sample from the outset. For example, in a recent survey of licensed anglers that focused on constraints to fishing in Texas, there was particular interest in constraints to female anglers. This is timely because of the recognition of a significant trend toward increased participation in recreational fishing by females (Snepenger and Ditton 1985). Also, there has been little previous research on gender differences in constraints to recreation participation (Henderson 1994; Shaw 1994). To better understand constraints experienced by female anglers, it was necessary to oversample females in addition to the sample of licensed anglers drawn for research purposes. With only 21% of the original sample being female, there would have been an insufficient number of female respondents without a stratified sampling design.

As fishery managers recognize they need to know more about other angler groups (e.g., Hispanic and African American anglers) and their fishing experiences and preferences, these groups will need to be oversampled accordingly. Even the USFWS (1993a) national survey has insufficient sample size to provide adequate attention to ethnic groups; Hispanics are currently aggregated within the "Other" category. As demographics change over the next 25 years and the implications for participation in recreational fishing become clear (Murdock et al. 1992), it may become necessary to learn more about African Americans, Hispanics, and other racial and ethnic groups in the angler population to provide support for new fisheries management programs and initiatives.

Once agencies recognize the need for a marketing perspective and for understanding the diverse clientele they serve, how do they integrate this need with the species-specific orientation of fisheries management efforts? This is difficult. Once a species-specific focus is adopted, constraints to sampling frames emerge; for example, there is no listing of licensed largemouth bass anglers unless a state requires a stamp to target, catch, or possess largemouth bass. Only with such a list of stamp purchasers for sampling purposes is it possible to investigate the within-group diversity of licensed largemouth bass anglers. Where such a list is not available, there are no suitable alternatives. The membership roster of the Bass Angler Sportsmen's Society (BASS), for example, is not likely to be representative of all largemouth bass anglers in the population and thus is of little use for sampling the diversity of this population. According to the USFWS (1993a), only 5% of the angler population 16 years or older paid dues to fishing membership organizations.

In lieu of a statewide sampling frame of largemouth bass anglers, managers may seek to investigate the largemouth bass angler market along with ongoing creel survey efforts. Although such efforts may be supportive of lake-specific management programs, they pose several problems. First, there is an avidity bias where more frequent users have a greater chance of being sampled. Also, results are casebound if a specific management regime (e.g., slot limit, mandatory catch and release) is in effect. If anglers seek fishing destinations that meet their particular needs, and if a specific management regime has a deterministic effect on the types of recreational fishing experiences afforded (see Driver et al. 1987, for an overview of the recreation opportunity spectrum concept), then it is unlikely a complete cross section of anglers will be intercepted at any particular lake. Likewise, results can be casebound depending upon travel distance considerations, the extent of the adjacent angler population, and other recreation demand considerations.

In the 1991 Statewide Survey of Texas Anglers funded by the Texas Parks and Wildlife Department (TPWD), anglers were asked to rank their species preferences (first, second, and third). This was done to learn more about freshwater anglers who prefer to target largemouth bass in Texas. Sample size (N = 10,000) was increased well beyond what was necessary to represent the population of licensed anglers in order to have a sufficient pool of largemouth bass anglers for follow-up research purposes.

This approach is not a panacea for gathering data from statewide populations of largemouth bass anglers but rather warrants additional scrutiny and evaluation. Whereas we were able to collect data of direct managerial utility (e.g., statewide distribution of largemouth bass fishing trips; preferences on management alternatives, regulations, and issues; and level of satisfaction with largemouth bass fishing), we are likewise coping with potential biases and design flaws. These include: (1) an assumption that all anglers had a species preference as stated, and if not, they would have volunteered they had no species preference, (2) a need to correct for nonresponse bias (more committed anglers were more likely to respond), and (3) an assumption that recreational fishing participation would continue beyond the initial survey contact. We are at a point in time with human dimensions research, however, where it is widely recognized that each survey approach has a unique set of design problems that must be understood and overcome (Brown 1991).

Over the past two decades, human dimensions research has demonstrated there is a diversity in preferences within most participant groups (Hendee 1974; Driver 1985). Previous work would suggest largemouth bass anglers vary in their reasons for fishing and definitions of fishing quality. To the extent they understand the composition and range of this diversity, managers can plan effectively to enhance the experiences of identifiable angler groups (not to be confused with constituency groups that represent the political interests of angler groups). Fishery managers may choose to continue to treat largemouth bass anglers as an aggregate in their planning and allocation efforts or, worse yet, they may assume all largemouth bass anglers have the same needs and preferences as those represented by politically powerful constituency groups. If they take the former approach, they are likely to manage for fishing outcomes that might please no one; if they embrace the latter, they will disproportionately represent the interests of some anglers and ignore the interests of the majority of largemouth bass anglers.

What constitutes a largemouth bass angler? Usually, this category would include anyone who has reported one or more days of largemouth bass fishing over a particular time period. This could include people who target largemouth bass exclusively, those who target largemouth bass as well as other species on each trip (or throughout the year), those who target largemouth bass but catch something else, and those recently converted to this pursuit but who have not yet caught one. Furthermore, in addition to differences due to their level of socialization in largemouth bass fishing, there are likely to be gender and racial or ethnic differences in angler participation patterns, motivations, attitudes, and management preferences that remain as yet unexplored. As a result of managerial indifference to some of these human dimensions concerns, and to the sampling strategies used previously to represent angler populations, most of the current literature on the human dimensions of recreational fishing pertains only to white males.

Management-Based Market Segmentation

Market segmentation is the process of partitioning potential clients into groups with similar characteristics that are likely to exhibit similar behaviors (Backman 1994). Whereas there are various approaches to segmentation, emphasis should be on measurement simplicity and on partitioning groups that are identifiable and accessible for management purposes (Kotler 1980). Previously, for example, recreationists have been segmented by personal characteristics (social and economic), cognitive measures (motivations or benefits sought), situations (specific times or places), or some combination thereof. Although studies that segment participants by demographic characteristics are useful if an agency seeks to respond to emerging trends, this approach provides little insight to the cognitive benefits sought by various participant segments. Alternately, whereas understanding motivations (or benefits sought) on a segmentation basis may be the most effective for predicting behavior (Haley 1985), it is difficult to access participants based on their motivational profiles. Also, previous work has shown that visitors to an area seek different benefits by situation (i.e., season and specific locale in the area; Backman 1994). Each of these market segmentation approaches may be appropriate depending on information needs and planned agency actions. Consequently, the aggregate of largemouth bass anglers can be partitioned variously to form subgroups where diversity is minimized to some extent. For example, they can be grouped by fishing frequency, residence location, gender, or by whether or not they participate in fishing tournaments. Although other approaches are available, these four are discussed using examples.

Segmenting by Fishing Frequencies

For years it has been collective wisdom that a small percentage of anglers account for a disproportionately large number of fish caught or days of fishing (effort). To better understand market segments based on avidity, the distribution of largemouth bass fishing days (in the previous 12 months, for example) can be viewed across quintiles of a statewide sample of resident anglers with a first preference for largemouth bass. In a statewide survey of largemouth bass anglers in Texas (1992), we observed that 20% of the angler sample (N = 940) accounted for 58% of the total largemouth bass fishing days in the previous year. Further, 40% of the angler sample accounted for 79% of the total largemouth bass fishing days. The remaining 60% of the angler sample accounted for the remaining 21% of largemouth bass fishing days. It remains to be seen if these quintile groups of anglers differ in personal characteristics, trip expenditures, fishing benefits sought, fishing attitudes, and management preferences. A better understanding of the most- and least-avid anglers would be useful to managers in that it would establish a range of angler capabilities, thinking, wants, and preferences. As fisheries agencies reach out to underserved constituencies, for example, perhaps agencies can help overcome the constraints of those who participate the least in fishing.

Segmenting by Residence Location

Although some may think there are too many people fishing already, new reservoir construction is often justified on the basis of recreation demand and the effect various recreationist groups have on state, regional, and local economies. To understand the economic impact of a reservoir fishery, for example, three market segments are often identified: (1) local resident anglers, (2) state resident anglers (but not locals), and (3) out-of-state anglers. In addition to economic impact differences, each angler group can pose different and often conflicting demands on fisheries management.

In an economic study of Lake Fork Reservoir (Texas) currently underway, in cooperation with TPWD and the Sabine River Authority of Texas, expenditures for nondurable purchases were separated for these three angler groups. It was assumed that money spent by local anglers to go fishing would have been spent for something else had they not gone fishing. From a state perspective, it was assumed that expenditures by local and nonlocal state residents do not have any economic consequences; there is simply a redistribution of monies within the state. Alternately, expenditures by nonlocal residents and out-of-state anglers are new monies which increase the respective local or state economic base and produce economic impacts (Milon and Thunberg 1993). This occurs as anglers transfer money to state and local businesses which in turn spend money to purchase goods and services; hence some level of economic impact is likely from fishing at Lake Fork Reservoir.

Market segmentation for economic impact analysis purposes poses challenges for sampling. As might be expected, most anglers intercepted by access point surveys were local and nonlocal state residents. To understand the economic impact of Lake Fork Reservoir on the state of Texas, it was necessary to increase the number of days sampled to intercept additional out-of-state anglers. In addition to expenditure data, anglers were probed on a variety of social and marketing descriptor variables to enable fishery agencies and local tourism interests to be more responsive to each segment.

Segmenting by Gender

In the 1992 statewide survey of largemouth bass anglers in Texas, only 141 (13%) of the respondents were female. Let us suppose that to support programs intended to socialize more women into outdoor activities (Thomas and Peterson 1993), more information was needed on female largemouth bass anglers than is currently available. This sample size would be insufficient to produce meaningful results; thus, it would be necessary to oversample female anglers in the previous statewide angler survey to acquire a sufficient sample size of female largemouth bass anglers. Only then would it be possible to test for gender differences in participation patterns, constraints, benefits sought from fishing, information sources used, attitudes toward fishing, and management preferences. As a result of greater understanding of female largemouth bass anglers, information programs and services can be better targeted to meet the needs of this particular angler segment.

TABLE 1.—Mean responses of Texas largemouth bass anglers to questions regarding attitudes towards recreational fishing. Scale developed by Graefe (1980) to understand four subdimensions of consumption: number of fish caught, type of fish caught, disposition of catch, and general orientation toward catching something. Data were collected as 1 = strongly disagree, 2 = disagree, 3 = neutral, 4 = agree, 5 = strongly agree and then averaged. Asterisks indicate chi-square significance at $P < 0.05$.

Statement	Tournament $N = 192$	Nontournament $N = 889$	X^2
The more fish I catch, the happier I am	3.62	3.44	
Trip may be successful even if no fish are caught	3.66	3.81	
I usually eat the fish I catch	2.53	3.44	*
Successful trip is one in which many fish are caught	2.95	2.94	
I would rather catch one or two big fish than 10 small fish	3.50	3.22	*
When I go fishing, I'm just as happy when I don't catch a fish	2.71	2.83	
It doesn't matter to me what type of fish I catch	2.55	3.07	*
The bigger the fish caught, the better the fishing trip	3.50	3.17	*
I'm just as happy if I don't keep the fish I catch	3.99	3.50	*
I like to fish where there are several kinds of fish to catch	3.38	3.79	*
I want to keep all the fish I catch	1.74	2.12	*
I catch fish for sport and pleasure rather than food	3.96	3.47	*
I'm just as happy if I release the fish I catch	3.99	3.46	*
I usually give away the fish I catch	1.95	2.18	*
I like to fish where I know I have a chance to catch a trophy fish	3.96	3.22	*

Segmenting by Tournament Participation

Largemouth bass anglers can be segmented according to whether or not they fished in tournaments. In the 1992 statewide survey of largemouth bass anglers in Texas, we found that 82% did not participate in fishing tournaments in the previous twelve months. Diversity in the Texas largemouth bass fishery was indicated by the large number of statistically significant differences between tournament and nontournament largemouth bass anglers in personal characteristics, participation variables, motivations or benefits sought from fishing, attitudes toward fishing, and management preferences (R. Riechers, G. Wilde, R. Ditton, M. Fisher, Texas Parks and Wildlife Department, unpublished data). Nontournament anglers were older, less likely to be members of fishing clubs, had lower self-reported skill levels, fished less frequently in freshwater, and had more years of fishing experience than tournament anglers. Furthermore, there were significant group differences on 11 of 17 motivation items. For example, nontournament anglers placed greater importance on fishing for family recreation and being able to obtain fish for eating. Not surprisingly, tournament anglers rated all catch-related motives higher than nontournament anglers. There were significant differences on 11 of 15 attitude items between tournament and nontournament largemouth bass anglers (Table 1). Tournament anglers agreed more strongly with statements that dealt with catching large (and trophy-sized) fish and catching and releasing fish; nontournament anglers agreed more strongly with statements that dealt with keeping and eating their catch and having a chance to fish for a variety of fish.

This segmentation approach would be useful to agency discussions on whether the experience needs of both angler groups were being met. First, it might be surprising to some that tournament anglers are a minority among the overall population of largemouth bass anglers. Second, results of this type of segmentation could help managers to better understand largemouth bass anglers in general vis-a-vis largemouth bass tournament anglers. Lastly, however, much diversity within the population of nontournament anglers remains to be understood through further segmentation.

The preceding four approaches to segmenting the largemouth bass market can help promote a better understanding of group diversity. Whereas most segmentation efforts are one-dimensional, they nevertheless provide managers with greater insights than available by just focusing on an "average" largemouth bass angler. In many cases, however, considerable diversity in characteristics, attitudes, and behaviors remains within identified segments. Although such segmentations provide useful information for natural resource managers, they do not explain or predict group differences in variables that are independent of the classification variable (Wilde and Ditton 1994).

A Conceptual Approach to Segmentation

The theory of recreation specialization provides an alternative means for understanding the diversity of largemouth bass anglers. Understanding anglers

TABLE 2.—Characteristics and types of participation in social worlds (adapted from Unruh 1979).

Characteristics	Subworld types: Strangers	Tourists	Regulars	Insiders
Orientation	Naive	Curiosity	Habituation	Identity
Experiences	Disorientation	Orientation	Integration	Creation
Relationships	Superficiality	Transiency	Familiarity	Intimacy
Commitment	Detachment	Entertainment	Attachment	Recruitment

in terms of varying levels of specialization stands in contrast with aggregate measures of central tendency or market segmentation approaches, which are usually problem-solving in orientation. Unlike with market-based segmentation, resultant groups based on level of specialization are not readily identifiable and accessible for management purposes. Bryan (1977, 1979) first proposed the concept of recreation specialization to explore the potential diversity of recreationists within an activity. He defined specialization as "a continuum of behavior from the general to the particular reflected by equipment and skills used in the sport and activity setting preferences" (Bryan 1977). He identified four types of anglers, with each having a unique place on the specialization continuum. At the lower end were occasional anglers, followed by generalists, technique specialists and, finally, at the upper end, technique and setting specialists. He suggested that the typology and location of anglers on the continuum were reflected in their frequency of participation, setting preferences, technique preferences, choice of equipment, importance of catch, social unit of participation, and resource management preferences.

Building on this work, Ditton et al. (1992) sought to develop a theory of recreation specialization based on a social worlds perspective. A social world is "an internally recognizable constellation of actors, organizations, events and practices which have coalesced into a perceived sphere of interest and involvement for participants" (Unruh 1979). The social world of largemouth bass fishing, for example, is larger than its groups or organizations; is not defined by formal boundaries, membership lists, or spatial territory; and lacks a powerful centralized authority structure. Like other recreational fishing social worlds, this social world has a social organization which is diffuse and amorphous in character.

Inevitably, social worlds segment into subworlds. This occurs as specialized interests develop within the social world, which act to differentiate members from each other (Kling and Gerson 1978). Subworlds can segment around various sources or conditions (Strauss 1984). These include:

1. spatial distinctions (e.g., particular lakes or reservoirs),
2. different management regimes that lead to increased opportunities to catch trophy size largemouth bass,
3. technology and skill (e.g., where new technologies can lead to new forms of largemouth bass fishing),
4. ideology (e.g., some largemouth bass anglers believe that catch and release is an appropriate norm while others do not, and
5. intersections with other social worlds to create a new fishing subworld (e.g., fishing for largemouth bass in walk-in access remote areas.

To categorize participants into social worlds and enhance conceptual understanding, Unruh (1979, 1980) proposed a typology of four generalized trans-situational social subworlds. His subworlds (strangers, tourists, regulars, and insiders) were based on their social proximity to knowledge about the social world and activities therein. These four subworlds were ordered along a theoretical dimension with four characteristics (orientation, experiences, relationships, and commitment, Table 2). Also, as Unruh (1979) cautions, this process may not be linear or inevitable, that is, we should not assume that everyone moves through the largemouth bass social world from stranger to insider. Understanding this process will require a longitudinal survey design that permits observations of a sample of largemouth bass anglers over an extended period.

In applying the concept of recreation specialization to the social world of largemouth bass fishing, it is reasonable to expect to find clusters of anglers differing in the social, behavioral, attitudinal, and other characteristics mentioned earlier (Ditton et al. 1992). Like others who have species preferences, largemouth bass anglers are a subset of a statewide population of anglers. They are perhaps slightly skewed toward the higher end of fishing specialization because they have species preferences. According to the USFWS (1993a), about 8% of the freshwater angler population did not have a species preference. In between the least and most specialized subworlds of largemouth bass anglers are an indeterminate number of subworlds with intermediate levels of specialization. Ditton et al. (1992) provide a series of propositions which can be used to link specialization with elements of largemouth bass fishing subworlds. As level of specialization in largemouth bass fishing increases:

1. the value of side bets (one's overall investment in the activity) will likely increase;
2. the centrality of that activity in a person's life will likely increase;
3. acceptance and support for the rules, norms, and procedures associated with the activity will likely increase;
4. the importance attached to equipment and the skillful use of that equipment will likely increase;
5. dependency on a specific resource will likely increase;
6. level of mediated interaction (number of media sources of fishing information) relative to that activity will likely increase;
7. the importance of fishing-specific elements of the fishing experience will decrease relative to nonactivity-specific elements of the experience.

These propositions remain to be tested to better understand subworld diversity within the largemouth bass social world.

Chipman and Helfrich (1988) focused on the recreational specialization of black bass anglers. They intercepted anglers as a part of a creel survey on two rivers in Virginia and followed up with a mail survey. The dominant fish targeted by anglers was smallmouth bass *M. dolomieu*. Specialization was defined using four dimensions of angler behavior: (1) fishery resource use, (2) experience, (3) investment, and (4) centrality of angling to lifestyle. Six relatively distinct types of anglers or angler groups were identified using cluster analysis representing low to high specialization. High specialization anglers were likely to cite resource-related motivations, rely on skill to catch fish, prefer to catch and release large fish, and favor restrictive management regulations. Less specialized anglers reported escape and family-oriented recreation as fishing benefits sought, placed more emphasis on luck to catch fish, were satisfied catching smaller fish, and favored less restrictive regulations. Whereas Chipman and Helfrich (1988) concluded more on-site studies are needed to standardize survey methodologies and to determine angler typologies among site-specific settings, Fisher (1993) suggests a diverse sample of anglers is needed to assess specialization, not a sample limited by a specific species and setting focus.

Other studies on recreation specialization (Bryan 1977; Gill 1980; Graefe 1980; Chipman and Helfrich 1988; McGurrin 1988; Hahn 1991; Fisher 1993) have enhanced understanding of the constituent elements of specialization and diversity among anglers. None, however, have resulted in practical tools that can be used readily by field personnel to identify different types of anglers at creel intercept interviews or public hearings. Currently, to establish an angler's level of specialization, it would be necessary to evaluate their responses to various questionnaire items. Because it is impossible to have anglers complete in-depth questionnaires or interviews for classification purposes when in the field, there is a need for a conceptually well-grounded surrogate measure or "marker" to establish each angler's specialization level. Wilde and Ditton (1994) divided largemouth bass anglers into two groups according to the specificity of responses to an open-ended mail survey question regarding preferred species in Texas waters. One group (N = 1,681) consisted of those who gave a general response (i.e., bass) and the other (N = 983) included those who gave a specific response (i.e., black bass, largemouth bass, or bigmouth bass). Wilde and Ditton (1994) found significant ($P < 0.05$) group differences in fishing motivations and attitudes, years of fishing experience, annual fishing frequency, and personal investment in fishing. These group differences were consistent with predictions based on the theory of recreation specialization (Ditton et al. 1992) and provide initial evidence that angler identification of specific species preferences is a useful surrogate for general level of fishing specialization. Although much conceptual and empirical work remains, the theory of recreation specialization has potential for providing a grounded understanding of angler group differences on many matters of managerial concern.

Discussion

Much has changed since Voiland and Duttweiler (1984) pondered the question, Where's the humanity? in response to a noticeable lack of human dimensions articles in fisheries journals. Not only has the number of human dimensions articles increased in the fisheries literature, but there is also a realization that this type of work is available in other scholarly and technical outlets. A more important question may be, How many human dimensions specialists are working in fisheries management today? In response to a recent survey conducted by the American Fisheries Society (AFS) Committee on the Human Dimensions of Recreational Fisheries, human dimensions fisheries specialists were identified in 39 of 52 (75%) reporting state and territorial fisheries agencies (G. Wilde, personal communication). Much has also changed since Bennett et al. (1978) asked state recreational fisheries management personnel to rank nine identified management objectives. Three yield-oriented goals (maximum sustainable yield, maximum catch, and establishing trophy fisheries) accounted for most of the goals ranked number one; only a small minority of respondents ranked angler-oriented goals. Furthermore, most respondents felt future management goals were not likely to be different. This recent AFS survey also showed the fisheries field is in the process of moving from indifference to human dimensions concerns to a status where many state agencies have recently conducted or are in the process of investing in statewide angler investigations. As fisheries management traditions change and as agency personnel recognize anglers as their customers, we can expect more states to study angler populations and do so on a more regular and comprehensive basis. This will surely include a marketing orientation with increased sample sizes for attention to particular market segments in support of specific management efforts or to resolve conflicts and allocation issues.

Surely, from experience, fisheries managers have recognized the diversity of angler clientele discussed earlier in this paper. However, it can be argued that they haven't had the educational background to transfer "what they know" into targeted programs and services (Kelso and Murphy 1988; Ditton and Fedler 1989). Consequently, it is understandable how we come to have the "one shoe fits all" approach to provision of fishing opportunities we often see today. Much more than simply a research perspective in support of management, fishery managers are able to provide various potential product offerings in response to expressed demands (Driver 1985). These products include much more than just largemouth bass, for example. They include "opportunities to use, appreciate, or benefit from the wildlife in particular ways" (Driver 1985), that is, the opportunity to participate in competitive largemouth bass fishing events, to fish for trophy-size largemouth bass, to catch largemouth bass among other species, to socialize new recruits into largemouth bass fishing, to participate in catch-and-release largemouth bass fishing, etc. Each of these products of management (or use opportunities) have differentiating attributes and provide different benefit packages. Each will be more or less attractive to particular angler segments. Furthermore, having largemouth bass available is essential, but the products provided include much more than the fish itself.

Having more sophisticated insight to the largemouth bass fishing market is one thing; using insight effectively to evaluate the current product mix and develop new fishing products is quite another. Fishery managers need to think more in terms of recreational fishing systems with careful attention to supply (e.g., fishery resources and use opportunities available) and demand (e.g., current participation as well as latent interest in recreational fishing) considerations. Such a system would include a statewide assemblage of interdependent fishing products that are accessible to major population centers. This will require much more attention to the planning process, whereby some of the following questions are answered: How do largemouth bass fishing demands vary statewide and in urban and rural areas? What types of fishery products (or use opportunities) are being demanded in each area? Can fishing products in fact be developed to meet these various demands? How much of each fishing product is sufficient? and Where does it need to be available? An approach for coping with the demographic changes underway in most states is to develop a wide range of fishing products in each region or within range of each major population cluster. It may not be necessary (or possible) to have the same product mix in each region, but this matter needs careful consideration. In addition to development of new reservoirs where needed, and new fishing products as necessary, efforts need to focus on creating management regimes at current reservoirs that provide an appropriate product mix of fishing opportunities so that overall largemouth bass angler fishing satisfaction increases.

As some largemouth bass angler market segments are perceived as getting more fishing opportunities than others (or more than they may be entitled to), or as the interests of some angler segments are ignored, freshwater fishery managers will be confronted by the equity (or fairness) issue. When managers develop and promote some fishing products to the exclusion of others, they are making allocation decisions. Loomis and Ditton (1993) discuss the basics of two conceptual approaches to equity and offer managers some specific suggestions for dealing with this topic. First, managers need to better understand their constituencies in order to deal with fairness issues. In particular, managers should not assume that largemouth bass anglers see the world as managers do. Second, they need to promote educational efforts to shape or correct individual and group perceptions of what's fair. Finally, constituent concerns about the fairness of an allocation decision may spawn concerns about the fairness of the entire fishery management process. Failure to deal effectively with fairness issues can seriously reduce the credibility of the fisheries agency. This alone with its various implications should provide plenty of incentive for agencies to better understand the various market segments within its recreational fishing constituency.

Acknowledgments

I thank my current and former Texas Parks and Wildlife Department colleagues, Phil Durocher, Robin Riechers, Gene R. Wilde, Mark R. Fisher, Gary C. Matlock, Maury Osborn and Steve Gutreuter, for their support and assistance in the design and execution of the freshwater and black bass angler surveys reported in this paper. Also, I am indebted to Kevin Hunt for his leadership and participation in my research program.

References

Babbie, E. 1989. The practice of social research, fifth edition. Wadsworth, Belmont, California.

Backman, S. J. 1994. Using a person-situation approach to market segmentation. Journal of Park and Recreation Administration 12(1):1–16.

Bennett, D. H., E. L. Hampton, and R. T. Lackey. 1978. Current and future management goals: implications for future management. Fisheries 3(1):10–15.

Brown, T. L. 1991. Use and abuse of mail surveys in fisheries management. Pages 255–261 *in* D. Guthrie and seven coeditors. Creel and angler surveys in fisheries management. American Fisheries Society Symposium 12.

Bryan, H. 1977. Leisure value systems and recreational specialization: the case of trout fishermen. Journal of Leisure Research 9:174–187.

Bryan, H. 1979. Conflict in the great outdoors. University of Alabama, Bureau of Public Administration, Sociological Studies 4, Tuscaloosa.

Chipman, B. D., and L. A. Helfrich. 1988. Recreational specializations and motivations of Virginia river anglers. North American Journal of Fisheries Management 8:390–398.

Ditton, R. B., and A. J. Fedler. 1989. Importance of fish consumption to sport fisherman: a reply to Matlock et al. Fisheries 14(4):4,6.

Ditton, R. B., D. K. Loomis, and S. Choi. 1992. Recreation specialization: re-conceptualization from a social worlds perspective. Journal of Leisure Research 24:33–51.

Driver, B. L. 1985. Specifying what is produced by management of wildlife by public agencies. Leisure Sciences 7:281–294.

Driver, B. L., P. J. Brown, G. H. Stankey, and T. G. Gregoire. 1987. The ROS planning system: evolution, basic concepts, and research needed. Leisure Sciences 9:201–212.

Fisher, M. R. 1993. The relationship between non-response bias and angler specialization. Doctoral dissertation. Texas A&M University, College Station.

Gill, A. D. 1980. The social circle of catfishermen: a contribution to the sociology of fishing. Master's thesis. Kansas State University, Manhattan.

Graefe, A. R. 1980. The relationship between level of participation and selected aspects of specialization in recreational fishing. Doctoral dissertation. Texas A&M University, College Station.

Hahn, J. 1991. Angler specialization: measurement of a key sociological concept and implications for fisheries management decisions. Pages 380–389 *in* D. Guthrie and seven coeditors. Creel and angler surveys in fisheries management. American Fisheries Society Symposium 12.

Haley, R. 1985. Developing effective communication strategies: a benefit segmentation approach. Wiley, New York.

Hendee, J. C. 1974. A multiple satisfaction approach to game management. Wildlife Society Bulletin 2:104–113.

Henderson, K. A. 1994. Broadening an understanding of women, gender, and leisure. Journal of Leisure Research 26:1–7.

Kelso, W. E., and B. R. Murphy. 1988. University fisheries programs in the United States: structure and needs. Fisheries 13(2):9–15.

Kling, R., and E. M. Gerson. 1978. Patterns of segmentations and interaction in the computing world. Symbolic Interaction 1:24–43.

Kotler, P. 1980. Marketing management. Prentice Hall, Englewood Cliffs, New Jersey.

Kutkuhn, J. 1994. Letter to the editor. Fisheries 19(7):40.

Loomis, D. K., and R. B. Ditton. 1993. Distributive justice in fisheries management. Fisheries 18(2):14–18.

McGurrin, J. M. 1988. Diversity in Gunpowder River trout anglers and implications for management. Proceedings of the Annual Conference Southeastern Association of Fish and Wildlife Agencies 40(1986): 111–116.

Milon, J. W., and E. Thunberg. 1993. A regional analysis of current and future Florida resident participation in marine recreational fishing. Florida Sea Grant College Program, Sea Grant Report 112, Gainesville.

Murdock, S. H., K. Backman, R. B. Ditton, and M. N. Hoque. 1992. Demographic change in the United States in the 1990's and the twenty-first century. Fisheries 17(2):6–13.

Salwasser, H., G. Contreras, M. Dombeck, and K. Siderits. 1989. A marketing approach to fish and wildlife program management. Transactions of the North American Wildlife and Natural Resources Conference 54: 261–270.

Shafer, E. L., Jr. 1969. The average camper who doesn't exist. U.S. Department of Agriculture Forest Service,

Northeast Forest Experiment Station, Research Paper NE-142, Upper Darby, Pennsylvania.

Shaw, S. M. 1994. Gender, leisure, and constraint: towards a framework for the analysis of women's leisure. Journal of Leisure Research 26:8–22.

Snepenger, D. J., and R. B. Ditton. 1985. A longitudinal analysis of nationwide hunting and fishing indicators: 1955–1980. Leisure Sciences 7:297–319.

Strauss, A. 1984. Social worlds and their segmentation processes. Studies in Symbolic Interaction 5:123–139.

Thomas, C. L., and T. A. Peterson. 1993. Becoming an outdoors-woman: concept and marketing. Transactions of the North American Wildlife and Natural Resources Conference 58:185–192.

Unruh, D. R. 1979. Characteristics and types of participation in social worlds. Symbolic Interaction 2:115–130.

Unruh, D. R. 1980. The nature of social worlds. Pacific Sociological Review 23:271–296.

USFWS (U.S. Fish and Wildlife Service). 1993a. National survey of fishing, hunting, and wildlife-associated recreation. U.S. Government Printing Office, Washington, DC.

USFWS (U.S. Fish and Wildlife Service). 1993b. National survey of fishing, hunting, and wildlife-associated recreation: Texas. U.S. Government Printing Office, Washington, DC.

Voiland, M. P., and M. W. Duttweiler. 1984. Where's the humanity? a challenge and opportunity for the fisheries community. Fisheries 9(4):10–12.

Wilde, G. R., and R. B. Ditton. 1994. A management-oriented approach to understanding diversity among largemouth bass anglers. North American Journal of Fisheries Management 14:34–40.

Witter, D. J., and C. E. Adams. 1994. Fish, wildlife, and public service: why resource processionals must market. Fisheries 19(9):20–21.

American Fisheries Society Symposium 16:145–151, 1996

A Qualitative Evaluation of Specialization among Crappie Anglers

MICHEAL S. ALLEN AND LEANDRO E. MIRANDA

National Biological Service
Mississippi Cooperative Fish and Wildlife Research Unit[1]
Post Office Drawer BX, Mississippi State, Mississippi 39762, USA

Abstract.—Recognizing the need to incorporate sociological factors into management decisions, fisheries managers have increased efforts to research angler opinions through mail and telephone surveys designed to yield primarily quantitative data. We evaluated perspectives of crappie *Pomoxis* spp. anglers through "field research," a social sciences survey technique that yields primarily qualitative data that can complement and improve the focus of quantitative surveys. Anglers were interviewed to determine their opinion of harvest restrictions, personal motivations for fishing, and resource management needs. Four groups of crappie anglers were identified: occasional anglers, generalists, springtime anglers, and specialists. The subgroups differed in fishing frequency and seasonality, fishing techniques, attitudes concerning harvest, and motivations. Catching fish to eat was a key motivation for the majority of crappie anglers, second only to relaxation. However, specialists enjoyed the challenge of crappie fishing and placed less emphasis on harvesting fish and more emphasis on harvest restrictions and conservation. Field studies such as this one provide insight into angler characteristics that cannot be obtained with rigidly structured data-collection surveys and may improve the focus of quantitative angler surveys.

The use of quantitative angler surveys designed to measure angler opinion has proliferated as fisheries managers recognize the importance of incorporating social science research into fisheries management decisions. Telephone, mail, and structured on-site surveys typically yield large sample sizes and quantitative data. Statistical analyses can then be used to separate anglers based on motivations (Chipman and Helfrich 1988; see review by Fedler and Ditton 1994), satisfactions (Graefe and Fedler 1986; Holland and Ditton 1992), or attitudes toward management practices (Duttweiler 1976; Hampton and Lackey 1976; Harris and Bergersen 1985), for example. Such quantitative studies are the primary method of social research currently used in fisheries management (Pollock et al. 1994).

A technique that may complement quantitative surveys of angler opinion is "field research" (Babbie 1986). Field research yields qualitative data through direct observation of the participants under study. The field researcher interacts directly with subjects under study to qualify motivations, satisfactions, or opinions. Unlike quantitative surveys, field research combines data collection, data analysis, and subsequent inferences in an inductive approach. Thus, field research is subjective. However, field research may improve quantitative studies by identifying segments of the angling population that can then be targeted using quantitative surveys.

Direct observation techniques similar to field research have been used to qualify groups of anglers targeting trout (Salmonidae; Bryan 1977, 1979) and catfish (Ictaluridae; Gill 1980). Bryan (1977, 1979) described a continuum of four trout angler specialization categories and hypothesized that groups differed in fishing methods, attitudes about fish harvest, and motivations for fishing. Quantitative studies followed Bryan's work and refined the specialization concept (Chipman and Helfrich 1988; Ditton et al. 1992). Similarly, Quinn (1992) described walleye *Stizostedion vitreum* angler subgroups from observations and questionnaire responses.

Black crappie *Pomoxis nigromaculatus* and white crappie *P. annularis* are among the most popular sport fishes in the United States. A national survey conducted by the U.S. Fish and Wildlife Service identified 9.2 million anglers as crappie anglers and estimated annual effort for crappies was 91 million angler-days (USDOI 1991). Crappies are important species in many reservoirs, often ranking either first or second in annual biomass harvested (Jenkins and Morais 1971; Campbell et al. 1978; Mitzner 1981). The advent of crappie clubs, crappie tournaments, and magazines about crappie fishing suggests that the popularity of these species is increasing. However, characteristics and ideologies of crappie anglers have not been formally detailed. We used the

[1]The Unit is jointly sponsored by the Mississippi Department of Wildlife, Fisheries and Parks; Mississippi State University; and the National Biological Service.

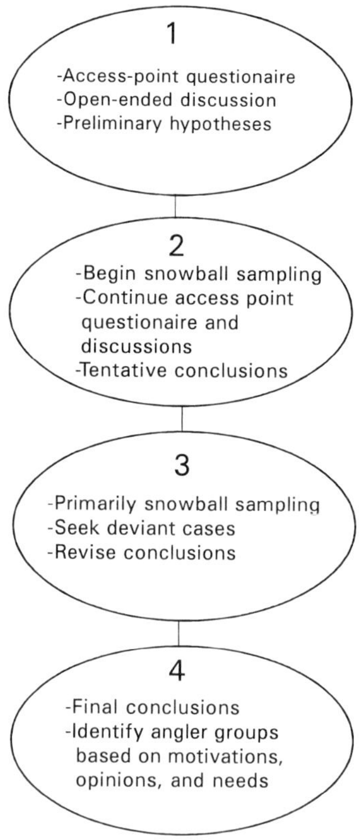

FIGURE 1.—An illustration of the dynamic methodology used in field research.

"field research" technique (Babbie 1986) to identify groups of crappie anglers, and discuss the use of field research to complement quantitative surveys.

Methods

Our field research approach began by contacting crappie anglers at a local reservoir access area to identify their fishing methods and motivations with a written questionnaire. When interacting with anglers, we identified ourselves as researchers and made no pretense of being actual participants. In addition to the written questionnaire, we asked some anglers to participate in an open-ended discussion. We identified additional contacts from referrals given by the access-area participants, and from the membership of a statewide crappie club (Figure 1). As suggested by Babbie (1986), throughout the study we modified our sampling methods as needed to identify the motivations and characteristics of participants.

Access-Area Interviews

The first portion of the study consisted of access-area interviews at Columbus Lake, Mississippi, during April and May 1993 (see number 1 on Figure 1). Columbus Lake is a 8,803-acre reservoir located on the Tennessee–Tombigbee Waterway just north of Columbus, Mississippi. The reservoir is surrounded by Clay, Lowndes, Monroe, and Oktibbeeha counties with a total population of approximately 155,000 people (J. W. Rush, Mississippi State University, unpublished data). In spring 1992, a 9-in minimum length limit for crappies was imposed on Columbus Lake with a bag limit of 30 fish per angler per day. We contacted anglers at a boat ramp 1.2 mi from the Columbus city limits. This ramp received heavy angling traffic for several reasons, including its extensive facilities for both shore and boat anglers, its proximity to the city, and it was the only public ramp providing access to the southern portion of the lake. Sampling was conducted during 8 d/month, and sample days were divided equally on randomly selected week and weekend days (strata) as recommended by Malvestuto et al. (1978). Each sampling day consisted of a 5-h period selected at random during either morning (0800–1300 hours) or afternoon (1300–1800 hours).

We approached anglers arriving and leaving the boat launch area, as well as shore anglers present during the 5-h sampling period. Anglers were asked what species of fish they sought on that trip. Those who stated they were fishing for crappies, "anything," or fished for crappies "sometimes" were asked to complete a two-page questionnaire designed to identify fishing frequency, motivations for fishing, attitudes toward harvest regulations, and angler age. Anglers fishing for other species who stated that they never fished for crappie were not asked to participate. Anglers usually completed the questionnaire themselves, but we offered to administer the questionnaire verbally to ensure participation of all anglers. Participants indicated the importance of attitude and motivation statements on a modified Likert-type scale with responses ranging from not important (1) to extremely important (5).

Open-Ended Discussion

Following completion of the two-page questionnaire, selected anglers were asked if they would "talk for a while" about fishing, either at the boat ramp or at another time and location. We sought anglers for detailed discussions using a quota sample procedure (Babbie 1986), which selected anglers to fill different participation categories. Categories included boat and shore anglers, male and female anglers, occasional versus frequent anglers, and categories of attitudes toward harvest restrictions identified through the written questionnaire. The interviewer selected anglers for open-ended discussions based on questionnaire responses in an attempt to include anglers from each of the categories identified above. Anglers who were interested but unable to participate in a discussion at that time were solicited for a later discussion either at their home or at a local restaurant within 2 weeks. Four anglers (4%) refused to complete the on-site questionnaire, and we had no refusals for the longer discussion.

The open-ended discussions were unstructured; we had a general plan of topics to be discussed, but no specific, ordered questions. Our goal was to make the participant feel comfortable so they would elaborate in their own words. Questions for discussion included: Do you prefer to catch crappie over other species, and why or why not? What is your opinion about the value of length and bag limits? Do you prefer to fish with live or artificial bait? Why do you fish? What is your preferred equipment (i.e., fishing rods, reels, electronics, boats)? and Do you fish all year? Questions were integrated into the conversation rather than asked sequentially, and participants were free to raise their own topics to discuss. Throughout the study, we modified questions when we felt a better approach had been identified from previous discussions. Brief notes were taken during each discussion and were enhanced with detailed notes within 24 h.

Snowball Sampling

After about 3 weeks of contacting anglers at the access-area, we added snowball sampling to the access-area interviews and discussions (see number 2 on Figure 1). Snowball sampling entails asking participants to recommend other individuals who may have introduced them into an activity or influenced them, interview those other individuals, then have them recommend others, and so on (Babbie 1986). Anglers identified through snowball sampling were contacted by phone and asked to meet in person as described above. There were no refusals. Before the discussion, anglers contacted through snowball sampling completed the same two-page questionnaire given to access-area participants.

Deviant Cases

Access-area interviews were discontinued at the end of May 1993 because the number of anglers seeking crappies had declined to about 1 or 2 per 5-h period, and because we saw a need to target deviant cases (see number 3 on Figure 1). Deviant cases (Babbie 1986) are subjects that hold different attitudes and behaviors from the majority of subjects. Thus, we sought anglers who held different motivations, satisfactions, and opinions than the majority of anglers we had previously encountered. Anglers who had returned multiple tags from crappies tagged during a concurrent tagging project on Columbus Lake (Brock 1994) were included in this group. We also contacted a statewide crappie club that held crappie fishing tournaments throughout Mississippi. Telephone discussions were conducted with three club members, and we attended a club tournament. At the tournament, we conducted discussions with four members of the club, but also observed and limited interaction with most of the 40 members present. We then concluded sampling and synthesized our conclusions (see number 4 on Figure 1).

Results

The two-page questionnaire was completed by 103 anglers. We conducted 47 discussions (Columbus Lake access area and crappie club combined), averaging 45 min each (range = 12–150 min). Based on responses to the short questionnaire and observations made during discussions, we identified four subgroups of crappie anglers. Thus, the questionnaire was helpful in establishing a basis for quota sampling for the open-ended discussions. Anglers who participated in discussions were classified into one of the four subgroups as the study progressed. However, the questionnaire alone was not sufficient to categorize those anglers into specialization groups because sample size was too small for statistical separation of groups and because the questionnaire did not provide the flexibility required to interactively pursue issues it identified. Nevertheless, responses to the questionnaire did provide partial information about attitudes toward harvest restrictions, importance of crappie fishing to each angler, motivations for fishing, and fishing frequency. Also, while administering the questionnaire we were able to make observations about the gear used by anglers. Moreover, because the questionnaire was initially administered to randomly-selected anglers, it allowed us to index the relative size of each group.

Group 1: Occasional Anglers

This group consisted of approximately 10% of anglers encountered at Columbus Lake. Anglers associated with this group fished infrequently (less than 1 d/month). They used simple tackle, usually consisting of a spin-cast rod. Occasional anglers were sometimes the spouse, child, or friend of a more frequent angler. Nevertheless, occasional anglers did not seek a particular species, would be satisfied catching "anything that will bite," and typically harvested their catch. Motivations for fishing by anglers in this group were based on companionship and relaxation. Spending time with fishing partner(s), experiencing the outdoors, or both, were primary motivations for this group. For example, occasional anglers often mentioned the experience of seeing wildlife as a satisfying part of their fishing trips. Anglers associated with this group perceived harvest regulations as beneficial but had no opinion about alternate regulations and were satisfied with those that existed.

Group 2: Generalists

Generalists consisted of both shore and boat anglers that fished more frequently than occasional anglers. Generalists were more abundant than occasional anglers and constituted approximately 20% of anglers at Columbus Lake. Some generalists fished as frequently as any anglers we encountered (5–6 d/week). Those that fished most frequently fished from shore and considered fishing for food as important as relaxation factors and enjoying nature. In contrast, generalists who fished less frequently (1–2 d/week) fished primarily for relaxation but harvested most of their catch and released only undersized fish. Generalists used simple tackle, either spin-cast rods or cane poles. Generalists stated they were fishing for "anything" but some indicated they preferred to catch crappies because of their quality as food. Thus, generalists were opportunistic anglers who would catch whatever they could. Generalists placed a greater emphasis on catching many fish regardless of size over catching few large ones. Most generalists were satisfied with existing harvest restrictions, but some wanted more liberal regulations and none preferred more restrictive regulations than those in effect. We encountered four generalists who preferred no regulations and observed two members of this group who kept fish below the minimum length limit.

Group 3: Springtime Anglers

This group included the majority (approximately 70%) of crappie anglers encountered at Columbus Lake. They targeted crappies each spring when the fish could be caught in shallow water and fished frequently during that period (3–4 d/week). During other times of the year these anglers fished for other species or did not fish often. Springtime anglers older than 60 years of age often mentioned that they had fished for other species (e.g., largemouth bass *Micropterus salmoides*) when they were younger but preferred to fish for crappies because the fishing techniques were less strenuous and more relaxing. Younger springtime anglers fished for other species also, but fished for crappies frequently in the spring because they were easy to catch and good to eat.

Members of this group used specialized gear compared to groups 1 and 2. They fished exclusively from a boat and used both live bait and artificial jigs. Almost all members used graphite rods about 12 ft long designed for crappie fishing. Each springtime angler had between three and five of these rods. Though these anglers fished in shallow water, their boats were often equipped with electronic depth sounders.

Anglers in this group were supportive of harvest restrictions and some preferred more strict length limits than the existing 9-in minimum length limit. They preferred catching fewer large fish over many small ones, and springtime anglers usually stated that crappie smaller than 10 in were not large enough to fillet (dress boneless), and thus, preferred larger fish. Nevertheless, springtime anglers kept legal-sized fish up to the 30-fish bag limit. They preferred to catch crappies over other species because of their quality as food and because they found them easier to catch than other species.

Relaxation appeared to be the most important reason this group fished, followed by catching fish to eat. Springtime anglers typically fished with family members or other springtime anglers; many of them knew one another, and we perceived that some of these anglers sought a limit to attain status among their peers. For example, anglers in this group often asked what other anglers had caught, or asked if a particular angler had come and gone and what they had caught. This phenomenon was not observed among anglers in groups 1 or 2.

We perceived the spring crappie season to be an important component of the yearly recreation of springtime anglers. Springtime angler effort began to decrease in early May, after which we encountered few springtime crappie anglers through access-area interviews. However, we continued to encounter other springtime anglers through snowball sampling.

Group 4: Crappie Specialists

These anglers preferred to fish for crappies year-round or nearly so and typically fished about 2–3 d/week. We classified nine anglers as specialists, but these were encountered only through tag returns, snowball sampling, and discussions with crappie club members. Anglers we met from the crappie club conformed to the specialist category. Because we encountered no specialists at the boat ramp, their numbers relative to the other angler groups was likely low.

Specialists chose to fish for crappies over other species because they considered them a challenge to find and catch, but not necessarily harvest, under a variety of conditions. Specialists placed less emphasis on catching fish to eat than did any other group. Factors such as the challenge of finding and catching crappies in offshore habitats and information exchange appeared more important to this group than to groups 1–3. For example, specialists often mentioned the thrill of finding a school of large crappies as an important reason they fished for these species. Specialists typically fished with other specialists rather than with family members, and they often had one or two individuals with whom they fished regularly. Members of this group fished exclusively from boats.

Specialists differed from springtime anglers in the techniques they used. Members of this group claimed to be proficient with the use of electronic depth sounders and many fished offshore or in deep (>10 ft) water during much of the year. Most specialists used rods and reels similar to those of springtime anglers, but specialists appeared more likely to use artificial jigs exclusively; some of them made their own jigs. Two specialist anglers we encountered used ultralight spinning reels in addition to long graphite rods for fishing deep water.

Specialists appeared motivated to learn more about fishing techniques and fisheries management than any other group. Several specialists requested scientific publications or other information about crappie management, and these anglers usually sought information about fishing techniques through interaction with peers and magazines. The crappie club published a quarterly newsletter that provided crappie fishing information to club members and was distributed to bait stores in Louisiana and Mississippi.

Specialists appeared interested in management strategies that would improve crappie size. Anglers in this group were supportive of length limits and most believed a 10-in minimum length limit to represent optimal management for most crappie fisheries. They harvested a large portion of their catch but typically released small fish that might be kept by groups 1–3. Thus, specialists considered catching a bag limit less important and catching large fish more important than did groups 1–3; we perceived that specialists considered catching trophy fish a stronger motivation than did groups 1–3. Crappie fishing appeared to be the primary year-round recreation to these anglers and an important component of their lives. Though specialist anglers constituted a small proportion of the angling population, they seemed highly interested in crappie management and more likely to react to changes in regulations with organized support or opposition than anglers in other groups.

Discussion

Groups of crappie anglers we qualified differed in their attitudes toward and expectations from fishery resources. Similar groups (i.e., occasional anglers, generalists, and specialists) have been documented among trout (Bryan 1977, 1979), catfish (Gill 1980), smallmouth bass *Micropterus dolomieu* (Chipman and Helfrich 1988), and walleye (Quinn 1992) anglers. However, no species-specific angler typologies have identified seasonal anglers such as the springtime anglers we encountered. Although the springtime angler group we identified may have been a local phenomenon, we believe they are widespread. Crappie fisheries may attract springtime anglers because the fish spawn inshore (Hansen 1965), are often abundant in lakes and reservoirs (Hill 1984; Miller et al. 1990; Mitzner 1991), and are more vulnerable to fishing during that period than during the remainder of the year (Siefert 1969; Markham et al. 1991). If springtime anglers represent a distinct group for crappie fisheries, they may substantially increase harvest during spring because they may be more numerous and fish more frequently than all other angler groups. Therefore, fisheries managers may be able to modify total harvest substantially by manipulating regulations during spring.

Specialist anglers were less numerous compared to other groups we identified; however, their numbers may increase in the future. Crappie tournaments, crappie clubs, and magazines devoted to crappie fishing are relatively recent phenomena that may facilitate such an increase. Bryan (1977:175) hypothesized that anglers tend to become more specialized over time and defined recreational specialization as "... a continuum of behavior from the general to the particular, reflected by equipment and skills used in the sport and activity setting preferences." Bryan (1977) hypothesized that as anglers become specialists, they maintain a network of communication through fishing clubs, magazines, and newsletters. Ditton et al. (1992) reconceptualized recreational specialization and provided evidence for a theory of recreational specialization. Ditton et al. (1992) found that highly specialized anglers were more dependent on mediated interaction (i.e., newspapers, magazines, television), placed more emphasis on attaining large or trophy fish, and considered activity-specific elements (i.e., catching fish to eat, catching fish to have a satisfying trip) less important than did less specialized anglers. Specialist anglers we encountered conformed well to these characteristics.

Our intent in this study was not to contribute to the development of a theory of recreational specialization. However, if anglers are likely to become more specialized over time, as hypothesized by Bryan (1977, 1979), we predict that the number of specialist crappie anglers will increase. Therefore, fisheries managers will probably encounter a more diverse group of crappie anglers in the future, with more anglers preferring size limits, bag limits, or other management strategies for crappie fisheries.

The angler groups we qualified may have been difficult to discern with a quantitative survey. Babbie (1986:260) noted that an advantage of field research is the validity of the conclusions and stated that "... being there" is a powerful technique for gaining insights into the nature of human affairs. We perceived differences among angler groups based on direct interactions with the anglers. For example, all groups we identified harvested a relatively large portion of their catch. However, springtime anglers often mentioned they preferred large crappies (>10 in) because of their quality as food, whereas specialists usually mentioned they sought large crappies for the experience of finding and catching them. We also noted a competition among some springtime anglers to catch a limit, a phenomenon that would be difficult to identify with quantitative surveys and would have implications for the use of bag limits. For example, managers may wish to reduce or liberalize bag limits on crappie fisheries depending on the crappie population structure and amount of springtime effort. We perceived that some springtime anglers older than 60 years of age fished for crappies because the techniques were less strenuous than those needed to catch other species; fisheries managers could alter management strategies for fisheries where a large proportion of anglers are older. Conversely, if the majority of anglers for a fishery were occasional anglers or generalists, harvest restrictions may be inappropriate if used solely to produce larger fish.

A disadvantage inherent to field research is that our conclusions are subjective. Pollock et al. (1994) noted that purposive sampling strategies (e.g., quota sampling, snowball sampling) include unknown biases and unknown precision. Admittedly, our conclusions are based on nonrandom, personal observations of anglers and may not represent the "typical" crappie angler. Our snowball sampling technique, contacting anglers through multiple tag returns and the crappie club were probably biased toward avid and skilled crappie anglers. However, purposive sampling allowed us to target and profile anglers who fished for crappies frequently and may have provided information that could improve the focus of future quantitative surveys. Another disadvantage of field research is that conclusions may differ among individual researchers, even with the same amount of observation. To eliminate this bias, we used only one interviewer. Babbie (1986) cautioned against the use of broad, descriptive statements in reporting results from field research because judgement may be linked to the observer. The field researcher should provide examples that clarify the inductive reasoning used (Babbie 1986), as we have attempted to show.

We conclude that field research may reveal angler characteristics that will not be identified in surveys, and provide managers with a more in-depth understanding of anglers. However, field research is subjective and somewhat time consuming, reducing the reliability of conclusions and the number of anglers surveyed. Therefore, field research should be used to improve the focus of quantitative surveys that provide the foundation for management decisions.

Acknowledgments

This research was funded by the Mississippi Department of Wildlife, Fisheries and Parks through Federal Aid in Sport Fish Restoration Act, Project F-105, and by the Mississippi Agriculture and Forestry Experiment Station. We especially thank D. Gill for his guidance during the implementation of the project. M. Colvin, D. Gabelhouse, K. Hodges, G. Hurst, D. Jackson, J. Reed, and H. Schramm provided helpful comments on previous drafts of this manuscript.

References

Babbie, E. R. 1986. The practice of social research. Wadsworth, Belmont, California.

Brock, R. 1994. Growth, fishing and natural mortalities of crappies in two Mississippi lakes. Master's thesis. Mississippi State University, Starkville.

Bryan, H. 1977. Leisure value systems and recreational specialization: the case of trout fishermen. Journal of Leisure Research 9:174–187.

Bryan, H. 1979. Conflict in the great outdoors. University of Alabama, Bureau of Public Administration, Sociological Studies 4, Tuscaloosa.

Campbell, W. J., E. Hayes, W. R. Chapman, and W. Seawell. 1978. Angling pressure and sport fish harvest in the predator-stocking-evaluation reservoirs. Proceedings of the Annual Conference Southeastern Association of Fish and Wildlife Agencies 30(1976): 114–119.

Chipman, B. D., and L. A. Helfrich. 1988. Recreational specializations and motivations of Virginia river anglers. North American Journal of Fisheries Management 8:390–398.

Ditton, R. B., D. K. Loomis, and S. Choi. 1992. Recreation specialization: re-conceptualization from a social worlds perspective. Journal of Leisure Research 24:33–51.

Duttweiler, M. W. 1976. Use of questionnaire surveys in forming fishery management policy. Transactions of the American Fisheries Society 105:232–239.

Fedler, A. J., and R. B. Ditton. 1994. Understanding angler motivations in fisheries management. Fisheries 19(4):6–13.

Gill, A. D. 1980. The social circle of catfishermen: a contribution to the sociology of fishing. Master's thesis. Kansas State University, Manhattan.

Graefe, A. R., and A. J. Fedler. 1986. Situational and subjective determinants of satisfaction in marine recreational angling. Leisure Sciences 8:275–295.

Hampton, E. L., and R. T. Lackey. 1976. Analysis of angler preferences and fisheries management objectives with implications for management. Proceedings of the Annual Conference Southeastern Association of Game and Fish Commissioners 29(1975):310–316.

Hansen, D. F. 1965. Further observations on nesting of the white crappie, *Pomoxis annularis*. Transactions of the American Fisheries Society 94:182–184.

Harris, C. C., and E. P. Bergersen. 1985. Survey on demand for sport fisheries: problems and potentialities for its use in fishery management planning. North American Journal of Fisheries Management 5:400–410.

Hill, K. R. 1984. Correlation of total and "angler-acceptable" crappie standing stocks with lake basin slopes and siltation indexes. North American Journal of Fisheries Management 4:350–354.

Holland, S. M., and R. B. Ditton. 1992. Fishing trip satisfaction: a typology of anglers. North American Journal of Fisheries Management 12:28–33.

Jenkins, R. M., and D. I. Morais. 1971. Reservoir sport fishing effort and harvest in relation to environmental variables. Pages 371–384 in G. E. Hall, editor. Reservoir fisheries and limnology. American Fisheries Society, Special Publication 8, Bethesda, Maryland.

Malvestuto, S. P., W. D. Davies, and W. L. Shelton. 1978. An evaluation of the roving creel survey with nonuniform probability sampling. Transactions of the American Fisheries Society 107:255–262.

Markham, J. L., D. L. Johnson, and R. W. Petering. 1991. White crappie summer movements and habitat use in Delaware Reservoir, Ohio. North American Journal of Fisheries Management 11:504–512.

Miller, S. J., D. D. Fox, L. A. Bull, and T. D. McCall. 1990. Population dynamics of black crappie in Lake Okeechobee, Florida, following suspension of commercial harvest. North American Journal of Fisheries Management 10:98–105.

Mitzner, L. 1981. Influence of floodwater storage on abundance of juvenile crappie and subsequent harvest at Lake Rathburn, Iowa. North American Journal of Fisheries Management 1:46–50.

Mitzner, L. 1991. Effect of environmental variables upon crappie young, year-class strength, and the sport fishery. North American Journal of Fisheries Management 11:534–542.

Pollock, K. H., C. M. Jones, and T. L. Brown. 1994. Angler survey methods and their applications in fisheries management. American Fisheries Society Special Publication 25.

Quinn, S. P. 1992. Angler perspectives on walleye management. North American Journal of Fisheries Management 12:367–378.

Siefert, R. E. 1969. Biology of the white crappie in Lewis and Clark Lake. U.S. Bureau of Sport Fisheries and Wildlife Technical Paper 22.

USDOI (U.S. Department of the Interior). 1991. 1991 national survey of fishing, hunting, and wildlife-associated recreation. U.S. Department of the Interior, Washington, DC.

American Fisheries Society Symposium 16:152–162, 1996

Trends in Regulatory and Voluntary Catch-and-Release Fishing

STEVE QUINN
In-Fisherman, Incorporated
2 In-Fisherman Drive, Brainerd, Minnesota 56401, USA

Abstract.—Regulations requiring the release of many fish species have increased dramatically. Fisheries management agencies have shifted from viewing harvest regulations as necessary evils to seeing them as popular mandates in the quest for high-quality fishing. Voluntary release of legally harvestable fish by anglers also seems to be increasing. This study documents trends in catch-and-release fishing including the increase in regulations for various fish species, the extent of voluntary release, reasons for increases in catch-and-release fishing, and agency promotion of catch and release. A survey of state agencies showed that during the last decade, catch-and-release regulations for trout (Salmonidae) have increased in 88% of the responding states, and for black bass *Micropterus* spp. in 86% of those states. Sixty-seven percent of states reported increases in regulations for walleye and sauger *Stizostedion* spp.; 52% for *Morone* spp.; 49% for crappies *Pomoxis* spp.; 48% for esocids *Esox* spp.; 39% for salmon; 19% for catfish (Ictaluridae); and 17% for sunfish *Lepomis* spp. Increases in voluntary release among states ranged from 95% for black bass to 16% for catfish and were similar in scope to increases in regulations. Regulations for catch and release increased greatly in some states, but voluntary release increased by smaller percentages. State agencies reported a wide array of biological and sociological reasons for increases in regulatory and voluntary catch and release. Despite the attention to regulations by agencies and great interest in this type of management by anglers, the published literature contains a paucity of case histories or sociological reports on the topic. It seems clear that the role of regulations in managing fish populations and providing recreation for the public will expand. Implementing regulations that will maintain ecosystem integrity successfully and meet society's needs will require greater communication among fisheries managers and between managers and the public about ecological principles and management strategies.

Behnke (1989) proposed that catch-and-release fishing was first associated with angling for anadromous Atlantic salmon *Salmo salar*, and attributed to Harold Trowbridge Pulsifer the statement "Do not be afraid to join the slowly growing fraternity of anglers whose password is 'we put them back alive' " published in *Outlook* magazine around 1919 (Pulsifer 1919). In comparing conservation aspects of angling and hunting in his 1936 book *The Modern Angler*, John Alden Knight (1936) wrote, "With the fisherman, however, the question of killing is nearly always optional." Fifty-seven years ago, salmon angler and outdoor writer Lee Wulff (1939) wrote, "Game fish are too valuable to be caught only once," which has become the rallying cry of catch-and-release proponents. Although articles in outdoor magazines often mentioned voluntary release of legally harvestable game fish in the early decades of the twentieth century, statewide harvest regulations typically included low minimum length limits and liberal creel limits through the 1940s, when regulations became even more liberal (Redmond 1986). An exception was the establishment of a trophy trout (Salmonidae) fishery on Spring Creek in Pennsylvania in 1934 through stocking and restrictive regulations (Behnke 1989).

During the 1970s, fisheries management agencies began to implement special regulations intended to raise catch rates and the proportion of large fish in the catch, at the expense of harvest. These policies were a response to documented overharvest of fisheries (Redmond 1974) and to angler requests for more opportunities for "quality fishing," generally defined as greater proportions of larger fish in the catch, along with satisfactory catch rates and aesthetic factors (Weithman and Anderson 1978). A symposium on catch-and-release fishing (Barnhart and Roelofs 1977) illustrated national interest in this topic and papers presented case histories that focused primarily on trout management (13 of 20 papers in the symposium). A decade later, a second catch-and-release symposium (Barnhart and Roelofs 1987) provided a broader focus, as 13 of 28 papers dealt with trout management but more case histories reported results of studies on black bass *Micropterus* spp., and other papers covered walleye *Stizostedion vitreum* (Fletcher 1987), bluegill *Lepomis macrochirus* (Burdick and Wydoski 1987), and crappies *Pomoxis* spp. (Childress 1987), as well as marine fish species.

Access to case histories that reported positive effects of release regulations led managers to begin testing more restrictive regulations on waters that held promise for this type of management. Although some evaluations of release regulations have been published in peer-reviewed journals, and others have appeared in agency reports, the extent of such regulations and their biological and sociological success remain largely undescribed.

Coincident with regulations that require release, anglers apparently have been voluntarily releasing increasing proportions of legally harvestable fish. Quinn (1989) noted that the extent of voluntary release has been poorly documented for three reasons: (1) harvest was the traditional measure of fishing success, (2) accurately measuring the extent of catch and release was difficult, and (3) some fisheries managers assumed that voluntary release had little effect on fisheries. Through the mid-1980s, the paucity of studies on survival of released fish (Ragan et al. 1986) may also have reduced confidence in the positive effects of voluntary as well as regulatory release. Since Quinn's (1989) study, many fisheries researchers have refined creel survey methods to count released as well as harvested fish, though older designs linger. Managers, through personal contact with anglers and angler groups, and through studies of angler attitudes, have become more aware of the extent of voluntary catch and release and the desire of many anglers for stricter limits. Hooking mortality studies generally have documented acceptably high survival rates for many fish species (Muoneke and Childress 1994). Despite these advances, documentation of voluntary catch-and-release rates and their effects remain scarce and are often hidden in descriptions of other aspects of fisheries. Studies of fishing tournaments have widely reported release rates following weigh-in (Schramm et al. 1991), but such releases typically are required by tournament organizers and thus are not truly voluntary. Tournament analyses are not included in this review.

Objectives of this study were to (1) compile information about trends in the extent of regulatory and voluntary catch and release in the United States for a variety of sport fish, (2) highlight case histories of the extent and effects of catch and release, and (3) monitor changes in attitude toward catch and release by management agencies and the fishing tackle industry.

Methods

Fisheries chiefs in 50 states were sent questionnaires concerning voluntary and regulatory catch and release of black bass, trout, esocids, walleye, sauger *Stizostedion vitreum*, *Morone* spp., catfish (Ictaluridae), sunfish *Lepomis* spp., and salmon. The survey asked whether catch-and-release regulations and voluntary catch and release had increased, decreased, or remained similar over the previous decade for each species group, the extent of change, and reasons for changes. The survey also asked whether case histories had been developed; space for comments was provided on the questionnaire. Agencies were contacted by phone if no response was received within about three months. In six cases, duplicate questionnaires were sent.

Percent of states reporting increases in regulations and voluntary catch and release were calculated for each species group. Percent increase of regulation (within states) was calculated as the percent increase in waters regulated. Percent increase of voluntary catch and release was estimated from managers observation and reports.

Regulation summaries, educational pamphlets, and agency press releases were compiled to monitor the extent to which fisheries management agencies have educated anglers about special regulations and promoted voluntary catch and release. A review was conducted of papers published in recent volumes of the *Proceedings of the Annual Conference Southeastern Association of Fish and Wildlife Agencies*, *Transactions of the American Fisheries Society*, *North American Journal of Fisheries Management*, *Fisheries*, and *Canadian Journal of Fisheries and Aquatic Sciences*, to document trends in publications about catch-and-release fishing.

Results

Fisheries chiefs or their appointees from 47 states returned questionnaires that were at least partially completed. One respondent indicated that the requested data were not available, and two states did not respond.

Concerning regulations that substantially limit harvest of desirable-size fish, agencies in only two states reported declines in catch-and-release regulations, one for black bass and one for trout. Catch-and-release regulations for trout and black bass showed large increases (88% for trout and 86% for black bass among states reporting; Table 1). More than half indicated increases in regulations for *Stizostedion* and *Morone* spp., while catch-and-release regulations for catfish and sunfish increased least (<20%; Table 1). Most new regulations protected specific size ranges of fish, but reductions in daily bag limits also were common. Catch-and-release seasons for black bass, reduced bag limits, or increased size limits during spring were also enacted (Quinn 1993).

TABLE 1.—Percent of state agencies reporting increases in regulatory and voluntary catch and release of various fish species groups, with mean and range of reported increases. Increases in regulations reflect the number of regulated waters. Increases in voluntary catch and release reflect documented or estimated increases in the proportion of legally harvestable fish that anglers release.

Species group	Increase in regulations (%)	Number reporting	Mean increase (%)	Range (%)[a]	Increase in voluntary release (%)	Number reporting	Mean increase (%)	Range (%)[a]
Black bass	86	42	284	2–3,100	95	42	38	2–250
Trout	88	40	107	1–1,600	77	39	28	1–75
Esocids	48	33	307	10–2,400	58	33	38	5–75
Stizostedion spp.	67	30	256	5–1,100	48	29	19	5–50
Morone spp.	52	27	242	5–1,000	45	29	19	5–25
Crappies	49	39	170	1–1,000	21	34	13	5–20
Salmon	39	18	556	1–3,200	40	20	22	5–50
Sunfish	17	36	26	1–100	19	32	8	5–10
Catfish	19	37	240	5–1,500	16	32	13	5–20

[a]Ranges represent the lowest and highest percent increases in regulations requiring catch and release and voluntary catch and release. An increase from 10 to 11 regulated lakes, for example, would represent a 10% increase whereas an increase from 1 to 43 would represent a 4,200% increase.

Percent increase in catch-and-release regulations over the previous decade varied greatly among states for all species, often from a few percent (representing, for example, an increase from 22 to 25 waters with catch-and-release regulations) to several thousand percent (increase from 1 to 11 regulated waters would represent a 1,000% increase). In a few cases, regulations for a species changed from no size or bag limit to statewide restrictions on size and a daily bag limit, not calculable as percents. In states where increases occurred, the number of regulated waters more than doubled (>100% increase) for all species except sunfish (Table 1).

Voluntary catch and release reportedly increased most widely for black bass (in 95% of states reporting), followed by increases for trout (77% of states), Esocids (58%), *Stizostedion* spp. (48%), and *Morone* spp. (45%; Table 1). Increases in voluntary release of sunfish (19%) and catfish (16%) were reported in the fewest states. The average percent of increase in voluntary catch and release across states during the decade also was highest for black bass and Esocids (both 38%), followed by trout (28%), salmon (22%), and *Stizostedion* spp. (19%; Table 1).

Agencies cited varied reasons for increases in catch-and-release regulations for various species. Responses included both biological and sociological justifications, and some that could be interpreted as a combination of both. For black bass, the most commonly cited biological reasons were suspected overharvest, particularly of larger fish (N = 11), documented overharvest (N = 8), management for quality and trophy fishing (N = 6), and adjustment of the population size structure (N = 4). Other primarily biological reasons cited for regulations were protection of spawners, contaminants in fish, new waters open to public fishing, low recruitment, high recruitment, habitat degradation, and control of bluegills. Angler preference was the most frequently cited sociological reason (N = 11) for increased regulation. Related reasons included political pressure and changing conservation ethics (N = 3 each), diversifying opportunity, and the influence of bass clubs (N = 1 each). Other sociological motives were encouraging a release philosophy among anglers, reducing angler confusion by standardizing regulations (N = 2 each), and creating consistency in regulations for border waters (N = 1).

Agencies increased catch-and-release regulations for trout for reasons similar to those identified for black bass, and angler preference was cited 20 times, far more frequently than any other reason. Documented overharvest and management for quality fishing (N = 7 each), and suspected overharvest (N = 5) were the next most frequently cited reasons. Reasons cited for trout but not for black bass included habitat degradation (N = 4), recovery of native subspecies (N = 2), youth fishing, improvement of catch rates, and promotion of year-round fishing.

In the 16 states where regulations for esocids increased, the most common reason cited was manipulation of populations to produce higher quality or trophy fisheries (N = 4). Attempts to maximize benefits of stocked fish, typically tiger muskellunge *Esox masquinongy* × *E. lucius*, suspected overharvest, and encouragement of a release ethic each were mentioned three times. Other reasons were documented overharvest, habitat degradation, and maintaining predator density to control common carp *Cyprinus carpio*. In the 20 states where regulations for walleye or sauger increased, enhancement of the size distribution for higher quality fishing was the predominant reason, cited by half the respondents. Five states reported documented overharvest and four suspected overharvest. Four cited restoration of weak populations and angler preferences, and two wished to foster voluntary release. Other reasons were to protect a rare stock, establish a fishery (saugeye *Stizostedion vitreum* × *S. canadense*), increase predation on panfish, increase spawning success, and match a neighboring state's regulation on a border water.

In the 14 states that reported increases in regulations that required substantial release of *Morone* spp., the most common reason was to restore populations of anadromous striped bass *M. saxatilis* on the Atlantic coast (N = 4). Desire for higher quality fisheries or increased benefit from stocked fish after they had grown larger, suspected overharvest (striped bass), and angler preference were second in frequency (N = 3 each), followed by documented overharvest and encouragement of voluntary release (N = 2 each). Other reasons were to reduce greed and to impose the same regulations for striped bass and their hybrids in order to reduce problems with identification by anglers.

Ten of the 19 states that reported increases in regulations for crappies reported that documented overharvest was the reason. Suspected overharvest and a need to improve size structures of populations were cited in five states each, and angler preference, political pressure, and better utilization of strong year-classes were each cited once. Five of seven states that reported increases in regulations for crappies cited a need to improve or maintain size structure as the reason. Four states cited documented overharvest as the reason. Protection of spawners, attempts to encourage voluntary release, and contaminant problems were cited once each. For sunfish species, three states indicated that regulation changes were intended to improve the size structure of populations and two cited attempts to encourage release as a reason; angler preference and spreading the harvest of large bluegill among anglers were cited once each. Five of seven states with increased regulations for salmon reported documented overharvest and four cited habitat degradation. Suspected overharvest and improving size structure were cited twice each and rebuilding of stocks, political pressure, and angler preference were cited once each.

Only two states reported decreases in regulations that required substantial release of legally harvestable game fish. In North Dakota, a regulation for black bass was removed after it failed to alter the population structure and release rates remained unchanged. In South Dakota, a trout regulation was removed when it didn't achieve desired results.

Evolution of a conservation or catch-and-release ethic was the most common reason cited (24 of 45 respondents) for increase in voluntary release of black bass. The second most common reason was the apparent success of regulations that required release (N = 7). Other reasons included influence of tournaments, angler concern with the health of the resource, and influence of fishing organizations and publications (N = 5 each); peer pressure, contaminant worries, and desire to improve fishing quality or trophy potential (N = 4 each); declining willingness to clean and cook fish by anglers or their spouses, and lack of tradition of black bass as a food fish (N = 2). One other reason cited was that replica mounts make better trophies than skin mounts (N = 2).

Reasons for voluntary release of trout also were dominated by a growing conservation ethic (N = 22), followed by prevalence and apparent success of stricter regulations (N = 8), and peer pressure and education efforts including those of agency Information and Education (IE) sections (N = 4 each). Other reasons cited included loss of cooking tradition, releasing fish is a "green thing" to do, enhancement of the fishing experience, desire to catch the same fish again, and a real or perceived decline in the fishery. For esocids, conservation ethics dominated (N = 14), followed by influence of regulations (N = 4), IE efforts (N = 3), and peer pressure (N = 2). Other reasons cited were desire to improve trophy fishing, organization influence, "green thing," perceived decline in fisheries, and predominance of fish of undesirably small size.

Concerning voluntary release of walleye and sauger, the conservation ethic was the most common response (N = 7), followed by the influence of regulations (N = 4); encouragement by angler organizations and IE efforts (N = 3 each), and peer pressure (N = 2). Tournament influence, desire for quality angling, perceived declines in fisheries, and concern for contaminants each were mentioned once. Nine respondents cited increasing conservation ethic as a reason for increase in voluntary release of *Morone* spp.; three mentioned effects of regulations; two mentioned desire for quality or trophy fisheries and contaminant concerns; and one mentioned peer pressure. Among the relatively few respondents who reported increases in voluntary release of crappies, sunfish, catfish, or salmon, the conservation ethic dominated for salmon and was cited in three of six responses for crappies, contamination was the most common response for catfish, and disinterest in cleaning and cooking was the most frequent for sunfish. South Dakota was the only state to note declines in voluntary release of any species group, reporting that catfish and *Morone* spp. were facing increased harvest as alternate species, presumably as regulations for more popular species became more strict.

Thirty-four of 47 state fisheries representatives reported that studies of the effects of regulations requiring substantial catch and release were available, including analyses of field sampling and creel survey data. Two states failed to note reports prepared by agency biologists, bringing the total to 36. Twenty-nine states have analyses of regulations for black bass, 14 for trout, 8 for walleye and sauger, 6 for crappies, 3 for esocids, 2 for sunfish and catfish, and 1 each for *Morone* spp., and salmon. Thirty-one agencies reported case histories of voluntary release of fishes, which included creel data, mail survey results, tag returns, and resort reports. Twenty-five reported on voluntary release of black bass, 14 on trout, 6 on esocids, 5 on walleye and sauger, 4 on crappies, 3 each on salmon, sunfish, and *Morone* spp., and 2 on sunfish and catfish.

Although many respondents indicated only whether case histories had been developed, others summarized data or included agency reports or technical papers. A selection of noteworthy summaries follows. Ohio employed a reservation system to reduce fishing pressure together with strict harvest regulations and maintained excellent bluegill size structure in lakes within a wildlife management area. Missouri biologists encouraged harvest of invading spotted bass *Micropterus punctulatus* to reduce competition for prey with native smallmouth bass *M. dolomieu* and incidence of hybridization. In Kentucky, regulations on spotted bass harvest have been liberalized because of their slow growth and high abundance relative to largemouth bass *M. salmoides*. The regulation successfully increased harvest of spotted bass while improving the size structure of largemouth bass populations (Buynak et al. 1991a). In Utah, regulations allow harvest of only one cutthroat trout *Oncorhynchus clarki* while allowing more rainbow trout *O.mykiss* to be taken. Such regulations allow anglers to harvest alternate species while protecting rarer or more valuable fish, but depend on angler identification of species. Utah also enacted a 20-in minimum length regulation for rainbow trout to require recycling of stocked trout. The high minimum length limit and increase in the size of stocked trout was deemed necessary due to high predation by cormorants *Phalacrocorax auritus* on small stocked fish. On the Housatonic River, Connecticut, a complete harvest ban due to PCB contamination resulted in more angler trips and more economic value from the fishery than was achieved with the earlier put-and-take management strategy. Catch rates of trout were 300–400% higher after comprehensive catch-and-release regulations were enacted. Oklahoma's 10-in minimum length limit and 15-fish daily bag limit for crappies increased the number of large fish but increased natural mortality of sublegal crappies to 76%, which was counterproductive to objectives (Boxrucker 1994). In Missouri, a population assessment of rock bass *Ambloplites rupestris* has led to a minimum length limit and reduced bag limit in a stream section.

Most case histories of voluntary release provided with this survey involved black bass, though some focused on trout or other fisheries. In Lake Winnepesaukee, New Hampshire, anglers voluntarily released over 85% of largemouth and smallmouth bass caught. In Connecticut, voluntary release of black bass has risen from 40% in 1986 to 60% in 1988 (Jacobs and Hyatt 1988), and to nearly 80% in the 1990s. In Maryland's portion of the Chesapeake Bay tidal rivers, 70–95% of black bass anglers released all bass. Virginia anglers voluntarily released 94% of black bass caught in Smith Mountain Lake and 88% caught in Kerr Reservoir. In Florida, voluntary release rates ranged from 8 to 85% in various lakes (Champeau and Thomas 1993). In Delaware, 70–90% of all black bass caught were released. In a small Iowa impoundment, creel surveys showed that since 1989, anglers have released nearly all largemouth bass caught (Hill 1992); however release rates for bluegills, crappies, and channel catfish *Ictalurus punctatus* were much lower. In Alabama, voluntary release rates of smallmouth bass in a reservoir fishery averaged 68%, which maintained the population in excellent condition, despite liberal statewide regulations (Reed et al. 1991). In several New York lakes, anglers released 56% of legal-sized largemouth bass but only 18% of legal-sized smallmouths (Green et al. 1986). Following implementation of a catch-and-release-only regulation for smallmouth bass in a section of Iowa's Maquoketa River, biomass, size structure, and angler catch rates increased, while total mortality declined, in comparison to control sections of the river. In Kansas, voluntary release of largemouth bass reduced growth rates due to intensified intraspecific competition.

In Nebraska, the 1% of avid anglers who participated in a walleye diary program took 38% of the total fish catch and released 70% of walleyes they caught, whereas the general public released only 7% of the walleyes they caught. In Kansas, diary cooperators released 73–80% of walleyes caught (Gabelhouse and Cox 1993). In Minnesota, voluntary release of walleyes over 20 in rose to nearly 50%. Release of walleyes under 20 in was not substantial, however.

In Nebraska, the percent of released trophy-sized fish submitted to the Master Angler Program has increased greatly in recent years, particularly for black bass, esocids, catfish, bluegills, and crappies. There has been little increase in the percent of released *Morone* spp., walleye, and trout. In-Fisherman's Master Angler Program, which invites entries from the United States and Canada, has charted an increase in the proportion of released fish from 21 to 66% in the last 10 years. The highest proportions of release entries have been for channel catfish, black bass, muskellunge, and lake trout *Salvelinus namaycush*.

Review of regulations summaries issued in all 50 states in 1994 showed that 29 included some endorsement of catch and release, ranging from instructions on minimizing injuries to fish to strong advocacies for voluntary reduction of harvest. States have commonly issued press releases to encourage voluntary release, and also offered detailed booklets and videotapes encouraging catch and release.

Among the five journals surveyed for papers concerning regulatory or voluntary catch and release, volumes 5 through 14 of *North American Journal of Fisheries Management* included 53 papers (6.4% of all papers) that focused on regulatory or voluntary catch and release. The *Proceedings of the Annual Conference Southeastern Association of Fish and Wildlife Agencies* published 27 papers (9.2% of all papers) on catch and release from 1985 to 1994. From 1985 through 1994, *Fisheries* published four papers that focused on catch and release, the *Canadian Journal of Fisheries and Aquatic Sciences* published three, and *Transactions of the American Fisheries Society* published one.

Although it could not be quantified, promotion of regulatory and voluntary catch and release in the sportfishing media increased greatly during the last decade, including promotion in magazines, television and radio broadcasts, club newsletters, and in seminars by expert anglers. More members of the fishing industry, including tackle and boat manufacturers and resorts, have emphasized catch and release in their advertisements and promotions. In recent years, manufacturers or promoters of other peripherally related products, including alcoholic beverages and automobiles, have included catch-and-release messages in their advertisements.

Discussion

Survey results indicate that state agencies have greatly expanded regulations that require anglers to release fish of edible and desirable size. Increases in regulations for black bass and trout continue a trend begun in earlier decades, but the scope of regulation has expanded, along with the expected benefits of regulation. Objectives of black bass management have changed from focus on harvest, such as Missouri's angler harvest goal of 0.25 black bass per hour (Novinger 1987) to higher catch rates of preferred-, memorable-, or trophy-sized bass as defined by Gabelhouse (1984a). Many new regulations have been enacted in response to angler preferences for catching more large fish, at the expense of total harvest. Agencies have assessed angler preferences with surveys and through less formal channels such as advisory councils, public meetings, petitions, and personal contacts. Harvest remains an important part of freshwater fishing for many anglers, however, and the importance of this motive may vary among regions and species specializations (Miranda and Frese 1989; Fedler and Ditton 1994). The proportion of harvest-oriented anglers seems smaller than it has ever been, and most anglers are satisfied with a smaller harvest than in the past (Leitch and Baltezore 1987; Hale et al. 1992). Growing numbers of specialized anglers also tend to favor more restrictive regulations than do casual anglers (Chipman and Helfrich 1988; Quinn 1992). Agencies have enacted stricter regulations in response to overharvest by anglers, which can occur quickly if fishing pressure is high (Redmond 1974). Agencies have also educated the public about fisheries problems to gain acceptance of stricter regulations (Behnke 1987), because regulations fail without high levels of compliance (Gigliotti and Taylor 1990; Van Vooren 1991).

Survey results indicate that regulations requiring greater release of walleye and sauger, striped bass, hybrid striped bass *Morone saxatilis* × *M. chrysops*, and white bass *M. chrysops* have increased in most states that manage these species, and regulations for esocids and crappies have become more stringent in almost half those states that manage for them. Esocids have been managed as trophy fish in many states (Wingate 1986), and anglers have been increasingly willing to release large fish (Graff 1986). Crappies have been widely regarded as difficult to manage due to stunting, extreme fluctuations in year-classes, and angler focus on harvest (Hooe 1991). The increase in catch-and-release regulations suggests increasing emphasis on trophy fisheries for esocids, particularly muskellunge *Esox masquinongy* (Dent 1986; Richards and Ramsell 1986), and greater understanding of population dynamics and the effects of harvest on crappie populations (Colvin 1991a, 1991b; Larson et al. 1991; Reed and Davies 1991; Webb and Ott 1991). Regulations for sunfish and catfish have increased least of all species groups examined, but any increase is notable as these fish have long been regarded as food fish, overabundant, or less valuable than trout or black bass. Angler interest in quality-sized sunfish and large or trophy-sized catfish, and recognition of angling effects on bluegills (Coble 1988; Olson and Cunningham 1989) apparently have spurred changes. The variety of reasons cited by state agencies for implementing regulations and motives for voluntary catch and release are interesting. The cursory nature of this paper precludes further examination of these reasons, which include documented, perceived, and suspected factors, but further investigation of reasons for more stringent regulations seems warranted.

Many states have documented dramatic changes in population structures and angler catch statistics in response to regulations for black bass and trout. In other cases, regulations have failed to improve fish population structures or angling statistics, apparently due to environmental variability, noncompliance by anglers, or adoption of regulations inappropriate for the ecosystem. Although some states have been testing and evaluating regulations for decades (trout management in Colorado, Wisconsin, and California, and black bass management in Missouri, for example), many more have only begun this process. The next decade will bring more knowledge and refinements of regulations that protect fish size groups, particularly of fishes other than black bass, trout, and muskellunge. Documented positive effects of total catch-and-release regulations for black bass (Weithman et al. 1977; Schneider and Juetten 1989; Porak et al. 1993) and trout (Deinstadt 1987; Orciari and Leonard 1990) have spurred expansion of total catch-and-release regulations for these species and others.

Despite the abundance of studies of catch-and-release regulations and great interest shown by anglers in improved fishing through these regulations, relatively few peer-reviewed papers have addressed these topics. In the last decade, the second symposium on catch-and-release fishing (Barnhart and Roelofs 1987), the *North American Journal of Fisheries Management* and the *Proceedings of the Annual Conference Southeastern Association of Fish and Wildlife Agencies* have published a total of 116 papers on catch-and-release fishing. Few additional refereed publications exist to guide managers and agencies working in this area. The relative scarcity of papers on catch and release could be due to the difficulty in determining cause and effect in natural systems (Buynak et al. 1991b) and the difficulty in adequately testing hypotheses. Also, recognition that fisheries regulations involve both biological and sociological components may have restricted their full evaluation by fisheries scientists, who traditionally have hesitated to address sociological issues (Bryan 1976; Bromley and Bryan 1982; Voiland and Duttweiler 1984; Ditton and Fedler 1989). Further studies of the complexities of regulations and their effects on fish populations and angler behavior clearly are needed (Loftus 1987). In the meantime, managers should consult the far larger gray literature (Collette 1990) for relevant case histories or guidelines, though readers should be careful about overextending the relevance of results.

Voluntary release of fish apparently has increased in addition to the increase in regulations (Table 1), though some responses to the questionnaire were based on observation and estimates rather than on rigorous quantification. Responses suggested that some new regulations have been implemented as a result of greater angler interest in catch and release and angler perceptions that catch-and-release regulations improve fishing. Managers often assume that more restrictive regulations will not harm populations, so management actions that are popular and do no harm may be implemented, although less vocal or less numerous anglers with harvest-oriented philosophies may subsequently voice opposition. Survey responses also suggest that some increases in voluntary release have been due to the increasing number of regulations that require release. Increasing acceptance and popularity of catch and release seems to stem from both agency and angler attitudes.

In some states, agencies have underestimated the strength of the catch-and-release ethic among specialized anglers, particularly those who pursue trout and black bass. Specialized trout anglers have requested that agencies ban or severely limit harvest on increasing numbers of waters, particularly those that harbor wild trout, so that fish abundance and catch rates will rise. Responses to this questionnaire suggest that trout anglers fishing stocked waters were more harvest oriented than those seeking wild fish, though there were exceptions. In black bass fisheries regulated by slot limits, declines in fish condition have been associated with unwillingness of anglers to harvest fish under the slot (Gabelhouse 1984b; McQuarrie 1995). In some cases, overabundant small largemouth bass have slowed population growth rates and hurt fisheries, coincident with the growing release ethic (Quinn 1993). Moreover, efforts by agencies to encourage anglers to harvest small black bass have drawn criticism from specialized bass anglers and angler groups (Dean 1994). Kumar (1994) reported that bass anglers in Kentucky opposed an age-growth study because it involved killing fish to remove otoliths.

Reported rates of voluntary release of species other than black bass, trout, and muskellunge are considerably lower, though they are rising in several states. Some fishes, including crappies, sunfish, and catfish, have traditionally been viewed more as food fish than sport fish by anglers and agencies. Yet enthusiasm for catching large specimens of all species has stimulated willingness to release large walleyes, northern pike *Esox lucius*, and catfish. Greater release rates of large fish are suggested by applications to angler award programs and have been documented for largemouth bass (Keefer 1988). Voluntary release of larger panfish, including crappies, sunfish, white perch *Morone americana*, and yellow perch *Perca flavescens* appears to be lower than for sport fish, though voluntary release of panfish appears to be increasing as anglers reduce their harvest expectations (Hale et al. 1992). Harvest of large panfish specimens and release of small ones apparently has deteriorated size structures of populations and angling quality in many waters (Goedde and Coble 1981; Coble 1988; Colvin 1991a), and popularization of this information in the outdoor press may have influenced angler behavior. Moreover, the success of catch-and-release regulations and voluntary release of traditional game fish may have spurred acceptance of similar management strategies for other fish.

In-Fisherman publications introduced the concept of selective harvest as the choice that must be made to harvest or release each legally harvestable fish caught (Stange 1989; Quinn and Reinert 1991). Lindner (1989) noted that harvest decisions should be based on fish species and size, status of the species in that body of water, and the angler's need for food. This philosophy strikes a balance between the tradition of harvesting fish for food and the realization that anglers must release most large fish, and all fish of some species in some bodies of water, to sustain good fishing. Selective harvest requires that anglers have some knowledge of aquatic productivity and population size structures of various fish species. Although many angler groups have embraced selective harvest and several state and federal agencies have adopted the term to describe their management strategies, or have recommended practicing selective harvest, some fervent catch-and-release advocates have viewed it as a step backward (Stange 1994).

Response to this survey indicates that there is a trend toward reduced daily bag limits set by fisheries management agencies. Despite evidence that bag limits have little effect on harvest (Redmond 1974; Snow 1982), agencies continue to manipulate them, apparently due to angler requests, and to the expectation that lower bag limits reduce the consumptive attitude of anglers or lower angler expectations (Rosen 1994). In addition, agencies have begun instituting further limits on the harvest of fish over a certain size, within the daily bag limit. This constitutes a bag limit on large fish, and biologists have questioned its effectiveness for walleyes (Schupp 1990) and northern pike (T. J. Goeman, Minnesota Department of Natural Resources, unpublished data). Yet this type of regulation also may encourage voluntary release of large fish (Schupp 1990).

Since recognition that the most effective fisheries management, from an ecological perspective, may require management of individual waters with regulations tailored to their fish populations, biologists have sought compromise between this fact and a need to keep regulations simple. Simple regulations may be easier for the public to read and understand, and easier to enforce, but may not provide optimum fishing in all waters. The survey revealed that states differ on their philosophies toward management of individual waters and how they balance idealism and pragmatism.

The involvement of sociology and biology in harvest regulations has perhaps never been so complete. A great challenge for fisheries managers into the twenty-first century will be to use angler input to craft regulations that will maintain fishing quality and ecosystem integrity, while sustaining recreational and other uses of fish habitat. This can only be accomplished with accelerated educational programs for anglers and the nonfishing public on ecological principles and management strategies. Involvement of conservation law enforcement agencies also will be crucial because officers can play a large role as educators as well as enforcers.

Acknowledgments

I thank the fisheries chiefs and their assistants in all states for providing the requested information and other helpful publications. Scott Lawrence helped compile and analyze responses. This study was supported by In-Fisherman, Incorporated.

References

Behnke, R. J. 1987. Catch-and-release—the last word. Pages 291–299 *in* Barnhart and Roelofs (1987).

Behnke, R. J. 1989. We're putting them back alive. Trout 30(3):48–61.

Barnhart, R. A., and T. D. Roelofs, editors. 1977. Catch-and-release fishing as a management tool. California Cooperative Fishery Research Unit, Arcata.

Barnhart, R. A., and T. D. Roelofs, editors. 1987. Catch-and-release fishing, a decade of experience. California Cooperative Fishery Research Unit, Arcata.

Boxrucker, J. 1994. Effect of a minimum length limit and daily creel reduction on the white crappie population structure and angler creel on Fort Supply Reservoir. Oklahoma Department of Wildlife Conservation, Federal Aid in Sport Fish Restoration, Project F-37-R, final report, Oklahoma City.

Bromley, P. T., and H. Bryan. 1982. The human dimension. Proceedings of the Annual Conference Southeastern Association of Fish and Wildlife Agencies 34(1980):530–535.

Bryan, H. 1976. The sociology of fishing: a review and critique. Pages 83–92 *in* R. H. Stroud and H. Clepper, editors. Marine recreational fisheries. Sport Fishing Institute, Washington, DC.

Burdick, B., and R. Wydoski. 1987. Effects of hooking mortality on a bluegill fishery in a western reservoir. Pages 187–196 *in* Barnhart and Roelofs (1987).

Buynak, G. L., L. F. Kornman, S. Surmont, and B. Mitchell. 1991a. Evaluation of a differential-harvest regulation for black bass in Cave Run Lake, Kentucky. North American Journal of Fisheries Management 11:277–284.

Buynak, G. L., W. N. McLemore, and B. Mitchell. 1991b. Changes in largemouth bass populations at Kentucky and Barkley lakes: environmental or regulatory response? North American Journal of Fisheries Management 11:285–292.

Champeau, T. R., and P. W. Thomas. 1993. Voluntary release of largemouth bass by Florida anglers. Proceedings of the Annual Conference Southeastern Association of Fish and Wildlife Agencies 45(1991): 317–322.

Childress, W. M. 1987. Catch-and-release mortality of white and black crappie. Pages 175–186 *in* Barnhart and Roelofs (1987).

Chipman, B. D., and L. A. Helfrich. 1988. Recreational specialization and motivations of Virginia river anglers. North American Journal of Fisheries Management 8:390–398.

Coble, D. W. 1988. Effects of angling on bluegill populations: management implications. North American Journal of Fisheries Management 8:277–283.

Collette, B. B. 1990. Problems with gray literature in fishery science. Pages 27–31 *in* J. Hunter, editor. Writing for fishery journals. American Fisheries Society, Bethesda, Maryland.

Colvin, M. A. 1991a. Population characteristics and angler harvest of white crappies in four large Missouri reservoirs. North American Journal of Fisheries Management 11:572–584.

Colvin, M. A. 1991b. Evaluation of minimum-size limits and reduced daily limits on the crappie populations and fisheries in five large Missouri reservoirs. North American Journal of Fisheries Management 11:585–597.

Dean, D. 1994. Of management, genetics, and the future. Honey Hole 10(5):18–19.

Deinstadt, J. M. 1987. California's use of catch-and-release angling regulations on trout waters. Pages 49–67 *in* Barnhart and Roelofs (1987).

Dent, R. J. 1986. A case history of voluntary catch and release of muskellunge with management implications for fishery programs. Pages 316–322 *in* Hall (1986).

Ditton, R. B., and A. J. Fedler. 1989. Importance of fish consumption to sport fisherman: a reply to *Matlock et al.* (1988). Fisheries 14(4):4,6.

Fedler, A. J., and R. B. Ditton. 1994. Understanding angler motivation in fisheries management. Fisheries 19(4):6–13.

Fletcher, D. H. 1987. Hooking mortality of caught-and-released walleye. Pages 170–174 *in* Barnhart and Roelofs (1987).

Gabelhouse, D. W., Jr. 1984a. A length-categorization system to assess fish stocks. North American Journal of Fisheries Management 4:273–285.

Gabelhouse, D. W., Jr. 1984b. An assessment of largemouth bass slot length limits in five Kansas lakes. Kansas Game and Fish Commission, Federal Aid in Fish Restoration, Project F-15-R, final report, Emporia.

Gabelhouse, D. W., Jr., and C. A. Cox. 1993. Results of the 1992 walleye angler diary program. Kansas Department of Wildlife and Parks, Federal Aid in Sport Fish Restoration, Project FW-9-P-11 & 12, final report, Emporia.

Gigliotti, L. M., and W. W. Taylor. 1990. The effect of illegal harvest on recreational fisheries. North American Journal of Fisheries Management 10:106–110.

Goedde, L. E., and D. W. Coble. 1981. Effects of angling on a previously fished and an unfished warmwater fish community in two Wisconsin lakes. Transactions of the American Fisheries Society 110:594–603.

Graff, D. R. 1986. Musky management—a changing perspective from past to present. Pages 195–198 *in* Hall (1986).

Green, D. M., B. J. Schonhoff, III, and W. D. Youngs. 1986. The New York state bass study 1977–1980, use of angler collected data to determine population dynamics. New York Department of Environmental

Conservation, Federal Aid in Sport Fish Restoration, Projects F-35-R and FA-5-R, final report, Albany.

Hale, R. S., M. Price, and E. Schneider. 1992. 1991 Kentucky angler survey. Kentucky Department of Fish and Wildlife Resources, Federal Aid in Fish Restoration, Project F-40, final report, Frankfort.

Hall, G. E., editor. 1986. Managing muskies. American Fisheries Society Special Publication 15.

Hill, K. 1992. Evaluation of catch rates and harvest rates as goals for an intensively managed fishery. Iowa Department of Natural Resources, Federal Aid in Sport Fish Restoration, Project F-90-R, completion report, Des Moines.

Hooe, M. L. 1991. Crappie biology and management. North American Journal of Fisheries Management 11:483–484.

Jacobs, R. P., and W. A. Hyatt. 1988. Statewide largemouth bass research and management. Connecticut Department of Environmental Protection, Federal Aid in Sport Fish Restoration, Project F-57-R5, final report, Hartford.

Keefer, L. C. 1988. A survey of the sport fisheries of four major reservoirs in southwest Georgia. Georgia Department of Natural Resources, Federal Aid in Sport Fish Restoration, Project F-28, final report, Atlanta.

Knight, J. A. 1936. The modern angler. Scribner, New York.

Kumar, J. C. 1994. Fish 'aging' method in Kentucky criticized. B.A.S.S. Times, April 1994:15.

Larson, S. C., B. Saul, and S. Schleiger. 1991. Exploitation and survival of black crappies in three Georgia reservoirs. North American Journal of Fisheries Management 11:604–613.

Leitch, J. A., and J. F. Baltezore. 1987. Attitudes of Minnesota anglers. Minnesota Department of Natural Resources, Division of Fish and Wildlife, St. Paul.

Lindner, A. 1989. Harvesting selectively in the 1990s. The In-Fisherman 84:7–8.

Loftus, K. H. 1987. Inadequate science transfer: an issue basic to effective fisheries management. Transactions of the American Fisheries Society 116:314–319.

McQuarrie, J. 1995. Catch and release: when it works, when it doesn't. Bassin' January 1995:55–58.

Miranda, L. E., and W. Frese. 1989. Mississippi angler survey. Mississippi Department of Wildlife, Fisheries, and Parks, Freshwater Fisheries Report 81, Jackson.

Muoneke, M. I., and W. M. Childress. 1994. Hooking mortality: a review for recreational fisheries. Reviews in Fisheries Science 2:123–156.

Novinger, G. D. 1987. Evaluation of a 15.0-inch minimum length limit on largemouth bass and spotted bass catches at Table Rock Lake, Missouri. North American Journal of Fisheries Management 7:260–272.

Olson, D. E., and P. K. Cunningham. 1989. Sport-fisheries trends shown by an annual Minnesota fishing contest over a 58-year period. North American Journal of Fisheries Management 9:287–297.

Orciari, R. D., and G. H. Leonard. 1990. Catch-and-release management of a trout stream contaminated with PCBs. North American Journal of Fisheries Management 10:315–329.

Porak, W. F., S. Crawford, D. J. Renfro, and J. Chadwick. 1993. Management of exploited largemouth bass populations. Florida Game and Fresh Water Fish Commission, Federal Aid in Sport Fish Restoration, Project F-24, completion report, Tallahassee.

Quinn, S. P. 1989. Recapture rates of voluntarily released largemouth bass. North American Journal of Fisheries Management 9:86–91.

Quinn, S. P. 1992. Angler perspectives on walleye management. North American Journal of Fisheries Management 12:367–378.

Quinn, S. 1993. Bass seasons, conservation measure or needless regulation? The In-Fisherman 18(6):30–38.

Quinn, S., and B. Reinert. 1991. Hidden dimensions in selective harvest. The In-Fisherman 100:38–52.

Ragan, J. A., T. Steinwand, and G. Van Eeckout. 1986. A synopsis of results from a questionnaire on muskellunge—resource agency survey. Pages 370–372 *in* Hall (1986).

Redmond, L. C. 1974. Prevention of overharvest of largemouth bass in Missouri impoundments. Pages 54–68 *in* J. L. Funk, editor. Symposium on overharvest and management of largemouth bass in small impoundments. American Fisheries Society, North Central Division, Special Publication 3, Bethesda, Maryland.

Redmond, L. C. 1986. The history and development of warmwater fish harvest regulations. Pages 186–195 *in* G. E. Hall and M. J. Van Den Avyle, editors. Reservoir fisheries management: strategies for the 80's. American Fisheries Society, Southern Division, Reservoir Committee, Bethesda, Maryland.

Reed, M. S., M. B. Bain, and K. C. Weathers. 1991. A quality smallmouth bass fishery: trophies and voluntary catch and release. Pages 151–157 *in* D. C. Jackson, editor. The first international smallmouth bass symposium. Mississippi Agricultural and Forestry Experiment Station, Mississippi State University.

Reed, J. R., and W. D. Davies. 1991. Population dynamics of black crappies and white crappies in Weiss Reservoir, Alabama: implications for the implementation of harvest restrictions. North American Journal of Fisheries Management 11:598–602.

Richards, K., and R. Ramsell. 1986. Quantifying the success of muskellunge catch and release programs: a summary of cooperative angler-tagging studies. Pages 309–315 *in* Hall (1986).

Rosen, R. A. 1994. To bag or not to bag. The In-Fisherman 19(6):26–28.

Schneider, J. C., and R. P. Juetten. 1989. An evaluation of special fishing regulations on Sylvania Lakes. Michigan Department of Natural Resources, Fisheries Technical Report 89-9, Ann Arbor.

Schramm, H. L., Jr., and nine coauthors. 1991. Sociological, economic, and biological aspects of competitive fishing. Fisheries 16(3):13–21.

Schupp, D. 1990. Factors keying effective catch and release regulations. Walleye In-Sider 1(1):14, 16–17.

Snow, H. E. 1982. Hypothetical effects of fishing regulations in Murphy Flowage, Wisconsin. Wisconsin Department of Natural Resources, Technical Bulletin 131, Madison.

Stange, D. 1989. Selective harvest—a step beyond catch and release. Walleye In-Sider 1(1):20.

Stange, D. 1994. When the enemy is among us. The In-Fisherman 19(6):9–10.

Van Vooren, A. 1991. Roles of biology and sociology in regulating bass harvest in Idaho. Pages 278–282 *in* Warmwater fisheries symposium I. U.S. Department of Agriculture General Technical Report RM-207, Fort Collins, Colorado.

Voiland, M. P., and M. W. Duttweiler. 1984. Where's the humanity? A challenge and opportunity for the fisheries community. Fisheries 9(4):12–12.

Webb, M. A., and R. A. Ott, Jr. 1991. Effects of length and bag limits on population structure and harvest of white crappies in three Texas reservoirs. North American Journal of Fisheries Management 11:614–622.

Weithman, A. S., and R. O. Anderson. 1978. A method of evaluating fishing quality. Fisheries 3(3):6–10.

Weithman, A. S., R. O. Anderson, and A. A. Ciuffa. 1977. Catch-and-release fishing for largemouth bass on Hunter's Lake—1963–1977. Pages 109–119 *in* R. A. Barnhart and T. D. Roelofs, editors. Catch and release fishing as a management tool. California Cooperative Fishery Research Unit, Arcata.

Wingate, P. J. 1986. Philosophy of muskellunge management. Pages 199–202 *in* Hall (1986).

Wulff, L. 1939. The handbook of freshwater fishing. Frederick A. Stokes Co., New York.

American Fisheries Society Symposium 16:163–171, 1996

Black Bass Movements after Release from Fishing Tournaments in Four New York Waters

Douglas L. Stang

New York State Department of Environmental Conservation, Bureau of Fisheries 50 Wolf Road, Albany, New York 12233, USA

David M. Green[1]

Cornell University
Biological Field Station, Department of Natural Resources 900 Shackelton Point Road, Bridgeport, New York 13030, USA

Rodger M. Klindt

New York State Department of Environmental Conservation, Bureau of Fisheries State Office Building, 317 Washington Street, Watertown, New York 13601, USA

Thomas L. Chiotti

New York State Department of Environmental Conservation, Bureau of Fisheries Fisher Avenue, P.O. Box 5170, Cortland, New York 13045, USA

William W. Miller

New York State Department of Environmental Conservation, Bureau of Fisheries Hudson Street, P.O. Box 220, Warrensburg, New York 12885, USA

Abstract.—We tracked largemouth bass *Micropterus salmoides* and smallmouth bass *M. dolomieu* using radiotelemetry to determine movements from weigh-in release sites after tournaments. Forty-two fish were tracked after five tournaments held on the St. Lawrence and Hudson rivers and Cayuga and Saratoga lakes. The fish displayed limited dispersal away from the release site and exhibited little tendency to return to their capture site. Only 8 of 42 fish returned to the vicinity of their capture site in 2–22 weeks. Six of 15 Hudson River largemouth bass moved away from their release site but did not return to their capture site, two fish returned to their capture location, and the remaining fish stayed near their release site. On Saratoga Lake, 56% of radio-tagged largemouth bass were harvested within 2 months. Although immediate mortality is low at New York black bass tournaments, effects on survival, growth, and reproduction that might result from relocating and concentrating a significant portion of a black bass population need to be considered. Until such potential impacts are evaluated for waters hosting numerous tournaments, fisheries managers should encourage tournament organizers to disperse fish after the tournament weigh-in, encourage "golden-rule" tournaments, or limit the amount of tournament activity originating from a particular site.

A comparison of surveys describing the status of competitive sportfishing conducted more than a decade apart (Shupp 1979; Schramm et al. 1991b) indicated at least a 20% increase in tournaments involving black bass *Micropterus* spp. and attests to the increasing popularity of competitive sportfishing. Black bass tournaments constitute approximately 78% of competitive fishing events (Schramm et al. 1991b). Associated with the increase in competitive fishing is growing concern about the effects of such events. Through a survey of inland and marine fisheries management agencies in North America, Schramm et al. (1991b) identified biological and administrative problems and benefits of competitive fishing. Stimulation of angling effort, concentration of angling effort, reduction in standing stocks, increased mortality, and relocation of fish were identified as the primary (perceived) biological problems associated with competitive fishing.

Effort has been directed at quantitating initial mortality (Holbrook 1975; Chapman and Fish 1985; Schramm et al. 1985) and delayed mortality (May 1973; Archer and Loyacano 1975; Moody 1975; Seidensticker 1975; Schramm et al. 1987) of black bass caught in live-release tournaments. In a study of 11 fishing tournaments on two Florida lakes, total mortality of largemouth bass *M. salmoides* ranged from 5.2 to 47.8% and averaged 26.7% (Schramm et al. 1987). Plumb et al. (1988) reported that mortality

[1]Present address: Post Office Box 1865, Richfield Springs, New York 13439, USA.

averaged 25% in several studies conducted in the 1970s.

Although most black bass survive live-release tournaments, many are transported considerable distances from their capture site before being released after the weigh-in or when "culled" for a larger individual. Relocation of fish can "alter their availability to anglers" and may affect prey and predator populations (Schramm et al. 1991a:17). Studies on movement and dispersal of smallmouth bass *M. dolomieu* and largemouth bass indicated limited dispersal of these fish away from a central release site after a tournament (Blake 1981; Healey 1990). Schramm et al. (1991a) indicated a need for further study of the effects of live release on the distribution of black bass.

Although an occasional small- to moderate-sized tournament may have negligible impacts on the size, structure, or distribution of the black bass population(s) in a water body, the impacts are potentially exacerbated on popular waters that host large numbers of tournaments administered from one or two locations. We document the movement and dispersal of tournament-relocated black bass and suggest interim strategies to alleviate potential effects of fish relocation pending further study.

Study Areas

We chose four popular black bass fishing waters to monitor post-tournament movement and dispersal from the weigh-in site. During 1988, the four study waters combined provided more than 500,000 angler days targeting black bass (Connelly et al. 1990).

The St. Lawrence River study area encompassed eastern Lake Ontario and the New York portion of the St. Lawrence River from its source to approximately 60 km downstream. The tournament weigh-in and black bass release site was located on French Creek, a tributary to the St. Lawrence River approximately 26 km downstream from Lake Ontario.

Cayuga Lake is a 17,236-ha natural lake located in central New York with a shoreline length of 143.4 km and mean and maximum water depths of 55 and 133 m, respectively. Because of the steeply sloped bottom over most of the lake, black bass habitat in Cayuga Lake is limited to the shoreline and the northern 2,673 ha. The tournament weigh-in and release site was on the west shore of Cayuga Lake, approximately 5 km south of the lake outlet.

Saratoga Lake is a 1,523-ha natural lake in northeastern New York with a shoreline length of 19.5 km and mean and maximum water depths of 7.6 and 29 m, respectively. The tournament weigh-in site was in Fish Creek, the outlet to Saratoga Lake, approximately 4.3 km downstream from the lake proper.

The Hudson River study area encompassed the upper 178 km of the tidal estuarine reach of the river (approximately 16,200 ha) and the tidal portions of four of its larger tributaries. The tournament weigh-in and release sites were located in Catskill Creek, a tributary to the Hudson River at River Kilometer (RK) 183 and 0.5 km north of the mouth of Catskill Creek on the river. Tournaments have been held on the Hudson River since at least the mid 1970s. Regional and national attention was drawn to the fishery when a major Bass Anglers Sportsman Society (BASS) tournament was held in 1984 that yielded an average of 3.3 black bass/d per contestant (mean total length >381 mm) and set a new BASS record for limit catches (39% of contestants). After 1984, the number of tournaments increased. By the late 1980s and early 1990s, tournaments were held at a minimum of 14 sites throughout the 178-km distance between Albany and Peekskill, RK 65–243.

Methods

Radiotelemetry procedures were similar for each study. Cooperating tournament anglers identified the capture location of each fish. Black bass were externally fitted with radio transmitters with individual frequencies in the 49- and 53-MHz bands. Dimensions of the transmitters were 5 × 2 × 1 cm and weight in water was 6–8 g. Depending upon the transmitter used, expected transmitter life was 2–12 months.

Transmitters were affixed to the fish after the tournament weigh-in. To attach the transmitters, stainless steel or teflon-coated wires were inserted through the back of the fish immediately below the soft dorsal in a procedure similar to that described by Winter (1977). At three of the study areas, fish were held in freshwater without anesthetic during the tagging procedure, but Hudson River largemouth bass were anesthetized with MS-222 before tagging. The tagging procedure took 3–5 min per fish, and tagged fish were subsequently placed in an aerated recovery tank or live car for 0.5–2 h before release.

All fish were released at weigh-in sites except at Saratoga Lake, where three fish were released into the lake proper approximately 4.5 km upstream from the weigh-in site. These three fish were re-

TABLE 1.—Number of largemouth bass (lmb) and smallmouth bass (smb) radio-tagged and tracked, mean (range in parentheses) distance from angler catch site to release site, release dates, and length of time locations of radio-tagged black bass were monitored to describe post-tournament dispersal and movement.

Study area	Number of fish monitored	Mean (range) distance from capture to release sites (km)	Release date(s)	Length of monitoring period (d)
St. Lawrence River	10 (5 smb, 5 lmb)	43.8 (3.2–90.2)	13 Sep 1989	282
Hudson River	2 (lmb)	18.3 (0.5–36)	29 Oct 1989	44
	2 (lmb)	46 (18–74)	4 Nov 1989	41
Cayuga Lake	10 (lmb)	7.1 (1–15.1)	12 Aug 1990	295
Hudson River	9 (lmb)	21.4 (18–36)	30 Sep 1990	222
	2 (lmb)	58 (37–79)	3 Oct 1990	219
Saratoga Lake	10 (lmb)	6.1 (0.7–11.9)	23 Jun 1991	151

leased into the lake proper so that differences in movement and dispersal between lake-released and outlet-released largemouth bass could be evaluated.

Movements were monitored with scanning receivers, directional yagi antennae, and directional tuned-loop antennae. Locations of individual fish were determined by boat for the first 2 weeks after release and by boat, truck, or aircraft thereafter.

Depending on study area and year, 4–11 black bass were tracked to characterize postrelease movements (Table 1). These fish were captured by anglers 0.5–90.2 km from their weigh-in and release sites and their locations were monitored for 41–295 d after release (Table 1). Eight of the 10 Saratoga Lake largemouth bass were caught in the lake proper, and 2 were caught in the lake outlet.

Information on black bass fishing tournaments on the Hudson River was obtained through attendance at tournaments by NYSDEC or Cornell University (CU) personnel beginning in 1984, or beginning in 1986, through a questionnaire mailed to organizations known or expected to hold tournaments. The information requested from the tournament organizers included organization identity, date(s), weigh-in location, number of contestants, number of limit catches, and numbers of largemouth and smallmouth basses weighed in. A follow-up letter was mailed to those organizations that did not respond to the first letter.

Attempts were made to identify the total number of tournaments held on the freshwater, tidal portion of the Hudson River by obtaining tournament-scheduling information from known tournament organizers. In 1989, CU biologists also visited as many popular tournament sites as possible on weekends during the open black bass season, from the third Saturday in June until 30 November.

Results

Movement of Relocated Fish

St. Lawrence River.—During the first 2 d after release, the radio-tagged fish did not make significant movements away from the release site. Smallmouth bass congregated near the confluence of French Creek and French Creek Bay 3 d after release. About 4–9 d after release, all the smallmouth bass moved from French Creek into French Creek Bay and the main body of the St. Lawrence River. After becoming established in deeper water in French Creek Bay, four of the five smallmouth bass exhibited little movement for the next 8–10 d. One smallmouth bass returned to its capture location 3.7 km downstream from the release site 4 d after release.

Four of the five largemouth bass remained within French Creek for the initial 14-d tracking period. One largemouth bass was located 19.6 km downstream from the release site 9 d after release and returned to within 2.4 km of the release site 48 d postrelease.

In general, the 10 fish moved very little after the initial dispersal period in September. Nine months after release, seven fish were located alive, one within 6.4 km of the release site and six others within 2.4 km of the release site. Final locations of fish were 18.6–79 km from their respective capture sites.

Cayuga Lake.—Upon release, the radio-tagged largemouth bass moved into macrophyte *Myriophyllum* spp. habitat less than 1 km offshore from the release site. Mean distance from the point of release 24 h after release was 0.3 km. Fish did not move substantially farther over the next 2 weeks. Mean distance from the release point for most fish increased to only 0.4–0.5 km. Two fish dispersed 1.4

and 2.1 km from the release point during the initial 14-d period.

From 29 to 46 d after release, mean distance from the release site increased from 1.1 to 1.3 km. Seventy-five days after the tournament, mean distance from the release site was 2.3 km. Distance from the release site and the location of tagged fish 76 d after release were compared to the distance to capture site for each fish to determine if tagged fish were "homing" back to their area of capture. Six of the nine fish with known capture locations moved closer to their capture site—a mean distance of 1.5 km—whereas three other fish moved away from their capture site a mean distance of 2.1 km. In general, however, the fish did not appear to exhibit a homing tendency because 2.5 months after release, the mean distance from the capture site (6.6 km) was equivalent to that on the release date (6.9 km).

Of the seven fish that were located 221 d after release, five were within 2 km of the release site and four were less than 1 km from the release site. Mean distance from the release site for the five largemouth bass located 277 and 295 d after release were 2 and 2.2 km, respectively. Only one fish returned to the vicinity of its capture site 1 km due east of the release site 221 d after release.

Saratoga Lake.—During the first 2 weeks after release, seven largemouth bass remained within 1 km of their release site, two others were less than 2 km from their release site, and one other moved 7.2 km from the release site to its capture site in 4 d. One radio-tagged largemouth bass was found dead by the telemetry tracking team 2 d after release and one fish was harvested 8 d after release.

From 15 to 36 d after release, mean distance from the release site increased from 1.9 to 2.4 km. Fish were harvested 23 and 24 d after release. By 151 d postrelease, mean distance from the release sites increased to 4.4 km. Two radio-tagged largemouth bass were harvested 43 and 96 d after release.

The two fish caught and released in the outlet returned to within 100 m of the location where they were caught. These fish returned 0.7 and 2.7 km to their capture sites 9 and 23 d after release.

Movement of lake-caught and released radio-tagged largemouth bass from the lake release site was not more pronounced than movement of lake-caught largemouth bass released at the weigh-in station. Thirty-six days after release, lake-caught and released fish averaged 0.8 km from the release site compared to 1.3 km (excluding one fish that moved >7 km) for lake-caught fish released at the outlet weigh-in station.

Hudson River 1989.—All four radio-tagged largemouth bass remained near the release point for approximately 1 month and two of these fish never left Catskill Creek during the 40-d tracking period. After 1 month, two fish moved out of Catskill Creek 5.5 and 25 km to the south, where they remained through the end of the tracking period (3 weeks). At the end of the tracking period, these fish were 12.5 and 49 km away from their respective capture sites. No fish returned to their capture site during the 40-d tracking period.

Hudson River 1990.—All radio-tagged largemouth bass released in Catskill Creek remained in the creek for the first 15 d after release, two fish remained in the release creek for the entire 176-d monitoring period, and one fish remained in the creek 36 d before radio contact with this fish was lost. One fish moved out of the creek after day 15 and moved at least 7.2 km downstream before returning to Catskill Creek by day 36, where it remained until day 46, when radio contact was lost. Other largemouth bass left Catskill Creek 26, 27, and 36 d after release and moved 5.4, 7.6, and 7.6 km downstream, respectively, where they remained for the balance of the tracking period. The final locations for these three largemouth bass were 11.7, 9.7, and 8.9 km, respectively, from their capture sites. Two other largemouth bass left Catskill Creek after 15 d, returned to the tributary where they were caught 18 km downstream 36 d postrelease, and remained within this creek for the rest of the monitoring period (140 d).

Two largemouth bass that were released directly into the Hudson River dispersed away from the release site. These fish moved 3 and 7.2 km upriver during the first 20 d postrelease before confining their movements to localized areas in the Hudson River for the balance of the tracking period.

Dispersal and return to catch sites.—Only 1 of 5 smallmouth bass and 7 of 37 largemouth bass returned to their capture sites after release at tournament weigh-in sites, including 2 largemouth bass whose capture sites were only 0.7 and 1 km from the release site (Table 2). Thirty-two black bass either remained at their release site or moved very short distances (<1 km) and remained for 15–30 d, and another did not leave the release site for 110 d. Another three black bass moved 1.5–3 km from the release site during the first 2 weeks but returned to the release site. Among the black bass that did not return to their capture sites, there were 11 fish that moved away from their release site at least 1 but less than 3 km, including one that moved more than 19 km from the release site and then returned to less

TABLE 2.—Number of largemouth bass (lmb) and smallmouth bass (smb) monitored via radiotelemetry, final distance from release site for black bass that did not return to their capture site, and length of time and distance traveled from release site to approximate capture site for black bass that returned to their capture location 1989–1991.

Study area	Number of black bass monitored	Final distance from release site for black bass that did not return to capture site	Black bass that returned to capture location
St. Lawrence River	8	3 smb < 2.4 km 3 lmb < 2.4 km 1 smb < 6.4 km	1 smb: 3.7 km in 4 d
Hudson River	4	2 lmb < 0.5 km 2 lmb from 5–25 km	None
Cayuga Lake	10	4 lmb < 1 km 3 lmb from 2.3–2.7 km 2 lmb from 3.6–3.9 km	1 lmb: 1 km in 221 d
Saratoga Lake		2 lmb from 2.7–2.9 km 3 lamb harvested	4 lmb: 7.2 km in 4 d; 0.7 km in 9 d (harvested); 2.6 km in 23 d (harvested); 7.4 km in 36–100 d
Hudson River	11	4 lmb < 1 km 5 lmb from 3–7.6 km	2 lmb: 18 km in 27–36 d

than 3 km from the release site, nine fish that moved at least 3 but less than 8 km, and one that moved 25 km. Another 10 largemouth bass remained at or very close (<1 km) to their release site (Table 2). Most largemouth (94%) and smallmouth bass (80%) that were relocated were displaced for more than 20 d from the site where they were captured by anglers.

Hudson River Fishery

The number of fish reported by tournament organizers or counted by fisheries biologists as brought in for weighing at Hudson River tournaments in 1984–1992 ranged from 973 to 3,581 largemouth bass and from 267 to 682 smallmouth bass (Table 3). These numbers were based on 1 to 27 tournaments each year.

Riverwide mark–recapture population estimates were made for largemouth bass 305 mm or longer (the minimum legal length) by Carlson (1992) for the winters of 1985–1986, 1986–1987, and 1987–1988, and by Green et al. (1993) for largemouth bass 279 mm or longer for the springs of 1989, 1990, and 1991 (Table 3). Largemouth bass that were 279 mm in March when the population estimates were made would reach the 305-mm minimum legal length by late June when tournaments began. Population size structure was similar during the 3 years

TABLE 3.—Angler effort at black bass tournaments, number of black bass entered for weighing at tournaments, and largemouth bass population size on the freshwater, tidal estuary of the Hudson River, New York, for 1984–1992. (Data summarized by P. J. Festa, New York State Department of Environmental Conservation.)

Variable	1984	1985	1986	1987	1988	1989	1990	1991	1992
Number of tournaments monitored or reported	1	3	15	13	13	27	15	15	17
Number of angler trips	798	563	1,946	1,443	1,129	1,607	1,704	1,132	679
Limit catches by contestants (%)	38.6	N/A[a]	16.0	6.7	15.1	11.9	4.2	4.6	17.5
Number of largemouth bass	2,609[b]	1,165[b]	3,581	1,443	1,423	2,330	1,534	1,200	973
Number of smallmouth bass	[c]	[c]	525	620	361	675	682	374	267
Estimated population size for largemouth bass[d]	N/A	N/A	30,200–45,200	21,000–42,300	29,500	21,954	14,503	14,230	N/A
Confidence interval[d]			N/A	N/A	±10,250	±2,844	±1,768	±1,762	
Percentage of largemouth bass population weighed-in at tournaments			7.9–11.9	3.4–6.9	4.8	10.6	10.6	8.4	
Tournament trips between 27 Sept and 7 Nov (%)			47.8	34.3	35.4	24.4	37.1	61.5	69.8

[a]N/A = not available.
[b]Includes both largemouth and smallmouth basses.
[c]Numbers are included in row above.
[d]Population estimates for 1987–1989, confidence interval of 95% (Carlson 1992); for 1989–1991, confidence interval of 90% (Green et al. 1993).

that populations were estimated by Carlson (1992); therefore, populations during the 1985–1986 to 1987–1988 period would have been approximately 10% larger when tournament fishing began. During 1986–1991, up to 11% of the riverwide population was weighed in either at tournaments where catch was reported to the NYSDEC or at tournaments monitored by fisheries biologists (Table 3).

Tournaments reported or monitored are only a portion of the total number of tournaments held on the Hudson River each year. Although it was possible to obtain schedules for most moderate-to-large-sized tournaments, it was apparent that tournaments (especially small tournaments, i.e., <30 boats or <60 anglers) were missed. Cornell University fisheries biologists identified 50 tournaments held in 1989, but the survey of sites was incomplete and it is unlikely that all tournaments were identified. Results were reported or monitored in 27 tournaments in 1988; schedules were obtained for 22 tournaments in 1990 and 30 in 1991, but only 15 of these were reported or monitored in each year. The minimum number of tournaments occurring annually on the freshwater, tidal portion of the Hudson River is on the order of at least 50 to 60 during the 24-week open season for black bass.

The average number of trips that occurred during the small tournaments for which there were records in 1986–1992 was 27 per tournament. If we assume that the tournaments from which catch records were not obtained were all small tournaments and that 55 tournaments occurred annually, then, using the mean number of largemouth bass weighed in per trip for the 1987–1992 period (1.2 largemouth bass per trip, Table 3), a number equivalent to 15–20% of the river's largemouth bass population was predicted to be weighed in and released at weigh-in sites annually.

An unknown number of black bass in addition to those released at weigh-in sites would also be transported in live wells and displaced. Smaller fish are culled for larger fish when an angler has a limit catch in the live well. Four to 39% of contestants (mean, 16.4%) in Hudson River tournaments weighed in a limit of fish (Table 3).

Discussion

Mortality of black bass caught and immediately released by anglers is usually negligible (Green et al. 1987). Immediate and delayed mortality of black bass caught during tournaments was a concern during the first few years of tournament activity and these sources of mortality have been evaluated. Most estimates of immediate and delayed mortality are less than 25% (May 1973; Wellborn and Barkley 1974; Archer and Loyacano 1975; Holbrook 1975; Moody 1975; Seidensticker 1975; Chapman and Fish 1985; Schramm et al. 1985, 1987; Plumb et al. 1988), except when tournaments are conducted during periods of high water temperature (Schramm et al. 1987; Plumb et al. 1988). Kwak and Henry (1995) found that tournament mortality was minimal compared with other sources of mortality in a Minnesota lake with two tournaments per 1,000 ha. Large numbers of black bass are relocated during tournaments at popular tournament waters and, in large waters, fish are relocated over great distances. Ridgway and Shuter (1994) found that smallmouth bass displaced greater distances from their home range tended to take longer to return.

The limited movement we observed (29% <1 km and 57% <3 km) was similar to that observed by Blake (1981) and Healey (1990), where 48% (smallmouth and largemouth basses) and 38% (largemouth bass), respectively, of tagged tournament-caught fish recaptured by anglers were less than 1.6 km from the weigh-in sites. In an Oklahoma lake, 49% of tagged largemouth bass recaptured by anglers were less than 0.8 km from the release site and 64% of the recaptures were less than 1 km from the release site up to 6 months postrelease (G. Gilliland, Oklahoma Department of Wildlife Conservation, personal communication). Blake (1981) reported that a substantial number of these fish were caught within 0.8 km of the release site 50 or more days after release, and Healey (1990) found that the average distance of capture from release site was 2 km in the first 40 d post-tournament. Smallmouth bass moved relatively farther than largemouth bass during these studies (Blake 1981; Healey 1990).

The relocation of black bass has potential significance only if substantial portions of populations are affected. At least a number of largemouth bass equal to 15–20% of the population of the Hudson River are relocated annually after being weighed and released. An unknown additional number of fish are also transported in live wells and relocated. In a 1978 BASS tournament on the St. Lawrence River, 23% of the black bass caught that were larger than 356 mm (minimum size for the tournament) were never weighed in (F. Panek, NYSDEC, unpublished data). There are no estimates of the number of smallmouth bass present in the Hudson River; however, 13–31% of the black bass weighed in were smallmouth bass. Because most tournaments are based on total weight of black bass entered for weighing, tournament anglers select for largemouth bass. The percentage of the Hudson River smallmouth bass population held and relocated is probably less than that of the largemouth bass population.

Handling of the largemouth bass population has been expressed as a percentage of the total adult population; however, some fish probably experience multiple handling annually, whereas others may not be caught. Hudson River largemouth bass are fully recruited to the fishery by the end of their third growing season and some are exposed to capture until about age 7. Largemouth bass older than age 7 are rare in the Hudson River. Over their life span, many fish experience repeated occurrences of being caught and released, and being caught, held, transported, and relocated. Annual survival rates were 0.45, 0.35, and 0.57 from 1988–1989, 1989–1990, and 1990–1991, respectively (Green et al. 1993), compared to a mean of 0.65 for 11 other New York waters (Green et al. 1986). Because a health advisory against eating black bass from the tidal Hudson River and release rates of 96% or more (Green and Jackson 1991), harvest of largemouth bass is negligible. Low survival rates of largemouth bass in the Hudson River suggest the population is subjected to stress.

Black bass populations in popular tournament waters throughout North America may experience similar amounts of capture and relocation. Tournaments occur on the Hudson River and Saratoga Lake at the annual rate of 3–4/1,000 ha. In popular waters that cover at least 1,000 ha, tournaments occur at rates of 1–4/1,000 ha in Alabama (Reeves and McHugh 1992; W. C. Reeves, M. J. Maceina, and A. S. Hendricks, Alabama Department of Conservation and Natural Resources, unpublished data), 13–56/1,000 ha in Connecticut (E. O'Donnell, Connecticut Department of Environmental Protection, personal communication), 1–40/1,000 ha in Florida (R. Watson, Florida Game and Fresh Water Fish Commission, personal communication), 13/1,000 ha in Missouri (Schramm et al. 1987), 4–16/1,000 ha in New Hampshire (K. Sprankle, New Hampshire Fish and Game Department, personal communication), and 1–16/1,000 ha in Texas (D. Strickland, Texas Parks and Wildlife, personal communication).

The behavior of largemouth bass contributes to their vulnerability to capture by anglers and probably to elevated levels of stress. In the Hudson River, dense concentrations of largemouth bass just before the onset of winter (Carlson 1992; Green et al. 1993) occur concurrent with the greatest amount of tournament effort and result in higher tournament catch rates during this period (1.74 fish/h during fall compared with 1.59 fish/h over the entire season). Sixty percent or more of the total tidal Hudson River largemouth bass population move into five winter refuges totaling approximately 118 ha. Densities of adult largemouth bass ranged from 200 to 464 fish/ha at two of these refuges (Green et al. 1993). Largemouth bass caught from these concentrations have only recently moved to these areas probably to avoid the current of the main river (Carlson 1992). When largemouth bass relocate out of these areas of winter concentration they are subjected to movement in the main stem of the river during periods of low water temperature. Sheehan et al. (1990) demonstrated that the ability of largemouth bass to maintain themselves in river channel current was greatly reduced as water temperature declined from 4 to 0°C.

Black bass are also highly vulnerable to capture by anglers when they move into shallow water to spawn (Philipp et al. 1994). The black bass fishing season is closed in New York during the period when most spawning occurs, and no tournaments are held during this period; however, in many states it is legal to fish for black bass during the spawning period. Spawning is a stressful period for black bass (Heidinger 1975). Relocation of fish at this time disrupts the spawning process, adds to the stress of spawning, and increases predation risk on young smallmouth bass (Kieffer et al. 1995).

Relocation of black bass at any time increases energy requirements and may increase vulnerability to capture and exploitation. In Saratoga Lake, 56% of the radio-tagged largemouth bass were harvested by anglers within 3–13 weeks after their release at the tournament weigh-in site. This exploitation rate was more than three times the average exploitation rate (17.5%) of 12 New York waters studied from 1978 to 1980 (Green, Schonhoff, and Youngs, unpublished data). Increased exploitation could occur because of artificially increased densities of legal-sized black bass. Large numbers of black bass are released at popular weigh-in sites on waters where large numbers of tournaments occur. Our study shows that many smallmouth and largemouth bass remain near their release site for at least 2–4 weeks, and anglers (tournament and otherwise) are aware of these concentrations, thereby increasing the potential for exploitation or, at a minimum, repeated capture.

Significant portions of black bass populations are caught, held, and relocated by tournament participants in popular tournament waters. Our findings suggest that tournament angling may affect black bass populations by redistributing the population and increasing exploitation. Other stress-related impacts on survival, growth, and reproduction that might result from relocating and concentrating a significant portion of a black bass population need to be addressed.

Recommendations

These initial studies of displaced black bass were done on popular tournament waters. Under such circumstances, radio-tagged black bass tracked during this study may have already been displaced fish, possibly even displaced several times. Studies of fish movements should be conducted at waters where little or no tournament activity has occurred to preclude the possibility of tracking movements of fish that may have already experienced multiple displacement.

Pending further study, tournament organizers should be encouraged to disperse fish away from the weigh-in site to minimize artificially high concentrations of black bass susceptible to exploitation, provided fish are transported at low density in aerated or oxygenated live wells. Because most state agencies (68%; R. Hartley, Massachusetts Division of Fish and Wildlife, personal communication) require permits for fishing tournaments or otherwise regulate them, dispersal of fish after tournaments may be a permit condition. Limits may be established on the number of tournaments originating from a single site or area within a water body. Fisheries managers should also encourage "golden rule" tournaments where participants measure, immediately release, and record their catch. Because hooking stress alone is not directly responsible for acute or delayed mortality (Gustaveson et al. 1991), this type of tournament alleviates tournament-induced mortality of black bass, relocation of fish, and concentration of black bass near the weigh-in site.

Acknowledgments

We thank the following people and organizations for their assistance with this study: P. J. Festa, New York State Department of Environmental Conservation (NYSDEC), for initiating the radiotelemetry and Hudson River tournament-monitoring projects and for reviewing the manuscript; S. B. Nack and D. L. Bunnell, Cornell University, who conducted the majority of the field studies and analysis of data for the work on the Hudson River; the Hudson River Foundation for Science and Environmental Research for funding much of the studies on the Hudson River; P. Hess, NYSDEC, for field assistance on Cayuga Lake; and J. LaPierre, NYSDEC, for field assistance on Saratoga Lake. The assistance of anglers and tournament organizers in providing fish for radiotelemetry is especially appreciated. J. M. Pitlo, Jr., L. E. Miranda, and two anonymous reviewers provided helpful comments on early drafts of the manuscript. This paper is contribution 179 of the Cornell University Biological Field Station.

References

Archer, D. L., and H. A. Loyacano. 1975. Initial and delayed mortalities of largemouth bass captured in the 1973 National Keowee B.A.S.S. tournament. Proceedings of the Annual Conference Southeastern Association of Game and Fish Commissioners 28(1974): 90–96.

Blake, L. M. 1981. Movement of tournament-caught and released bass. New York Fish and Game Journal 28:115–117.

Carlson, D. M. 1992. Importance of wintering refugia to the largemouth bass fishery in the Hudson River estuary. Journal of Freshwater Ecology 7:173–180.

Chapman, P., and W. V. Fish. 1985. Largemouth bass catch results in Florida. Proceedings of the Annual Conference Southeastern Association of Fish and Wildlife Agencies 37(1983):495–505.

Connelly, N. A., T. L. Brown, and B. A. Knuth. 1990. New York statewide angler survey 1988. New York State Department of Environmental Conservation, Albany.

Green, D. M., and J. Jackson. 1991. Characterization of angler activity on the Hudson River Estuary. Pages VII-1 to VII-52 *in* E. A. Blair and J. R. Waldman, editors. Polgar fellowship reports of the Hudson River Estuarine Research Reserve Program, Hudson River Foundation for Science and Environmental Research, New York, and the New York State Department of Environmental Conservation, New Paltz.

Green, D. M., S. E. Landsberger, S. B. Nack, D. Bunnell, and J. L. Forney. 1993. Abundance and winter distribution of Hudson River black bass. Hudson River Foundation for Science and Environmental Research (Grant 001/88B & 009/91A), New York.

Green, D. M., B. J. Schonhoff, III, and W. D. Youngs. 1986. The New York state bass study 1977–1980: use of angler collected data to determine population dynamics. New York State Department of Environmental Conservation, Albany.

Green, D. M., B. J. Schonhoff, III, and W. D. Youngs. 1987. Evaluation of hooking mortality of smallmouth and largemouth bass. Pages 229–240 *in* R. A. Barnhart and T. D. Roelofs, editors. Catch and release fishing, a decade of experience. A national sportfishing symposium. Humboldt State University, Arcata, California.

Gustaveson, W. A., W. S. Wydoski, and G. A. Wedemeyer. 1991. Physiological response of largemouth bass to angling stress. Transactions of the American Fisheries Society 120:629–636.

Healey, T. P. 1990. Movement and survival of tournament-caught black bass at Shasta Lake. California Fish and Game 76:36–42.

Heidinger, R. C. 1975. Life history and biology of largemouth bass. Pages 11–20 *in* H. Clepper, editor. Black bass biology and management. Sport Fishing Institute, Washington, DC.

Holbrook, J. A., II. 1975. Bass fishing tournaments. Pages 408–415 *in* H. Clepper, editor. Black bass biology and management. Sport Fishing Institute, Washington, DC.

Kieffer, J. D., M. R. Kubacki, F. J. S. Phelan, D. P. Philipp, and B. L. Tufts. 1995. Effects of catch-and-release angling on nesting male smallmouth bass. Transactions of the American Fisheries Society 124:70–76.

Kwak, T. J., and M. G. Henry. 1995. Largemouth bass mortality and related causal factors during live-release fishing tournaments on a large Minnesota lake. North American Journal of Fisheries Management 15:621–630.

May, B. E. 1973. Evaluation of large scale release programs with special reference to bass fishing tournaments. Proceedings of the Annual Conference Southeastern Association of Game and Fish Commissioners 26(1972):325–334.

Moody, H. L. 1975. Tournament catch of largemouth bass from St. Johns River, Florida. Proceedings of the Annual Conference Southeastern Game and Fish Commissioners 28(1974):73–82.

Philipp, D. P., C. A. Toline, B. F. Philipp, and F. J. S. Phelan. 1994. The impact of pre-season catch-and-release angling on the reproductive success on bass. Pages 7–16 *in* S. J. Kerr and R. Cholmondeley, editors. Bass management in Ontario workshop proceedings. Queens University Biological Station, Ontario.

Plumb, J. A., J. M. Grizzle, and W. A. Rogers. 1988. Survival of caught and released largemouth bass after containment in live wells. North American Journal of Fisheries Management 8:325–328.

Reeves, W. C., and J. J. McHugh. 1992. B.A.I.T. bass anglers information team 1991 annual report. Alabama Department of Conservation and Natural Resources, Montgomery, Alabama.

Ridgway, M., and B. J. Shuter. 1994. Home range fidelity by displaced smallmouth bass in Lake Opeongo. Pages 33–41 *in* S. J. Kerr and R. Cholmondeley, editors. Bass management in Ontario workshop proceedings. Queens University Biological Station, Ontario.

Schramm, H. L., Jr., P. J. Haydt, and N. A. Bruno. 1985. Survival of tournament-caught largemouth bass in two Florida lakes. North American Journal of Fisheries Management 5:606–611.

Schramm, H. L., Jr., P. J. Haydt, and K. M. Portier. 1987. Evaluation of prerelease, postrelease, and total mortality of largemouth bass caught during tournaments in two Florida lakes. North American Journal of Fisheries Management 7:394–402.

Schramm, H. L., Jr., and nine coauthors. 1991a. Sociological, economic, and biological aspects of competitive fishing. Fisheries 16(3):13–21.

Schramm, H. L., Jr., and seven coauthors. 1991b. The status of competitive sport fishing in North America. Fisheries 16(3):4–12.

Seidensticker, E. P. 1975. Mortality of largemouth bass from two tournaments utilizing the "Don't kill your catch" program. Proceedings of the Annual Conference Southeastern Association of Game and Fish Commissioners 28(1974):83–86.

Sheehan, R. J., and six coauthors. 1990. Winter habitat requirements and overwintering of riverine fishes. Southern Illinois University, Fisheries Research Laboratory, Federal Aid in Fish Restoration F-79-R, final report, Carbondale.

Shupp, B. D. 1979. 1978 status of bass fishing tournaments in the United States. Fisheries 4(6):11–19.

Wellborn, T. L., Jr., and J. H. Barkley. 1974. Study on the survival of tournament released bass on Ross R. Barnett Reservoir, April, 1973. Proceedings of the Annual Conference Southeastern Association of Game and Fish Commissioners 27(1973):512–519.

Winter, J. D. 1977. Summer home range movements and habitat use by four largemouth bass in Mary Lake, Minnesota. Transactions of the American Fisheries Society 106:323–330.

American Fisheries Society Symposium 16:172–175, 1996

The Role of Angler Organizations in Fisheries Management

JERRY DEAN
Honey Hole, Post Office Box 9027, Fort Worth, Texas 76147, USA

Abstract.—Because fishing organizations include concerned and outspoken anglers, they may be perceived as adversaries to resource management agencies. However, explanation and justification of agency actions prior to changes in policies and regulations will facilitate angler support and reduce conflict. I believe that public hearings have traditionally been ineffective in facilitating communication between agencies and angler groups because often they are held not to learn about anglers's views on a given issue, but instead to allow agencies to convey decisions that have already been made. Agency personnel should work to understand public priorities and match management and research emphases and resource allocation to meet those priorities. Angler organizations serve to stimulate honest communication between anglers and managing agencies so that benefits to the agency, anglers, and the resource can be optimized.

Honey Hole is a regional fishing organization in the southwestern United States which publishes a magazine that has a circulation of over 40,000. Honey Hole also produces a weekly television program reaching more than 15 million households, a weekly taped one-hour radio program that airs throughout Texas, and a weekly one-hour live question-and-answer radio program. The radio program invites listeners to comment on recreational fishing issues; the lively discussion reflects a wide range of opinions. The communications industry, of which Honey Hole is a part, has an important role in educating anglers. Communication between the Honey Hole organization and anglers is a given; we speak to them directly on a daily basis. We hear and understand their wants and needs.

Honey Hole maintains a focus on black bass *Micropterus* spp. fishing in Texas, and the organization's audience is made up primarily of Texas anglers, and others who wish to use Texas' vast system of reservoirs. Thus, the information presented and the questions asked here pertain mainly to Texas, but also apply to other geographical and political provinces.

Black bass have no concern for state and national borders and boundaries. Fish do not realize that they are living in waters "owned" by Texas, Oklahoma, or Oregon for that matter. Neither do most anglers care where the head of their state fisheries agency was raised or went to school. What they do care about is how and why agency officials regulate their chosen sport.

Fisheries managers recognize the importance of regulations in management of fish populations and maintenance of balance in fish communities. Fisheries managers have been trained to recognize problems affecting fish. They often function as problem solvers, and their expertise allows them to formulate ideas and plans to improve, not just maintain, our nation's fisheries. However, what some fisheries managers perhaps lack is a true understanding of how to manage the users of the resources. The idea that agencies also manage anglers as well as fish is very offensive to some of those anglers, regardless of its basis in fact.

The increasingly broad acceptance of catch and release by anglers has contributed to the abundance of large, trophy-size fish available to anglers. This single example of anglers' acceptance of fisheries management is significant; it proves beyond doubt that some anglers accept policies that may improve their odds of enjoying outings, whether during competition with other anglers or simple outings with their family and friends. Whether by regulation or voluntary change in behavior, anglers have begun to assume a larger role in fisheries management than ever before. For this reason alone, it is essential that fisheries management agencies keep anglers informed and thoroughly listen to their needs and wants. Agencies and anglers must reach a point at which each can respect the other's position.

The Need for Communication: A Texas Example

The following example illustrates the need for two-way communication. In 1994, Texas Parks and Wildlife Department (TPWD) had produced an excess of approximately one million largemouth bass fingerlings of the Florida strain *M. salmoides floridanus*. This was after stocking goals for public lakes had been achieved. The decision was made to place equal numbers of fingerlings in four lakes. Someone at headquarters had decided it was a great time to invite the governor and release the fish in front of the public and the media. Perhaps because it was an election year, the governor's staff jumped at the opportunity. Like most agencies facing cutbacks on funds and resources, TPWD saw this affair as a positive opportunity; they could release excess fish and show the public what a great job they were doing. What angler would complain about more Florida largemouth bass being stocked into the reservoirs?

I was one who did complain. I was disturbed because of the release locations. The governor's press secretary informed me one quarter million fingerlings were going into Sam Rayburn Reservoir. This reservoir supports one of the most productive black bass fisheries in our state, and numerous other reservoirs across Texas had not been stocked in 10–20 years. Immediately, I contacted the chief of fisheries for an explanation. When queried, the officer explained that although fisheries surveys of Sam Rayburn Reservoir showed it to be one of the top reservoirs in the state for numbers of largemouth bass *M. salmoides*, it had a relatively low percentage of Florida largemouth bass genes. This lake had not been stocked with Florida largemouth bass since 1987.

That one telephone call which resulted in such a brief explanation settled what easily could have been a controversy. Communication is important, and the message here is twofold. This example demonstrates the importance of asking questions, rather than assuming the negative and reacting to it. I would like to think it also impressed upon management agency staff the importance of fully explaining their actions, rather than assuming an attitude of "we know what's good for the public and the fisheries."

Too often those in control have a tendency to unwittingly use their authority ambiguously. Anglers and other constituents of fisheries managers may be justified in doubting or rejecting new policies, even sound policies, if no effort is made to explain the rationale behind them.

The Need to Speak Out

Honey Hole's most actively involved members can be found in its Bass Club Affiliation, which comprises over 250 Texas clubs with over 9,500 individual members, and as such it is one of the largest angler groups in the southwestern United States. While Honey Hole staffers and editors do interact with the fisheries management agency directly, they recognize that they cannot represent all anglers. Honey Hole supports angler involvement in the fisheries management process, informs anglers, and encourages them to be responsive on their own. Anglers must realize that they cannot join an organization and expect someone else to fight for their rights. It is unrealistic for any one group or organization to be an instrument for all anglers. Individual rights, powers of speech, and actions must be continually sought for true majority representation. Groups and organizations working towards specific legislation are helpful in bringing to the public's attention the need for changes and new ideas and proposals. Fisheries management issues in each state must, or at least should, meet public approval and citizens should work toward an equal and balanced relationship with government agencies. Individuals should support those organizations that seek their own specific goals. At the same time, anglers need to voice their personal opinions when an agency enacts policies that appear to oppose their goals.

Angler Opinion

To what extent should angler opinions influence fisheries managers? Whose fishery is it? Government agencies regulating fisheries and wildlife must be accountable to those who pay for and use the resource. What anglers desire and what fisheries managers can provide sometimes differ greatly. Without the power to protest or support fisheries managers, the public cannot feel that the resource is truly "public domain."

Fisheries managers understand why resource management is important; on the other hand, bureaucrats appear at times to use "management" as an excuse to pursue personal initiatives. For example, agencies make promises for certain reservoirs to secure approval of a new regulation, or may use tournament permits to gain control of sportfishing activities. These conditional offerings nearly always hold some "ifs, ands, or buts." Where is the line over which state or federal power should not cross? Should anglers have a vote?

Establishing Communications

It is my opinion that public hearings are ineffective, and that they consist mostly of rhetoric. There seems to be a lack of desire on the part of management agencies to truly understand or care about citizens' complaints. Most state agencies hold some type of public hearings on at least a yearly basis. It has been my experience that less than adequate time is allotted at these meetings to allow participants to voice their concerns and opinions. Many public hearings are held at inconvenient times, thereby limiting the number of participants. Public hearings, in my opinion, are not held to hear complaints in an effort to understand, but instead are held so the agency can announce the decisions already made.

Another example from Texas is implementation of the statewide five-fish, 14-in minimum length black bass regulation in 1986. When the state fisheries agency instituted this regulation, the public outcry was fierce. However, after the fisheries managers explained the reasoning for the regulation, and anglers adjusted to this new idea, their initial rejection changed to acceptance. Acceptance occurred because the agency had successfully communicated the potential results of their proposal.

Public involvement in management decisions could be improved by (1) a questionnaire that is filled out when a fishing license is purchased, and (2) an official vote at the time of proposed regulations affecting anglers. If agencies can show just cause and just rewards for proposals, they should have little difficulty persuading the public to support their decisions.

Only when agencies are willing to listen can a true dialogue be opened between agencies and anglers. It has been my experience that state agencies listen only to those who will agree with and support them most of the time. However, the agency must listen to all people, including those who oppose their policies. I believe that any fisheries agency in this country that truly makes an effort to listen to anglers, or to explain management activities that initially seem unfavorable to anglers will find it much easier to gain angler support. Explanations would likely yield the results (i.e., compliance) agencies seek, and agencies might soon find themselves besieged by volunteers willing to support, and even assist, with labor-intensive field projects, for example.

Experienced fisheries managers seem to have a better grasp of public opinion than their less experienced counterparts. Through their work, they have developed a better understanding of fisheries resources and of anglers. The public respects fisheries biologists, and believes that they sincerely care about the resource. However, biologists need to better understand the public's needs and perceptions. Some resource studies are viewed as wastes of time and money and some projects are viewed as simply serving biologists' personal interests rather than real public needs. The economic times we live in are such that public funding cannot be wasted on projects the public neither wants nor needs. The public should have the means to influence expenditures on major projects.

In June 1993, I attended the Sports Fishing Institute's National Recreational Fisheries Summit. For me, it was a mind-opening event, and I was extremely impressed with some of the difficulties facing management agencies. But, I also left with a foreboding feeling about the future. At the summit Dr. L. Nielsen reported on an informal survey which revealed that a majority of fisheries and wildlife students had never hunted nor fished, and had no desire to do so. These students will be making decisions and judgements that will affect the future of wildlife and fisheries resources. Frankly, it scares me that tomorrow's managers may not have the fundamental experience needed to relate to a major resource user group.

Delivering the Message

Honey Hole has aggressively sought information on the management of Texas' black bass resources. It is our goal to represent black bass anglers. We respect the opinions of other anglers, whether they agree or disagree with ours. At the same time, there are still the "fence sitters" who need prodding to express their own positions. The facts need to be relayed honestly to the public. It is the role of the media to stimulate honest and open communication between management agencies and the public.

Regardless of the questions asked, if the answers do not suit the public, the public will not accept those answers. In most cases, however, fisheries agencies can alleviate problems through simple, honest answers, even when the public does not like the answers. Trust is just as important among agencies, the media, and the public, as it is in any relationship. Regardless of content or subject matter, the media carries messages to both the public and the agencies who serve them. This stimulates honesty and open communication, whereas the acceptance of the status quo hinders communication on fisheries management issues.

Without stimulation and education, anglers and the agencies would ignore and overlook each other. By open and direct communication (and confrontation if necessary), angler organizations such as Honey Hole prompt regulatory agencies to review their policies and proposals, and prompt the public to review its activities and options, or lack thereof. Silence and acceptance of whatever comes can be the downfall of all concerned.

Consensus and Working Together—Can it be Done?

A consensus can be reached only if you get the attention of those with whom you must achieve agreement, and working together can be accomplished only when all parties seek to do so. Despite the aggressive positions of some angler organizations, including Honey Hole, sportfishing groups are willing to work with fisheries agencies if they foster a "give-and-take" environment. For example, give-and-take may mean anglers learn to live with some unwanted regulations in exchange for favorably perceived regulations on other issues.

Angler organizations, including Honey Hole, are usually more willing to work with the agency than the agency may realize. Fisheries officials, in my view, generally look to supporters who can be easily persuaded and who are perceived as more likely to "go along with" agency ideas and proposals. This may provide short-term success but present long-term problems. Agencies must come to realize that neither the general public nor organized anglers are trying to usurp their authority. They simply desire to be treated fairly as required by agency mandate.

Agencies depend on the public for financial and political support. Yet, at times, agencies show a tendency to ignore suggestions from the public or reject them outright. Communication can benefit resource management by ensuring that all involved parties participate in decisions and all facts are thoroughly considered. When open communication between resource management agencies and the public is improved, public trust for agency activities and decisions will increase and our "adversarial" role will be less necessary.

The agency, the users, and the resource will benefit.

Facilities Management

American Fisheries Society Symposium 16:179–185, 1996

Balancing Recreational User Demands and Conflicts on Multiple Use Public Waters

WILLIAM W. JONES

Indiana University, School of Public & Environmental Affairs
Bloomington, Indiana 47405, USA

Abstract.—Recreational demand for public freshwater lakes and reservoirs is increasing nationally and is especially high in areas where there are few natural lakes. Although most public reservoirs have been constructed for flood control or water supply, they are used heavily for fishing, boating, swimming, and other activities. Because these reservoirs and the adjacent public property may represent the only locally available public recreational lands, additional uses (e.g., hiking, picnicking, aesthetics) are equally important. These activities cannot all be optimized on a single water body without conflicts. When conflicts over use arise, or when heavy recreational demand threatens the very health of the public resource, restrictions on individual uses, such as space and time zoning, may be required. Balancing competing uses of water resources is a major challenge facing resource managers today. Satisfactory resolution is possible through long-range planning and careful management.

The demand for recreation, and especially water-based recreation, escalated following the Second World War as increased affluence, new highway construction, and widespread automobile ownership gave Americans the means to recreate. In addition, the shortening of the average work week to five days gave them the time to recreate. Increasing human population and the development of new technologies have led to rapid increases in recreational uses of lakes in the United States.

Recreational Demand

National Trends

The great demand for aquatic recreation in the United States is reflected in market surveys which show increased participation in these activities. National Recreation Surveys show that between 1965 and 1983 participation in powerboating increased by 1.5%, water skiing by 3%, fishing by 5%, sailing by 4%, and canoeing by 5% (Kelly 1987). Also experiencing growth were related recreational uses which can be enhanced by aquatic settings, such as camping (15%) and birdwatching (6%). Future projections for the above recreational activities suggest steady growth (Kelly 1987).

The importance of water to noncontact aquatic recreation is illustrated in the results of the latest National Survey of Fishing, Hunting, and Wildlife-Associated Recreation. In 1991, 20.9 million anglers in the United States fished in freshwater lakes (excluding the Great Lakes) and reservoirs larger than 4 ha and a nearly equal number (19.2 million people) traveled to lakes or streams to observe, photograph, or feed wildlife (U.S. Fish and Wildlife Service and U.S. Bureau of the Census 1993).

Aquatic recreation and tourism generate substantial revenues for local economies. Travel and tourism on 3,800-ha Big Sandy Lake in Minnesota are estimated to generate direct consumer purchases of US$277 per lake hectare per year. Lakes in Georgia generated nearly $324 per hectare in fishing-related expenditures alone in 1991 (Williams 1994). The total contribution of fishing to Indiana's economy is estimated as $736.6 million annually (IDNR 1994)

Monroe Reservoir, Indiana

The effects of meeting recreational demand can be illustrated by a case study of Monroe Reservoir. Monroe Reservoir, in south–central Indiana, is the largest inland water body in the state. The reservoir, constructed by the U.S. Army Corps of Engineers (USCOE) in 1965, has a surface area of 4,330 ha and a mean depth of 4.5 m at normal pool. Monroe Reservoir was constructed for the joint purposes of flood control, water supply, and low flow augmentation (USCOE 1966).

Monroe Reservoir supports the largest water recreation base in Indiana. Six state recreation areas, one federal recreation area, a wilderness area, two state wildlife refuges, three campgrounds and ten boat ramps provide ample public access and recreational opportunities. Estimated visitorship has increased from 1 million in 1983 to 1.3 million in 1992. Peak visitorship occurred in 1989 with 1.5 million lake users (USCOE 1993). Visitors seek a variety of recreational activities when they visit Monroe Reservoir; sightseeing, fishing, swimming, picnicking, and boating are the most common activities (Table 1). The recreational activities that had the greatest increase in participation at Monroe

TABLE 1.—Participation in various recreational activities, by number of visitors to Monroe Reservoir, Indiana, 1983–1992. Source: USCOE (1993).

Activity	1983	1985	1987	1989	1990	1991	1992
Camping	125,525	118,077	138,303	154,443	141,554	144,907	129,847
Picnicking	193,241	188,818	243,645	280,990	237,896	236,818	217,337
Boating	183,840	187,981	234,242	263,800	158,260	128,417	211,879
Fishing	246,317	232,024	324,044	425,739	339,085	335,503	314,352
Hunting	12,814	11,901	16,542	12,005	11,999	12,911	12,822
Sightseeing	234,310	233,072	300,857	351,353	337,848	254,977	395,410
Skiing	77,617	70,141	85,252	91,357	87,391	90,180	82,440
Swimming	267,775	270,144	308,295	328,138	214,252	223,803	294,322
Other	100,640	128,375	141,283	140,003	144,350	122,493	123,568
Total[a]	1,035,688	1,069,469	1,332,017	1,525,239	1,386,415	1,398,798	1,344,843

[a]Many visitors reported involvement in several recreational activities. Therefore, the yearly sums of all activities are greater than the total visitors for each year.

Reservoir between 1983 and 1992 were sightseeing (68.8%), fishing (27.6%), other (22.8%), boating (15.0%), and picnicking (12.5%).

In 1992 we surveyed 179 lake users on two summer Saturdays at two boat ramps, one swimming area, and a campground on Monroe Reservoir. Our survey results (Table 2) corroborate the lake user data collected by the USCOE. Respondents to our survey felt that their favorite recreational activities in the reservoir had been adversely affected by deteriorating water quality, and they supported management activities to improve the recreational experience. Additionally, our survey showed that users were willing to travel great distances to recreate at their favorite reservoir. The users surveyed came from 28 Indiana counties throughout the state. Those who traveled more than 160 km averaged four visits to Monroe Reservoir per year. Those from nearby counties averaged more than 20 visits per year.

Recreational Limitations

The ability of a water body to support a given number of users and a variety of uses has been referred to as its carrying capacity (Kusler 1972). Carrying capacity differs for each water body. Factors affecting a reservoir's carrying capacity include physical and ecological constraints that can be identified scientifically (Figure 1). However, the acceptable limits of environmental disturbance and activity which users will tolerate requires more subjective evaluation.

Physical Constraints

Lake basin morphometry may limit the potential capacity for recreation on some reservoirs. For example, in Vermont, with only a few exceptions, no lakes smaller than 30 ha are used for high-speed boating (Bulmer and Garrison 1994). Small, sheltered lakes might likewise be unsuitable for sail boating due to wind shielding. Steep shorelines support little fish habitat and may be a safety hazard for swimmers. On the other hand, a shallow lake with extensive macrophyte development in the littoral zone restricts the amount of open water available for boating, interferes with swimming, restricts fishing access and may promote stunted fish populations.

Ecological Constraints

Water quality characteristics may also affect reservoir recreational uses (Figure 1). Reservoirs with

TABLE 2.—Public use of Monroe Reservoir and opinions of 179 users surveyed during summer 1992.

Reservoir use	Participation (%)	Uses adversely affected by deteriorating water quality (%)	Which uses should be improved by management (%)
Swimming	75.1	52.4	64.6
Fishing	57.1	26.5	31.2
Lake beauty	55.0	25.4	28.6
Power boating	45.5	20.6	12.2
Wildlife observation	42.3	16.9	20.1

Recreational Activities	Small size (< 30 ha)	Shallow depth	Steep shoreline	Extensive littoral community	Oligotrophy	Eutrophy
Fishing	○	◐	◐	◐	●	○
Motor boating	●	●	○	◐	○	○
Water skiing	●	●	○	◐	○	◐
Sailing	●	○	○	◐	○	○
Canoeing	◐	○	○	◐	○	◐
Swimming	○	◐	◐	◐	○	●
Aesthetic enjoyment	○	○	◐	◐	○	●

● Major constraint to recreational activity
◐ Moderate constraint to recreational activity
○ Minor or no constraint to recreational activity

FIGURE 1.—Physical and ecological lake conditions affecting aquatic recreation.

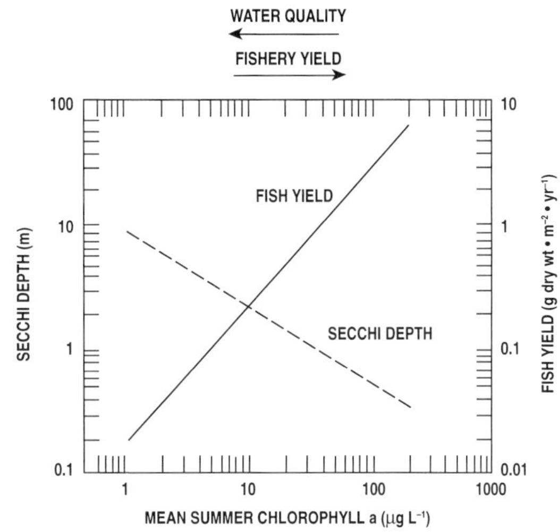

FIGURE 2.—Relationships of Secchi transparency and fish yield to water quality as measured by chlorophyll-*a* concentrations. (Adapted from Olem and Flock 1990.)

high transparency and low algal productivity have aesthetic appeal and are preferred by many for swimming and other body contact activities. For example, assessments of swimmability and aesthetics by lake users in Minnesota and Vermont were positively correlated with Secchi disk transparency (Smeltzer and Heiskary 1990). However, the reduced nutrient availability in such lakes limits fish biomass (Yurk and Ney 1989).

Conversely, as nutrient and algal concentrations increase, transparency decreases and fish productivity increases (Figure 2). The causative sequence is: greater phosphorus leads to increased algal production, which leads to increased trophic level response, which leads to increased fish biomass. Numerous studies have found statistically significant relationships between increased fish biomass and total phosphorus (Yurk and Ney 1989; Ney et al. 1990; Bayne et al. 1994), and between increased fish biomass and chlorophyll *a* (Jones and Hoyer 1982).

The nutrient–fish biomass relationship is so strong and the importance of fishing as a recreational activity is so great that several states recommend or use nutrient additions to stimulate algal growth, enhance fish production and improve fishing opportunities in unproductive waters (IDOC 1984; Lehman 1989). Management plans developed by the U.S. Forest Service have emphasized the need for improved fishing opportunities in unproductive ponds in Wayne–Hoosier National Forest, Indiana (Pfister and Jones 1992).

As nutrient concentrations and fish biomass increase, other conditions (color, transparency, odors) may become objectionable to reservoir users seeking aesthetic enjoyment or body contact recreation. Excessive nutrient enrichment can result in eutrophic or hypereutrophic lake conditions. Low hypolimnetic dissolved oxygen concentrations associated with eutrophic lakes may result in summerkill or winterkill of fish (Bennett 1983; Baker et al. 1993) or may favor rough fish over sport fish (Bayne et al. 1994).

Ecologically, we cannot have it all. A clear, unproductive reservoir cannot support many fish, and a eutrophic reservoir which supports fish cannot provide the aesthetic enjoyment that many users demand.

Operational Limitations

In lakes and especially reservoirs, which were constructed for special purposes other than recreation, it is common for constraints to be placed on recreational uses. Gasoline motors may be prohibited on waters serving as drinking water supplies to prevent contamination by gasoline products. Swimming may also be prohibited in drinking water sup-

Activity suffering impact

Recreational Activity	Fishing	Motor boating	Water skiing	Sailing	Canoeing	Swimming	Sunbathing	Aesthetic enjoyment
Fishing	—	◐	◐	○	○	○	○	○
Motor boating	●	—	◐	◐	◐	●	◐	●
Water skiing	●	◐	—	◐	◐	●	◐	●
Sailing	○	◐	◐	—	○	◐	○	○
Canoeing	○	◐	◐	○	—	○	○	○
Swimming	◐	◐	◐	○	○	—	○	◐
Sunbathing	○	○	○	○	○	○	—	◐
Aesthetic enjoyment	○	○	○	○	○	○	○	—

● Major effect on impacted activity
◐ Moderate effect on impacted activity
○ Minor or no effect on impacted activity

FIGURE 3.—A recreational use compatibility matrix.

plies out of concern for bacterial contamination (Jones et al. 1984; Forest 1985).

Other operational activities that can affect recreational uses include deliberate water level drawdowns for flood control storage or for consumptive water uses. These can lower water levels to the extent that boat launching, navigation, or beach use become problematic.

The implementation of certain reservoir management programs may also require temporary limitations on recreational use. For example, body contact recreation and fish consumption are restricted following herbicide applications for macrophyte control. Boat traffic is restricted during dredging operations because dredges are anchored by cables attached to shore. Fishing may be prohibited on reservoirs where the fisheries have been recently renovated to allow time for recovery.

Recreational Incompatibility

The ability of a water body to accommodate a given number of users and mixed recreational uses depends on the compatibility of those uses. Some aquatic-based recreational uses are incompatible with others. For example, motorboating cannot safely occur in swimming areas nor can motorboating and fishing successfully co-exist. Potential conflicts between recreational uses can be illustrated by a compatibility matrix (Jones 1988). Motorboating and waterskiing are the most incompatible with other lake uses (Figure 3). Note that fishing boats, while motoring to the fishing site, will have the same incompatibilities as motorboats.

Characteristics of motorboats that cause this high degree of incompatibility include high speeds, loud engine noise, and boat wakes. The high speeds at which many modern motorboats travel are inherently unsafe and may cause additional safety problems on crowded lakes. The noise generated by motorized watercraft may also be objectionable to other reservoir users. Loud noise levels from motorboats and personal watercraft were the most frequent complaint of other users at a Rhode Island lake (Wagner 1994). Sounds generated by motorboats with larger than 50 hp motors were above background sound levels at a distance of up to 250 m away and above annoying sound levels (75 dB) from 40 to 50 m away.

Wakes and turbulence generated by motorboats may erode unprotected shorelines (Liddle and Scorgie 1980; Wagner 1991), resuspend sediments and phosphorus from the reservoir bottom (Yousef et al. 1978), and disrupt smaller or less stable watercraft. For example, turbulence generated by a 75-hp motorboat can resuspend clay-sized particles from sediments as deep as 2.5 m (Yousef et al. 1978). Resuspended sediments and phosphorus decrease transparency and may stimulate algal growth, consequences which are detrimental to other recreational uses.

When asked to identify potential problems at Monroe Reservoir, respondents to our informal survey named unsafe boaters and overcrowding as the major problems (Table 3). This suggests that more recreational value at Monroe Reservoir is lost due to the impact of other recreationalists than is lost from the impact of more traditional reservoir problems such as excessive nutrient additions or sedimentation. This phenomenon was explored in a recent paper by Klessig (1994).

TABLE 3.—Public perception of Monroe Reservoir problems: opinions of 179 users surveyed during summer 1992.

Potential lake problem	Users indicating this as a major problem (%)[a]
Unsafe boaters	53.4
Overcrowding	42.9
Water clarity	31.7
Rooted aquatic weeds	25.4
Sedimentation	22.7

[a]Survey respondents could select more than one problem.

Management Options

The high demand for recreation along with the physical and ecological limits of the resource dictate that controls must be placed on several recreational uses, despite the ideal of free access and use of public waters. At 18-ha Dow's Lake, Ottawa, Canada, recreation demand was so high that up to 170 boats were reported using this lake at one time (Jaakson 1984). In such cases, resource damage and conflicts among recreational uses are inevitable without the use of controls.

Reservoir use controls and limits must be a part of a carefully developed management plan. Goals for such a plan must be fashioned in a clear, logical manner and should consider (1) resource limitations; (2) recreational demand; (3) recreational use compatibility; (4) alternative siting; and (5) health, safety, and liability issues. Water and adjacent land-based uses, such as hiking and birdwatching, should be included in this evaluation. However, the need to provide for diverse uses should not diminish the quality of the recreational experience, nor does diversity of use imply equal use for all activities. This approach was used successfully in developing a management plan for Griffy Lake, Indiana (Jones 1988). The Griffy Lake plan, which included upland areas, remains a viable, working plan 10 years after it was adopted.

Once goals are established, strategies can be developed to achieve them. Reservoir use restrictions and time and space zoning are among the more common options available for managing aquatic recreation.

Use Restrictions

Among the most common use restrictions are boat speed limits. Speed limits are used on small reservoirs to insure safety, to maintain a more tranquil setting, or to reduce boat-induced waves and the shoreline erosion they cause. Thus, in Indiana there is a 16 km/h speed limit on all public lakes smaller than 120 ha (Indiana Revised Code, Title 14: Recreation and Land Management).

Examples of three other reservoir use restrictions follow:

1. Gasoline motors are often prohibited on small reservoirs within fish and wildlife areas to reduce disruption of nesting waterfowl or other animals.
2. Swimming may be prohibited in reservoirs where supervision is not possible or where poor water quality creates a health risk.
3. Fishing seasons and limits are imposed to protect spawning fish, to manage fish population size structure, and to maintain a healthy fishery.

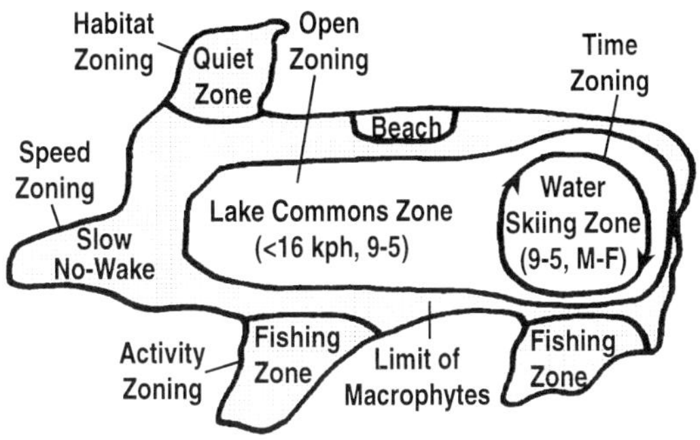

FIGURE 4.—Zoning of lake activities on a hypothetical lake by time and space. (Adapted from Engel 1989a.)

In some cases, designated uses may be established for entire lakes or reservoirs (Engel 1989a). For example, shallow lakes can be designated for fishing or waterfowl hunting, and deep basins could support motorboating and waterskiing. This approach is most feasible in a region with numerous diverse lakes or reservoirs. Vermont has established the Use of Public Waters Policy which designates permitted lake uses on some 283 public lakes according to four lake use types: wilderness–solitude, nonmotorized, low-speed motorized, and high-speed motorized (Bulmer and Garrison 1994).

Space Zoning

Lakes and reservoirs under intense recreational pressure can be spatially zoned to partition use (Engel 1989a, 1989b). Activity zoning in large, diverse reservoirs allocates space for all allowed water uses (Figure 4). Open-water areas can be reserved for motorboats and skiing. A shoreline area can be designated for swimming. Shallow bays can be designated as quiet zones for fish or waterfowl habitat where only minimal disturbance is allowed. In these areas, motors may be prohibited or limited to idle speed. For example, the shallow, eastern half of Monroe Reservoir is designated as an idle zone to promote fishing, fish and waterfowl habitat, and to prevent resuspension of sediments. Another form of spatial zoning in Indiana is the state law which prohibits boat speeds greater than 16 km/h within 50 m of shoreline on all public lakes (Ind Rev Code, Title 14: Recreation and Land Management).

Each of the zoned areas should be separated from other areas by clearly designated boundaries and buffers. Aquatic vegetation can be used to separate adjacent zones or slow approaching boats. Harvested channels can serve as boat corridors. Zoning regulations should be posted at reservoir access points to inform users.

Time Zoning

Time zoning can be applied successfully in small reservoirs where spatial zoning is not feasible (Engel 1989a). For example, water skiing may be allowed during certain mid-day hours or on alternate days to avoid conflicts with fishing or quiet enjoyment of the reservoir. At 44-ha Great Pond on Cape Cod, Massachusetts, motorized watercraft are not permitted on odd numbered days to allow time for windsurfers, canoeists, and anglers to enjoy themselves without interference from motorboats (Wagner 1994). Time zoning was an integral component of the management plan for Dow's Lake, where incompatible lake uses were scheduled at different times of the week (Jaakson 1984).

Another form of time zoning is a night speed limit for boats. In Indiana, boats may travel no faster than 16 km/h between sunset and sunrise (Ind Rev Code, Title 14: Recreation and Land Management).

Conclusion

A reservoir cannot be all things to all people. The demand for aquatic recreation is sufficiently high that conflicts among recreational uses and ultimately, damage to the resource will result. With heavy recreational use, we run the risk that recreation will have a greater impact on reservoirs than will nutrients or other traditional problems. Limits on individual rights to use public waters must be accepted in order to provide for the greater public good: protection of the resource. By understanding the physical and ecological limitations of reservoirs, and by considering the inherent potential for conflict among reservoir uses, resource managers and recreation planners can effectively manage multiple-use reservoirs.

References

Baker, J. P., H. Olem, C. S. Creager, M. D. Marcus, and B. R. Parkhurst. 1993. Fish and fisheries management in lakes and reservoirs. U.S. Environmental Protection Agency (EPA 841-R-93-002). Terrene Institute and U.S. Environmental Protection Agency, Washington, DC.

Bayne, D. R., M. J. Maceina, and W. C. Reeves. 1994. Zooplankton, fish and sport fishing quality among four Alabama and Georgia reservoirs of varying trophic status. Lake and Reservoir Management 8:153–163.

Bennett, G. W. 1983. Management of lakes and ponds, second edition. Robert Krieger Publishing, Malabar, Florida.

Bulmer, S. K., and V. Garrison. 1994. Statewide lakes and ponds recreation management in Vermont. LakeLine 14:29–31.

Engel, S. 1989a. The restructuring of littoral zones. Lake and Reservoir Management 2:235–242.

Engel, S. 1989b. Lake use planning in local efforts to manage lakes. Pages 101–1016 *in* J. Taggart, editor. Enhancing states' lake management programs. Northeast Illinois Planning Commission and North American Lake Management Society, Chicago.

Forest, H. S. 1985. Hemlock Lake, New York: interrelationships among ecology, engineering, and political structure. Lake and Reservoir Management 1:9–14.

IDOC (Illinois Department of Conservation). 1984. Management of small lakes and ponds in Illinois, first edition. IDOC, Springfield.

IDNR (Indiana Department of Natural Resources). 1994. How federal funds benefit Indiana's anglers. Focus Newsletter, March/April 1994.

Jaakson, R. 1984. Recreation planning for a small urban lake. Town Planning Review 56:90–111.

Jones, W. W. 1988. Balancing competing uses for water resources—a Griffy Lake example. Lake and Reservoir Management 4:73–80.

Jones, J. R., and M. V. Hoyer. 1982. Sportfish harvest predicted by summer chlorophyll-a concentrations in midwestern lakes and reservoirs. Transactions of the American Fisheries Society 111:176–179.

Jones, W. W., and seven coauthors. 1984. Griffy Lake long-range use and management plan. Indiana University, School of Public and Environmental Affairs, Environmental Systems Application Center, (ESAC-84-02), Bloomington.

Kelly, J. R. 1987. Recreation trends toward the year 2000. Management Learning Laboratories, Ltd., Champaign, Illinois.

Klessig, L. 1994. New loadings to lakes. LakeLine 14:49.

Kusler, J. A. 1972. Carrying capacity controls for recreation water uses. University of Wisconsin Press, Madison.

Lehman, L. L. 1989. Use of liquid fertilizer on small infertile impoundments, 1989 fish management report. Indiana Department of Natural Resources, Division of Fish and Wildlife, Indianapolis.

Liddle, M. J., and A. R. A. Scorgie. 1980. The effects of recreation on freshwater plants and animals: a review. Biological Conservation 17:183–206.

Ney, J. J., C. M. Moore, M. S. Tisa, J. J. Yurk, and R. J. Neves. 1990. Factors affecting the sport fishery in a multiple-use Virginia reservoir. Lake and Reservoir Management 6:21–32.

Olem, H., and G. Flock, editors. 1990. Lake and reservoir restoration guidance manual, second edition. Prepared by North American Lake Management Society for U.S. Environmental Protection Agency (EPA 440/4-90-006), Washington, DC.

Pfister, M. A., and W. W. Jones. 1992. Pond management

studies in Hoosier National Forest. Indiana University, School of Public and Environmental Affairs, Bloomington.

Smeltzer, E., and S. A. Heiskary. 1990. Analysis and applications of lake user survey data. Lake and Reservoir Management 6:109–118.

USCOE (U.S. Army Corps of Engineers). 1966. Wabash River Basin, Indiana, reservoir regulation plan—appendix C: Monroe Reservoir. USCOE, Louisville, Kentucky.

USCOE (U.S. Army Corps of Engineers). 1993. Visitorship records—files. USCOE, Bloomington, Indiana.

U.S. Fish and Wildlife Service, and U.S. Bureau of the Census. 1993. 1991 national survey of fishing, hunting, and wildlife-associated recreation. U.S. Government Printing Office, Washington, DC.

Wagner, K. J. 1991. Assessing impacts of motorized watercraft on lakes: issues and perceptions. Pages 77–94 *in* G. Flock, editor. Enhancing the states' lake management programs. U.S. Environmental Protection Agency, Washington, DC.

Wagner, K. J. 1994. Of hammocks and horsepower: the noise issue at lakes. LakeLine 14:24–28.

Williams, V. 1994. What's a lake worth? LakeLine 14: 37–39.

Yousef, Y. A., W. M. McLellon, R. H. Fagan, H. H. Zebuth, and C. Larabee. 1978. Mixing effects due to boating activities in shallow lakes. U.S. Department of the Interior Technical Report (ESEI 78-10), Washington, DC.

Yurk, J. J., and J. J. Ney. 1989. Phosphorus-fish community biomass relationships in southern Appalachian reservoirs: can lakes be too clean for fish? Lake and Reservoir Management 5:83–90.

American Fisheries Society Symposium 16:186–189, 1996

U.S. Army Corps of Engineers Natural Resource Management Program

DAVID J. WAHUS

U.S. Army Corps of Engineers, Natural Resource Management Branch
20 Massachusetts Avenue, NW, Washington, DC 20314-1000, USA

Abstract.—The U.S. Army Corps of Engineers (COE) is responsible for the management of diverse natural resources on over 11 million acres of land and water. The COE plays an important role in the nation's economy by providing recreation opportunities. The COE stewardship responsibilities include managing recreation opportunities, fisheries resources, shoreline and associated conflicts, an aquatic plant control program, and a fish transportation program at dams in the Pacific Northwest.

The U.S. Army Corps of Engineers (COE) has become increasingly involved in outdoor recreation in recent decades. There were few legal provisions that mandated attention to recreational or environmental values during the early decades of the century when reservoir construction projects were first initiated. Although the Flood Control Act of 1944 (16 U.S.C. § 460d) authorized the COE to include recreational features at reservoir projects, from the 1940s through the 1960s the COE operated in basically a custodial mode at its projects, that is, improvising parks at road ends where people gathered, mowing grass, cleaning restrooms, and collecting garbage. This mode of operation was largely due to administrative and funding reasons, and not to a lack of concern. During the early 1970s, the COE hiring policies expanded to include professional resource managers, biologists, and park rangers to manage the resources in and surrounding the reservoir projects. One primary goal of COE managers is to provide the public with quality outdoor recreation experiences without adversely impacting those resources. Recreation has since become the most visible mission of the COE, and in 1994 the COE officially adopted a mission statement that reflects the significance of its natural resources management program and its impact on the nation as follows.

> The Army Corps of Engineers is the steward of lands and waters at Corps water resource projects. Its Natural Resources Management Mission is to manage and conserve those natural resources, consistent with ecosystem management principles, while providing quality public outdoor recreation experiences to serve the needs of present and future generations.
>
> In all aspects of natural and cultural resources management, the Corps promotes awareness of environmental values and adheres to sound environmental stewardship, protection, compliance and restoration practices. The Corps manages for long-term public access to, and use of, the natural resources in cooperation with other Federal, State, and local agencies as well as the private sector.
>
> The Corps integrates the management of diverse natural resource components such as fish, wildlife, forest, wetlands, grasslands, soils, air, and water with the provision of public recreation opportunities. The Corps conserves natural resources and provides public recreation opportunities that contribute to the quality of American life.

Recreation Opportunities

The U.S. Army Corps of Engineers is responsible for 463 water resource development projects throughout the United States. Managing almost 12 million acres of land and water with 4,300 developed recreation areas and almost 41,000 miles of shoreline, the COE makes significant contributions to the nation's recreational opportunities. Over 80% of COE reservoirs are within a 1-h drive of a major metropolitan area (U.S. Army Corps of Engineers 1994).

Corps of Engineers recreational areas host more than 380 million annual visits; over 25 million people (10% of the 1992 U.S. population) visit a COE project at least once each year. The COE hosts almost 30% of the outdoor recreation or tourism activities occurring on federal lands, making the COE the nation's largest provider of water-related recreation and the second largest federal provider of outdoor recreation, surpassed only by the U.S. Forest Service (Figure 1). This activity takes place on just 2% of the nation's federal land base, using less than 9% of the total federal funds expended for recreation (National Park Service 1992).

Visitors to COE reservoirs spend significant amounts of money on goods and services, thereby contributing to the local and national economies. The COE recreation program is an important part of the U.S. travel and tourism industry, which is the second largest service industry in the country. Recreation activities at COE reservoirs account for

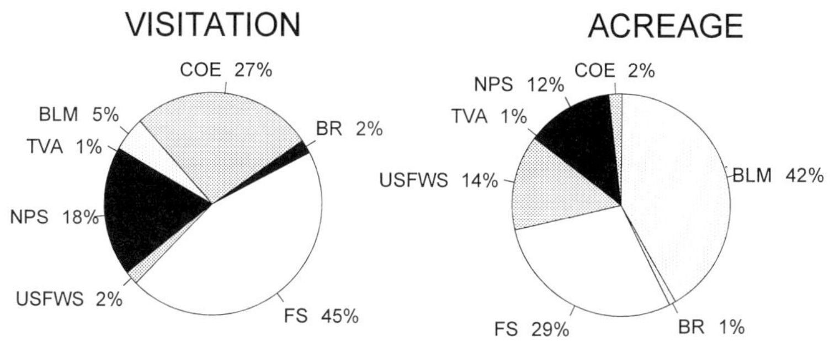

FIGURE 1.—Visitation and acreage distribution of sites managed by Bureau of Land Management (BLM), Bureau of Reclamation (BR), Corps of Engineers (COE), Forest Service (FS), National Park Service (NPS), Tennessee Valley Authority (TVA), and U.S. Fish and Wildlife Service (USFWS).

over 1.4% of the direct sales in this US$200 billion industry. In 1991, visitors to COE reservoirs spent over $10 billion on goods and services related to recreation (Jackson et al. 1994). The direct and indirect effects of this economic activity resulted in $12.4 billion in employee income and 617,000 full and part-time nonfederal jobs with an average annual salary of $18,300. These jobs represented 0.4% of nonfederal employee income and 0.5% of the employment in the United States. With a Fiscal Year 1994 budget of $196 million, the COE recreation program expends less than $325 per job. Although analyses employing indirect effects may overstate overall economic activity, they give some sense of the value of the COE recreation program.

An example of the local economic impact made by COE projects was illustrated in studies by the Oklahoma Department of Wildlife Conservation. Study results indicated that fishing activities on Lake Texoma contributed $25.6 million to the local economy (Schorr et al. 1995). Researchers reported that anglers spent more than 1.3 million hours fishing at Texoma in 1990. This benefitted the local economy and resulted in 417 jobs directly associated with Lake Texoma angler expenditures.

Boating and fishing are two very popular recreation activities available at many COE projects. The COE operates over 3,500 boat-launching ramps, including more than 6,500 launching lanes, and has 536 marina facilities under lease at 178 reservoir projects. Fishing activities account for 26% of total visits to COE projects. For example, Sam Rayburn Lake near Jasper, Texas, annually hosts the McDonald's fishing tournament, attracting over 7,000 boats which are assessed an entry fee of $200 each. Over 70 major tournaments sponsored by the Bass Anglers Sportmen's Society have been held at COE projects since 1967.

Recreational Fisheries Management

Many COE projects provide outstanding fishing opportunities. Record catches have been documented in many of the reservoirs managed by the COE. Many of these reservoirs are stocked in cooperation with local fish and game agencies who have primary responsibility for managing the local fisheries resources.

Fish hatcheries have been built by the COE at many project sites, but are managed by other federal and state agencies. For example, the Bonneville Fish Hatchery is operated by the Oregon Department of Fish and Wildlife and is the oldest hatchery in the state. Spring Creek Hatchery is operated by the U.S. Fish and Wildlife Service. Also, in cooperation with the Tennessee Wildlife Resources Agency and the U.S. Fish and Wildlife Service, the COE signed an agreement that calls for the expansion of the 24-year-old Dale Hollow National Fish Hatchery at Dale Hollow Lake in Tennessee. The expansion could mean an additional 100,000 pounds of trout each year to anglers.

Millions of fish have been stocked by the Missouri Department of Conservation at Harry S. Truman Lake to supplement the existing fish population. Over 8,000 acres of timber left standing by the COE at Truman Lake prior to inundation continue to provide superior aquatic habitat. Habitat management remains a high priority at many COE reservoirs.

Special fishing events are also becoming increasingly common. The annual Jay Cees-sponsored "Fishing Has No Boundaries" event at Lake Oahe in South Dakota is attended by over 100 disabled persons. Each angler is accompanied by a friend or family member. Several hundred volunteers provide boats, prepare the picnic site, cook and serve meals, clean and package fish, and clean up afterwards.

The COE, along with other federal, state, and private partners, actively support National Fishing Week in celebration of America's recreational fisheries and to promote environmental stewardship and fishing education. Tailwater, reservoir, and river fishing opportunities abound in COE reservoirs and waterways.

Other Water-Related Recreation

Boating remains a popular activity at all COE reservoirs, and many regional and national regattas are held at the projects. Table Rock Lake offers scuba divers interesting underwater rock bluffs to explore. The Bonneville Lock and Dam project on the Columbia River is a world premier windsurfing site with several professional meets held annually. Canoeing opportunities on COE projects range from calm waters to rapids. An annual canoe slalom is held each October downstream from East Sidney Lake in New York. Releases from Youghiogheny River Lake in Pennsylvania allow some of the best white water canoeing in the East.

During the 1960s and 1970s, boating usage and the subsequent development and use of marina facilities at COE reservoirs continued to grow. Boat-related activities account for 52% of yearly visitation at COE projects.

An issue linked to boating is the COE concern for reservoir carrying capacity. During the last three decades the popularity of boating, waterskiing, fishing, and other water-based activities has dramatically increased. Activities such as windsurfing and personal watercraft add new dimensions of complexity to managing the water resources available for recreation at COE projects. To better understand these issues and conflicts, the COE is studying reservoir carrying capacity with the goal of providing higher quality water-based recreation opportunities (Chilman et al. 1995; Titre and Vogel 1995).

Shoreline Management

During the early days of water resources development, recreation visitation and the emphasis on environmental quality were not as intense as they are today. There was thought to be more land, water, and shoreline than would ever be needed for public use. As public use of the reservoirs began to increase, conflicts developed between private and public use of COE lands. By the late 1960s, policies were developed which prohibited private, exclusive use, while authorizing certain private floating recreation facilities on COE projects. Public use had grown to such proportions on COE land that managers, rangers, biologists, and professionals from other disciplines were hired to aid in the proper management of COE resources. These professionals soon realized the inequities associated with private use of public lands and initiated new policies to prohibit additional exclusive use. These policies stressed maximum storage of boats and related equipment at commercial concessions and provided for shoreline use permits without a fee.

Shoreline management is one of the programs being used to manage federal lands to insure their availability to all members of the public. During the mid-1970s, COE shoreline management policy expanded to include minor land and vegetation modification. This policy required shoreline management plans for projects with existing private shoreline use and encouraged the use of community docks. For the first time, fees were required for the issuance of permits for these activities. This program has remained essentially unchanged, with the exception of changes in fee structures.

The policy governing the shoreline management program is to protect and manage shorelines in a manner that promotes safe and healthy use by the public, while maintaining environmental safeguards. Balance is the key objective between permitted private use and resource protection for general public use. Almost 55,000 shoreline permits have been issued at 87 projects. More than 35,000 of these permits are for private or community boat-docking facilities.

In addition to the many activities and structures managed by the shoreline management program on lands surrounding the water, all agricultural leases on COE managed lands require compliance with the Environmental Protection Agency's label requirements for herbicides and pesticides to protect water quality and fish and wildlife habitat.

Aquatic Plant Management

The Corps of Engineers has also played a role in the management of aquatic plants since the late nineteenth century. This program was initiated in response to expanding populations of water hyacinth *Eichhornia* spp. impeding commercial river traffic. The Removal of Aquatic Plant Growths Project was included in the Rivers and Harbors Act of 1899 (33 U.S.C. §§ 401, 403) and limited activities to Florida, Alabama, Mississippi, Louisiana, and Texas. Expansion of the program was recommended by a review committee in 1945, and enacted into law by the Rivers and Harbors Act of 1958 (33 U.S.C. § 610). The law authorized a comprehensive program for control and programmatic eradication of water hyacinth, alligator weed *Alternanthera philoxeroides*, and other obnoxious aquatic plant growth in eight southern states. The aquatic plant control program is now a nationwide program. Through this program, the COE cooperates with other federal and nonfederal agencies when projects benefit the combined interests of navigation, flood control, drainage, agriculture, fish and wildlife conservation, and public health. The costs of current aquatic plant control operations are shared: 50% from federal sources and 50% from nonfederal sources.

Fisheries Concerns

The COE has also been involved in recent years with issues surrounding anadromous fish in the Columbia and Snake river systems in the Pacific Northwest. Twelve hydroelectric dams create physical obstructions to salmon migration. Adults returning to their spawning grounds must negotiate fish ladders. Juveniles not diverted into bypass systems or spilled over the top of the dams pass through the power turbines, which can reduce the survival of the fish. The dams also change habitat through the creation of reservoirs resulting in further potential reductions in recruitment of young fish. The Columbia and Snake river flows have been augmented throughout the spring and summer by controlled releases of water stored upstream to aid in water temperature regulation and downstream movement of juvenile salmon. Water has been released from some dams to divert juvenile fish from the powerhouses.

The COE fish transportation program involves loading juvenile fish onto barges or trucks and transporting them below Bonneville Lock and Dam. A number of improvements have been made to fish passage facilities at dams, and COE biologists monitor adult and juvenile fish to provide survival data for the National Marine Fisheries Service and state fisheries agencies. These studies continue to provide information to agencies with a variety of fisheries concerns.

Water levels are regulated to improve fisheries habitat where it is consistent with other project demands. Excellent trout fisheries are present below several COE dams such as Broken Bow and Hartwell. Dissolved oxygen levels and temperature are monitored and adjusted where possible to reduce potential impacts to a fishery.

The COE is moving towards an ecosystem management approach. This approach, with a watershed focus, seeks to restore the structure or function of ecosystems, recognizing that all components are interrelated. The COE works closely with state and federal fisheries, wildlife, and natural resources management agencies to contribute to improving the total ecosystem.

Conclusion

The COE outdoor recreation and natural resource management mission continues to evolve in response to public demand, policy, and environmental philosophy. Today, the agency is a major provider of recreation opportunities across the nation. Corps of Engineers facilities continue to complement the many state, local, and private recreation facilities operated at COE reservoirs. The COE outdoor recreation and conservation stewardship programs have improved relations between the COE and the public at large.

Current trends indicate a continued and growing demand for water-based recreational opportunities at COE reservoirs. These opportunities make COE outdoor recreation and natural resource missions comparable to other congressionally authorized purposes. The COE will continue to focus on providing the public with cost effective, quality services and facilities.

References

Chilman, K. C., J. J. Vogel, and J. P. Titre. 1995. Recreational carrying capacity decision-making and planning for water-based recreation. U.S. Army Corps of Engineers Waterways Experiment Station, Instructional report, Vicksburg, Mississippi.

Jackson, R. S., D. J. Stynes, and D. B. Propst, D. B. 1994. An assessment of the national economic effects of the U.S. Army Corps of Engineers recreation program. U.S. Army Corps of Engineers Waterways Experiment Station, Miscellaneous Paper R-94-2, Vicksburg, Mississippi.

National Park Service. 1992. Federal recreation fee report to Congress. National Park Service, Washington, DC.

Schorr, M. S., J. Sah, D. F. Schreiner, M. R. Meador, and L. G. Hill. 1995. Regional economic impact of the Lake Texoma (Oklahoma–Texas) striped bass fishery. Fisheries 20(5):14–18.

Titre, J. P., and J. J. Vogel. 1995. Carrying capacity study of Table Rock Lake. U.S. Army Corps of Engineer Waterways Experiment Station, Technical Report, Vicksburg, Mississippi.

U.S. Army Corps of Engineers. 1994. Natural resources management information system. U.S. Army Corps of Engineers, Washington, DC.

American Fisheries Society Symposium 16:190–199, 1996

Environmental Considerations in Licensing Hydropower Projects: Policies and Practices at the Federal Energy Regulatory Commission

JENNIFER HILL

Federal Energy Regulatory Commission, Office of Hydropower Licensing
888 First Street, N.E., Washington, DC 20426, USA

Abstract.—The Federal Energy Regulatory Commission (FERC) has the responsibility, when deliberating the licensure of a hydropower project, to consider all aspects of the public interest and to license only those projects that are consistent with the best comprehensive use of the waterway. Amendments to the Federal Power Act, pursuant to the passage of the Electric Consumers Protection Act in 1986, require FERC to give equal consideration to environmental resources and energy conservation, as well as developmental values such as power production. The responsibility of determining the proper balance between the development of hydropower as a renewable energy source and environmental protection rests with FERC. Additional mandatory conditions may be imposed by certain resource agencies. The most common issues FERC considers in its environmental impact documents for hydropower projects include aesthetics, recreation, water quality and quantity, fisheries, vegetation and wildlife, soils, geology, and historical and archeological resources. The applicant for a hydropower license is required by FERC to file study results or other information that facilitates assessment of the potential environmental impacts associated with project development. The staff of FERC decides on the need for and scope of environmental studies, based on the potential for adverse impacts and the costs associated with the studies. Often, there is a tradeoff between limiting study costs and the precision or certainty associated with the study results. Early communication and cooperation among the applicants, agencies, and other entities greatly facilitates the licensing process.

The Federal Energy Regulatory Commission (FERC) issues licenses or exemptions from licensing for nonfederal hydropower projects pursuant to the Federal Power Act (FPA; 16 U.S.C. §§ 791a–825r). Under the "comprehensive development" standard of Section 10(a) of the FPA, FERC has responsibility to consider all aspects of the public interest in using a waterway and to license only those projects that are consistent with the best comprehensive use of the waterway. Licensing decisions are made by FERC after the preparation of an environmental assessment or environmental impact statement in compliance with the Council on Environmental Quality's regulations for implementing the National Environmental Policy Act (NEPA; 42 U.S.C. §§ 4321–4370d). The NEPA document (environmental assessment or environmental impact statement) considers both adverse and beneficial environmental impacts and balances environmental protection, mitigation, and enhancement with other project benefits if FERC decides to license the project (FERC 1991a).

In 1986, the Electric Consumers Protection Act (16 U.S.C. §§ 791a–825h) was passed, revising the FPA. This act requires that FERC, in deciding whether to issue a license, shall, in addition to considering power and development purposes, give "equal consideration" to energy conservation, and the protection, mitigation of damage to, and enhancement of, fish and wildlife (included related spawning grounds and habitat), the protection of recreation opportunities, and the preservation of other aspects of environmental quality (Section 4[e] of the FPA).

Competing uses must be carefully weighed by FERC to determine the best comprehensive development of the waterway. Competing uses may be, for example, power generation versus some environmental concern, such as flows over waterfalls for aesthetic enhancement and mitigation. In contrast, there may be competing environmental concerns, such as flow needs for whitewater boating versus fish habitat. Sometimes one action may have ramifications for several different environmental resources. For example, fish passage may increase recreational fishing for inland waters. However, if the fish are contaminated, fish passage could be detrimental to humans or other species, such as the bald eagle *Haliaeetus leucocephalus*, that consume the fish. This is an issue regarding fish passage at hydropower projects on tributaries to the Great Lakes.

Section 10(a) of the FPA also requires FERC to consider a project's consistency with federal and state comprehensive plans for improving, developing, or conserving waterways. Such plans may include a state restoration plan for a certain fish species, a management plan for a national forest, or any other federal or state agency plan that is filed with FERC and accepted as a comprehensive plan. The FERC considers the consistency of a project with goals outlined in comprehensive plans and weighs such plans heavily in any licensing decision.

Section 10(j) of the FPA requires FERC to issue hydropower licenses that are consistent with recommendations pursuant to the Fish and Wildlife Coordination Act (16 U.S.C. §§ 661–666c) if such recommendations are consistent with the FPA and other applicable law. The National Marine Fisheries Service, the U.S. Fish and Wildlife Service, and state fish and wildlife agencies may make such recommendations. For projects affecting tribal lands or waters, the appropriate councils must be consulted and their concerns addressed.

Fish and wildlife agencies have various mandatory conditioning authorities. Pursuant to FPA Section 18, the Secretary of the Interior or the Secretary of Commerce may require upstream and downstream fishways. Appropriately prescribed fishway conditions must be included by FERC in any license issued, whether or not FERC agrees that providing fishways are in the public interest.

Section 401 of the Clean Water Act (33 U.S.C. §§ 1251–1376) grants the U.S. Environmental Protection Agency the authority to issue water quality certification. The Environmental Protection Agency has delegated authority to issue the certification to the state in which the hydropower project is or would be located. No license can be issued by FERC without certification from the state, or waiver thereof. Valid water quality certification conditions are incorporated into any license issued. If the certification is neither issued nor denied within 1 year from the date of request, FERC considers it to be waived.

If a proposed or existing project occupies a U.S. reservation, then the federal administering agency of those lands can issue conditions, under FPA Section 4(e), that must be incorporated into the license. Commonly, the U.S. Forest Service, the Department of the Interior, and the U.S. Army Corps of Engineers provide such mandatory conditions for projects on lands they administer.

If the proposed or existing project affects areas purchased with money provided by grant under the Land and Water Conservation Fund Act (16 U.S.C. §§ 460d, 460l-4–460l-11), FERC is required to consult with the National Park Service, which may require in-kind replacement or compensation. Areas protected under this act may include, for example, parks with shoreline access for recreation, scenic overlooks, or sanitary facilities. Likewise, areas may be protected by designation as national wild and scenic rivers. Section 7(a) of the Wild and Scenic Rivers Act (16 U.S.C. §§ 1271–1287) prohibits FERC from authorizing construction of a new project on or that would directly affect rivers that have been so designated, or are being studied for possible designation. However, FERC can and has relicensed some existing projects on wild and scenic rivers when the original project licensing preceded the designation.

The Endangered Species Act (16 U.S.C. §§ 1531–1544) requires FERC to consult with the Secretary of the Interior and the Secretary of Commerce on federally listed and proposed species and the designated and proposed critical habitat that may be affected by any licensing action taken (FERC 1993). For projects affecting natural resources within a state's coastal zone, conditions may also be incorporated into licenses for compliance with the Coastal Zone Management Act (16 U.S.C. §§ 1451–1464).

Types of Hydropower Projects

Environmental impacts associated with hydropower projects vary considerably, depending on site-specific environmental characteristics and the size and type of the hydropower project. The size of the project itself may not be a good indicator of the potential environmental impacts; small projects operating at high head may use small quantities of water, but have substantial environmental impacts. Therefore, potential impacts must be evaluated on a case-by-case basis. It is also important to understand the differences in hydropower types to address the kinds of impacts that may occur.

There are two types of hydropower projects, pumped storage and conventional. Pumped storage hydropower projects use bodies of water at two different elevations. Water flows to the lower body of water by gravity, generating power during periods of peak electrical use (e.g., daylight hours). During off-peak hours (e.g., evening and night hours), the turbine generators are reversed and water is pumped back up to the upper body with electrical energy from fossil or nuclear plants. If both the upper and lower bodies of water are distinct reservoirs, the pumped storage is considered closed. Conversely, an open pumped storage system would typically have a dammed river as either the upper or lower water body. Although FERC has authorized the development of both open and closed system projects, most current original license applications for pumped storage projects are for closed systems, which generally have fewer environmental impacts. Pumped storage projects typically operate under high head (e.g., 100–700 m), generating 500–2,000 MW.

Conventional projects, which use a dam in a waterway, typically operate with a low head (1–100 m), and generate from a few kilowatts to several thousand megawatts for large projects. The powerhouse is usually located at the dam. These projects may operate in a run-of-river mode, where outflow from the project is equal to inflow. Conversely, a daily peaking project drafts the reservoir to generate more during the day and refills the reservoir during the night, thus causing reservoir fluctuations. Such daily peaking operations cause high flows downstream from the project during the day and low flows downstream during the night.

To increase head for electrical or hydromechanical generation, the developer may construct the powerhouse downstream from the dam, diverting flow from a section of river known as a bypassed reach. This reach has reduced and typically more uniform flows than would occur naturally. Most current applications for an original license propose run-of-river conventional projects with little or no bypassed reach.

The Hydropower Licensing Process

Original licenses for projects generally expire in 30 to 50 years. An application for relicense must be filed with FERC at least 2 years before the current license expires. Under FERC's regulations, applicants must follow a three-stage prefiling consultation process with resource agencies. Early in the process, the applicant prepares an initial consultation package of information regarding the proposed project and meets with the agencies, interested nongovernmental organizations (NGOs) such as Trout Unlimited and American Rivers, any affected tribes, and the general public.

The resource agencies identify environmental concerns and may request studies to address those concerns. The applicant consults with the agencies on the studies needed and conducts the agreed-upon studies. Should the applicant and agencies disagree on the studies needed, either entity may file a request for dispute resolution with FERC staff. The applicant interacts with agencies, NGOs, affected tribes, and the general public, to develop a draft, and then a final application that addresses multiple uses of the waterway.

When the application is filed, FERC issues a notice inviting agencies, tribes, NGOs, or the general public to request additional studies. The applicant may respond to the study requests. The application is then reviewed by FERC for compliance with its regulations, and FERC may either find it deficient and dismiss it, request correction of deficiencies, or accept the application with or without a request for additional information.

After any deficiencies are corrected, FERC issues a public notice soliciting any protests or motions to intervene in the proceeding. Intervenors receive pertinent documents, can participate in project meetings, and can appeal FERC's order. After all additional information is filed and deemed adequate, FERC issues another public notice stating that it is ready to prepare an environmental analysis and requesting comments, recommendations, terms and conditions, or prescriptions, as appropriate, from agencies, affected tribes, NGOs, and the general public.

The FERC staff holds public scoping meetings to help define the issues and reasonable alternatives to the proposed project, which will be analyzed in the NEPA document. It then prepares a draft environmental assessment or environmental impact statement in which the applicant's proposal, agencies' recommendations, and other reasonable alternatives are evaluated. The determination of whether to prepare an environmental impact statement or an environmental assessment depends on a number of factors, including the potential for the project to cause significant adverse impacts, the level of complexity, the potential for cumulative effects, the level of controversy, and the number of pending actions covered in the NEPA document. The NEPA document may serve as a multipurpose document particularly if the project affects interests of other federal entities such as the U.S. Bureau of Land Management, the U.S. Army Corps of Engineers, the U.S. Forest Service, or the military.

If FERC determines that fish and wildlife recommendations of the resource agencies are inconsistent with the purposes of the FPA or other applicable law, it attempts to resolve the inconsistencies through a negotiation process. If unresolved, FERC must explain why its recommendations represent the best comprehensive use of the waterway and public interest, and must also show how they address the identified impact.

The Council on Environmental Quality's (1978) Regulations for implementation of NEPA allow agencies to cooperate in the preparation of environmental NEPA documents and to use such documents to satisfy each of the agency's NEPA requirements. The decision to allow another agency to be a "cooperating agency" is at the discretion of the lead agency, which, in the case of nonfederal hydropower licensing actions, is FERC.

The final action in the licensing process is issuance of a license order. The order may deny a license or include license conditions for project construction and operation, as well as environmental protection, mitigation, and enhancement measures identified and analyzed in the NEPA document. Guidance on following the complex licensing and relicensing regulations are available from FERC (e.g., FERC 1990, 1991b).

After a license is issued the project is monitored by FERC to ensure compliance with project license conditions. Investigations are conducted in response to complaints received throughout the term of a 30–50 year license. Enforcement action, at the discretion of FERC, may include halting project construction or operation, license revocation, and civil penalties up to US$10,000 per day. Additional mitigation for damages caused by noncompliance with a license may also be required by FERC.

Environmental Concerns for Hydropower Projects

Aesthetic resources

Reservoir fluctuations can cause low water and bank erosion that is unsightly, which is of particular concern if there is recreational use of the impoundment. Diversion of water into a hydropower facility may alter the sight and sound of water flowing over falls and in the river channel, and have associated adverse effects on recreational use in the vicinity. Blending the appearance of project structures into the surroundings may also be a consideration in licensing.

Recreation

Based on existing and future supply and demand, FERC must determine the types and number of recreational facilities to be required in a license. Parks, picnic areas, scenic trails, boat docks, fishing piers, and associated parking facilities may be required at hydropower projects. In areas that have poor water quality or public safety concerns, the licensee may be required to restrict certain recreation uses. Recreation access and facilities for disabled persons must be considered pursuant to FERC's regulations and the Americans with Disabilities Act (Hayes 1992; 42 U.S.C. § 12101). Most large projects provide educational programs about power and the environment, including fish viewing facilities if the project has fish passage provisions.

Whitewater boating is common in riverine sections downstream from peaking hydropower projects where flows during generation can exceed natural flows in the river. Often, NGOs representing whitewater recreationists request releases of water from the reservoir to augment downstream flows and create recreational opportunities that would not otherwise exist. In contrast, owners of shoreline property and reservoir recreationists typically request high and stable reservoir water levels for consistent access to boat launches, docks, and other access points for fishing, canoeing, row boating, motorboating, and waterskiing. These issues have been raised in the relicensing proceedings for the North Georgia Project. The water management needs for these and other competing uses must be balanced by FERC.

Water quality

Hydropower projects may alter dissolved oxygen (DO) concentrations in the water by diverting water from falling over a dam, waterfalls, or rapids, where the water would be oxygenated. The diverted water flows through turbines, which prevents it from mixing with air. Outflow water has about the same DO concentration and temperature upon leaving the turbines as at the intake location in the reservoir. Water quality may be degraded by the cumulative effect of reduced aeration at hydropower projects in combination with other factors such as domestic and industrial waste discharges, which can lower the assimilative capacity. As a condition of the license, FERC often requires the maintenance of state standards for DO and temperature at the site.

If reservoirs thermally stratify, the depth of the intake may have a profound effect on downstream water quality. Many reservoirs, particularly deeper ones, naturally stratify by temperature and DO concentration during the summer. Surface photosynthetic activity and wave action lead to oxygenation of surface waters. In contrast, decomposition of biotic material on the cooler lake bottom depletes oxygen. Thus, during the summer, the surface water is warmer and more oxygenated than the deeper water. Because the density of water is greatest at about 39°F, the opposite thermal stratification may occur in the winter.

Withdrawal of water from a specific depth, and the loss of aeration at the hydropower facility, alters the DO level and temperature of water in the project tailrace from natural levels. These effects are evaluated by FERC with respect to the temperature and DO concentration requirements of fishes (e.g., Brungs and Jones 1977; Chapman 1986) and potential impacts on other resources. Selective withdrawal and aeration techniques may be considered to maintain water quality (Pastorok et al. 1982; Waide 1986; Bohac and Ruane 1990). Artificial destratification and localized mixing may also be considered to enhance hypolimnion DO (Waide 1986).

Damming a river can change sediment load, channel morphology, and water quality characteristics, particularly if there is a series of impoundments. Reservoirs have accelerated eutrophication rates, and different nutrient cycling characteristics when compared to free-flowing rivers. Further, the increased surface area contributes to a cumulative warming trend downstream, which may be less suitable for some or all coldwater fishes (McMichael and Kaya 1991). In contrast, releases of warmer water in the winter may decrease the chance of frazil ice formation downstream from a project (FERC 1992b).

Water quality and habitat changes associated with the construction and operation of hydropower can alter the invertebrate and fish communities (Walburg et al. 1983). Fish and invertebrates may suffer from gas bubble disease when exposed to water supersaturated with gas. At hydropower projects, the cause is normally from spilling water, although air vented into turbines may also cause gas bubble disease (Weitkamp and Katz 1980). Spillway deflectors may be effective in reducing the level of gas supersaturation at hydropower projects (Weitkamp and Katz 1980).

Contaminants may be a concern at hydropower facilities (Waide 1986). Mercury, polychlorinated biphenyls (PCBs), and organochlorine pesticides are the most common contaminant concerns at hydropower projects. Hydropower facilities generally do not contribute contaminants to waterways. However, paper mills or other factories (that historically may have used the waterway for transportation and energy), agricultural runoff, airborne fallout, and other sources may cumulatively affect contaminant loading in waterways.

Contaminants accumulate in the sediments, and may resuspend if disturbed by construction activities or water level fluctuations from peaking operations or project maintenance. Fish can take up the contaminants directly from the water or from their food (Ray and Treiff 1980). Because of biomagnification and bioconcentration, piscivorous sport fishes and some benthic feeding fishes may contain high levels of contaminants and be unfit for human consumption. The presence of contamination at a site may affect FERC's decisions regarding recreational improvements or fish passage.

Fisheries

Dams, altered flows, and water quality changes associated with hydropower projects can disrupt or halt fish passage. Altered flows can delay fish passage by attracting fish to the project tailrace, rather than to a fishway. Inadequate flow releases from a dam can also alter fish passage if a bypassed reach does not provide passage between the tailrace and the opening of the fish passage facility. Upstream fishways, such as pool and weir, Denil, vertical slot, Alaska steep pass and, fish lifts or elevators, are commonly used to facilitate upstream fish movement (Bell 1986). Downstream fish passage is commonly accomplished by a guidance device leading fish to an opening (e.g., bar racks angled towards a sluiceway), traveling screens, and the high velocity Eicher screen developed recently (EPRI 1992c). Behavioral guidance devices such as strobe lights or sound deterrents show promise in guiding fish into passage facilities and away from project works (e.g., EPRI 1992b). When evaluating whether or not to provide fish passage at a project, FERC must consider the costs and benefits of such measures. The benefits may be questionable if, for example, passage would provide access beyond the native range and promote competition with native species (Li 1975).

Fish mortality from impingement on trash racks (used to protect turbine runners from mechanical damage) and turbine entrainment is an issue at many hydropower facilities (EPRI 1992a; FERC 1995). Turbine type and site-specific characteristics can be used to determine the mortality from turbine entrainment (Boreman 1977; Rochester et al. 1984). For anadromous fishes, sophisticated fish bypass systems (e.g., traveling screens) are often employed to move the fish downstream efficiently, as well as to minimize entrainment. For resident fishes exhibiting substantial entrainment losses, entrainment may be reduced by using trash racks with small spacing or other avoidance systems, sometimes associated with fish bypass facilities (Stone and Webster Engineering Corporation 1986; FERC 1995). However, care must be taken in designing trash rack or screening systems because, if the velocity in front of the trash racks is higher than the fish's swimming speed, fish can be impinged on the racks or screens.

Reservoirs provide benefits other than hydropower generation, including flood control, water supply, and recreational opportunities. Tailwaters tend to provide good fishing opportunities (Walburg et al. 1983). However, projects that operate in a peaking mode generally have more potential adverse impacts than run-of-river projects. Low flows during peaking operations can strand fish downstream of the tailrace and high flows can displace fish downstream. High and fluctuating flows scour the substrate and resident benthic invertebrates, thus reducing available food for fish. Such flow regimes also reduce instream cover, exacerbating the effects of high flows on fish, as well as displacing or stranding fish eggs and older life stages (Rochester et al. 1984). High flows may alter normal home range movements, particularly if there are no velocity refuges (Hill and Grossman 1987). Instream structures may be used to create velocity barriers and provide fish habitat (Walburg et al. 1983). Flow rates affect the velocity, depth, substrate composition, water quality, and other instream habitat variables. Changes in water velocity, as well as other variables, may alter the energetic benefits of stream habitat for some water column fishes (Hill and Grossman 1993). Instream flow incremental methodology and other flow or standard setting techniques may be used to set minimum flows in a bypassed reach, and minimum and maximum flows in the tailrace.

The fluctuating flows may also cause bank erosion, which can contribute to sedimentation, which adversely affects fish and molluscs, including some threatened and endangered species. Salmonids are especially sensitive to sedimentation because the small particles may fill interstitial spaces in the gravel and suffocate eggs. Bank scouring may also limit the establishment of riparian vegetation, thereby reducing stream shading, and inputs of forage (e.g., from terrestrial insects) and large woody debris.

Water level fluctuation of reservoirs at peaking hydropower projects causes some destruction of the very productive shallow zone along the shoreline and may strand fish, particularly eggs (Rochester et al. 1984), or cause a guarding adult to abandon its nest. Nest abandonment can be a problem with black bass *Micropterus* spp., which spawn in shallow areas of the reservoirs. Once a guarding male abandons the nest, the eggs are vulnerable to predation (Neves 1975). Rapidly fluctuating water levels in the impoundment may increase the likelihood of fish stranding or nest abandonment, and FERC considers reservoir elevation fluctuation limits and maximum ramping rates (associated with the rate of increasing or decreasing generation) for peaking facilities to alleviate adverse effects upstream and downstream from the project. However, if power generation is needed during critical periods of energy use, restrictive ramping rates and generation may decrease or eliminate the project's ability to meet peak or load energy demands. Operation restrictions can cause the loss of significant economic benefits associated with a project's ability to provide dependable capacity when needed in the power grid.

Hydropower operators may draw down the reservoir in the winter to provide downstream flood control benefits, greater hydroelectric generation in the winter when it may be most needed, and greater utilization of spring runoff for generation. Without the drawdown, much of the spring runoff flows may be spilled. The concentration of fishes during the drawdown may lead to overpredation, thus destabilizing the predator–prey relationships in some systems. However, if forage fish tend to be overpopulated, the drawdown may have a beneficial effect by allowing for increased predation rates and enhancing the sport fishery (e.g., Reil 1965; Lewis 1967). Depending on the timing of the drawdown and refill, drawdowns may adversely affect fish spawning. Typically, FERC requires environmental studies to address such impacts.

Vegetation and Wildlife

Clearing for project development and transmission line placement requires defoliation and disruption of wildlife habitat. Injury from abrasion or electrocution may result when raptors fly into the transmission lines (Avery 1978). When there is potential for adverse impact, licensees are required to adhere to the Raptor Research Foundation's suggested practices for raptor protection (Olendorff et al. 1981). The Commission must also consider potential adverse impacts on vegetation and wildlife habitat that may occur during construction of the project or recreation facilities.

Reservoir fluctuations expose and may freeze aquatic vegetation and benthos (Paterson and Fernando 1969). Exposure can reduce wetlands and associated waterfowl breeding and feeding habitat. Winter drawdowns may freeze burrowing amphibians, reduce habitat and food availability for beaver and muskrat, and may disrupt waterfowl nesting. There may be circumstances, however, for which drawdowns or other measures are desirable for the control of certain aquatic vegetation (Pieterse and

Murphy 1990). Fluctuating flows downstream also can cause loss of wetlands and soil destabilization. Changes in the sediment loading, water quality, and flow regime downstream from the dam may affect riverine vegetation.

Biodiversity concerns related to vegetation and wildlife, as well as fisheries, are addressed by FERC in accordance with Council on Environmental Quality's guidance on this issue. Federally listed species and associated critical habitat are a priority for protection and enhancement, as are other targeted biodiversity concerns outlined in resource agencies' comprehensive plans. Large contiguous tracts of land in a watershed are sometimes associated with hydropower facilities; these areas have been protected from development and may offer important habitat (Russo 1994). Loss of a project license may have ramifications for the biota on these lands if these holdings were sold to developers.

Soils and Geology

Typically, FERC requires a description of the soils and geological resources in the project area, which may include an assessment of erosion sites. Particular attention must be given to soils and geology when considering licenses for new construction. In addition, peaking operations may accelerate erosion in the reservoir and downstream from the project. Altered operations, including a reduction in water surface elevation fluctuations and a reduction in flow fluctuations downstream, and structural or vegetative stabilization of the banks may be used to ameliorate adverse impacts (FERC 1992a).

Historical and Archeological

All applicants are required by FERC to consult with their respective state historic preservation officer concerning effects on properties (e.g., archeological sites, historic structures and buildings) which are listed on or are eligible for listing on the National Register of Historic Places. The National Park Service must also be consulted regarding historical and archeological resources.

Future Directions in Hydropower Licensing

Currently, FERC regulates over 1,600 licensed or exempted hydropower projects having a combined capacity of about 52 GW of electric power. This represents about 54% of the total energy generated by hydropower in the United States.

Most license applications currently before FERC are for relicensing existing projects. In 1993, 157 licenses of generally small capacity projects (totalling 1.94 GW of energy) expired, prompting a flood of relicense applications in 1991. Thirty-eight licenses (totalling 1.14 GW of energy) are scheduled to expire between 1994 and 1999. Between the years 2000 and 2010, 221 project licenses will expire. These projects have a combined capacity of 23.12 GW (nearly half of all hydropower capacity under FERC's jurisdiction).

Cooperative approaches are encouraged by FERC in the licensing process. The Energy Policy Act of 1992, which amended the FPA, allows a third party or an applicant to prepare the environmental document. In the third party process, FERC staff establish the scope of work and procedures and supervise the contractor's drafting of the NEPA document. When an applicant prepares a draft environmental assessment to file with their license application, FERC staff act in an advisory role only. For instance, Georgia Power Company developed an applicant-prepared environmental assessment for relicensing the Sinclair hydropower project. Other companies have begun using this approach, as well as third party contracting, in their relicensing efforts.

The development of settlement agreements among applicants, agencies, and interested parties is promoted by FERC (Emery 1995). Such agreements are likely to be substantively adopted by FERC if supported by evidence that they are consistent with the best comprehensive use of the waterway and in the public interest. This approach was used by Consumers Power Company for the relicensing of 11 projects located on three Michigan rivers.

The use of settlement agreements, applicant-prepared NEPA documents, and third-party contracting provides early agreement on the need for and design of studies, and potential environmental alternatives (Russo and Narins 1993). Early application of the NEPA process expedites licensing. Also, close coordination and open communication among applicants, agencies, tribes, NGOs, and the general public, in a cooperative rather than adversarial role, is likely to encourage better decision making on alternatives and mitigation and enhancement measures. In contrast, an approach that encourages or produces adversarial relationships may cost ratepayers significant funds for litigation, with no environmental benefit.

Challenges

All decisions regarding study, mitigation, and enhancement needs of a project require a certain amount of calculated risk. To spend extraordinary amounts of money with the intent of getting absolute precision for impact definition is unnecessary. If study, mitigation, or enhancement funds are spent in excess of that which is dictated by the potential impact to the resources, then the public ultimately suffers through increased electrical rates. Further, we are entering a time when electrical markets are becoming increasingly competitive (FERC Order Number 888, *Federal Register* 61 [10 May 1996] p. 21,540). A hydropower project may be able to compete only if it can produce energy at a lower cost than alternate energy sources. Therefore, if resource agencies insist on "cadillac" versions of studies and mitigation and enhancement, they could ultimately force the loss of hydropower generation in exchange for alternate energy sources, which are typically oil, coal, or natural gas (Russo and Narins 1993).

The excavation of fossil fuels may have adverse effects on biota. Mining can contribute to acid mine drainage, human health impacts, vegetation and associated wildlife losses, erosion, and stream sedimentation. Drilling causes habitat disturbances and vulnerability to oil spills. The transportation of fossil fuels increases automotive emissions, habitat loss from construction of pipelines, and vulnerability to oil spills.

The production of energy from fossil fuels also has a set of adverse impacts. Coal-fired generation plants or other plants requiring cooling water may cause impingement and entrainment of fish, and thermal pollution to the waterway (Hansen et al. 1977). The burning of fossil fuels releases airborne emissions of sulfur dioxide and nitrogen oxides (and other chemicals), which return as acidic deposition (Haines 1981). Areas with granitic bedrock and soils with poor buffering capacity show the greatest adverse impacts to biota from acidic deposition (Altshuller and Linthurst 1984). Fish and other aquatic biota may be adversely affected by changes in ambient acidity (Haines 1981). Some sport fishes, such as smallmouth bass *M. dolomieu*, are particularly vulnerable to adverse effects from acid exposure (e.g., Hill et al. 1988; Hill 1989). Also, there may be synergistic effects from acid deposition, including increased leaching and toxicity of metals such as aluminum (Haines 1981). Burning of fossil fuels releases carbon dioxide, which may contribute to global warming and ozone depletion (Fisher 1989). Combustion of fossil fuels also releases trace elements of mercury, which is toxic to fish and their consumers (Haines 1981).

Other common energy resources also have adverse impacts. Nuclear wastes and use of cooling water are of concern at nuclear plants. Wind power turbines may kill large birds of prey that use the wind for lift; they may also create interference with electromagnetic transmission and have adverse aesthetic impacts (Pimental et al. 1994). Biomass energy systems require large tracts of land, cause air pollution, and may threaten biodiversity (Pimental et al. 1994). Solar energy is not cost competitive for most developed areas. Current technology has some adverse impacts associated with the use of toxic chemicals in the manufacture of the photovoltaic cells (Pimental et al. 1994).

Hydropower can offer a relatively inexpensive source of dependable capacity, which reduces the possibilities of blackouts or other short-term energy shortages. Hydropower generation produces no atmospheric pollution, and is consistent with national policy to reduce global warming and increase the use of renewable resources. By comparison to fossil fuels, hydropower energy typically can have relatively low environmental impacts and contributes to a base of sustainable and renewable energy in the United States (Russo 1994). Decisions regarding hydropower must be weighed in light of the alternative energy impacts.

Acknowledgements

Helpful input and reviews were provided by Gerhard Hill, Gladys Hill, David Boergers, John Clements, Eddie Crouse, Kelly Fargo, Mary Golato, James Griffin, Patti Leppert–Slack, Christopher Metcalf, John Novak, Mark Pawlowski, Thomas Russo, and Richard Takacs. The views expressed in this paper are those of the author and do not represent the official views and policies of the Federal Energy Regulatory Commission.

References

Altshuller, A. P., and R. A. Linthurst, editors. 1984. The acidic deposition phenomenon and its effects, volume II. Effects sciences. U.S. Environmental Protection Agency (EPA-600/8-83-016), Washington, DC.

Avery, M. L., editor. 1978. Impacts of transmission lines on birds in flight: proceedings of a workshop. U.S. Fish and Wildlife Service (FWS/OBS-78/48), Washington, DC.

Bell, M. C. 1986. Fisheries handbook of engineering requirements and biological criteria. Fish passage development and evaluation program. U.S. Army Corps of Engineers, Portland, Oregon.

Bohac, C. E., and R. J. Ruane. 1990. Solving the dissolved oxygen problem. Hydro Review 9:62–72.

Boreman, J. 1977. Topics briefs: fish and wildlife resources and electric generation, no. 1.—impacts of power plant intake velocities on fish. U.S. Fish and Wildlife Service (FWS/OBS-76/20.1), Office of Biological Services, Washington, DC.

Brungs, W. A., and B. R. Jones. 1977. Temperature criteria for freshwater fish: protocol and procedures. U.S. Environmental Protection Agency (EPA-600/3-77-061), Environmental Research Laboratory, Office of Research and Development, Duluth, Minnesota.

Chapman, G. 1986. Ambient water quality criteria for dissolved oxygen. U.S. Environmental Protection Agency (EPA-440/5-86-003), Environmental Research Laboratories, Office of Research and Development, Duluth, Minnesota.

Council on Environmental Quality. 1978. Regulations for implementing the procedural provisions of the National Environmental Policy Act. Code of Federal Regulations, Title 40, Parts 1500–1508. U.S. Government Printing Office, Washington, DC.

Emery, L. 1995. Do settlement agreements act as a catalyst for receiving a license? Pages 633–642 *in* J. J. Cassidy, editor. Proceedings of the international conference on hydropower. American Society of Civil Engineers, New York.

FERC (Federal Energy Regulatory Commission). 1990. Hydroelectric project relicensing handbook. Federal Energy Regulatory Commission, Office of Hydropower Licensing, Washington, DC.

FERC (Federal Energy Regulatory Commission). 1991a. Evaluating relicense proposals at the Federal Energy Regulatory Commission. Federal Energy Regulatory Commission, Office of Hydropower Licensing (Paper DPR-2), Washington, DC.

FERC (Federal Energy Regulatory Commission). 1991b. Hydroelectric project licensing handbook. Federal Energy Regulatory Commission, Office of Hydropower Licensing, Washington, DC.

FERC (Federal Energy Regulatory Commission). 1992a. Evaluating geologic and soils issues at the Federal Energy Regulatory Commission. Federal Energy Regulatory Commission, Office of Hydropower Licensing (Paper DPR-3), Washington, DC.

FERC (Federal Energy Regulatory Commission). 1992b. Report on stream ice processes. Federal Energy Regulatory Commission, Office of Hydropower Licensing (Paper DPR-5), Washington, DC.

FERC (Federal Energy Regulatory Commission). 1993. Hydropower licensing and endangered species: procedures for complying with the Endangered Species Act. Federal Energy Regulatory Commission, Office of Hydropower Licensing (Paper DPR-7), Washington, DC.

FERC (Federal Energy Regulatory Commission). 1995. Preliminary assessment of fish entrainment at hydropower projects—a report on studies and protective measures. Federal Energy Regulatory Commission, Office of Hydropower Licensing (Paper DPR-10), Washington, DC.

Fisher, A. 1989. Global warming, part one. Playing dice with earth's climate. Popular Science 235(2):51–58.

Haines, T. A. 1981. Acidic precipitation and its consequences for aquatic ecosystems: a review. Transactions of the American Fisheries Society 110:669–707.

Hansen, C. H., J. R. White, and H. W. Li. 1977. Entrapment and impingement of fishes by power plant cooling water intakes: an overview. U.S. National Marine Fisheries Service Marine Fisheries Review 39(10):7–17.

Hayes, D. 1992. Becoming familiar with the Americans with Disabilities Act. Hydro Review 11:107–108.

Hill, J. 1989. Analysis of six foraging behaviors as toxicity indicators, using juvenile smallmouth bass exposed to low environmental pH. Archives of Environmental Contamination and Toxicology 18:895–899.

Hill, J., R. E. Foley, V. S. Blazer, R. G. Werner, and J. E. Gannon. 1988. Effects of acidic water on young-of-the-year smallmouth bass (*Micropterus dolomieui*). Environmental Biology of Fishes 21:223–229.

Hill, J., and G. D. Grossman. 1987. Home range estimates for three North American stream fishes. Copeia (1987):376–380.

Hill, J., and G. D. Grossman. 1993. An energetic model of microhabitat use for rainbow trout and rosyside dace. Ecology 74:685–698.

Lewis, W. M. 1967. Predation as a factor in fish populations. Pages 386–390 *in* Reservoir fisheries resources symposium. American Fisheries Society, Southern Division, Reservoir Committee, Bethesda, Maryland.

Li, H. 1975. Competition and coexistence in stream fish. Pages 19–30 *in* P. B. Moyle and D. L. Koch, editors. Symposium on trout/non-game fish relationships in streams. Center of Water Resources Research, Desert Research Institute, University of Nevada System, Miscellaneous Report 17, Reno.

McMichael, G. A., and C. M. Kaya. 1991. Relations among stream temperature, angling success for rainbow trout, and fisherman satisfaction. North American Journal of Fisheries Management 11:190–199.

Neves, R. J. 1975. Factors affecting fry production of smallmouth bass (*Micropterus dolomieui*) in South Branch Lake, Maine. Transactions of the American Fisheries Society 104:83–87.

Olendorff, R. R., A. D. Miller, and R. N. Lehman. 1981. Suggested practices for raptor protection on power lines: the state of the art in 1981. Raptor Research Report 4, Raptor Research Foundation, St. Paul, Minnesota.

Pastorok, R. A., M. W. Lorenzen, and T. C. Ginn. 1982. Environmental and water quality operational studies. Environmental aspects of artificial aeration and oxygenation of reservoirs: a review of theory, techniques, and experiences. U.S. Army Corps of Engineers Waterways Experiment Station (Technical Report E-82-3), Vicksburg, Mississippi.

Paterson, C. G., and C. H. Fernando. 1969. The macroinvertebrate colonization of a small reservoir in eastern Canada. Internationale Vereinigung fur Theoretische und Angewandte Limnologie, Verhandlungen 17:126–136.

Pieterse, A., and K. Murphy, editors. 1990. Aquatic

weeds. The ecology and management of nuisance aquatic vegetation. Oxford University Press, New York.

Pimental, D., and eleven coauthors. 1994. Renewable energy: economic and environmental issues. BioScience 44:536–547.

Ray, S. M., and N. M. Treiff. 1980. Bioaccumulation of anthropogenic toxins in the ecosystem. Pages 93–120 *in* N. M. Trieff, editor. Environment and health. Ann Arbor Science Publishers, Inc., Ann Arbor, Michigan.

Reil, A. D. 1965. The control of an overpopulation of yellow perch in Bow Lake, Strafford, New Hampshire. Progressive Fish-Culturist 27:37–41.

Rochester, H., Jr., T. Lloyd, and M. Farr. 1984. Physical impacts of small-scale hydroelectric facilities and their effects on fish and wildlife. U.S. Fish and Wildlife Service (FWS/OBS-84-19) Office of Biological Services, Washington, DC.

Russo, T. N. 1994. Making hydropower sustainable. Hydropower and Dams 1(6):126–131.

Russo, T. N., and M. J. Narins. 1993. Effects of U.S. national energy legislation on future U.S. and international hydro. International Water Power and Dam Construction. August 1993:28–31

Stone and Webster Engineering Corporation. 1986. Assessment of downstream migrant fish protection technologies for hydroelectric application. Electric Power Research Institute, Palo Alto, California.

Stone & Webster Environmental Services. 1992a. Fish entrainment and turbine mortality review and guidelines. Final report to Electric Power Research Institute TR-101231. EPRI, Palo Alto, California.

Stone & Webster Environmental Services. 1992b. Evaluation of strobe lights for fish diversion at the York Haven Hydroelectric Project. Final report to Electric Power Research Institute TR-101703. EPRI, Palo Alto, California.

Stone & Webster Environmental Services. 1992c. Evaluation of the Eicher Screen at Elwha Dam: 1990 and 1991 test results. Final report to Electric Power Research Institute TR-101704. EPRI, Palo Alto, California.

Waide, J. B. 1986. General guidelines for monitoring contaminants in reservoirs. U.S. Army Corps of Engineers, Waterways Experiment Station (Instruction Report E-86–1), Vicksburg, Mississippi.

Walburg, C. H., J. F. Novotny, K. E. Jacobs, and W. D. Swink. 1983. Effects of reservoir releases on water quality, macroinvertebrates, and fish in tailwaters: field study results. U.S. Army Corps of Engineers, Waterways Experiment Station (Technical Report E-83-6), Vicksburg, Mississippi.

Weitkamp. D. E., and M. Katz. 1980. A review of dissolved gas supersaturation literature. Transactions of the American Fisheries Society 109:659–702.

American Fisheries Society Symposium 16:200–203, 1996

A Utility's Perspective on Life After the Electric Consumers Protection Act

DUANE HARRELL
Duke Power Company, Environmental Center
13339 Hagers Ferry Road, Huntersville, North Carolina 28078, USA

Abstract.—The relicensing of hydropower projects throughout the United States changed dramatically with the enactment of the Electric Consumers Protection Act (ECPA) of 1986. The resulting three-stage consultation process both encouraged and mandated agency and licensee interaction. As an owner and operator of 25 conventional and two pumped-storage hydropower projects in the Piedmont North and South Carolina, Duke Power Company has become extensively involved in hydropower relicensing. Duke Power Company's initial relicensing efforts began in 1980, and relicensing studies have been conducted for a total of six hydropower projects. Studies designed and conducted in coordination with state and federal resource agencies have included evaluations of reservoir and tailwater water quality and quantity, reservoir and tailwater fish abundance and diversity, fish entrainment and mortality, fisheries habitat in bypassed stream reaches, navigation, and recreation. The post-ECPA process has been a learning experience for Duke Power Company and the resource agencies. An essential key to success is a relicensing strategy that begins early, establishes open communication and good working relationships with the resource agencies, employs genuine negotiation and conflict resolution, and strives to achieve settlement agreements on major issues before involvement by the U.S. Federal Energy Regulatory Commission.

Duke Power Company (DPC) is a private, investor-owned utility located in the Piedmont of North and South Carolina. Duke Power Company's generating system consists of 8 coal-fired, 3 nuclear, and 27 hydroelectric stations, with a combined total generating capacity of approximately 18,000 MW. The hydroelectric system consists of 25 conventional and 2 pumped-storage projects ranging in capacity from 610 KW to 1,065 MW. The total capacity of DPC's hydroelectric system is 2,682 MW. Most of DPC's hydropower projects were built in the early to mid-1900s and licensed for a period of 50 years. The company is now entering an extended period of hydropower relicensing which involves relicensing one project for which the license expired in 1985 and five projects for which licenses expired in 1993. Plans are being developed for an additional watershed project with 13 hydropower developments, the license for which expires in 2008.

Duke Power Company's hydropower relicensing efforts began in 1980 with environmental studies for the 15-MW Buzzards Roost Hydroelectric Project, South Carolina, scheduled for relicensing in February 1985. Since 1980, relicensing studies were conducted for five additional hydropower projects scheduled for relicensing in December 1993. Typical studies designed and conducted in coordination with various state and federal resource agencies included evaluations of reservoir and tailwater water quality and quantity, fish entrainment and mortality, fisheries habitat in bypassed stream reaches, navigation, and recreation.

Background

To properly understand the current relicensing process facing licensees and resource agencies, it is imperative to understand the environmental and regulatory climate during the 1970s and early 1980s, prior to passage of the Electric Consumers Protection Act (ECPA; 16 U.S.C. §§ 791a–825h) of 1986. Although legislation for the protection of environmental resources during relicensing did exist in the form of legislation such as the Federal Power Act (FPA; 16 U.S.C. §§ 791a–825r), the Fish and Wildlife Coordination Act (FWCA; 16 U.S.C. §§ 661–667e), and the National Environmental Policy Act (NEPA; 42 U.S.C. §§ 4321–4361), clear definitions of the specific roles of the various players in the process were lacking.

During the energy crisis of the mid 1970s, a renewed emphasis was placed on the development of alternative and renewable energy sources. In 1978, the U.S. Congress enacted the Public Utilities Regulatory Policy Act (PURPA) as an incentive for the development of alternative energy. Through economic incentives and a guaranteed market, small hydropower development became an attractive alternative. Although there was an emphasis on the importance of project-related environmental impact assessment through the FWCA and NEPA processes, the overall national concern was focused on the need for energy. This fact and the resulting resurgence in hydropower development heightened the existing concerns by state and federal fish and wildlife agencies over the adequacy of the environmental review process and the resulting consideration of nonpower values of waterways. During the early 1980s, increasing pressure on Congress by state and federal resource agencies and environmental lobbyists contributed to the passage of the ECPA.

The process of relicensing hydropower projects throughout the United States changed dramatically with the enactment of the ECPA. The ECPA has probably had more impact on hydropower licensing and relicensing and the hydropower industry, in general, than any other hydropower-related legislation. The ECPA amendments to the relicensing provisions of the FPA necessitated a revision of the Federal Energy Regulatory Commission (FERC) regulations governing the hydropower relicensing process. Specifically, section 3 of the ECPA (Environmental Consideration in Licensing), amended the FPA to require FERC to give equal consideration to power and developmental purposes and preserving environmental quality, including the "protection, mitigation of damage to, and enhancement of, fish and wildlife (including related spawning grounds and habitat), and the protection of recreational opportunities. . . ." Additionally, the ECPA added section 10(j) to the FPA to require that each license issued include conditions based on recommendations received, pursuant to the FWCA, from the National Marine Fisheries Service, the U.S. Fish and Wildlife Service, and state fish and wildlife agencies, unless adoption of such recommendations is inconsistent with applicable law.

One major implication of the ECPA was the need for more extensive and better consultation between resource agencies and applicants. This was a need that had already been identified by FERC, based on deficiencies noted in the avalanche of hydropower license applications received by FERC in the early 1980s following enactment of PURPA. Prompted by these deficiencies and the need to revise FERC regulations to incorporate amendments to the relicensing provisions of the FPA made by the ECPA, FERC issued a notice of proposed rulemaking in May 1988. The final rule was issued on 17 May 1989, and the final relicensing rules became effective 3 July 1989. The new relicensing rules outlined a prefiling, three-stage consultation process between the licensee and the resource agencies. The rules were designed to facilitate the relicensing process by encouraging the resolution of issues and the identification of necessary scientific studies and mitigation options prior to the filing of the relicense application.

The three-stage process is an extensive consultation process that both encourages and mandates agency and licensee interaction prior to the filing of the license application by the licensee. The first stage consists of the preparation of an initial consultation package by the licensee that describes the project, project operations, and the project environment. A joint meeting with the various resource agencies is held to review and discuss this package. The resource agencies prepare written comments, including recommendations for necessary studies, and these comments are included with the initial consultation package filed with FERC. The second stage of consultation consists of the completion of all necessary studies and the preparation of the draft license application. The third stage consists of filing the final license application with FERC and all consulted agencies. This filing must occur at least 24 months prior to the expiration of the current license. Under the new procedure, some of the greatest challenges will be those faced by FERC. All licensee and agency data must be reviewed by FERC for preparation of the required NEPA documents. As the final authority in this process, FERC must then balance power and nonpower interests in determining the terms and conditions to be included in the new license for a project.

Before passage of the ECPA, the agency and licensee consultation process was vague and in some cases essentially nonexistent. Historically, relationships between licensee and resource agencies often were strained and positions relative to many relicensing issues were divergent. Until the relicensing class of 1993, many licensees and resource agencies had little or no experience with the relicensing process. Most existing licenses were 50-year licenses, and at the time of their issuance, the primary societal concerns were jobs and energy availability. Over the course of the license period, policies, attitudes, and values had undergone extensive changes. Resource agencies with concerns about current project operations see relicensing as an opportunity to address the environmental concerns and project-related impacts that they have been unable to challenge under the current license. The additional legislative and regulatory authority afforded to these agencies under the ECPA further emphasize the importance of current relicensings, especially those in the class of 1993. The decisions made relative to the class of 1993 will serve as the benchmark for all future relicense proceedings.

Attitudes

What real world experiences has DPC had under the new relicensing procedure, and what have we learned through the process? First of all, the increased interaction built into the process has resulted in new opportunities and challenges for agencies and licensees. To take advantage of these opportunities and successfully deal with the challenges, a critical first step is a change in attitudes. Licensees must understand that status quo will probably not be acceptable. Resource agencies must understand that they will probably not get everything that they ask for. Both parties must understand the need to negotiate, and that FERC will attempt to equitably balance the needs of all concerns. These may be difficult concepts to embrace, given that historical relicensing proceedings essentially have been conducted in the legal arena, but their acceptance is essential for success.

Barriers

To plan a successful relicensing strategy, one must first understand some of the barriers inherent in the process. One major barrier relates to inexperience with relicensing in general, and with the three-stage consultation process, specifically. Because of the lengthy license periods for hydropower projects (up to 50 years) prior to 1993, most of the key players in the process had never been involved in relicensing proceedings. Additionally, the procedural changes necessitated by the ECPA radically changed the existing relicensing process. Consequently, agencies and license applicants are still on the ascending portion of the relicensing learning curve.

Another barrier to the process deals with posturing and role determination. Where does everyone fit into this new process, and what are the boundaries of their power and authority? As with any new process, each player is trying to establish their territory. There are underlying concerns of licensees and agencies relative to compromise or concession, due to the possible establishment of a precedent that may affect future projects. Additionally, resource agencies are under intense scrutiny by proponents of environmental protection.

Finally, there are the divergent accountability issues that govern the operations of licensees and resource agencies. Resource agencies are the custodians of public resources and, as such, focus primarily on environmental issues and user interests. They answer directly to the public for their performance in protecting public interests. On the other hand, licensees must balance multiple issues of environmental stewardship, power needs, and financial accountability to customers and investors. Because of their differing responsibilities and perspectives, it can be difficult for licensees and resource agencies to relate to each other's sometimes mutually exclusive needs.

A Successful Strategy

As mentioned earlier, an understanding of the barriers to the relicensing process is essential in designing and executing a successful strategy. Although awareness of the barriers will not solve the problems they create, a basic understanding of and respect for the different perspectives and objectives of licensees and resource agencies is required by all participants if the negotiation process is to be successful.

Obviously a good relicensing strategy consists of much more than attitudinal changes and an understanding of the barriers involved. What are the key elements of a good relicensing strategy? Foremost, a licensee can never start too early. Under FERC regulations, a current licensee must file a Notice of Intent to relicense a project no sooner than 5.5 years or later than 5 years before expiration of the current license. Initiation of consultation with resource agencies, however, can begin anytime, providing that the three-stage consultation requirements are met in time to file the license application at least 24 months prior to the expiration of the current license. There are no procedures or rules that instruct licensees when the process must be begun. Because the relicensing process is highly susceptible to pitfalls, a licensee must ensure that the schedule is sufficient to accommodate unforeseen delays. The more complex the project, the longer the process will take. A timeline should be established that includes both regulatory and self-imposed milestones and deadlines to ensure that the process remains on schedule.

Another key element of a successful relicensing strategy is a proactive approach, which indicates the licensee's interest in agency concerns and the resolution of those concerns. This cooperative initiative can set the tone for the entire relicensing process and is critical to successful negotiation. An important proactive step is licensee-initiated, informal meetings with the resource agencies to identify all potential participants and areas of concern. Early identification allows the licensee to determine the best approach to collect and analyze additional baseline data to address relicensing issues. Ideally, these early steps involving supplemental data collection are completed prior to initiation of the formal, first-stage consultation process.

A good relicensing plan must also be built around good communication. No matter how early the process starts or how proactive the licensee is, if good, ongoing communications with the resource agencies and FERC have not been established, the relicensing process is sure to be an arduous and frustrating one. Although the three-stage relicensing process is predicated upon communication among all parties, it is often easy for licensees and agencies alike to assume that everyone is on the same plane of understanding. With the multitude of issues and opinions possible in the relicensing procedure, universal understanding seldom occurs. Accordingly, a good rule of communication is to assume nothing. Licensees should make sure that what they think is understood intuitively, is stated and clarified verbally, and is documented in writing.

By nature, the relicensing process is conducted with a certain level of bias on the part of both licensees and resource agencies. Although there is a shared concern for the environment, each party has differing responsibilities and is accountable to different constituents. This leads to the fourth key element of a good relicensing strategy. That element is the objective consideration of and appreciation for the other party's perspective and position. Considering the issues in an objective manner may not directly result in a resolution of conflict, but it can streamline the negotiation process. This approach also results in a good understanding by both licensee and agency personnel of the needs of the other party. An important distinction that should be made at this point is the difference between "needs" and "wants". In the relicensing arena, "wants" are seldom a realistic expectation in the context of balancing power and nonpower values. On the other hand, "needs" represent the real world, acceptable conditions that can be negotiated to achieve a successful outcome for all parties. The "needs" of both licensees and agencies must become the focus of the relicensing negotiations.

Finally, a good relicensing strategy must begin and end with a willingness to negotiate. The entire three-stage consultation process is geared toward negotiation. Obviously, an ideal relicensing scenario would be one in which the licensee and resource agencies have reached agreements on all major issues prior to the actual filing of the license application with FERC. With the multitude of issues most projects deal with, this scenario may be more the exception than the rule. Even so, the important point to remember is that each resolved issue brings the licensee one step closer to a new license, and can significantly shorten a historically long process. To this end, FERC encourages licensees to seek settlement agreements with the resource agencies to resolve many issues before they reach FERC. Because of their direct involvement in determining the outcome, many licensees are finding this approach preferable to the conventional approach where FERC decides these issues.

Duke Power Company has successfully used the settlement agreement approach for three of its five current relicensings. Legal instream flow settlement agreements have been signed with the South Carolina Department of Natural Resources and the U.S. Fish and Wildlife Service for the Ninety-Nine Islands and Gaston Shoals Projects on the Broad River, and the Saluda Project on the Saluda River, all in South Carolina. Although instream flow is not the only major relicensing issue at these projects, it was a major issue of contention at each project. Other major issues such as entrainment mortality have not yet been resolved, but are being considered for settlement agreement negotiations. Although not resolved through formal settlement agreement documents, DPC agreed to recreation enhancements at all five projects, as requested by the resource agencies.

The licensing process since enactment of the ECPA has truly been a learning experience for DPC and the associated resource agencies. In dealing with barriers, licensees must design a proactive relicensing strategy that begins early, establishes good lines of communication and working relationships with the resource agencies and FERC, involves genuine negotiation and conflict resolution, and strives to achieve settlement agreements on major issues prior to FERC involvement. The ultimate objective of the process must be a "win–win" scenario that benefits all parties involved.

Has DPC done everything right and been totally successful in relicensing? No. Just like all other licensees, DPC is still learning. The lessons and experience gained through the class of 1993 relicensings will be invaluable as DPC looks ahead. Success in the future is predicated upon learning from our mistakes and taking advantage of that wisdom. Just as the challenges of relicensing are many, so are the opportunities.

American Fisheries Society Symposium 16:204–209, 1996

Shared Management of Fish Resources on Developed Rivers: A Balance of Needs and Resources

JOHN E. NAVARRO AND DENNIS J. MCCAULEY
Great Lakes Environmental Center
739 Hastings Street, Traverse City, Michigan 49686, USA

ANDREW R. BLYSTRA
Thunder Bay Power Company
13561 West Bay Shore, Traverse City, Michigan 49684, USA

Abstract.—As part of Federal Energy Regulatory Commission requirements for relicensing of hydroelectric facilities, studies were conducted to determine the impact of four hydroelectric and two water storage facilities on the Thunder Bay River fish resources in northeast Michigan. These studies were focused on instream flows needed to protect fish habitat below water storage facilities, downstream passage of fish at hydroelectric and water storage facilities, and effectiveness of fish diversion at hydroelectric facilities. Flows necessary to maximize available habitat varied by species and life stage, but agency-proposed minimum flows provided baseline habitat during critical low flows. For fish passing through facilities, species composition was site-specific, most fish were small, diel and seasonal patterns were species specific, and overall acute turbine mortality was 8% (and varied by species, fish size, and turbine type). No acute mortality was associated with reservoir spill. Fish diversion techniques were successful, but effectiveness depended on the diversion technique used, species, and time of day. These results also demonstrated that each facility was unique in its effect on fish resources, but potential negative effects could be mitigated using an operational plan that incorporated fish management. Consequently, for hydroelectric or water storage facilities, we suggest a site-specific comprehensive water resource and operational plan that manages for energy production, and is consistent with natural resource agency management goals.

Hydroelectric and water storage facilities are an important component of developed rivers. Historically, in the Great Lakes region of the United States, many low-head impoundments were developed to facilitate logging, various mill operations, and other industrial projects. Permanent changes to the riverine ecosystem occur after impoundment when the river is converted from a lotic to a lentic system. Resident riverine species must migrate to more suitable habitat, modify their behavior in the lentic system, or perish. The hydrology and aquatic communities of the river downstream also are altered. In addition to the initial changes caused by impoundment, the daily operation of hydroelectric facilities affects the aquatic environment. Variability of water releases can have a direct impact on the downstream aquatic community. In addition, water level fluctuations in the upstream impoundment can influence fish communities, particularly as they influence spawning activity.

The operation of a hydroelectric or water storage facility can transport fish downstream through turbines or via reservoir spill. Downstream movement of fish is likely whenever water is discharged from a project. This is often viewed as a direct loss of impoundment resources, but can be a potential gain to the river resources if fish survive passage. If mortality from passage is high, or if movement of important sport fish from the impoundment is considered excessive, then a fish diversion program may be implemented to exclude fish entering turbine intakes or water control structures. Diversion may be behavioral, mechanical, or a combination of both, and the optimal design is generally site-specific.

We have previously conducted studies on four hydroelectric and two water storage facilities on the Thunder Bay River in northeast Michigan to determine their effect on fish resources. In this paper, we review results of four studies dealing with instream flow at water storage facilities (Navarro et al. 1994), turbine passage at hydroelectric facilities (Navarro et al. 1996), downstream passage of fish in reservoir spill at water storage facilities (Navarro and McCauley 1993), and fish diversion at a hydroelectric facility (McCauley et al. 1996). Collectively, these studies demonstrated that each facility has a unique effect on the fish resources of the Thunder Bay River. Specifically, instream flow, downstream fish passage, and fish diversion studies were reviewed and used to develop water resource management and operational strategies. These strategies emphasized an integrated approach, recognizing the need for sound environmental and operational planning, and involved federal, state, and local agencies, project owners, and private citizens.

Study Site

The Thunder Bay River system in northeast Michigan drains a watershed of approximately 2,200 km^2. Four hydroelectric facilities and their associated impoundments are located on the main branch of the river and two water storage reservoirs are located on two tributaries. The four hydroelectric facilities were constructed in the early 1900s, and have both vertical and horizontal shaft turbines, ranging in capacity from 220 to 2,800 KW. The dams range in height from 4 to 12 m. All four projects are owned and operated by the Thunder Bay Power Company (TBPC) and are operated in a run-of-river mode using an operating water fluctuation band of ± 0.15 m.

The two water storage reservoirs are also owned and operated by the TBPC and are located on the Upper South Branch (Upper South Dam) and the Lower South Branch (Hubbard Dam) of the river. Upper South Dam was constructed in 1930 and has a head height of 4 m, while Hubbard Dam was constructed in 1890 and has a head height of 2 m. These reservoirs are not operated in a run-of-river mode, but instead provide supplemental water to power-producing facilities downstream. The dams are similar in construction, and water flow is direct from the reservoir to the tailwater, without any turbines or screening devices. Impoundment water level management at these facilities is practiced according to legally binding agreements between the local lake associations and the TBPC.

Instream Flow

Because the two water storage impoundments are not operated as run-of-river facilities, the Michigan Department of Natural Resources (MDNR) requested that minimum flows be established using instream flow incremental methodology (IFIM). Using IFIM, we determined the effect of various flows on habitat available for four life stages (fry, juvenile, adult, and spawning adult) of smallmouth bass *Micropterus dolomieu*, northern pike *Esox lucius*, and white sucker *Catostomus commersoni*. Smallmouth bass and northern pike were selected because they are representative sport fish; white sucker was selected because of its importance as a prey species. The objective of the instream flow studies was to determine appropriate water release flows below these impoundments. The agency-proposed minimum flows were 0.85 m^3/s below Upper South Dam and 0.43 m^3/s below Hubbard Dam.

Cross sectional transects were used to represent typical habitat below Upper South and Hubbard Dams. Each transect was surveyed to determine stream width, distance between transects and water surface elevation. Water flow from the dams was manipulated to achieve low, medium, and high flows. Water velocity, water depth, and substrate type were recorded at 1-ft intervals across each transect during each flow regime. The physical habitat simulation system (PHABSIM) models, as described by Milhous et al. (1989), were used to calculate total habitat and available habitat for the four life stages of each species at the three measured flows, as well as at six other simulated water flows. The PHABSIM models use a hydraulic component (HEC2 hydrologic simulation model, U.S. Army Corps of Engineers) and a biological component (HABTAV habitat simulation model, PHABSIM) to determine available habitat. Habitat suitability criteria used in modeling available habitat were developed by the U.S. Fish and Wildlife Service and modified by the MDNR.

Flows that optimized habitat availability varied by species and life stage. At all sites, available habitat for juvenile, adult, and spawning adult smallmouth bass increased as flow increased, but decreased with increasing flow for fry. For northern pike, available habitat for fry, juvenile, and spawning adults decreased as flow increased, but increased with increasing flow for adults. While available habitat for juvenile and adult white sucker increased with flow, the situation for fry and spawning adults was more complex, decreasing for fry with flows up to 2m^3/sec, then increasing, and increasing for spawning adults with flows up to 2m^3/sec, then decreasing. Life stages with less than 5% available habitat at the agency-recommended minimum flows were adult and spawning adult smallmouth bass on both rivers, northern pike fry on the lower south branch and adult northern pike on both rivers, and spawning adult white sucker on the lower south branch. Total habitat changed little with increased flow, and even the highest flows created only a modest increase in total habitat.

Downstream Fish Passage

Given concern about fish movement through the four hydroelectric facilities and the two water storage facilities on the Thunder Bay River, we conducted studies to assess the following characteristics of downstream fish passage: (1) community characteristics, (2) diel distribution at the hydroelectric facilities, (3) seasonal distribution at all facilities, and (4) acute mortality.

Turbine passage studies were conducted from March 1991 through February 1992, at Ninth Street and Hillman dams, and May 1992 through April 1993, at Norway Point and Four Mile dams. These studies were conducted monthly during winter (November–March) and biweekly during spring, summer, and fall (April–October). Tailrace nets were custom designed to fit the discharge opening of one turbine at each dam. Net capture efficiency was determined by placing 50 white suckers (10–25 cm) into the turbine at Four Mile Dam. Ninety percent of those fish were recovered; consequently, a 90% efficiency was assumed for all facilities because net designs were similar. Fish were collected from the catch container every 2–4 h over the 72-h sample period. At Hubbard Dam, a tailrace net was used to sample reservoir spill from April through October 1992. Because of unique structural characteristics at Upper South Dam, electrofishing was used in conjunction with a block net to sample fish in the reservoir spill from May–September 1991 and April 1992. One 24-h sample was taken each month at each water storage facility. For all fish passage studies, fish were removed from the net, placed in a live well, and immediately processed. Fish were not observed over an extended time period to determine latent mortality, so only acute mortality was measured.

The number of fish collected and species composition varied across facilities, but most fish were small. Species-specific diel and seasonal patterns were observed and appeared related to species-specific behavior. Diel patterns in turbine passage were likely related to feeding behavior as has been observed for salmon in the western United States (Raemhid et al. 1985; Ransom and Steig 1994) and several species in the Midwest (EPRI 1992). Increased movement in the spring appeared related to water temperature, in response to spawning activity, rather than water flow. This is contrary to studies that have found a correlation between water flow and fish passage (Walburg 1971; Ransom and Steig 1994). Increased movement in the fall may reflect a seasonal redistribution of juvenile fish as noted in a recent review of turbine passage studies (EPRI 1992). Fish behavior is the key variable when addressing diel and seasonal patterns in passage because movement through the facilities indicate behavioral responses to time of day, water temperature, and water flow, as opposed to the concept that fish were pulled involuntarily through the facilities.

In a review of turbine passage studies conducted mainly in Wisconsin and Michigan, mortality rates varied greatly depending on species, fish size, and turbine type (EPRI 1992). In our studies, turbine mortality was species-specific, depending on fish durability and size (decreasing with durability and size). Lower mortality for more durable fish was expected, but higher mortality for smaller fish was contrary to what would be expected, based on previous work (Gloss and Wahl 1983; Cada 1990; EPRI 1992). Handling stress may have caused increased mortality of smaller fish, as noted by Mathur et al. (1994). Turbine mortality at Four Mile Dam (horizontal shaft turbine) was significantly greater than at Norway Point Dam (vertical shaft turbine). The horizontal turbine has smaller wicket openings and more closely spaced runners which are more likely to cause damage and mortality to fish. This relationship between turbine design and the probability of a blade strike was addressed by Cada (1990). In contrast, acute mortality at the two water storage facilities was zero, which is less than the 3% mortality observed in reservoir spill for salmonid smolts (Schoeneman et al. 1961; Bell and Delacy 1972).

Fish Diversion

Downstream movement of fish through turbine passage and reservoir spill, and mortality associated with turbine passage is a concern on developed rivers. It should not be the intention of resource agencies to completely prevent downstream movement of fish from water control structures, but to ensure that those fish that do move are passed safely. If fish can be diverted from turbine intakes through bypass structures, or held within the impoundment when downstream passage is considered excessive, then potentially adverse effects of water control facilities may be reduced. Fish diversion can reduce adverse effects to fish resources due to project operation. Consequently, we determined the effectiveness of strobe lighting and air bubble curtains to deter fish from the turbine intake at Four Mile Dam in May 1994.

The air bubble curtain consisted of perforated tubing (perforation every 8 cm) secured along the bottom of the reservoir approximately 1 m upstream of the water intakes, and provided a curtain of air that completely encompassed the front of the intake area. A single strobe light was secured at the top center of each of the six water intake openings, flashing 60 times a minute. Each treatment and a control period were tested independently and an effectiveness index (EPRI 1989) was calculated to determine the ability of each treatment to deter fish from entering the turbine intake during different periods of the day: dawn (0300–0900 hours), day (6900–1500 hours), dusk (1500–2100 hours), and night (2100–0300 hours). This index value represents the percent effectiveness of the deterrent device when compared to the control period (no treatment). Fish were collected and analyzed using the same method employed for the turbine passage studies.

Deterrent effectiveness varied across species and time of day. In general, the deterrents were more effective during dusk and night and less effective during dawn and day. Bullhead *Ameiurus* spp. was the only group of fish that consistently passed through the turbines when deterrents were operating. Because of diel patterns in turbine passage, only selected time periods could be analyzed for rock bass *Ambloplites rupestris* and black bass *Micropterus* spp.. Rock bass were effectively deterred during both dusk and night by both deterrents, and black bass were effectively deterred during dusk by both deterrents. The deterrents used in this study were rudimentary and were not as effective as others have demonstrated (Erotheck 1975; Patrick et al. 1985; EPRI 1989; Martin and Sullivan 1992; Nemeth and Anderson 1992). Further study using commercially manufactured equipment was conducted in 1995 and these results showed an 81% reduction in turbine passage using the air and strobe treatments in combination (McCauley et al. 1996).

Discussion

Results from these studies demonstrate that the facilities differ in their effects on fish resources, but potential negative effects can be mitigated using an operational plan that incorporates fish management. Although the projects are relatively close to each other geographically, structural differences among sites may explain differences in effect and damage to fish resources. For any hydroelectric or water storage facility, there should be a site-specific comprehensive water resource and operational plan that manages for energy production, and is consistent with state natural resource agency management goals. A willingness to act on any recommendations set forth by the plan is necessary (Somach 1990), and compromise is paramount because not all competing users can get all that they want. Two principal issues were addressed in these studies: (1) protection of instream flow versus maintenance of reservoir water levels, and (2) downstream fish passage and the use of fish diversion to minimize negative impacts.

The relatively simple concept of defining a minimum water flow standard was confounded by (1) the difficulty of managing instream flows for the needs of different species and life stages and the misconception that maintaining a single minimum flow is adequate to preserve healthy aquatic biota, and (2) the limitation of the impoundment to produce the needed instream flow under natural hydrologic conditions without violating water level criteria in the impoundments. It is evident that many variables must be incorporated when choosing an instream flow standard that not only sufficiently protects aquatic biota, but is also achievable.

Orth (1987) stresses that IFIM is a process and not simply a model. He suggests additional areas of interest that need to be considered in the IFIM process, including assessment of (1) the prey base (both fish and invertebrates) because it is an integral component of the food chain, (2) the effect of instream flows on feeding and predator avoidance, given that lower flows will increase crowding and may increase both competition and susceptibility to predation, (3) temperature and water quality data, and the potential for reaching threshold limits, and (4) a wider array of species to protect ecological integrity. Leonard and Orth (1988) proposed the use of habitat use guilds (i.e., grouping fish by habitat needs) so that a single target species can be analyzed to represent a wider array of species in the IFIM process.

Aside from the inclusion of white sucker as a prey species, we did not address these additional factors, or use the habitat use guild approach in our IFIM study. However, we were able to determine the available habitat for agency-selected species and life stages at the proposed agency flows. With few exceptions, these minimum flows protected at least 90 m^2 of available habitat for each 300 m stream reach. The exception was adult northern pike which had no available habitat on the Lower South Branch at the proposed minimum flow. Flows in excess of 2 m^3/s, which exceed the storage capacity of the impoundment, would be necessary to achieve only modest gains in available habitat for adult northern pike. Available habitat is also limited for adult northern pike on the Upper South Branch; as a consequence, it would be counter productive to manage flow in either of these systems for northern pike.

At the proposed flows on both river systems, white sucker have substantially more available habitat than smallmouth bass. Thus, minimum flows would have a greater impact on smallmouth bass, making them the most logical target species on which to concentrate instream flow analysis. At the proposed minimum flows, the percentage of total habitat available for all life stages of smallmouth bass ranged from 2 to 7%. Doubling the proposed minimum flows on both systems would increase available habitat by as much as 50% for certain life stages, but increased available habitat still would constitute a relatively small percentage (≤10%) of the total habitat. This is an indication that available habitat for smallmouth bass on both systems is limited by other factors; as a result, smallmouth bass cannot be effectively managed with flow alone. As a possible alternative, habitat improvements may be the best management option for increasing available habitat for smallmouth bass.

It is important to note that these systems will not be maintained at a constant minimum flow, a phenomenon noted by Stalnaker (1990) in the water-limited western United States, but will generally follow the natural hydrologic cycle that allows flows to exceed the minimum. However, if the minimum flow is used over a period of time, the habitat available during these flows will be the limiting factor ultimately controlling the carrying capacity in these systems. Consequently, selection of adequate minimum flow is critical. During dry periods, a problem could arise if violation of impoundment water levels were necessary to maintain minimum flows. Analysis of historic data to determine the impoundment's normal storage capacity, as well as the storage capacity during dry and wet years, is necessary to determine if a potential conflict exists between maintenance of impoundment water levels and maintenance of adequate instream flows. Because one of these criteria may be violated under some circumstances, guidelines will be required to resolve potential conflicts.

Movement of fish from impoundments (both water storage and hydroelectric) and the effect of turbine mortality on fish resources has been assessed. Although the negative aspects of fish displacement and turbine mortality is considered low, further studies were conducted using commercially manufactured equipment to determine the effectiveness of air curtain and strobe lighting deterrent techniques in reducing turbine passage at the hydroelectric facilities. We also examined diel and seasonal variation in fish passing through the turbines to help maximize fish deterrent effectiveness. These techniques could also be used at the water storage facilities if deemed necessary, although in our case there was no mortality associated with fish escape, and the displacement of fish through reservoir spill in these facilities was not considered detrimental to the viability of the impoundment fisheries.

In some instances, aquatic resources are viewed by power companies as a liability. Alternatively, resource agencies and the public sometimes feel that facility owners should pay the entire cost of environmental protection, even when these costs exceed the project's worth. Cost-benefit analysis is a critical component of the licensing process (Bearzi 1991), and this process should stress a reasoned approach to mitigation with well-defined goals and objectives (Hunt and Hunt 1994). If this approach is not used, the competitiveness of hydroelectric power may be jeopardized (Lagassa 1994). There will be circumstances when project retirement is the best alternative because the benefits derived from project operation are exceeded by its negative impacts on the resources or the cost of mitigating them. Developers need to understand the importance of aquatic resources, and resource agencies need to understand the intended purpose of these projects (i.e., power production or water storage) and the associated operational and financial constraints.

Dams are relatively permanent structures, many of which have been in place for nearly 100 years, and natural resource management will typically be limited to the environment created by the dam. Consequently, a comprehensive river management plan should be directed toward protecting and enhancing the existing resources. The successful management of natural resources on any river system necessarily involves a partnership encompassing resource agencies, power companies, and private citizens, and promotes a shared stewardship of the river and its natural resources. The integration of the expertise and interests of all the parties should provide a clear and concise set of goals and objectives for the successful long-term management of natural resources.

Acknowledgments

We thank Howard Energy and the Thunder Bay Power Company for their support of this research. We also thank both the Thunder Bay Power Company operations and the staff from the Great Lakes Environmental Center for their assistance in the field work.

References

Bearzi, J. A. 1991. The delicate balance of power and nonpower interests in the nation's rivers. Rivers 2:326–332.

Bell, M. C., and A. C. Delacy. 1972. A compendium on the survival of fish passing through spillways and conduits. U.S. Army Corps of Engineers, North Pacific Division, Portland, Oregon.

Cada, G. F. 1990. A review of studies relating to the effects of propeller-type turbine passage on fish early life stages. North American Journal of Fisheries Management 10:418–426.

EPRI (Electric Power Research Institute). 1989. Field testing of behavioral barriers for fish exclusion at cooling-water intake systems. EPRI GS-6246, Palo Alto, California.

EPRI (Electric Power Research Institute). 1992. Fish entrainment and turbine mortality review and guidelines. EPRI TR-101231, Palo Alto, California.

Erotheck, L. M. 1975. Evaluation of an air curtain as a fish deterrent device at the Prairie Island nuclear generating plant cooling water intake. Annual report. Prairie Island Nuclear Generating Plant, Environmental Monitoring and Ecological Studies Program, volume III.

Gloss, S. P., and J. R. Wahl. 1983. Mortality of juvenile salmonids passing through Ossberger crossflow turbines at small-scale hydroelectric sites. Transactions of the American Fisheries Society 112:194–200.

Hunt, R., and J. M. Hunt. 1994. An emerging entitlement program. Independent Energy 24(4):84–87.

Lagassa, G. 1994. Fighting for economic survival. Independent Energy 24(4):76–84.

Leonard, M. L., and D. J. Orth. 1988. Use of habitat guilds of fishes to determine instream flow requirements. North American Journal of Fisheries Management 8:399–409.

Martin, P. D., and C. W. Sullivan. 1992. Guiding American shad with strobe lighting. Hydro Review 11(4): 52–58.

Mathur, D., P. G. Heisey, and D. A. Robinson. 1994. Turbine passage mortality of juvenile American shad at a low-head hydroelectric dam. Transactions of the American Fisheries Society 123:108–111.

McCauley, D. J., L. M. Montuori, J. E. Navarro, and A. R. Blystra. 1996. Using strobe lights, air bubble curtains for cost-effective fish diversion. Hydro Review 15(2): 42–51.

Milhous, R. T., M. A. Updike, and D. M. Schneider. 1989. Physical habitat simulation system reference manual, version II. U.S. Fish and Wildlife Service, Instream Flow Information Paper 26, Biological Report 89(16), Washington, DC.

Navarro, J. E., and D. J. McCauley. 1993. Fish escapement from two storage reservoirs in Michigan. Rivers 4:36–47.

Navarro, J. E., D. J. McCauley, and A. R. Blystra. 1994. Instream flow incremental methodology (IFIM) for modeling fish habitat. Pages 1–22 *in* W. James, editor. Current practices in modelling the management of stormwater impacts. Lewis Publishers, Boca Raton, Florida.

Navarro, J. E., D. J. McCauley, and A. R. Blystra. 1996. Turbine passage at four low-head hydroelectric facilities in northeast Michigan. North American Journal of Fisheries Management 16:182–191.

Nemeth, R. S., and J. J. Anderson. 1992. Response of juvenile coho and chinook salmon to strobe and mercury vapor lights. North American Journal of Fisheries Management 12:684–692.

Orth, D. J. 1987. Ecological consideration in the development and application of instream flow-habitat models. Regulated Rivers 1:171–181.

Patrick, P. H., A. E. Christie, D. Sager, C. Hocutt, and J. Stauffer, Jr. 1985. Responses of fish to a strobe light/air-bubble barrier. Fisheries Research 3:157–172.

Raemhid, G. A., R. Nason, and S. Hays. 1985. Hydroacoustic studies of downstream migrating salmon at hydropower dams: two case studies. Pages 244–250 *in* F. W. Olson, R. G. White, and R. H. Hamre, editors. Symposium on small hydropower and fisheries. American Fisheries Society, Western Division and Bioengineering Section, Bethesda, Maryland.

Ransom, B. H., and T. W. Steig. 1994. Using hydroacoustics to monitor fish at hydroelectric dams. Lake and Reservoir Management 9:170–172.

Schoeneman, D. E., R. T. Pressey, and C. O. Junge. 1961. Mortalities of downstream migrant salmon at McNary Dam. Transactions of the American Fisheries Society 90:58–72.

Somach, S. L. 1990. The American river decision: balancing instream protection with other competing beneficial uses. Rivers 1:251–263.

Stalnaker, C. B. 1990. Minimum flow is a myth. Pages 31–33 *in* M. B. Bain, editors. Ecology and assessment of warmwater streams: workshop synopsis. U.S. Fish and Wildlife Service, Biological Report 90(5), Washington, DC.

Walburg, C. H. 1971. Loss of young fish in reservoir discharge and year-class survival, Lewis and Clark Lake, Missouri River. Pages 441–448 *in* G. E. Hall, editor. Reservoir fisheries and limnology. American Fisheries Society Special Publication 8.

American Fisheries Society Symposium 16:210–214, 1996

Strategies for Improved Sportfishing Access on U.S. Army Corps of Engineers Reservoirs

H. Roger Hamilton and R. Scott Jackson

U.S. Army Corps of Engineers, Waterways Experiment Station
3909 Halls Ferry Road, Vicksburg, Mississippi 39180-0699, USA

Abstract.—Large lakes and artificial reservoirs are important resources for providing sportfishing opportunities. The U.S. Army Corps of Engineers (USCOE) is a significant provider of such opportunities through its system of reservoirs located throughout the United States. Although USCOE lakes were constructed primarily for purposes other than fishing, such as flood control, hydroelectric power production, and water supply, recreational activities including sportfishing, has emerged as a dominant use of these lakes. In 1993, more than 372 million visits were recorded at USCOE-managed lakes, and visitors spent more than US$10 billion to engage in recreation at these sites. Effective management of USCOE lakes requires evaluation of competing and changing demands. To ensure a balanced response to these demands, while meeting more stringent environmental requirements, the operation of USCOE lakes must continuously be reevaluated because of changes in land use, population patterns, and environmental laws. We discuss the dynamics of planning, development, and management of large, multipurpose projects, with emphases on providing public access for sportfishing and on the roles of fisheries scientists and outdoor recreation planners in these activities.

Some of the best fishing in the United States is found in large lakes, most of which are artificial impoundments constructed for reasons other than fisheries development. As reservoirs continue to support higher levels of recreational fishing, in some cases management strategies will have to be developed to optimize fishing opportunities while maintaining the original operational objectives. The original purposes for which the projects were authorized should be reexamined at older reservoirs to determine if they remain valid many years after impoundment.

Many U.S. reservoirs were built and are operated by the U.S. Army Corps of Engineers (USCOE). The USCOE currently has jurisdiction over 463 reservoirs in 43 states (USCOE 1993). These reservoirs are the focus of this discussion on sportfishing access, but the principles are applicable to other reservoirs and to smaller lakes, regardless of jurisdiction.

Reservoirs have been constructed throughout the country for a wide range of purposes, including flood control, hydroelectric power production, navigation, water supply, irrigation, recreation, and fish and wildlife enhancement. Although a single purpose often drives the demand for construction, the USCOE investigates and evaluates all other needs that might be satisfied by construction of a large dam to determine if projected benefits are greater than funds expended while minimizing effects on the environment. Federal legislation that governs authorization and appropriation of funds for water projects under USCOE jurisdiction requires that no single-purpose water resource project be built. Thus, a need for municipal and industrial water supply might be met by a federally authorized project that includes recreation or flood control along with water supply if those purposes are justifiable.

Before construction of a dam, long planning periods make it possible for alternative plans to be developed and thoroughly analyzed, benefits and costs to be optimized, and all legal requirements to be met. Thorough public input is required in the planning process. An environmental impact statement is required for significant actions. Archaeological surveys are conducted, and fish and wildlife mitigation plans are prepared. Unfortunately, fish and wildlife and cultural resource considerations were mandated only after passage of the National Environmental Policy Act (42 U.S.C. §§ 4321–4361) in 1970, after most USCOE projects were developed.

No new large reservoirs are likely to be constructed in the United States. Economic and environmental constraints, accelerated public resistance to large projects that substantially modify the landscape, and occupation of the most desirable sites by existing dams limit expansion of fishing opportunities. Public fishing can, however, be enhanced by modifying existing reservoirs—principally through structural modification and implementation of new management or operational strategies. Recreational anglers and other recreation users can benefit greatly from such improvements, and the benefits are best realized when user groups actively participate in the modification process.

Study Area

The USCOE reservoirs comprise some 4.7 million hectares of land and water. At normal pool elevations, about one-half of the total area is inundated and the other half is adjacent dry land. Total length of shoreline at normal pool elevation for all lakes (65,401 km) is more than triple the total length of the coastline of the entire United States (Hoffman 1988; USCOE 1992).

Although many reservoirs were originally developed in rural areas, urban spread and transportation improvements have resulted in the close proximity of a large number of Americans to these valuable resources. About 80% of USCOE lakes are within 80 km of major metropolitan areas. Ninety-four percent of these lakes are within a 2-h drive of the nation's population centers (USCOE 1990).

The USCOE is the nation's leading provider of water-based outdoor recreation. In 1993, more than 372 million visits were recorded at USCOE managed reservoirs. Twenty-five percent of these visits were by anglers, and 20% were engaged in other recreation activities that required boating access to the lakes (USCOE 1993). An estimated 25 million individuals visit a USCOE reservoir at least once each year (USCOE 1990).

Reservoirs managed by the USCOE are an important economic development factor in many regions of the country. In 1991, visitors spent more than US$10 billion at USCOE projects, resulting in $2.9 billion in employee income and 180,000 jobs. Secondary effects of this visitor spending accounted for an additional $9.5 billion in employee income and 437,000 jobs (Jackson et al. 1994).

U.S. Army Corps of Engineers Policy

Fishing access is a very important aspect of reservoirs, but, as noted, fishing is almost never the driving force in reservoir development. Dams are usually built for flood control, navigation, or some other conventional purposes. Although sportfishing and other recreational uses are important benefits, they have often been viewed by engineers as "add-ons."

Legal authority does exist for proponents of sportfishing and other forms of recreation to demand equitable treatment in planning, formulation, design, and management of USCOE projects. Outdoor recreation and fish and wildlife management were authorized in the Flood Control Act of 1944 (16 U.S.C. § 460b). As early as 1952, the USCOE policy stated that "Land and water areas of civil works projects under jurisdiction of the Corps of Engineers will be administered and managed so as to obtain maximum sustained public benefits from conservation and use of their natural resources . . . " (USCOE 1952:125). The policy further states that " . . . full consideration will be given to all collateral uses of land and water areas including, but not limited to, public park and recreational use, wildlife management, farming, grazing, and industrial uses and adequate provision will be made for soil conservation, forestry, mosquito control, sanitation, control of pollution arising in project areas, health and safety of the visiting public and similar activities relating to administration and management of project lands and waters in the public interest."

In the Federal Water Project Recreation Act of 1965 (16 U.S.C. § 460-1-12), recreation and fish and wildlife were afforded project purpose status commensurate with conventional purposes such as navigation and flood control. The same public law requires that not less than 50% of the development costs for recreation and fish and wildlife enhancement be provided by a nonfederal public agency. All operation, maintenance, and replacement costs must be assumed by a nonfederal sponsor as well.

The challenge is to maximize fishing opportunities within the scope of the authorized purposes of the project. Fish and wildlife enhancement and recreation, although legally equal in status to other purposes, have historically been subjected to administrative policies that make them lower in priority compared with other project purposes (USCOE 1969; Crafts 1970; CZRC 1975; Verburg 1975; Jahn et al. 1984).

Site Design

Access to the water for boating and pedestrian angling may require work on the landward side of the shoreline. Adequate access and circulation roads, ample parking, and a means of storing or launching boats are basic requirements.

Site selection is based on several physical and social criteria. Obvious physical characteristics that are important are sustained water depth (throughout the recreational season) and access to both land and water. Access on the landward side should be easily available from major population centers or transportation corridors. Water access should be to attractive, reliable fish habitat. Bank slope, erosion rates, scenic quality, availability of utilities such as electricity, and ample parking are also important.

Development of new recreation facilities at USCOE reservoirs requires cost-sharing with a nonfederal public agency, which has a large voice in siting boat ramps and other facilities. Many USCOE projects have more than one cost-sharing partner. Because marinas are leased to private entrepreneurs, market studies are necessary to assure investors that the project is profitable and provides adequate public service. A delicate balance is required for the correct level of competition in these situations.

Safety is always a prime consideration in site design. The USCOE design criteria include specific safety requirements such as adequate lighting, control and directional signs, buoys, and security measures. Access roads to launching ramps are designed to require a deliberate turn from the approach onto the ramp. Where economically feasible to install and maintain, courtesy boat docks are provided at boat ramps. The docks are sized to assure safe use and access for disabled persons (USCOE 1987).

Access is not ideal in many places because some reservoirs are quite old, and project land was acquired with little or no regard for public recreation use. Some of the reservoirs designed for flood control or navigation have only a very narrow ribbon of publicly owned land around them. Subsequent subdivision of adjacent land by private developers often precludes public access to USCOE reservoirs except at a few designated areas. Access is not always near the best fishing areas, and in some cases is difficult to reach from major roads. Initial development or restudy of any public reservoir with sportfishing potential must include input from professional fisheries biologists and outdoor recreation planners, who carefully examine and recommend adequate fishing access.

In addition to biological data concerning the fishery and aquatic habitat, other information is needed for well-informed planning and management. Analysis of fishing trends and transportation networks, information about the origin and destination of visitors, and analysis of landscape features are important components of these studies. Physical and social carrying capacity of the lake should be determined for fishing and other water-based recreation activities. Physical carrying capacity is a measure of public use that the natural resources can support without deterioration. Social carrying capacity refers to the visitor's comfort level while engaged in recreation. Types and sizes of boats to be used, number of boats expected on peak days, and other pertinent information will help determine location of launching ramps, number of ramps, number of lanes at each ramp, size of parking lots, and land-based facilities needed. It is possible to exert some control over the number of boats on a reservoir by controlling the extent of land-based facilities, such as size of ramps and size of parking lots at the ramps.

Bank and wading anglers are often important user groups at reservoirs. Pedestrian fishing often requires little more than road access, parking facilities, and a trail or path to favorite fishing areas. Tailwaters downstream of large dams, particularly those that generate hydropower, are often congested and can be quite dangerous. It is critical that dams be designed to have safe areas for fishing and observation. Rapid rises in water levels are common when generators are engaged, and many anglers are endangered because they are not able to escape the turbulent discharge. Warning signs, flashing lights, sirens and other audible warning signals, and proactive public information campaigns have not prevented several anglers from losing their lives each year below USCOE dams. Parking and access controls can be good management tools, but if the fishing is consistently good in these areas, limiting public use can be difficult. Special attention should be paid to these areas in all phases of fishing-access planning and management.

Data for use in these analyses are often collected through a variety of survey techniques, including face-to-face interviews, telephone interviews, and mail-back questionnaires. Public meetings, workshops, and seminars are used with news releases and a variety of techniques to solicit public input to the decision process and keep all parties as informed as possible.

Special facilities are needed to accommodate physically impaired anglers. Approximately 43 million Americans have one or more physical or mental disabilities, and this number is increasing as the population ages. The Americans with Disabilities Act of 1990 (42 U.S.C. § 12101) specifies measures required to provide appropriate access for disabled people to public facilities. Public entities are required to provide readily accessible and usable facilities for individuals with disabilities, including wheelchair access, when constructing new facilities or altering existing facilities. Projections of types of disabilities and numbers of users from those populations are important considerations in spending limited funds wisely to ensure that all of the public has high-quality fishing experiences.

Boating access can take several forms. Marinas are in the business of storing boats and providing supplies and services for their customers. Boats are commonly stored in slips in the water or at dry storage facilities located nearby. Launching facilities are required for trailered or cartop boats. Provisions must be made for parking the boat trailer and the towing vehicle while anglers enjoy their sport. Restrooms, fish-cleaning stations, trash disposal containers, and other land-based facilities enhance a launching area.

Landowners who have developed property adjacent to USCOE reservoirs require special consideration. About 50,000 private boat houses, launching ramps, picnic areas, gardens, landscaped and mowed areas, and other developments have been permitted at 100 USCOE reservoirs (USCOE 1990). Private, exclusive use of public lands at USCOE projects did not receive much opposition during the 1950s and early 1960s. On the contrary, this kind of activity was generally acceptable because it provided additional use of the resources. To reflect changing attitudes and increasing demands for limited resources, USCOE policy was revised and now prohibits such activity except where it was previously allowed (USCOE 1974). The goal is to phase out existing private facilities. Where private development is permitted, boating and pedestrian access are provided for the adjacent landowners and their family and friends. Access by the general public is greatly restricted and sometimes prohibited.

Federal design criteria are usually more stringent than those of state or local agencies. Because federal legislation requires that costs for recreation and fish and wildlife enhancement be shared by a nonfederal sponsor, the cost-sharing partner also becomes a planning partner. The existence of stricter design criteria often means increased construction costs. Representatives of the federal sector must be able to negotiate with their partners and make compromises without endangering public safety.

Management

Some large multipurpose lakes are subject to dramatic water-level fluctuations. Water is released in large amounts, for example, to augment downstream flows for navigation or increase storage capacity for water supply. The amount of water level fluctuation varies widely depending on operating procedures and surrounding topography. Lake Seminole in the Georgia–Alabama–Florida tri-state area retains a fairly constant pool elevation. Conversely, some western projects located in mountainous areas routinely have great water-level changes. Dworshak Lake, Idaho, can fluctuate as much as 50 m to supply navigation water to the Columbia River system. Large lakes in flat terrain such as Oklahoma or the Dakotas experience pool drawdowns that leave the shoreline at normal pool levels miles from the lower lake elevation. When this occurs, boat access is nearly impossible and marina operators are out of business.

Biologists and recreation planners need to anticipate these situations and work with agencies that have jurisdiction over the project to optimize fishing access during these periods. Long ramps that extend under the water in normal times but continue to accommodate boat launching when the pool drops are often the answer. This technique has been used at several reservoirs, including Lake Sakakawea, North Dakota, and Lake Texoma, Oklahoma and Texas. Alternate access areas in times of low pool or alternate fishing opportunities also must be investigated.

Monitoring is needed to determine effectiveness of planning. Projections of public use, demand for all project outputs, mitigation of fish and wildlife impacts, and other important features are made during project planning. Seldom are adequate monitoring plans devised and implemented to determine if the projections are true. Only a few projects have ongoing monitoring programs. Lake Aquilla and Ray Roberts Lake, both in Texas, have pre- and postimpoundment data available and data collection is conducted at 3- to 5-year intervals. All too often the implications of management decisions affecting large reservoirs are not well understood.

Periodic monitoring should consider use-levels, trends in activities and types of equipment, proximity of access to fishing activity, conflicts with other uses of the resources, and a host of parameters designed to gauge the effectiveness of fishing access management. Where nonfederal partners participate in project management, they must have an equitable voice in formulating and implementing monitoring plans.

As management evolves and continues throughout the life of a project, new questions arise; some are peculiar to a single project or a region, whereas others are very general in nature. Research is usually needed to solve short- and long-term problems.

Large reservoirs developed for various purposes are quite complex. Provision of adequate access for recreational anglers can be equally complex. Consider fishing access as an organism, one of many, which, in combination with many other organisms and other physical and biological attributes, constitutes an ecosystem, the reservoir. Using this analogy, one can see that fishing access is important unto itself and is also a vital part of a larger domain.

References

Crafts, E. C. 1970. How to meet public recreation needs at Corps of Engineers reservoirs. U.S. Army Corps of Engineers, Office, Chief of Engineers, Washington, DC.

CZRC (Coastal Zone Resources Corporation). 1975. Study of land use for recreation and fish and wildlife enhancement. Coastal Zone Resources Corporation, Wilmington, North Carolina.

Hoffman, M. S. 1988. The world almanac and book of facts. Scripps Howard Company, New York.

Jackson, R. S., D. J. Stynes, and D. B. Propst. 1994. An assessment of the national economic effects of the U.S. Army Corps of Engineers recreation program. U.S. Army Corps of Engineers Waterways Experiment Station, Miscellaneous Paper R-94-2, Vicksburg, Mississippi.

Jahn, L. R., W. C. Cook, and J. D. Hughes, Jr. 1984. An evaluation of U.S. Army natural resource management programs on selected military installations and civil works projects. Office, Secretary of the Army, Washington, DC.

USCOE (U.S. Army Corps of Engineers). 1952. Orders and regulations. U.S. Army Corps of Engineers, Office, Chief of Engineers, Washington, DC.

USCOE (U.S. Army Corps of Engineers). 1969. Corps of Engineers resources managers. U.S. Army Corps of Engineers, Office, Chief of Engineers, Washington, DC.

USCOE (U.S. Army Corps of Engineers). 1974. Shoreline management at Corps of Engineers civil works projects. Engineer Regulation 1130-2-406. U.S. Army Corps of Engineers, Office, Chief of Engineers, Washington, DC.

USCOE (U.S. Army Corps of Engineers). 1987. Recreation Planning and Design Criteria, Engineer Manual 1110-1-400. U.S. Army Corps of Engineers, Office, Chief of Engineers, Washington, DC.

USCOE (U.S. Army Corps of Engineers). 1990. U.S. Army Corps of Engineers recreation study: a plan prepared for the Assistant Secretary of the Army (Civil Works). U.S. Army Corps of Engineers, Office, Chief of Engineers, Washington, DC.

USCOE (U.S. Army Corps of Engineers). 1992. Nationwide statistical data, 1992. U.S. Army Corps of Engineers, Office, Chief of Engineers, Washington, DC.

USCOE (U.S. Army Corps of Engineers). 1993. Natural resource management system. U.S. Army Corps of Engineers, Office, Chief of Engineers, Washington, DC.

Verburg, E. A. 1975. The U.S. Army Corps of Engineers recreation development program: a comparative perspective of equity related to policies and program outcomes. U.S. Army Corps of Engineers IWR (Institute for Water Resources) Paper 75-P6, Washington, DC.

American Fisheries Society Symposium 16:215–220, 1996

On-Site Weight Certification Facilities: A Service to Anglers and a Tool for Managers

RICHARD A. OTT, JR.

Texas Parks and Wildlife Department
11942 FM 848, Tyler, Texas 75707–9657, USA

MARK A. WEBB

Texas Parks and Wildlife Department
1004 East 26th Street, Bryan, Texas 77803, USA

Abstract.—On-site weight certification facilities (weigh stations) were constructed on two Texas reservoirs to provide anglers the opportunity to weigh fish for personal gratification or official certification in angler recognition programs, and to provide managers with information not routinely collected during standard surveys. When modified catch-and-release-only regulations for largemouth bass *Micropterus salmoides* (allowing temporary retention of fish 22 inches and longer for transport to the weigh station) were enacted, these stations provided anglers the only legal method to obtain certified weights. Where harvest is permitted, weigh stations can provide anglers with accurate weight information, increase the probability released fish will survive, and provide management agencies with information on large fish which are not reliably sampled. Where harvest is not permitted, weigh stations provide anglers certified weights for angler recognition programs, and provide agencies a method of assessing the results of their management efforts. By combining weigh stations with angler recognition programs and traditional bulletin boards, it may be possible to create lakeside information transfer centers, thereby minimizing the time lag for information exchange between fisheries managers and anglers.

In their review of the importance of goals, objectives, and values in fisheries management Barber and Taylor (1990:365) wrote "throughout the history of fisheries management, goals have been rarely stated or have been stated in generalized terms of best or wise use with no supporting objective statement." They further stated that "objectives operationally support goals and are measurable, verifiable statements of intermediate tasks that must be accomplished for goal attainment." However, measurement, or the ability to verify attainment of goals, is often difficult. This is particularly apparent in the development of "quality" fisheries, where even the definition of quality is in question. Anderson (1975), Weithman and Anderson (1978), and Weithman and Katti (1979) attempted to describe quality fisheries based on angler catch rates combined with the catch of larger-than-average fish as a trophy component. Quinn (1987) further suggested fish of memorable size represent an important fisheries resource which far outweighs their small numerical representation in populations.

Because trophy-sized fish (particularly largemouth bass, *Micropterus salmoides*) exist in relatively low numbers in most populations, they are difficult to sample (Cofer 1995; Horton and Gilliland 1995). Quinn (1987) proposed the use of angler recognition programs (ARPs) as a method of assessing trophy fisheries. In these programs, an angler must confirm fish weight and length and fill out proper documentation to enter the fish for trophy consideration. At the time of Quinn's publication, 31 state agencies reported active ARPs but few, if any, used the information for much more than public relations. Forshage et al. (1991) analyzed information collected through the Texas Parks and Wildlife Department (TPWD) ARP to document annual increases in mean weight of fish submitted and improvements in fishing quality for largemouth bass and striped bass *Morone saxatilis*, and related these directly to stocking programs. Horton and Gilliland (1995) described monitoring trophy ($\geq$ 8 lb, $\geq$ 22.6 in) largemouth bass in Oklahoma using a taxidermist network. Entry of the fish in the Oklahoma ARP provided managers documentation of the catch along with the address and telephone number of the angler. By interviewing anglers, managers obtained names of taxidermists holding the fish who then provided liver samples from trophy largemouth bass for genetic analyses. These analyses enabled managers to directly measure the contribution of Florida largemouth bass *M. s. floridanus* introductions to producing a trophy fishery. However, in all three cases, researchers stated higher levels of angler participation in ARPs would have improved data acquisition.

Several factors influence an angler's decision to participate in ARPs: the angler must have caught a fish of significant size, must be aware of the program, and must have access to necessary entry forms. Because, in the ARP administered by TPWD, three of the four major recognition categories (State Record, Water Body Record, and Big Fish) require certified weights for submission (A. A. Forshage and J. Knight, TPWD, unpublished data), anglers must also have access to a scale with a level of accuracy acceptable to the agency administering the program. In most cases, this forces the angler to leave the lake, reducing the probability the caught fish will be released alive. In the current climate of voluntary or mandatory catch-and-release fisheries, anglers may be unwilling or unable to obtain certified weights and fill out applications, depriving managers of valuable information. The objective of our paper is to describe the utility of on-site weight certification facilities (weigh stations) as a strategy to increase participation in ARPs, thereby providing both a service to anglers and a method for collecting information on quality and trophy fisheries.

Weigh Station Scenarios

Scenario 1

Purtis Creek State Park Lake is a 355-acre reservoir contained within the boundaries of Purtis Creek State Park. The reservoir is located in Henderson and Van Zandt counties within a 1-h drive from Dallas, Texas. The reservoir was impounded in 1984 and stocked with Florida largemouth bass at 90 per acre in 1985. The reservoir was opened to public fishing in November 1988 under a regulation requiring immediate release of largemouth bass. This regulation produced a very popular largemouth bass fishery with high catch rates and a high proportion of large fish (Storey and Ott 1994), and by 1991 anglers began reporting release of largemouth bass in excess of 15 lb. Although these reports were anecdotal, they fueled managers' concerns that valuable management information was being lost and the potential existed for state-record-size fish to be caught that could not be legally certified in the ARP. To relieve these concerns, the Purtis Creek State Park weigh station was constructed in August 1992. To allow anglers to legally retain largemouth bass for weight certification, the regulation was amended to read "... catch-and-release-only, except that any largemouth bass 22 inches or greater in length may be retained in a live well or other aerated holding device and immediately transported to the Purtis Creek State Park weigh station. After weighing, the bass must be released immediately back into the lake" (TPWD 1995:26).

The weigh station is located at the end of a courtesy dock across the lake from the only boat ramp on the reservoir. This location provides good access to anglers with fish in live wells, is most convenient to electrical power, and offers the greatest security from clandestine removal of large fish contained in live wells. The station is a fully enclosed structure 4 ft^2 $\times$ 8 ft tall equipped with certified scales, a measuring board, ARP entry forms, and a citizens band radio. The station is kept closed and locked when not in use, and anglers must contact park staff for access. This provides security for all equipment and documents and prevents entry of muddaubers (*Sceliphron* sp.) whose nests can incapacitate the scale mechanism. Instructions posted on the exterior of the weigh station advise anglers of staff hours and the protocol for submitting documentation. Existing Water Body Records for all species (including those for which harvest is optional) are provided along with minimum lengths or weights of species for qualification in the ARP. Anglers are instructed to insure largemouth bass are at least 22 in long, using the measuring board permanently mounted on the weigh station exterior, before contacting park staff. Anglers are also advised to transfer the fish from the live well to a submersible cage (as described in Ott and Storey 1993) attached to the courtesy dock. The two-way citizens band radio is provided to alert park staff of the capture of qualifying fish. When park staff arrive, they conduct the official weighing and measuring of fish, fill out ARP entry forms, and insure largemouth bass are released.

Scenario 2

Gibbons Creek Reservoir is a 2,500-acre power generation cooling reservoir owned by the Texas Municipal Power Authority (TMPA). The reservoir is located in Grimes County within a 1-h drive from Houston, Texas. The reservoir was impounded in 1981 and stocked with Florida largemouth bass at 49 per acre in 1981 and Florida largemouth bass $\times$ largemouth bass at 102 per acre in 1985. The reservoir was opened to public fishing in March 1985 under a 14–18 in slot length limit and a three-fish daily bag limit. This regulation was amended in 1988 to a 14–21 in slot length limit and a three-fish daily bag limit. Directed angling effort for largemouth bass was high (62 h/acre during the first 3

TABLE 1.—Numerical distribution (%) of species entered for angler recognition program (ARP) awards from Purtis Creek State Park Lake (March 1989–December 1994) and Gibbons Creek Reservoir (March 1985–December 1994), and species weighed and measured but not entered in the ARP at the entrance station at Gibbons Creek Reservoir (September 1989–December 1993). Entries include, but are not limited to, those submitted at the Purtis Creek State Park and Gibbons Creek Park weigh stations.

Species	Purtis Creek State Park Lake ARP entries	Gibbons Creek Reservoir ARP entries	Gibbons Creek Reservoir entrance station measurements
Common carp *Cyprinus carpio*	0.4		
Smallmouth buffalo *Ictiobus bubalus*			0.2
Blue catfish *Ictalurus furcatus*		3.7	11.5
Channel catfish *Ictalurus punctatus*	3.5	0.5	0.2
Flathead catfish *Pylodictis olivaris*		1.4	6.9
Blackstripe topminnow *Fundulus notatus*	0.4		
Bluegill *Lepomis macrochirus*	2.6		
Redear sunfish *Lepomis microlophus*	0.9		
Largemouth bass *Micropterus salmoides*	87.9	90.2	79.8
White crappie *Pomoxis annularis*	4.3	1.4	0.6
Black crappie *Pomoxis nigromaculatus*		1.4	
Freshwater drum *Aplodinotus grunniens*			0.2
Blue tilapia *Tilapia aurea*		1.4	0.6

months after the reservoir was opened). In 1989 the popular fishery that developed prompted TMPA to provide certified scales and a measuring board for angler use at the entrance station (approximately 1 mi from the reservoir). This facility provided anglers the opportunity to weigh and measure harvestable-sized fish for personal gratification or for entry in the Texas ARP. Although TMPA maintained a list of lengths and weights, as well as photographs of trophy fish, they did not provide anglers with ARP entry forms or encourage participation in the program.

Scenario 3

By 1990 managers began to observe what appeared to be overharvest of legal-sized largemouth bass smaller than the protected slot size range. This resulted in fewer fish entering the protected slot size and threatened the quality of the fishery. By 1992, managers encouraged by the success at Purtis Creek State Park Lake, decided to invoke a catch-and-release-only regulation for largemouth bass at Gibbons Creek Reservoir. On 1 September 1993 the regulation was amended to read "... catch-and-release-only, except that any largemouth bass 22 inches or greater in length may be retained in a live well or other aerated holding device and immediately transported to the Gibbons Creek Park weigh station. After weighing, the fish must be released immediately back into the lake" (TPWD 1995:26). To increase the probability that released fish would survive, an existing pavilion was converted into a lakeside open air weigh station. The new Gibbons Creek Park weigh station went into operation in December 1993 and is available to anglers any time the reservoir is open to fishing; assistance by park staff is not required. Scales, a measuring board, and ARP forms are available for angler use. The presence of a witness is required to certify fish weights, in accordance with ARP rules (Forshage and Knight, unpublished data). This service is usually provided by an angling companion. In cases where the angler is alone, TMPA park staff may assist, but anglers are responsible for notifying staff. This may require driving to the entrance station or using a portable telephone if no staff are present in the boat ramp area. Anglers are required to maintain fish in live wells until weighed. To encourage angler participation, sample ARP certificates and award pins as well as a list of all Gibbons Creek Reservoir Water Body Records are posted at the weigh station along with instructions for submission.

Observations on Use of Weigh Stations

Largemouth bass was the species most frequently weighed (Table 1) and represented the majority of ARP entries. Large catfishes (*Ictalurus* spp. and *Pylodictis olivaris*) were often weighed but not necessarily submitted as ARP entries. Panfish entries included crappies *Pomoxis* spp., as well as sunfishes *Lepomis* spp. and included the State Record bluegill *L. macrochirus* caught at Purtis Creek State Park Lake in 1993. The most unusual entry was a Water Body Record blackstripe topminnow *Fundulus notatus* caught at Purtis Creek State Park Lake.

At Purtis Creek State Park Lake, prior to con-

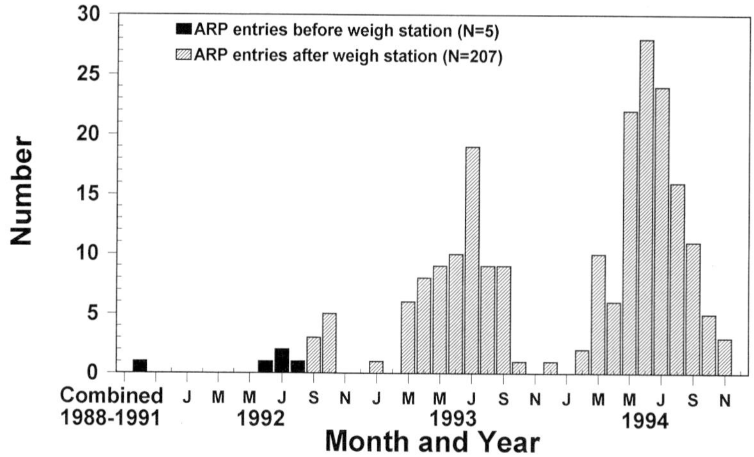

FIGURE 1.—Monthly angler recognition program (ARP) entries for largemouth bass from Purtis Creek State Park Lake before and after construction of the Purtis Creek State Park weigh station. The Purtis Creek State Park weigh station became operational 1 September 1992.

struction of the weigh station and modification of the catch-and-release regulation, only five largemouth bass had been submitted as ARP entries (Figure 1). Although anglers could submit Lunker-Catch-and-Release ARP entries based on fish length alone, only one entry was submitted in this category prior to weigh station construction. Entries in the ARP increased immediately following weigh station construction. Eight entries were submitted in fall 1992, 72 entries during 1993, and 127 during 1994. Anglers continued to submit Lunker-Catch-and-Release entries after construction of the weigh station; however, most anglers took advantage of the ability to certify fish weight.

Results at Gibbons Creek Reservoir further illustrate the utility of on-site weigh stations. From September 1989 (when TMPA began keeping records of largemouth bass weights and lengths obtained at the entrance station) through December 1993 (when the Gibbons Creek Park weigh station was completed), 390 largemouth bass 21 in or longer were weighed and measured (Figure 2). However, only six of these were entered in the ARP. An increase in the number of Gibbons Creek Reservoir fish entered in the ARP began in January 1993 prior to completion of the Gibbons Creek Park weigh station; however, most were weighed at off-site locations; only three of the 1993 entries were weighed at the TMPA entrance station. In the first full year following completion of the Gibbons Creek Park weigh station, 100 ARP entries were recorded there. This is equivalent to the average number of largemouth bass weighed per year at the TMPA entrance station, but the information obtained has greater utility to managers due to standardization with statewide program data.

Discussion

When harvest is prohibited, on-site weigh stations provide anglers the opportunity to obtain certified weights necessary for ARP entry. At Purtis Creek State Park Lake, ARP entries in the first 2 months following construction of the weigh station exceeded entries for the previous 4 years. In the first 2 years of operation, the Purtis Creek State Park weigh station provided information on approximately 200 fish 22 in or longer caught and released in this 355-acre reservoir. In addition, anglers began bringing fish from other area reservoirs to the Purtis Creek State Park weigh station for certification, indicating the potential for these weigh stations to act as regional ARP certification facilities.

When harvest is optional, on-site weigh stations can be used successfully to promote ARPs. At Gibbons Creek Reservoir, ARP entries from the entrance station increased markedly 10 months prior to completion of the weigh station, presumably due to advanced publicity about station development and pending regulation changes. Nevertheless, some anglers who used the scales at the TMPA entrance station did not enter these fish in the ARP, and some anglers took their fish to other locations for certification and entry. It is possible local busi-

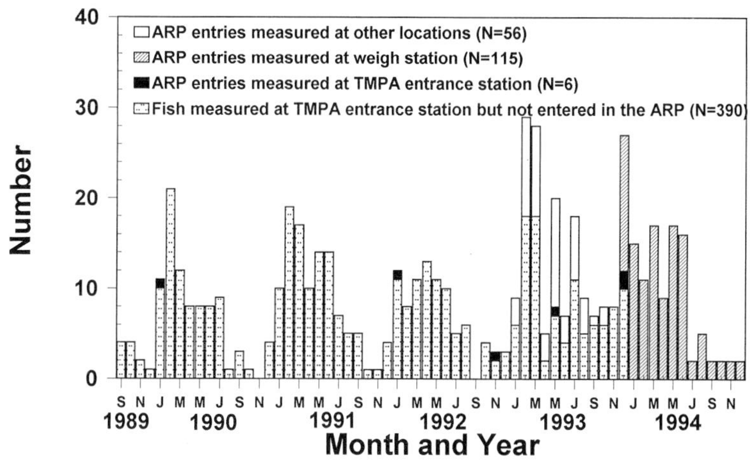

FIGURE 2.—Monthly angler recognition program (ARP) entries for largemouth bass from Gibbons Creek Reservoir before and after construction of the Gibbons Creek Park weigh station compared to numbers of largemouth bass measured at the entrance station but not submitted for ARP awards. The Gibbons Creek Park weigh station became operational 15 December 1993.

nesses maintained promotional information regarding the ARP and provided anglers access to certified scales, entry forms, and assistance. Change in harvest regulation at Gibbons Creek Reservoir prevented harvest after 1 September 1993; therefore, we do not know how many anglers would have continued to use off-site weigh stations in the absence of the regulation. We suspect anglers would have chosen to use the on-site weigh stations preferentially even if harvest had remained optional, because on-site facilities are less disruptive to the recreation experience. Also, providing the opportunity to weigh fish and receive official recognition on-site may influence an angler's decision to harvest fish because no extra travel is required for release of the fish.

Although the accessibility of on-site weigh stations is an advantage to anglers, managers must think of equipment security when considering station design and operation. Vandalism has been rare at Purtis Creek Park or Gibbons Creek Park; however, a digital scale was damaged at Gibbons Creek Park, evidently from placing a very heavy object on the scale pan. Although the act was likely more mischievous than malicious, it does indicate vandalism could be a problem if weigh stations are located in unsecured areas.

Both the Purtis Creek State Park and the Gibbons Creek Park weigh stations have attracted angler interest as information centers as well as weigh stations. Typically, agencies transmit their information to anglers using written reports, news releases, bulletin boards, and radio or television broadcasts. Feedback from anglers often takes the form of comments during public hearings, creel surveys, or outreach events. These traditional information loops may fail to fully meet angler expectations in a society accustomed to immediate answers and a constant flow of information. Weigh stations not only provide anglers immediate information regarding their catch and allow them to compare their catch to existing records, but also provide a mechanism for managers to obtain data from anglers through volunteer creel surveys, open-ended comments, and ARP data. As technology advances, weigh stations have the potential to become interactive with real-time computer applications, allowing anglers access to ARP and other agency databases, as well as providing feedback from the agency regarding fish population sampling, stocking, angling techniques, regulations, etc. With advances in telecommunications, weigh stations may ultimately facilitate dialog between anglers at the reservoir and off-site managers, thereby reducing recall bias, and insuring the best possible input while fishing experiences are still fresh in the minds of anglers.

The weigh stations at Purtis Creek Park and Gibbons Creek Park have been well regarded by anglers as a mechanism for direct feedback about reservoir management efforts. By providing this type of facility in conjunction with restrictive regulations, agencies make it possible for anglers to directly participate in management, thereby forming a partnership that is fundamental in meeting the increasing demands on our fisheries resources.

Acknowledgments

We thank Sportsman's Extravaganza for funding the Purtis Creek State Park weigh station, and Texas Municipal Power Authority for funding the Gibbons Creek Park weigh station. We also thank Richard W. Luebke for his timely editorial assistance. This project was partially funded under Federal Aid in Fish Restoration Act, Grant F-30-R to TPWD.

References

Anderson, R. O. 1975. Factors influencing the quality of largemouth bass fishing. Pages 183–194 *in* R. H. Stroud and H. Clepper, editors. Black bass biology and management. Sport Fishing Institute, Washington, DC.

Barber, W. E., and J. T. Taylor. 1990. The importance of goals, objectives, and values in the fisheries management process and organization: a review. North American Journal of Fisheries Management 10:365–373.

Cofer, L. M. 1995. Evaluation of a trophy bass length limit on Lake Fuqua, Oklahoma. Proceedings of the Annual Conference Southeastern Association of Fish and Wildlife Agencies 47(1993):702–710.

Forshage, A. A., P. P. Durocher, M. A. Webb, and D. G. Lewis. 1991. Management application of angler recognition program data. Proceedings of the Annual Conference Southeastern Association of Fish and Wildlife Agencies 43(1989):36–40.

Horton, R. A., and E. R. Gilliland. 1995. Monitoring trophy largemouth bass in Oklahoma using a taxidermist network. Proceedings of the Annual Conference Southeastern Association of Fish and Wildlife Agencies 47(1993):679–685.

Ott, R. A., and K. W. Storey. 1993. Hooking mortality of channel catfish by trotline. Proceedings of the Annual Conference Southeastern Association of Fish and Wildlife Agencies 45(1991):399–406.

Quinn, S. P. 1987. The status and usefulness of angler recognition programs in the United States. Fisheries 12(2):10–16.

Storey, K. W., and R. A. Ott, Jr. 1994. The largemouth bass catch-and-release regulation at a new reservoir, Purtis Creek State Park Lake, Texas. Proceedings of the Annual Conference Southeastern Association of Fish and Wildlife Agencies 46(1992):377–386.

TPWD (Texas Parks and Wildlife Department). 1995. Texas fishing guide. TPWD, PWD BK N3000-302 (6/95), Austin.

Weithman, A. S., and R. O. Anderson. 1978. A method of evaluating fishing quality. Fisheries 3(3):6–10.

Weithman, A. S., and S. K. Katti. 1979. Testing of fishing quality indices. Transactions of the American Fisheries Society 108:320–325.

Socioeconomics and Optimum Yield

American Fisheries Society Symposium 16:223–235, 1996

Minimum Sociocultural and Economic Data Requirements for Optimum Yield Management of Reservoir Fisheries

Michael D. Hudgins[1] and Stephen P. Malvestuto

Fishery Information Management Systems, Inc.
Post Office Box 3607, Auburn, Alabama 36831-3607, USA

Abstract.—Numerous socioeconomic variables come into play when evaluating recreational fisheries. To provide a pragmatic and optimum yield approach for including socioeconomic aspects into reservoir fisheries management, we suggest a series of sociocultural and economic variables that reservoir managers should gather in concert with traditional fish harvest and effort data. The vehicle for data gathering is a modified creel survey interview. The sociocultural data we cover are used to describe anglers, measure their satisfaction levels, and evaluate information about the reservoir that is available to the public. The economic data we present are used to evaluate trip expenditures, investments in durable equipment, consumer surplus, and other values of willingness to pay. Minimally, we suggest that reservoir managers describe anglers by several sociodemographic and activity categories, measure satisfaction associated with several use characteristics, determine the importance of consumption of wild-caught fish, estimate trip expenditures and consumer surplus, and evaluate printed information about the reservoir. With these data in hand, reservoir managers will be able to evaluate and document the array of quantitative and qualitative yields that reservoirs generate for angling groups and communities.

Imagine this conversation:

"Why are sociocultural and economic data important to reservoir fisheries managers?"

"Because without that information, managers can't account for the many benefits that reservoirs provide to users and nearby communities."

"But why are these benefits important to me as a reservoir manager?"

"Because we live in a time of shrinking budget dollars and increased water use, and if we can't document the value of the resources and our management efforts to users, our bosses, and politicians, we may be out of a job!"

"But I'm a biologist. What do I know about economics or sociology?"

"It will take some initiative on your part, but you can learn the basics well enough to apply them to your work!"

This discussion reflects the political and day-to-day issues that force many fisheries managers to explore the applied aspects of sociology and economics for the first time. Fisheries biologists typically take one introductory course each in sociology and economics as undergraduates; at the graduate level, coursework in these areas disappears (Kelso and Murphy 1988). The transition from introductory courses to applied fieldwork in these areas is difficult to make, but it is manageable with some basic guidance; for example, see Babbie (1979), American Fisheries Society (1987), Guthrie et al. (1991), and Pollock et al. (1994). These sources, however, lack a holistic approach and do not provide a methodological or decision-making framework for fisheries managers.

Malvestuto and Hudgins (1996) present a reconceptualization of optimum yield (OY) as a new paradigm for recreational fisheries management data gathering and decision making (Roedel 1975). The four information categories used for OY decision making are ecosystem health, human health, sociocultural, and economic accounts (Malvestuto 1989). These accounts encompass the array of benefits that can be derived from a fishery by individuals and communities. We conceptualize the ecosystem health account to be the foundation from which all benefits to people and communities arise (Figure 1). Use of, and returns from, the resource are implied by the two-way arrows in Figure 1.

As human use of a fishery increases over time, ecosystem integrity will decrease (the dotted line in Figure 2). Loss of integrity may be indicated by loss of water quality or quantity, destabilization of aquatic and terrestrial populations, increased water pollution, or bioaccumulation of contaminants; however, a variety of social and economic benefits (Barber and Taylor 1990) are generated as more and more people use a fishery. Use can come in many forms, such as power generation, navigation, water storage, harvest of aquatic organisms, and land- and water-based recreation. Each of these uses generates monetary and sociocultural benefits

[1]Present address: Huntingdon College, 1500 E. Fairview Avenue, Montgomery, Alabama 36106-2148, USA.

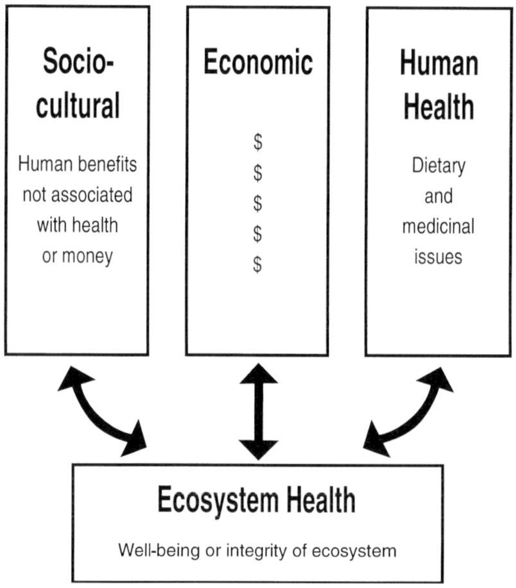

FIGURE 1.—The interdependent relationships among the four optimum-yield accounts, with the ecosystem health account being the primary source that promotes all yields to anglers.

that can be quantified as "yields." Conceptually, these amalgamated "yields" (the parabolic curve in Figure 2) will reach a maximum at some level of human use, at M, and then begin to subside. If a fisheries manager is able to evaluate the biological status of a resource, in concert with sociocultural, human health, and economic "yields" across increasing levels of reservoir use, then it may be possible to conceptualize a level of use, at O, that provides sustained "yields" with minimum degradation of ecosystem health. The essential conceptual shift illustrated in Figures 1 and 2 is that traditional indicators of fishery yield, such as catch and harvest data, are viewed only as indicators of ecosystem health. We do not believe that an OY assessment of a reservoir will ever generate a graph such as that in Figure 2. The main purpose of our approach is to place an appropriate array of data in the hands of decision makers so that a common database exists for OY discussions.

Because reservoirs tend to be multipurpose aquatic resources, fisheries biologists should familiarize themselves with basic sociological and economic data as they relate to OY reservoir use and benefits, or yields, to people. Initially, managers need to develop an understanding of how the outdoor environment, fish populations, and sport and subsistence anglers function together at a reservoir and its environs to generate a variety of yields. This can be accomplished through multidimensional investigations that are applied in a statistically valid manner (Malvestuto et al. 1978; Malvestuto 1983, 1991; Pollock et al. 1994). That is, the resource manager must devise a well-rounded sampling program that generates a balanced array of resource-specific data over several years and must develop sampling designs for randomized data collection.

In this paper, we assume that anglers will be the user group investigated via modified creel survey interviews obtained using either roving or access point techniques. Creel surveys provide a convenient and cost-effective method of collecting OY information because they are routinely used and the questions suggested can be covered in a 10- to 15-min interview with anglers, not including evaluation of harvest (see appendix). Individuals performing these OY interviews should be well briefed to reduce interviewer bias (Hudgins and Malvestuto 1985). If the larger issue of multiple-use assessment of reservoirs is to be addressed, the data collection framework presented here should be expanded to include other reservoir users, an inventory of facilities at the reservoir, and an evaluation of the information available to the public. Eventually, managers will need this additional data to examine issues of environmental health versus benefits to society and evaluate OY trade-offs among user groups.

With this background, we consider the fundamental question: What are the minimum sociocultural and economic data that need to be collected from anglers so that biologists can evaluate and eventually manage reservoir fisheries from an OY point of view? To place this question in a holistic, OY context, we briefly discuss the ecosystem health and human health accounts, we then specify the sociocultural and economic data that should be collected to evaluate anglers from an OY perspective. We will not cover mail or telephone surveys, even though they have been used successfully to obtain certain types of sociocultural and economic data for fisheries management (Guthrie et al. 1991; Pollock et al. 1994). The holistic approach we advocate here requires on-site interviews of anglers so that real-time information is gathered on actual fishing trips across the four OY accounts. All numbered questions referred to in this paper can be found in the appendix.

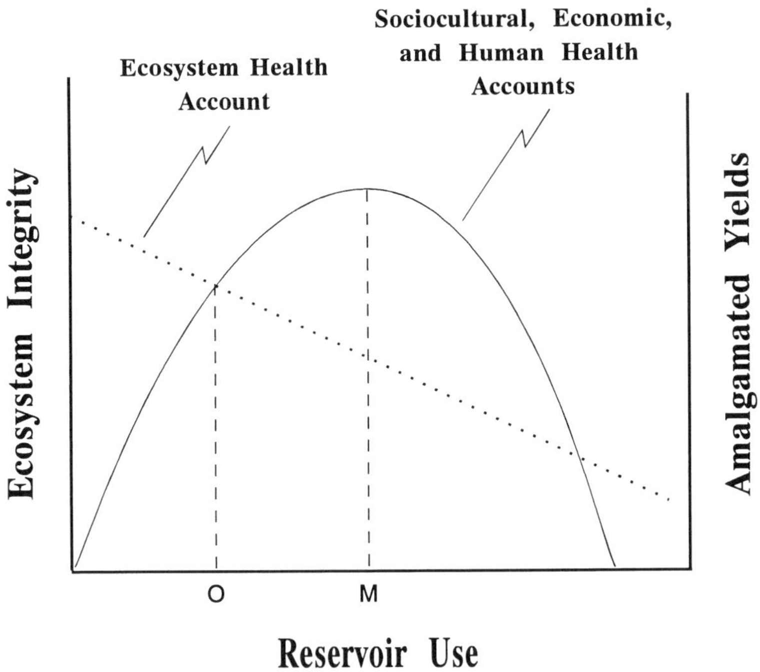

FIGURE 2.—A modified version of the theoretical optimum-yield model (Malvestuto and Hudgins 1996) that integrates reservoir use, ecosystem integrity (dotted line), and amalgamated yields to resource users (parabolic curve); M = level of use associated with maximum yield, O = level of use associated with optimum yield.

The Ecosystem Health Account

In this account we are primarily concerned with measuring variables that provide an indication of the health or condition of a defined ecosystem or an indication of the performance of particular biological populations within the ecosystem. For reservoir fisheries management, it is the health of the fish community that is crucial to providing benefits to people through fishing activities. Data that characterize the performance of fish populations as they respond to environmental changes, and to harvest by anglers, can provide indications of stressed populations that are in need of remedial action by managers to restore equilibrium.

The traditional data on fish harvest collected or generated via creel surveys (e.g., total harvest, harvest rates, catch rates, species composition, length-frequency distributions, condition factors, etc.) are indicators of fish population performance and thus of ecosystem health. Again, OY requires a critical reorientation for fisheries managers. From an OY perspective, harvest rates, species, and sizes of fish in the creel are not yields in and of themselves; rather, they generate sociocultural, economic, and human health yields. That is, anglers are willing to spend money for the opportunity to catch fish, and the fish they harvest provide personal satisfaction, as well as food for the dinner table. Traditional harvest information is solely a measure of how angling exploitation affects the fish community and thus other components of the ecosystem.

For more details on monitoring ecosystems and fish populations, refer to U.S. Environmental Protection Agency's rapid-bioassessment protocols (EPA 1989a), Adams et al. (1993), texts on traditional fishery assessment methods (e.g., Nielsen and Johnson [1983] and Schreck and Moyle [1990]), and papers on environmental assessment in this volume.

The Human Health Account

The human health account deals with health issues associated with eating wild-caught fish. Fish can provide a nutritional source of food for anglers and their families; however, fish flesh can be a source of contaminants potentially harmful to people (Puffer et al. 1982). In both cases, a key index of the importance of fish to human health is the per capita consumption rate (grams of fish eaten per day). This value is a key component of human health risk assessment formulas for determining allowable discharges of harmful chemicals into public waters (EPA 1989b) and is also the basis for assessing the nutritional contribution of wild-caught fish to the diets of reservoir angler populations. Questions 14 and 15 ask anglers for the information required to calculate per capita consumption of wild-caught fish over a 1-month survey period; interviews are done on-site (EPA 1992; FIMS 1994). If a fish consumption advisory is in place at a reservoir, then additional questions regarding cleaning and cooking methods and number of people consuming wild-caught fish at family meals should be added to the interview.

As an alternative to empirical per-capita-consumption calculation methods (Ebert 1996; Meredith and Malvestuto 1996, both this volume), a crude measure of the value of fish as food for family meals can be determined by simply asking anglers to rate the importance of fish as a food source for their families. This question might be: "How important are wild-caught fish as a source of food for your family? Not Important, Slightly Important, Moderately Important, Very Important, Extremely Important" (Question 13). For additional information on fish consumption studies, refer to Cunningham et al. (1994).

The Sociocultural Account

The sociocultural account includes data that describe users and measure user benefits. Three general categories of data describe users: standard demographics (Murdock et al. 1992a, 1992b), angling descriptions (Phippen and Bergersen 1987), and opinions (Dawson and Wilkins 1981; Gigliotti and Peyton 1993). To measure the sociocultural benefits to users, we suggest four basic areas of investigation: satisfactions derived through perceptions of ecosystem health, through facilities used, through fishing success, and through the availability of information about the resource. These satisfactions should be viewed as yields that are generated, either directly or indirectly, by the angling experience (McMichael and Kaya 1991; Holland and Ditton 1992; Spencer and Spangler 1992; Spencer 1993). Another yield or benefit potentially provided to an angler by the fishery is employment; that is, an angler's income may depend primarily on the viability of the fishery, either through employment in or ownership of a business that depends on the fishery for its revenues. Question 22 makes it easier to identify anglers in this category.

Descriptions of anglers and other sociocultural data may be obtained as nominal, ordinal, or ratio measurements. Nominal descriptions are those based on names (e.g., bass anglers versus catfish anglers). Ordinal measurements are based on a series of words or numbers that imply a rating scale (e.g., poor, fair, good, and excellent) or a ranking scale (e.g., 0–10). Ratio measurements are numeric values that have known intervals between the numbers, and changes in values logically can be expressed as ratios or percentages (e.g., dollars, angler-hours, or numbers of fish). Nominal and ordinal measurements can be evaluated via nonparametric statistical analyses (Conover 1971; Carl 1982). Ratio measures can be evaluated using standard nonparametric or parametric statistical methods (Marascuilo and McSweeney 1977; Steel and Torrie 1980). The interview schedule, or scripted questionnaire, in the appendix includes all three types of measurements. Many of the questions have been field tested, but the entire instrument has not.

Describing Users

Descriptions of anglers should be based on easily determined characteristics or on categories generated through the interview process (Moring 1985). These variables will allow fisheries managers to define many subpopulations within the total angler population sampled and enable the manager to partition and evaluate resource use, fish harvest, opinions, and economic impacts by all groups described.

The five standard demographic variables of ethnic background, gender, age, income, and residence are valuable angler descriptors that are used in many fishery investigations for market segmentation. The first three of these variables, listed before Question 1, can be collected without verbal exchange if crude categorizations are acceptable. The importance of ethnic background is often debated, but this information will allow resource use to be partitioned so that minority input and participation can be evaluated. Also, ethnic categories have proven to be useful when evaluating "populations at risk" in fish consumption studies. Gender and age data provide a description of the angling public so that facilities (Hudgins 1984) and information initiatives can be evaluated (Schramm and Dennis 1993). For example, fishing sites that draw a large number of women and children should have additional security measures in place (e.g., better lighting) and a variety of land-based, family-oriented recreational opportunities (Howard 1979). Also, signs and pamphlets about the fishery should be produced with age, ethnic, and gender diversity in mind.

Standard demographic information also can be used for basic sociocultural research. For example, residence data (Question 1) will allow tourism researchers or natural-resource economists to determine if fishery use is local or extends to other parts of the state or other regions of the country (Matlock 1983). Household income (Question 23) determines to a large degree how much anglers are willing to spend for the opportunity to fish at a particular site. Therefore, incomes of anglers (Question 23) and family food needs (Question 13) can be investigated relative to species preferences (Question 10) and fish consumption rates (Questions 14 and 15).

These are just a few examples of how data from a scripted questionnaire can be used to understand clientele and plan OY management actions. Collectively, the angler description data obtained in the appendix will allow managers to define specific sociodemographic subpopulations and evaluate subpopulation behaviors and perceptions (i.e., their expenditures, satisfactions, preferences, and opinions). Through these data, managers will be able to focus on particular angler groups and behaviors in different ways to enhance yields to society in an OY framework. These data can also be used by local chambers of commerce, tourism agencies, economists, and recreation planners in marketing campaigns that promote the fishery and its environs.

Two angling descriptors can be entered on the interview form before the interview: whether fishing is taking place from bank or boat and the fishing location of bank anglers. The first descriptor is useful to managers because bank and boat anglers tend to have different attitudes regarding access development, fishing regulations, and fish harvest (Hudgins 1984). Also, fishing location puts anglers into categories that generally relate to trip expenditures, that is, bank anglers tend to spend less per day than boat anglers. The parks and facilities used by bank anglers will allow managers to pinpoint sites where bank anglers tend to cluster. This information can be used in concert with facility inventory data, data from Question 6 (trip description), and data from Question 7 (number of trip days) to determine if existing facilities for bank anglers are adequate or if new facilities are needed at various locations. Question 2, along with Questions 6 and 7, will allow similar information to be generated for boat anglers.

Number in the fishing party (found at the top of the appendix) may or may not be ascertained visually. For instance, one person in the group may not be actively fishing at the time of interview, but has been and will be fishing. Thus, verbal query is usually necessary for an accurate count. Number in the angling party becomes critical when calculating per-angler or per-trip values based on questions that are asked of the entire fishing party. For example, harvest and trip expenditure data usually are easily and more accurately collected from each member of the entire party, rather than from only one angler in the party. For management purposes, sizes of angling groups across other descriptors, such as bank and boat, are required when evaluating carrying capacity (Bryan 1982; Schoolmaster 1986) and use of shoreline facilities (Gold 1980). In concert with Questions 6 (trip description) and 7 (trip duration), number in the angling party will help managers develop profiles of use based on size, type, and duration of social events supported by the resource, which can be taken as indicators of "family and community values" associated with resource use. In addition, economic values can be partitioned relative to these events.

Other descriptors enable managers to partition anglers based on fishing behavior. Questions 3 and 4 allow the length of the fishing trip to be estimated if anglers are interviewed while they are still fishing. Trip lengths, based on interviews of anglers leaving the water at access points, are calculated using Question 3 (time fishing began) and the time of the interview (found at the top). The hours during the day spent at the site will allow managers to establish peak times of facility use (e.g., boat ramps and parking spaces) which, in concert with data on accesses, parks used, and facility satisfaction (Question 17), will help managers plan facility development.

For angling-research purposes, number of angling days per year at the reservoir (Question 5) indicates the avidity of anglers (Teirney and Richardson 1992). Question 12 (knowledge of the resource) is a proxy for avidity and an indicator of interest in the resource. Data from these questions allow relative use to be evaluated across angling groups and community distances (Question 1). When considered along with Question 6 (tournament anglers), Question 10 (targeted fish group or species), and Question 11 (perceived skill), avidity information can help to build a profile of angler specialization similar to those presented by Bryan (1977), Loomis and Ditton (1987), Chipman and Helfrich (1988), and Wilde and Ditton (1994).

The fish group or species of fish that individual anglers primarily seek (Question 10) is a traditional angling descriptor. This classification is often used to establish "intended catch and effort" statistics for a fishery. Describing anglers by the fish they intend to harvest allows managers to develop another series of angler profiles and provides an indication of fishes that are highly desired. These data will indicate the fish populations that are most heavily affected by angling and thus in need of management action.

The last category of angling descriptors is opinions (Questions 20 and 21). Questioning should be customized for site-specific issues, for example, creel length limits, habitat restoration, or new facility development. Opinion questions can be presented using a rating or ranking scale that explores the level of satisfaction anglers experience when faced with an angling regulation, for example, on a scale of 0–10 where 0 is "Not At All Satisfied" and 10 is "Extremely Satisfied" (Matlock et al. 1991). These data are valuable when assessing public response to management initiatives. Of course, interviews should be conducted before an initiative is implemented in order to gauge angler sentiment. A preliminary question might be: "If our fish population data indicate that a 9-inch length limit on crappie would improve crappie fishing, would you support this regulation? Yes, No, and Why?"

Measuring Benefits to Users

The primary sociocultural benefits derived by resource users are measured as levels of satisfaction associated with various aspects of the resource and resource use. Levels of satisfaction can be measured using the scale presented for opinions, namely, 0–10 where 0 is "Not At All Satisfied" and 10 is "Extremely Satisfied." For research purposes, it is best to investigate the range of satisfactions found (i.e., from "Not At All Satisfied" to "Extremely Satisfied") and the reasons for all levels of satisfaction; however, for management purposes, it is most critical to know why users experience average or less-than-average satisfaction (i.e., 5 or lower on the 0–10 scale) and what can be done specifically to enhance the resource to increase user satisfaction. The follow-up question "Why?" after a satisfaction question, is termed an open-end question. The variety of responses from open-end questions will have to be categorized (coded) after data gathering begins. This categorization process can be somewhat laborious, but the information gleaned from open-end questions will be invaluable when considering the many circumstances that can reduce levels of satisfaction. Undoubtedly, some aspects of satisfaction will not be within the manager's control.

When asking people if they are satisfied with the ecological health of the reservoir (Question 16), the interviewer must be able to explain the concept of health as it relates to aquatic ecosystems. The addition of the sentence "A very healthy reservoir is one that is free of pollution, destruction of habitat, and other negative human impacts" should provide the appropriate frame for responses. If anglers cannot relate to this explanation, then an alternative that reflects local circumstances will be required. The large blank after Question 16 is the space for the answer to the open-end question: "What is the most important thing that can be done here to increase your satisfaction with this reservoir's ecological health?" Data from this question can be used to determine anglers' knowledge about human impacts on the reservoir. This type of inquiry could lead to the development of information packets that explain issues raised by anglers.

Shoreline facilities and their upkeep are always sources of intense conversation when anglers are asked to discuss reservoir use and management. Question 17 allows managers to link facility satisfaction ratings to specific facilities used. For example, family outings at a reservoir (data from Question 6, along with age and gender data) may occur most frequently at access areas where restroom facilities and picnic tables are available. If users wanted more places for family outings, or if managers needed to orient them to under-utilized perimeter parks, then adding these facilities to other sites and publicizing the additions could increase satisfaction, provide more family-oriented parks, and shift use patterns.

Ratings or rankings of fishing success have been presented in a variety of formats (e.g., poor, fair, good, or excellent). Question 18 asks anglers to indicate their levels of satisfaction with fishing success using a 0–10 scale. Even though it may not be possible for a reservoir manager to improve catch per unit of effort (CPUE) or harvest per unit of effort (HPUE), satisfaction with angling is certainly a fundamental reason why anglers return to a reservoir year after year. It is assumed that if anglers who travel a long distance (Question 1) respond to Question 18 with a 5 or below, then they may not return. Fewer returns might mean fewer travel dollars in the local economy. The fundamental issues here are the mobility of anglers and their perceptions of success (Hudgins and Davies 1984). For OY purposes, Question 18 should be investigated across a number of reservoir and angler types, and particularly in relation to values of CPUE and HPUE.

The last category of satisfactions involves those

satisfactions derived through the availability of information about an aquatic resource (Question 19). By asking this question, we assume that anglers value information, derive some level of satisfaction from reading that information, and use it to enhance their angling experience. In other words, resource-specific information provides benefits to people while they are on and off the reservoir. This information is usually available to the public in the form of pamphlets, information packets, newspapers, magazines, television shows, and videotapes. The topics that may be covered in these publications are directions to accesses, access descriptions, the names and numbers of local merchants and emergency personnel, resource characteristics (e.g., bottom contour maps, unique habitats, fish species present), and harvest characteristics. As with the satisfactions discussed previously, satisfaction with the availability of information about a resource can be measured using a 0–10 scale. The follow-up question might be, "What is the most important kind of information that could be provided to you to increase your satisfaction with your fishing trip to this reservoir?"

The issue of information availability requires that the reservoir manager assemble, catalog, and evaluate all information available to the public (i.e., brochures, maps, travel guides, posters, etc.). Because these materials come from a variety of sources, probably no one has looked at them as one body of information. Evaluation of this material probably will uncover omissions and outdated information that should be updated for the angling public.

The Economic Account

The economic account measures the monetary value of the resource (Cauvin 1980; Matlock 1986; Sorg and Loomis 1986; Guthrie et al. 1991; Pollock et al. 1994). Valid economic information can be collected on-site by face-to-face interviews with anglers (Milton 1991; Storey and Allen 1993). There are four primary economic issues relevant to OY management of reservoirs: (1) trip expenditures; (2) investments in durable equipment; (3) consumer surplus (CS), a type of willingness to pay (WTP) (Swanson and McCollum 1991); and (4) other WTP values. Trip expenditures and investments in durable equipment are most relevant to measuring benefits to the business community. The minimum economic variables for evaluation of angling are trip expenditures and CS. We think that investments in durable equipment and other WTP values are beyond those minimally required for OY assessment of a reservoir; however, a cursory review is presented.

Investments in durable fishing equipment represent fixed costs for items that can be used over a number of years (Gordon et al. 1973). It is likely that the purchase of durable equipment, in particular, might have taken place at stores away from the locality of the resource, and thus would be more important for regional or state economic-impact assessment (Reichers and Fedler 1996, this volume). In addition, the estimation of investments in durable equipment can be particularly difficult because data are required on when, and for how much, each item was purchased, as well as the proportion of time each item is used at the survey site. Getting this detailed information can be time consuming and potentially fraught with recall bias. Thus, we do not recommend the collection of cost data for durable equipment unless it is very pertinent to the objectives of the reservoir manager.

Other WTP values (item 4 above) are theoretical dollar values that anglers would be willing to pay for aspects of the fishery not associated with their daily use of the resource. For example, how much would anglers be willing to pay simply to know the ecosystem would continue to be healthy or be available to future generations for fishing? An outline and discussion of these values were covered by Ozuna and Stoll (1991). In general, these WTP values are rather abstract and theoretically can be associated with both users and nonusers of resources. We do not consider these values to be a necessary component of minimum OY data requirements.

Trip expenditures are variable trip costs incurred for items such as gas, food, picnic supplies, lodging, and bait (Question 8). These dollar values are primarily a measure of benefits derived by the business community serving the fishery (Martin et al. 1982; Guthrie and Peterman 1988; Johnson et al. 1995). Trip expenditures also are a part of the total value that anglers place on the opportunity to fish at a particular site. Anglers certainly value the opportunity to fish at least as much as the money they are willing to spend to travel to and from the site and engage in recreational activity at the site; however, money actually spent during the trip goes to businesses that provide supplies and services associated with fishing. Trip expenditures are more likely to occur in the locality of the resource than are expenditures on durable equipment items and are therefore more closely linked to management actions at the resource. Trip expenditures are thus a measure of benefits to the local business community and also are part of the economic value of the recreational site to the user. Expenses for trips of more than one day can be converted to daily values for statistical expansions along with other creel survey data (Questions 7 and 8).

Consumer surplus measures the amount of money over and above trip expenses that users would be willing to pay to use the resource (Palm and Malvestuto 1983). Theoretically, CS is the amount of money that a resource owner would be able to collect by providing fishing opportunities at a particular site, as if users were required to pay a daily fee to go fishing. This amount of money is the theoretical profit (not realized) associated with the fishery and is directly comparable to profits of other economic enterprises. Thus, by estimating CS of anglers, the value of the sport fishery can be compared to other economic enterprises in the region. In addition, effects of management actions on this WTP value can be evaluated.

The most direct way to estimate CS for anglers is to ask them how much money, over and above the money they have spent on their fishing trip, they would be willing to pay for the opportunity to fish. This approach is termed the contingency valuation method, or CVM (Reichers and Fedler 1996). There are various ways of posing questions to users to estimate CS. We recommend the following simple series: First, "How much money will you spend during this fishing trip on the following items?" (Question 8). The angler should give a series of dollar values for the listed items that the interviewer will have to add together to get total expenditures for the trip (X dollars). Second, "How much more than "X dollars" would you be willing to pay to fish here today?" The amount of money that the user spent during that day or trip is used as a reference point, or threshold, to enable the user to think about how much more money he or she would be willing to pay to use the resource that day. This amount of money is the angler's CS for that particular day at the site. The CS question could be phrased as the "maximum you would be willing to pay" if maximum values are required or may be presented as a "bid" question (Reichers and Fedler 1996). We feel that our modest approach to determination of CS is reasonable, both in the interest of interview time (particularly for roving surveys) and given the difficulty people seem to have when asked about hypothetical values.

It is important to note that another approach for estimating CS is the travel cost method, or TCM (Reichers and Fedler 1996). With anglers, however, we have found that many critical assumptions in the TCM are not met. In addition, the quantitative procedures for the TCM are cumbersome and involve relatively advanced mathematical skills, such as nonlinear and multiple regression, and integration. We recommend using contingent valuation rather than a travel cost approach.

Realization of Optimum-Yield Management

This paper has presented an overview of the sociocultural and economic data requirements for OY assessment of reservoir fisheries. Minimally, we suggest that reservoir managers describe anglers in several different ways, measure the satisfaction levels of anglers, collect information on the consumption of wild-caught fish, gather data on trip expenditures and consumer surplus, and evaluate the information that is available to the public about the reservoir. Over time, the proposed OY assessment of recreational reservoir fisheries will provide managers with a better understanding of the biological and physicochemical state of the reservoir, the benefits and yields derived by user groups, and the market and nonmarket values generated by reservoir use. Evaluation of the OY information across reservoirs, drainages, and regions will (1) help fisheries managers evaluate options for enhancement of the recreational experience through exposure to a variety of angler satisfaction, preference, and opinion data; (2) provide documentation in political and budget discussions for the expenditures and WTP values generated by reservoir fisheries use; (3) allow researchers to investigate the development of an OY index; and (4) identify areas where physical improvements can be made that maintain or increase yields to individuals and communities, while still protecting the health and performance of sport fish populations.

References

Adams, S. M., A. M. Brown, and R. W. Goede. 1993. A quantitative health assessment index for rapid evaluation of fish condition in the field. Transactions of the American Fisheries Society 122:63–73.

American Fisheries Society. 1987. Social assessment of fisheries resources. Transactions of the American Fisheries Society 116(3).

Babbie, E. R. 1979. The practice of social research. Wadsworth, Belmont, California.

Barber, W. E., and J. N. Taylor. 1990. The importance of goals, objectives, and values in the fisheries management process and organization: a review. North American Journal of Fisheries Management 10:365–373.

Bryan, H. 1977. Leisure value systems and recreational specialization: the case of trout fishermen. Journal of Leisure Research 9:174–187.

Bryan, H. 1982. A social science perspective for managing recreational conflict. Pages 15–22 *in* F. E. Carlton and R. H. Stroud, editors. Marine recreational fisheries. Sport Fishing Institute, Washington, DC.

Carl, L. M. 1982. Social impacts of a stream reclamation project on urban anglers. North American Journal of Fisheries Management 2:164–170.

Cauvin, D. M. 1980. The valuation of recreational fisheries. Canadian Journal of Fisheries and Aquatic Sciences 38:145–154.

Chipman, B. D., and L. A. Helfrich. 1988. Recreational specializations and motivations of Virginia river anglers. North American Journal of Fisheries Management 8:390–398.

Conover, W. J. 1971. Practical nonparametric statistics. Wiley, New York.

Cunningham, P. A., S. L. Smith, J. P. Tippett, and A. Greene. 1994. A national fish consumption advisory data base: a step toward consistency. Fisheries 19(5): 14–23.

Dawson, C. P., and B. T. Wilkins. 1981. Motivations of New York and Virginia marine boat anglers and their preferences for potential fishing constraints. North American Journal of Fisheries Management 1:151–158.

Ebert, E. S. 1996. Fish consumption and human health: developing partnerships between risk assessors and resource managers. Pages 261–270 *in* L. E. Miranda and D. R. DeVries, editors. Multidimensional approaches to reservoir fisheries management. American Fisheries Society Symposium 16.

EPA (U.S. Environmental Protection Agency). 1989a. Rapid bioassessment protocols for use in streams and rivers. U.S. Environmental Protection Agency, Office of Water Regulations and Standards, Publication WH-553, Washington, DC.

EPA (U.S. Environmental Protection Agency). 1989b. Assessing human health risks from chemically contaminated fish and shellfish: a guidance manual. U.S. Environmental Protection Agency, Office of Water Regulations and Standards, Document 503/8-89-002, Washington, DC.

EPA (U.S. Environmental Protection Agency). 1992. Consumption surveys for fish and shellfish: a review and analysis of survey methods. U.S. Environmental Protection Agency, Office of Water Regulation and Standards, Document 822/R-92-001, Washington, DC.

FIMS (Fishery Information Management Systems). 1994. Estimation of daily per capita freshwater fish consumption of Alabama anglers. Final report for the Alabama Department of Environmental Management, Montgomery.

Gigliotti, L. M., and R. B. Peyton. 1993. Values and behaviors of trout anglers, and their attitudes toward fishery management, relative to membership in fishing organizations: a Michigan case study. North American Journal of Fisheries Management 13:492–501.

Gold, S. M. 1980. Recreation planning and design. McGraw-Hill, New York.

Gordon, D., D. W. Chapman, and T. C. Bjornn. 1973. Economic evaluation of sport fisheries—what do they mean? Transactions of the American Fisheries Society 102:293–311.

Guthrie, I. C., and R. M. Peterman. 1988. Economic evaluation of lake enrichment strategies for British Columbia sockeye salmon. North American Journal of Fisheries Management 8:442–454.

Guthrie, D., and seven coeditors. 1991. Creel and angler surveys in fisheries management. American Fisheries Society Symposium 12.

Holland, S. M., and R. B. Ditton. 1992. Fishing trip satisfaction: a typology of anglers. North American Journal of Fisheries Management 12:28–33.

Howard, I. C. 1979. Opinions, preferences, satisfactions, and importance of women anglers in Massachusetts. Fisheries 4(6):32–34.

Hudgins, M. D. 1984. Structure of the angling experience. Transactions of the American Fisheries Society 113:750–759.

Hudgins, M. D., and W. D. Davies. 1984. Probability angling: a recreational fishery management strategy. North American Journal of Fisheries Management 4:431–439.

Hudgins, M. D., and S. P. Malvestuto. 1985. An evaluation of factors affecting creel clerk performance. Proceedings of the Annual Conference Southeastern Association of Fish and Wildlife Agencies 36(1982): 252–263.

Johnson, D. M., R. J. Behnke, D. A. Harpman, and R. G. Walsh. 1995. Economic benefits and costs of stocking catchable rainbow trout: a synthesis of economic analysis in Colorado. North American Journal of Fisheries Management 15:26–32.

Kelso, W. E., and B. R. Murphy. 1988. University fisheries programs in the United States: structure and needs. Fisheries 13(2):9–15.

Loomis, D. K., and R. B. Ditton. 1987. Analysis of motive and participation differences between saltwater sport and tournament fishermen. North American Journal of Fisheries Management 7:482–487.

Malvestuto, S. P. 1983. Sampling the recreational fishery. Pages 400–423 *in* L. A. Nielsen and D. L. Johnson, editors. Fisheries techniques. American Fisheries Society, Bethesda, Maryland.

Malvestuto, S. P. 1989. Sociological perspectives on large river management: a framework for the application of optimum yield. Pages 589–599 *in* D. P. Dodge, editor. Proceedings of the international large river symposium. Canadian Special Publication of Fisheries and Aquatic Sciences 106.

Malvestuto, S. P. 1991. The customization of recreational fishery surveys for management purposes in the United States. Pages 201–213 *in* I. G. Cowx, editor. Catch effort sampling strategies: their application in freshwater fisheries management. Fishing News Books, Oxford, UK.

Malvestuto, S. P., W. D. Davies, and W. L. Shelton. 1978. An evaluation of the roving creel survey with nonuniform probability sampling. Transactions of the American Fisheries Society 107:255–262.

Malvestuto, S. P., and M. D. Hudgins. 1996. Optimum

yield for recreational fisheries management. Fisheries 21(6):6–7.

Marascuilo, L. A., and M. McSweeney. 1977. Nonparametric and distribution-free methods for the social sciences. Brooks/Cole Publishing Company, Monterey, California.

Martin, W. E., F. H. Bollman, and R. L. Gum. 1982. Economic value of Lake Mead fishery. Fisheries 7(6): 20–24.

Matlock, G. C. 1983. Angler origin and their catch retention rate while fishing on seven Texas bay systems. North American Journal of Fisheries Management 3:438–444.

Matlock, G. C. 1986. Estimating the direct market economic impact of sport angling for red drum in Texas. North American Journal of Fisheries Management 6:490–493.

Matlock, G. C., H. R. Osburn, R. K. Riechers, and R. B. Ditton. 1991. Comparison of response scales for measuring angler satisfaction. Pages 413–422 *in* Guthrie et al. (1991).

McMichael, G. A., and C. M. Kaya. 1991. Relations among stream temperature, angling success for rainbow trout and brown trout, and fisherman satisfaction. North American Journal of Fisheries Management 11:190–199.

Meredith, E. K., and S. P. Malvestuto. 1996. Evaluation of two on-site survey methods for determining daily per capita freshwater fish consumption by anglers. Pages 271–276 *in* L. E. Miranda and D. R. DeVries, editors. Multidimensional approaches to reservoir fisheries management. American Fisheries Society Symposium 16.

Milton, J. W. 1991. Measuring the economic value of anglers' kept and released catches. North American Journal of Fisheries Management 11:185–189.

Moring, J. R. 1985. Formulating simple angler profiles to better manage "catchable" trout fisheries. North American Journal of Fisheries Management 5:413–414.

Murdock, S. H., K. Backman, R. B. Ditton, M. N. Hoque, and D. Ellis. 1992a. Demographic change in the United States in the 1990s and the twenty-first century: implications for fisheries management. Fisheries 17(2):6–13.

Murdock, S. H., K. Backman, R. B. Ditton, M. N. Hoque, and D. Ellis. 1992b. Implications of future demographic change for participation in fishing in Texas. North American Journal of Fisheries Management 12:548–558.

Nielsen, L. A., and D. L. Johnson, editors. 1983. Fisheries techniques. American Fisheries Society, Bethesda, Maryland.

Ozuna, T., Jr., and J. R. Stoll. 1991. The significance of data collection and economic methods in estimating nonmarket resource values. Pages 328–335 *in* Guthrie et al. (1991).

Palm, R. C., and S. P. Malvestuto. 1983. Relationships between economic benefit and sport-fishing effort on West Point Reservoir, Alabama–Georgia. Transactions of the American Fisheries Society 112:71–78.

Phippen, K. W., and E. P. Bergersen. 1987. Angling definitions and their effects on the accuracy of count-interview creel survey harvest estimates. North American Journal of Fisheries Management 7:488–492.

Pollock, K. H., C. M. Jones, and T. L. Brown. 1994. Angler survey methods and their application in fisheries management. American Fisheries Society Special Publication 25.

Puffer, H. W., M. J. Duda, and S. P. Azen. 1982. Potential health hazards from consumption of fish caught in polluted coastal waters of Los Angeles County. North American Journal of Fisheries Management 2:74–79.

Riechers, R. K., and A. J. Fedler. 1996. An overview of economic impact and value of recreational fisheries. Pages 245–250 *in* L. E. Miranda and D. R. DeVries, editors. Multidimensional approaches to reservoir fisheries management. American Fisheries Society Symposium 16.

Roedel, P. M., editor. 1975. Optimum sustainable yield as a concept in fisheries management. American Fisheries Society Special Publication 9.

Schoolmaster, F. A. 1986. Bank- and float-angler perceptions of use levels on the Madison River, Montana. North American Journal of Fisheries Management 6:430–438.

Schramm, H. L., Jr., and J. A. Dennis. 1993. Characteristics and perceptions of users and nonusers of an urban fishery program in Lubbock, Texas. North American Journal of Fisheries Management 13:210–216.

Schreck, C. B., and P. B. Moyle, editors. 1990. Methods for fish biology. American Fisheries Society, Bethesda, Maryland.

Sorg, C. F., and J. B. Loomis. 1986. Economic values of Idaho sport fisheries with an update on valuation techniques. North American Journal of Fisheries Management 6:494–503.

Spencer, P. D. 1993. Factors influencing satisfactions of anglers on Lake Miltona, Minnesota. North American Journal of Fisheries Management 13:201–209.

Spencer, P. D., and G. R. Spangler. 1992. Effect that providing information has on angler expectations and satisfactions. North American Journal of Fisheries Management 12:379–385.

Steel, R. G. D., and J. H. Torrie. 1980. Principles and procedures of statistics: a biometrical approach, second edition. McGraw-Hill, New York.

Storey, D. A., and P. G. Allen. 1993. Economic impact of marine recreational fishing in Massachusetts. North American Journal of Fisheries Management 13:698–708.

Swanson, C. S., and D. W. McCollum. 1991. Application of economics to recreational fisheries management: an overview. Pages 299–315 *in* Guthrie et al. (1991).

Teirney, L. D., and J. Richardson. 1992. Attributes that characterize angling rivers of importance in New Zealand, based on angler use and perceptions. North American Journal of Fisheries Management 12:693–702.

Wilde, G. R., and R. B. Ditton. 1994. A management-oriented approach to understanding diversity among largemouth bass angler. North American Journal of Fisheries Management 14:34–40.

Appendix: Interview Form for Minimum Optimum Yield Data Requirements

System:______________________ Date:____/____/____(Month/Day/Year)

Sample No._____ Time of Interview: ____:____

Number in fishing party: _____ Bank ____ Boat ____

For Bank Anglers, Park or Facility Used:______________________

Ethnic Background: **(Enter number of people in each blank)**

____ (**B**) Afro-American ____ (**H**) Hispanic-American
____ (**A**) Asian-American ____ (**N**) Native-American
____ (**C**) Euro-American ____ (**O**) Other

Age Categories: **(Enter number of people in category)**

MALES:

<10 ___, 10/19 ___, 20/29 ___, 30/39 ___, 40/49___, 50/59 ___, ≥60___

FEMALES:

<10 ___, 10/19 ___, 20/29 ___, 30/39 ___, 40/49___, 50/59 ___, ≥60___

1. Where do you live?

   Town:______________ County:____________ State:______

2. If boat angler, where did you launch your boat today?_________

3. At what time did you begin to fish today? _____:______

   (for access survey, use time of interview to calculate effort)

4. At what time do you think that you will finish fishing today? ______: _______ (ask only during roving surveys).

5. How many days per year do you fish here? _______ days

6. Would you describe your fishing trip today as...Individual _____, a Family outing _____ , a group of Friends _____ , participating in a Tournament _____ , or a sponsored group (church, scouts, civitan, etc.) _____?

7. Is this a 1-day trip? Yes _____ No _____ (If No, then ask) What are the total number of days in this trip? _______ days

8. How much money will you spend during this fishing trip on the following items?

Gas for Car and Boat	Food Beverages Ice	Picnic Supplies	Lodging	Equipment Rental	Activity Supplies
$	$	$	$	$	$

9. You have spent about $______________ [**Sum of Q#8/number of trip days in Q#7**] to fish here today. How much more than $__________ [**Sum of Q#8/number of trip days**] would you be willing to pay to fish here today?

   $________________

10. What species of fish are you primarily fishing for today? ________

11. How skilled of an angler do you consider yourself to be?

    Beginner_____ Average _____ Better than Average _____

12. Using a scale of Not, Slightly, Moderately, Very, or Extremely, how knowledgeable are you about fishing at this lake? ______

13. Using a scale of Not, Slightly, Moderately, Very, or Extremely, how important are wild-caught fish as a source of food for your family? _______

14. How many family meals of fish caught from this reservoir have you eaten over the last month? _________ meals

15. [***Give respondent visual aid***] How many servings of fish, like the 4-ounce one illustrated, would you normally eat at a family meal where wild-caught fish are served? ____

For the next 6 questions, please respond using a 0 to 10 scale, where 0 is Not At All Satisfied and 10 is Extremely Satisfied. [If person gives an answer that is 5 or below, then ask: "What is the most important thing that can be done here to increase your satisfaction with ...... (**the topic of the question**)?]

16. How satisfied are you with this reservoir's ecological health? A very healthy reservoir is one that is free of pollution, destruction of habitat, and other negative human impacts. _______
    ____

17. How satisfied are you with the facilities and upkeep of this site (bank anglers) or the site you launched from (boat anglers)?_____

18. How satisfied are you with your fishing success today? ______

19. How satisfied are you with the information available about this reservoir? _______

20. How satisfied are you with the current 16-inch length limit on largemouth bass? __________

21. How satisfied are you with the current 9-inch length limit on crappie? _________

22. Is your employment dependent on the viability of this fishery?

Yes ______ No _______

23. What is your annual household income? [***Give respondent visual aid to choose category***]

A: LESS THAN $5,000 PER YEAR, B: $5,000 TO $9,999, C: $10,000 TO $14,999, D: $15,000 TO $19,999, E: $20,000 TO $24,999, F: $25,000 TO $34,999, G: $35,000 TO $49,999, H: $50,000 TO $74,999, I: $75,000 TO $100,000, J: OVER $100,000

Harvest Data

Species	Length (cm)	Number		Species	Length (cm)	Number

American Fisheries Society Symposium 16:236–244, 1996

Using Survey Research in Support of Fisheries Management: The 1994 Texas Statewide Angler Survey

KEVIN M. HUNT AND ROBERT B. DITTON
Texas A&M University, Department of Wildlife and Fisheries Sciences
210 Nagle Hall, College Station, Texas 77843–2258

Abstract.—Despite widespread recognition of the importance of human dimensions research in fisheries, fisheries managers and field personnel need greater understanding of the general approaches to angler surveys and specific methods for conducting them. Instead of focusing on study results, we developed the background for how and why the most recent Texas statewide angler survey was completed. From our 10 years of cooperative experience with the Texas Parks and Wildlife Department, we highlight the need for developing relationships between fisheries managers and social science researchers, and we discuss survey objectives, question development, sample size considerations, appropriate sampling frames for angler studies, other pertinent design issues, and strategies for using survey results in fisheries management. Finally, we make recommendations on how state fisheries agencies can most effectively gather and report human dimensions information.

Much has been written on the human dimensions of fisheries, why and how research on the topic is done, its importance in understanding fisheries, and its use in decision making (Hendee 1974; Smith 1983; Driver 1985; Brown 1987; Essig and Holliday 1991). However, the general approaches and specific methods used in angler surveys have received little attention, especially regarding examples and applications for fisheries managers and field personnel who seek to understand their clientele but may not be formally trained in human dimensions inquiry.

Most of the angler information available to fisheries biologists comes from creel survey interviews, public hearings, and chance encounters during management activities. Creel surveys, which focus on collecting catch and effort data, provide few additional opportunities for ascertaining anglers' fishing activity or preferences. For example, managers have little basis for knowing the level of angler support on a particular issue, such as support for a reservoir's slot-length limit on largemouth bass *Micropterus salmoides*. Although it would be beneficial to conduct expanded creel surveys or follow-up mail surveys of anglers intercepted for individual waterbodies, budgetary constraints limit such surveys (Pollock et. al 1994). Statewide or regional angler survey designs may be the only means by which managers can quantify the anecdotal information they receive from anglers and special interest groups.

This paper does not report results from a traditional data-based fisheries study, but rather offers insights gained from our experience with angler surveys in Texas. After introducing the Texas statewide angler survey and noting general types of human dimensions inquiry, we review survey objectives, sampling issues, rationale for data collected, methods of measurement, and important sources of bias associated with angler surveys. We then discuss the dissemination of results from angler surveys and emphasize the importance of building partnerships for conducting human dimensions work. We do not attempt to cover all aspects of angler surveys but rather emphasize key points. If more detail is desired, it can be found in Smith (1983), Malvestuto (1983), and Pollock et al. (1994), who provide thoughtful reviews of the steps involved in creel and angler surveys. We use our experience from the 1994 Texas Statewide Angler Survey to identify key issues and describe human dimensions research in a practical sense. By doing so, expectations for such inquiry may be clarified for fisheries managers.

Evolution of the Texas Statewide Angler Survey

Statewide angler surveys in Texas have evolved over time. The first survey was a household interview study conducted in 1957 by Belden Associates, a Dallas-based market research firm, to learn more about saltwater-angler fishing habits and characteristics. The study was never released by the then Texas Game and Fish Commission because of perceived inaccuracies in reported catch data. No other angler surveys were completed in Texas until 1986, when a survey of saltwater anglers was possible because a saltwater sportfishing stamp authorized by the Texas Legislature in 1985 provided a means to access this angler group (Table 1). In 1987, the file of general fishing license receipts was sampled

TABLE 1.—Surveys conducted by the Texas A&M University Human Dimensions of Fisheries Laboratory in conjunction with the Texas Parks and Wildlife Department 1986–1994.

Sampling frame	Year of study	Sample size	Effective response rate[a]
Saltwater stamp holders	1986	6,371	71.8
General fishing license holders	1987	10,657	64.7
Saltwater stamp holders	1988	7,847	66.3
Saltwater stamp holders	1989	3,455	66.5
General fishing license holders	1989	4,111	66.4
General fishing license holders	1989	10,001	65.0
General fishing license holders	1990	9,981	62.0
Largemouth bass anglers[b]	1992	1,965	69.9
Spotted seatrout anglers[c]	1993	1,597	65.6
General fishing license holders	1994	10,000	61.5

[a]Effective response rate = returned usable + returned non-usable × 100/sample size − nondeliverables

[b]Anglers with stated preference for largemouth bass in 1990 survey.

[c]Anglers with stated preference for spotted seatrout in 1990 survey.

manually and those selected were sent a mail questionnaire about their freshwater fishing activities. In 1989, in response to vendor demands to reduce the time devoted to license sales, anglers were required to provide driver's license numbers when purchasing licenses. With cross-referencing between the Texas Parks and Wildlife Department (TPWD) and Texas Department of Public Safety computer files, angler samples could be derived more conveniently. For the first time, anglers sampled were questioned about their participation in both freshwater and saltwater fishing. Beginning in 1991, the angler survey was conducted every third year, follow-up studies of anglers according to their preferences for specific species were conducted in the intervening years. Numerous forces internal and external to TPWD fueled this evolution in data needs, sampling frames, and survey approaches. Furthermore, angler surveys became institutionalized over this period with implications for both management agency and university partners. Agency personnel learned to participate fully in the design of angler surveys to ensure that results would be relevant to management, and we were in a position to learn from our mistakes, improve the survey process, and to make surveys more time and cost efficient.

Types of Human Dimensions Inquiry

Fisheries professionals are confronted with important questions about resource users. Unless an agency wants to rely on anecdotal information, it is necessary to have valid and reliable data. The information needed from users can be "broad" for long-range planning purposes; "comprehensive" for short-range planning decisions, commitment of resources, and establishment of goals and objectives at the program level; or "focused" for more immediate action decisions and implementation of activities (Brown 1987).

There are also various kinds of human dimensions inquiry: surrogate biology, audience profiles, administrative justification of programs, user preferences, and an integrated management perspective (Brown 1987). Surrogate biology refers to the conduct of creel surveys, where staff members question anglers about their catch and effort. It is important to remember that anglers provide this information orally. This is the first experience most agencies have with human dimensions research. Audience profiles include the traditional data gathering on demographic characteristics of clientele. As funding becomes scarce, it is important that agencies evaluate their various program offerings in an effort to determine which ones can be justified for continuation and increased funding. Instead of believing clientele have the same preferences as the agency, it is useful to determine the extent of support and opposition to proposed management regulations. As state agencies continue to conduct human dimensions studies, they can develop a trend perspective in each of the aforementioned areas. Finally, with an integrated management perspective, human dimensions data and understanding are considered simultaneously with biological, economic, and policy matters to make recommendations to fisheries decision makers.

Survey Components

Study Objectives

The most critical step in any survey is formulating a clear set of objectives (Johnson and Nielsen 1983), which are needed at the beginning of the study to have a basis for the remaining steps. Without objectives, a survey lacks focus. Those who conduct the survey and those who use it need to be reminded constantly of the reasons for the survey and the particular management questions that need answering.

Stating objectives is not easy because various individuals, often with different agendas, are involved in the process. Objectives should nevertheless focus on what managers need to know about their anglers. The objectives will vary depending on agency and researcher experience in conducting surveys. In 1986, our objectives were broad; we sought baseline data on angler demographics, extent of participation, and general motivations for fishing; however, as each subsequent survey was completed, new issues arose and objectives were modified to focus more on agency needs. Initially, for example, angler support for management tools was not a main focus of our surveys, but it was a major concern in the 1994 survey. We even presented anglers with management tools not currently in use. We questioned whether angler responses to general questions in the survey would be consistent when asked in a species-specific context and consequently sought to obtain data from anglers fishing for particular species. Because we have repeated measures over time, we tried to keep questions consistent so that trend analysis would be possible. Specifically, the objectives of the 1994 Texas Statewide Angler Survey were

1. to establish the extent and distribution of fishing activity by the statewide population of anglers—variables of concern included days of participation in farm ponds, lakes and reservoirs, rivers, saltwater bays, and the Gulf of Mexico, and years of participation in fresh and salt water;
2. to describe the characteristics, general fishing habits, attitudes, motivations, and expenditures of anglers in Texas—variables of concern included demographic information (age, gender, income, race, and ethnicity), club membership, tournament participation, and expenditures on a "typical" fishing trip. Also, scale items (statements to which anglers respond on an agree–disagree, support–oppose, or importance continuum) were included for measuring motivations for fishing, consumptive attitudes, constraints to fishing, and support for both freshwater and saltwater management tools;
3. to give special attention to measures of species preference, inasmuch as this was the basis for follow-up survey research in intervening years—variables of concern included an open-ended identification of species preference in fresh or salt water; and
4. to explore trends and managerial implications of the statewide data collection and make comparative analyses with other available data (i.e., previous statewide survey results conducted by TPWD and federal fisheries agencies).

Population to be Sampled and Sampling Frame

Angler surveys can target various populations of concern. For human dimensions research, the target population may include the general population of all anglers at the state level, the population of licensed anglers (e.g., anglers >16 and <65 years in Texas), or anglers fishing at a particular location. Among licensed anglers, the target population may focus on one or more subsets: freshwater anglers, saltwater anglers, or fishing stamp holders. With a local focus, if information is desired about anglers who fish a particular reservoir, licensed anglers within a particular set of zip code areas may be targeted.

General population surveys obtain information from licensed and unlicensed anglers and can be useful for determining the proportion of anglers in the entire population or in a particular area. This type of information is collected routinely in the National Survey of Fishing, Hunting and Wildlife Associated Recreation Studies conducted by the U.S. Fish and Wildlife Service (1955–present). This approach would probably not be cost effective at the state level but should not be ruled out. For example, an agency trying to identify ways to increase participation needs to know why people do not fish or buy a fishing license and what would help facilitate their participation. A general population survey also provides the only means of ascertaining agency image and policy preferences beyond those of licensed anglers.

Licensed anglers (those supporting most state agencies through license sales and excise taxes on fishing equipment) are considered by many agencies as their most important clientele because they are the ones who pay the bills. For this reason, and because an alternative household survey would be more costly, the 1994 Texas Statewide Angler Survey focused on resident licensed anglers. Therefore, the target population included all holders of a resident recreational fishing license. These included anglers who held (1) a resident fishing license, (2) a resident combination hunting and fishing license, (3) a temporary (14-d) resident fishing license, or (4) a special resident fishing license (including those who were blind, disabled, or held a resident commercial fishing license).

Before selecting the sample, the population of licensed anglers was divided into sampling units so that every angler in the population was defined as one unit. Although it would appear that this division had already been accomplished with separate computer files of anglers who purchased one of the four types of resident fishing licenses, it was often not the case. Anglers who purchased a 14-d license more than once in the same year would be in the computer files twice, for example, and they would thus have a higher probability of being selected in a random sample (see Perdue and Ditton [1983] for a discussion of sampling bias associated with duplicate listings and the effects of that bias on survey results). Additionally, some licensed anglers may have been recorded more than once due to data entry errors. Duplications were checked before sample selection by sorting the listing by driver license number and checking for duplicates.

When we were assured that every licensed angler had an equal chance of being selected for the survey, the listing became the sampling frame for the study. The sample pulled from this frame was representative of that frame only, that is, results from the sample were generalizable only to the population of resident licensed anglers in Texas. Results are not representative of all anglers in Texas, because anglers younger than 17 years or older than 64 years were not sampled (they were not required to purchase a fishing license).

Data to be Collected

Study objectives guide data collection from survey participants. Once study variables are decided upon, it is necessary to verify that the data to be collected are relevant to the objectives of the survey and that no essential questions have been omitted (Cochran 1977). At this point, restraint is needed to avoid "nice to know" questions that have no value or direct link to project objectives.

The 1994 Texas Statewide Angler Survey consisted of management-related and conceptually based questions. Management-related questions were driven by specific decision-making needs (i.e., who is participating when and where). For most of these questions, previous research may give little indication of what to expect. Conversely, conceptually based questions build on previous human dimensions research and hence expected results can be hypothesized and can suggest which variables to include and how questions should be worded. Without an adequate knowledge of previous human dimensions literature, staff may be tempted to look at relationships among all variables included in the study.

Degree of Precision Desired

The results of sample surveys are always subject to some uncertainty because of errors in sampling and measurement (Cochran 1977). To minimize this uncertainty, particularly to reduce the confidence intervals around estimates, larger sample sizes are required. Determining sample size depends on what estimates are most critical for the study. For the 1994 Texas Statewide Angler Survey, we wanted to make estimates of the proportion of licensed anglers who prefer to catch largemouth bass, and we wanted this to be estimated with an SE of not more than 1% at $\alpha = 0.05$. From previous studies, we knew that the true proportion of licensed anglers who prefer to catch largemouth bass was between 0.45 and 0.65. We estimated minimum sample size to be:

$$n_o = \frac{t^2 p\,(1 - p)}{d^2} = \frac{1.96^2\,(0.5)\,(0.5)}{(0.0196)^2} = 2{,}500; \quad (1)$$

t = tabular t value for $\alpha = 0.05$;
p = estimated proportion of largemouth bass anglers;
d = margin of error = $t(\text{SE})$ = 1.96×0.01 = precision.

In this particular case, the finite population correction (Cochran 1977) was negligible and thus could be ignored because the population size was large. Anticipating a 70% response rate as per Dillman (1978), we calculated that we needed a final sample size of at least 3,572. If we had desired a lower level of precision (e.g., SE = 0.02 was acceptable), the sample size could have been reduced. Amidst budgetary constraints, it is imperative that researchers provide convincing evidence that the higher costs of increased sample size are necessary in terms of achieving greater precision of estimates (Cochran 1977).

Further considerations are necessary if the researcher wants to do follow-up surveys of respondents. One of the objectives for the 1994 Texas Statewide Angler Survey was to conduct a follow-up survey of largemouth bass anglers (as identified by respondent's stated species preference in the statewide survey). Currently, without a sampling frame for licensed largemouth bass anglers, there is no alternative means of accessing this angler group. Having sufficient sample size for the species-specific follow-up angler survey also needed to be considered in determining sample size for the statewide survey.

A sample of 10,000 was sufficient for the degree of precision desired in the 1994 Texas Statewide Angler Survey; it allowed sufficient response rates for conducting follow-up surveys of anglers fishing for largemouth bass and spotted seatrout *Cynoscion nebulosus* in years 2 and 3, respectively. Obviously, this approach would work best with a species that receives extensive fishing effort, such as largemouth bass (in Texas, 58% of licensed anglers listed largemouth bass as their preferred target species). With less-preferred species, sample size would need to be greater and hence may be inefficient from a sampling perspective.

Methods of Measurement

How data are collected depends on a variety of factors, including cost, desired sample size, the amount and type of data to be collected, and the capabilities (computer and logistic) of those who will conduct the survey. Survey data can be collected by personal interview, telephone interview, or self-administered mail questionnaire. Advantages and disadvantages of these methods are discussed in detail by Dillman (1978), Babbie (1982), and Bailey (1987) and reviewed with regard to angler surveys by Brown (1991), Essig and Holliday (1991), and Weithman (1991). In general, personal interviews are the most expensive to conduct, and mail surveys are the least expensive (Bailey 1987). For the 1994 Texas Statewide Angler Survey, a mail questionnaire was chosen because of the large sample size needed. Personal interviews were not feasible because the geographic distribution of anglers was too great; interviewing 10,000 anglers would not be cost effective. Creel clerks come in contact with many anglers, but the original purpose of a creel survey—collecting catch and effort data—leaves limited time for additional questions. Telephone interviews would have greatly increased the cost of the survey because of the amount of information desired and, in some cases, would not have been appropriate for the types of information sought; however, it may be necessary to shift to telephone interviews in the future to overcome potential problems of nonresponse as the Hispanic population in Texas increases (Murdock et al. 1992).

The 1994 Texas Statewide Angler Survey was conducted using self-administered mail questionnaires, following the total design method (TDM) presented by Dillman (1978). This methodology was followed with few exceptions, which were made only after deliberation and evaluation. For example, a certified mailing was not sent to anglers who did not respond to previous mailings because some anglers may have to travel great distances to their post office. Also, instead of using personal signatures, we have experimented with computer-scanned signatures and have found no significant differences in response rate.

The TDM works because of persistence, a personal approach to respondents, and attention to detail. License files usually reveal first name, last name, and address for each angler. Therefore, it is unnecessary to address the potential respondent as "Dear Angler" in the cover letter. Dear "first name" or Dear Mr. or Mrs. "last name" (when only an initial for first name was known) were used as the salutation, along with a standard inside address. Additionally, mailing labels were not used on envelopes; instead, addresses were printed on envelopes using a laser printer to simulate personal communication.

These steps required that first name, middle name or initial, last name, address, and other pertinent information each have a separate "field" in the data file. These fields were then merged with the data file to produce customized letters and envelopes. This merging technique is a built-in feature of word-processing software such as WordPerfect® or Microsoft Word®. Development of merging capabilities took extra planning, computer work, and time, but this attention to detail differentiated the survey mailings from "junk mail" and encouraged angler response.

The body of the opening letter also affects an angler's willingness to respond to surveys. In addition to personalization, the potential respondents need to know why the survey is being done, the importance of their answers to the success of the survey, how their answers will help improve fisheries management, and that their responses will be confidential. See Dillman (1978) for additional details about these aspects and about ways to increase response rates.

Some researchers use no particular survey methodology when conducting angler surveys and may send only one mailing and a postcard. Not sending follow-up surveys usually results in a 25–35% response rate (Dillman 1978). Low response rates reduce confidence in the reliability of estimates, which is why persistence is important. As per Dillman (1978), we sent a postcard reminder 1 week after the original mailing to all members of the sample, and we sent second and third mailings (with replacement questionnaires) to those who had not responded to previous correspondence. This resulted in increased survey response rates, ranging from a low of 61% to a high of almost 72% (Table 1). Such high response rates give confidence that results are reliable and generalizable to the target population of concern.

Errors in Measurement

Despite high response rates, potential sources of sampling and nonsampling error affect survey estimates. Sampling errors occur when the population is nonrepresentative, when inaccessibility prevents randomness in selection, or when sample size is insufficient. Nonsampling errors include response bias (recall, digit, and prestige), nonresponse bias, and errors associated with editing, coding, analyzing, and interpreting data. Nonsampling biases result from improper question wording and order, salience of the study to the participant, and social desirability of the study (Filion 1980; Brown 1991). Comments will be made on the three biases that we think are the most important: sampling bias, recall bias, and nonresponse bias.

First, samples need to be selected in a manner that makes them representative of the target population (Filion 1980). Although fishing license files in Texas are currently computerized, this was not always the case. Before 1989, it was necessary to manually "pull" our samples from more than 250 boxes of license receipts sorted by month and vendor. At the time, we used a systematic random sample (every 10th license receipt) obtained from randomly selected vendors; however, vendor files varied in size and some files were incomplete or had illegible addresses and replacements had to be selected (i.e., the next legible license receipt). The ability to address sampling bias was greatly hindered by the way the license files were maintained at the time. Computerization of license files allowed for a variety of checks to be made and improved our ability to minimize sampling bias.

Second, recall bias occurs when the angler cannot accurately recall if an event happened, or its frequency, time, and place (Filion 1980). Hilgard and Atkinson (1967) indicated an inverse relationship existed between an individual's ability to retain information and the time elapsed since the occurrence of the event. In angler studies, recall bias is most evident in measures of days of participation and expenditures. Evidence of this for our surveys (which typically have a 12-month recall period) can be seen in digit bias of some responses. Digit bias occurs when anglers cannot specifically recall how many days they fished last year or how much they spent on fishing-related goods and services, and thus they respond with guesses, which usually end in a five or zero (Tarrant and Manfredo 1993). Reducing the recall period from 12 to 3 months reduces recall bias problems (Fisher et al. 1991) and possibly associated digit bias. This reduction may be impractical for large statewide surveys because conducting the survey four times per year would increase costs. Because many of the questions asked do not suffer from recall problems, efforts need to be made to find alternative means of making corrections for data subject to recall bias. For all surveys, however, an effort should be made to keep the recall period as short as possible.

Third, it is almost impossible to obtain replies from every person in the survey, regardless of the data collection method used or how well the survey is designed (Filion 1980; Fisher 1996). Consequently, inferences made about the target population will be subject to nonresponse bias (Bethlehem and Kersten 1985; Fisher 1996). As a general trend among all types of mailed surveys, nonresponse has been increasing since the 1950s (Steeh 1981). Thompson and Hubert (1990) referenced "low and generally declining response rates" to mailed surveys in Wyoming, and, likewise, we have observed a general decline in angler response rates over the past several years in Texas (Table 1).

Unit nonresponse occurs when no information is collected from an angler because they refuse to participate, cannot be contacted, or are unable to respond to the survey (e.g., language barrier; Fisher 1996). Unit nonresponse is difficult to correct for because all that may be known about the angler may be location of residence and other external variables such as age, gender, or race (provided license files incorporate such information). Unit nonresponse is compounded because early respondents are not representative of nonrespondents or even later respondents (Brown et al. 1981; Choi et al. 1992). Before the 1994 Texas Statewide Angler Survey, nonresponse bias was investigated by contacting a sample of nonrespondents by phone. Each individual contacted was asked 10 basic questions that could be cross-checked with respondents. Generally, nonrespondents were less avid participants and not very involved or invested in fishing (Hunt et al. 1990). Weighted means for respondents and nonrespondents were used to correct estimates for annual days of fishing participation. After correction, the mean days of fishing participation in Texas decreased from 24 to 19 d/year. This adjusted estimate was more consistent with estimates made for Texas (mean = 17 d/year) from the National Survey of Fishing, Hunting and Wildlife Associated Recreation (USFWS 1985).

The process of follow-up telephone surveys to nonrespondents, however, was time consuming because phone numbers had to be solicited from directory assistance and often numerous calls were required to reach the angler. For logistical reasons and budgetary constraints, sample size (N = 300) was small for the total number of nonrespondents (N = 3,163). As a result of these limitations, Fisher (1996) developed a method to adjust for nonresponse without making phone calls. Adjustments for nonresponse were calculated using the method of response propensity stratification (Little 1986). Response probabilities and weights were then determined using logistic regression, and weights were used to calculate adjusted survey results. For further details, see Fisher (1996).

Many biases associated with surveys can be minimized by careful attention to pretesting the survey instrument. Because the same or similar questions have been asked for 10 years, previous studies served as a partial pretest for the 1994 Texas Statewide Angler Survey. Furthermore, the questionnaire was pretested with a subset of the angler sample. Responses (or lack thereof) and comments from anglers alerted us to problems in question wording and order. For example, some anglers viewed questions regarding age and income as "sensitive," and therefore these questions were moved to the back of the questionnaire, where they would least likely affect other responses. Also, bias can be minimized by paying attention to the time of year the survey is conducted. For example, we avoid sending questionnaires during the summer, and prefer to survey immediately after summer (1 September–15 November) or after the New Year (1 January–15 March). These periods are consistent with findings by Brown et al. (1989) regarding optimal times of the year to conduct angler surveys.

The matter of errors and subsequent biases is important. Although it may be impossible to correct for all biases in survey research, biases should be recognized and corrected where possible. Unaccounted for or unreported biases are likely to lead to incorrect estimates (no matter how large the sample size) and a loss of credibility for the researcher.

Transition

Data collection, coding, data entry, and analysis are mentally taxing on all involved because they happen in a short period of time. Therefore, there may not be much time to review and to think about the data collected, although observations and ideas are usually filed away. We call this the transition from "collection mode" to "creative mode." Whereas study objectives determine how the data will be used for reporting purposes, other timely management issues arise that may require the data set to be segmented into various groups, such as reservoir and river anglers, club and nonclub members, tournament and nontournament anglers, or gender, race, and ethnic segments. For example, a segmentation of freshwater tournament and nontournament anglers was used by TPWD to (1) understand the proportions in each angler group, (2) evaluate the extent of delivery of services to each group, and (3) ascertain unmet needs and future program directions.

Public Dissemination of Survey Information

Paying attention to the methodological and analytical requirements of angler survey research is one thing; publicizing and implementing the results effectively is another. Feedback of information to managers, respondents, the public, and the scientific community is important. Until recently, we provided TPWD with a data set and management data report, the only two requirements for agency projects (e.g., Ditton et al. 1988, 1989; Hunt et al. 1990). After a rigorous review process, reports were distributed to agency personnel and commissioners. The reports rarely went to the media or to anglers (respondents or otherwise), thus nobody really knew much about the survey except agency administrators and managers.

Recently, we expanded our dissemination efforts. Instead of providing management reports strictly in support of data files, we now provide TPWD with executive summaries and "user-friendly" reports that concentrate on key points related to study objectives. Technical appendices are available to those interested in probing the data further. A series of oral presentations have been made to management staff, department advisory committees, and to the Texas Parks and Wildlife Commission as requested.

Feedback is provided to survey respondents in the form of executive summaries or newsletters before public dissemination of results. This is done to thank anglers for their time and effort and to let them know how the information they provided was used before they hear about it through media outlets. In addition, we make an effort to respond to questions written by respondents on the questionnaire.

Public dissemination of data is provided in various forms. User-friendly reports, newsletters, and press releases are developed in cooperation with agency personnel for distribution to outdoor writers. Reports and newsletters are distributed to field offices where they can be available to the public. The TPWD news releases have included a series titled "What fishermen and women tell us," which builds on themes and issues identified by respondents in volunteered comments. This allows the agency to answer important questions and clarify misconceptions.

Finally, the research community can benefit from the results of survey efforts. Brown (1991) noted that data from statewide surveys usually are not published in scientific literature because they are typically used for monitoring and assessment rather than to answer basic research questions; however, data collected from statewide surveys in Texas have been used to contribute to the human dimensions technical literature on practical management issues (e.g., Wilde and Ditton 1991; Riechers et al. 1993; Wilde and Ditton 1994), methodology (e.g., Choi et al. 1992; Fisher 1996), and conceptual development (e.g., Ditton et al. 1990). This is an important part of human dimensions research, because those wishing to conduct angler research can benefit from previous results and the successes and failures of other researchers.

Developing Partnerships

There are several ways for state agencies to gather human dimensions information through statewide surveys: (1) develop an in-house staff capability, (2) work cooperatively with interested university social science researchers to develop agency capabilities, (3) use competitive bidding to request proposals for survey research from academic researchers and consulting firms, and (4) hire a private consultant on a sole-source basis to provide necessary support to the agency or complete surveys as needed. There are pros and cons to each of these alternatives. Whereas many agencies probably cannot afford to develop an in-house survey capability at this time, there are also problems with seeking external services: (1) there may be insufficient agency expertise to write an adequate scope of work for a statewide angler survey, (2) university researchers are not likely to be interested in what they might view as monitoring projects, (3) the costs of gearing up for a single project (or one every 5–10 years) may increase survey costs to prohibitive levels, (4) continuity may be lost as consultants change over time, (5) contractors may not be interested in how agency personnel can use results to help improve fisheries management, and (6) because of (5), the relationship with contractors usually ceases with delivery of a final report and payment for services.

Work in Texas over the past 10 years resulted from a close and continuing partnership between TPWD management and research personnel and university researchers. The partnership included (1) identification of an agency coordinator, (2) joint agency and researcher responsibilities throughout each project, (3) numerous joint meetings to review survey questions and results with more agency personnel included over time, (4) identification and elaboration of critical management issues by the fisheries chief and support staff, (5) extensive efforts to enhance two-way communication, (6) the possibility for university researchers to collect data for basic research questions in addition to a management focus, (7) joint publications and presentations of survey results, and (8) continuing efforts beyond what was specified in agency contracts.

This partnership approach has helped to develop or enhance (1) support for an understanding of human dimensions research among TPWD personnel in support of fisheries management, (2) greater understanding and appreciation for management issues among university personnel, (3) basic infrastructure and capability in TPWD to use survey research in decision making, (4) survey research innovations that have enhanced the quality of survey efforts or reduced costs (i.e., various quality control efforts and use of bar codes to automate survey processing), and (5) graduate student education and entry to the professional world of fisheries management. As long as the partnership meets the needs of both managers and agency researchers and contributes to fisheries management in Texas, it is likely to continue. To the extent that we continue to assist agency decision making on critical issues, the partnership will endure.

Acknowledgments

We thank our current and former TPWD colleagues Phil Durocher, Robin Riechers, Mark Fisher, Gene Wilde, Ken Kurzawski, Gary Matlock, Maury Osborn, and Steve Gutreuter for their support and assistance in survey development and implementation. Also, we thank the anonymous reviewers for their thoughtful comments that helped improve earlier versions of this paper.

References

Babbie, E. R. 1982. Social research for consumers. Wadsworth, Belmont, California.

Bailey, K. D. 1987. Methods of social research. The Free Press, New York.

Bethlehem, J. G., and H. M. P. Kersten. 1985. On the treatment of nonresponse in sample surveys. Journal of Official Statistics 4:251–260.

Brown, T. L. 1987. Typology of human dimensions information needed for Great Lakes sport-fisheries management. Transactions of the American Fisheries Society 116:320–324.

Brown, T. L. 1991. Use and abuse of mail surveys in fisheries management. Pages 255–261 *in* Guthrie et al. (1991).

Brown, T. L., C. P. Dawson, D. L. Hustin, and D. J. Decker. 1981. Comments on the importance of late respondent and nonrespondent data from mail surveys. Journal of Leisure Research 13:76–79.

Brown, T. L., D. L. Decker, and N. A. Connelly. 1989. Response to mail surveys on resource-based recreation topics: a behavioral model and an empirical analysis. Leisure Sciences 11:99–110.

Choi, S., R. B. Ditton, and G. C. Matlock. 1992. Homogeneity across mail surveys waves: a replicated study. Journal of Leisure Research 24:79–85.

Cochran, W. G. 1977. Sampling techniques. Wiley, New York.

Dillman, D. A. 1978. Mail and telephone surveys: the total design method. Wiley-Interscience, New York.

Ditton, R. B., and five coauthors. 1989. Trends in demographics, participation, attitudes, expenditures, and management preferences of Texas freshwater anglers, 1987–88. Human Dimensions of Fisheries Research Laboratory Technical Report HD-601. Texas A&M University, College Station.

Ditton, R. B., D. K. Loomis, and S. Choi. 1990. Recreation specialization: reconceptualization from a social worlds perspective. Journal of Leisure Research 24:33–51.

Ditton, R. B., and six coauthors. 1988. Demographics, participation, attitudes, expenditures, and management preferences of Texas freshwater anglers, 1987. Texas Parks and Wildlife Department, Management Data Series Report, Austin.

Driver, B. L. 1985. Specifying what is produced by management of wildlife by public agencies. Leisure Sciences 7:281–295.

Essig, R. J., and M. C. Holliday. 1991. Development of a recreational fishing survey: the marine recreational fishery statistics survey case study. Pages 245–254 *in* Guthrie et al. (1991).

Filion, F. 1980. Human surveys. Pages 441–454 *in* Wildlife management techniques manual, 4th edition (revised). The Wildlife Society, Washington, DC.

Fisher, M. R. 1996. Estimating the effect of nonresponse bias on angler surveys. Transactions of the American Fisheries Society 125:118–126.

Fisher, W. L., A. E. Grambsch, D. L. Eisenhower, and D. R. Morganstein. 1991. Length of recall period and accuracy of estimates from the national survey of fishing, hunting and wildlife-associated recreation. Pages 255–261 *in* Guthrie et al. (1991).

Guthrie, D., and seven coeditors. 1991. Creel and angler surveys in fisheries management. American Fisheries Society Symposium 12.

Hendee, J. C. 1974. A multiple-satisfactions approach to game management. Wildlife Society Bulletin 2:104–113.

Hilgard, E. R., and R. C. Atkinson. 1967. Remembering and forgetting. Pages 314–335 *in* E. R. Hilgard and R. C. Atkinson, editors. Introduction to psychology, 4th edition. Harcourt, Brace and World, New York.

Hunt, K. M., and five coauthors. 1990. Demographics, participation, attitudes, and management preferences of Texas anglers, 1989. Human Dimensions of Fisheries Research Laboratory Technical Report HD-603. Texas A&M University, College Station.

Johnson, D. L., and L. A. Nielsen. 1983. Sampling considerations. Pages 1–21 *in* L. A. Nielsen and D. L. Johnson, editors. Fisheries techniques. American Fisheries Society, Bethesda, Maryland.

Little, R. 1986. Survey response adjustments for estimates of means. International Statistical Review 54: 139–157.

Malvestuto, S. P. 1983. Sampling the recreational fishery. Pages 397–419 *in* L. A. Nielsen and D. L. Johnson, editors. Fisheries techniques. American Fisheries Society, Bethesda, Maryland.

Murdock, S. H., K. Backman, R. B. Ditton, M. N. Hoque, and D. Ellis. 1992. The implications of future demographic change for participation in fishing in Texas. North American Journal of Fisheries Management 2:548–558.

Perdue, R. R., and R. B. Ditton. 1983. Sampling from registration files: the problem of duplicate listings. Journal of Leisure Research 15:95–99.

Pollock, K. H., C. M. Jones, and T. L. Brown. 1994. Angler survey methods and their application in fisheries management. American Fisheries Society Special Publication 25.

Riechers, R., G. R. Wilde, R. B. Ditton, and K. M. Hunt. 1993. Freshwater and saltwater anglers: a comparative analysis of differences in attitudes toward management tools. Proceedings of the Annual Conference Southeastern Association of Fish and Wildlife Agencies 45(1991):246–253.

Smith, C. L. 1983. Evaluating human factors. Pages 431–445 *in* L. A. Nielsen and D. L. Johnson, editors. Fisheries techniques. American Fisheries Society, Bethesda, Maryland.

Steeh, C. G. 1981. Trends in nonresponse rates, 1952–1979. Public Opinion Quarterly 45:40–57.

Tarrant, M. A., and M. J. Manfredo. 1993. Digit preferences, recall bias, and non-response bias in self reports of angler participation. Leisure Sciences 15: 231–238.

Thompson, T., and W. A. Hubert. 1990. Influence of survey method on estimates of statewide fishing activity. North American Journal of Fisheries Management 10:111–113.

USFWS (U.S. Fish and Wildlife Service). 1985. Survey of fishing, hunting, and wildlife-associated recreation. U.S. Government Printing Office, Washington, DC.

Weithman, A. S. 1991. Telephone survey preferred in collecting angler data statewide. Pages 271–280 *in* Guthrie et al. (1991).

Wilde, G. R, and R. B. Ditton. 1991. Diversity among anglers in support of fishery management tools. Pages 329–335 *in* J. L. Cooper and R. H. Hamre, editors. Warmwater fisheries symposium. U.S. Forest Service, Rocky Mountain Forest and Range Experiment Station, General Technical Report RM-207, Fort Collins, Colorado.

Wilde, G. R., and R. B. Ditton. 1994. A management-oriented approach to understanding diversity among largemouth bass anglers. North American Journal of Fisheries Management 14:34–40.

American Fisheries Society Symposium 16:245–250, 1996

An Overview of Economic Impact and Value of Recreational Fisheries

ROBIN K. RIECHERS
Texas Parks and Wildlife Department, Coastal Fisheries Division
4200 Smith School Road, Austin, Texas 78744, USA

ANTHONY J. FEDLER
University of Florida, Department of Recreation, Parks and Tourism
Gainesville, Florida 32653, USA

Abstract.—Determination of the value of natural resources is becoming increasingly important in resource management. Specifically, this information may be used by the public, policy analysts, and public officials in support or opposition of funding and construction of projects, to obtain or maintain funding for specific programs, or in cost-benefit analyses and damage assessments. Valuation of natural resources is necessary to justify funding allocations and other decisions. Direct expenditures for recreational activities and their associated economic impacts are often used as a measure of the economic importance of a resource. Economic value, or the value that society places on a resource, is also an important measure which is often used. Because direct markets for sport fishing experiences do not exist, indirect methods for economic valuation have been derived. We discuss economic impact and economic value in the context of reservoir fisheries management, propose guidelines for evaluating economic impact, and describe two methods for determining economic value.

Natural resource managers, from the local biologist to the administrator, are increasingly being asked to determine the value of the resources they are charged with managing. The public, policy analysts, and public officials may use several means of determining the value of a recreational activity such as fishing at a reservoir (Weithman 1986). This information may be needed in support or opposition of funding and construction of projects, to obtain or maintain funding for specific programs, or for cost-benefit analyses and damage assessments.

Because the recreational use of fisheries is not a marketed good, and because users do not pay directly for the good, there is no established market price. The lack of a direct market for recreational use of a resource does not mean that it has no value, but rather that direct measurement of its value is not possible (Rockland 1985). The economic importance of recreational activities often is measured through direct expenditures made by users in pursuit of those activities, and their resulting economic impact on the local, regional, or state economies (Hushak et al. 1986; Hushak 1987; Propst and Gavrilis 1987). This requires identifying monetary expenditures associated with resource use and tracing the impact of those expenditures through the economy.

Unfortunately, expenditure and economic impact data do not fully account for the economic value of a fishery. Although related, the concepts of economic value and economic impact have different meanings and measurements. Economic impact is a measure of the economic activity generated by use of the resource, whereas economic value is the total economic worth or value that society places on the resource. Economic value of a resource includes not only the gross expenditures of users, but the total amount they would be willing to pay for continued use of the resource. In this paper we discuss economic impact and economic value, and the methods used to determine them.

Economic Impact

Expenditures by recreational anglers include everything anglers spend to use fisheries resources at local, state, and national levels. Changes in regional economic output, income, and employment associated with changes in expenditures by recreational anglers do not measure the total benefits of a recreational fishery. Instead, they represent accounting measures of transactions and income payments which give an indication of the economic activity generated in a local area or region by a particular activity (i.e., fishing). Expenditures represent costs to anglers, and are not a measure of economic benefits. They do not measure the loss of value to anglers that would result from loss of fishing opportunities, or the gain in value from expanded or new fishing opportunities. They do represent the expenditures by anglers at present levels of participation in the activity.

Recreational anglers purchase things such as

tackle boxes, rods, reels, boats, ice, fuel, and groceries for their fishing activities. These expenditures represent revenues and jobs generated in the study area. Surveys can be used to estimate the expenditures made by anglers, and also to ascertain where those expenditures are made so they can be assigned to the proper location (i.e., local, state, or national level). Once angler expenditures are measured, economic impacts can be calculated.

There are basically three types of economic impacts: (1) direct impacts, which are the purchases made by recreational anglers; (2) indirect impacts, which are the purchases made by businesses to produce the goods or services demanded by anglers; and (3) induced impacts, which are the purchases of goods and services by households receiving wages from businesses producing direct or indirect goods and services. For example, the purchase of a fishing rod is an example of a direct impact. The associated indirect impact would include purchases of graphite, paint, guides, reel seats, and other items by the rod manufacturer. Further indirect impacts might include the purchase of pigments by the paint supplier and aluminum by the guide supplier. Induced impacts would include all purchases by households with the wages paid to employees of the retailer, rod manufacturer, and paint, graphite, guide, pigment and aluminum manufacturers in the proportion they contributed to the final product. Total economic impact is the summation of these three levels of impact. The total economic impact divided by the direct impact is called the multiplier and reflects the number of times the initial expenditure circulates through the economy.

The economic effect of any activity may seem to be an infinite loop as initial expenditures are spent and respent by suppliers and households; however, there are several factors that will limit the scope of this respending effect. First, expenditures are limited to the impacts on a specific region (i.e., the expenditures and the resulting impacts are confined to a prescribed geographic area). In the case of a fishing rod, the suppliers of paint, graphite, and aluminum are probably not located in the same region as the rod manufacturer. Therefore, the money paid to those suppliers for the essential goods and services represents expenditure leakage, because the money spent leaks out of the local economy. Second, households, which are the source of the induced impacts, do not respend all of their wages. A portion of their wages may be saved and a portion may be withheld in taxes. Also, businesses do not spend a dollar in inputs for every dollar they receive. These expenditure leakages and portioning provide a finite limit to the effects of the initial expenditures.

Obtaining the requisite expenditure data for analysis is often the most difficult part of conducting an impact study. There are five general steps that should be followed when conducting economic impact studies.

1. Define the scope of the study. For instance, the scope of the study may be the economic activity of fishing for a particular species at a specific reservoir, for example, striped bass *Morone saxatilis* at Lake Texoma (Schorr et al. 1995). This is defining the participants by activity and the scope of the study area or local economy. It may be a group of counties or a region surrounding a reservoir or an entire state or country, but a boundary or specific area is set so that when expenditures occur outside this area or monies leak out of the area they are not used to calculate impact in the area.
2. Collect data from anglers. Personal interviews or mail surveys may be used. Sampling and questionnaire design should be considered and appropriate steps should be taken to ensure reliable data collection (Dillman 1978; Nachmias and Nachmias 1987; Thomson 1991).
3. Categorize expenditure data. Expenditure categories must be defined, and data must be grouped by type of expenditure, such as lodging expenses, restaurant expenses, fishing tackle, ice, and groceries. The groupings are necessary because the individual multipliers and leakages out of the local economy may differ for each group.
4. Determine the amount of expenditures within the impact study region. To determine the impact on the local economy, the expenditures must be portioned according to where the expenditure occurred. The questionnaire may be designed to portion the expense, or other data may be used to support assumptions on where expenditures occur (e.g., previous survey information on distance travelled may be used to proxy the portion of expenditures in another area).
5. Estimate economic impact. Using the expenditure data and the appropriate multipliers, the total economic impact of the activity can be estimated. Models such as RIMS II (Beemiller et al. 1986) and IMPLAN (Alward et al. 1989) are available for estimating economic impacts. Alternatively the state or country where the study is being conducted may have developed an input–output model which can be used to determine

the economic impacts of expenditures in certain sectors of the economy.

Economic Value

Economic value of a resource is the total worth or value society places on the resource. To be comprehensive, societal assessment of resource values should include valuation by nonusers of the resource as well as users. Nonusers may never visit or use the resource, but still may value it. The value that the nonuser may place on the resource may be an option, existence, or bequest value (Randall and Stoll 1983; Brookshire et al. 1986; Peterson and Sorg 1987). Option value is an individual's willingness to pay to retain the option of using the resource in the future. Existence value is the value of knowing that the resource exists and that others have the opportunity to use the resource. Bequest value is the value associated with the existence of the resource for future generations.

User values may take one of three forms: consumptive use, nonconsumptive use, and indirect use (Brookshire et al. 1982; Randall and Stoll 1983; Bishop et al. 1987). The total consumptive use value of a fishery resource includes both the expenditures currently used to gain access to the fishery, and the additional dollars anglers would be willing to spend to continue their activities associated with the fishery. On-site use includes consumptive and nonconsumptive activities associated with the fishery. Indirect use may include activities performed away from the site, including reading about a particular place, species, or special activity at the recreational location (Stoll et al. 1987).

Because the rights to public recreational fisheries are not bought and sold in the traditional market sense, indirect assessment methods are necessary. Nonmarket valuation techniques have been developed to determine the value of these resources. The travel cost method (TCM) and the contingent valuation method (CVM) are two of the most accepted and widely used techniques (Clawson and Knetsch 1966; Burt and Brewer 1971; Fisher 1986; Mitchell and Carson 1989). Both methods are used to estimate consumer surplus which is the amount over the actual cost or expenditure to use a resource that the user would have been willing to pay to continue to use the resource (Rockland 1985).

Travel Cost Method

The TCM uses actual participant (angler) behavior as a measure of worth or one's willingness to pay for the recreational activity. A demand curve is derived for the number of trips (quantity) and the cost of the trips (price). Specifically, costs are incurred to travel and participate in an activity at a recreation site. The basic functional form of the demand curve is depicted in Figure 1, and assumes that the number of trips to a recreational site (e.g., fishing for striped bass at Lake Texoma) will decrease as the price of each trip increases. The functional form of the travel cost model can be expressed simply as:

$$Q = f(\text{TC}, \text{S}, \text{A}, \text{E})$$

where:

Q = quantity demanded, defined as visits to the interview site in the past year;
TC = cost of round-trip travel to the site (including a measure of the value of travel time to the site;
S = price of substitute sites;
A = quality attributes of the site;
E = socioeconomic characteristics of angler.

The basic model may include quantity demanded (Q) at a specific price (TC). More complex models may include the price of substitute sites (S), quality attributes of the site (A), and the socioeconomic characteristics of those using the site (E). Both (S) and (A) from the model above allows for the incorporation of more complex behavior from the recreationist. Specifically, it allows for the choice of a site to be affected by the amenities (quality attributes) at that site and by the relative price of substitute sites.

Two common models are used: the individual model and the zonal model. In the individual model the dependent variable is visits per individual and the independent variables are characteristics of each individual. This model requires intensive data collection. In the zonal model the dependent variable is visits per capita from the origin zone and the independent variables are the attributes for the population within that zone, thus data on the behavior of individuals are not required. Using the average distance a person from each zone traveled to reach the site, the cost per mile traveled, and the number of visits to the site, an estimated demand curve is derived in aggregate for that zone.

For both models, travel expenditures (TC) include the actual expenditures per trip to the site, and the opportunity cost for travel, representing wages that potentially could have been earned during travel time (Cicchetti 1970; Cesario 1976). This opportunity cost is often estimated using some minimum wage level and an average driving speed.

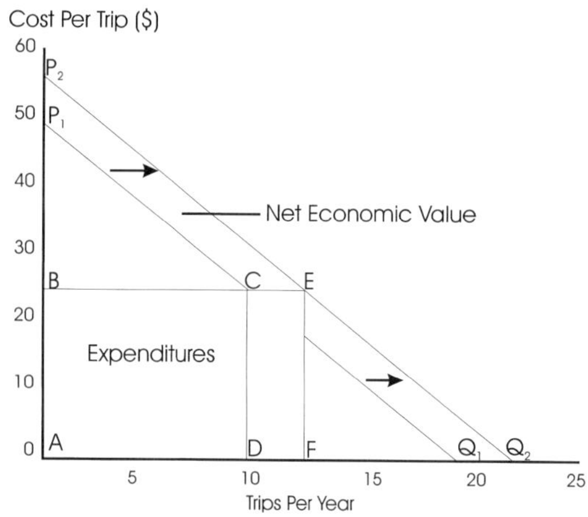

FIGURE 1.—Shifts in angler demand curve for fishing trips. A shift in the angler demand curve from (P_1-Q_1) to (P_2-Q_2) represents a shift caused by a quality change (e.g., increased harvest rates). The marginal change associated with this shift in demand for expenditures is D, C, E, F as compared to the expenditures before the shift A, B, C, D. The marginal change in the net economic value or the difference in what the angler is willing to pay versus what the angler currently pays is represented by the area P_1, P_2, E, C.

Because the TCM relies on actual behavior to develop a demand curve, the value obtained for the activity or use of the resource is limited only to users of the resource and will not include any nonuse values. Limitations associated with data collection and reliability of data on individual behavior must be considered (Dillman 1978; Nachmias and Nachmias 1987; Thomson 1991). Limitations also may occur in creating the demand curve when: (1) the travel distances among visitors to the site are very similar, (2) multiple activities occur during a trip and respondents may be unable to partition their expenses associated with each activity, and (3) minimal changes in the quality of the recreational experience actually may not alter trip behavior, but significantly alter the individual's valuation of the recreational experience.

Contingent Valuation Method

The CVM takes an alternate approach to determining a market value for a resource. The user is asked hypothetically what he or she would be willing to pay for a specific good or service, or for a particular change in the good or service. Users might be asked about their willingness to pay for an increased bag limit at a specific lake, or what they would be willing to pay to continue to fish at that reservoir or lake.

Because a market is being defined through a survey mechanism (personal interview or mail survey) both users and nonusers can be solicited to obtain the value of the resource or the value of the marginal change which would result from some management action. Issues of sampling, questionnaire design, and the effects of these on the functional form of the analysis must also be considered (Dillman 1978; Nachmias and Nachmias 1987; Mitchell and Carson 1989; Ozuna and Stoll 1991).

The CVM elicitation procedure can be broken into five basic components: (1) survey technique, (2) description of policy impacts, (3) method of provision, (4) payment vehicle, and (5) bid elicitation (Hoehn and Krieger 1988). Survey techniques may include personal interviews, or mail or telephone surveys. The questionnaire and the scenario presented for valuation may include detailed descriptions or other explanatory tools, such as graphs, diagrams, or photographs (Brookshire et al. 1979) to help the respondent understand how they will be affected by the scenario being presented. The presentation medium leads directly to the description of policy impacts, whereby it is essential that policies be clearly described and that the survey is not biased or misleading concerning the actual alternatives being described.

Both the method in which the payment may be administered (component 4) and the agency or group charged with managing the additional payment (component 3) may affect an individual's willingness to pay. Attempts should be made to minimize the personal bias that might be associated with a particular agency or type of payment (e.g., taxes or user fees) when designing the questionnaire. Many studies have been conducted to test various bid elicitation procedures, sample size requirements, and the range of values (component 5), which should be presented and how these factors affect the final results (Hoehn and Randall 1987; Mitchell and Carson 1989).

The CVM has at least three primary advantages: (1) on-site participation is not required, (2) presentation of several scenarios and subsequent data analysis is not difficult, and (3) both user and nonuser values can be obtained. Disadvantages of the CVM include the expertise required for survey design and administration, and the fact that the results are based on implied, not actual, behavior.

Marginal Value

Whereas measuring the total value of a resource is useful, most situations requiring economic valuation (e.g., management activities, construction projects, marketing) are concerned with the valuation of a marginal change (Swanson and McCollum 1991). For example, a management change that may cause a quality shift (e.g., increased harvest rates) may shift the demand curve, resulting in a change in the net economic value (Figure 1). The marginal change as compared to the total value concept requires that the value of recreational fishing be placed in a particular context or under particular management regimes (e.g., bag limit, habitat restoration, fish stocking) (Freeman 1979; Peterson and Cordell 1991).

For example, based on the original demand curve (Figure 1) represented by the line P_1 to Q_1, at a cost of US$50 per trip the angler would not take any fishing trips, but if the angler paid nothing per trip the angler would take 20 trips per year. At $25 per trip the angler would choose to take 10 trips per year. Therefore, under current conditions, the angler's willingness to pay or the total value the angler places on participation is $375 (area A, P_1, C, D in Figure 1). The angler's actual expense is $250 (area A, B, C, D). The difference between total value and actual expenditures would be $125 (area B, P_1, C), and is referred to as the consumer surplus of the angler. If a quality shift causes the demand curve to shift to P_2 and Q_2, an individual would choose either to take more trips per year for the same cost per trip, or to take the same number of trips and pay a greater amount per trip. The incremental change or net economic value associated with the management action or cause of the demand shift is the area P_1, P_2, E, C, depicted in Figure 1.

Conclusion

The methods presented here are applicable to all fisheries in a reservoir, entire state or regional fisheries, or specific angler groups. Both economic impact and economic value can provide valuable information to the resource manager. Many factors must be considered when designing and executing these types of studies. These considerations should include biases in sampling, survey methods, questionnaire design, and analytical techniques, and the effects that any of these may have on the final results and interpretation.

These types of studies have at least three implications for management. First, fisheries management agencies, environmental organizations, and constituency groups can use economic impact and economic valuation data to support arguments in favor of resource conservation. Second, estimates derived from the methods presented here can be used to demonstrate potential economic impacts and user costs associated with proposed development projects that would affect fisheries resources and habitats deleteriously. Third, these estimates can be used in support of resource enhancement efforts (i.e., stocking and habitat protection) because consumer surplus would be expected to increase. In every one of these situations, efforts should be made to ensure that economic information is widely disseminated and understood by the public.

The importance of knowing the economic impact and value of a fishery should be apparent to all levels of resource managers. Presented by itself, economic information is useful to decision makers, but when considered in a multidimensional framework with other social and biological information, economic information takes on even greater importance. Better management decisions, both short- and long-term, can be made when fisheries are examined multidimensionally. For instance, valuation information combined with production potential estimates and participation data may lead to one stocking location being chosen over another, or may lead to a change in the way a natural resource agency markets a recreational activity or reservoir in a particular area.

References

Alward, G., and five coauthors. 1989. Micro IMPLAN software manual. Colorado State University, Fort Collins.

Beemiller, R. M., Z. O. Ambargis, and H. L. Friedenberg. 1986. Regional multipliers: a user handbook for the regional input–output modeling system (RIMS II). U.S. Department of Commerce, Bureau of Economic Analysis, Washington, DC.

Bishop R. C., K. J. Boyle, and M. P. Welsh. 1987. Toward total economic valuation of Great Lakes fishery resources. Transactions of the American Fisheries Society 116:339–345.

Brookshire, D. S., L. S. Eubanks, and C. F. Sorg. 1986. Existence values and normative economics: implications for valuing water resources. Water Resources Research 22:1509–1518.

Brookshire, D. S., M. Thayer, W. Schulze, and R. C. D'Arge. 1982. Valuing public goods: a comparison of survey and hedonic approaches. American Economic Review 72:165–177.

Brookshire, D. S., and six coauthors. 1979. Methodological experiments in valuing resources. Phase I interim report to the U.S. Fish and Wildlife Service, Washington, DC.

Burt, O. R., and D. Brewer. 1971. Evaluation of net social benefits from outdoor recreation. Econometrica 39:813–827.

Cesario, F. 1976. Value of time in recreation benefit studies. Land Economics 52:32–41.

Cicchetti, C. J. 1970. Time bias in recreation benefit estimates. Water Resources Research 6:700–704.

Clawson, M., and J. L. Knetsch. 1966. Economics of outdoor recreation. Resources for the Future, Inc., Johns Hopkins University Press, Baltimore, Maryland.

Dillman, D. A. 1978. Mail and telephone surveys: the total design method. Wiley, New York.

Fisher, W. L. 1986. Measuring the economic value of sport fishing. Sport Fishing Institute economics program technical report III. Sport Fishing Institute, Washington, DC.

Freeman, M., III. 1979. The benefits of environmental improvements. Johns Hopkins University Press, Baltimore, Maryland.

Guthrie, D., and seven coeditors. 1991. Creel and angler surveys in fisheries management. American Fisheries Society Symposium 12.

Hoehn J. P., and D. Krieger. 1988. Methods for valuing environmental change. Michigan State University, Department of Agricultural Economics, Staff Paper 88–30, East Lansing.

Hoehn, J. P., and A. Randall. 1987. A satisfactory benefit-cost indicator from contingent valuation. Journal of Environmental Economics and Management 14: 226–247.

Hushak, L. J. 1987. Use of input–output analysis in fisheries assessment. Transactions of the American Fisheries Society 116:441–449.

Hushak, L. J., G. W. Morse, and K. K. Apraku. 1986. Regional impacts of fishery allocation to sport and commercial interests: a case study of Ohio's portion of Lake Erie. North American Journal of Fisheries Management 4:472–480.

Mitchell, R. C., and R. T. Carson. 1989. Using surveys to value public goods: the contingent valuation method. Resources for the Future, Inc., Johns Hopkins University Press, Baltimore, Maryland. Nachmias, D., and C. Nachmias. 1987. Research methods in the social sciences, 3rd edition. St. Martin's Press, New York.

Ozuna, T., Jr., and J. R. Stoll. 1991. The significance of data collection and econometric methods in estimating nonmarket resource values. Pages 328–335 *in* Guthrie et al. (1991).

Peterson, G. L., and H. K. Cordell. 1991. Estimating recreation demand functions: experience with the public area recreation visitor survey. Pages 316–327 *in* Guthrie et al. (1991).

Peterson, G. L., and C. F. Sorg. 1987. Toward the measurement of total economic value. U.S. Forest Service General Technical Report RM-148.

Propst, D. B., and D. G. Gavrilis. 1987. Role of economic impact assessment procedures in fisheries management. Transactions of the American Fisheries Society 116:450–460.

Randall, A., and J. R. Stoll. 1983. Existence value in a total valuation framework. Pages 265–274 *in* R. D. Rowe and L. G. Chestnut, editors. Managing air quality and scenic resources at national parks and wilderness areas. Westview Press, Boulder, Colorado.

Rockland, D. B. 1985. The economic benefits of a fishery resource: a practical guide. Sport Fishing Institute economics program technical report I. Sport Fishing Institute, Washington, DC.

Schorr M. S., J. Sah, D. F. Schreiner, M. R. Meador, and L. G. Hill. 1995. Regional economic impact of the Lake Texoma (Oklahoma–Texas) striped bass fishery. Fisheries 20(5):14–18.

Stoll, J. R., J. B. Loomis, and J. C. Bergstrom. 1987. A framework for identifying economic benefits and beneficiaries of outdoor recreation. Policy Studies Review 7:443–452.

Swanson, C. G., and D. W. McCollum. 1991. Application of economics to recreational fisheries management: an overview. Pages 299–315 *in* Guthrie et al. (1991).

Thomson, C. J. 1991. Effects of the avidity bias on survey estimates of fishing effort and economic value. Pages 356–366 *in* Guthrie et al. (1991).

Weithman, A. S. 1986. Measuring the value and benefits of reservoir fisheries programs. Pages 11–17 *in* G. E. Hall and M. J. Van Den Avyle, editors. Reservoir fisheries management: strategies for the 80's. American Fisheries Society, Southern Division, Reservoir Committee, Bethesda, Maryland.

American Fisheries Society Symposium 16:251–260, 1996

Sustaining Reservoir Fisheries and Angler Benefits through Comprehensive Management Analysis

RICHARD A. COLE

New Mexico State University, Department of Fishery and Wildlife Sciences
Box 30003, Department 4901, Las Cruces, New Mexico 88003, USA

FRANK A. WARD

New Mexico State University, Department of Agricultural Economics and Business
Box 30003, Department 3169, Las Cruces, New Mexico 88003, USA

Abstract.—Anglers decide to fish at a reservoir by comparing the expected qualities and travel costs of several alternate sites. These choices include but are not necessarily limited to reservoirs. Nonetheless, in enacting management policy for a reservoir, fishery managers typically ignore the effects of management actions at other reservoirs. We tested the hypothesis that management actions at alternate reservoirs have no significant effect on angler use and economic benefits derived from management decisions for a target reservoir. This hypothesis was tested with a recently developed version of the comprehensive management model RIOFISH, which enables its users to contrast single- and multiple-site analyses of angler use and economic benefits. Our results showed that a single-site analysis incorrectly assumes that the effects of management decisions made at a target reservoir are independent from the effects of management actions at other reservoirs. Also, we found that a single-site analysis significantly overestimates the impact of management on use and benefits at a target reservoir when similar management occurs at other reservoirs. Fisheries managers can improve total angler benefits at a lower cost by distributing sustainable opportunities across all reservoirs in a managed fishery. Multiple-site models provide a mechanism for agencies with different jurisdictions and missions to be more effective in coordinating management for increased public benefit.

The growing dimensions of sport fishery management goals and policy analysis (Harville 1991; Nielsen 1993) require more comprehensive approaches to management systems. In keeping with those trends, contemporary management theory has replaced a management principle focused on maximum sustainable fish yield with the more comprehensive, angler-oriented principle of optimum sustainable yield (OSY) of diverse fishery benefits (Roedel 1975). Weithman (1993) identified the effects of site quality and substitute sites as important considerations for OSY-guided management at inland fisheries. Based on the OSY principle, effective management strategy needs to consider interactions among diverse attributes within and among reservoir and other fishing sites. Yet, recent reviews reveal the continuing single-site emphasis placed on management and research at large lakes and reservoirs (Hayes et al. 1993; Summerfelt 1993).

Single-site focus on management effectiveness is caused partly by the complexity of regional management administration. Numerous government agencies and other institutions typically contribute to determining the quality of fishing sites at regional scales. Thus, one important question for reservoir management is the degree to which reservoirs can be managed effectively as isolated management units.

The objective of this paper is to show how management effectiveness of large-reservoir systems, as measured by angler use and economic benefit, can be significantly improved by decisions made across numerous reservoirs that serve as fishing substitutes. We show that a single-reservoir management analysis substantially overestimates angler benefits compared with results of a more comprehensive multiple-reservoir management analysis. Independent reservoir-by-reservoir management leads to distribution of management resources among all sites which produces less total angler benefit than can be sustained with a multiple-reservoir system analysis.

Recent technical advances have linked fishery site management more comprehensively to angler benefit, revealing that many quality attributes of recreational fishing, including the availability and quality of substitute fishing sites, contribute to the willingness of anglers to pay for fishing (Weithman 1993). Recreationists base their choices of recreational sites on the qualities and costs associated with traveling to alternative sites (Hansen 1986; Cole and Ward 1994).

As Weithman (1993:163) notes for anglers, "If

expenses exceed perceived value, people might change location, species, or method, or could stop fishing altogether and substitute other activities." Many dissatisfied reservoir anglers would rather switch reservoirs than stop fishing entirely. Resultant changes in fishing intensity among reservoirs may alter the angler–resource interaction enough to require a redistribution of management effort to compensate for the redistribution of fishing effort and sustain fishing quality. Stocking, harvest regulation, and other management tactics may need to be intensified at some reservoirs and curtailed at others to sustain the benefit-maximizing distribution of angler use across all reservoirs in a managed fishing system. Angler switching among substitute sites affects estimates of angler-fishing effort on fish stocks, compensatory management needs, and the benefits realized from management.

The potential for managing reservoir fisheries more cost effectively across numerous reservoirs and administrative jurisdictions has been improved by developing outcome-oriented models with multiple-site perspectives. Several multiple-site, recreational-benefits studies (Burt and Brewer 1971; Cichetti et al. 1976; Knetsch et al. 1978; Ward 1989) preceded development of a statewide policy analysis model, RIOFISH, which simulates angler-use adjustments and resulting fishery effects among sites after changes induced by management and natural events (Cole et al. 1990a; Cole et al. 1995a).

In this paper, we demonstrate the utility of RIOFISH for analyzing the influence of multiple-reservoir management interactions on sustained management effectiveness. Our results cause us to question the assumption that large reservoirs can be treated as independent fisheries. In most respects relevant to this assumption, New Mexico recreational reservoir fisheries are representative of reservoir fisheries elsewhere. New Mexico fishing sites are, however, less densely distributed than sites in wetter regions, a condition that increases the probability that large sites can be managed independently without much loss in cost effectiveness. For that reason, New Mexico is an excellent location for conservatively testing assumptions about site management independence.

Methods

Structural Overview

Cole et al. (1995a, 1995b) describe RIOFISH's mathematical structure and use. The model RIOFISH is a DOS-based numerical simulation of a comprehensive management system for public sport fisheries in New Mexico. For modeling purposes, a comprehensive sport-fishery management system is defined by the interaction of management policies with resources, publics, institutions, and ecological events. The RIOFISH model simulates the status encountered by fisheries managers who inherit resource conditions and revenues. With these "inputs," managers are expected to provide fishing opportunities with beneficial outcomes to anglers, given management constraints. Combined with independent estimates of management costs, RIOFISH can be used to estimate management efficiency from comparison of angler benefit and management cost. Using RIOFISH one can easily change variables controlled by management or by outside forces, such as watershed runoff and angler preferences. Earlier versions of RIOFISH (Cole et al. 1986, 1990a, 1990b) were modified to simulate the interaction of hydrology, ecology, and angler use and benefits among the 132 sites where more than 90% of the state sport fishing occurs. Model iterations can be executed for periods as long as a decade.

The RIOFISH model integrates the most recent nonmarket resource valuation theory. It also employs techniques that predict effects of fisheries management on ecosystem process. Model users are provided with an array of options from which to select management actions, which can then be assembled into strategies for policy analysis. Users also may modify certain critical dimensions of the planning environment, including water runoff rates, regional human population numbers and distribution, the fraction of people who fish, and travel costs. Among the more important considerations, RIOFISH estimates stock abundance, fishery-use rates, and angler benefits resulting from management actions selected by the model user.

The three components of RIOFISH may be used independently or interactively (Figure 1). When integrated, RIOFISH links hydrologic, biologic, and economic components of ecosystem and fishery system process with fisheries management decisions. Run independently, fixed inputs in each component take the place of outputs supplied by other components. For example, the economics model can be run independently of the hydrology and biology models. Stream and reservoir hydrology are linked within river basins and semimonthly hydrologic changes based on U.S. Geological Survey data are transmitted throughout the basin. The RIOFISH user may choose either historic simulation (1975–1992) or synthetic forecasting flows. These flow regimes define seven hydrologic conditions represent-

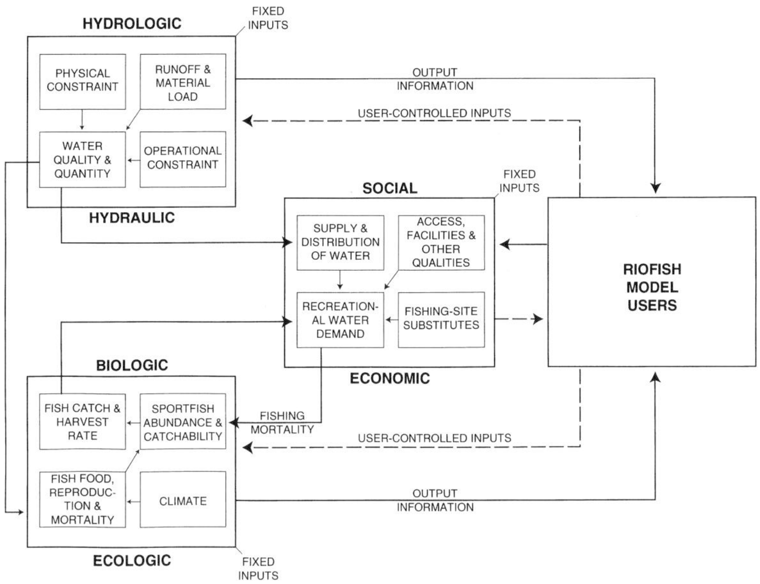

FIGURE 1.—Output and input relationships among the three main model components and model users of RIOFISH. Many of the model inputs can be changed by model users, but each model component also has fixed inputs. Each component has numerous internal elements symbolized here by category.

ing extreme-low to extreme-high flows. Synthetic flows can be assembled for a future time interval in any desired sequence to forecast effects of management actions on fishing benefits that would emerge under that forecast flow regime.

The hydrologic regime selected by the RIOFISH user influences concentrations of nutrients, light transmission, flushing rates, and sediment size, which, in combination with inputs of solar radiation and air temperature, determine aquatic community process. Important elements in community process include ecosystem primary production, loading of allochthonous organic matter, and partitioning of ecological resources. Estimates of fish population density, fish size distribution, and weight–length relationships are needed and provided in a default state, which can be changed by RIOFISH users. Energy and mass–balance principles and transformation efficiencies are used to convert total organic load and forage production to fish production and biomass. Fish production is partitioned within trophic levels among population life stages by use of growth, reproduction, natural mortality, and fishing mortality coefficients that are all changeable by the user. These factors vary in effect, depending on specific population tolerances to environmental conditions that apply equally to naturally recruited, stocked, and introduced fish.

The economic component is a multiple-site, travel cost demand model (Hansen 1986; Rosenthal et al. 1986; Ward and Loomis 1986). It was estimated from a multiple-regression analysis of fishing travel costs and site qualities encountered from 1988 to 1989 in New Mexico. Data on fishing demands, fish catch, harvest rates, and average size of harvested fish were obtained from statewide monthly telephone surveys of New Mexico anglers. Other information on site quality was gathered onsite. The complete data set consisted of 9,504 observations, including 132 sites, 9 zones-of-origin, and 8 sequential seasons. The system of estimated demands is consistent with choice theory, in which total fishing expenditures exhaust the observed angler fishing budget. We specified a constant elasticity-of-substitution model of angler preferences with site quality characteristics (Cole et al. 1995b). The relationship among site qualities, fish populations, and anglers form a fishery system. This system is specified to attract anglers to an array of discrete sites (e.g., reservoirs) depending on fishing quality at all sites, demographic factors, and travel costs from each origin to all sites in the system. Factors affecting demand include driving distance, gasoline price, gas mileage, other vehicle operating costs, site entry fee, and value of travel time. Angler demand (fishing effort) in RIOFISH is determined by angler expenditures and costs (price). Angler expenditure is measured by travel costs per trip, multiplied by average observed number of trips per site, summed over all sites. This economic model is described in more detail in Cole and Ward (1994) and Cole et al. (1995a).

Among the numerous quality factors that influence angler use, demand, and benefit, many are manageable and accessible to RIOFISH users. These include number of fish caught per day, number of fish harvested per day, average weight of fish harvested, water surface area, amount of aquatic vegetation, and water turbidity. Also included are number of boat ramps, site closure status, percentage shoreline access, availability of drinking water, number of campsites, and number of toilets (Cole and Ward 1994; Cole et al. 1995a). Certain unmanageable site qualities that also affect results include seasonal precipitation, air temperature, site location in a designated national forest, and site elevation. Although crowding per se was not explicitly included, factors that are sensitive to crowding were included, such as ramps, toilets, and reservoir size.

Total statewide angler use and benefits are predicted for management actions implemented simultaneously or sequentially at any desired combination of all 132 sites in RIOFISH. Parameter estimates from the demand equations are used in specifying an angler benefits function. The benefits function is based on the theoretically correct expenditure function calculated from demand equations consistent with underlying angler preference (e.g., Hanemann and Morey 1992).

Because many factors in addition to fish density attract anglers, site use in RIOFISH often exceeds the capacity of an unmanaged site to sustain fish yield naturally. Feedback between the economics and biology models simulates the impact of angler demands on fish production and biomass. As modeled anglers substitute one site for another in response to actions by managers or nature, they redistribute their demand and their impact on the fisheries. Thus, they affect the sustainability of future catch and harvest and the need for additional management adjustments. Through angler adjustments to fishery change, fishery system boundaries in RIOFISH go beyond watershed or other ecosystem boundaries.

Model Scenario Development

For this paper, a series of model-run scenarios were developed to test the null hypothesis that single- and multiple-site analyses of the benefits derived from management result in the same estimated use, benefit, and indicated management. Individual RIOFISH simulations, referred to here as scenarios, were run in pairs (with and without management action) and differences between scenario outputs were calculated to estimate management effectiveness. Except where otherwise specified, we constrained model hydrology to average flow, to unchanged human population from 1992, and to status quo management at RIOFISH sites other than those included in the examples.

The first set of scenarios was designed to show how RIOFISH behaves under highly constrained, single-reservoir management. A single tactical modification (management action) was made for each scenario and output results were contrasted back to the reference condition (without the management action) defined in Table 1. The second set of scenarios in Table 2 illustrates how management at one reservoir (different from the reservoir in Table 1) is affected by the number of other reservoirs included in the system of substitute reservoirs when site qualities are virtually the same. A third set of scenarios was designed to show the effect of changed fishery catch and harvest on benefits derived from example tactics, proposed for the first model experiments. Scenarios also were designed to show the interactions that exist between reservoir facilities management and fisheries management by stocking at two coldwater reservoirs improved by facilities addition. Other scenarios were analyzed to show the effect of independent analysis of proposed warmwater reservoirs on estimating benefits of management, when it is commensurate among substitute reservoirs.

Results

Single- and Multiple-Reservoir Output Behavior

A popular warmwater reservoir was first targeted to model response of angler use and benefits to common management tactics exposed to single-site analysis (Table 1). For the scenarios shown in Table 1, management actions were implemented adding one tactic at a time to an otherwise constant background condition at the target reservoir. The background conditions represented minimal site development. Tactical effort was varied over a range of conditions encountered at New Mexico reservoirs. Consistent with economic theory and common sense, improvement of angler catch and harvest success increased angler use and benefit at decreasing rates. Angler use nearly tripled as fish harvest increased from 0.5 to 4.0/d. Adding boat ramps created the greatest angler use and value

TABLE 1.—Incremental gain in angler-use rates (angler-days) and fishing benefits (US$) for a popular warmwater reservoir managed to demonstrate the responses to changes in tactics, using the management model RIOFISH. For each management scenario, background conditions were held constant except for the indicated tactic. Background conditions were 1,000 ha surface area, 1.0 fish caught/d and 0.5 fish kept/d, 150 g fish mean weight, 5% shore access by vehicle, one boat ramp, four toilets, and 10 campsites.

Output category	Level of management (tactical-input scenarios)		
	Low	Medium	High
Site size (ha)	2,000	4,000	8,000
Angler-days	5,281	6,416	7,069
Benefits	94,000	115,000	127,000
Fish harvest and catch/d	1.0 and 2.0	2.0 and 4.0	4.0 and 8.0
Angler-days	6,982	11,183	15,127
Benefits	126,000	205,000	291,000
Fish size (g)	300	600	1,200
Angler-days	5,281	5,541	5,651
Benefits	94,000	99,000	101,000
Ramps	2	5	10
Angler-days	11,532	40,027	78,951
Benefits	211,000	768,000	1,555,000
Toilets	8	16	32
Angler-days	8,078	13,395	17,241
Benefits	146,000	247,000	320,000
Campsites	10	20	40
Angler-days	5,991	6,640	6,695
Benefits	107,000	119,000	125,000
Shore access by road (%)	10	20	40
Angler-days	5,281	5,312	5,319
Benefits	94,000	95,000	95,000

increase, whereas increasing the size of fish created the smallest angler use and value increase. Similar behavior occurred when other reservoirs were chosen, although the numerical impact varied among reservoirs, depending on their location relative to the distribution of angler populations in the state. For this single-site analysis, assuming constant conditions other than the tactical change, benefits and impacts increased proportionally to increases in angler use after site improvement.

The effects of management actions at a given reservoir depended strongly on what actions were implemented at substitute reservoirs (Table 2). Output estimates changed when several identical substitute reservoirs various distances away from the target reservoir were included in the analysis and managed identically to the target reservoir. Angler-days expected from management at the first reservoir decreased as management investment increased at substitute reservoirs. Management actions at substitute reservoirs strongly affected the predicted use and benefit expected from management planned at the target reservoir. Estimated management effectiveness decreased at the target reservoir as more management was implemented at other reservoirs.

The incremental impact on management effectiveness at the target reservoir varied among substitute reservoirs, depending on management level, quality, and location of each substitute reservoir. Because quality and management level were held constant in these scenarios, resource distribution with respect to angler distribution was the critical factor determining benefits of a management action. Integrating substitute reservoirs into the analysis resulted in a changing estimate of value per angler-day derived from decisions. As more man-

TABLE 2.—Angler use and fishing benefits (US$) from selected management actions at a reservoir treated as part of single-, two-, and five-site systems. Analyses were conducted at selected New Mexico reservoirs. All sites had equal site qualities, except for varying travel costs and the examined tactical changes. Background conditions were as described in Table 1, except different reservoirs were chosen for analysis.

Tactical input (scenario)	Level of management		
	One site	Two sites	Five sites
Site size[a]			
Angler-days	8,634	8,147	7,636
Benefits	230,000	201,000	167,000
Fish harvest and catch[b]			
Angler-days	24,049	22,321	20,495
Benefits	644,000	540,000	422,000
Fish mean size[c]			
Angler-days	1,689	1,605	1,517
Benefits	105,000	40,000	26,000
Boat ramps[d]			
Angler-days	71,053	62,765	54,521
Benefits	1,985,000	1,417,000	938,000
Toilets[e]			
Angler-days	54,304	22,549	20,699
Benefits	710,000	551,000	426,000
Campsites[f]			
Angler-days	30,503	20,262	5,222
Benefits	168,000	139,000	105,000
Access (%)[g]			
Angler-days	24,582	257	235
Benefits	8,000	7,000	5,000

[a]Site size was increased from 1,000 to 4,000 ha only at the managed reservoir.

[b]Fish harvest/d was increased from 0.5 to 2.0 and catch/d was increased from 1.0 to 4.0 only at the managed reservoir.

[c]Fish mean weight increased from 150 to 600 g only at the managed reservoir.

[d]Concrete boat ramps were increased from one to four only at the managed reservoir.

[e]Toilets were increased from 16 to 32 only at the managed reservoir.

[f]Campsites were increased from 10 to 40 only at the managed reservoir.

[g]Percentage road access was increased from 5 to 20% only at the managed reservoir.

agement actions were implemented at substitute reservoirs, estimated benefit per angler-day at the target reservoir decreased.

Existing fish catch and harvest quality influenced the benefits accrued from managing other quality factors (Figure 2). Also, the effectiveness of supplying reservoir access or amenities was influenced by whether the analysis was a single- or multiple-site analysis. The benefits derived from the placement of ramps and amenities were influenced by the specific catch and harvest realized. Estimated use, value, and impact are affected when management at a secondary reservoir is accounted for in the analysis.

Policy analysis outcomes (Table 3) indicated that fish management decisions interact with facility management to alter conclusions about management effectiveness. Adding fish, facilities, and access at two minimally managed coldwater reservoirs increased angler use despite reduced harvest and catch rates under the defined conditions (Table 3). When access and facilities were limited, as in the first scenario in Table 3, the stocking rate needed to sustain a harvest rate of 0.5 fish/h resulted in a low return to creel (5.0%) and benefits far less than the cost of stocking at US$1.00/fish stocked. Adding ramps, facilities, and road access (second scenario in Table 3) greatly increased the return to creel as a consequence of angler attraction to the sites and improved the net benefits from fish stocking while simultaneously reducing mean fish density and mean fish harvest rate.

Depressed harvest rate indicated a stocking adjustment was justified. For the third scenario in Table 3, the two reservoirs were stocked at rates that would generate the 0.5 fish/h harvest rate targeted by many agencies; this action resulted in a small increase in benefit and a marginal benefit to cost ratio greater than 1.0. Angler use and benefits were increased even more, however, by redistributing the total number of fish stocked in the two reservoirs so as to generate lower harvest rates in one reservoir than in the other. The interactions resulting from differences in size and facilities at multiple reservoirs caused the improved cost effectiveness from harvest rates tailored for each reservoir within the system of substitute reservoirs. Results like these lead us to conclude that fisheries management agencies seeking greater cost effectiveness need to tailor reservoir management objectives to each reservoir based on existing or proposed conditions at all substitute reservoirs, as well as at the reservoir in question.

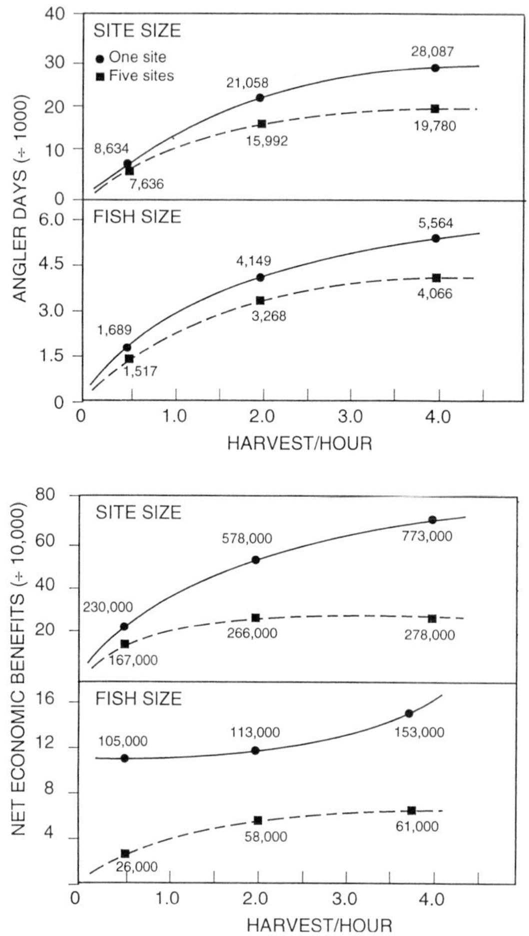

FIGURE 2.—The effect of background daily fish harvest and catch on angler use and benefit (US$) from selected management actions implemented at one reservoir using single- (solid line) versus multiple-site (dashed line) analysis. Background conditions are those used for Tables 1 and 2. At the managed reservoir, site size (surface area) was increased from 1,000 to 4,000 ha and fish size (mean weight) was increased from 150 to 600 g.

Comprehensive Management

When a boat ramp was added to one reservoir (Cochiti), and no other actions were taken at other reservoirs (first scenario in Table 4), increased use, benefits, and impact on days of use exceeded estimates that included planned management at other reservoirs shown in subsequent scenarios. Placing additional ramps at other reservoirs managed by the U.S. Army Corps of Engineers (Santa Rosa and Abiquiu) lowered estimated use and benefits from actions implemented at Cochiti Reservoir. Based

TABLE 3.—Incremental gain in resident angler benefit, total angler use, and effects on fish population harvest as a consequence of stocking fish, and adding boat ramps and toilets at two coldwater reservoirs (A and B). The reference condition created at the two reservoirs included no gamefish (only cyprinids and crayfish), no toilets, 5% road access, no camp sites, no boat ramps, and water surface areas of 938 ha (reservoir A) and 2,292 ha (reservoir B). Catchable size (>230 mm) fish were stocked initially at a rate of 37.5 fish/ha.

Tactical input (scenario)	Fish (number/ha) A	Fish (number/ha) B	Harvest (fish/h) A	Harvest (fish/h) B	Return to creel (%)	Annual angler use (days)	Annual benefits (US$1000s)
Stocking alone[a]	20.6	23.8	0.5	0.4	5.0	3,383	64
Facilities[b]	9.9	9.6	0.2	0.2	54.4	88,956	1,719
Stock to harvest[c]	22.2	27.3	0.5	0.5	59.3	255,959	2,251
Stock to benefit[d]	29.3	24.2	0.7	0.5	59.3	256,104	2,299

[a]121,125 trout were stocked in total: 35,175 at A and 85,950 at B.
[b]121,125 trout were stocked as in the first scenario above, but two ramps, 10 toilets, 100 campsites, and 15% road access were added at site A; and four ramps, 20 toilets, 200 campsites, and 30% road access were added at site B.
[c]431,735 trout were stocked in total: 87,935 in A and 343,800 in B; facilities were in place as in the second scenario above.
[d]431,735 trout were stocked in total: 131,050 in A and 300,685 in B; facilities were in place as in the second and third scenarios above.

on these three reservoirs alone, the state fisheries management agency might conclude that substitute-site interaction was too low to justify multiple-site analysis; however, recognizing the proposed placement of another ramp at another reservoir (Elephant Butte), managed by a different agency (the U.S. Bureau of Reclamation), decreased the estimated benefits generated by a ramp added at Cochiti Reservoir by more than 10%.

Over a 5-year planning period, facilities probably will be added at a number of reservoirs managed by the two federal agencies, as indicated in the remaining scenarios of Table 4, especially the last one. The estimated annual value to anglers resulting from building a ramp at Cochiti Reservoir is predicted to be halved as a consequence of adding boat ramps at other substitute reservoirs. Angler use is predicted to be reduced by 20%, a change that could affect anticipated biological management effectiveness. For example, with boat ramps elsewhere, there is less need for boat ramps and fish stocking at Cochiti Reservoir.

TABLE 4.—Predicted effect of management activity at other warmwater reservoirs on annual use (angler-days) developed at Cochiti Reservoir, change of angler use at all other reservoirs, and benefits derived from building a boat ramp at Cochiti Reservoir, assuming otherwise constant conditions among other reservoirs in New Mexico.

Tactical input (scenario): one ramp added at Cochiti Reservoir and . . .	Management output: Angler-days at Cochiti Reservoir	Decreased days of use at other reservoirs	Statewide benefits (US$)
No other action taken	26,313	5,763	489,000
One ramp at Santa Rosa	26,148	5,658	488,000
One ramp at Abiquiu	25,931	5,538	479,000
One ramp at Elephant Butte	25,054	5,399	434,000
One ramp at three other reservoirs[a]	24,561	5,217	410,000
Two ramps at Elephant Butte	3,836	5,095	337,000
Numerous ramps[b]	21,088	2,779	257,000

[a]Abiquiu, Santa Rosa, and Elephant Butte.
[b]One ramp at Santa Rosa, Ute, Caballo, and Bluewater; two ramps each at Elephant Butte and Navajo.

Jemez Canyon Reservoir is presently (1995) closed to fishing by the surrounding landholders, but negotiations in-progress may open access to the public. The RIOFISH model predicted that angler use of Jemez Canyon Reservoir would increase dramatically if access is negotiated and the reservoir is developed for recreational use (Table 5); however, most angler use would not be new but rather would be diverted from other reservoirs. Of the 52,641 angler-days predicted for an opened and developed reservoir, only 9,387 are expected to be new angler-days. The modeled decision to open Jemez Canyon Reservoir reduced fishing effort at substitute reservoirs, requiring biological or other management adjustments to sustain fishing benefits.

The estimated annual benefits gained from opening and developing Jemez Canyon Reservoir are influenced by management decisions made for other reservoirs. If no other management changes are expected, benefits derived from developing Jemez Canyon Reservoir are significantly larger than if improvements are installed at substitute reservoirs. Depending on the extent of management applied elsewhere, the average annual benefits may be cut by two-thirds.

Timing is important. If it takes 5 years to develop Jemez Canyon Reservoir before benefits are derived, development elsewhere may reduce benefit at Jemez Canyon Reservoir to the point where opening the reservoir is no longer economically beneficial to anglers. On the other hand, opening Jemez

TABLE 5.—The effect of other management activities on the annual use (angler-days) and value generated by opening a new reservoir site at Jemez Canyon Reservoir without any other change in the system (first scenario) and with simultaneous change in ramp number at other sites in the system (all other scenarios).

Tactical input (scenario)	Management output			
	New site use	Cochiti use change	Statewide use change	Statewide benefits (US$)
New site opened	52,641	−3,050	9,387	651,000
Cochiti ramp	41,758	−2,330	6,901	503,000
Ramps at two sites[a]	41,487	−2,297	6,886	494,000
Ramp at three sites[b]	39,118	−2,047	6,260	430,000
Three ramps[c]	37,067	−1,845	5,698	382,000
Nine ramps[d]	30,844	−1,354	5,118	249,000

[a]Santa Rosa and Cochiti.
[b]Cochiti, Santa Rosa, and Elephant Butte.
[c]Two ramps at Elephant Butte and one ramp at Cochiti.
[d]One ramp each at Santa Rosa, Cochiti, Bluewater, and Caballo; two ramps each at Elephant Butte and Navajo.

Canyon Reservoir reduces the values gained by management at other reservoirs.

Discussion

Significance of Site Substitution Effects

Effective pursuit of optimum sustainable yield requires comprehensive management of diverse opportunities over a system of substitute reservoirs. Our results show that angler willingness to substitute among reservoirs is high when reservoir qualities change under the conditions that existed at the time of model development. Reservoir substitution is an important factor determining management effectiveness, even in a dry state where water bodies are relatively far apart. As anglers adjust their fishing activity toward improved reservoirs, the fishing effort changes—enough in many instances to warrant adjustments in regulations, stocking, and other biological management. Although we focused on reservoirs in this study, site substitution can occur among reservoirs, natural lakes, rivers, and other site types, depending on angler preferences.

Restricting use and benefits analysis to a single-site perspective overestimates anticipated reservoir use and benefit of management at a given site. Over short planning periods of 5–10 years, the sum of proposed management changes across substitute reservoirs markedly alters the use and benefits expected at any one reservoir within the system.

We tested our hypothesis with RIOFISH in a constrained management environment with static human dimensions. Our general conclusion about interactions across substitute reservoirs remains unchanged, however, when demographic change is incorporated into the analysis. Inputs to RIOFISH can be modified easily to simulate changes in human population number, distribution, and fraction who fish, for example, to assess management effectiveness with more versatility.

Our results also indicate that an average benefit per angler-day approach to estimating benefits obtained from individual site management may substantially misrepresent estimates of statewide benefits gained from the proposed management. Effective comprehensive planning to give maximum total net benefit considers management proposals at all substitute reservoirs.

Our research shows that multiple-reservoir analysis limited to the boundaries of an agency jurisdiction overestimates effectiveness of recreational management at those reservoirs. It also affects reservoir use, resource status, and benefits of management by agencies outside its jurisdiction. In New Mexico, at least five federal and two state agencies control recreational opportunities that can alter the estimated effectiveness of angler-oriented management. Most agencies emphasize recreational improvements at locations within their management jurisdiction; however, they rarely consider the effects of those decisions on management effectiveness outside that jurisdiction, except where such effects are obvious (e.g., reservoir hydrology interactions). Developing improved policy analytical tools with geographical as well as functional comprehensiveness could contribute to improved interagency coordination.

Interagency Management and Research Implications

Federal and state land- and water-management agencies typically concern themselves with the benefit of all water-based recreationists, not just anglers. Therefore, facilities influencing angler benefits might not be placed optimally for anglers. State fisheries management agencies are responsible most specifically to anglers and for related regulation, stocking, and other biological management at public reservoirs, whether or not the reservoirs are also managed by other federal or state agencies. In addition to biological responsibilities, state fishery management agencies collect federal aid revenues that must be used to develop boat ramps and attendant facilities. The most cost-effective provision of fishing opportunities usually requires the state fishery agencies to anticipate the decisions of other agencies.

Our results indicate that the effectiveness of management actions planned by a fisheries management agency to maximize statewide angler benefit is conditioned by improvements planned by other agencies and demographic changes influencing recreational demand. In New Mexico, as in many locations, decisions about the most cost-effective placements of roads, ramps, amenities, and other improvements have been based primarily on past facility use at reservoirs within each agency's jurisdiction. Typically, improvements are made at locations where pressure on existing facilities is highest within the agency jurisdiction and usually with limited concern for management effectiveness outside the agency jurisdiction. Results presented here for New Mexico indicate that locally considered decisions, such as reservoir openings, boat ramp construction, and other provisions, can limit angler use and benefit derived from management at numerous reservoirs beyond agency jurisdiction. These parochial decisions can substantially change angler fishing intensity, resource status, need for follow-up management, and regional management effectiveness. Truly comprehensive management systems transcend parochial decision making and facilitate more beneficial management by integrating analysis across jurisdictional areas.

Fisheries managers can use RIOFISH to analyze the optimum placement of access and facilities for angler benefit, and to communicate that analysis to other agencies providing recreational facilities. It also allows fisheries managers to systematically plan stocking and regulation adjustments that will most benefit anglers, given the planned management actions of other agencies. Of course, RIOFISH requires application and testing with measures of angler use and benefits to confirm the projected results and to provide insights into improvements according to principles of scientifically responsive management (Walters 1986).

One of the more important limitations is the restriction of RIOFISH domain to angler use and benefits. Because RIOFISH does not include all water-based recreational use and benefits, the total benefits of multiple-use facilities are understated. Because land- and water-management agencies are mandated to serve a larger recreational need than that of anglers alone, the extent to which anglers get a proportional share of the benefits remains equivocal without more comprehensive tools for policy analysis. Decisions made as a result of RIOFISH analysis are made to benefit resident anglers and therefore also benefit the state. Out-of-state residents may also benefit coincidentally but are viewed more for the income they add to the state. Federal agencies have more comprehensive mandates, which require interstate benefits analysis in regional models. Nevertheless, comprehensive management system models such as RIOFISH offer a resource of considerable value for optimally sustaining fisheries resources and angler benefits.

References

Burt, O., and D. Brewer. 1971. Evaluation of net social benefits from outdoor recreation. Econometrica 39: 813–827.

Cichetti, C. J., A. Fisher, and K. Smith. 1976. An economic evaluation of a generalized consumer surplus measure: the mineral king controversy. Econometrica 44:1259–1276.

Cole, R. A., and five coauthors. 1990a. RIOFISH, a fishery management planning model for New Mexico reservoirs. New Mexico Water Resources Research Institute Report Number 252, New Mexico State University, Las Cruces.

Cole, R. A., K. A. Green-Hammond, F. A. Ward, T. J. Ward, and R. A. Deitner. 1995b. User's guide for RIOFISH: a comprehensive management system model for New Mexico sportfisheries. New Mexico Water Resources Research Institute Technical Completion Report 292, New Mexico State University, Las Cruces.

Cole, R. A., and six coauthors. 1995a. RIOFISH: a statewide comprehensive management system model for New Mexico sportfisheries. New Mexico Water Resources Research Institute Technical Completion Report 291, New Mexico State University, Las Cruces.

Cole, R. A., and F. A. Ward. 1994. Optimum fishery management policy: angler opportunity versus angler benefit. North American Journal of Fisheries Management 14:22–33.

Cole, R., T. Ward, F. Ward, and R. Deitner. 1986. A simulation model for managing fisheries in reservoirs of the Rio Grande of New Mexico. Pages 18–28 *in* G. E. Hall and M. M. Van den Avyle, editors. Reservoir fisheries management strategies for the 80's. American Fisheries Society, Southern Division, Reservoir Committee, Bethesda, Maryland.

Cole, R. A., F. A. Ward, T. J. Ward, and R. M. Wilson. 1990b. Development of an interdisciplinary model for water and fishery management. Water Resources Bulletin 26:597–609.

Hanemann, M., and E. Morey. 1992. Separability, partial demand system and consumer's surplus. Journal of Environmental Economics and Management 22:241–258.

Hansen, W. J. 1986. Valuating the recreational use of fisheries resource sites with the travel cost method. Technical Report V, Sport Fishing Institute, Washington, DC.

Harville, J. P., editor. 1991. Proceedings of the North American fisheries leadership workshop. American Fisheries Society, Bethesda, Maryland.

Hayes, D. B., W. W. Taylor, and E. L. Mills. 1993. Natural lakes and large impoundments. Pages 493–515 *in* Kohler and Hubert (1993).

Knetsch, J. L., R. E. Brown, and W. J. Hansen. 1976. Estimating expected use and value of recreation sites. Pages 103–115 *in* C. E. Gearing, W. W. Swat, and T. Var, editors. Planning for tourism development: quantitative approaches. Praeger, New York.

Kohler, C. C., and W. A. Hubert, editors. 1993. Inland fisheries management in North America. American Fisheries Society, Bethesda, Maryland.

Nielsen, L. A. 1993. History of inland fisheries management in North America. Pages 3–30 *in* Kohler and Hubert (1993).

Roedel, P. M. 1975. Optimum sustainable yield as a concept in fisheries management. American Fisheries Society Special Publication 9, Bethesda, Maryland.

Rosenthal, D. H., D. M. Donnelly, M. B. Shiffhauer, and G. E. Brink. 1986. User's guide to RMTCM: software for travel cost analysis. U.S. Forest Service General Technical Report RM-132.

Summerfelt, R. C. 1993. Lake and reservoir habitat management. Pages 231–261 *in* Kohler and Hubert (1993).

Walters, C. 1986. Adaptive management of renewable resources. Macmillan, New York.

Ward, F. A. 1989. Effectively managing spatially competing water uses: new evidence from a regional recreation demand model. Journal of Regional Science 29:229–246.

Ward, F. A., and J. B. Loomis. 1986. The travel cost demand model as an environmental policy assessment tool: a review of literature. Western Journal of Agricultural Economics 11:164–178.

Weithman, S. W. 1993. Socioeconomic benefits of fisheries. Pages 159–177 *in* Kohler and Hubert (1993).

American Fisheries Society Symposium 16:261–270, 1996

Fish Consumption and Human Health: Developing Partnerships between Risk Assessors and Resource Managers

ELLEN SMITH EBERT

ChemRisk®—A Division of McLaren/Hart
1685 Congress Street, Portland, Maine 04102, USA

Abstract.—Until recently, fisheries managers have been primarily concerned with maintaining the quality and extent of fishery benefits and healthy populations of fish. Although fish consumption has long been promoted as a healthy behavior, it is apparent that through consumption humans may be exposed to chemicals that accumulate in fish tissues. Public health officials are often faced with the need to evaluate health risks associated with consumption of fish from individual fisheries and may, as a result, require additional information about the behavior of anglers who use those fisheries. Factors that affect potential health risks due to chemical contamination include the species, age, and size of fish being harvested for consumption, frequency of fish meals, parts of the fish being consumed, concentrations of contaminants in the edible tissues, and demographic characteristics of the angler population. Although information concerning preferred species, catch and harvest rates, and sizes and ages of fish caught and harvested from managed fisheries are often readily available, current data collection methodologies generally fall short of providing all of the information necessary for risk assessors to evaluate consumption behaviors and resulting health risks. In this paper I discuss ways in which fisheries data are used to estimate human health risks and recommend that fisheries managers use modified creel survey forms to collect site-specific data on harvested fish and consumption patterns of anglers in order to meet the dual needs of resource management and the protection of public health.

Consumption of fish has long been promoted as a healthy dietary practice. Diets rich in fish oils reduce plasma cholesterol and triglyceride concentrations and probably contribute to reduced risk of cardiovascular disease (Fehily et al. 1983; Fisher et al. 1986; Kantha 1987; Hornstra 1988; Simonsen and Nordoy 1989; van Houwelingen et al. 1989; Wahlqvist et al. 1989). Given this association, several authors (Brown 1986; Hanninen and Agren 1989) have recommended an overall increase in fish consumption.

Although fish do contribute a beneficial, low-fat source of high-quality protein, if fish tissues have accumulated toxic substances, consumption of fish may lead to health risks. This increased risk can be particularly problematic for recreationally caught fish that may not be monitored for toxic contaminants. Because many types of contaminants, including certain chlorinated organic compounds and metals, present no physical evidence of contamination, do not adversely affect fish at low environmental concentrations, and cannot be identified without laboratory analysis, anglers and their families may be unaware that they are being exposed to hazardous materials by consuming fish.

Factors that affect the potential for health risks caused by chemical contamination include the species, age, and size of fish being consumed, the frequency and amount of fish consumed at meals, the parts of the fish being consumed, and the concentrations of contaminants in those tissues; the age and gender of the anglers who consume those fish also contribute to variation in risk. Although some data do exist concerning fish consumption by anglers (Pierce et al. 1981; Puffer et al. 1981; Fiore et al. 1989; Connelly et al. 1992; Ebert et al. 1993), these data have limited value in predicting angler behavior at specific water bodies. Factors such as access, productivity, seasonality, and angler demographics all influence the rate of fish consumption in different locations. In the absence of site-specific information, many risk assessors rely on default standard values that are based on dietary patterns of the U.S. population and are not necessarily representative of anglers who use a particular water body. For example, most U.S. population studies include consumption of commercially available marine and freshwater fish and shellfish and would not be relevant or applicable to a freshwater reservoir that is only fished recreationally (Ebert et al. 1994). The result of using default assumptions is that actual human health risks may be substantially overestimated or underestimated and lead to inappropriate regulatory control.

In this paper I discuss the evaluation of human health effects from contaminated fish. I provide an overview of the variables that are considered in such an evaluation and discuss the ways in which fisheries managers can effectively collect data to ensure that characterization of consumption from their fisheries will be appropriately based on site-specific characteristics.

The Risk Assessment Process

The risk assessment process generally includes deriving an estimate of the level of contaminant intake (dose) and comparing that intake to levels associated with certain health effects. Ultimately, risk assessment results are used by environmental regulators who need to understand, as clearly as possible, who may be at risk and what their potential for exposure may be (EPA 1992).

For many years, risk assessors have lacked needed data on fish consumption and have had to rely on default assumptions. The public and regulatory agencies are increasingly recognizing that these assumptions may inappropriately underestimate exposures for certain populations and overestimate exposures for others (EPA 1989a, 1989c). Thus, in recent years there has been an attempt to broaden the database on fish consumption and collect site-specific data to make possible more precise estimates of health risks (EPA 1989a). This effort will ensure that public health will be protected and that limited economic resources are being directed toward areas of greatest need.

When attempting to estimate potential health risks associated with consumption of contaminated fish, the following general equation is used (EPA 1989b):

$$\text{Intake} = (\text{CF})(\text{CR})(\text{FI})(\text{EF})(\text{ED})(1/\text{AT})(1/\text{BW}); \tag{1}$$

intake = average daily intake (mg/kg · day);
CF = contaminant concentration in the fish (mg/g);
CR = fish consumption rate (g/day);
FI = fraction ingested from contaminated source (unitless);
EF = frequency of consumption (d/year);
ED = duration of exposure (years);
AT = averaging period (d); and
BW = body weight (kg).

Each of these variables is discussed below.

Contaminant Concentration in Fish (CF)

Sometimes, when attempting to monitor levels of contaminants in water bodies of interest, whole-body tissue concentrations in young-of-year or rough fish (e.g., sucker and carp) are measured. Although whole-body concentrations in smaller fish of various species are appropriate for considering potential hazards for fish-eating birds and mammals, humans typically consume only the muscle meat of larger gamefish or panfish rather than the whole fish. The most appropriate data for human health considerations are those concentrations measured on a wet-weight basis in the portions, sizes, and species of fish commonly consumed by the anglers of interest.

Rate of Fish Consumption (CR)

Numerous survey methods may be used to collect fish consumption data, including mail and telephone recall surveys, food diary studies, and creel surveys. Any of these methods can be effective in collecting information, but each presents unique problems when characterizing long-term consumption from a single fishery (Ebert et al. 1994). First, in any survey that relies on random sampling or the sampling of licensed anglers, it can be difficult to identify an adequately large sample of the population that uses a specific fishery. Second, although recall surveys may provide a good understanding of long-term behavior, they are subject to recall bias that may contribute to uncertainties in any estimates derived from them. Shorter term recall, dietary, or creel surveys can minimize recall bias and provide good data on short-term behavior, but their use is limited when one attempts to derive long-term estimates, because a number of assumptions about long-term behavior must be made. In addition, creel surveys tend to oversample more frequent anglers, resulting in overestimated rates of consumption (Puffer et al. 1981; Price et al. 1994). Finally, although long-term diary studies of the target population are effective, they tend to be costly and are subject to attrition.

The creel survey is a method that has long been used by fisheries managers for the collection of data about specific fisheries. Because creel survey methods are well understood by managers, and because they are focused on the fishery and population of interest, they may provide the best means of collecting information about consumption from an individual fishery—assuming they are well designed and pieces of information necessary for long-term extrapolation from short-term data are collected.

Fish consumption can be estimated from a creel survey by two methods. The first involves asking anglers to indicate how often they consume fish from the fishery and to recall the average size of those fish meals. This method can be problematic because anglers may have some difficulty recalling exactly how many meals they consume during the year; they may forget, for example, that the frequency may vary substantially over the year because of seasonal variations in the availability of preferred fish, angling activity, and success rates. If anglers fish multiple water bodies, it may be difficult for them to recall the percentage of meals that came from the water body of interest. Finally, the size of the portions consumed are likely to be variable, depending on the sizes and types of fish harvested for consumption and the number of people consuming, making it difficult for anglers to identify a reliable single-portion size.

The second method of estimating consumption is based on the size and species of fish harvested for consumption at the time of interview (Puffer et al. 1981; Landolt et al. 1985, 1987; Ebert et al. 1994). In this method, fish weights and lengths and species are recorded by the creel clerk, and the angler is asked to indicate which fish are to be eaten, what parts are to be eaten, and how many people are to share in the consumption (Figure 1). Based on this, "per trip" fish consumption estimates can be derived. These estimates can then be extrapolated by risk assessors from information concerning frequency of trips and seasonal variations to derive an annualized rate of consumption. Although these estimates provide strong data on each trip, they are subject to prediction bias in the long-term extrapolation.

Most reported fish consumption rates have been annualized; that is, if an angler eats 10 meals per year that are 227 g each, the consumption rate is reported as 6.2 g/d (10 meals/365 d × 227 g/meal). This type of consumption rate is most commonly used when a risk assessor is evaluating health effects resulting from long-term exposures (i.e., cancer, chronic reproductive effects, etc.). When a chemical contaminant presents a risk to human health as a result of short-term exposure, however, annualized rates of consumption may not be appropriate. The estimate should be based instead on the average number of meals eaten during the period of time being considered. For example, if evaluating an acute toxicant, it may be necessary to evaluate effects resulting from a single meal occurring on a single day. In this case, daily estimates in the range of 100–200 g, based on meal size, may be appropriate (EPA 1994).

Fraction of Fish Ingested From a Contaminated Source (FI)

Because anglers tend to consume fish from more than one fishery, and not all fisheries are contaminated, it is important to determine what fraction of the consumed fish is contaminated. This is generally not an issue when conducting a creel survey for a single body of water, since creel survey questions will be limited to that water body. If other survey methods, such as mail, telephone, or diary studies are conducted, however, it is important to differentiate among fish consumed from contaminated and uncontaminated sources. Although a single fishery may include a combination of contaminated and uncontaminated fish, this variability should be characterized in a well-designed fish-sampling program and does not generally need to be considered separately in the risk assessment.

Frequency of Fish Consumption (EF)

Because consumption rate is generally annualized when evaluating potential long-term health effects, it is usually assumed that the frequency of consumption is 365 days per year. This changes if subchronic (short-term) or acute health effects are being evaluated. In cases where annualized consumption rates are not used, the frequency should be reported as the number of days on which fish were consumed during the exposure period.

Duration of Exposure (ED)

Duration of exposure is generally considered to be 30 years for chronic health effects. It was once common to use a 70-year exposure duration (EPA 1989c), but EPA and many state agencies changed their positions after recognizing that individuals do not engage in the same activities throughout their lives. This is particularly true when one is evaluating a single, recreational fishery. Although some anglers may live near and fish a single fishery throughout their lives, most probably do not. Census statistics on population mobility can be useful in determining the likely duration of exposure in a given area.

Averaging Time (AT)

The length of time over which exposures are averaged, the averaging time, varies, depending upon the health effects being evaluated. For cancer, the averaging time is generally assumed to be a lifetime of 70 years (25,550 d) (EPA 1989c). For other chronic health effects, it equals the number of years of exposure times 365 d. Shorter term averaging times, as little as a few days for compounds that may have acute health effects, are sometimes used (EPA 1994).

GENERAL CREEL CENSUS

Location ________________ Date ___________ Day of Week ________ Clerk ______ Weather _____

Census Method: Stationary Vehicle Census Started ________ Census Ended ________ Left Census ________ to ________

River Reach ________________________ Boat/Shore______ Party Size _____ Age Of Individual Interviewed ______

Trip Over?___ Repeat Contact? ___ Before Interviewed? _______ Interview Refused ___________ Target Species _________, ________, _________, _________, _________

Resident? _____ How Long? _______ Zip Code __________ Time Started _________ Time Now _________

How many years have you fished here?_________

	Jan	Feb	Mar	Apr	May	Jun	Jul	Aug	Sep	Oct	Nov	Dec
What Months Of The Year Do You Fish Here?												

	Jan	Feb	Mar	Apr	May	Jun	Jul	Aug	Sep	Oct	Nov	Dec
How Frequently do You Fish Here During These Months (times/month)												

Date of Last Trip: _________ How Many Hours Did You Fish? _______ Did You Catch Anything? ________________

Did You Eat Any Of The Fish You Caught? __________Do you normally consume the fish you harvest here (y/n)?_______

Today's Catch

Species	No. Caught	No. Released	Why Released

Comments:

Successful Anglers with Harvest Only

Species	Weight (grams)	Field or Dressed	Length (mm)	To be Cons. by Household?	# Persons to Consume (Ad/Ch)	Parts To Be Eaten*	How Prepared**	Other Use	Comments

Put all numbers that apply

* 1. Entire 2. Fillet 3. Skin 4. Liver 5. Roe 6. Head 7. Broth 8. Other (specify)

** 1. Raw 2. Baked 3. Broiled 4. Grilled 5. Pan fried 6. Boiled 7. Poached 8. Deep fried 9. Skin on 10. Smoked 11. Other (specify)

FIGURE 1.—Sample of expanded, general, creel census form. Abbreviations are y = yes, n = no, No. = number, Con. = consumed, Ad = adults, and Ch = children.

TABLE 1.—Different goals of fisheries managers and risk assessors.

Fisheries managers	Risk assessors
How much fishing pressure is occurring?	Who, if anyone, is using the fishery for food?
Is the habitat healthy?	Are there any subpopulations of anglers who may, for cultural or ethnic reasons, have consumption behaviors that differ from those of the general angler population?
Are resources sustainable?	What types and sizes of fish are exposed anglers harvesting for consumption?
What are catch and harvest rates?	How large are their fish meals?
What types and sizes of fish are being targeted and harvested?	How often do they eat fish and how does their consumption activity vary over the year?
Are fish stocks healthy and balanced?	What parts of the fish do they eat?
What are the stocking needs, if any?	What are the levels of contamination in those parts of the fish?
	How do they prepare the fish that they eat (pan fry, bake, grill, etc.)?
	How many years have they eaten from this fishery?
	How many persons will consume harvested fish?

Body Weight (BW)

There are extensive data available concerning the ranges of body weights for males and females of different age groups within the U.S. population. These age–weight distributions for the U.S. population are well presented in the Environmental Protection Agency's (EPA) *Exposure Factors Handbook* (EPA 1989c). They can be considered representative of the general angler population, assuming there are a variety of individuals using the fishery. These distributions may not be appropriate if the fishery is being used by a subpopulation for whom the general U.S. weight distributions may not be applicable.

Different Goals of Fisheries Managers and Risk Assessors

Fisheries managers typically focus on maintaining a high-quality fishery through maintaining high-quality habitat, healthy and well-balanced fish populations, and fishing pressure compatible with available resources. Toward this end, creel surveys are often used to determine the level of angling effort, targeted species, success rates, sizes of fish caught and harvested, and the socioeconomic characteristics of the angler population (Malvestuto 1983; Matlock 1991).

Although much of the information now being collected by fisheries managers is also of interest to risk assessors, generally the level of detail is not sufficient to meet risk assessment needs. Creel surveys document short-term behavior of the user population, and from them managers can estimate the overall use of the fishery; however, data collected and subsequently reported provide little information about variations in long-term behavior among different individuals. For example, data collected for fisheries management purposes may indicate that annual effort on a particular body of water is 10,000 angler-days. Such a finding may be sufficient to determine fishing pressure and adequacy of recreational services; however, risk assessors are interested in the number of trips by individual anglers to that site because the number of trips may have a substantial impact on the number of meals eaten. Lack of information on a "per angler" basis forces risk assessors to make assumptions about individual behavior. In this example, a risk assessor could assume that 10,000 anglers fish 1 d/year, that 1,000 anglers fish 10 d/year, or that 100 anglers fish 100 d/year. If it is assumed that the number of meals consumed is proportional to the number of fishing trips, use of any one of these assumptions would result in substantially different estimates of consumption than would be derived using any other; substantial differences would then be seen in estimates of potential risk. This problem is further compounded by the fact that angler activity and consumption behaviors are highly skewed, with most anglers reporting very low levels of activity and consumption and a very small percentage of anglers reporting high rates of activity and consumption (Pierce et al. 1981; Puffer et al. 1981; Landolt et al. 1987, 1985; Fiore et al. 1989; Cox et al. 1990; Connelly et al. 1992; Ebert et al. 1993, 1994, 1996; C. Cox and coworkers, Ontario Ministry of the Environment, unpublished data). Thus, it is important when evaluating potential exposure from a fishery to have some understanding of the variability in angler behaviors.

To model the potential for exposure via fish consumption, risk assessors must understand the characteristics and behaviors of anglers who fish in a particular location (Table 1). These additional data will help define the population as a whole and provide an understanding of the range of variation within that population. That understanding will, in turn, help to determine who, if anyone, is at risk, and estimate what their potential for exposure may be.

Bridging the Gaps

To improve the quality of data on which risk assessors rely for determining the potential for health risks, fisheries managers can provide information on variability in success rates, effort, and consumption over the season. This information will help risk assessors to understand the behavior of anglers who consume fish from particular types of fisheries in different regions of the United States. Below are descriptions of additional angler behaviors that can be measured by fisheries managers and subsequently used to improve risk estimates.

Frequency of Fishing Trips

For evaluating chronic effects of toxic chemicals, annualized consumption rates need to be developed. Thus, in addition to knowing how much fish is consumed at a single meal, it is important to determine how often an individual eats the fish. Collecting this information is generally straightforward when determining rates of consumption of commercially available fish, which are generally available throughout the year. It becomes more problematic when determining the frequency of meals of sport-caught fish, because of variabilities in fishing effort, success rates, availability of target species, and seasonality.

It is helpful if the variability of fishing and consumption frequency from an individual fishery can be identified. Although anglers often are asked to indicate the date of their most recent previous trip, long-term variability of fishing frequency is not determined. For example, anglers who are interviewed during a fishing vacation may report that they are fishing daily, but this frequency may only be applicable for a 1- or 2-week vacation period, and they may not fish in that area at all during the remainder of the fishing season. Although a daily fishing frequency may be representative at the time of the interview, it may substantially under- or overestimate effort during the remainder of the year. Thus, to extrapolate to long-term consumption rates, it is important to understand how the frequency varies throughout the season by asking surveyed anglers to identify the months of the year that they fish and to report their typical fishing and consumption frequencies during those months.

Fraction of Fishing Trips Resulting in Consumption

The fraction of trips resulting in consumption of fish can be determined by asking anglers to indicate how often they consume fish that they have harvested from the fishery. Although it is useful to ask how many sport-caught fish meals they consume during a specified period of time, it is important to determine what percentage of fishing trips result in consumption of the fish so that corrections can be made for seasonal variability in effort and consumption. In addition, if anglers have not harvested fish for consumption at trip end or on the previous trip, it is important to determine whether this is because they have been unsuccessful at catching target species or because they do not, as a matter of practice, take fish home for consumption. This information is useful for determining the size of the potentially exposed population.

Number of Persons to Share in Consumption

When consumption estimates are to be based on harvested fish, it is important to understand how many individuals will be sharing the fish to determine likely portion sizes. Anglers can easily be asked how many persons are likely to share the harvested fish. It is also advisable to ask them to indicate, if possible, how many of these consumers are adults and how many are children to characterize the demographics of the potentially exposed population.

Fish Tissue Concentrations

For evaluating human health, it is most appropriate to measure contaminant levels in the species and the parts of the fish that are actually consumed, which is usually the muscle meat but may include other parts as well. Chemical analysis of the relevant fish parts can be conducted, a strategy that results in meaningful health risk estimates.

To determine whether chemical contamination is a problem at a given site, regulatory agencies often collect information on whole fish or on species (e.g., carp and suckers) that are known to accumulate higher levels of certain contaminants (Cunningham et al. 1990). Unless contaminant concentrations are available for edible parts or species of fish that are likely to be consumed, agencies may mistakenly use these concentrations to assess potential effects on public health.

Fish tissue concentrations are generally sampled by regulatory agencies or potentially responsible parties (groups to whom potential environmental liabilities have been assigned) in preparation for a risk assessment. Fisheries managers may not, as a rule, be responsible for collecting these data, but there may be circumstances when they are involved in these efforts or they may be asked to help design such a program based on their knowledge and familiarity with a fishery. Fisheries managers must therefore understand the types of data that are needed for risk assessment.

If creel surveys are used to collect information on the types, sizes, and parts of fish that are eaten, fisheries managers can provide valuable data to agencies or groups that have been charged with collecting and analyzing fish-sampling data for the purpose of evaluating public health risks. If some anglers using the fishery eat unusual parts of the fish (i.e., the viscera, liver, etc.), that information should be passed along to these groups so that chemical analysis of the appropriate parts of the fish will be conducted. It is better to report wet-weight concentrations than dry-weight concentrations because fish consumption rates are derived on a wet-weight basis. Finally, when submitting fish samples for analysis, it is important to specify the use of low detection limits to ensure that low concentrations in fish tissues will be detected.

Edible Portion of the Fish

After determining the species, size, and parts of fish to be consumed and the number of individuals to share the fish, it becomes necessary to determine the mass of the fish that is consumed (edible portion) to derive a "per trip" consumption estimate. It is generally assumed that the muscle meat, representing 30% of the total mass of the fish (EPA 1989a), is the only part of the fish that is consumed, but creel survey data may indicate otherwise. For example, a recent survey of Chippewa Indians in Wisconsin indicates that they commonly eat the livers of walleye *Stizostedion vitreum* and use walleye heads in soups (Morgan et al. 1996). Thus, certain angler subgroups may eat larger portions of fish and have a higher potential for exposure than the typical sport angler. Because the parts consumed may vary, depending on species, it is important to collect species-specific information.

It may also be worthwhile, during routine fish collection activities associated with the management of the fishery, to characterize the edible portions of the most frequently consumed fish. We have found that the EPA's recommended value of 30% is a reasonable default estimate, but there is a fair amount of variability, caused by differences among species and fisheries. As part of the creel surveys for the West Branch Penobscot River and the Saco River in Maine, we undertook an analysis of edible portions (ChemRisk, unpublished data). In that study, smallmouth bass *Micropterus dolomieu* were collected from both rivers, and white perch *Morone americana* and landlocked salmon *Salmo salar* were collected from the West Branch. Although we found that the mean fillet portions of smallmouth bass from both rivers (29%) were consistent with the EPA's default value, the edible portions of white perch were lower (mean of 24%) and the fillet portions of landlocked salmon were higher (mean of 37%). Thus, if anglers are predominately eating one species of fish, exposure may be over- or underestimated using the EPA's value. Studies of this type can be undertaken easily in the field.

Duration of Exposure

In risk assessment, it is usually assumed that anglers maintain the same fishing effort and consumption activity throughout their lives. This probably is not the case, particularly for a single fishery, because of changing work and family status, changing residences, and conflicting priorities. The duration of fishing at a single fishery can be adjusted by collecting information on the number of years the angler has fished from a particular location.

Demographics of the Angler Population

It is important to collect information on the demographics of the population that uses the fishery. Most long-term recall surveys have identified the "typical" recreational angler as a middle-class, Caucasian male between the ages of 30 and 40 (Pierce et al. 1981; Puffer et al. 1981; NYSDEC 1990; Ebert et al. 1993), and have characterized the fishing and consumption behavior of these anglers. At some fisheries the angler population may differ substantially from this "typical" population described above. For example, urban fisheries may have a much higher number of non-Caucasian anglers who may, for cultural reasons, eat different species or different parts of the fish or who may prepare the fish differently (Landolt et al. 1985, 1987). In addition, certain fisheries may be used by a subpopulation of anglers who rely on the fishery as a major source of food for themselves and their families (CRIFC 1994). Because these types of anglers have not yet been well characterized, it is important to understand whether their consumption differs substantially from that of the general angler population. Thus, it would be useful for fisheries managers to identify such populations of users and to evaluate and compare behaviors of these groups with those of the general, recreational angler population.

Although it is possible and preferable to collect

detailed information from individual anglers, these anglers may be unwilling to report on certain demographic characteristics. Often anglers will be willing to provide their ages; if not, ages can be estimated. Gender can be easily recorded. Anglers may be unwilling to share information pertaining to ethnic background or income level. Although not critical to risk assessment, such information clarifies characteristics of the angler population. Towns or zip codes can be compared with local census statistics to provide another type of understanding of the demographics of the surveyed angler population.

When evaluating risks to anglers, a common assumption is that a subsistence population may exist in a given area, that is, there may be individuals who rely on the fishery as a major source of food. Anglers who fish frequently and eat large amounts of fish should be characterized during a creel survey as long as the survey design matches their behavior in terms of locations, seasons, and times of day fished. If the survey has a sound experimental design, these anglers will be adequately represented.

Preparation Methods

Methods of preparation and cooking can be important considerations when measuring certain types of chemical contaminants. For example, chemical losses of PCBs have been shown to vary among the different cooking methods, ranging from 15% for microwaving and poaching to 60% for frying (Sherer and Price 1993). Other contaminants (e.g., mercury) have been shown to be unaffected by cooking methods (Morgan et al. 1996). It is worthwhile to determine how the anglers prepare and cook their fish. Surveys have demonstrated that pan frying is the most common method of cooking most harvested fish (Puffer et al. 1981; NYSDEC 1990; Connelly et al. 1992; Ebert et al. 1993; P. C. West, J. M. Fly, R. Marans, and F. Larken, University of Michigan, unpublished data), but cooking method may vary from species to species.

Management Tools for the Protection of Public Health

Many fisheries management tools used to protect the resource can also be applied in the protection of public health. Data collected in creel surveys can be invaluable in developing risk management strategies. If fish inhabiting a given fishery have the potential to affect public health adversely, discharges can be controlled, size and bag limits can be implemented, seasons can be adjusted, consumption bans or advisories can be posted to warn anglers of the potential risks of eating fish, and more stringent water quality standards can be considered. Because contaminants can accumulate in different species of fish at different rates, depending on their trophic level, lipid content, and feeding behavior, it is important to present anglers with specific information concerning the types of fish that can and cannot be eaten, the size limitations of fish to be consumed, and the ways in which anglers can reduce their exposure to contaminants in consumed fish (i.e., trimming, cooking method, parts eaten, etc.).

Alternatively, it may be worthwhile to consider use of catch-and-release fisheries. Several studies have indicated that, in many cases, anglers prefer to practice catch-and-release because of the quality and number of fish available and the preservation of the resource (Barry 1988; Boyle et al. 1990; ChemRisk, unpublished data).

Ultimately, data collected by fisheries managers will be most helpful to risk assessors if these data are more readily available. It is likely that substantial data sets exist that would help to expand the database on fish consumption, but few data have been published. As a result, risk assessors may have difficulty identifying, obtaining, or using them. If fisheries managers can publish more of their work (Matlock 1991), these data can be more easily obtained and utilized.

Conclusions

Lack of adequate data on consumption behavior of anglers has led to the use of default estimates of consumption for estimating risks to potentially exposed populations. In some cases, in the absence of site-specific data, extremely high consumption levels have been assumed to occur in order to ensure that public health is being protected. For example, estimates as high as 180 g/d (nearly one half-pound fish per day, 365 days per year) have been proposed for use (EPA 1989a; McCormack and Cleverly 1990), despite the fact that in most fisheries, consumption at that level is not occurring regularly (Fiore et al. 1989; Cox et al. 1990; Connelly et al. 1992; Ebert et al. 1993; Cox and coworkers, unpublished data). Also, lower consumption estimates have been assumed in many regulatory actions, without adequate site-specific consideration of potential subpopulations that may consume substantially greater amounts.

Although the goals of fisheries managers and risk assessors are not identical, many of their data collection needs are similar. Through the use of ongoing creel surveys of angler pressure, fisheries managers can collect information on the sizes, species, and amounts of fish being consumed by anglers and the frequency of consumption. This information can then be used by risk assessors to evaluate potential human health risks, if such an analysis should become necessary. In addition, the collection and publication of additional data obtained in creel surveys will greatly expand the database on fish consumption from individual fisheries. These data can then be made available to risk assessors and regulators interested in identifying data on usage and consumption specific to the region or body of water.

If a creel survey is used, the cost of collecting these additional data is minimal because it involves only modifications to the survey intake form and some additional staff time during the interview process. Because these additional data may not be useful to fisheries managers, there is no need to undertake more analyses; rather this information may be collected, preserved, and made available at a later date, should risk assessment become necessary.

Site-specific fish sampling is another necessary component of risk assessment. Although this sampling can be expensive, it is generally conducted by the regulatory agency or the potentially responsible parties, and should not add to costs for the fisheries managers. It is important, however, that information on the types, parts, and sizes of fish being consumed by local anglers be provided to those individuals involved in fish sampling so that the most relevant tissue data on which to base the risk assessment will result.

By working together, fisheries managers, regulators, and risk assessors can collect the types of data that allow for meaningful evaluation of the fishery from both a resource and a public health perspective. Fisheries managers, who are the individuals most knowledgeable about regional and site-specific fishing behavior, need to become involved in the regulatory and risk assessment processes. With the cooperation of public health officials and fisheries managers, more accurate evaluations of public health risks can be developed, valuable resources can be preserved and protected, and monies and regulatory focus can be appropriately directed to the greatest needs.

References

Barry, T. 1988. A creel and economic survey of the Housatonic River. Connecticut Department of Environmental Protection, Bureau of Fisheries, Final report, Hartford.

Boyle, K. J., R. K. Roper, and S. D. Reiling. 1990. Highlights from the 1988 survey of open water fishing in Maine. University of Maine, Department of Agricultural and Resource Economics (ARE 416), Orono.

Brown, W. V. 1990. Dietary recommendations to prevent coronary heart disease. Annals of the New York Academy of Sciences 598:376–388.

Connelly, N. A., B. A. Knuth, and C. A. Bisogni. 1992. Effects of the health advisory changes on fishing habits and fish consumption in New York sport fisheries. Cornell University, Department of Natural Resources, Human Dimensions Research Unit, (New York Sea Grant Institute Project R/FHD-2-PD), Ithaca.

Cox, C., A. Vaillancourt, and A. F. Johnson. 1990. The results of the 1989 "Guide to eating Ontario sport fish" questionnaire. Ontario Ministry of the Environment, Water Resources Branch (ISBN 0-7729-7084-X), Ontario, Canada.

CRIFC (Columbia River Inter-Tribal Fish Commission). 1994. A fish consumption survey of the Umatilla, Nez Perce, Yakama, and Warm Springs tribes of the Columbia River basin. Columbia River Inter-Tribal Fish Commission, Technical Report 94-3, Portland, Oregon.

Cunningham, P. A., J. M. McCarthy, and D. Zeitlin. 1990. Results of the 1989 census of state fish consumption advisory programs. U.S. Environmental Protection Agency, Office of Water Regulations and Standards, Washington, DC.

Ebert, E., P. S. Price, and R. E. Keenan. 1994. Selection of fish consumption estimates for use in the regulatory process. Journal of Exposure Analysis and Environmental Epidemiology 4:373–393.

Ebert, E. S., N. W. Harrington, K. J. Boyle, J. W. Knight, and R. E. Keenan. 1993. Estimating consumption of freshwater fish among Maine anglers. North American Journal of Fisheries Management 13:737–745.

Ebert, E. S., S. H. Su, T. J. Barry, and M. N. Gray. 1996. Estimated rates of fish consumption by anglers participating in the Connecticut Housatonic River creel survey. North American Journal of Fisheries Management 16:81–89.

EPA (U.S. Environmental Protection Agency). 1989a. Assessing human health risks from chemically contaminated fish and shellfish: a guidance manual. U.S. Environmental Protection Agency, Office of Marine and Estuarine Protection, Office of Water Regulations and Standards (EPA-503/8-89-002), Washington, DC.

EPA (U.S. Environmental Protection Agency). 1989b. Risk assessment guidance for superfund; volume I: human health evaluation manual (part A)—interim final. U.S. Environmental Protection Agency, Office of Emergency and Remedial Response (EPA/540/1-89-002), Washington, DC.

EPA (U.S. Environmental Protection Agency). 1989c. Exposure factors handbook. U.S. Environmental Protection Agency, Office of Health and Environmental Assessment (EPA/600/8-89/043), Washington, DC.

EPA (U.S. Environmental Protection Agency). 1992. Guidelines for exposure assessment. U.S. Environmental Protection Agency, 57 Federal Register 104: 2288–22938, Washington, DC.

EPA (U.S. Environmental Protection Agency). 1994. Guidance for assessing chemical contaminant data for use in fish advisories. U.S. Environmental Protection Agency, Office of Water Regulations and Standards (EPA 823-B-94-004), Washington, DC.

Fehily, A. M., M. L. Burr, K. M. Phillips, and N. M. Deadman. 1983. The effect of fatty fish on plasma lipid and lipoprotein concentrations. American Journal of Clinical Nutrition 38:349–351.

Fiore, B. J., H. A. Anderson, L. P. Hanrahan, L. J. Olson, and W. G. Sonzogni. 1989. Sport fish consumption and body burden levels of chlorinated hydrocarbons: a study of Wisconsin anglers. Archives of Environmental Health 44:82–88.

Fisher, M., P. H. Levine, and B. H. Weiner. 1986. The potential clinical benefits of fish consumption. Archives of Internal Medicine 146:2322–2323.

Hanninen, O., and J. J. Agren. 1989. Effects of moderate freshwater fish diet on lipid metabolism of Finnish students. Journal of Internal Medicine 225:77–81.

Hornstra, G. 1988. The significance of fish and fish-oil enriched food for the prevention and therapy of ischemic cardiovascular disease. Pages 151–235 *in* M. A. Crawford and A. J. Vergroesen, editors. The role of fat in human nutrition, volume II. Academic Press, London. (Not seen, cited *in* van Houwelingen et al. 1989.)

Kantha, S. S. 1987. Dietary effects of fish oil on human health: a review of recent studies. Yale Journal of Biology and Medicine 60:37–44. (Not seen, cited *in* Wahlqvist et al. 1989.)

Landolt, M. L., and five coauthors. 1985. Final report: potential toxicant exposure among consumers of recreationally caught fish from urban embayments of Puget Sound. NOAA (National Oceanic and Atmospheric Administration) Technical Memorandum OMA 33, Rockville, Maryland.

Landolt, M., and five coauthors. 1987. Final report: potential toxicant exposure among consumers of recreationally caught fish from urban embayments of Puget Sound. NOAA (National Oceanic and Atmospheric Administration) Technical Memorandum NOS (National Ocean Service) OMA 33, Rockville, Maryland.

Malvestuto, S. P. 1983. Sampling the recreational fishery. Pages 397–419 *in* L. A. Nielsen and D. L. Johnson, editors. Fisheries techniques. American Fisheries Society, Bethesda, Maryland.

Matlock, G. C. 1991. Overview: use of surveys in decision making. Pages 1–4 *in* D. Guthrie and seven coeditors. Creel and angler surveys in fisheries management. American Fisheries Society Symposium 12.

McCormack C., and D. Cleverly. 1990. Analysis of the potential populations at risk from the consumption of freshwater fish caught near paper mills. U.S. Environmental Protection Agency, Office of Policy Planning and Evaluation, Office of Research and Development, Washington, DC.

Morgan, J. N., M. R. Berry, and R. L. Graves. 1996. Effects of Native American cooking practices on total mercury concentrations in walleye. Journal of Exposure Analysis and Environmental Epidemiology 6(4).

NYSDEC (New York State Department of Environmental Conservation). 1990. New York statewide angler survey 1988. New York State Department of Environmental Conservation, Division of Fish and Wildlife, Albany.

Pierce, D., D. Noviello, and S. Rogers. 1981. Commencement Bay seafood consumption study. Tacoma–Pierce County Health Department, Seattle.

Price, P. S., S. H. Su, and M. N. Gray. 1994. The effect of sampling bias on estimates of angler consumption rates in creel surveys. Journal of Exposure Analysis and Environmental Epidemiology 4:355–372.

Puffer, H. W., S. P. Azen, M. J. Duda, and D. R. Young. 1981. Consumption rates of potentially hazardous marine fish caught in the metropolitan Los Angeles area. University of Southern California, School of Medicine for Environmental Research Laboratory, Grant R 807 120010, Los Angeles.

Sherer, R. A., and P. S. Price. 1993. The effect of cooking processes on PCB levels in edible fish tissue. Quality Assurance: Good Practice, Regulation, and Law 2:396–407.

Simonsen, T., and A. Nordoy. 1989. Ischemic heart disease, serum lipids and platelets in Norwegian populations with traditionally low or high fish consumption. Journal of Internal Medicine 225:83–89.

van Houwelingen, A. C., G. Hornstra, D. Kromhout, and C. de Lezenne Coulander. 1989. Habitual fish consumption, fatty acids of serum phospholipids and platelet function. Atherosclerosis 75:157–165.

Wahlqvist, M. L., C. S. Lo, and K. A. Myers. 1989. Fish intake and arterial wall characteristics in healthy people and diabetic patients. Lancet 2:944–946.

American Fisheries Society Symposium 16:271–276, 1996

Evaluation of Two On-Site Survey Methods for Determining Daily per Capita Freshwater Fish Consumption by Anglers

EARL K. MEREDITH AND STEPHEN P. MALVESTUTO

Fishery Information Management Systems, Inc., Post Office Box 3607, Auburn, Alabama 36830, USA

Abstract.—A statewide freshwater fish consumption survey was conducted from August 1992 to July 1993 to estimate daily per capita consumption of freshwater fish harvested by Alabama anglers. Completed trip interviews of anglers were collected at 29 fishing access points in Alabama. Estimates of daily per capita consumption were derived from each interview by two methods. One method used the actual harvest of fish that anglers indicated would be consumed at the next meal (harvest method). This method incorporated the dress-out percentage of the fish and the number of people who would consume those fish. The second method utilized a recall approach in which the number of servings of fish typically eaten by the angler at a meal was determined by using a visual representation of a typical 113-g (4-oz) serving (serving method). Both methods used the number of fish meals eaten by the angler in the previous month as a component of the per capita consumption equation. Annual estimates of daily per capita consumption were 33 g/d (N = 563, SE = 3.3) for the harvest method and 30 g/d (N = 1,303, SE = 1.8) for the serving method. Although the two methods corroborated one another, the serving method provided a more precise estimate because angler harvest estimates typically are highly variable and because the serving method yielded a larger sample size. Depending on survey objectives, one method may be better suited over the other, and various alternatives are discussed.

Findings on cardiovascular and other health benefits associated with fish oils have sparked increased interest in, and consumption of, fish for nutritional purposes (Kromhout et al. 1985; Bonna et al. 1992). According to the National Fisheries Institute, Americans consumed 22% more fish in 1995 than they did in 1985, and according to *The Aquaculture News* in 1992, annual fish consumption by the general public will increase from the present 7 kg (19 g/d) per capita to 9 kg (25 g/d) by the year 2000. However, there could be disadvantages in increasing fish consumption (Puffer et al. 1982; Humphrey 1987; Foran et al. 1989; Svensson et al. 1991). Ingestion of freshwater fish is considered to be a significant avenue for human exposure to certain toxic compounds in the environment (Reinert et al. 1991; Rifkin and LaKind 1991).

United States federal and state agencies and the general public are concerned with the potential human health risks from consuming chemically contaminated fish. Federal rules define guidelines and criteria for states to use in assessing increased lifetime cancer risks associated with the consumption of freshwater fish containing chemical contaminants (EPA 1989, 1992). Maximum tolerable concentrations of contaminants have been established by many states. When regular sampling and testing of fish tissue reveals concentrations of contaminants that exceed established levels, action is taken to protect the public. Several states have set standards calculated from national averages of per capita fish consumption by the general public. However, different standards may be necessary for protection of local populations of fish consumers. Anglers and their families may face increased health risks because they tend to consume more fish and to eat fish from the same location repeatedly (Reinert et al. 1991).

The U.S. Environmental Protection Agency (EPA) used 6.5 g/d as an estimate of consumption of commercially and recreationally harvested fish and shellfish from estuarine and freshwater environments to develop water quality criteria based on human health guidelines (EPA 1989). Further, the EPA suggested that consumption rates for portions of the U.S. population (e.g., by region, age, race, and gender) may vary from 6 to 100 g/d (Suta 1978; SRI 1980; Puffer et al. 1982). It was determined that approximately 0.1% of the U.S. population consumes 165 g/d of commercially harvested fish and shellfish (EPA 1989). For a "reasonable worst case" scenario, based on individuals consuming fish as their sole source of protein (i.e., a fish consumption rate equivalent to normal consumption of red meat, poultry, fish, and shellfish combined) the EPA (1989) recommended that 180 g/d be used. Individual states, however, are responsible for setting their own fish consumption rates for risk assessments purposes and for determining when fish consumption advisories are needed to protect the general public or specific subpopulations when specified contamination levels are exceeded.

Several states have conducted fish consumption studies as part of their evaluation and assessment of potential additional risks associated with consuming freshwater fish from specific sites. These site-specific data have been collected using a variety of methods. The EPA (1992) reviewed methods for fish and shellfish consumption surveys. Five approaches for collection of fish consumption data were identified: (1) recall information (what people remember) collected through telephone surveys, (2) recall information collected by in-person (face-to-face) interviews, (3) recall information collected via self-administered mail questionnaires, (4) information collected from angler diaries, and (5) information obtained from angler creel surveys. Each method has advantages and disadvantages, but the absence of empirically derived statistical comparisons among these methods makes it difficult for agencies to choose the appropriate method for determining site-specific fish consumption rates of special populations.

Of the five methods listed above, face-to-face interviews and creel surveys can be conducted on site, and creel surveys normally incorporate face-to-face interviews. The main distinction between the two is that the collection of fish harvest data usually is a primary objective of creel surveys, which allows fish consumption rates to be estimated from actual measurements of fish harvested, rather than on a respondent's estimate of average amounts of fish eaten over some recall period (typically a year).

We compared two methods of estimating freshwater fish consumption rates among Alabama anglers. Data for comparisons were obtained during face-to-face interviews in conjunction with an access point creel survey. This approach allowed personnel to observe actual harvests and to ask anglers to recall their prior fish consumption. More specifically, the harvest method estimated consumption rates by basing meal sizes on the observed harvest; the recall method based meal sizes on a visual representation of a typical 113-g (4-oz) serving of fish, which helped anglers recall their average meal sizes when fish was served. Potentially, each interview provided information for two independent estimates of daily per capita fish consumption for the angler.

Methods

Survey Design

Twenty-nine locations throughout Alabama were selected for sampling. These locations included 23 tailwater sites and six reservoir sites representing 11 river drainages. Tailwater areas at many of Alabama's dams afford easy public access to rivers for anglers. These sites were chosen because the primary objective of the survey was to determine fish consumption rates for river anglers.

The survey was conducted from August 1992 to July 1993. Sampling days were selected within each of four seasonal blocks defined as: fall (1 August to 30 November), winter (1 December to 20 February), spring (21 February to 8 May), and summer (9 May to 30 July). This sampling design allowed estimation of daily per capita freshwater fish consumption rates for each season.

A clustered sampling approach was used. Each study site was surveyed once, for two-consecutive days, within each of the four seasonal blocks. Sampling was conducted from sunrise to sunset on the days chosen at each site. The 2-d sampling units were either Friday and Saturday or Sunday and Monday, and they were randomly assigned to study sites within each season. This approach helped minimize day type bias and maximize the potential for intercepting the most anglers. Samples were thus clustered spatially and temporally to coincide with peak fishing activity.

Anglers were intercepted and interviewed at the completion of their fishing trips (copies of the 4-page questionnaire are available from the authors). This intercept method (completed trip) assured that all fish harvested were enumerated. As many active anglers as possible were interviewed at each site on sample days. After an interview was administered, total length (mm) and weight (g) were measured by species for all harvested fish. Weight was measured within ± 0.5 g for fish under 50 g, within ± 5 g for fish over 50 g but under 2.5 kg, and ± 100 g for fish over 2.5 kg but under 10 kg. Heavier fish were measured to ± 1 kg up to 50 kg.

Estimation of Daily Fish Consumption Rates

Two methods were used to estimate per capita daily fish consumption (C_{daily}). One method used anglers who had harvested fish in their possession. These anglers were asked if they planned to eat their fish. If they answered "yes," then C_{daily} was calculated for that interview by using the harvested fish that would be eaten at the next meal as specified by the interviewee. These fish were weighed, and weights were reduced by species-specific dress-out percentages determined for various sizes of fish and dressing methods in an ancillary study. The total dress-out weight per interview was divided by the number of people who would be consuming those fish as indicated by the interviewee. This provided an estimate of meal size in grams based on angler harvests by a technique we call the "harvest method."

All anglers interviewed, with or without harvested fish, were asked the number of standard (113-g) fish servings they would typically eat at a meal when fish was served. Only anglers indicating they consumed fish from the study sites were included in the analysis. A serving was determined by equating the entire surface (palm side) of the flat, open hand to a single 113-g (4-oz) serving. This gave anglers a visual frame of reference for the serving size being addressed. We termed this method of determining meal size the "serving method." An ancillary study was conducted to equate palm size to serving weight of fish. We found that over the fillet thicknesses and palm sizes tested, the mean weights for hand sized fillets ranged from 85 to 156 g, so the 113-g equivalent seemed reasonable.

The calculation used to determine C_{daily} from interviews using the harvest method was

$$C_{daily} = \frac{\left(\frac{\Sigma \, Wt_{dressed}}{q}\right) m}{30}; \quad (1)$$

$\Sigma Wt_{dressed}$ = the sum of the dressed weight of fish harvested that would be eaten for a meal;
q = the number of people who would eat those fish at the meal;
m = the number of meals eaten at which fish was served during the last month;
30 = average number of days in a month to convert to a daily rate.

Equation (1) was applied to both the number of meals eaten in the past month consisting of fish caught at a study site (site meals), and the number of meals eaten in the past month consisting of fish caught at a sample site plus those caught at other lakes and rivers in Alabama (all meals), excluding farm ponds. We assumed that meal sizes based on fish harvested from the study sites would represent meal sizes composed of fish harvested from other lakes and rivers.

The calculation used to determine C_{daily} by the serving method from interviews was

$$C_{daily} = \frac{(g)\,(s)\,(m)}{30}; \quad (2)$$

g = weight per serving (113 g);
s = number of servings (palm side of the open hand) eaten at a meal.

Equation (2), like equation (1), was applied to sample site meals and to all meals including fish caught from Alabama lakes and rivers.

Mean values of C_{daily} were calculated on a seasonal basis ($C_{seasonal}$) by pooling the interviews from all sample sites within each season. This estimate was considered to be a seasonal statewide average. Variances and relative standard errors (RSE; 100·SE/mean) were calculated by taking a weighted variance across seasonal time periods. Weighting factors were determined by the fraction of the total number of interviews taken each season ($W1$), and by the fraction of the total year represented by each season ($W2$). Annual consumption (C_{annual}) was calculated as

$$C_{annual} = \Sigma \, (W_{seasonal}) \, (C_{seasonal}); \quad (3)$$

$W_{seasonal}$ = the weighting factor for a particular season
SE_{annual} = $\Sigma(W1)(W2)/\Sigma(W1)(W2)$, where the summation is over the four seasons.

The SE of C_{annual} was calculated as:

$$SE_{annual} = (\Sigma \, (W_{seasonal})^2 \, (V_{seasonal}))^{0.5}; \quad (4)$$

$V_{seasonal}$ = variance of $C_{seasonal}$.

Statistical Comparisons

The Statistical Analysis System (SAS Institute 1992) was used for data manipulation and statistical analyses. Analysis of variance (ANOVA) with a Student–Newman–Keul's multiple-comparison test was used to determine differences among seasonal consumption rates. A Student's t-test for two independent samples was used to compare mean values of C_{daily} between all observations from the harvest and serving methods. A paired t-test was used to compare estimates made by the serving and harvest methods for those interviews that yielded data for both methods.

Sample sizes needed for the estimation of consumption rates within ±15% of the mean with 90% confidence were calculated for each estimation technique by

$$N = \frac{t^2 CV^2}{D^2}; \quad (5)$$

N = required sample size;
t = Student's tabular t ($\alpha = 0.10$);
CV = coefficient of variation (SD/mean);
D = the half-width of the confidence interval expressed as a percentage (15%).

Results

Sample Sizes

Altogether, 1,586 interviews were obtained during the study period: 363 in fall, 224 in winter, 511 in spring, and 488 in summer. The number of interviewed anglers who indicated that they had eaten fish from the sample sites was 1,303, or 83% of all interviews taken. All of these interviews were used for calculations of mean estimates of C_{daily} based on the serving method. There were 563 interviews of anglers (33%) with harvested fish who indicated that they would consume their fish; all were included in calculations of C_{daily} based on the harvest method. From equation (5), the estimated sample sizes required to produce 90% confidence intervals of ±15% around annual means of C_{daily} (C_{annual}) were 752 interviews for the harvest method (which thus did not meet the criterion) and 456 for the serving method (which did).

Annual Estimates of Daily per Capita Fish Consumption

The harvest method estimates of C_{annual} were 32.6 g/d (N = 563, RSE = 10%) for meals eaten from study sites only, and 43.1 g/d (N = 563, RSE = 9%) for meals eaten from study sites plus those from other lakes and rivers in Alabama. The serving method estimates of C_{annual} were 30.3 g/d (N = 1,303, RSE = 6%) for meals eaten from study sites only, and 45.8 g/d (N = 1,311, RSE = 5%) for meals eaten from study sites plus those from other lakes and rivers in Alabama. There was no significant difference ($P > 0.05$) between the harvest method and serving method estimates of C_{annual}, whether C_{annual} was based on only study site meals or on all meals. The paired t-test between the estimates derived with harvest and serving methods also resulted in no significant difference between methods. There was a significant difference ($P < 0.05$) between estimates of C_{annual} based on site meals and all meals with both harvest and serving calculation methods.

Seasonal Estimates of Daily per Capita Fish Consumption

Estimates of $C_{seasonal}$ were consistently lowest in spring across all categories analyzed (Table 1). The highest consumption rates were during the summer. Three of the estimates of $C_{seasonal}$ based on the harvest method, for both site meals and all meals, had RSEs greater than 20%; two of these occurred in winter, when the sample size was smallest (N = 56). An RSE of 20% implies that the 95% confidence interval will be approximately 40% of the mean, which is too large for risk assessment purposes. In general, the precision of $C_{seasonal}$ was better when the serving method was used because all RSEs were 12% or less.

TABLE 1.—Seasonal estimates of daily freshwater fish consumption ($C_{seasonal}$, g/d) by Alabama anglers (August 1992–July 1993). Estimates from the harvest and serving methods for site meals and all meals are given; N is sample size used to estimate mean consumption rates, and relative standard error (RSE) is 100 · SE/mean.

	Meals of fish caught on site		Meals of fish caught throughout Alabama	
Season	N	Mean (RSE)	N	Mean (RSE)
Harvest method				
Fall	130	29.7 (18)	130	43.4 (18)
Winter	56	26.2 (28)	56	34.2 (24)
Spring	185	21.5 (21)	185	29.3 (17)
Summer	192	46.7 (17)	192	57.0 (15)
Serving method				
Fall	303	32.0 (12)	303	49.4 (11)
Winter	177	30.8 (11)	177	43.9 (10)
Spring	411	20.5 (9)	414	33.6 (8)
Summer	412	36.4 (10)	417	53.0 (10)

No statistical differences ($P > 0.05$) were found among $C_{seasonal}$ values estimated by the harvest method, even though summer estimates across site meals and all meals generally were 1.5 to 2 times greater than estimates for other seasons (Table 1). Serving method estimates of $C_{seasonal}$ were statistically different between the spring and summer seasons for both site meals and all meals. No other seasonal differences were found. The seasonal contrasts in mean daily consumption rates between the spring and summer seasons partially reflect seasonal differences in the frequencies of meals consumed, which was lowest in spring (3.9 meals/month) and highest during the summer (4.8 meals/month). Note that the same estimates of seasonal fish consumption frequencies (meals/month) were used for both the harvest and serving methods.

Discussion

Both methods for the estimation of per capita freshwater fish consumption yielded similar rates, however, their precision differed. The serving method had greater precision (i.e., lower RSE) than the harvest method. Typically, high variances are associated with harvest rate estimates determined from traditional creel surveys (Crone and Malvestuto 1991). The inherent variability associated with values of C_{daily}, based on meal sizes derived from fish harvested by anglers in this survey, was greater than that based on meal sizes derived from the number of servings typically eaten. The coefficient of variation (100·SD/mean) was 249% across values of C_{daily} for the harvest method and 194% for the serving method. This difference was at least partially due to the discrete nature of angler responses to the serving method survey, which were generally whole numbers (e.g., 1, 2, 3, 4, etc.), and which translated into fixed gram equivalents (equation 2). Thus, the serving method likely masked true variability in meal sizes, but the harvest method likely magnified this variability, because meal sizes were derived from weights of whole fish prior to their preparation for consumption.

Sample sizes for harvest method estimates were lower than those for serving method estimates. The number of anglers indicating they had eaten fish from sample sites, but who had not harvested enough fish that day to make a family meal, yielded a sample size for the harvest method that was approximately 60% smaller than the serving method sample. A distinct advantage of the serving method is that it allowed mean per capita consumption to be estimated even when no fish had been harvested, which resulted in larger sample size and greater precision. In a relative sense, the variability associated with C_{annual} based on the serving method (RSE = 6%) was 40% less than that based on the harvest method (RSE = 10%). It should be noted that the smaller sample sizes obtained with the harvest method were caused not only by lack of harvest, but also by some anglers who were unwilling to wait for their fish to be measured during attempted interviews.

Given the target criteria of 90% confidence intervals of ±15% for estimation of annual per capita freshwater fish consumption, the number of interviews (sample size) obtained in this study for the harvest method would need to be increased by 33%, to approximately 750 anglers (equation 5). The number of interviews obtained for the serving method was more than adequate and could be decreased by approximately 65%, to 456 anglers.

The primary advantage of using the harvest method was that species and sizes of fish consumed by anglers interviewed could be determined by direct observation. This is an important consideration, because certain species and sizes of fish have greater potential for bioaccumulation of toxic substances.

Conclusions

We chose the creel survey approach for estimating freshwater fish consumption rates in Alabama because the survey objectives called for estimates from a specific subpopulation of people in Alabama (primarily anglers fishing rivers), and for identification of species and sizes of fish consumed. Additionally, it was believed that direct observation of fish to be consumed would provide more accurate estimates of meal sizes than more indirect methods, such as using visual representations of serving sizes.

By incorporation of a visual serving size method into the creel survey, we were able to contrast estimates of serving size, and thus consumption rate, based on direct observation of fish in angler creels with those based on a visual representation of serving size. We found that estimates of annual and seasonal per capita fish consumption were very similar regardless of the method used to estimate meal sizes. An unexpected result, however, was that the sample sizes obtained with the serving method were much higher than those obtained with the harvest method, which had a direct bearing on the relative precisions of the two methods. Another distinct advantage of the serving method was the simplification of field procedures, because enumeration and identification of harvested fish were not necessary. However, if species and sizes of fish consumed are primary survey objectives, then there is no substitute for recording fish harvested. In this study, there were not enough interviews of anglers with harvested fish to meet reasonable precision requirements with the harvest method; thus, sampling intensity, and overall survey costs, would have to be increased beyond the effort reported here to obtain acceptable per capita fish consumption values using this method.

When fish consumption surveys are being planned, consideration should be given to the kind of data required and the most efficient (cost effective) means of collecting that information. Recall telephone or mail survey methods may be better when the general population is to be surveyed. However, when specific subpopulations or high-risk groups are to be targeted, the advantages of on-site surveys, in which harvested fish are measured or serving size recall techniques are used, may outweigh the increased cost. The use of more than one technique within a survey may be advisable for corroboration or follow-up purposes. Traditional recall biases encountered in face-to-face surveys in which short time frames are used are less of a concern than the accuracy of the visual representation of serving size (Chu et al. 1992). Further work is needed to establish standard visual representations of serving sizes for accurate estimation of consumption rates. When harvested fish are used to determine meal sizes, it is imperative that specific fish to be consumed at a single meal be identified, including portions of large fish that would provide more than one meal. Additionally, accurate dress-out percentages must be applied to whole fish based on species, size and dress-out method.

Finally, it should be pointed out that bioaccumulation of toxic substances is related to exposure over time. This process is cumulative in nature, and it is very difficult to determine chronic toxicity over entire life spans of individuals, given changes in exposure rates, biological characteristics, and locations of people over time. This problem can be addressed only via costly and time-consuming longitudinal studies of fish-consuming individuals. For assessment of human health risk from contaminants in fish, short-term cross section studies of high-risk populations, by means of survey approaches similar to those discussed above, likely will play the major role.

References

Bonna, K. H., K. S. Bjerve, and A. Nordoy. 1992. Habitual fish consumption, plasma phospholipid fatty acids, and serum lipids: the Tromso study. American Journal of Clinical Nutrition 55:1126–34.

Chu, A., and five coauthors. 1992. Measuring the recall error in self-reported fishing and hunting activities. Journal of Official Statistics 5:13–39.

Crone, P. R., and S. P. Malvestuto. 1991. Comparison of five estimators of fishing success from creel survey data on three Alabama reservoirs. Pages 61–66 *in* D. Guthrie and seven coeditors. Creel and angler surveys in fisheries management. American Fisheries Society Symposium 12.

EPA (U.S. Environmental Protection Agency). 1989. Assessing human health risks from chemically contaminated fish and shellfish: a guidance manual. U.S. EPA, Office of Marine and Estuarine Protection, Office of Water Regulations and Standards, EPA-503/8-89-002, Washington, DC.

EPA (U.S. Environmental Protection Agency). 1992. Consumption surveys for fish and shellfish: a review and analysis of survey methods. U.S. EPA, Office of Water Regulations and Standards, EPA-822/R-92-001, Washington, DC.

Foran, J. A., B. S. Glenn, and W. Silverman. 1989. Increased fish consumption may be risky: letter to the editor. Journal of the American Medical Association 262:28.

Humphrey, H. E. B. 1987. The human population—an ultimate receptor for aquatic contaminants. Hydrobiologia 149:75–80.

Kromhout D., E. B. Bosschieter, and C. D. Coulander. 1985. The inverse relation between fish consumption and 20-year mortality from heart disease. New England Journal of Medicine 312:1205–1209.

Puffer, H. W., M. J. Duda, and S. P. Azen. 1982. Potential health hazards from consumption of fish caught in polluted coastal waters of Los Angeles County. North American Journal of Fisheries Management 2:74–79.

Reinert, R. E., B. A. Knuth, M. A. Kamrin, and Q. J. Stober. 1991. Risk assessment, risk management, and fish consumption advisories in the United States. Fisheries 16(6):5–12.

Rifkin E., and J. LaKind. 1991. Dioxin bioaccumulation: key to a sound risk assessment methodology. Journal of Toxicology and Environmental Health 33:103–112.

SAS Institute. 1992. The statistical analysis system, edition 6.03 release. SAS Institute, Cary, North Carolina.

SRI (Stanford Research Institute). 1980. Seafood consumption data analysis. Final report. Prepared by SRI International for U.S. Environmental Protection Agency, Office of Water Regulations and Standards, Washington, DC.

Suta, B. E. 1978. Human exposures to mirex and kepone. U.S. Environmental Protection Agency, EPA-600/1-78-045, Washington, DC.

Svensson, B. C., and five coauthors. 1991. Exposure to dioxins and dibenzofurans through the consumption of fish. New England Journal of Medicine 324:8–12.

American Fisheries Society Symposium 16:277–282, 1996

Factors Affecting Angler Fishing Success Ratings at Fifteen Texas Reservoirs, 1984–1993

MICHAEL E. COSTELLO[1] AND ROBERT K. BETSILL[2]

Texas Parks and Wildlife Department, Heart of the Hills Research Station HCR-7, Box 62, Ingram, Texas 78025, USA

Abstract.—We used categorical models to explore relationships between fishing success ratings and catch-related elements (species, numbers, and sizes of fish caught) for anglers targeting largemouth bass *Micropterus salmoides*, channel catfish *Ictalurus punctatus*, white bass *Morone chrysops*, and crappies *Pomoxis* spp. at 15 Texas reservoirs. Success ratings differed among angler groups and reservoirs. Multivariate models explained significant amounts of variation in fishing success ratings in 51 of 54 possible angler group and reservoir combinations. Number of target fish caught was the most influential factor for fishing success ratings with largemouth bass and white bass anglers. Number of target fish caught, number of target fish harvested, and total weight of harvested target fish were most often significantly associated with success ratings of channel catfish and crappie anglers. Nontarget species were important to all angler groups. Catch-per-unit-effort descriptors were less useful than those based on measures of total catch. Single-variable predictive models were developed for 15 angler group and reservoir combinations.

The management concept of optimum sustainable yield recognizes fishing as a multidimensional experience, with multiple satisfactions derived from multiple determinants (Hampton and Lackey 1976). Both catch-related factors (i.e., species, numbers, and sizes of fish caught) and recreational factors not related to catch influence angler perception of the quality of a fishing experience (Hudgins 1984). Motivations not related to catch may be more important in defining overall trip satisfaction than catch-related motivations (Holland and Ditton 1992; Fedler and Ditton 1994); however, motivations related to the catch are consequential (Graefe and Fedler 1986; Ditton et al. 1991; Green et al. 1991). Angler satisfaction with the consumptive elements of a fishing experience (defined in this paper as fishing success) can be distinguished by the angler from overall trip satisfaction (Weithman and Katti 1979; Hudgins 1984), and the two are not necessarily correlated (Hudgins and Davies 1984). Research on angler satisfaction has focused primarily on motivations and overall trip satisfaction (e.g., Vaske et al. 1982; Graefe and Fedler 1986; Holland and Ditton 1992), and only to a lesser extent on catch-related satisfaction (Hudgins and Davies 1984; McMichael and Kaya 1991; Spencer 1993).

Of the factors that influence angler satisfaction, catch-related elements are most amenable to influence by fisheries managers. The ability to manipulate fisheries in areas other than stock structure and abundance is limited. Although recognized as important (Hampton and Lackey 1976; Hudgins 1984), recreational components such as relaxation, weather, and companionship, often are not realistic management considerations.

To be effective, natural resource managers must identify components that contribute to satisfaction and evaluate management through measures of angler contentment (Holland and Ditton 1992). Catching fish is an important element of a fishing experience for Texas freshwater anglers (Ditton et al. 1991). Aspects of the catch were the most important factors influencing trip satisfaction of Texas saltwater anglers (Green et al. 1991). However, quantitative associations between fishing satisfaction and catch-related components are lacking. Identifying these associations would elucidate how anglers rate their fishing success, and to what extent angler expectations are being met in Texas reservoirs.

As a part of routine fisheries monitoring in Texas, the Texas Parks and Wildlife Department conducts creel surveys to observe the species, numbers, and sizes of fish caught and harvested by anglers. Creel clerks also ask anglers to rate their fishing success. In this paper, we attempt to (1) identify components of the catch that significantly affect fishing success ratings; (2) assess the relative importance of those components; and (3) develop single-descriptor linear models based on a catch-related variable to predict fishing success ratings.

[1]Present address: Ohio Department of Natural Resources, Division of Wildlife, Inland Fisheries Research Unit, 10517 Canal Road SE, Hebron, Ohio 43025, USA.

[2]To whom correspondence should be addressed.

Methods

Creel Surveys

Data from annual creel surveys were available for 15 Texas reservoirs from 1984 through 1993. Both access point and roving surveys were used; therefore, interviews included incomplete and completed trips. Sampling effort averaged 36 d/year, but varied among reservoirs and years. Normally, daily survey periods were randomly selected 6-h intervals between sunrise and sunset. Creel clerks interviewed one angler from each party. Information collected from interviews included (1) species sought; (2) total directed party effort; (3) numbers, lengths, and weights of harvested fish; and (4) numbers of released fish. Anglers were asked "How would you rate today's fishing success compared with fishing at other lakes that you have fished: very good, good, average, poor, or very poor?" Responses were recorded on a 5-point scale, from 5 (very good) to 1 (very poor). We assumed anglers could differentiate between fishing satisfaction and overall trip satisfaction (Weithman and Katti 1979). We divided catch and harvest by directed effort to obtain species-specific catch and harvest rates for each party. Our analyses included only those parties targeting largemouth bass *Micropterus salmoides*, channel catfish *Ictalurus punctatus*, white bass *Morone chrysops*, or crappies *Pomoxis* spp. We defined angler groups by their targeted species.

Analysis

We chose 15 catch-related, situational variables (i.e., directly measurable characteristics of a fishing experience; Graefe and Fedler 1986) and two class variables that we hypothesized were associated with fishing success ratings (Table 1). We treated fishing success rating as an ordinal response and used categorical data models (CATMOD and LOGISTIC procedures, SAS 1992) in our analyses.

Associations between target species (SOUGHT), reservoir (LAKE), and fishing success were tested using a mean response analysis of variance (ANOVA) model. Both main effects of these variables and their interaction were significant. Accordingly, all further analyses were separated by target species and reservoir. Four study reservoirs did not have white bass populations, and two reservoirs did not have crappie populations. We developed 54 distinct angler group and reservoir combinations from 46,542 interviews.

We assessed the relationship between fishing success and the situational variables using stepwise, cumulative-logit regression. The significance level

TABLE 1.—Descriptors of angler catch used in analysis of fishing success ratings.

Variable	Description
Class variables	
SOUGHT	identified target species
LAKE	reservoir at which the interview occurred
Situational variables	
CATCH	total number of target fish caught
HARVEST	total number of target fish kept
YIELD	total weight (lb) of harvested target fish
AVEWGT	mean weight (lb) of harvested target fish
AVETL	mean total length (in) of harvested target fish
RELEASE	number of target fish caught and released
CPUEt	catch rate of target fish (N/h)
CPUEh	harvest rate of target fish (N/h)
CPUEyld	harvest rate of target fish (lb/h)
CPUErel	release rate of target fish (N/h)
OTHERCAT	number of nontarget fish caught
OTHERHAR	number of nontarget fish harvested
OTHERYLD	total weight (lb) of nontarget fish harvested
OTHERREL	number of nontarget fish caught and released
EFFORT	party directed effort (h) towards target species

(α) for entry and retention in the model was 0.05. For each variable, the highest 10% of values were set equal to the 90th percentile to conform to the proportional odds assumption of the model (Agresti 1990). Model construction was decided by the magnitude of Wald chi-square statistics (SAS 1992). The variable having the highest Wald statistic entered the model first, and was interpreted as having accounted for more variation in fishing success ratings than other variables in the model.

Additionally, the initial variable fit in the multivariate model (i.e., the variable to which the most model variation was attributed) was regressed independently against fishing success ratings using a mean response model (CATMOD procedure; SAS 1992). Our purpose was to identify individual catch descriptors having significant linear relationships with anglers' perceived fishing success. A mean response model was chosen over a logistic model because of ease of interpretation of model parameters. Again, the highest 10% of values were set equal to the 90th percentile for each descriptor, and used the ADDCELL = 0.005 option (SAS 1992) to eliminate cell counts of zero. Significance levels were 0.05 for all analyses.

Results

Both target species (SOUGHT; Wald chi-square = 33.7, $P < 0.001$) and reservoir (LAKE; Wald chi-square = 788.2, $P < 0.001$), and their interaction (Wald chi-square = 198.3, $P < 0.001$) were significantly associated with fishing success

TABLE 2.—Number of reservoirs for which catch descriptors were significant in a stepwise, cumulative-logit regression model, for each of four angler groups. The number of reservoirs for which the descriptor was the first variable fitted (i.e., the variable having the highest Wald chi-square statistic, and accounting for the most variation) into the multivariate model is given in parentheses. The number of significant models for each angler group is given as N. Catch descriptors are defined in Table 1.

	Angler group			
Catch descriptor	Largemouth bass $N = 15$	Channel catfish $N = 15$	White bass[a] $N = 10$	Crappies[b] $N = 11$
CATCH	15 (12)	6 (3)	7 (6)	4 (3)
HARVEST	4 (1)	4 (2)	0 (0)	2 (2)
YIELD	5 (1)	5 (5)	1 (1)	4 (4)
AVEWGT	2 (0)	1 (0)	1 (1)	1 (0)
AVETL	3 (0)	2 (0)	2 (0)	2 (0)
RELEASE	4 (0)	1 (0)	1 (0)	0 (0)
CPUEt	4 (0)	2 (0)	1 (0)	5 (1)
CPUEh	1 (0)	1 (1)	3 (2)	1 (0)
CPUEyld	4 (0)	3 (0)	1 (0)	2 (0)
CPUErel	3 (0)	2 (0)	0 (0)	0 (0)
OTHERCAT	13 (1)	5 (1)	5 (0)	3 (0)
OTHERHAR	7 (0)	4 (0)	0 (0)	4 (1)
OTHERYLD	2 (0)	6 (3)	6 (0)	3 (0)
OTHERREL	2 (0)	1 (0)	2 (0)	2 (0)
EFFORT	5 (0)	3 (0)	5 (0)	4 (0)

[a]Only 11 reservoirs had sufficient sample sizes of white bass angler interviews.
[b]Only 13 reservoirs had sufficient sample sizes of crappie angler interviews.

ratings in the factorial ANOVA model. The effect of LAKE accounted for most of the variation explained by the model, followed by the SOUGHT*LAKE interaction.

Significant Catch Components

We constructed significant multivariate models for 51 of the 54 possible angler group and reservoir combinations (Table 2). Each catch descriptor explained a significant amount of variation in at least one angler group and reservoir model. All regression coefficients were positive, implying higher fishing success ratings were associated with increased levels of catch descriptors. Across all angler groups, CATCH and OTHERCAT were included in the most models; AVEWGT and CPUErel were included in the fewest models; and CATCH, YIELD, and HARVEST most often accounted for the greatest amount of variation in success ratings in the models. Although generalizations across angler groups were useful, we were more interested in patterns across reservoirs within angler groups.

Largemouth bass anglers.—In all 15 reservoirs CATCH was significant; OTHERCAT was significant in 13 reservoirs (Table 2). All other descriptors were significant in at least one reservoir. The most significant source of model variation in fishing success ratings in 12 reservoirs was CATCH. In the remaining three reservoirs, HARVEST, YIELD, and OTHERCAT were the most important descriptors.

Channel catfish anglers.—There was greater diversity among reservoirs in the factors associated with fishing success ratings of channel catfish anglers than of largemouth bass anglers (Table 2). No one descriptor was significant in a majority of the reservoirs. In six reservoirs CATCH and OTHERYLD were significant; YIELD and OTHERCAT were significant in five reservoirs. In five reservoirs YIELD was the most significant descriptor; CATCH, HARVEST, CPUEh, OTHERCAT, and OTHERYLD were the most significant variables in the other reservoirs.

White bass anglers.—In a majority of the reservoirs CATCH and OTHERYLD were significant (Table 2); HARVEST, CPUErel, and OTHERHAR were not significant in any reservoir. The most significant variable in 6 of 10 reservoirs was CATCH. The most significant sources of model variation in the remaining reservoirs were YIELD, AVEWGT, and CPUEh.

Crappie anglers.—No descriptor was significant in a majority of the reservoirs (Table 2). In five reservoirs CPUEt was significant. All other descriptors except RELEASE and CPUErel were significant in at least one reservoir. The most significant variable in four reservoirs was YIELD; CATCH, HARVEST, CPUEt, and OTHERHAR were the most important variables in the remaining reservoirs.

Predictive Models

Multivariate models were useful in determining which catch components were associated with fishing success ratings, and the relative importance of those components across reservoirs. However, a simpler model would be preferred for prediction purposes. Single-predictor models allow easy interpretation of regression coefficients, and avoid problems of multicollinearity. So, for each of the significant multivariate models, we determined whether the most significant descriptor (i.e., the variable first entered in the model) could stand alone as a predictor of fishing success.

From the 51 significant angler group and reservoir multivariate models, only 15 significant single-descriptor models were developed (Table 3). Of the 12 reservoirs for which CATCH was the most sig-

TABLE 3.—Linear regression parameters ($\hat{y} = \beta_0 + \beta_1 x$) and standard errors for significant single-descriptor, mean response models, where $\hat{y}$ = predicted fishing success rating and x = catch descriptor. *P*-values are for tests of the null hypothesis that model residual variation is equal to zero. Table rows represent individual reservoirs. Number of interviews from the angler group and reservoir combination is given as *N*.

Angler group	Descriptor	β_0 (SE)	β_1 (SE)	P[a]	N
Largemouth bass	CATCH	2.1827 (0.0300)	0.2134 (0.0119)	0.0877	1,410
	CATCH	2.4582 (0.0309)	0.1394 (0.0117)	0.3095	1,395
	CATCH	1.9576 (0.0393)	0.1938 (0.0123)	0.0557	547
	CATCH	2.2446 (0.0519)	0.2529 (0.0267)	0.2526	454
	CATCH	2.2595 (0.0575)	0.1116 (0.0135)	0.3140	393
	OTHERCAT	2.2542 (0.0605)	0.2001 (0.0278)	0.2867	247
Channel catfish	HARVEST	2.0939 (0.0193)	0.3672 (0.0165)	0.5072	2,566
	HARVEST	1.9967 (0.0686)	0.1155 (0.0080)	0.2442	305
White bass	CPUEh	2.4152 (0.0830)	0.5664 (0.0679)	0.0515	180
Crappies	CATCH	2.4799 (0.0973)	0.1227 (0.0427)	0.0510	121
	CATCH	2.1796 (0.0437)	0.1317 (0.0129)	0.4063	588
	HARVEST	2.3356 (0.0285)	0.1672 (0.0085)	0.3230	1,251
	HARVEST	2.0962 (0.0460)	0.2525 (0.0168)	0.4370	379
	CPUEt	1.5770 (0.1124)	1.3902 (0.2923)	0.3440	35
	OTHERHAR	2.0911 (0.5634)	0.5634 (0.1286)	0.7873	215

[a] $P > 0.05$ indicates that the null hypothesis should not be rejected; the residual variation in the model is not different from zero, and the model adequately fits the data.

nificant factor for largemouth bass anglers, five could be adequately modeled using just CATCH. Among channel catfish anglers, HARVEST by itself was adequate in explaining variation in fishing success ratings in two reservoirs. Only one significant model was generated for the white bass angler group. For anglers seeking crappies, 6 of 13 reservoirs were successfully modeled with a single descriptor. Despite YIELD being the most significant source of variation in at least one reservoir for all angler groups, it failed as a single predictor of fishing success ratings for all angler groups in all reservoirs. Within angler groups, regression coefficients for individual descriptors varied by reservoir.

Discussion

Fishing success ratings were influenced by species sought and reservoir. Spencer (1993) also found significant variation in fishing success ratings among anglers targeting different species. Because angler group motivations differ (Hudgins 1984; Fedler and Ditton 1994), and motivations affect desired outcomes (Fedler and Ditton 1994), these results are not unexpected. However, reservoir effects are not as easily understood. Variation in fishing success responses, within angler groups, might be attributable to differences in experience or fishing strategies among anglers seeking the same species. Bryan (1977) called this the "specialization continuum" and pointed out that anglers may have different motivations, even though they seek the same species. Another possible source of variation might have been anglers failing to adjust their expectations to match a reservoir's fishery potential (i.e., catch rates, size structure). The wording of the fishing success question, "... compared with fishing at other lakes ...," would directly contribute to this type of variability. Our findings differ from those of Hudgins and Davies (1984), who found fishing success ratings were similar among reservoirs because anglers adjusted their fishing success criteria by reservoir.

Our approach assumed angler motivations did not change over the study period. This was probably simplistic because motivations may have been affected by changes in fish populations, harvest regulations, or angler ethics (e.g., catch-and-release) during the study period. Our results should be viewed as the average conditions during this 10-year period.

It appeared that anglers in our study could discern fishing success quality from overall trip quality, which is not always the case (McMichael and Kaya 1991). Variation caused by anglers incorporating noncatch components into fishing success ratings would have been included in the error terms of our models. Using only catch descriptors, we were able to successfully model fishing success ratings for 51 out of 54 possible angler group and reservoir combinations.

Overall, CATCH, YIELD, and HARVEST appeared to be the most important variables contributing to fishing success ratings. Factors associated with the total number of fish caught (CATCH and OTHERCAT) were most often incorporated into the multivariate models, suggesting total catch (targeted and nontargeted) was very important to anglers. Our findings are consistent with others who have found that satisfaction increases with increases in harvest (Vaske et al. 1982; Hudgins and Davies 1984; Spencer 1993), total weight of harvested fish (Hudgins and Davies 1984), and catch rates (McMichael and Kaya 1991). Rate descriptors (CPUEt, CPUEh, CPUErel, CPUEyld) generally were not useful in our models. Angler satisfaction was more closely related to achieving a certain quantity of catch, and not the effort required to amass that quantity. Catch components associated with fish size (AVETL, AVEWGT) were rarely included in multivariate models. This was unexpected, especially for the largemouth bass angler group, whom we assumed might prefer larger fish. However, those descriptors were measured for harvested fish only; the sizes of released fish were not measured. Lengths or weights of released fish might help explain variation in fishing success ratings for anglers who practice catch-and-release fishing.

Differences among angler groups have implications for management strategies. Factors related to total number of fish caught were important to both largemouth bass and white bass anglers, suggesting that regulations maximizing catch would be well-received by these angler groups at most reservoirs. Within largemouth bass anglers, the consistency of CATCH as the most significant factor in 12 of 15 reservoirs shows a statewide management policy might be acceptable. The variability within the channel catfish and crappie angler groups, where several different factors (YIELD, CATCH, HARVEST) were most significant depending on the reservoir, suggests a need for an individual reservoir management approach for these species. Emphasis on numbers of fish caught, harvested, or total weight of harvested fish might be appropriate, depending on the reservoir. The importance of nontargeted catch for all angler groups emphasizes the need for managers to consider the entire sport fish community when developing management strategies.

Multivariate models tended to fit the data better than single-variable models, which implies that fishing success ratings were dependent on several aspects of the catch. The low number of significant single-variable models, and the low P values obtained with goodness-of-fit tests, caution against indiscriminate use of linear models. However, in cases where models were significant, success ratings could be predicted for given levels of a catch component. Variation in success ratings for largemouth bass anglers at five reservoirs was explained by CATCH. Observed mean success ratings of largemouth bass anglers at these five reservoirs ranged from 2.3 to 2.7, and CATCH ranged from 1.7 to 3.1 fish per party. Using the regression equations from Table 3, we predicted that a CATCH of 3.4 to 7.5 fish per party was necessary to obtain mean success ratings of above "average" (3.1 or higher).

Managers who wish to maximize fishing success ratings should find utility in our study. Undoubtedly, motivations vary among and within angler groups and change over time. However, if continuous sampling includes a broad cross-section of anglers, models like those developed here can identify appropriate management actions, and can help evaluate those actions.

Acknowledgments

We thank R. W. Luebke for his assistance during manuscript development. T. M. Steeger and three anonymous reviewers provided helpful recommendations on early drafts of the manuscript. Partial funding for this project was provided by the Federal Aid in Sport Fish Restoration Project F-30-R of the Texas Parks and Wildlife Department.

References

Agresti, A. 1990. Categorical data analysis. Wiley, New York.

Bryan, H. 1977. Leisure values systems and recreational specialization: the case of the trout fisherman. Journal of Leisure Research 9:174–187.

Ditton, R. B., and six coauthors. 1991. Demographics, participation, attitudes, expenditures, and management preferences of Texas freshwater anglers, 1987. Texas Parks and Wildlife Department, Management Data Series 66, Austin.

Fedler, A. J., and R. B. Ditton. 1994. Understanding angler motivations in fisheries management. Fisheries 19(4):6–13.

Graefe, A. R., and A. J. Fedler. 1986. Situational and subjective determinants of satisfaction in marine recreational fishing. Leisure Sciences 8:275–295.

Green, L. M., H. R. Osburn, G. C. Matlock, and R. B. Ditton. 1991. Development of a social and economic questionnaire for on-site interviews of marine recreational anglers in Texas. Pages 406–412 *in* D. Guthrie and seven coeditors. Creel and angler surveys in fisheries management. American Fisheries Society Symposium 12.

Hampton, E. L., and R. T. Lackey. 1976. Analysis of angler preferences and fisheries management objectives with implications for management. Proceedings of the Annual Conference Southeastern Association of Game and Fish Commissioners 29(1975):310–316.

Holland, S. M., and R. B. Ditton. 1992. Fishing trip satisfaction: a typology of anglers. North American Journal of Fisheries Management 12:28–33.

Hudgins, M. D. 1984. Structure of the angling experience. Transactions of the American Fisheries Society 113:750–759.

Hudgins, M. D., and W. D. Davies. 1984. Probability angling: a recreational fishery management strategy. North American Journal of Fisheries Management 4:431–439.

McMichael, G. A., and C. M. Kaya. 1991. Relations among stream temperature, angling success for rainbow trout and brown trout, and fisherman satisfaction. North American Journal of Fisheries Management 11:191–199.

SAS (SAS Institute). 1992. Categorical data analysis course notes. SAS Institute, Cary, North Carolina.

Spencer, P. D. 1993. Factors influencing satisfaction of anglers on Lake Miltona, Minnesota. North American Journal of Fisheries Management 13: 201–209.

Vaske, J. J., M. P. Donnelly, T. A. Heberlein, and B. Shelby. 1982. Differences in reported satisfaction ratings by consumptive and nonconsumptive recreationalists. Journal of Leisure Research 14:195–206.

Weithman, A. S., and S. K. Katti. 1979. Testing of fishing quality indices. Transactions of the American Fisheries Society 108:320–325.

Habitat Management

American Fisheries Society Symposium 16:285–295, 1996

Oligotrophication and Its Discontents: Effects of Reduced Nutrient Loading on Reservoir Fisheries

JOHN J. NEY

Virginia Polytechnic Institute and State University
Department of Fisheries and Wildlife Sciences, Blacksburg, Virginia 24061, USA

Abstract.—Oligotrophication is the reversal of the eutrophication process and can occur in reservoirs as the result of nutrient trapping in upstream impoundments or of operation of advanced waste treatment (AWT) facilities on inflowing rivers. I examined the response of reservoir fisheries to oligotrophication using case studies and regression analysis of the relationship between concentration of phosphorus, the principal limiting nutrient, and fishery productivity. In Smith Mountain Lake, Virginia, and Beaver Lake, Arkansas (upstream AWTs), and in Lake Mead, Nevada (upstream impoundment), oligotrophication was accompanied by declines of more than 50% of standing stock of planktivorous forage fishes and reductions in growth, standing stock, or harvest of piscivorous sport fishes. Both phytoplankton and total fish productivity were highly correlated (r = 0.7–0.9) with total phosphorus concentration in lakes and reservoirs, increasing linearly over a wide range of concentrations. Eutrophication status, in terms of cholorophyll *a* and water transparency, is achieved in temperate storage reservoirs when total phosphorus concentration exceeds 40 μg/L, but sport fish biomass probably does not peak at less than 100 μg/L, providing the basis for conflicts among reservoir user groups. Remedial measures to restore reservoir sport fisheries following oligotrophication include species introductions and in-lake fertilization, both of which are unlikely to be successful. Prevention of oligotrophication through top-down manipulation of the food web has the potential to promote both "clean" lakes and good fishing, but will probably not be effective in large eutrophic impoundments. A more promising approach is to prevent undesirable oligotrophication through informed decision making. Fisheries scientists first must collaborate with limnologists to predict the impacts of proposed nutrient loading reductions on reservoir fisheries, then act as educators and advocates for their resource.

The limnological term "eutrophication" has become synonymous in the public consciousness with water quality degradation. Eutrophication is the process through which aquatic plant and other biomass of a waterbody increases (Lee and Jones 1991), most often as the result of nutrient additions. Because the buildup of plant (algae and macrophyte) biomass is considered undesirable for aesthetic and recreational uses (Heiskary and Walker 1988; Harper 1992), the United States and many other countries have undertaken large programs to reduce nutrient concentrations in surface waters (Moore and Thornton 1988). These well-intentioned efforts generally ignore nonplant biomass, under the perception that "cleaner" water will benefit all aquatic life forms, including fish. However, community energetics dictates otherwise: the biomass of fish at or near the top of the trophic pyramid is highly dependent on the amount of primary production at the base (Lindemann 1942). Consequently, expected outcomes of reduced nutrient loading should include not only less algae and clearer water, but also lower total fish biomass and, to the degree that sport fishes are affected, poorer fishing and lower angler satisfaction (Yurk and Ney 1989).

Phosphorus is usually the limiting nutrient for primary production in surface waters and the focus of nutrient reduction programs. Decreases in water column phosphorus concentrations may reduce biological productivity at all trophic levels; this reversal of the eutrophication process is oligotrophication. Where the process is induced by human actions, the term "cultural oligotrophication" properly may be applied.

Reservoirs are particularly susceptible to experiencing large and rapid reductions in phosphorus and other nutrient concentrations because of their dependence on upstream water quality. The two principal causes of oligotrophication of reservoirs are point-source nutrient abatement programs on tributary rivers and construction of upstream impoundments. Both actions have been widely implemented, but the consequences to the fisheries of downstream reservoirs have seldom been considered in advance or documented afterwards.

I reviewed case histories of reservoir fisheries responses to phosphorus reductions from both nutrient abatement programs and upstream dams, examined the relation of phosphorus to water quality and fisheries productivity, and critiqued alternatives

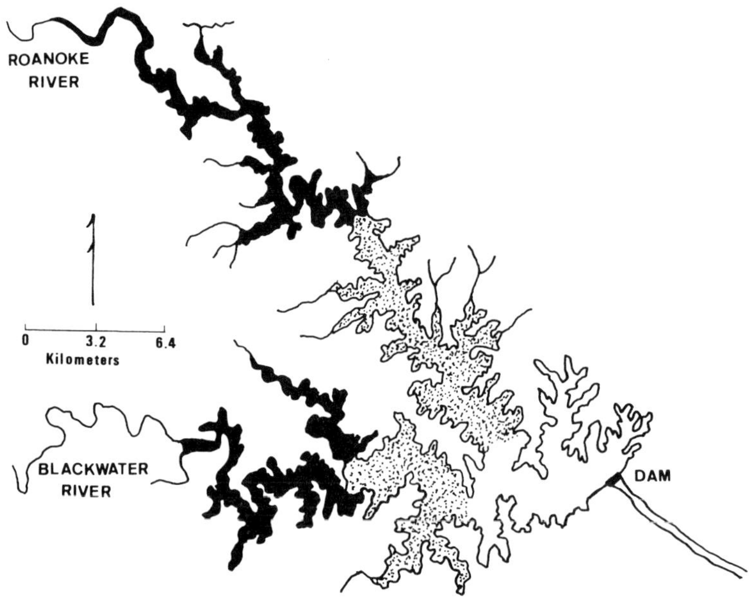

FIGURE 1.—Smith Mountain Lake, Virginia. Stippled and darkened areas experienced summer hypolimnetic oxygen deficits in 1974, before nutrient abatement. Only darkened portions had such deficits during 1984–1994.

for prevention as well as remediation of the consequences of oligotrophication.

Case Studies

Response to Nutrient Abatement Programs

Smith Mountain Lake, Virginia, is an 8,300-ha hydroelectric impoundment on the Roanoke River (Figure 1). This highly dendritic reservoir was impounded in 1963 and soon developed a premier fishery for black bass *Micropterus* spp. and stocked striped bass *Morone saxatilis*. The forage fish resource was dominated by alewife *Alosa pseudoharengus* and gizzard shad *Dorosoma cepedianum*. In response to complaints about algal blooms in the upper section of the reservoir, an advanced waste treatment plant (AWT) began operation in the city of Roanoke, 40 km upstream, in 1975. Yurk and Ney (1989) compared mean annual epilimnetic total phosphorus concentrations with estimates of fish standing stock (kg/ha) obtained from cove rotenone samples over the 12-year period 1973–1984. Total phosphorus concentration declined 79% while total fish standing stock dropped 76%. Standing stock of planktivorous fishes (primarily gizzard shad) was highly correlated (r = 0.81) with total phosphorus, but biomass of piscivorous sport fish was not (r = 0.01). However, the annual growth rate (in total length) of black bass and striped bass declined about 12% between 1973 and 1984 (Ney et al. 1988).

The apparent lack of relationship between phosphorus concentration and piscivore standing stock may be spurious. Standing stock of piscivores, particularly pelagic fishes such as striped bass, is poorly estimated by cove rotenone sampling (Davies and Shelton 1983); creel survey data indicate that striped bass harvest in Smith Mountain Lake was 65% lower in 1984 than in 1977 (Hart 1978; Whitehurst 1987). However, oligotrophication did expand potential summer habitat for striped bass. By 1984, the percentage of Smith Mountain Lake (by surface area) experiencing summer hypolimnetic oxygen deficits contracted by 60%, doubling the suitable habitat (≤24°C, ≥5 mg/L dissolved oxygen) for this coolwater species (Figure 1). The trade-off between less food and more habitat might have partially mitigated nutrient abatement impacts on the sportfish resource in this case. Between 1984 and 1994, the phosphorus–fishery relationship stabilized in Smith Mountain Lake; total phosphorus concentration averaged about 20 μg/L and total fish standing stock fluctuated around 370 kg/ha. Summer habitat for coolwater fish also stabilized (Figure 1).

Beaver Lake, Arkansas, has experienced a more recent, but similar, response to an upstream nutrient abatement program. Beaver Lake is an 11,000-ha flood control hydroelectric impoundment on the White River, which historically provided about 60% of the phosphorus load to the lake (Aggus 1985). In 1987, an AWT went into service in the upstream city of Fayetteville, reducing phosphorus load to Beaver Lake by about 50%. Before-and-after comparisons of mean annual standing stock, as described by cove rotenone sampling for the 6-year periods 1982–1987 and 1988–1993 showed reductions of 55% in planktivorous clupeids and 9% in piscivorous sport fishes (primarily black bass and ictalurids). Possible changes in sportfish growth rate and condition as consequences of the shad decline have not been investigated, and long-term effects of the phosphorus loading reduction on the fishery have not been extrapolated. Further, there is some indication that the AWT-induced reduction may eventually be replaced with phosphorus from other sources (M. Bivin, Arkansas Game and Fish Commission, personal communication).

The phosphorus–fish biomass relationship has been poorly documented for most other reservoir systems with upstream AWTs. However, anecdotal evidence of fisheries responses is abundant and usually negative in terms of post-AWT changes in biomass, growth, or harvest. Further substantiation of the impact of phosphorus abatement programs on fishery productivity is provided by the ongoing saga in Lake Ontario (Great Lakes Fishery Commission 1992). A basinwide nutrient abatement program has resulted in a 25% reduction in the total phosphorus concentration in offshore Lake Ontario since the early 1980s. Although water transparency has increased 20%, zooplankton production has dropped 50%, as has the forage fish biomass (rainbow smelt *Osmerus mordax* and alewife). Weight and condition of stocked salmonids has begun to decrease, and dead and dying salmon are collected in trawl catches. State and provincial fisheries agencies reduced annual salmonid stockings by 50% in 1993 to adjust predator abundance to the lower prey supply (Jude and Leach 1993). Despite a potentially large economic impact on the sport fishery, the Lake Ontario nutrient abatement program will continue due to a "strong international commitment to water quality enhancement" (Great Lakes Fishery Commission 1992).

Effects of Upstream Impoundments

Most large river systems in the United States host series of impoundments, including those built primarily for navigation. In the Southeast, about 75% of major (>500 ha) impoundments are arranged in series (Soballe et al. 1992). Because it functions as a nutrient and sediment trap, each new upstream reservoir has the potential to diminish the fertility of existing downstream impoundments. The degree of oligotrophication will be situation specific, depending on watershed nutrient load, water retention time, reservoir morphometry, and depth of the discharge port (epilimnetic or hypolimnetic).

Lake Mead, Nevada–Arizona, is perhaps the best documented example of fisheries response to nutrient deprivation following upstream dam construction. Lake Mead is a 66,000-ha storage reservoir on the Colorado River impounded in 1935. A highly successful fishery for largemouth bass *Micropterus salmoides* soon developed. Threadfin shad *Dorosoma petenense* was introduced as forage in the mid 1950s. Glen Canyon Dam, 450 km upstream of Lake Mead, was closed in 1963, creating Lake Powell. The new reservoir retained up to 90% of the phosphorus load in the Colorado River (Gloss et al. 1981), and much of the remainder entered Lake Mead as a density current deeper than the photic zone (Prentki and Paulson 1983). The new dam also substantially altered the annual flow regime of the Colorado River: spring floods were replaced with 2–3-m drawdowns in Lake Mead (Evans and Paulson 1983). Annual harvest of largemouth bass declined 50% by 1969, apparently due to poor reproduction resulting from the spring drawdowns (Romero and Allan 1975). Striped bass and rainbow trout *Oncorhynchus mykiss* were introduced in 1969 as limnetic predators of threadfin shad. The trout fishery soon collapsed under predation by striped bass, which established a large reproducing population (Baker and Paulson 1983). Between 1975 and 1980, the threadfin shad population also crashed, in response to both intense predation by striped bass and declining primary productivity (Paulson 1994). Adult striped bass soon became emaciated and experienced die-offs, but managed to reproduce successfully. Lake Mead has shifted from mesotrophic to oligotrophic status (<10 μg/L total phosphorus), and the fishery is now dominated by small striped bass in poor condition (Paulson 1994). Although it is not possible to precisely partition the effects of species introductions and reduced lake fertility, it is evident that the fishery of Lake Mead has suffered from oligotrophication (Kimmel et al. 1990).

In particular situations, upstream impoundments may enhance primary production (and perhaps fishery productivity) of downstream reservoirs by increasing the availability of nutrients during the growing season. Elser and Kimmel (1985) theorized

TABLE 1.—Predictive relationships between measures of plant and fish productivity in lakes and reservoirs, as determined from single-variable regression models.

Independent variable	Dependent variable	Data set (N)	Percent of variation explained (r^2)	Source
Gross photosynthesis	Total fish yield	Indian lakes (15)	82	Melack (1976)
Phytoplankton standing stock	Total fish yield	Natural lakes, northern hemisphere (19)	84	Oglesby (1977)
Gross photosynthesis	Total fish yield	Chinese lakes and ponds (18)	76	Liang et al. (1981)
Chlorophyll *a*	Sport fish yield	Midwestern U.S. lakes and reservoirs (25)	83	Jones and Hoyer (1982)
Primary production	Total fish production	Cosmopolitan lakes (19)	67	Downing et al. (1990)

that deep-discharge storage reservoirs, which release nutrients downstream in late summer, stimulate algal production more than unregulated river discharge, which carries peak nutrient loads in the winter–spring periods.

Phosphorus, Fisheries, and Water Quality

Enhancement of water quality is the principal goal of nutrient abatement programs and may also be considered a benefit of upstream impoundment. As applied to lakes and reservoirs, water quality and lake trophic status are usually defined in terms of phytoplankton production and water transparency, most often measured as chlorophyll *a* (mg/m^3) and Secchi disk disappearance depth (Carlson 1977; Moore and Thornton 1988). Dense growths of aquatic macrophytes are sometimes perceived as a water quality problem in shallow, relatively stable reservoirs, but excessive algae and resultant low transparency are considered a more ubiquitous threat to aesthetic and recreational uses (Heiskary and Walker 1988; Klessig et al. 1988).

Primary production in most lentic waters is limited by phosphorus (Jones and Bachmann 1976; USEPA 1993), and the correlation between total phosphorus concentration and chlorophyll *a* is strongly positive (Canfield and Bachmann 1981; Maceina et al. 1996, this volume). Regressions of data from several sets of lakes and reservoirs show that the relationship between primary production and various measures of fishery productivity is also strongly positive; differences in the former explain most of the variation in the latter (Table 1). Consequently, it is not surprising that total phosphorus concentration has been an excellent predictor of the standing stock, yield, and production of fish in lakes and reservoirs (Table 2). Because phosphorus promotes both algal and fish production, it would seem that target concentrations for phosphorus management might differ substantially between anglers and other users. A logical first step in optimal phosphorus management in multiple-use waters is to answer two questions: (1) how low must the concentration be to avoid undesirable algal production?; and (2) how high must the concentration be to sustain good fishing? Some effort has been directed to answer the first question, but the phosphorus–fishery issue has received little quantitative attention. What follows is an interpretative synthesis of the state of knowledge regarding desirable phosphorus concentrations in reservoirs.

Phosphorus Concentration and Water Quality

There is general agreement among limnologists that total phosphorus concentrations of 20–30 $\mu g/L$ are associated with eutrophic conditions in natural lakes (Carlson 1977; Rast and Lee 1978; Wetzel 1983). Eutrophic lakes are characterized by high photosynthetic activity, low (<2.0 m) transparency, and development of anoxic conditions in the hy-

TABLE 2.—Relationship between total phosphorus concentration ($\mu g/L$) as the independent variable and various measures of fish production in lakes and reservoirs.

Dependent variable	Data set (N)	Percent of variation explained (r^2)	Source
Total yield	North American lakes (21)	84	Hanson and Leggett (1982)
Sport fish yield	Midwestern U.S. lakes and reservoirs (21)	52	Jones and Hoyer (1982)
Total standing stock	Southern Appalachian (U.S.) reservoirs (21)	84	Ney et al. (1990)
Piscivore standing stock	Southern Appalachian (U.S.) reservoirs (11)	51	Ney et al. (1990)
Total fish production	Cosmopolitan lakes (14)	67	Downing et al. (1990)

polimnion (USEPA 1993). However, there is a great deal of variability among lakes in the phosphorus–phytoplankton relationship. In the National Eutrophication Survey (NES) of 894 U.S. lakes and reservoirs, only 40% of lakes with total phosphorus of 40 μg/L were considered eutrophic; at 60 μg/L, the eutrophic percentage climbed to sixty (Walker 1988).

User perceptions ultimately determine whether the phosphorus concentration of a water body is acceptable. Heiskary and Walker (1988) surveyed users of 99 Minnesota natural lakes, finding that impairment of physical appearance and swimming use occurred in only 25% at 40 μg/L total phosphorus and 50% at 60 μg/L.

Chlorophyll-*a* concentrations tend to be lower in reservoirs than in natural lakes (Soballe et al. 1992) because higher turbidity and flushing rates may limit the ability of phosphorus to stimulate phytoplankton production (Maceina et al. 1996). In the southeastern United States, storage reservoirs on tributary streams have higher chlorophyll-*a* levels at particular phosphorus concentrations than do mainstem, run-of-the-river impoundments (Soballe et al. 1992). Algal production is particularly low in reservoirs with retention times less than 7 d, the approximate phytoplankton doubling rate (Kimmel et al. 1990). In a comparative analysis of the NES data set, Canfield and Bachmann (1981) found that reservoirs tended to have substantially lower chlorophyll-*a* levels than natural lakes at the same phosphorus concentrations. Interpretation of their scatter diagram indicates that to produce 10.0 mg/m^3 of chlorophyll *a*, indicative of eutrophic status, the average natural lake would require 30 μg/L total phosphorus, whereas the average reservoir would need 40 μg/L.

In summary, excessive algal production is defined by user perception, and the phosphorus concentrations causing it vary with the characteristics of the water body (morphometry, chemistry, and hydrology). A conservative, "not to exceed" target for storage reservoirs appears to be 40 μg/L total phosphorus. Because this concentration is a whole-lake average, localized nuisance algae problems still may occur, particularly near nutrient point sources. For highly turbid or rapidly flushing systems, higher phosphorus concentrations may produce no objectionable results.

The concentration of total phosphorus in U.S. reservoirs ranges from less than 10 to more than 100 μg/L (Dodd et al. 1988; Yurk and Ney 1989; Soballe et al. 1992). Because of their higher ratios of drainage to surface areas and their dependence on river nutrient load, reservoirs are more frequently eutrophic than are natural lakes. Nutrient abatement programs can be expected to affect the biological productivity of more reservoirs in the future.

Phosphorus Concentration and Fishery Productivity

Regression analysis has also been used to determine how fish standing stock and production vary with total phosphorus concentration (Table 2). The strength of these relations (r^2) and their shape (i.e., linear, asymptotic, or parabolic) can be examined to develop a first-order estimate of how much phosphorus is required to maximize fishery productivity.

Only one of these empirical analyses has focused exclusively on reservoirs. Ney et al. (1990) examined the relationship between fish standing stock and a variety of potential predictors in a set of 21 southern Appalachian reservoirs for which fishery and water chemistry information was available for similar lengths of time (±2 years). These reservoirs varied greatly in surface area (1,700–132,000 ha), retention time (4–438 d), and total standing stock (77–2,321 kg/ha). Total phosphorus was the best predictor of fish standing stock ($r^2 = 0.84$; Figure 2), followed by Secchi disk depth (negative slope; $r^2 = 0.42$) and $\log_{10}$chlorophyll *a* ($r^2 = 0.31$). Retention time and the morphoedaphic index (MEI, total dissolved solids/mean depth) had no predictive power ($P > 0.3$ for each variable) in this data set ($r^2 = 0.01$ and 0.14, respectively). Fish standing stock increased linearly over the range of total phosphorus concentrations (8–81 μg/L) in the southern Appalachian reservoirs, suggesting that maximum fish biomass would occur at higher phosphorus concentrations.

Downing et al. (1990) reported a linear increase in total annual fish production (kg/ha) through 100 μg/L total phosphorus for a data set of 13 natural, northern hemisphere lakes. However, inclusion of fish production in a hypereutrophic African lake (9,850 μg/L total phosphorus) produced a curvilinear regression, suggesting that fish production will peak somewhere between 100 and 1,000 μg/L total phosphorus in temperate waters.

Total fish standing stock or total fish production may not be indicative of sportfishing potential of reservoirs because sport and food fishes usually account for less than half of that total (USEPA 1993). Three studies have examined the relationship between total phosphorus and sport or food fish productivity. For the southern Appalachian reservoirs data set, Yurk and Ney (1989) found that

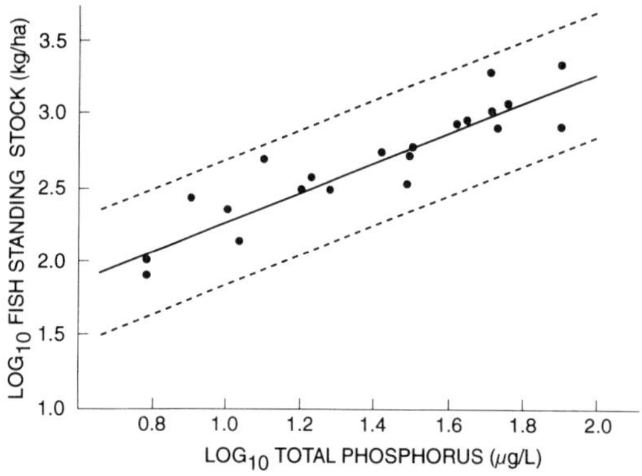

FIGURE 2.—Regression of $\log_{10}$ total fish standing stock (FSS) versus $\log_{10}$ total phosphorus (TP) concentration for 21 southern Appalachian reservoirs. $\text{Log}_{10}\text{FSS} = 1.24 + 1.02\log_{10}\text{TP}$; $r^2 = 0.84$. Dashed lines are 90% predictive limits (from Ney et al. 1990).

piscivore (largely sport fish) standing stock increased linearly over the range of total phosphorus concentrations, however, the relationship was not as strong ($r^2 = 0.51$) as for planktivores ($r^2 = 0.84$).

Jones and Hoyer (1982) reported that annual sportfish harvest increased linearly with total phosphorus concentration over the range of 15–90 μg/L in 25 midwestern U.S. lakes and reservoirs ($r^2 = 0.52$). In a study of 21 north temperate natural lakes, Hanson and Leggett (1982) found that long-term sport and commercial annual harvests increased with total phosphorus concentrations up to 500 μg/L ($r^2 = 0.84$). However, only five of the 21 lakes had phosphorus concentrations greater than 100 μg/L.

Within the sport fish complex, individual species are likely to respond differently to varying levels of lake fertility. Schupp and Wilson (1993) compared their knowledge of the relative abundance of several sport fish species in northern U.S. natural lakes to trophic status indicators. Extrapolation of their data indicates that relative abundance of lake trout *Salvelinus namaycush* and walleye *Stizostedion vitreum* peaks at less than 10 μg/L and at 25 μg/L total phosphorus, respectively. Black crappies *Pomoxis nigromaculatus* are most abundant at 70 μg/L total phosphorus, but white crappie *P. annularis* populations do best at more than 100 μg/L total phosphorus. In a study of four large Alabama reservoirs, Bayne et al. (1994) found that largemouth bass growth and harvest were substantially greater in eutrophic systems (50–100 μg/L total phosphorus) than in a mesotrophic (10 μg/L total phosphorus) counterpart. Using an expanded data base of 32 Alabama impoundments, Maceina et al. (1996) confirmed that growth rates and condition of both largemouth bass and crappies were higher in eutrophic than in oligo-mesotrophic reservoirs, but that angler catch rates did not differ significantly with trophic status. They suggested that good fishing and publicly acceptable water clarity (120 cm Secchi disk transparency) might be compatible in Alabama impoundments, although the catch of trophy black bass and crappies would decline. Maceina et al. (1996) developed linear regressions to predict chlorophyll-*a* concentration from total phosphorus level and water clarity from chlorophyll-*a* concentration. In combination, these regressions predict transparency of 120 cm at 13 μg/L chlorophyll *a* and 46 μg/L total phosphorus.

The phosphorus–fishery regressions in the multilake data set (Table 2) failed to clearly identify those total phosphorus concentrations that promote maximum productivity of either the total fish assemblage or the sport fish component, in part because few data points with total phosphorus exceeding 100 μg/L were included. The species-specific analyses above did not consider standing stock or yield as the dependent variable. Clearly, more detailed examinations of the phosphorus–sport fish relationship are warranted (Carline 1986; Lee and Jones 1991).

It is intuitive that both total fish and sport fish productivity in reservoirs will peak at some specific phosphorus concentration as the trade-off between more food and less habitat (through hypolimnetic oxygen depletion or excessive vegetative cover) becomes acute. The sport fish maximum will likely occur at a substantially lower total phosphorus concentration than the total fish maximum. Sport fishes, especially coolwater and coldwater species, are relatively demanding in habitat and reproductive requirements, as the Smith Mountain Lake case study illustrates. Opportunistic, eurytolerant, nonsport species are likely to replace stenotolerant fishes, maintaining or increasing total fish biomass over a range of increasing fertility.

Considered collectively, the available evidence indicates that maximum biomass of sport fish will occur at phosphorus concentrations above 100 μg/L, which will stimulate eutrophic conditions in many reservoirs (Figure 3). Potential for user conflicts over lake trophic status will remain high, and fisheries managers will need to seek innovative means to cope with the threat as well as the realities of induced oligotrophication.

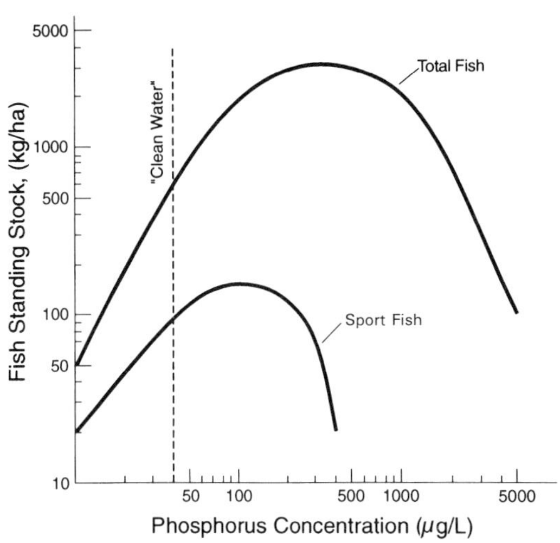

FIGURE 3.—Generalized relation of total and sport fish standing stock to total phosphorus concentration and trophic status in temperate-latitude reservoirs. Standing stock values are representative of southeastern U.S. reservoirs to 100 μg/L total phosphorus. Standing stocks at higher phosphorus concentrations are hypothetical.

Meeting the Challenge of Oligotrophication

Declines in fishery productivity following construction of upstream impoundments or AWTs have usually come as a surprise to both anglers and the general public; reservoir fisheries managers have not been vocal prophets or advocates for their resource. If the challenge of oligotrophication is to be met, fisheries scientists need to be involved at the education and decision-making stages long before construction begins; remedial measures to restore fishery productivity (species manipulations; in-lake fertilization) are likely to achieve limited success at best.

Remedial Measures

Stocking salmonids or coolwater sport fishes to use newly created pelagic habitat may succeed in diversifying a reservoir fishery, but stocking will not restore its productive capacity. There are no precedents by which to evaluate the success of post-oligotrophication species manipulations on angler satisfaction. Fertilizing a large reservoir is a massive undertaking and could arouse the ire of those responsible for nutrient abatement. The only large-scale fertilization experiment reported to date, the Lake Mead Nutrient Enhancement Project, was expensive, ineffective, and controversial (Vaux and Paulson 1990). In May and June 1987, 1,000 volunteers in 300 boats added 76,000 L of ammonium polyphospate to an 8,000-ha arm of Lake Mead to raise the phosphorus concentration and stimulate plankton production, thereby increasing reproductive success and abundance of threadfin shad. Similar phosphorus additions were made in 1988 and 1989, but the effects on plankton abundance were small and transitory; threadfin shad abundance did not increase perceptibly. The Colorado River is a major drinking water supply for 20 million people, and opposition from downstream users caused the experiment to be terminated, despite negligible impacts on downstream drinking water quality (Vaux et al. 1995).

Preventive Measures

Fisheries managers must be proactive, becoming involved when the first consideration of an upstream impoundment is announced or the first complaints are made about deteriorating water quality. In the latter situation, it may be possible to treat the symptom of eutrophication, excessive algae, through food web management without resorting to nutrient abatement. For either situation, the potential change in nutrient loading and its impact on fishery productivity can be predicted. An informed dialogue among reservoir user groups can then develop before the decision-making stage.

Food web management.—Also known as biomanipulation or the trophic cascade, top-down food web management has great appeal to lake and reservoir managers because it promises both good fishing and clear water (Ney 1993). The central premise of top-down management is that planktivorous fishes affect algal abundance through intense predation on zooplankton. Predation pressure on zooplankton can be relieved by the addition of piscivorous sport fishes to prey on the planktivores. Large, efficiently grazing zooplankton then reappear to crop phytoplankton, improving water clarity (Carpenter et al. 1985). A number of case studies in natural lakes, ranging from small New England ponds to Lake Michigan, provide evidence that top-down food web manipulations do sometimes succeed in improving water clarity (Hayes et al. 1993). However, it appears unlikely that increasing the abundance of piscivorous sport fishes will in itself overcome the effects of bottom-up algal productivity due to high nutrient loading. An ambitious experiment in 4,000-ha, eutrophic Lake Mendota, Wisconsin, demonstrated the limitations of top-down management (Kitchell 1992). Between 1987 and 1989, fingerling walleye and northern pike *Esox*

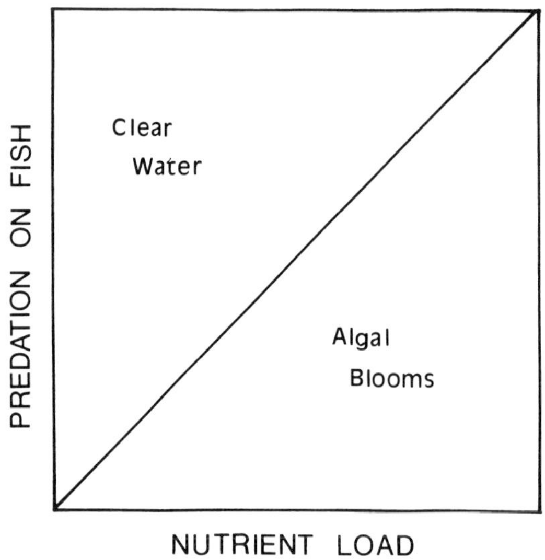

FIGURE 4.—Opposing effects of predation on zooplankton-feeding fish and nutrient loading on the potential for algal blooms in reservoirs and lakes. (Adapted from Kitchell and Carpenter 1992.)

lucius were stocked at an annual combined rate of 140/ha to reduce the abundance of planktivorous cisco *Coregonus artedi* and yellow perch *Perca flavescens*. In succeeding years, yellow perch and cisco year-class strength declined, but summer algal blooms still occurred in wet, high-runoff years. In reviewing the Lake Mendota experiment, Kitchell and Carpenter (1992) concluded that top-down and bottom-up processes interact to affect algal production and water clarity over years and among water bodies (Figure 4).

The potential for top-down control of algal blooms in reservoirs will be limited also by the characteristics of the forage fish resource and the ability to manipulate piscivore abundance (USEPA 1993). Gizzard shad are the dominant forage species in many U.S. reservoirs and rapidly grow too large to be vulnerable to predators (Noble 1981). Gizzard shad also feed on both phytoplankton and zooplankton, complicating the food web and energy flow pathways. In most situations, economic constraints will dictate that an effective increase in predation by piscivores can happen only through the establishment of self-sustaining populations, which may not be easily manipulated. The explosion of the striped bass population and subsequent crash of rainbow trout and threadfin shad in Lake Mead provides a cautionary example (Paulson 1994). In summary, predator–prey manipulations must still be considered as a highly experimental option for preventing reservoir oligotrophication.

Informed decision-making.—The consequences of proposed nutrient abatement programs for reservoir fisheries can be predicted and used in the decision-making process to determine how much, if any, reduction in point source nutrient loading to impose. Limnologists have developed accurate models for projecting water column phosphorus concentration from annual phosphorus load (reviewed by Walker 1988). Prediction of sport fish standing stock from water column phosphorus concentration is less certain, but first-order approximations are clearly possible (see preceding section).

Prediction of impacts of a potential upstream impoundment on the fishery of a downstream reservoir is more complicated but may offer an opportunity to recommend beneficial alterations in dam design and operation. Downstream changes in nutrient loading will depend on outlet depth, annual discharge regime, retention time, reservoir morphometry, and sediment transport as well as incoming nutrient load. Limnologists must be consulted to estimate how much of the nutrient load will reach the dam and at what depth. A deep discharge will not effectively transport nutrients downstream during the growing season unless nutrients accumulate in an anoxic hypolimnion adjacent to the dam and summer releases are substantial and consistent (Soballe et al. 1992). If the temperature of the upstream discharge is lower than that of the surface water in the downstream impoundment, it may travel through the impoundment as a deep density current, inaccessible to phytoplankton (Kimmel et al. 1990). Limnological projections, coupled with phosphorus loading models and phosphorus concentration–fish standing stock regressions, should permit the development of realistic fishery response scenarios under different operation and design alternatives.

Prediction of impacts on fisheries associated with reduced nutrient loading must be made known early in the planning process to receive thorough consideration. Public education and involvement, not only of anglers but of other user groups, is also essential; most people will be surprised that clearer water may not be an unmitigated blessing. Fisheries managers also have an obligation to argue for their resource. Conflicts among reservoir user groups can be expected, and mediation techniques are increasingly used for their resolution (McMullin and Nielsen 1991). The end result may not be an outright victory, but rather a compromise between water clarity and fish production (see Maceina et al. 1996). In the short history of reservoir oligotrophication, even a compromise must be considered an achievement.

Conclusions

The reversal of the eutrophication process can have deleterious effects on reservoir fisheries, as the case histories and strength of the phosphorus–fishery relation indicate. However, the nature and magnitude of the response of the sport fish component of the reservoir fish assemblage to oligotrophication cannot be precisely predicted in any particular reservoir until the database is greatly expanded. Concurrent nutrient, water quality, and fishery productivity (standing stock, harvest, or production) measurements must be made and compared over many years for before-and-after assessment of the response to reduced fertility. These data also must be developed on a same-year basis for a large number of impoundments to facilitate empirical prediction of how reservoir characteristics (biology, morphometry, hydrology, and chemistry) influence the oligotrophication response.

Uncertainties in predictive precision do not absolve reservoir fisheries managers from treating potential reductions in nutrient loading as they would any other environmental impact. First, they must assess the present status of their fisheries: are growth and abundance high or limited by habitat or food deficiencies? Next, they must consult with limnologists to determine how reduced nutrient load will affect water column nutrient concentration. Finally, reservoir managers should compare their fisheries to those in other impoundments in the region which have similar as well as lower nutrient levels. From this analysis, a first-order assessment of the impact of pending oligotrophication can be developed, an assessment which is probably as accurate as those now possible for a host of other perturbations.

Reservoir fisheries managers have not collected nutrient and water quality data routinely, nor have they coordinated sampling with limnologists who do. This oversight is due, in part, to a lack of appreciation of the importance of nutrients to fisheries productivity, an ignorance shared with the general public and policy makers. In the United States, the Clean Lakes Program initiated pursuant to Public Law 92–500 has provided a major impetus for nutrient abatement. Although this program has been beneficial in the treatment of hypereutrophic waters, its objectives are often narrowly focused on enhancement of aesthetic and recreational uses of lakes with little consideration of fisheries (Lee and Jones 1991). Fisheries scientists have been conspicuously absent in the formulation of nutrient abatement policy, relinquishing their advisory role to limnologists. A collaborative effort must be initiated between fisheries scientists and limnologists to first educate themselves and then the public and policy makers regarding all the consequences of nutrient control programs. Only then can the issue of eutrophication versus oligotrophication be resolved satisfactorily on a case-by-case basis.

Acknowledgments

I am indebted to the following individuals for freely providing information used in the development of this paper: David Bayne, Auburn University; Mike Bivin and Robert M. Jenkins, Arkansas Game and Fish Commission; M. C. Duval, Virginia Department of Game and Inland Fisheries; Larry Paulson, West Lakes; Gene Ploskey, U.S. Army Corps of Engineers; Clifford Schneider, New York State Department of Environmental Conservation; Peter Vaux, University of Nevada–Las Vegas; and Larry Willis, Virginia Department of Environmental Quality. Trent Sutton provided graphical assistance and, as always, Carolyn Linkous processed the manuscript.

References

Aggus, L. R. 1985. A strategy for managing water quality in Beaver Lake. Report to the Arkansas Water Resources Research Center, Fayetteville.

Baker, J. R., and L. J. Paulson. 1983. The effects of limited food availability on the striped bass fishery in Lake Mead. Pages 551–561 *in* V. D. Adams and V. A. Lamarra, editors. Aquatic resources management of the Colorado River ecosystem. Ann Arbor Science Publishers, Ann Arbor, Michigan.

Bayne, D. R., M. J. Maceina, and W. C. Reeves. 1994. Zooplankton, fish, and sport fishing quality among four Alabama and Georgia reservoirs of varying trophic status. Lake and Reservoir Management 8:153–163.

Canfield, D. E., and R. W. Bachmann. 1981. Prediction of total phosphorus concentrations, chlorophyll *a* and Secchi disk depths in natural and artificial lakes. Canadian Journal of Fisheries and Aquatic Sciences 38:414–423.

Carline, R. F. 1986. Indices as predictors of fish community traits. Pages 45–56 *in* G. E. Hall and M. J. Van Den Avyle, editors. Reservoir fisheries management: strategies for the 80's. American Fisheries Society, Southern Division, Reservoir Committee, Bethesda, Maryland.

Carlson, R. E. 1977. A trophic status index for lakes. Limnology and Oceanography 22:361–369.

Carpenter, S. R., J. F. Kitchell, and J. R. Hodgson. 1985.

Cascading trophic interactions and lake productivity. BioScience 35:634–639.

Davies, W. D., and W. L. Shelton. 1983. Sampling with toxicants. Pages 191–213 *in* L. A. Nielsen and D. L. Johnson, editors. Fisheries techniques. American Fisheries Society, Bethesda, Maryland.

Dodd, R. C., J. F. Smith, and J. D. Vogt. 1988. The development of a phosphorus management strategy for two Piedmont reservoirs. Lake and Reservoir Management 4:243–252.

Downing, J. A., C. Plante, and S. Lalonde. 1990. Fish production correlated with primary productivity, not the morphoedaphic index. Canadian Journal of Fisheries and Aquatic Sciences 47:1929–1936.

Elser, J. J., and B. L. Kimmel. 1985. Nutrient availability for phytoplankton production in multiple-impoundment series. Canadian Journal of Fisheries and Aquatic Sciences 42:1359–1370.

Evans, T. D., and L. J. Paulson. 1983. The influence of Lake Powell on the suspended sediment-phosphorus dynamics of the Colorado inflow to Lake Mead. Pages 57–68 *in* V. D. Adams and V. A. Lamarra, editors. Aquatic resources management of the Colorado River ecosystem. Ann Arbor Science Publishers, Ann Arbor, Michigan.

Gloss, S. P., R. C. Reynolds, Jr., L. M. Mayer, and D. E. Kidd. 1981. Reservoir influences on salinity and nutrient flux in the arid Colorado River basin. Pages 1618–1629 *in* H. G. Stefan, editor. Symposium on surface water impoundments. American Society of Civil Engineering, Minneapolis, Minnesota.

Great Lakes Fishery Commission. 1992. Status of the Lake Ontario pelagic fish community and related ecosystem in 1992. Great Lakes Fishery Commission, Kingston, Ontario.

Hanson, J. M., and W. C. Leggett. 1982. Empirical prediction of fish biomass and yield. Canadian Journal of Fisheries and Aquatic Sciences 39:257–263.

Harper, D. 1992. Eutrophication of freshwaters: principles, problems, and restoration. Chapman and Hall, London.

Hart, L. G. 1978. Smith Mountain Reservoir research study, 1973–1976. Completion report to the Virginia Commission of Game and Inland Fisheries, Richmond.

Hayes, D. B., W. W. Taylor, and E. E. Mills. 1993. Natural lakes and impoundments. Pages 493–515 *in* C. C. Kohler and W. A. Hubert, editors. Inland fisheries management in North America. American Fisheries Society, Bethesda, Maryland.

Heiskary, S. A., and W. W. Walker. 1988. Developing phosphorus criteria for Minnesota Lakes. Lake and Reservoir Management 4:1–9.

Jones, J. R., and R. W. Bachmann. 1976. Prediction of phosphorus and chlorophyll levels in lakes. Journal of the Water Pollution Control Federation 48:2176–2182.

Jones, J. R., and M. V. Hoyer. 1982. Sportfish harvest predicted by summer chlorophyll-a concentrations in midwestern lakes and reservoirs. Transactions of the American Fisheries Society 111:176–179.

Jude, D. J., and J. Leach. 1993. The Great Lakes fisheries. Pages 517–551 *in* C. C. Kohler and W. A. Hubert, editors. Inland fisheries management in North America. American Fisheries Society, Bethesda, Maryland.

Kimmel, B. L., O. T. Lind, and L. J. Paulson. 1990. Reservoir primary production. Pages 133–194 *in* K. W. Thornton, B. L. Kimmel, and F. E. Payne, editors. Reservoir limnology: ecological perspectives. Wiley, New York.

Kitchell, J. F., editor. 1992. Food web management: a case study of Lake Mendota. Springer-Verlag, New York.

Kitchell, J. F., and S. R. Carpenter. 1992. Summary: accomplishments and new directions for food web management in Lake Mendota. Pages 539–543 *in* J. F. Kitchell, editor. Food web management: a case study of Lake Mendota. Springer-Verlag, New York.

Klessig, L., R. Wedepohl, and D. Knauer. 1988. Problem identification. Pages 3-1–3-29 *in* L. Moore and K. Thornton, editors. Lake and reservoir restoration guidance manual. U.S. Environmental Protection Agency Publication (EPA 44015-88-002), Washington, DC.

Lee, G. F., and R. A. Jones. 1991. Effects of eutrophication on fisheries. Reviews in Aquatic Sciences 5:287–305.

Liang, Y., J. M. Melack, and J. Wang. 1981. Primary production and fish yields in Chinese ponds and lakes. Transactions of the American Fisheries Society 110:346–350.

Lindemann, R. L. 1942. The trophic-dynamic aspect of ecology. Ecology 23:399–418.

Maceina, M. J., and five coauthors. 1996. Compatibility between water clarity and quality black bass and crappie fisheries in Alabama. Pages 296–305 *in* L. E. Miranda and D. R. DeVries, editors. Multidimensional approaches to reservoir fisheries management. American Fisheries Society Symposium 16.

McMullin, S. L., and L. A. Nielsen. 1991. Resolution of natural resource allocation conflicts through effective public involvement. Policy Studies Journal 19:553–559.

Melack, J. M. 1976. Primary productivity and fish yield in tropical lakes. Transactions of the American Fisheries Society 105:575–580.

Moore, L., and K. Thornton, editors. 1988. Lake and reservoir restoration guidance manual. U.S. Environmental Protection Agency Publication (EPA 440/5-88-002), Washington, DC.

Ney, J. J. 1993. Top-down management of water quality: clear lakes plus better fishing? Lake Line 13(4):16–17.

Ney, J. J., C. M. Moore, M. S. Tisa, J. J. Yurk, and R. J. Neves. 1988. The Smith Mountain Lake fishery: factors affecting major sport and forage fish populations. Report to the Virginia Department of Game and Inland Fisheries, Richmond.

Ney, J. J., C. M. Moore, M. S. Tisa, J. J. Yurk, and R. J. Neves. 1990. Factors affecting the sport fishery in a multiple-use Virginia reservoir. Lake and Reservoir Management 6:21–32.

Noble, R. L. 1981. Management of forage fishes in impoundments of the southern United States. Transactions of the American Fisheries Society 110:738–750.

Oglesby, R. T. 1977. Relationships of fish yield to lake phytoplankton standing crop, production and morphoedaphic factor. Journal of the Fisheries Research Board of Canada 34:2271–2279.

Paulson, L. J. 1994. Effect of water diversions in the Virgin River on the sport fisheries in the Overton arm of Lake Mead. Report to the U.S. National Park Service, Boulder City, Nevada.

Prentki, R. T., and L. J. Paulson. 1983. Historical patterns of phytoplankton productivity in Lake Mead. Pages 105–123 *in* V. D. Adams and V. A. Lamarra, editors. Aquatic resources management of the Colorado River ecosystem. Ann Arbor Science Publishers, Ann Arbor, Michigan.

Rast, W., and G. F. Lee. 1978. Summary analysis of the North American (U.S. portion) OECD eutrophication project: nutrient loading-lake response relationships and trophic state indices. U.S. Environmental Protection Agency Publication (EPA-60013-78-008), Corvallis, Oregon.

Romero, J., and R. C. Allan. 1975. Underwater observation of largemouth bass spawning and survival in Lake Mead. Pages 104–112 *in* R. H. Stroud and H. Clepper, editors. Black bass biology and management. Sport Fishing Institute, Washington, DC.

Schupp, D., and B. Wilson. 1993. Developing lake goals for water quality and fisheries. Lake Line 13(4):18–21.

Soballe, D. M., B. L. Kimmel, R. H. Kennedy, and R. F. Gaugush. 1992. Reservoirs. Pages 421–474 *in* C. T. Hackney, S. M. Adams, and W. H. Martin, editors. Biodiversity of the southeastern United States: aquatic communities. Wiley, New York.

Vaux, P. D., and L. J. Paulson. 1990. Lake Mead nutrient enhancement project. Final report to the Nevada Department of Wildlife, Las Vegas.

Vaux, P. D., L. J. Paulson, R. P. Axler, and S. Leavitt. 1995. The water quality implications of artificially fertilizing a large desert reservoir for fisheries enhancement. Water Environmental Research 67:189–200.

USEPA (U.S. Environmental Protection Agency). 1993. Fish and fisheries management in lakes and reservoirs. U.S. Environmental Protection Agency Publication (EPA-841-R-93-002), Washington, DC.

Walker, W. W. 1988. Predicting lake water quality. Pages 4-1–4-23 *in* L. Moore and K. Thornton, editors. The lake and reservoir restoration guidance manual. U.S. Environmental Protection Agency Publication (EPA 440/5-88-002), Washington, DC.

Wetzel, R. G. 1983. Limnology. Saunders, Philadelphia.

Whitehurst, D. K. 1987. Reservoir investigations: Smith Mountain Lake, 1986–1987. Report to the Virginia Commission of Game and Inland Fisheries, Richmond.

Yurk, J. J., and J. J. Ney. 1989. Phosphorus–fish community biomass relationships in southern Appalachian reservoirs: can lakes be too clean for fish? Lake and Reservoir Management 5:83–90.

American Fisheries Society Symposium 16:296–305, 1996

Compatibility between Water Clarity and Quality Black Bass and Crappie Fisheries in Alabama

MICHAEL J. MACEINA, DAVID R. BAYNE, AND A. SCOTT HENDRICKS[1]
Auburn University, Department of Fisheries and Allied Aquacultures
Alabama Agricultural Experiment Station, Auburn University, Alabama 36849, USA

WILLIAM C. REEVES
Alabama Department of Conservation and Natural Resources
64 North Union Street, Montgomery, Alabama 36130, USA

WILLIAM P. BLACK[2] AND VICTOR J. DICENZO[3]
Auburn University, Department of Fisheries and Allied Aquacultures
Alabama Agricultural Experiment Station, Auburn University, Alabama 36849, USA

Abstract.—The reversal of cultural eutrophication of reservoirs can be deleterious to warmwater sport fisheries because it can reduce fish production and biomass. However, this process has resulted in greater water transparency and improved aesthetics for nonangling recreational users. In Alabama, we found angler catch rates of black bass *Micropterus* spp. and crappie *Pomoxis* spp. did not vary between oligomesotrophic (chlorophyll $a \leq 7$ mg/m^3) and eutrophic (chlorophyll $a \geq 8$ mg/m^3) reservoirs. However, the mean weight of angler-caught black bass and catch rates of memorable-size black bass (≥2.27 kg) were higher in eutrophic than in oligomesotrophic reservoirs. Consonant with these observations, crappie and black bass growth and condition were usually greater in eutrophic reservoirs. However, within the range of eutrophic reservoirs, sport fishery and population characteristics did not change with higher chlorophyll-*a* concentrations. Based on empirical relations in 32 major impoundments in Alabama, chlorophyll-*a* concentrations greater than 15 mg/m^3 would generally result in water transparencies less than 120 cm, which may be less appealing to nonangling reservoir users. We propose that in southern U.S. reservoirs, reductions in chlorophyll-*a* concentrations to 10–15 mg/m^3 will not necessarily be detrimental to black bass and crappie fisheries, and will likely improve water clarity.

The reversal of eutrophication and improvement of water clarity in reservoirs can be deleterious to warmwater sport fisheries (Axler et al. 1988; Yurk and Ney 1989; Ney et al. 1990; Ney 1996, this volume) as reductions in fish biomass and production can occur. However, this process has resulted in greater water transparency and can increase satisfaction among nonangling reservoir users (Smeltzer and Heiskary 1990). Although the quantity of phytoplankton is inversely related to water transparency in reservoirs, nonalgal suspended particulate matter is also an important factor regulating water clarity in these systems (Canfield and Bachmann 1981; Hoyer and Jones 1983; Soballe and Kimmel 1987). Phosphorus and nitrogen are positive correlates of phytoplankton in reservoirs and natural lakes (Soballe and Kimmel 1987), and are usually considered the primary nutrients limiting algal production. Thus, reduction of anthropogenic nutrient loading by control of point or nonpoint sources, or by construction of an upstream reservoir that traps phosphorus, can result in a decline in primary production. Consequently, a reduction in predatory sport fish can occur by limiting production at the bottom of the trophic web (McQueen et al. 1986).

Oglesby (1977) and Jones and Hoyer (1982) found that algal production was positively correlated to fish yields in natural lakes and reservoirs. Axler et al. (1988) attributed the decline in the US$100 million per year fisheries for largemouth bass *Micropterus salmoides* and striped bass *Morone saxatilis* in Lake Mead, Nevada–Arizona, to advanced nutrient removal in wastewater and the construction of Lake Powell upstream. To increase phytoplankton production and improve fishing in Lake Mead, 202 t of liquid ammonium polyphosphate were applied to one arm of the reservoir in 1987, resulting in only a short-term increase in algal production (Axler et al. 1988). Thus, costs were incurred to reduce and subsequently add nutrients

[1]Present address: Georgia Power Company, 5131 Maner Road, Smyrna, Georgia 30080, USA.
[2]Present address: Florida Marine Research Institute, 3 Jackson Street, Fort Walton Beach, Florida 32548, USA.
[3]Present address: Virginia Department of Game and Inland Fisheries, 2206 South Main Street, Suite C, Blacksburg, Virginia 24060, USA.

to this reservoir. Reductions in anthropogenic loading were partially responsible for a decline in the sport fishery in Smith Mountain Lake, Virginia (Ney et al. 1990). Bayne et al. (1994a) reported that population characteristics of crappie *Pomoxis* spp. and black bass *Micropterus* spp. differed between a mesotrophic and three eutrophic reservoirs in Alabama and Georgia. Attributes of these populations were similar in the three eutrophic reservoirs, however, a nearly threefold difference in chlorophyll-*a* concentrations existed among them. Based on their observations of three reservoirs, Bayne et al. (1994a) proposed that a reduction in trophic state from hypereutrophic to the low range of eutrophic may not be detrimental to crappie and black bass fisheries.

We further tested the original hypothesis proposed by Bayne et al. (1994a) by examining the relationships among water quality, crappie, and black bass data from nearly all major (>400 ha) Alabama impoundments. Specific objectives were to describe and relate factors that influenced the relationship between water clarity and phytoplankton. Also, we used angler catch rates, size of fish caught by anglers, and growth and body condition of crappies and black bass as indicators of population status, and related these variables to chlorophyll-*a* concentrations, trophic state, and water clarity. These two fish groups usually account for more than 70% of the angler effort on Alabama reservoirs. We assessed the potential impacts of oligotrophication on these reservoir fisheries and attempted to determine target levels of primary production that provide both quality fisheries and satisfactory water clarity for other recreational users of reservoirs.

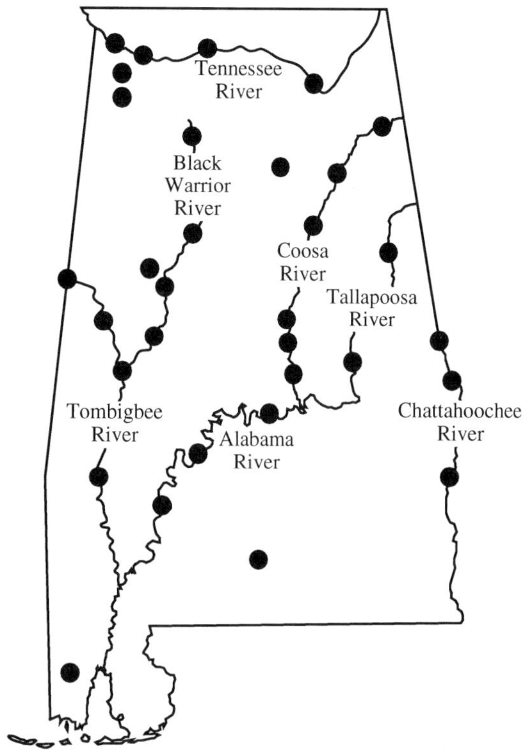

FIGURE 1.—Location of the 32 reservoirs examined in this study.

Methods

Water quality data for 32 reservoirs (Figure 1) in Alabama were obtained from unpublished records provided by the Tennessee Valley Authority, the Alabama Department of Environmental Management, Auburn University, Bayne et al. (1990, 1994a), and Hendricks et al. (1995). Phytoplankton abundance was determined by measuring chlorophyll-*a* concentrations. Based on the criteria of Forsberg and Ryding (1980) for determining trophic status, reservoirs were classified as either oligomesotrophic if chlorophyll-*a* concentrations were 7 mg/m^3 or less and eutrophic if chlorophyll-*a* levels were 8 mg/m^3 or more. Chlorophyll-*a* concentrations were corrected for phaeophytin from the equations in APHA (1985). Secchi disk transparency was measured with a standard 20-cm black-and-white disk. Sample size included 1,394 paired observations for Secchi disk transparency and chlorophyll *a*. The database also included some total phosphorus, total nitrogen, and total suspended solid measurements. Data to compute mean depth and hydraulic retention time (volume/discharge) were provided by the U.S. Army Corps of Engineers, Alabama Power Co., Tennessee Valley Authority, Georgia Power Co., and local municipalities.

Crappie harvest rates, catch rates (fish harvested + fish released), and average size harvested were determined from access point and roving creel surveys (Table 1). From 1989 to 1993, creel surveys were conducted from 1 March to 30 May, corresponding to the peak in angler effort for crappie in Alabama. Individual crappies from each reservoir and time period were placed in 25-mm size groups and average total length (TL) was extrapolated by multiplying the number of fish in each size group by the mid-point length, summing these values, then dividing this sum by the total number of fish mea-

TABLE 1.—Summary statistics used to assess crappie and black bass in Alabama reservoirs.

Variable	Mean	Range	SD	N
Angler catch rate of crappie (N/h)	1.09	0.10–2.91	0.70	23
Angler harvest rate of crappie (N/h)	0.64	0.09–1.27	0.35	20
Angler catch of black bass (N/hr)	0.24	0.12–0.31	0.04	27
Average total length (TL) of crappie harvested by anglers (mm)	265	234–292	17	19
Average weight of angler-caught black bass (kg)	0.71	0.57–0.82	0.06	27
Angler effort to catch black bass ≥2.27 kg (h)	356	136–1,298	277	28
Time for crappie to reach 229 mm TL (years)	2.30	1.57–3.01	0.39	32
Time for crappie to reach 254 mm TL (years)	2.81	2.00–4.04	0.48	31
Time for largemouth bass to reach 304 mm TL (years)	2.47	1.89–3.21	0.31	32
Time for largemouth bass to reach 381 mm TL (years)	3.79	3.15–4.89	0.40	31
Time for largemouth bass to reach 508 mm TL (years)	7.45	6.52–10.21	0.67	28
Time for spotted bass to reach 304 mm TL (years)	2.61	2.25–2.97	0.22	11
Time for spotted bass to reach 432 mm TL (years)	5.20	4.32–6.97	0.75	11
Relative weight (W_r) for stock-size black crappie	88	79–102	7	15
W_r for quality-size black crappie	89	80–104	8	11
W_r for preferred-size black crappie	93	81–103	8	7
W_r for stock-size white crappie	89	81–98	6	19
W_r for quality-size white crappie	93	81–105	7	19
W_r for preferred-size white crappie	95	86–115	8	14
W_r for stock-size largemouth bass	90	80–102	7	32
W_r for quality-size largemouth bass	93	81–108	8	32
W_r for preferred-size largemouth bass	95	83–112	8	31
W_r for stock-size spotted bass	88	77–97	8	11
W_r for quality-size spotted bass	90	77–102	9	11
W_r for preferred-size spotted bass	91	75–108	10	11

sured. Differences in year-class strength caused large variation in crappie catch rates and average lengths of harvested crappies from year to year; multiple observations for these variables were therefore made at each reservoir and compared with a single mean for chlorophyll *a*, a figure derived from annual means for each reservoir sampled. To assess harvest success, only reservoirs with no length limits were used, and the daily bag limit was consistent among reservoirs.

Black bass catch rates were obtained from a database compiled from reports provided between 1986 and 1992 by tournament anglers at 27 reservoirs where 304-mm TL limits were imposed (Hendricks et al. 1995). Catch statistics for black bass were summed and pooled for the entire sample period representing 2,183 tournaments and 596,407 h of angling (Table 1). Also, mean weight and the angler effort to catch a large (≥2.27 kg) black bass were determined for all reservoirs. Angler effort to catch a large black bass was computed for an additional reservoir that had a 406-mm TL limit on black bass. These black bass fisheries included largemouth bass and spotted bass *M. punctulatus*, except for two fisheries in the Tennessee River that also included smallmouth bass *M. dolomieu*.

From 1990 to 1992 in October and November, 5,451 black crappie *P. nigromaculatus* and white crappie *P. annularis* were collected in Indiana trap nets (20 × 0.91 m, 13-mm stretch mesh). Age-1 and older largemouth bass (N = 5,997) and spotted bass (N = 2,449) were collected with electrofishing gear from 15 March to 15 May between 1989 and 1993. Crappies and black bass were measured for total length (mm) and weight (g). Sagittal otoliths were removed and aged according to procedures described by Maceina and Betsill (1987) and Hoyer et al. (1985). Mean lengths at age and von Bertalanffy (1938) growth parameters were estimated in 9 reservoirs for black crappie, 8 for white crappie, 31 for largemouth bass, and 11 for spotted bass. For crappies, we used von Bertalanffy growth models to estimate the time (years) to reach Alabama's current legal minimum size (229 mm TL) and preferred size (254 mm TL; Table 1). Differences in growth rates were not evident between crappie species within reservoirs (Black 1994), hence growth rates were pooled for both species when reservoir-wide comparisons were made. Similarly for largemouth bass, we computed the time needed to reach quality (304 mm), preferred (381 mm), and memorable (508 mm) total lengths, and for spotted bass, the time to reach the minimum tournament size (304 mm) and memorable (432 mm) total length (Table 1).

Body condition for black crappie and white crappie was computed from the relative weight (W_r) equations of Neumann and Murphy (1991). Largemouth bass W_r was computed from the equation published by Anderson and Gutreuter (1983). For

TABLE 2.—Summary statistics for annual mean values for limnological variables in Alabama reservoirs.

Variable	Mean	Range	SD	*N*
Surface area (ha)	6,709	443–27,976	7,195	32
Mean depth (m)	7.2	1.6–20.0	4.0	32
Secchi disk transparency (cm)	164	50–467	100	32
Chlorophyll *a* (mg/m^3)	10	2–27	6	32
Retention time (d)	56	1–435	96	32
Total phosphorus (mg/m^3)	36	5–90	23	19
Total nitrogen (mg/m^3)	610	290–1,860	370	18
Total suspended solids (mg/L)	6.9	0.6–20.5	5.3	31

spotted bass, W_r was computed from the standard weight (W_s) equation (C. Guy, Kansas Cooperative Fish and Wildlife Research Unit, unpublished analysis): $\log_{10} W_s = -5.358 + 3.201(\log_{10}\text{TL})$. Relative weights were computed according to structural length groups (Table 1) described by Anderson and Gutreuter (1983). Higher W_r values are generally associated with better physiological condition in fish.

For each reservoir and year, we determined mean monthly (April–September) values for Secchi disk transparency, chlorophyll *a*, total suspended solids, total phosphorus, and total nitrogen, and described relationships among these variables using correlation and regression analyses. In addition, mean monthly retention time (volume/discharge) was computed to determine its influence on chlorophyll-*a* concentrations. Because crappie and black bass are long-lived (7–10 years), their response to water quality would accumulate over the long term. Thus, a mean of average annual chlorophyll-*a* concentration values was computed for each reservoir to compare to fish data. Also, means of annual mean values for the other limnological variables defined earlier were computed to characterize limnological variation among Alabama reservoirs.

Results

Water Quality Relationships in Alabama

Water quality among the major impoundments in Alabama was diverse (Table 2). Water clarity ranged from poor to excellent (0.5–4.7 m). Trophic state varied from oligotrophic to eutrophic, and total phosphorus and total nitrogen varied according to trophic state. Retention time varied over two orders of magnitude.

Higher chlorophyll-*a* concentrations or algal biomass were associated with reduced water clarity for 292 monthly means. A nonlinear inverse relationship (Figure 2) was evident between Secchi disk

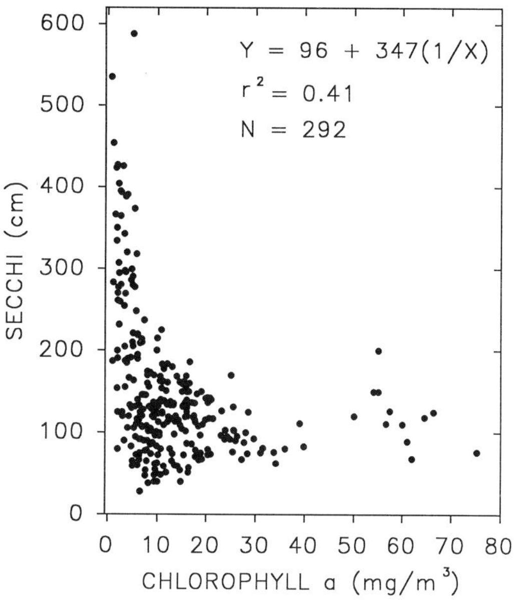

FIGURE 2.—The relationship between mean monthly values for Secchi disk transparency (Y) and chlorophyll *a* (X) in Alabama reservoirs.

transparency (SDT, cm) and chlorophyll-*a* concentrations (CHLA, mg/m^3);

$$\text{SDT} = 96 + 347(1/\text{CHLA}). \tag{1}$$

Chlorophyll *a* explained 41% ($r^2 = 0.41$) of the variation in Secchi disk transparency. Log_{10}-transformed values for total suspended solids (TSS) were also negatively correlated ($r = -0.74, P < 0.01; N = 234$) to Secchi disk transparency for mean monthly data. The multiple regression equation was:

$$\text{SDT} = 256 + 151(1/\text{CHLA}) - 171(\log_{10}\text{TSS}), \tag{2}$$

and this equation explained 64% ($r^2 = 0.64$) of the variation in water clarity. Chlorophyll *a* and total suspended solids were positively correlated ($r = 0.42$, $P < 0.01$; $N = 227$) as phytoplankton algae comprised some of the particulate matter measured in total suspended solids. Thus, some multicollinearity occurred between the two independent variables in equation (2) (multicollinearity can mask the importance of a single independent variable). Nevertheless, total suspended solids explained an additional 23% of the variation in Secchi disk transparency above that of chlorophyll *a*. In addition to phytoplankton, total suspended solids include nonalgal particles and these were also associated with a decline in water clarity.

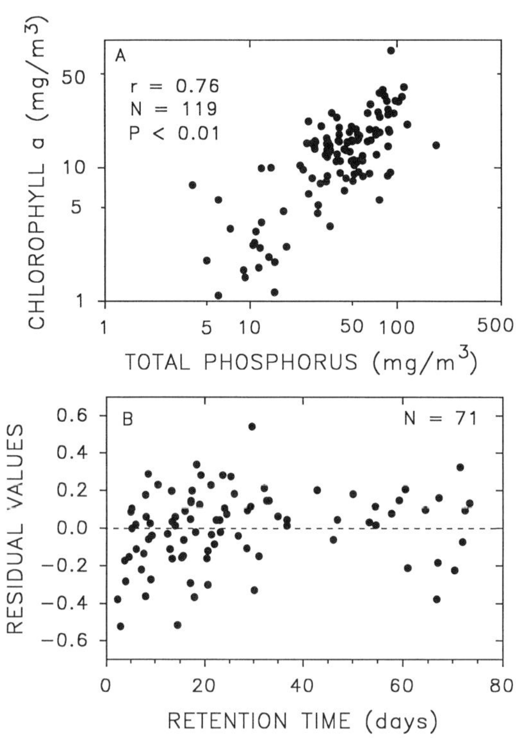

FIGURE 3.—Plot of the total phosphorus–chlorophyll-*a* relationship for mean monthly values (A). Residual values from the $\log_{10}$ chlorophyll *a*-to-$\log_{10}$ total phosphorus regression plotted against retention time for transitional (≤75 d retention time) reservoirs (B).

For a subset of 18 reservoirs that contained both total phosphorus and total nitrogen data were determined simultaneously, total phosphorus was a better correlate ($r = 0.71$, $P < 0.01$) of chlorophyll *a* than total nitrogen ($r = 0.18$, $P = 0.13$). For 118 mean monthly observations among a larger database of 19 reservoirs, total phosphorus (TP) explained 58% of the variation in chlorophyll-*a* concentrations (Figure 3a):

$$\log_{10}\text{CHLA} = -0.26 + 0.82(\log_{10}\text{TP}). \quad (3)$$

We examined chlorophyll-*a* relationships for transitional reservoirs, defined as reservoirs with a retention time (RET) of 75 d or less (Soballe and Kimmel 1987). In this smaller database, retention time and total phosphorus were positive and significant regressors ($P < 0.01$) of chlorophyll *a* ($R^2 = 0.50$):

$$\log_{10}\text{CHLA} = -0.27 + 0.70(\log_{10}\text{TP}) + 0.20(\log_{10}\text{RET}). \quad (4)$$

Retention explained an additional 7% of the variation in algal biomass after 43% of the variation in algal biomass was accounted for by total phosphorus.

For transitional reservoirs, a plot of residual values against retention time for chlorophyll *a* estimated from total phosphorus (from equation 3) indicated an inflection point at about 35 d, and a flattening of the relationship at longer retention times (Figure 3b). This analysis suggested that greater algal production for a given concentration of phosphorus could be expected at retention times of about 35 d or more given that nonalgal turbidity did not limit light and inhibit phytoplankton growth. In these transitional reservoirs, retention time was negatively correlated to total suspended solids ($r = -0.45$, $P < 0.01$), but not correlated to chlorophyll *a* ($r = 0.15$, $P = 0.14$) indicating that concentrations of nonalgal suspended particles were higher in faster flushing reservoirs. Retention time was a significant regressor of chlorophyll *a* in equation (4) only after the contribution of total phosphorus was accounted for.

Accounting for the effects of total phosphorus and retention time, mean depth (Z_x in m) was a significant negative regressor ($P < 0.01$) of chlorophyll *a* and adding this term to equation (4) we computed:

$$\log_{10}\text{CHLA} = 0.10 + 0.57(\log_{10}\text{TP}) + 0.22(\log_{10}\text{RET}) - 0.27(\log_{10}Z_x) \quad (5)$$

The coefficient of determination of equation (5) was 53%; thus, mean depth contributed an additional 3% to the total variation above that of total phosphorus and retention time. Although retention time and mean depth were weakly correlated ($r = -0.36$, $P < 0.01$), variance inflation factors and condition numbers were within acceptable ranges and multicollinearity was not too severe (Montgomery and Peck 1982). Thus, we conclude that for a given level of phosphorus in transitional reservoirs, algal biomass was lower in deeper, faster flushing reservoirs than in shallower reservoirs with longer retention time.

Water Quality–Sport Fishing Quality Interactions

Crappie catch and harvest rates averaged 1.05 and 0.60 fish/h in eutrophic reservoirs and were similar (*t*-test < 0.6, $P > 0.5$) to rates of 1.24 and 0.71 fish/h, respectively, in oligomesotrophic reservoirs. No relationships were evident between chlorophyll-*a* concentration and angler catch rates nor harvest rates of crappies (Figure 4). In addition, black bass catch rates were not correlated to chlo-

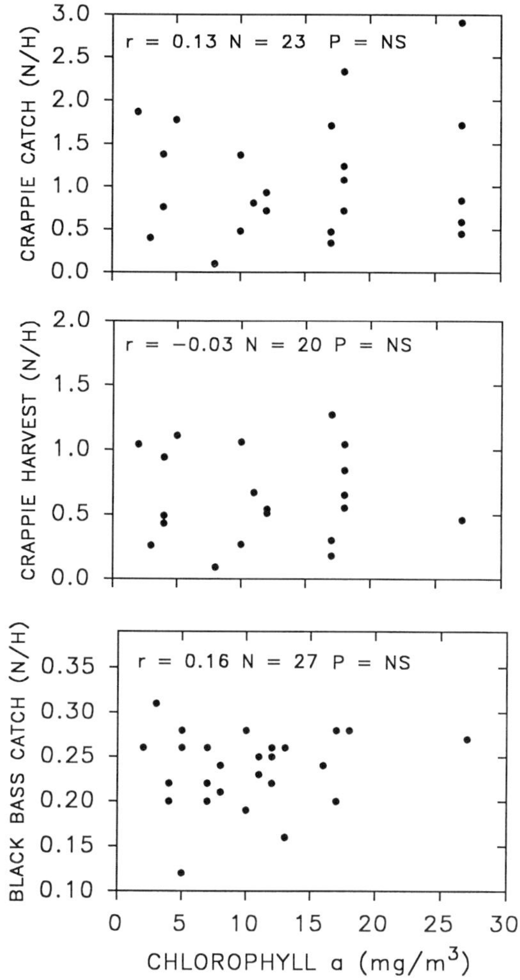

FIGURE 4.—Bivariate plots of crappie catch rates, crappie harvest rates, and black bass catch rates against mean of annual mean chlorophyll-*a* concentrations.

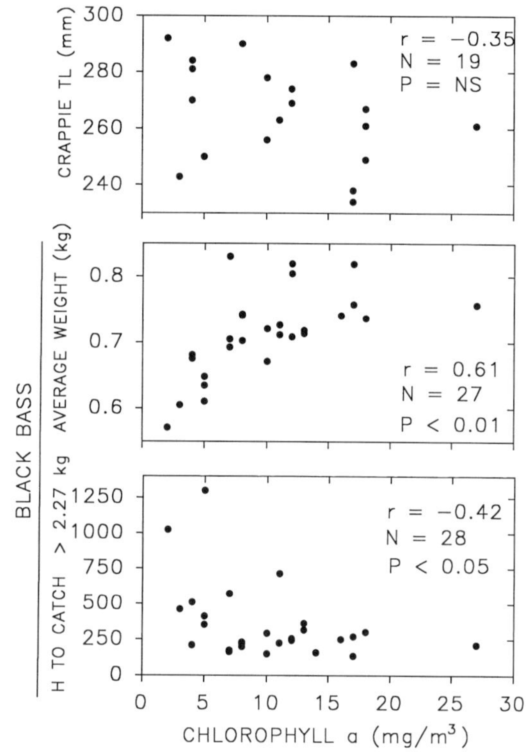

FIGURE 5.—Plots of total length of crappie (A) average weight (B) and the effort to catch (C) a black bass 2.27 kg or more by tournament anglers versus mean of annual mean chlorophyll-*a* concentrations.

rophyll *a* (Figure 4). Catch rates of black bass averaged 0.24 fish/h in eutrophic reservoirs and were similar (t-test = 0.41, $P > 0.5$) to the average 0.23 fish/h caught in oligomesotrophic reservoirs.

Size of crappies harvested by anglers was not related to chlorophyll-*a* concentrations ($r = -0.35$, $P > 0.10$; Figure 5a). Mean lengths of crappies harvested was 263 mm in eutrophic reservoirs and was similar ($t = -0.78$, $P = 0.4$) to 270 mm in oligomesotrophic reservoirs.

Although catch rates of black bass did not differ between trophic states of reservoirs, mean weights were higher and the amount of angler effort to catch a large (≥2.27 kg) black bass was less in eutrophic than in oligomesotrophic reservoirs. The mean weight of black bass longer than 304 mm TL caught in tournaments was 0.74 kg and was significantly greater (t-test = 3.03, $P < 0.01$) than 0.67 kg in oligomesotrophic reservoirs. Similarly, the amount of effort to catch a large black bass was 518 h in oligomesotrophic reservoirs, about twice as great ($t = 2.08$, $P = 0.05$) as 266 h required in eutrophic reservoirs. Mean weight of black bass caught and the amount of effort to catch a large black bass were correlated to chlorophyll *a* (Figure 5). However, in the eutrophic range, a flattening of these relationships were evident as chlorophyll *a* exceeded 8 mg/m³.

Crappie and Black Bass Growth and Condition

In eutrophic reservoirs, crappie reached 229 mm in 2.24 years and 254 mm in 2.73 years; these growing periods were significantly ($t > 2.5$, $P < 0.05$) less than the 2.73 years and 3.34 years for crappie to reach these same sizes, respectively, in oligomesotrophic reservoirs. However, faster crappie growth rates ($P > 0.5$) were not evident when chlorophyll-*a* levels were equal to or greater than 8 mg/m^3.

No relations were evident between W_r values for the three size groups of black crappie and chlorophyll *a* ($r = -0.04$–0.43, P > 0.10). Also, W_r values were similar between eutrophic and oligomesotrophic reservoirs. For white crappie, W_r for stock, quality, and preferred size groups increased over the entire range of chlorophyll-*a* levels ($r = 0.41$–0.75, $P < 0.10$).

The time for largemouth bass to reach 304, 381, and 508 mm TL was weakly, but negatively correlated to chlorophyll *a* (Figure 6), indicating faster growth occurred as trophic state increased. The time required for largemouth bass to reach 304, 381, and 508 mm TL was 2.37, 3.63, and 7.25 years, respectively, in eutrophic reservoirs which was less ($t = 1.67$–3.38, $P < 0.10$) than 2.63, 4.07, and 7.80 years, respectively, in oligomesotrophic reservoirs. Like for crappie, largemouth bass growth rates did not increase in eutrophic reservoirs at higher chlorophyll-*a* concentrations ($r = -0.18$–-0.22, $P > 0.4$). For the entire range of data, the asymptotic models did not significantly improve fit compared to the linear models.

Relative weights of three size groups of largemouth bass were better related than growth rates to chlorophyll-*a* concentrations (Figure 7). At chlorophyll-*a* concentrations greater than 7 mg/m^3, increases in condition were not evident for stock- and quality-size fish. Second-degree polynomial equation terms were significant ($P < 0.05$) for asymptotic relationships, but only for quality-size fish did the fit of W_r data improve significantly ($P < 0.10$) compared to the simple linear regression. Thus, higher chlorophyll-*a* concentrations within the eutrophic range did not confer higher W_r values. A weak correlation ($r = 0.43$, $P < 0.10$) was evident between W_r for preferred-size fish and chlorophyll *a* in eutrophic reservoirs.

Spotted bass grew faster and displayed better body condition in eutrophic than in oligomesotrophic reservoirs. Spotted bass reached 304 mm in 2.43 years in eutrophic reservoirs and 2.77 years in oligomesotrophic reservoirs ($t = 2.14$, $P < 0.10$); they reached 432 mm in 4.74 years in eutrophic reservoirs and 5.58 years in oligomesotrophic reservoirs ($t = 3.97$, $P < 0.01$). Relative weight values for stock-, quality-, and preferred-size groups were all higher ($t > 3.9$, $P < 0.01$) in eutrophic than in oligomesotrophic reservoirs. The relationship between increasing algal biomass and spotted bass growth and condition could not be adequately inferred because sample size was low (i.e., only 11 of the 32 major impoundments in Alabama contained substantial spotted bass fisheries and only 5 were in the eutrophic range).

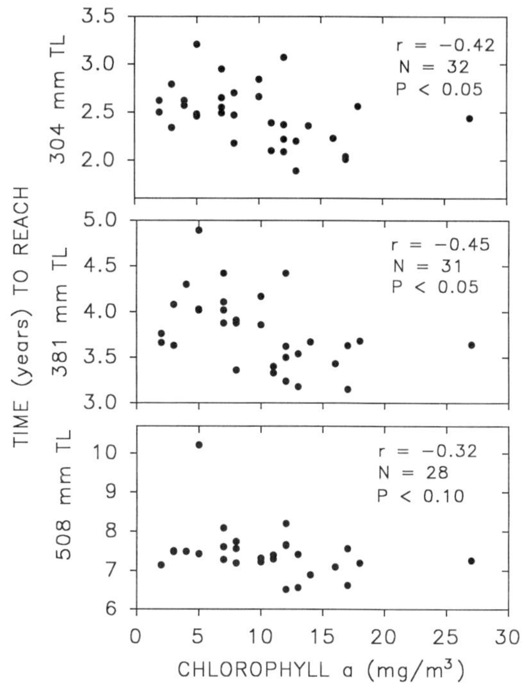

FIGURE 6.—Time in years for largemouth bass to obtain 304, 381, and 508 mm TL versus chlorophyll *a*.

Implications for Multiple Recreational Use

In Alabama, a reduction in chlorophyll *a* will increase the probability of greater water clarity and improve the aesthetic value among all recreational users. The National Academy of Sciences and National Academy of Engineering (1973) established a minimum criterion of 120 cm water clarity for safe swimming. If we assume that 120 cm visibility in the water is acceptable to most recreational users of Alabama reservoirs, then from predictive equation (1) this visibility equates to a chlorophyll-*a* concentration of 14 mg/m^3, which is still eutrophic. Attempts to reduce point or nonpoint nutrient sources will improve water clarity, if a nutrient–algal biomass relationship exists.

Algal biomass appeared as a slightly more important determinant of Secchi disk transparency in Alabama reservoirs compared to reservoirs located primarily throughout the southern and midwestern

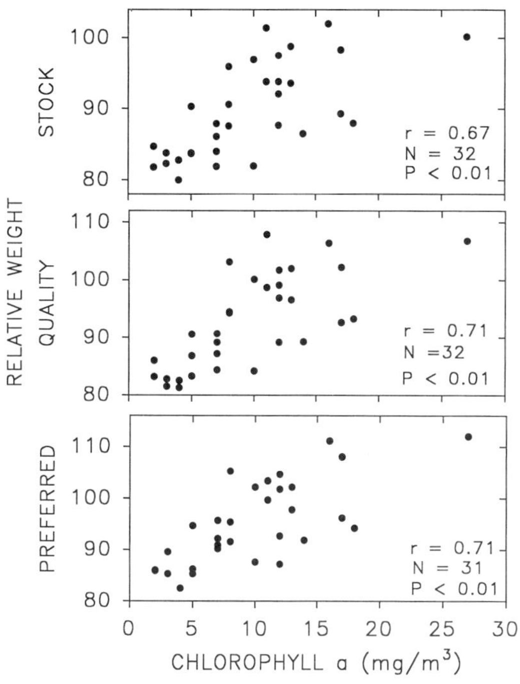

FIGURE 7.—Relative weight of stock, quality, and preferred sizes of largemouth bass versus chlorophyll *a*.

United States. In these regions, algal levels explained 12 to 42% of the variation in Secchi disk transparency (Canfield and Bachmann 1981; Hoyer and Jones 1983; Soballe and Kimmel 1987; Jones and Knowlton 1993) compared to 41% in Alabama. Much of Alabama is forested and agriculture is not as predominant as in the midwestern United States. Alluvial and clay soils are more common in these other regions, which likely results in waters with higher levels and persistence of nonalgal particulates.

The amount of chlorophyll-*a* variation explained by phosphorus in Alabama was similar to other regions. Total phosphorus accounted for 32 to 74% of the variation in chlorophyll *a* in other U.S. reservoirs (Canfield and Bachmann 1981; Hoyer and Jones 1983; Soballe and Kimmel 1987; Jones and Knowlton 1993) compared to 58% in Alabama. Poor fits to the chlorophyll *a*–phosphorus relationship in other regions were primarily due to high levels of inorganic suspended solids, which contained higher phosphorus concentrations but which also reduced water transparency, limiting the amount of light available for algal growth.

When allocating funds, reservoir managers need to consider retention time and nonalgal turbidity concentrations before embarking on any point or nonpoint nutrient reduction program. Our analyses indicated that a greater probability of reduction in algal biomass from nutrient control will occur in shallow reservoirs with retentions of 35 d or more than in deep, fast-flushing systems. Soballe and Kimmel (1987) also found total phosphorus, retention time, and mean depth explained about half the variation in chlorophyll-*a* concentrations in transitional reservoirs in the United States. Although empirically derived estimates of critical retention time for full algal production vary, previous reservoir investigations have consistently found this time was less than 60–100 d (reviewed by Soballe and Kimmel 1987). In fast-flushing reservoirs, the full production of phytoplankton algae is not realized because of time limitation. In addition, levels of nonalgal turbidity are likely higher in reservoirs with short-resident time, limiting light to algal production. Both these phenomena were indirectly observed in our analyses and in U.S. Army Corps of Engineers reservoirs in the southeastern United States (Kennedy and Gaugush 1988).

Variation in chlorophyll-*a* concentrations from low to highly eutrophic did not appear detrimental to quality crappie and black bass fisheries in Alabama. We found angler catch rates of crappie and black bass were similar for reservoirs ranging from oligomesotrophic to eutrophic. However, size of fish caught, growth rates, and condition were generally higher in eutrophic than in oligomesotrophic reservoirs. Thus, a reduction from eutrophic to oligomesotrophic will likely result in anglers catching smaller fish.

Increased biomass of plankton algae in reservoirs raises total fish standing stock via the planktivore (i.e., shad *Dorosoma* spp.) community, but the efficiency of transfer of this food (energy) source to piscivorous sport fish appears to decline in reservoirs of higher trophic status. In support of this, Yurk and Ney (1989) found that total phosphorus explained 84% of the variation in planktivore biomass, but explained only 51% of the piscivorous sport fish biomass in southern Appalachian reservoirs. In Smith Mountain Lake, Virginia, total phosphorus was positively correlated to biomass of gizzard shad *D. cepedianum*, but not to biomass of four piscivorous sport fish (Yurk and Ney 1989). Jenkins's (1982) parabolic plots of the morphoedaphic index (MEI) versus total fish and sport fish biomass showed decreasing efficiency of trophic transfer to sport fish as the percent composition of sport fish decline at higher MEI values. Increasing trophic state was associated with an increase in rough fish biomass (primarily gizzard shad), and a decline in sport fish abundance in Florida lakes (Kautz 1982).

Ney (1996) suggested that sport fish biomass would probably peak at total phosphorus concentrations of about 100 mg/m^3. Equation (3) suggests that about 50 mg/m^3 total phosphorus results in about 14 mg/m^3 chlorophyll-*a* concentration, which is at the low range of eutrophy. This level of algal biomass increases the probability of water clarity equal to or greater than 120 cm, but does not appear to sacrifice black bass and crappie fisheries in Alabama.

Contrary to our results, which indicated an asymptotic relationship between trophic state and crappie and black bass fisheries and population characteristics, other investigators have found fish yield and production increased linearly with algal biomass, primary production, total phosphorus, and the MEI (Melack 1976; Oglesby 1977; Matuszek 1978; Hanson and Leggett 1982; Jones and Hoyer 1982). However, our study was different because we examined only two major fish groups. Total recreational and commercial yield data were not collected in Alabama. Our data included only reservoirs, whereas lake data, reservoir data, or both were used in other studies. Trophic state response by fisheries for other species in Alabama may be different than for crappie and black bass fisheries.

Reservoir managers, limnologists, and fishery biologists should recognize and attempt to assess the impact of any planned oligotrophication project. Limnological investigations similar to the U.S. Environmental Protection Agency Diagnostic-Feasibility Studies should be conducted to determine factors that limit algal production in reservoirs. For example, nutrient reduction in fast-flushing reservoirs with high nonalgal turbidity may not result in greater water clarity, which is an important goal of any oligotrophication project. Currently, attempts are being made to reduce phosphorus loading in West Point Reservoir in Alabama–Georgia; the goal is to maintain an average chlorophyll-*a* concentration of 15–20 mg/m^3 (Bayne et al. 1994b). Based on our analyses, this level of algal biomass should continue to support desirable crappie and black bass fisheries. However, an extensive fishery for hybrid striped bass *Morone saxatilis* × *M. chrysops* exists and the impact of oligotrophication on this fishery is unknown, but shad populations are expected to decline. In Smith Mountain Lake, Virginia, management recommendations following oligotrophication included a reduction in the stocking rates of nonreproducing predatory game fish (striped bass and walleye *Stizostedion vitreum*; Ney et al. 1990). Thus, oligotrophication and subsequent reduction in prey fish may be detrimental to the success of stocking predator sport fish, other than crappie and black bass, and this impact should be assessed for each reservoir.

Our data indicated that reduction of chlorophyll-*a* concentrations in eutrophic reservoirs to 10–15 mg/m^3 will likely result in greater water clarity and improved aesthetics for other recreational users without adverse effects on quality black bass and crappie fisheries. If a shift in trophic state occurs from eutrophic to oligomesotrophic, catch rates of crappies and black bass will likely not change, but in some instances smaller and less robust fish will be caught, and the effort needed to catch a large-size black bass will increase. Thus, fisheries management approaches may need to be modified, and new expectations for the fishery should be communicated to the public.

Acknowledgments

We thank the many biologists from the Alabama Department of Conservation and Natural Resources and Auburn University who collected and processed much of the large data set presented in this paper. Alabama Power Company, the U.S. Army Corps of Engineers, the Tennessee Valley Authority, and the Alabama Department of Environmental Management provided reservoir hydrologic and water quality data. We thank the black bass fishing clubs of Alabama for recording and providing tournament data. W. Davies, R. Pasch, J. Putman, S. Szedlmayer, and two anonymous reviewers assisted to improve this paper. This study was funded, in part, through Federal Aid in Sport Fish Restoration Projects F-38 and F-40. This paper is journal number 8-944917 of the Alabama Agricultural Experiment Station.

References

Anderson, R. O., and S. J. Gutreuter. 1983. Length, weight, and associated structural indices. Pages 283–300 *in* L. A. Nielsen and D. L. Johnson, editors. Fisheries techniques. American Fisheries Society, Bethesda, Maryland.

APHA (American Public Health Association). 1985. Standard methods for the examination of water and wastewater, 16th edition. APHA, Washington, DC.

Axler, R. L., L. Paulson, P. Vaux, P. Sollberger, and D. H. Baepler. 1988. Fish aid—the Lake Mead fertilization project. Lake and Reservoir Management 4:125–135.

Bayne, D. R., M. J. Maceina, and W. C. Reeves. 1994a. Zooplankton, fish and sport fishing quality among four Alabama and Georgia reservoirs of varying trophic status. Lake and Reservoir Management 8:153–163.

Bayne, D. R., W. C. Seesock, C. E. Webber, and J. A. McGuire. 1990. Cultural eutrophication of West Point Lake—a 10-year study. Hydrobiologia 199:143–156.

Bayne, D. R., and six coauthors. 1994b. West Point Lake, phase I diagnostic/feasibility study. Final report to Georgia Department of Natural Resources, Atlanta, and Alabama Department of Environmental Management, Montgomery.

Black, W. P. 1994. Evaluation of factors affecting growth of black crappie and white crappie in Alabama. Master's thesis. Auburn University, Auburn, Alabama.

Canfield, D. E., Jr., and R. W. Bachmann. 1981. Prediction of total phosphorus concentrations, chlorophyll *a*, and Secchi depths in natural and artificial lakes. Canadian Journal of Fisheries and Aquatic Sciences 38:414–423.

Forsberg, C., and S. O. Ryding. 1980. Eutrophication parameters and trophic state indices in 30 Swedish waste-receiving lakes. Archives fur Hydrobiologie 89: 189–207.

Hanson, J. M., and W. C. Leggett. 1982. Empirical prediction of fish biomass and yield. Canadian Journal of Fisheries and Aquatic Sciences 39:257–263.

Hendricks, A. S., M. J. Maceina, and W. C. Reeves. 1995. Abiotic and biotic factors related to black bass fishing quality in Alabama. Lake and Reservoir Management 11:47–56.

Hoyer, M. V., and J. R. Jones. 1983. Factors affecting the relation between phosphorus and chlorophyll *a* in midwestern reservoirs. Canadian Journal of Fisheries and Aquatic Sciences 40:192–199.

Hoyer, M. V., J. V. Shireman, and M. J. Maceina. 1985. Use of otoliths to determine age and growth of largemouth bass in Florida. Transactions of the American Fisheries Society 114:307–309.

Jenkins, R. M. 1982. The morphoedaphic index and reservoir fish production. Transactions of the American Fisheries Society 111:133–140.

Jones, J. R., and M. V. Hoyer. 1982. Sportfish harvest predicted by summer chlorophyll-*a* concentrations in midwestern lakes and reservoirs. Transactions of the American Fisheries Society 111:176–179.

Jones, J. R., and M. K. Knowlton. 1993. Limnology of Missouri reservoirs: an analysis of regional patterns. Lake and Reservoir Management 8:17–30.

Kautz, R. S. 1982. Effects of eutrophication on the fish communities of Florida lakes. Proceedings of the Annual Conference Southeastern Association of Fish and Wildlife Agencies 34(1980):67–80.

Kennedy, R. H., and R. F. Gaugush. 1988. Assessment of water quality in Corps of Engineers reservoirs. Lake and Reservoir Management 4:253–260.

Maceina, M. J., and R. K. Betsill. 1987. Verification and use of whole otoliths to age white crappie. Pages 267–278 *in* R. C. Summerfelt and G. E. Hall, editors. Age and growth of fish. Iowa State University Press, Ames.

Matuszek, J. E. 1978. Empirical predictions of fish yields of large North American lakes. Transactions of the American Fisheries Society 107:385–394.

McQueen, D. J., J. R. Post, and E. L. Mills. 1986. Trophic relationships in freshwater pelagic ecosystems. Canadian Journal of Fisheries and Aquatic Sciences 43:1571–1581.

Melack, J. M. 1976. Primary productivity and fish yields in tropical lakes. Transactions of the American Fisheries Society 105:575–580.

Montgomery, D. C., and E. A. Peck. 1982. Introduction to linear regression analysis. Wiley, New York.

National Academy of Sciences and National Academy of Engineering. 1973. Water quality criteria. U.S. Government Printing Office, Washington, DC.

Neumann, R. M, and B. R. Murphy. 1991. Evaluation of the relative weight (W_r) index for assessment of white crappie and black crappie populations. North American Journal of Fisheries Management 11:543–555.

Ney, J. J. 1996. Oligotrophication and its discontents: effects of reduced nutrient loading on reservoir fisheries. Pages 285–295 *in* L. E. Miranda and D. R. DeVries, editors. Multidimensional approaches to reservoir fisheries management. American Fisheries Society Symposium 16.

Ney, J. J., C. M. Moore, M. S. Tisa, J. J. Yurk, and R. J. Neves. 1990. Factors affecting the sport fishery in a multiple-use Virginia reservoir. Lake and Reservoir Management 6:21–32.

Oglesby, R. T. 1977. Relationship of fish yield to lake phytoplankton standing crop, production, and morphoedaphic factors. Journal of the Fisheries Research Board of Canada 34:2271–2279.

Smeltzer, E., and S. A. Heiskary. 1990. Analysis and applications of lake user data. Lake and Reservoir Management 6:109–118.

Soballe, D. M., and B. L. Kimmel. 1987. A large-scale comparison of factors influencing phytoplankton abundance in rivers, lakes, and impoundments. Ecology 68:1943–1954.

von Bertalanffy, L. 1938. A quantitative theory of organic growth. Human Biology 10:181–213.

Yurk, J. J., and J. J. Ney. 1989. Phosphorus–fish community biomass relationships in southern Appalachian reservoirs: can lakes be too clean for fish? Lake and Reservoir Management 5:83–90.

American Fisheries Society Symposium 16:306–314, 1996

The Compleat Bass: Habitat Use Patterns of all Stages of the Life Cycle of Largemouth Bass

CYNTHIA ANNETT

University of Kansas, Department of Systematics and Ecology
Lawrence, Kansas 66045-2106, USA

JODEE HUNT[1]

University of Kansas, Department of Systematics and Ecology
Lawrence, Kansas 66045-2106, USA

ERIC D. DIBBLE[2]

U.S. Army Corps of Engineers, Waterways Experiment Station
Vicksburg, Mississippi 39180, USA

Abstract.—Predicting how changes in habitat availability will affect the population dynamics of managed warmwater fishes is difficult, in part because previous studies have used large spatial scales to categorize habitat and assess fish populations. However, the mechanisms that affect reproductive success or survival occur at the scale of individual fish and single microhabitat patches, a scale that is often lacking in the sport fish literature. Using underwater surveys, we investigated how swim-up fry, juvenile, nonnesting adult, and nesting adult largemouth bass *Micropterus salmoides* use qualitatively different microhabitat patches in the littoral zone of a small impoundment. Divers surveyed 1,368 microhabitat patches in 144 belt transects, where 549 nonnesting adults and 345 juveniles were observed. Physical characteristics of 270 largemouth bass nests also were measured, and observations were made on habitat use of 31 schools of free-swimming fry. Juveniles preferred complex physical structure, especially patches with combinations of leaf packs, aquatic vegetation, and woody debris, and avoided patches lacking structure. Nonnesting adults were strongly associated with patches containing vegetation, and were the only members of the population found regularly in areas lacking physical structure. Nonnesting adults and juveniles were more common than expected in patches containing silt. Nesting males used habitat patches differently than either juveniles or nonnesting adults, and occurred most frequently near large, simple, woody structure, and only rarely near complex structure or on sites lacking physical structure. Nesting males seldom used patches containing silt and preferred gravel or gravel and cobble substratum. Immobile embryos and weak-swimming fry were constrained by the habitat choices of nesting adults. The nonrandom, life-stage specific nature of largemouth bass distribution suggests that managers should assess and manipulate the abundance and interspersion of qualitatively different habitat patches, paying attention to the fact that a single kind of habitat manipulation (e.g., addition of artificial reefs composed of complex structure) may be inadequate for the needs of all stages.

Black bass *Micropterus* spp. use a wide variety of habitats in the littoral zone of warmwater reservoirs. These habitats contain patches of qualitatively different physical structure that vary in spatial complexity. The composition (scale and complexity of physical structure within individual patches), size (length, area, or volume of a patch), and arrangement (degree of similarity of adjacent patches) of microhabitat patches can have important impacts on population dynamics. Vegetation in the littoral zone, for example, provides important refugia and foraging sites for juvenile fish (Werner et al. 1977; Crowder and Cooper 1982). Nesting bass show strong habitat selectivity in lakes (Hunt 1995), and quality of nesting habitat may affect survival of embryos and fry (Raffetto et al. 1990). High densities of juvenile fishes, which serve both as prey for adult bass and predators on bass embryos and fry, concentrate in nearshore habitats, especially in microhabitats containing brush and vegetation (Keast et al. 1978; Rainwater and Houser 1982; Johnson and Lynch 1992). These studies all indicate that the specific nature, as well as availability, of physical structure in littoral habitats influences the behavior and survival of bass.

Although availability of structurally complex habitats is known to affect year-class strength of bass in reservoirs (Aggus and Elliott 1975; Miranda et al.

---

[1]Current address: Grand Valley State University, Department of Biology, Allendale, Michigan 49401, USA.

[2]Current address: Mississippi State University, Department of Wildlife and Fisheries, Mississippi State, Mississippi 39762, USA.

1984; Means and Miranda 1991), predicting how changes in habitat availability will influence population dynamics is difficult because existing studies of warmwater impoundments assess habitat effects on too large a spatial scale. For example, cove rotenone samples (Davies and Shelton 1983), boat electrofishing (Reynolds 1983), and shoreline seining collect fish over an area that incorporates more than one microhabitat patch. Combining qualitatively different microhabitats into a single sample obliterates some of the complexity inherent in aquatic systems. Using these techniques it is impossible to identify whether a particular fish occurred in association with a single patch, or whether the fish was driven into a patch by the approach of a boat (c.f., Dibble 1991), seine, or diffusing toxicants. These techniques also do not allow determination of what the fish was doing in the patch (e.g., guarding a nest or school of young, foraging, or resting). Data on catch-per-unit effort from such studies are typically used to estimate population size over a large area (Davies and Shelton 1983; Reynolds 1983), but these do not provide information on the density of individuals at smaller scales such as single microhabitat patches.

Data concerning how fish in warmwater lakes and impoundments use qualitatively different microhabitats are rare, making it difficult for fisheries biologists to manipulate habitat to achieve management goals (Summerfelt 1993). Detailed studies of the ways fish use microhabitat patches are the first step in determining the consequences of habitat choice for survival, growth, and reproductive success, all components of population dynamics. To meet this need, a different perspective is required.

We view the littoral zone of a clear warmwater reservoir as a landscape, and investigate which patches within this landscape are used by largemouth bass *Micropterus salmoides* during each stage of their life cycle. We use diver transects (Helfman 1983) to observe directly how habitat use differs among swim-up fry, juveniles, nonnesting adults, and nesting males within the same study area. Such data are necessary to determine the degree of landscape complexity required to accommodate individuals differing in developmental stage or reproductive condition, and allows managers to identify critical habitats that may restrict population growth.

Methods

Study Site

Wedington Reservoir (Washington Co., Arkansas), is a 41-ha impoundment within the Illinois River basin. Its watershed (approximately 1,050 ha) is wooded and sheltered from wind. The reservoir possesses a distinct littoral zone containing a variety of substrata and physical structures (Table 1). Water clarity, measured with a Secchi disk, ranged from 2 to 4 m. Water levels did not fluctuate within seasons during this study. Habitat and location of bass and nests were measured throughout the lake; however, a 2.2-ha cove was surveyed intensively for this research (Dibble 1993; Hunt 1995).

TABLE 1.—Categories of habitat recorded during surveys of fish distribution and nest site associations. Physical structure categories are similar to those designated by Gorman and Karr (1978) and Todd and Rabeni (1989) and substrata categories are designated according to a modified Wentworth scale (Bovee and Milhouse 1978).

Category	Description or particle diameter
Habitat structure	
Horizontal log	Logs, felled trees, large branches (>150 mm diameter)
Vertical log	Stumps, inundated trees (standing)
Woody debris	Sticks, brush, small branches (<150 mm diameter)
Leaf packs	Accumulations of leaves (>0.5 m)
Aquatic vegetation	Attached vascular hydrophytes
Boulder or rock piles	Boulders, rocks, concrete blocks
Substrata	
Silt	0.1–5.0 mm
Gravel	5.0–65.0 mm
Cobble	65.0–256.0 mm
Boulder	>256.0 mm

Quantifying Abundances of Fish and Nearshore Habitat

All data in this study were collected by divers, using either snorkeling equipment or scuba, depending upon water depth (for descriptions of methods see Northcote and Wilkie 1963; Keast and Harker 1977; Helfman 1983; Dibble 1991, 1993). Largemouth bass density and characteristics of the physical habitat were measured simultaneously by two divers on randomly placed belt transects that were 2 m wide by 100 m long (Dibble 1991, 1993). Transects were aligned parallel to the shoreline along consistent depths from 0.5 to 3.0 m. To ensure adequate visibility, we limited dives to days with a minimum Secchi disk transparency reading of 2 m.

Distribution and abundance of juvenile centrarchids and nonnesting adult largemouth bass were measured by the first of the two divers. This diver swam slightly ahead of the second and recorded on underwater checksheets the species, number, and length category (TL) of all fish observed, as well as the depth, substratum, and physical structure of the habitat. We determined fish length by comparison to reference marks on the diver's clipboard. We defined juvenile largemouth bass as those shorter than 150 mm. Age-1 largemouth bass in Arkansas range from 124 to 262 mm TL (Carlander 1977); therefore, our cutoff point is highly conservative and our "juvenile" category should not include any reproductively mature individuals. Largemouth bass 150 mm or longer that were not defending or remaining near nest sites were designated as nonnesting adults.

The second diver swam with a 100-m fiberglass tape and measured distances along the transect where each microhabitat patch began and ended. Combinations of different physical structure and substrata intercepted along the belt transects were considered patches. Patch size was measured as the linear distance over which the patch was encountered. Patches lacking physical structure were categorized according to substratum alone. For example, on a single 100-m transect, divers might encounter a first patch composed of gravel alone for the interval 0–4 m, a second patch composed of silt and aquatic vegetation (4–6.5 m), followed by a third patch composed of gravel and woody debris (6.5–8.7 m). Divers would then quantify the remainder of the transect in a similar manner.

Data on nonnesting adults and juveniles were collected June–October 1990, March–October 1991, and March–August 1992. Details on seasonality of habitat use can be found in Dibble (1993).

Quantifying Habitat of Nesting Adults and Dispersing Fry

We searched the littoral zone for nests using transects placed on shorelines facing each direction and incorporating shallow to steep slopes. Divers intensively surveyed the littoral zone at 0–3 m depth for nests. When nest sites were detected, they were marked with stakes or weighted flags bearing unique numbers. We then recorded depth, substratum, physical structure within 1 m of the nest, and the distance to the nearest neighboring nest. Nest searches were conducted from early April, before any spawning occurred, until mid-May, when schools of fry had dispersed and no new nests were observed. Nest density was calculated by combining data for all nests on each transect, regardless of date of initiation.

During nest searches, divers monitored presence and developmental stage of broods, as embryos (unhatched eggs), yolk sac fry (hatched fry with muscle contractions), or swim-up fry (able to swim above the substratum), as well as presence or absence of a guarding male. When a school of swim-up fry was encountered, we recorded its position above the substratum, association with physical structure, location relative to the nearest nest, and presence or absence of a guarding male.

TABLE 2.—Mean patch length recorded during diver surveys by category, with standard deviation.

Structure category	Mean length (m)	Standard deviation
Leaf packs	6.0	4.8
Vertical log	2.0	1.1
Boulder or rock piles	5.5	2.8
Aquatic vegetation	4.2	3.6
Woody debris	4.7	3.4
Horizontal log	3.5	1.7

Results

Transect Surveys: Habitat Patches

We surveyed 144 transects (15,598 m^2) in Wedington Reservoir, and categorized and measured 1,368 habitat patches. Of these patches, 62% contained one or more varieties of physical structure, and 38% contained no physical structure. Wedington Reservoir's littoral zone is fragmented into many small patches (Table 2). The mean length of microhabitat patches, by category, ranged only from 2.0 to 6.0 m (mean = 3.9 m for all categories combined). The longest patch recorded was 12.2 m.

Patches containing no structure were the most common (38%, Table 3), followed by patches containing aquatic vegetation (10%), woody debris (9%), leaf packs (9%), horizontal logs (7%), and horizontal logs combined with woody debris (5%). No other structure category accounted for more than 3% of the patches. Patches with substrata of gravel, and gravel mixed with other particles, accounted for 66% of the total (Table 4).

Transect Surveys: Juveniles and Nonnesting Adults

Three hundred patches (24% of all patches surveyed) contained largemouth bass. Of the patches occupied by largemouth bass, 87% contained one or more varieties of physical structure, while only 13% (39 patches) contained no structure. This is significantly different than expected if largemouth bass were distributed randomly between patches containing structure and patches lacking structure (chi-square test, $P < 0.001$).

We observed 549 nonnesting adult and 345 juve-

TABLE 3.—Availability of physical structure patch types (%) and associations of juvenile, nonnesting, and nesting adult largemouth bass *Micropterus salmoides* with structure types in Wedington Reservoir, Washington Co., Arkansas. Data are expressed as percentages of total patches, individuals, or nest sites sampled. Categories containing more than two structures were rare and are not included. Categories containing two structure types that were rarely used (<2%) by any life stage are not included. The three life stages differed significantly in their use of patches (chi-square test for homogeneity based on numbers of individuals, $P < 0.001$). Structure categories are defined in Table 1.

Structure category	Available habitat (N = 1,368 patches)	Juveniles (N = 345)	Nonnesting adults (N = 549)	Nesting adults (N = 270)
Horizontal log	7	8	7	46
Vertical log	2	0	1	4
Boulder or rock piles	2	8	<1	9
Aquatic vegetation	10	24	30	4
Woody debris	9	13	10	7
Leaf packs	9	11	4	0
Horizontal log and vertical log	1	0	0	3
Horizontal log and boulder or rock piles	1	0	<1	3
Horizontal log and aquatic vegetation	1	0	2	2
Horizontal log and woody debris	5	4	6	10
Aquatic vegetation and woody debris	1	3	2	2
Aquatic vegetation and leaf packs	2	6	2	0
Leaf packs and woody debris	3	11	2	0
No structure	38	0	19	6
Other combinations	9	12	15	5

nile largemouth bass. Nonnesting adult largemouth bass were most frequently observed in association with aquatic vegetation (30% of observed individuals; Table 3). Patches without structure were the second most commonly used (19% of observed nonnesting adults), followed by patches containing woody debris (10%), horizontal logs (7%), horizontal logs combined with woody debris (6%), and leaf packs (4%). Fewer than 3% of the nonnesting adults were found in each of the other patch categories.

If both structure and substratum are considered simultaneously, patches characterized by aquatic vegetation alone as structure and silt as substratum had the highest association of nonnesting adults of any patch category (17%). Aquatic vegetation with gravel and cobble substratum was the second most common variety of patch used by adults (8% of individuals).

Juvenile largemouth bass were most commonly observed in patches with complex physical structure, with 72% of individuals in patches containing aquatic vegetation, leaf packs, or woody debris, either alone or in combination with other varieties of physical structure. Aquatic vegetation was the single most commonly used category (24% of observed individuals; Table 3). Juveniles also used simple structure, such as boulder or rock piles (8%) and horizontal logs (8%). No juveniles were observed in patches lacking structure.

If both substratum and physical structure are considered simultaneously, patches of aquatic vegetation on silt substratum were used most often by juveniles (16% of observed individuals). Juveniles

TABLE 4.—Comparison of substrate availability to association of largemouth bass *Micropterus salmoides* with substrate categories in Wedington Reservoir, Washington Co., Arkansas. Data are expressed as percentages of total patches, individuals, or nest sites sampled. The three life stages differed significantly in patch use (chi-square test for homogeneity based on number of individuals, $P < 0.001$).

Substratum	Percent availability (N = 1,368)	Juveniles (N = 345)	Nonnesting adults (N = 548)	Nesting adults (N = 270)
Silt	15	26	26	3
Silt and gravel	14	9	13	5
Gravel	8	13	7	52
Silt, gravel, and cobble	3	<1	4	0
Gravel and cobble	41	35	41	34
Cobble	8	4	3	2
Boulder	<1	0	<1	3
Other	11	13	7	2

were also relatively common in patches of woody debris overlying gravel and cobble substratum (10% of individuals), boulder patches containing cobbles (8%), and aquatic vegetation with cobbles (6%).

Both adults and juveniles were observed in silty microhabitats more often than expected by habitat availability (Table 4; chi-square test, $P < 0.001$). All other substrata were used in a similar fashion by nonnesting adults and juveniles and use did not deviate from availability in the habitat. Regardless of the varieties of structure and substratum present, both nonnesting adults (93%) and juveniles (100%) used intermediate depths more often than either the upper or lower water column.

Nesting Adults and Dispersing Fry

Nesting was initiated between April 3 and April 22 during the study (1991–1993). Clean, empty nests, vigorously defended by territorial males, were usually observed several days before spawning occurred. Males continued to guard territories until schools of free-swimming fry dispersed well away from their nests. Although nesting was generally quite synchronous, reproduction persisted for as long as 5 weeks, and terminated between late April and mid-May.

We measured physical characteristics of 270 largemouth bass nests (Table 3). Of these, 73% were associated with a single category of physical structure, 22% were associated with two or more categories of physical structure, and only 6% had no associated structure. This pattern of association was significantly different than that expected based on the observed distribution of structure (chi-square test, $P < 0.001$). Largemouth bass selected nest sites near large, simple, woody structure (i.e., horizontal logs) more often than all other categories of structure combined (125 nests were associated with horizontal logs alone and 56 additional nests were associated with combinations of horizontal logs and other structural categories; Table 3). Although thin layers of abscised leaves covered much of the undisturbed substratum, no nests were associated with leaf packs (see Table 1 for habitat descriptions).

Gravel and gravel mixes were the dominant substrata of nests (Table 4). Fully 86% of all nests were constructed on gravel or gravel mixed with cobble. Largemouth bass also constructed nests on coarser substrata; 7% of nests were located on cobble, boulders, stumps, and mixtures of cobble and boulders. Few nests were located on silty substrata or silt mixed with other substrata (8% of nests, total), and this avoidance was statistically significant (chi-square test, $P < 0.001$).

Male largemouth bass cleared nest depressions of leaves, sticks, silt, and other debris prior to spawning, and maintained nests while actively guarding embryos and yolk sac fry. As a result, nest depressions were pale colored and visually distinct from surrounding, dark colored substrata. Embryos were readily distinguished from the gravel or cobble substrata of most nests. Yolk sac fry were difficult to detect visually from distances longer than 50 cm, therefore, divers examined all nests closely. Free-swimming fry were less cryptic than yolk sac fry and experienced divers could follow them as they swam in the water column.

Thirty-one schools of free-swimming fry were observed. Of these, multiple observations (at least 24 h apart) of free-swimming schools were recorded for 14 broods. Initially, schools of mobile fry hovered directly above the substratum of their natal nests. Within 24 h of becoming mobile, however, schools of fry were located away from, but within 1 m, of nest clearings. Fry hovered less than 0.5 m above the substratum or just above the nearest structure. Schools located within 2–3 m of nest sites were often found near or within clumps of branches or logs (47% of observations). We often noticed schools first in open water within a few meters of sizeable structure such as felled trees. Males guarded these schools tenaciously and aggressively. While males attacked potential threats (sunfish or divers), schools of fry swam into the nearby structure, where they were much more difficult to see.

Repeated observations of fry schools allowed us to map patterns of school movement over time. Schools observed for several consecutive days were found in different locations each day, but always within a few meters of the natal nest site. Schools were most frequently observed in depths similar to those of their nest sites, a result of school movements parallel to the shoreline (50% of observations), but schools were also observed in deeper water offshore of nest sites (33% of observations), and in shallow water inshore from nest sites (17% of observations). Thus, for any single brood, we could not predict the direction in which a school might move.

Comparison of Life Stages

A chi-square test for homogeneity was used to determine whether juvenile, nonnesting adult, and nesting adult largemouth bass were distributed among patches in a similar fashion. Statistical tests were conducted on the actual number of individuals observed. Significant differences were found in use of patches with regard to both physical structure ($P < 0.001$; Table 3) and substrata ($P < 0.001$; Table 4). All three stages show a different pattern of habitat use.

Nesting males used both horizontal logs and boulder patches much more than expected (almost three times the expected value), while juveniles and nonnesting adults used these patches less than expected assuming homogeneous distribution of the three stages. Nesting adults used aquatic vegetation less, and nonnesting adults used aquatic vegetation more frequently than expected; use of aquatic vegetation by juveniles was similar to the expected value. Patches lacking physical structure were used more by nonnesting adults, less by nesting adults, and not at all by juveniles. Leaf packs were used more often by juveniles and less by nesting adults. Patches with silt substratum were used more often and gravel less often by both juveniles and nonnesting adults, exactly the opposite pattern seen in patch use by nesting adults.

Discussion

In this study we have demonstrated that largemouth bass live in a landscape with many qualitatively different microhabitats, and that they are distributed in a complex pattern over the landscape. No single microhabitat was preferred by all life stages. Nearly half of all nesting males were located in patches containing only logs, yet abundances of juveniles and nonnesting adults were only moderate where logs were the sole structure present. Aquatic vegetation isolated from other physical structure, which accounted for only 10% of all patches, was the microhabitat most frequently used by both juveniles and nonnesting adults. In contrast, less than 5% of nesting adults were associated with aquatic vegetation alone, and nests were seldom associated with vegetated habitat, even when horizontal logs or woody debris were also present. The only life stage commonly found in patches lacking structure was nonnesting adults.

Habitat choices of mobile fry appeared to be restricted by the availability of habitat features within a few meters of nest sites. Our observations of schools of free-swimming fry suggest that newly dispersed schools returned to their natal nest sites during the evening (J. Hunt, University of Kansas, unpublished data). Fry schools then moved to new locations each day, possibly to avoid detection by predators or to find new foraging sites. They did not, however, move more than a few meters from the nest until they were sufficiently developed to swim in a sustained fashion. When approached by divers, fry retreated into structure, suggesting that physical structure provides an important refuge against predators (J. Hunt and C. Annett, University of Kansas, unpublished data). Nest site selection by males, therefore, was important not only to embryo survival in the nest, but also to the survival of swim-up fry.

There is, however, a possible cost in nesting too close to complex structure. Immobile embryos and very young fry are vulnerable to vertebrate predators such as sunfish *Lepomis* spp., which can quickly decimate broods in the absence of a guarding parent, and can even deplete guarded broods (Eipper 1975; Colgan and Brown 1988). Juvenile sunfish were abundant in shallow portions of the littoral zone, especially where complex structure such as leaf litter, woody debris, and aquatic vegetation was present (Dibble 1993). Nests juxtaposed to complex structure, where juvenile sunfish were abundant, may be especially vulnerable to sunfish predation.

Nonnesting adults used both complex structure (aquatic vegetation, woody debris) and areas lacking structure. Densities of sunfish (*Lepomis megalotis*, *L. macrochirus*, and *L. cyanellus*) in Wedington Reservoir were highest in patches containing structure (Dibble 1993), suggesting that adult largemouth bass may have used these patches when searching for prey. In contrast, patches lacking structure may act as paths through which largemouth bass can more easily move when swimming between food patches (Kilgore et al. 1989). Therefore, this pattern of habitat use may arise from the foraging behavior of largemouth bass.

Aquatic vegetation, leaf packs, and woody debris may have been used by juvenile largemouth bass as refugia from predators. Many investigators have suggested that spatially complex habitat may be important for survival of smaller fish (Werner et al. 1977; Miranda et al. 1984; Gutreuter and Anderson 1985; Schmitt and Holbrook 1985; Mittelbach 1988). Complex structure reduces predation risk (Savino and Stein 1982; Gotceitas and Colgan 1987). Complex structure also may act as substrata for invertebrates, increasing foraging opportunities for juvenile largemouth bass. Whether use of vegetated habitats increases or decreases growth rates and survival, however, depends on complex interactions among predation risk, food availability, and competitor density (Werner and Gilliam 1984; Mittelbach 1986, 1988).

The ontogenetic shifts in habitat use demonstrated in this paper are crucial to understanding largemouth bass ecology. Low availability of critical habitat for any single life stage could ultimately limit growth of entire populations. Similar effects have been documented for other centrarchids. For example, populations of both sunfish and crappies *Pomoxis* spp. increase with abundance of aquatic vegetation (Shireman et al. 1981), and decrease with reduced vegetative coverage (Forester and Lawrence 1978; Noble 1981). Seasonal senescence of aquatic macrophytes decreases abundance of invertebrates, thereby depressing fish abundance (Whitfield 1984). Similarly, largemouth bass production increases with abundance of aquatic vegetation and decreases with reduced plant abundance, probably because of the benefits of aquatic vegetation to juveniles and young-of-year (Haller and Sutton 1975; Durocher et al. 1984). Smallmouth bass *M. dolomieu* populations increase with the addition of logs to lakes lacking physical structure in the littoral zone (Hoff 1991).

Habitat availability may vary greatly among reservoirs within the same ecoregion (e.g., Wedington and Beaver Reservoirs in northwest Arkansas, Dibble and Annett, unpublished data); such variation could explain differences in population and community dynamics among water bodies. In addition, habitat availability may vary over time within reservoirs. Common examples of temporally variable phenomena include seasonal growth and senescence of aquatic vegetation (Whitfield 1984; Dibble 1993; Dibble and Annett, unpublished data), varying water levels in fluctuating reservoirs, and successional changes in aging reservoirs (e.g., decomposition of inundated vegetation after impoundment).

The nonrandom, lifestage-specific nature of largemouth bass distribution has strong implications for reservoir managers who wish to enhance largemouth bass populations. Managers traditionally improve habitat for black basses by adding complex structure to the littoral zone, for example, brush piles, tire reefs (Prince et al. 1975; Vogele and Rainwater 1975; Brouha and von Geldern 1979; Wege and Anderson 1979). Increasing complex structure increases the habitat type used most often by juveniles and nonnesting adults (Dibble 1993), but does not increase the habitat preferred by nesting largemouth bass (Hunt 1995). Bass may nest near complex structure in some reservoirs simply because it is the only structure available. This possibility has not been considered in evaluations of artificial reefs (e.g., Bull Shoals Reservoir; Vogele and Rainwater 1975). In Wedington Reservoir, Hunt (1995) found that nesting males prefer simple structure over either complex structure or no structure, a pattern strongly supported by the data presented here.

We suggest that managers assess and, where necessary, manipulate the abundance and interspersion of qualitatively different microhabitat patches to provide for the changing requirements of developing fry, juveniles, nonnesting adults, and nesting adults. Specifically, large, simple structure should be provided (e.g., submerged logs, hinged trees) as well as patches of complex structure (e.g., vegetation, brush piles). Simple and complex structure patches should be interspersed, but sufficiently spaced so that nesting males can effectively guard against vertebrate predators. Simple structure should be located on gravel or gravel mixes, where available, to create preferred nesting habitat (Hunt 1995). Finally, patches of both simple and complex structure should be provided in many locations within the littoral zone (i.e., construct many small reefs in different locations rather than a few, large reefs concentrated in one area).

Managing for a range of habitat variables has also been suggested in order to diversify the physical environment experienced by members of species-rich warmwater fish assemblages. This has been recommended as one of the most efficient ways of maintaining both sport fish populations and community biodiversity at high levels (Rabeni 1993). Because warmwater reservoirs are embedded within warmwater stream systems, and have many species in common, it makes intuitive sense that fisheries biologists face similar management problems in the two systems.

Acknowledgements

We thank E. Kupriyanova, K. Sigler, C. Timmerman, and K. Walits for their assistance with field work. Comments of R. Pierotti improved previous drafts of the manuscript. The Arkansas Game and Fish Commission, especially M. Armstrong and the U.S. Forest Service, especially K. England, generously helped with logistics for work in Wedington Reservoir. Special thanks to the Arkansas Game and Fish Commission (project F-54) and the Arkansas Bass Association for partial funding of this research. Facilities were provided by the Arkansas Cooperative Fish and Wildlife Research Unit.

References

Aggus, L. R., and G. V. Elliott. 1975. Effects of food and cover on year-class strength of large-mouth bass in

Bull Shoals Lake. Pages 317–322 *in* R. H. Stroud and H. Clepper, editors. Black bass biology and management. Sport Fishing Institute, Washington, DC.
Bovee, K. D., and R. Milhouse. 1978. Hydraulic simulation in instream flow studies: theory and techniques. Instream Flow Informational Paper 5, U.S. Fish and Wildlife Service, FWS/OBS-78/33, Washington, DC.
Brouha, P., and C. E. von Geldern. 1979. Habitat manipulation for centrarchid production in western reservoirs. Pages 11–18 in D. L. Johnson and R. A. Stein, editors. Response of fish to habitat structure in standing water. American Fisheries Society, North Central Division, Special Publication 6.
Carlander, K. D. 1977. Handbook of freshwater fishery biology volume 2. Iowa State University Press, Ames.
Colgan, P. W., and J. A. Brown. 1988. Dynamics of nest defense by male centrarchid fish. Behavioural Processes 17:17–26.
Crowder, L. B., and W. E. Cooper. 1982. Habitat structural complexity and the interaction between bluegills and their prey. Ecology 63:1802–1813.
Davies, W. D., and W. L. Shelton. 1983. Sampling with toxicants. Pages 199–213 *in* L. A. Nielsen and D. L. Johnson, editors. Fisheries techniques. American Fisheries Society, Bethesda, Maryland.
Dibble, E. D. 1991. A comparison of diving and rotenone methods for determining relative abundance of fish. Transactions of the American Fisheries Society 120: 663–666.
Dibble, E. D. 1993. A patch dynamics study of habitat use by juvenile centrachids in an Ozark reservoir. Doctoral dissertation. University of Arkansas, Fayetteville.
Durocher, P. P., W. C. Provine, and J. E. Kraai. 1984. Relationship between abundance of largemouth bass and submerged vegetation in Texas reservoirs. North American Journal of Management 4:84–89.
Eipper, A. W. 1975. Environmental influences on the mortality of bass embryos and larvae. Pages 295–305 *in* R. H. Stroud and H. Clepper, editors. Black bass biology and management. Sport Fishing Institute, Washington, DC.
Forester, T. S., and J. M. Lawrence. 1978. Effects of grass carp and carp on populations of bluegill and largemouth bass in ponds. Transactions of the American Fisheries Society 107:172–175.
Gorman, O. T., and J. R. Karr. 1978. Habitat structure and stream fish communities. Ecology 59:507–515.
Gotceitas, V., and P. Colgan. 1987. Selection between densities of artificial vegetation by young bluegills avoiding predation. Transactions of the American Fisheries Society 116:40–49.
Gutreuter, S. J., and R. O. Anderson. 1985. Selection between densities of artificial vegetation by young bluegills avoiding predation. Transactions of the American Fisheries Society 114:417–327.
Haller, W. T., and D. L. Sutton. 1975. Community structure and competition between hydrilla and vallisneria. Hyacinth Control Journal 13:48–50.
Helfman, G. S. 1983. Underwater methods. Pages 349–369 *in* L. S. Nielsen and D. L. Johnson, editors. Fisheries techniques. American Fisheries Society, Bethesda, Maryland.
Hoff, M. H. 1991. Effects of increased nesting cover on nesting and reproduction of smallmouth bass in northern Wisconsin lakes. Pages 39–43 *in* D. C. Jackson, editor. Proceedings of the first international smallmouth bass symposium. Mississippi Agricultural and Forestry Experimental Station, Mississippi State University, Mississippi State.
Hunt, J. 1995. Reproductive ecology of largemouth bass: habitat manipulation, habitat selection and parental care. Doctoral dissertation. University of Kansas, Lawrence.
Johnson, D. L., and W. E. Lynch, Jr. 1992. Panfish use of and angler success at evergreen tree, brush, and stake-bed structures. North American Journal of Fisheries Management 12:222–229.
Keast, A., and J. Harker. 1977. Strip counts as means of determining densities and habitat utilization patterns in lake fishes. Environmental Biology of Fishes 1:181–188.
Keast, A., J. Harker, and D. Turnbull. 1978. Nearshore fish habitat utilization and species association in Lake Opinicon (Ontario, Canada). Environmental Biology of Fishes 3:173–184.
Kilgore, K. J., R. P. Morgan, II, and N. B. Rybicki. 1989. Distribution and abundance of fishes associated with submersed aquatic plants in the Potomac River. North American Journal of Fisheries Management 9:101–111.
Means, K. O., and L. E. Miranda. 1991. Variability in abundance of age-0 centrarchids among littoral habitats of flood control reservoirs in Mississippi. North American Journal of Fisheries Management 11:298–304.
Miranda, L. E., W. L. Shelton, and T. D. Bryce. 1984. Effects of water level manipulation on abundance, mortality, and growth of young-of-year largemouth bass in West Point reservoir, Alabama–Georgia. North American Journal of Fisheries Management 4:314–320.
Mittelbach, G. G. 1986. Predator-mediated habitat use: some consequences for species interactions. Environmental Biology of Fishes 16:159–169.
Mittelbach, G. G. 1988. Competition among refuging sunfishes and effects of fish density on littoral zone invertebrates. Ecology 69:614–623.
Noble, R. L. 1981. Management of forage fishes in impoundments of the southern United States. Transactions of the American Fisheries Society 110:738–750.
Northcote, T. G., and D. W. Wilkie. 1963. Underwater census of stream fish populations. Transaction of the American Fisheries Society 92:146–151.
Prince, E. D., R. F. Raleigh, and R. V. Corning. 1975. Artificial reefs and centrarchid basses. Pages 498–505 *in* R. H. Stroud and H. Clepper, editors. Black bass biology and management. Sport Fishing Institute, Washington, DC.
Rabeni, C. F. 1993. Warmwater streams. Pages 427–443 *in* C. C. Kohler and W. A. Hubert, editors. Inland fisheries management in North America. American Fisheries Society, Bethesda, Maryland.

Raffetto, N. S., J. R. Baylis, and S. L. Serns. 1990. Complete estimates of reproductive success in a closed population of smallmouth bass (*Micropterus dolomieu*). Ecology 71:1523–1535.

Rainwater, W. C., and A. Houser. 1982. Species composition and biomass of fish in selected coves in Beaver Lake, Arkansas, during the first 18 years of impoundment (1963–1980). North American Journal of Fisheries Management 2:316–325.

Reynolds, J. B. 1983. Electrofishing. Pages 147–163 *in* L. A. Nielson and D. L. Johnson, editors. Fisheries techniques. American Fisheries Society, Bethesda, Maryland.

Savino, J. F., and R. A. Stein. 1982. Predator–prey interaction between largemouth bass and bluegills as influenced by simulated, submersed vegetation. Transactions of the American Fisheries Society 11:255–266.

Schmitt, R. J., and S. J. Holbrook. 1985. Patch selection by juvenile black surfperch (Embiotocidae) under variable risk: interactive influence of food quality and structural complexity. Journal of Experimental Marine Biology and Ecology 85:269–285.

Shireman, J. V., D. E. Colle, and D. F. DuRant. 1981. Efficiency of rotenone sampling with large and small block net in vegetated and open water habitat. Transactions of the American Fisheries Society 110:77–80.

Summerfelt, R. C. 1993. Lake and reservoir habitat management. Pages 231–261 *in* C. C. Kohler and W. A. Hubert, editors. Inland fisheries management in North America. American Fisheries Society, Bethesda, Maryland.

Todd, B. L., and C. F. Rabeni. 1989. Movement and habitat use by stream-dwelling smallmouth bass. Transactions of the American Fisheries Society 118: 229–242.

Vogele, L. E., and W. C. Rainwater. 1975. Use of brush shelters as cover by spawning black basses (*Micropterus*) in Bull Shoals Reservoir. Transactions of the American Fisheries Society 104:264–269.

Wege, G. J., and R. O. Anderson. 1979. Influence of artificial structure on largemouth bass and bluegills in small ponds. Pages 59–69 *in* D. L. Johnson and R. A. Stein, editors. Response of fish to habitat structure in standing water. American Fisheries Society, North Central Division, Special Publication 6, Bethesda, Maryland.

Werner, E. E., and five coauthors. 1977. Habitat partitioning in a freshwater fish community. Journal of the Fisheries Research Board of Canada 34:360–370.

Werner, E. E., and J. F. Gilliam. 1984. The ontogenetic niche and species interactions in size-structured populations. Annual Review of Ecology and Systematics 15:393–425.

Whitfield, A. K. 1984. The effects of prolonged aquatic macrophyte senescence on the biology of the dominant fish species in a southern African coastal lake. Estuarine and Coastal Shelf Science 18:315–329.

American Fisheries Society Symposium 16:315–323, 1996

Application of Geographic Information Systems in Fisheries: Habitat Use by Northern Pike and Largemouth Bass

KEVIN B. ROGERS AND ERIC P. BERGERSEN

Colorado Cooperative Fish and Wildlife Research Unit[1]
Colorado State University, Room 201 Wagar Building
Fort Collins, Colorado 80523, USA

Abstract.—The display and analysis of spatial information has been greatly enhanced with geographic information systems (GIS), which combine layers of geographic features with an associated database of attributes to form a composite map. This map database can then answer questions regarding habitat feature locations, temporal changes, spatial patterns, and can also be used for system modeling. The broad applicability of GIS has spawned an extensive array of systems with a vast spectrum of abilities. The process and precision of acquiring spatial information has been greatly enhanced by other technologies such as remote sensing and global positioning systems. With personal computers becoming increasingly powerful and affordable, GIS has become an attractive tool for facilitating analysis of spatial questions in fisheries research and management. Two simple GIS applications evaluating use of vegetation and shoreline by northern pike *Esox lucius* and largemouth bass *Micropterus salmoides* in two Colorado reservoirs are presented.

The ability to define attributes of geographical elements and their corresponding location has been a common goal of human societies for a long time. The first maps were created before the first alphabet (Marble 1987). With scientific information doubling every 5 years (Naisbitt 1984), assimilation of information must be facilitated with systems that can analyze, store, and present data in a tangible form (Star and Estes 1990). Geographic information systems (GIS) have this ability, greatly enhancing the display and analysis of coordinate-based spatial information.

Natural resource management has become increasingly complex over time. Resource managers are now charged with managing ecosystems, maintaining biodiversity, and preserving natural resources in the face of increasing human populations. Traditional analysis methods are unable to coordinate the vast information needed for this management approach. Fortunately, GIS is a tool designed to facilitate tackling both large- and small-scale questions. GIS allows for rapid assessment of data, visual output, and identification of solutions to management problems. With GIS technology, resource managers can analyze systems with spatial scales ranging from microhabitats to regional landscapes, allowing them to recognize landscape level ecological processes (Sample 1994b). The broad applicability of GIS along with plummeting hardware and software costs have made it one of the fastest growing areas of the computer industry (Korte 1994).

What are Geographic Information Systems?

Geographic information systems are a valuable tool for capturing, storing, manipulating, analyzing, and displaying geographically referenced information. A user is able to address questions on the location and condition of geographic features, trends, patterns, and even to model future conditions (ESRI 1991). Unlike computer aided design (CAD) systems, the ability to link spatial and descriptive data makes GIS powerful. Because of the synthesis between mapping and database systems, GIS can be thought of as a higher order map.

The foundation of GIS lies in the storage of spatially referenced information in map layers (Figure 1). Each layer stores information about a particular land feature (e.g., rivers, cover types, topography, land use). Themes need not be restricted to a single layer (Laurini and Thompson 1992). Temperature profiles can be generated with multiple planes representing, for example, incremental lake depths. Temporal changes in a given geographic feature can be depicted on consecutive layers. Geographic information systems must have four basic components (Marble 1987; Ashdown and Schaller 1990).

1. Data entry: the ability to incorporate a diverse array of spatial information into the system.
2. Data storage and retrieval: the ability to store

[1]Cooperators are the National Biological Service, the Colorado Division of Wildlife, and Colorado State University.

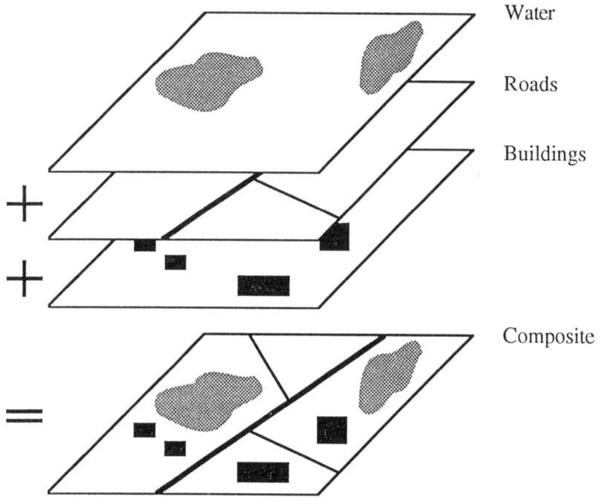

FIGURE 1.—Thematic map layers in GIS form a composite on which spatial analyses can be performed.

large quantities of data, with rapid access to facilitate efficient updating.

3. Data manipulation and analysis: the ability to yield responses to data queries (this is the foundation of any GIS). Modeling capabilities make it possible to predict alternative outcomes.
4. Data reporting system: the ability to display results of the analysis.

These four components provide resource managers the ability to inventory and monitor resources, plan site-specific and regional management, and analyze policy alternatives (Green 1994).

Some Concerns

Though GIS can be extremely powerful, it is just a tool. Resource management problems will not simply go away now that GIS is available (Moore 1994). Setting up GIS requires a substantial commitment of time and money, and using GIS effectively often requires significant training, which can scare potential users away.

Potential error associated with each map layer is another concern that is often overlooked. When many layers containing unquantified data are combined to form a composite, error may be compounded (Allen 1994). Precision and accuracy of GIS data and output must therefore be considered. Users must understand the probability of human error and be capable of making wise judgements and interpretations of GIS output (Laurini and Thompson 1992).

Global Positioning Systems

Base map precision has been greatly enhanced by another newly available technology known as global positioning systems (GPS). Precise location estimates anywhere on the planet can be generated by interrogating—with a GPS receiver—at least 3 of 24 Department of Defense satellites currently orbiting the earth (Hurn 1989). These hand-held devices determine the users' position in either latitude–longitude, loran, or universal transverse mercator (UTM) coordinate systems. Like other sophisticated electronic instruments, costs of GPS hardware have fallen dramatically in the last few years. Inexpensive units with ±100-m accuracy could be purchased for under US$300 in 1995 (Magellan, San Dimas, California). These units are less accurate than they could be because the Department of Defense scrambles satellite signals at undisclosed intervals, which results in "selective availability" of position data. Error introduced by selective availability and other sources (Hurn 1989) is easily eliminated by setting up a stationary base at a known nearby location (Figure 2). Error is recorded on the stationary GPS unit and can be subtracted from relevant field data, a process called differential correction. Correction factors are also available from commercial radio stations in larger population centers (J. Vallazza, Ohio State University, personal communication). Though a monthly subscription and a radio receiver must be purchased, this can be an economical alternative to purchasing a second GPS unit. Accuracy of less than ±5 m should be expected when the differential correction method is used. More expensive units allow the user to register attributes to positional coordinates in the field. These attributed data files can be downloaded directly into GIS, which minimizes transcription error and greatly enhances efficiency.

Setting up Geographic Information Systems

With the variety of GIS applications available, it is imperative that goals and objectives be established before equipment is purchased. This minimizes unnecessary expenditures and provides a standard against which to measure progress. Simple questions often can be addressed with inexpensive software packages that run on existing computer hardware. Other systems may require a considerable financial investment for workstation grade computers, large storage facilities, and complex software, as well as money for database management and user training.

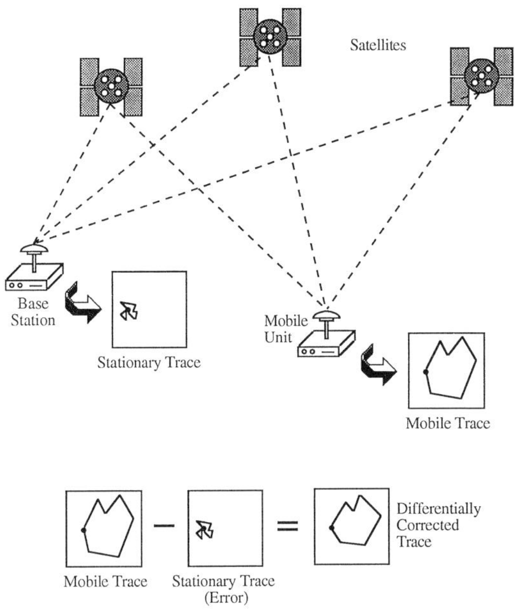

FIGURE 2.—Positional information with better than ±5-m accuracy can be achieved with global positioning systems (GPS) using differential correction. A base GPS is set up at a known stationary location. Deviations in position recorded at this location (stationary trace) can be attributed to error in the system. This error confounds the positional information collected from a mobile field unit (mobile trace). By subtracting error realized in the stationary trace from the mobile trace, a differentially corrected trace of true locations can be obtained.

There is a tremendous array of GIS software available (Table 1), ranging from public domain programs to commercial systems costing over $20,000. Inexpensive software is typically designed to support individual needs, and often lacks sophisticated data analysis and graphics capabilities. These systems are generally user friendly, in contrast to more sophisticated ones that often require significant training to operate. Software packages may be functionally similar in terms of the analyses they perform, yet the resulting graphics can be vastly different. If publication quality displays are essential, more expensive systems may be required.

Perhaps the most expensive and labor intensive portion of GIS involves entering spatial information into the computer (Korte 1994). A relational database can be built by scanning existing maps or by entering map features with digitizing tablets, then manually attributing descriptions to these features. Information can also be imported from spreadsheet or ASCII files. Remote sensing from satellites or computer based acquisition systems with GPS links can also be used to compile data. Thematic map data are often available electronically from outside sources such as the U.S. Geological Survey and the U.S. Bureau of the Census on tapes, CD ROMs, and recently over the Internet. Careful research to assess data availability is important and cost effective.

TABLE 1.—A sampling of some readily available software packages for geographic information systems with manufacturer and telephone number, data structure, available operating platforms (WIN = Windows, MAC = Macintosh), and the base price (US$) in 1995. Prices can be highly variable depending on the modules purchased. Educational institutions can often obtain substantial discounts from the basic list price.

Software	Company	Telephone	Data structure		Platform				Base price (1995, US$)
			Raster	Vector	UNIX	DOS	WIN	MAC	
ARC/INFO	ESRI	(909) 793–2853	•	•	•	•			18,000[a]
ArcView	ESRI	(909) 793–2853	•	•	•	•	•	•	995
Atlas GIS	Strategic Mapping	(408) 970–9600		•		•	•		495
FMS	Eagle Point	(415) 381–1750		•	•	•	•		6,995
GDS	Graphic Data Systems	(303) 741–8484	•	•	•		•		28,800[a]
GENAMAP	Genasys	(970) 226–3283	•	•	•				12,000[a]
Geo/SQL	Geo/SQL	(303) 940–9266	•	•	•	•	•		16,000[a,b]
GRASS	GRASS Information	(509) 963–1504	•		•			•	990
IDRISI	Clark University	(508) 793–7526	•			•	•		720
InfoCAD	Digital Matrix Services	(305) 445–6100	•	•	•		•		12,500[a]
MapII	Wiley	(212) 850–6000	•					•	125
MapiX	Delta Data Systems	(601) 799–1813	•	•			•		2,495
MGE	Intergraph	(205) 730–2000	•	•	•	•	•		13,000[a,b]
MIPS	MicroImages	(402) 477–9554	•	•	•		•	•	10,000[b]
TerraSoft	Essential Planning Systems	(604) 652–8895	•	•		•			14,000

[a]Highly variable.
[b]Personal computer versions are substantially less expensive.

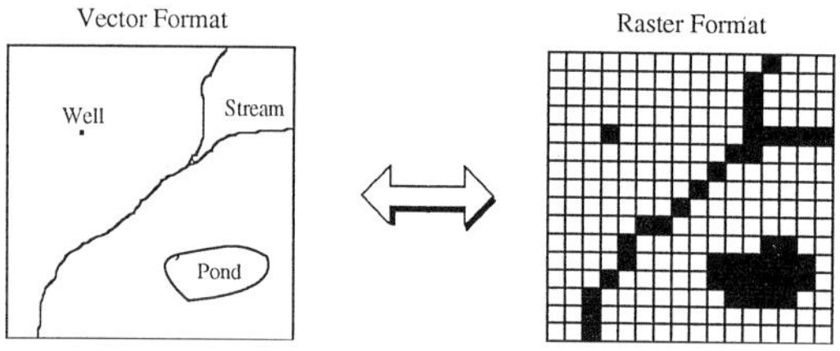

FIGURE 3.—The vector format can represent geographic features accurately using points, lines, and polygons. Grid cells used in the raster equivalent have a finite ability to specify location and the resulting map is less attractive.

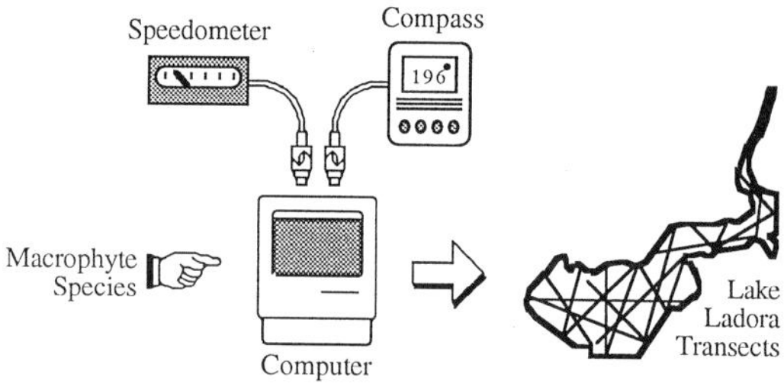

FIGURE 4.—Using our customized Macintosh-based acquisition system, the user can enter macrophyte species identity while running multiple transects of a lake. Input from an electronic compass and speedometer allow the system to add corresponding universal transverse mercator coordinates to plant locations. This information is then compiled in a format accessible to a geographic information system for interpolation.

Data Structures: Raster versus Vector

Geographically referenced spatial information can be organized either in raster format (grid cells) or vector format, that use points, lines, and polygons to characterize information (Figure 3). It is important to consider the appropriate data structure for specific user needs when setting up GIS. Raster systems are inexpensive, relatively easy to use, have simple data structures, are compatible with remotely sensed data, and are powerful spatial modeling tools. Unfortunately, the raster format provides a limited ability to specify location, and precision is a function of pixel size (Laurini and Thompson 1992). The method for determining a cell attribute is important because only one attribute per pixel can be designated in the raster system. Also, the resulting maps are not as attractive as those generated on vector systems. Vector GIS provides accurate graphics and compact data structures, unlike raster system files which can require more memory. Vector systems, however, tend to be expensive and are more difficult to combine with remotely sensed data. Although most natural resources issues can be addressed with raster-based packages, vector systems have some advantages. Some software manufacturers have made the choice of software easier by producing packages that can synthesize both formats. A simple example using each data structure follows.

Raster Format

The strong association between largemouth bass *Micropterus salmoides* and underwater structure has been well documented (Johnson et al. 1988; Smith and Orth 1991; Walters et al. 1991). Northern pike *Esox lucius*, although not as closely associated with structure, prefer vegetated littoral areas (Diana et al. 1977; Cook and Bergersen 1988). Because aquatic macrophytes make up the bulk of structural relief in 25-ha Lake Ladora on the Rocky Mountain Arsenal, Colorado, we investigated fish use of the vegetation with the raster-based GIS MapII (Wiley, New York).

A 10 × 10-m grid cell vegetation map for Lake Ladora was generated with a customized acquisition system (Figure 4) on board a 5-m johnboat. Twenty-four transects were run: vegetation was identified continuously. The dominant cover type in each 10 × 10-m block was determined and assigned to the cell. Vegetation data for the remaining lake cells were generated by interpolating the 24 transects across the entire lake basin in MapII using octants and the closest four points to produce a vegetation map (Figure 5).

Four adult northern pike and six adult largemouth bass were implanted with coded ultrasonic transmitters (Hart and Summerfelt 1975) and located weekly in the summer of 1992 in Lake Ladora. Fish were located with a directional hydrophone and receiver (Sonotronics, Tucson, Arizona) from a canoe. Locations were converted to UTM coordinates with the aid of an aerial photograph inscribed with a meter grid to facilitate data analysis and transfer to GIS.

Northern pike and largemouth bass binary (presence–absence) location maps were combined with vegetation coverage, and macrophyte use was subjected to Ivlev's electivity index (Ivlev 1961) to evaluate selection (Figure 6). The analysis of selection was repeated with a logistic regression (SAS Institute, Cary, North Carolina) to assess statistical significance of these trends (Rogers 1994). Both northern pike and largemouth bass avoided stands of

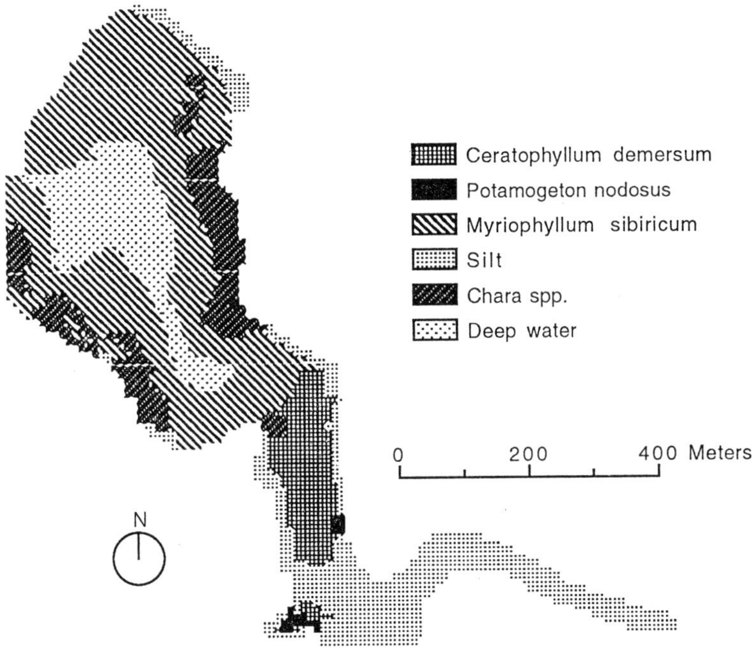

FIGURE 5.—Cover types were identified and entered into a computer-based acquisition system while running 24 transects on Lake Ladora. The dominant cover type for each 10 × 10-m grid cell was calculated. The resulting transect map was then interpolated using a mask of the lake perimeter.

Myriophyllum sibiricum, *Chara* spp., and deep water. Largemouth bass favored areas dominated by *Potamogeton nodosus*, whereas northern pike favored the silt bottomed arm of the lake ($P < 0.05$).

Our results were not surprising given the biology of largemouth bass (Heidinger 1975). We hypothesize that largemouth bass favored *P. nodosus* because they are ambush predators as adults, and the growth form of this macrophyte creates a canopy under which they can hide. Shade makes it difficult for forage species to see under the canopy and the view for largemouth bass maintaining position there is enhanced (Helfman 1981). The other plants (*Chara* spp., *M. sibiricum*, and *Ceratophyllum demersum*) grow in much denser mats that may not be navigable by large predators. Dense mats of vegetation were characteristic of the main body of Lake Ladora and perhaps explain why few fish were found there.

The results for northern pike were uncharacteristic. It is generally accepted that esocids are intimately associated with vegetation (Bregazzi and Kennedy 1980; Chapman and Mackay 1983; Cook and Bergersen 1988), yet our fish clearly selected for silt ($P = 0.001$). However, vegetation maps express only one aspect of the environment. Lake Ladora is filled with waters drawn from the bottom of Lower Derby Lake, an impoundment further up the drainage. Although the narrow inlet of Lake Ladora is one of the shallowest portions of the lake, it is also the coolest in the heat of summer. Surface water temperatures along the Front Range in Colorado may exceed 26°C. Northern pike are considered a coolwater species (Casselman 1978), and may select silt bottomed habitat due to thermal preference rather than cover type.

Despite the strong significance of the cover type parameters included in these models, they may not completely explain why largemouth bass and northern pike select the habitats they do. Although one patch of *P. nodosus* was heavily visited by largemouth bass, the remaining patches yielded no largemouth bass records. Northern pike were associated heavily with the silt-bottomed arm of Lake Ladora, yet they were never recorded in deep water habitats also dominated by silt. Such details are readily apparent with GIS but not with traditional regression models. We conclude that factors other than cover type apparently influence fish location. Addition of other habitat variables (e.g., depth, temperature, vegetation density, forage density, water clarity) as

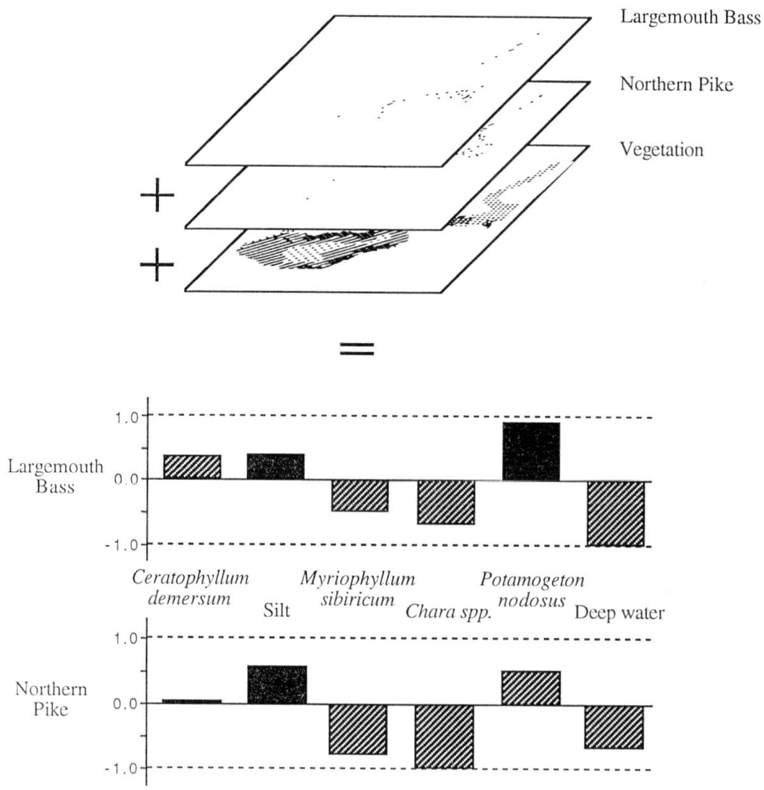

FIGURE 6.—Coverages depicting weekly telemetered largemouth bass and northern pike locations in Lake Ladora in the summer of 1992 were combined with a vegetation map. Selection for a given cover type by largemouth bass and northern pike was assessed using Ivlev's electivity index. Bars above zero indicate positive selection; those below zero indicate negative selection. Solid bars represent significant positive selection ($P < 0.001$) by logistic regression.

separate map layers in GIS would, perhaps, increase the predictive value of these models of fish location.

Vector Format

Although selection of nearshore habitat by northern pike has not been evaluated, several authors have suggested the importance of this habitat to largemouth bass (Winter 1977; Doerzbacher 1980; Betsill et al. 1988). We monitored monthly locations of 19 northern pike (1.20–4.28 kg) and 20 largemouth bass (0.94–2.39 kg) from June 1991 through August 1993 in Lakes Ladora and Lower Derby (25 ha each) on the Rocky Mountain Arsenal, Colorado.

Analysis of nearshore habitat preference is best evaluated with vector-based GIS because it allows measurement of precise distances from fish to shore. A raster analysis with very small pixels would be adequate, but map layers would require prohibitively large amounts of computer memory. The perimeter of the lakes were digitized into layers in ARC/INFO (Environmental Systems Research Institute, Redlands, California) and a 20-m buffer zone around the shoreline was established. The resulting layers were then combined with eight map layers that included fish locations by species and season. Locations in March, April, and May were defined as spring; in June, July, and August as summer; in September, October, and November as fall; and in December, January, and February as winter. The GIS was then queried for the proportion of fish locations falling within the nearshore buffer zone and the relative availability of this habitat (Figure 7). A chi-square analysis was used to determine if northern pike and largemouth bass selected for nearshore habitats.

Nearshore habitat proved to be important for largemouth bass but not for northern pike during most of the year. Largemouth bass favored near-

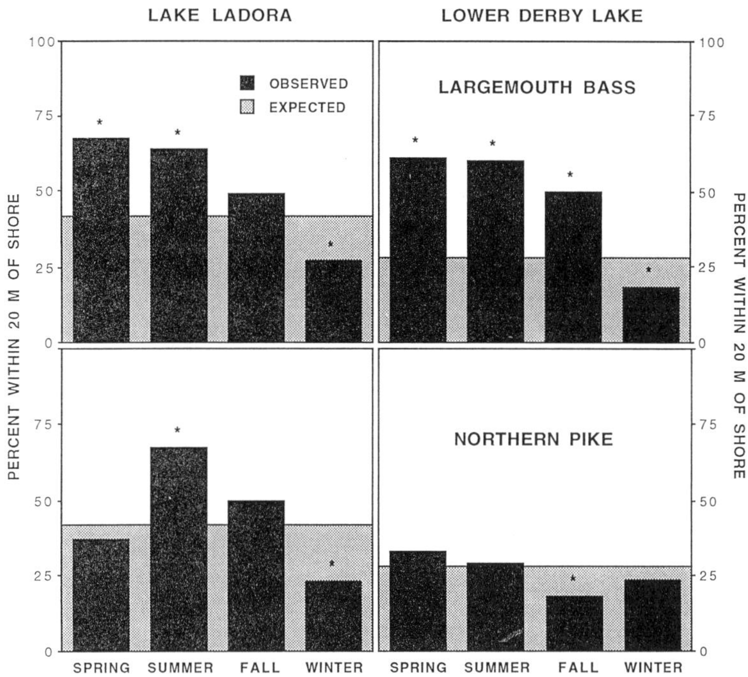

FIGURE 7.—Seasonal selection for habitat within 20 m of shore by largemouth bass and northern pike. The relative abundance of nearshore habitat (expected) is shown in gray, while the black bars (observed) represent the percent of fish locations falling within that 20-m zone. Significant differences between expected and observed ($P < 0.05$ by chi-square analysis) are marked with an asterisk.

shore habitat in both lakes during spring, summer, and fall (Figure 7); approximately 60% of all locations were within 20 m of the shoreline in both lakes. Northern pike exhibited similar shoreline selection only during summer and only in Lake Ladora. Both species moved off shore in winter.

Nearshore cover appears to be an important component of largemouth bass habitat (Doerzbacher 1980; Betsill et al. 1988). Winter (1977) reported that 94% of his locations were in water less than 3 m deep. Our analysis indicated that largemouth bass were located within 20 m of shore more often than would occur by chance ($P < 0.05$ by chi-square analysis). Largemouth bass are associated with structurally complex habitats (Prince and Maughan 1979), and it is the littoral zone that provides the most complex habitat. Lower Derby Lake lacks extensive structure, so relief and cover provided by surrounding cattails *Typha latifolia* appear to be important. Macrophytes extend beyond our 20-m buffer in Lake Ladora, yet prime habitat still seems to be adjacent to shore.

Northern pike did not rely on nearshore habitat any more than would be expected by chance (Figure 7). However, Lake Ladora northern pike did associate with nearshore areas in summer. Cool waters flowing into the narrow arm of Lake Ladora from the bottom of Lower Derby Lake may explain these results. Although this channel is less than 20 m wide, it is probably not the shoreline proximity that draws northern pike here. Headrick and Carline (1993) found that northern pike occupied the coolest available water when surface temperatures exceeded 25°C. This may explain why these fish favored the narrow inlet to Lake Ladora when summer water temperatures in the remainder of the lake exceed 26°C. Both species were found less frequently near shore than could be expected by chance in the winter months, moving off shore, presumably in search of deeper, warmer water.

Discussion

Although these examples deal with spatial questions on a small scale, the scale of possible analyses are limited only by software and hardware capabilities. Landscape level analyses are presumably where resource management agencies will derive the most benefit from GIS. Potential uses are diverse, and can integrate human dimensions with physical and biological considerations. Examples include the rapid isolation of optimal stocking protocols (Noble et al. 1994), an accounting framework for managing biological diversity (Giles and Nielsen 1992), siting fishery access points (Giles and Nielsen 1992), and defining potential sites for mitigation.

An obvious use of GIS is predictive mapping of the spatial distributions of important fish species (Giles and Nielsen 1992). Suitable habitat for reintroduction can be readily isolated with GIS technology. The search for potential native greenback cutthroat trout *Oncorhynchus clarki* reintroduction sites in Colorado would have been greatly facilitated with GIS because all criteria used could have been easily mapped into a spatial system (high elevation, above a barrier to migration, good water quality, and available spawning habitat). The search for additional populations of endangered fishes can also be aided by highlighting regions that share characteristics with the type of habitat (Giles and Nielsen 1992).

The modeling potential of GIS can also be tapped to facilitate environmental impact assessments. Sediment transport into a drainage from a potential clearcut can be predicted by integrating slope, soil type, soil permeability, and vegetative cover (Laurini and Thompson 1992). Resulting changes in probable spawning and nursery habitat for endangered, anadromous species can then be anticipated (Giles and Nielsen 1992).

The need for spatial analysis clearly exists in fisheries management. The ability to compile and interpret diverse resource interactions will make GIS an indispensable tool for informed decision making. Increasing complexity in natural resource management is cumbersome when traditional methods are used. New technologies such as GIS will provide for efficient, effective resource management in the future.

Acknowledgments

Research was funded by the National Biological Service. We thank Chris Rogers for lending his programming expertise. Our gratitude is also extended to Leslie Ager, Elise Irwin, Greg Langer, and two anonymous reviewers for their helpful comments on earlier versions of the manuscript.

References

Allen, C. D. 1994. Ecological perspective: linking ecology, GIS, and remote sensing to ecosystem management. Pages 111–139 *in* Sample (1994a).

Ashdown, M., and J. Schaller. 1990. Geographic information systems and their application in MAB projects, ecosystem research and environmental monitoring. UNESCO (United Nations Educational, Scientific and Cultural Organization), Bonn, Germany.

Betsill, R. K., R. L. Noble, and W. H. Neill. 1988. Distribution and habitat selection of telemetered northern and Florida largemouth bass in two small Texas impoundments. Proceedings of the Annual Conference Southeastern Association of Fish and Wildlife Agencies 40(1986):275–286.

Bregazzi, P. R., and C. R. Kennedy. 1980. The biology of pike, *Esox lucius* L., in a southern eutrophic lake. Journal of Fish Biology 17:91–112.

Casselman, J. M. 1978. Effects of environmental factors on growth, survival, activity, and exploitation of northern pike. Pages 114–128 *in* R. L. Kendall, editor. Selected coolwater fishes of North America. American Fisheries Society Special Publication 11.

Chapman, C. A., and W. C. Mackay. 1983. Versatility in habitat use by a top aquatic predator, *Esox lucius* L. Journal of Fish Biology 25:109–115.

Cook, M. F., and E. P. Bergersen. 1988. Movements, habitat selection, and activity periods of northern pike in Eleven Mile Reservoir, Colorado. Transactions of the American Fisheries Society 117:495–502.

Diana, J. S., W. C. Mackay, and M. Ehrman. 1977. Movements and habitat preference of northern pike (*Esox lucius*) in Lac Ste. Anne, Alberta. Transactions of the American Fisheries Society 106:560–565.

Doerzbacher, J. F. 1980. Movement and home range of largemouth bass (*Micropterus salmoides*) in relation to water quality of the Atchafalaya River basin, Louisiana. Master's thesis. Louisiana State University, Baton Rouge.

ESRI (Environmental Systems Research Institute, Inc.). 1991. Understanding GIS: the ARC/INFO method. ESRI, Redlands, California.

Giles, R. H., and L. A. Nielsen. 1992. The uses of geographic information systems in fisheries. Pages 81–94 *in* R. H. Stroud, editor. Fisheries management and watershed development. American Fisheries Society Symposium 13.

Green, K. 1994. The potential and limitations of remote sensing and GIS in providing ecological information. Pages 327–336 *in* Sample (1994a).

Hart, L. G., and R. C. Summerfelt. 1975. Surgical procedures for implanting ultrasonic transmitters into flathead catfish (*Pylodictus olivaris*). Transactions of the American Fisheries Society 104:56–59.

Headrick, M. R., and R. F. Carline. 1993. Restricted summer habitat and growth of northern pike in two southern Ohio impoundments. Transactions of the American Fisheries Society 122:228–236.

Heidinger, R. C. 1975. Life history and biology of the largemouth bass. Pages 11–20 *in* R. H. Stroud and H. Clepper, editors. Black bass biology and management, Sport Fishing Institute, Washington DC.

Helfman, G. S. 1981. The advantage to fishes of hovering in shade. Copeia 1981:392–400.

Hurn, J. 1989. GPS: a guide to the next utility. Trimble Navigation Ltd., Sunnyvale, California.

Ivlev, V. S. 1961. Eksperimental´naia ekologiia pitaniza ryb. (Experimental ecology of the feeding of fishes. Translated from Russian by Douglas Scott), Yale University Press. New Haven, Connecticut.

Johnson, D. J., R. A. Beaumier, and W. E. Lynch. 1988. Selection of habitat structure interstice size by bluegills and largemouth bass in ponds. Transactions of the American Fisheries Society 117:171–179.

Korte, G. B. 1994. GIS book, 3rd edition. Onword Press, Santa Fe, New Mexico.

Laurini, R., and D. Thompson. 1992. Fundamentals of spatial information systems. Academic Press, San Diego, California.

Marble, D. F. 1987. Geographic information systems: an overview. Pages 2–8 *in* W. J. Ripple, editor. Geographic information systems for resource management: a compendium. American Society for Photogrammetry and Remote Sensing and the American Congress on Surveying and Mapping, Falls Church, Virginia.

Moore, M. M. 1994. Analysis and mapping of late-successional forests in the American southwest. Pages 108–110 *in* Sample (1994a).

Naisbitt, J. 1984. Megatrends. Warner Books, New York.

Noble, R. L., J. R. Jackson, E. R. Irwin, J. M. Phillips, and T. N. Churchill. 1994. Reservoirs as landscapes: implications for fish stocking programs. Transactions of the North American Wildlife and Natural Resources Conference 59:281–288.

Prince, E. D., and O. E. Maughan. 1979. Telemetric observations of largemouth bass near underwater structures in Smith Mountain Lake, Virginia. Pages 26–32 *in* D. L. Johnson and R. A. Stein, editors. Response of fish to habitat structure in standing water. American Fisheries Society, North Central Division, Special Publication 6, Bethesda, Maryland.

Rogers, K. B. 1994. Movement patterns of largemouth bass and northern pike on the Rocky Mountain Arsenal. Master's thesis. Colorado State University, Fort Collins.

Sample, V. A., editor. 1994a. Remote sensing and GIS in ecosystem management. Island Press, Washington, DC.

Sample, V. A. 1994b. Realizing the potential of remote sensing and GIS in ecosystem management planning, analysis, and policymaking. Pages 346–352 *in* Sample (1994a).

Smith, S. M., and D. J. Orth. 1991. Distributions of largemouth bass in relation to submerged aquatic vegetation in Flat Top Lake, West Virginia. Proceedings of the Annual Conference Southeastern Association of Fish and Wildlife Agencies 44(1990):36–44.

Star, J., and J. Estes. 1990. Geographic information systems: an introduction. Prentice Hall, Englewood Cliffs, New Jersey.

Walters, D. A., W. E. Lynch, and D. L. Johnson. 1991. How depth and interstice size of artificial structures influence fish attraction. North American Journal of Fisheries Management 11:319–329.

Winter, J. D. 1977. Summer home range movements and habitat use by four largemouth bass in Mary Lake, Minnesota. Transactions of the American Fisheries Society 106:323–330.

American Fisheries Society Symposium 16:324–331, 1996

Effects of Reservoir Drawdown on Littoral Habitat: Assessment with On-Site Measures and Geographic Information Systems

ELISE R. IRWIN[1] AND RICHARD L. NOBLE

North Carolina State University, Department of Zoology
Box 7617, Raleigh, North Carolina 27695, USA

Abstract.—Effects of reservoir drawdown on littoral habitats in two reservoir embayments were quantified using on-site measures and remotely sensed data entered into geographic information systems (GIS). Habitat characteristics (slope, substratum, and cover) were measured at three water levels (65.8, 64.6, and 64.0 m above mean sea level) at 15 10 × 2-m fixed sites in Little Beaver Creek (LBC) and White Oak Creek (WOC) embayments of B. E. Jordan Reservoir, North Carolina. Additional fixed sites were analyzed to more precisely characterize habitats in each embayment. Differences in habitat between the two embayments were detected with sample sizes of 55–57 sites. Habitat databases that included bathymetric and dominant substratum data were incorporated into GIS. Both GIS and on-site measures indicated that slope characteristics remained stable as water level declined. The GIS indicated that LBC was steeper than WOC, whereas on-site measures did not detect these differences. Significant decreases in total cover with decreasing water levels were detected in both embayments via on-site habitat measures. Incorporation of cover into GIS at the embayment scale was not feasible. Amounts of gravel substratum decreased in both embayments as water level declined. The GIS allowed large-scale assessment of changes in size and configuration of habitat patches, whereas on-site measures provided finer resolution of habitat characteristics. Because some littoral habitats support higher densities of fish, differences in habitat on embayment-wide scales should be considered when managing water levels. Geographic information systems are powerful tools that provide reservoir managers the ability to visually and analytically assess habitat data.

Landscape ecology principles can be applied to reservoir systems to increase understanding of the processes, scale, and effectiveness of management scenarios (Noble et al. 1994). Many reservoirs exhibit complex dendritic patterns, characteristic of the river and stream valleys that are inundated when they are formed. The complexities of the watershed can produce water bodies that are more spatially heterogeneous than most lake systems (Lodge et al. 1988). The measurement of landscape features in spatially complex lentic environments is a new approach toward understanding functional linkages in these systems (Meals and Miranda 1991; Dibble 1993; Beauchamp et al. 1994; Irwin 1994).

Studies of the effects of spatial heterogeneity on reservoir fish populations have largely focused on pelagic fish species, larval fish, and zooplankton on lakewide scales (Jenkins 1967, 1982; Van Den Avyle et al. 1982; Siler et al. 1986; Tisa and Ney 1991). These studies have shown consistent relationships between abiotic parameters and some aspects of reservoir fish populations on broad scales, but investigations at smaller scales have seldom been conducted. Recent studies have related littoral fish abundance to nearshore habitat mosaics (Meals and Miranda 1991; Dibble 1993; Irwin 1994). Matching scale of spatial heterogeneity with behavior of littoral-dwelling fishes is critical to understanding habitat needs of reservoir fishes and landscape approaches should be applicable (Cullinan and Thomas 1992; Noble et al. 1994).

The primary impetus for our work came from a long-term study of age-0 largemouth bass *Micropterus salmoides* in B. E. Jordan Reservoir, North Carolina, where consistent trends in abundance occurred among four bays over a 6 year period (Phillips 1993). Jackson et al. (1991) hypothesized that those differences in abundance were a function of difference in embayment habitats. Because age-0 largemouth bass exhibit strong site fidelity in that system (Copeland and Noble 1994), size and arrangement of habitats may affect distribution patterns. Irwin (1994) reported a positive relationship between abundance of age-0 largemouth bass and the distribution of steeply sloping, rocky habitats. Therefore, the landscape mosaic may regulate aspects of largemouth bass ecology. Quantification of landscape features on the proper scales should increase our ability to manage fish populations and fisheries.

Water level fluctuation in freshwater systems acts as disturbance in the landscape. Most studies of the

[1]Present address: Alabama Cooperative Fish and Wildlife Research Unit, 331 Funchess Hall, Auburn University, Alabama, 36849-5414, USA.

effects of water level fluctuation on fish populations have focused on recruitment processes, linking spawning success to water level variation (Zalewski et al. 1990; Kohler et al. 1993). However, the extent of littoral habitat is directly related to water level (Ploskey 1986). Because water level fluctuation affects littoral habitats disproportionately, attempts to relate abundance of fish based on habitat must consider both shoreline morphology and water level.

Geographic information systems (GIS) provide the opportunity to build spatially explicit databases that incorporate many variables, but these systems have been used only to a limited extent by fisheries managers (Meaden and Kapetsky 1991; Giles and Nielsen 1992). Current GIS technology, where data are collected with remote sensing techniques, can be used to model response of habitat area and arrangement to water level fluctuations in reservoirs. However, measurements taken on site may provide managers with adequate data. We used two methods—on-site field measurements and GIS—for littoral habitat assessment; our objectives were to quantify changes in habitat resulting from declines in water level using both methods and to compare the utility of the two techniques.

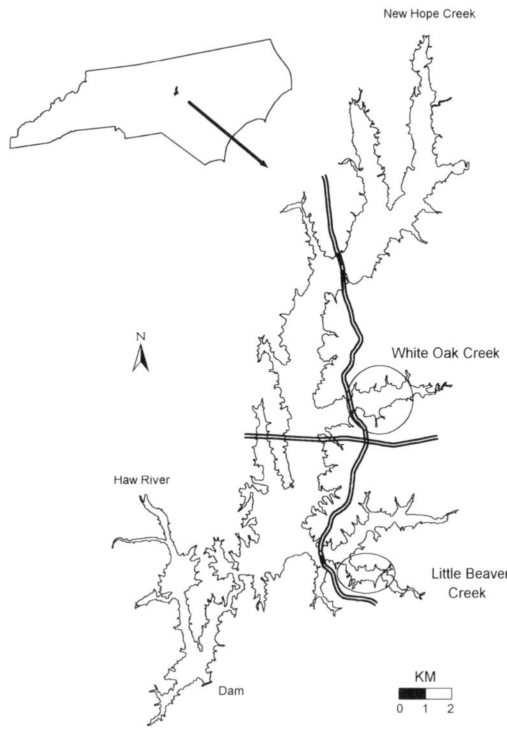

FIGURE 1.—Map of B. E. Jordan Reservoir showing the location of the reservoir in North Carolina (inset) and of the two study embayments (circled).

Study Site

B. E. Jordan Reservoir (Jordan Lake) is a 5,720-ha impoundment in the piedmont region of North Carolina (Figure 1). The dam began operating in 1981 on the Haw River below its confluence with the New Hope Creek. The primary use of the reservoir is flood control; other designated uses include fish and wildlife habitat, recreation, and water supply. Jordan Lake is shallow (mean depth = 5 m), turbid (Secchi disk transparency ≈0.5 m), and eutrophic (NCNRCD 1989). Annual variations in water level (usually ± 2 m) have prevented establishment of aquatic macrophytes. During the 2 years of our study, lake levels ranged from 64.0 to 68.6 m above mean sea level (msl, full pool = 65.8).

Little Beaver Creek (LBC) and White Oak Creek (WOC) embayments in Jordan Lake were studied (Figure 1). Surface area of LBC embayment is 64 ha at full pool with a shoreline length of 8.0 km; surface area of WOC at full pool is 172 ha with a shoreline length of 10.2 km (Table 1). Little Beaver Creek embayment supports higher densities of age-0 largemouth bass than WOC embayment (Jackson et. al 1991; Phillips 1993). Habitats within the two embayments are representative of those in other embayments of the reservoir.

Methods

On-Site Measurements

Habitat was quantified at three water levels, 65.8, 65.2, and 64.0 m msl in both WOC and LBC. Three 10 × 2 m habitat sites were systematically located in each of five areas in each embayment (*N* = 15 per

TABLE 1.—Summary of morphometrics for Little Beaver Creek and White Oak Creek embayments (from Atlas*GIS). Littoral area represents the area between the 65.8 and 64.6 m msl bathymetric contours and is reported to illustrate differences between the embayments.

	Little Beaver Creek		White Oak Creek	
Morphometric	65.8 m	64.6 m	65.8 m	64.6 m
Surface area (ha)	64.0	52.0	172.3	149.0
Shoreline length (km)	8.0	6.9	10.2	8.5
Littoral area (ha)	14.8		23.2	

bay) selected to correspond with fish sampling sites (Irwin 1994).

At each habitat site five transects were established at 2-m intervals perpendicular to shore. Substratum was quantified visually using a modified Wentworth scale (Bovee 1982). Cover composition (% of total) and total cover (% area covered) were estimated visually at two locations (one <0.5 and one >0.5 m from waterline) along each transect. Area of estimation included 0.5 m on both sides of transect for both locations. Cover components were small woody debris, cobble and boulder, stumps and standing timber, tree roots, grass hummocks, shrubs, and undercut banks. Depth was recorded at 0.5-m intervals along the transects for slope calculations. Habitat sites were marked with flagging and described relative to landmarks so that they could be located again for reassessment at the other water levels. Additional sites (42 in WOC; 40 in LBC) were systematically placed around the shoreline in both embayments to assess embayment-wide habitat at full pool only.

Habitat data were pooled at each site to provide a site mean for each variable. A one-way ANOVA (α = 0.05) was used to test water level effects on habitat components and differences between habitat characteristics in LBC and WOC. Tukey's multiple range test was used to identify differences among water levels.

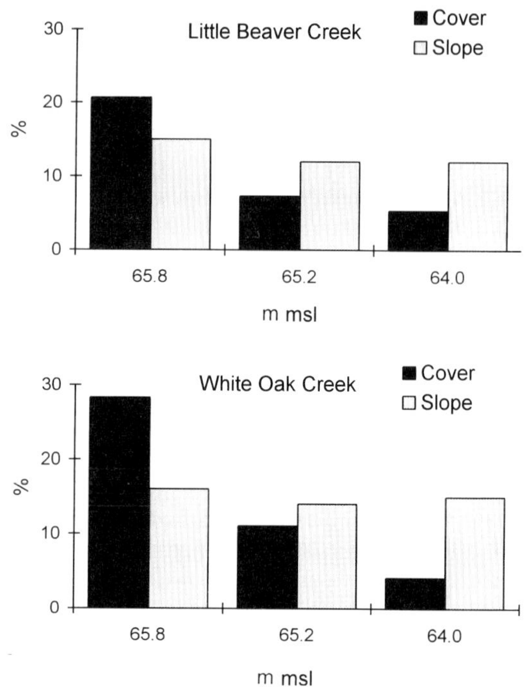

FIGURE 2.—Cover and slope characteristics from the 15 fixed habitat sites at three water levels (65.8, 65.2, and 64.0 m msl) in Little Beaver Creek and White Oak Creek. Significant declines in cover were detected when water level decreased from 65.8 to 65.2 m msl in both embayments (ANOVA, $P < 0.01$).

Geographic Information Systems Database

A digital elevation database of WOC and LBC was developed by digitizing two bathymetric isopleths (65.8 and 64.6 m msl) from aerial photographs of the reservoir into a vector GIS (Atlas*GIS, Strategic Mapping Inc., Sunnyvale, California). Aerial photographs from two sources were used: Chatham County Tax Assessors Office (map scale 1:400; 65.8 m msl) and Soil Conservation Service (map scale 1:660; 64.6 m msl). Substratum data were recorded for both embayments by video taping the entire shoreline when lake levels were low (2 m msl below full pool). Audio data were recorded while video data were collected. This procedure enabled us to visually quantify patches of dominant substratum in the field and locate them relative to shoreline features (e.g., point of land or the back of cove), but notation of cover variables was not feasible at this resolution. Dominant substratum type was identified and transferred to paper maps while viewing video tapes (with audio enabled) using a video cassette recorder. Substratum data were then digitized into the GIS database.

A raster-based software package (IDRISI, Clark University, Worcester, Massachusetts) was then used to develop a digital elevation model (DEM); using the INTERCON module with a cell size of 3 m^2. Substratum layers were imported from Atlas*GIS to the DEM. Slope data were calculated for individual cells using the SURFACE module. Also, the final database was used to model changes in macrohabitat (e.g., gravel patches; Irwin 1994) at various water levels.

Results

Fixed Site Measurements

Habitat data assessed at the 15 fixed sites and three water levels indicated that percent total cover decreased significantly in both LBC and WOC as water level declined from 65.8 to 65.2 m msl ($P <$ 0.01) but did not decrease as water level declined from 65.2 to 64.0 m msl (Figure 2). In contrast, slope characteristics did not change as water level decreased. Although LBC and WOC differed sig-

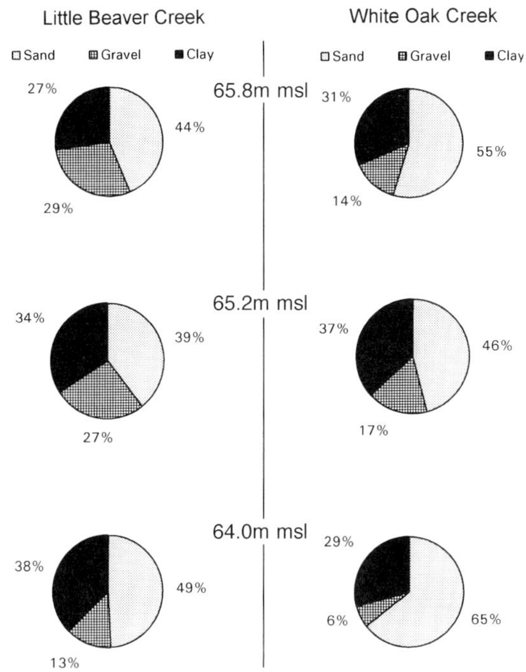

FIGURE 3.—Substratum characteristics from the 15 fixed habitat sites at three water levels (65.8, 65.2, and 64.0 m msl) in Little Beaver Creek and White Oak Creek. Significant declines in gravel were detected when water level decreased from 65.8 to 64.0 m msl in LBC (ANOVA, $P = 0.10$).

nificantly ($P < 0.10$) in substratum characteristics at full pool (65.8 m msl) as assessed at the 15 fixed sites, both showed a similar strong trend in reduction of gravel with declining water level (Figure 3). All substratum proportions changed little from 65.8 to 65.2 m msl, but from 65.2 to 64.0 m msl, percentage of gravel declined by over half in each embayment. The reduction in gravel from 65.8 to 64.0 m msl was significant ($P = 0.10$) in LBC.

Habitat data obtained from 55 to 57 sites (15 fixed sites and 42 or 40 sites selected systematically throughout shores of each embayment) indicated similar substrate composition at full pool to that estimated at the 15 fixed sites in the five areas (Table 2). Composition of substrata differed significantly ($P < 0.05$) between LBC and WOC. Based on the 55–57 site analysis, LBC had relatively more gravel (mean 21.0%; SE = 3.3) and clay (mean 37.8%; SE = 3.8) than WOC (gravel mean 11.6%, SE = 2.6; clay mean 27.6%, SE = 3.0) and WOC had more sand (mean 60.9%; SE = 3.8) than LBC (mean 41.1%; SE = 3.7). Slope and total cover were similar between embayments at full pool.

TABLE 2.—Estimates of percent substrate composition at full pool (65.8 m msl) in Little Beaver Creek and White Oak Creek embayments, based on 15 sites (3 sites in each of 5 areas in each embayment) and 55–57 sites systematically distributed along the total shoreline.

	Little Beaver Creek		White Oak Creek	
Substrate	15	55	15	57
Gravel (%)	29	21	14	12
Clay (%)	27	38	31	28
Sand (%)	44	41	55	60

Results obtained from the 55–57 systematically selected sites were not markedly different from those obtained from the 15 sites located within five selected areas in each embayment. Irwin (1994), by random selection of sample sizes from 5 to 100 sites in LBC, found that definition of habitat characteristics was not highly sensitive to the number of sites sampled, either in terms of means or variance.

Geographic Information Systems Database

Total littoral shoreline area in LBC between the 65.8 and 64.6 m msl bathymetric contours was 14.8 ha (Table 1). Substratum composition in LBC was 7.3 ha sand (49.3%), 5.3 ha clay (35.8%), and 2.2 ha gravel (15.0%) (Figure 4). The WOC littoral shoreline area was 23.2 ha (Table 1.). Substratum composition in WOC was 21.6 ha sand (93.1%), 0.8 ha clay (3.4%), and 0.8 ha gravel (3.4 %) (Figure 5). Slope ranged from 1.0 to 29.0% in both LBC and WOC and mean slope was 8.1 and 5.5 in the two embayments, respectively (Figure 6). Gravel at different water levels ranged from 11.5% at 64.9 m msl to 16.4% at 65.2 m msl in LBC and from 0.8% at 64.6 m msl to 5.7% at 65.5 m msl in WOC (Figure 7). Decreases in water level affected the size and arrangement of gravel patches. The summation of the distance between patches along a 770-m section of shoreline with a high percentage of gravel increased from 170 to 540 m and total area of gravel patches decreased from 0.55 to 0.028 ha when water level decreased from 65.8 m msl to 64.6 m msl (Figure 8).

Discussion

Small decreases in water level in Jordan Lake significantly decreased cover and changed substratum composition in LBC and WOC embayments. Several recent studies on littoral zone habitats in lakes have documented changes in littoral zone cover and substratum composition as lake levels decline. Beauchamp et al. (1994) estimated that

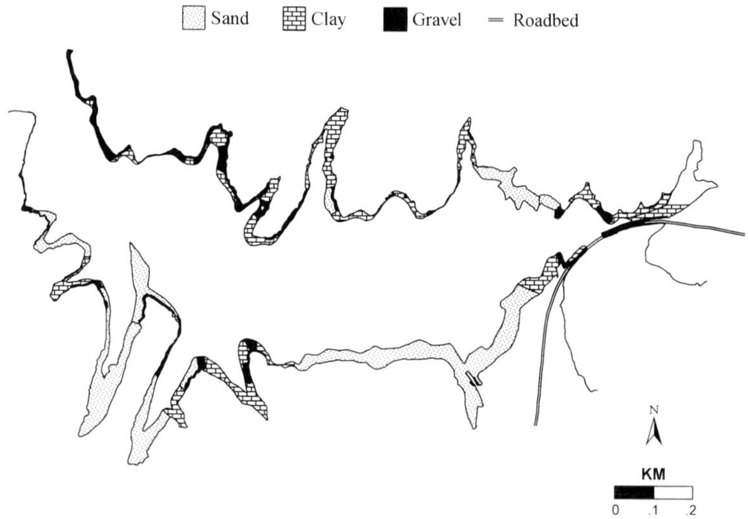

FIGURE 4.—Atlas*GIS map of Little Beaver Creek indicating substratum composition in the littoral area (65.8–64.6 m msl).

20% of rocky substratum, important as cover for small native fishes, was exposed during a drought that lowered Lake Tahoe's water level by 2 m. Gasith and Gafny (1990) described changes in substratum composition, from rocky to sandy, in Lake Kinneret, Israel, with a 4-m decrease in water level. Dibble (1993) mapped patches of habitat in a reservoir cove and reported that gravel substrata were located in shallow water relative to sandy substrata; he hypothesized that a decline in water level would

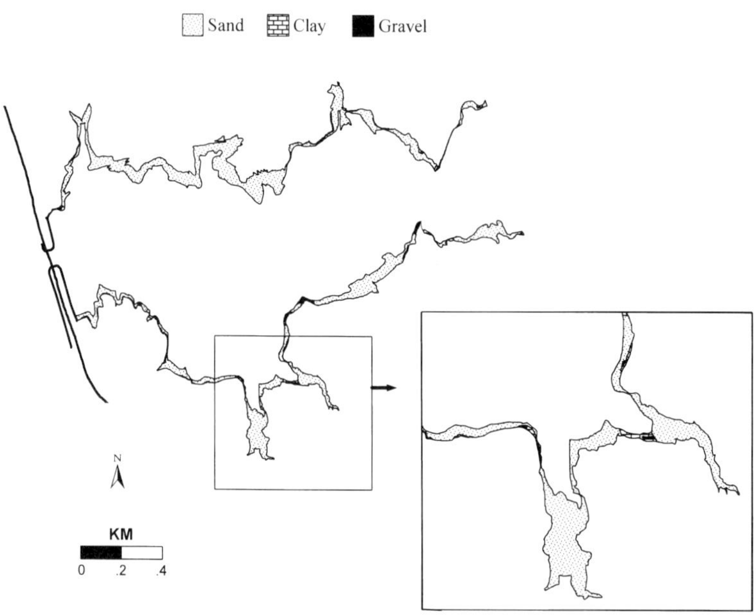

FIGURE 5.—Atlas*GIS map of White Oak Creek indicating substratum composition in the littoral area (65.8–64.6 m msl). The inset is provided for better visual resolution.

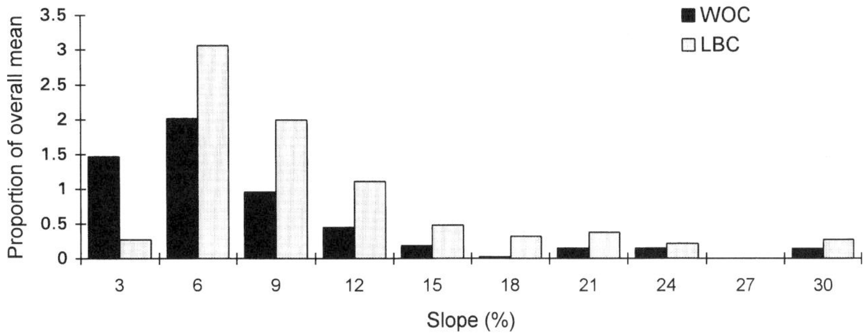

FIGURE 6.—Slope characteristics for Little Beaver Creek (LBC) and White Oak Creek (WOC) measured with IDRISI software. Mean slope was 8.1% and 5.5% for LBC and WOC, respectively. Values are represented as a proportion of the overall mean for relative scaling.

affect gravel habitat, which he found supported high densities of age-0 largemouth bass. Because habitat is a primary factor regulating the distribution and abundance of age-0 largemouth bass (Irwin 1994), changes in habitat with water level fluctuations may impact populations of this important species.

Both the GIS and the embayment-wide on-site assessment provided similar data, but the on-site assessment provided finer resolution. For example, components of cover were easily included in the on-site assessment, while inclusion in the GIS would have been very labor intensive. Also, because a minor decrease in water level significantly decreased cover in the littoral zone of both embayments, small scale measures of cover would provide better data than an embayment scale GIS database. However, some changes in habitat were not detected with the small scale, on-site measures. Patches of gravel decreased in size and became fragmented as water level receded. The changes in patch arrangement and size were more easily analyzed using the GIS. Habitat loss and fragmentation may be significant to fish by decreasing embayment carrying capacity or by forcing fish into marginal habitats.

Accurate on-site habitat measures could be obtained by sampling as little as 1% of the shoreline. Parameters estimated by measuring with instrumentation are less subject to investigator bias than those estimated visually (Block et al. 1987). Only one investigator (E. Irwin) measured habitat parameters; thus observer bias, if any, should have been consistent. Neither precision nor accuracy were particularly sensitive to sample sizes over the range of 15–57 used in this study. A sample size of 55–57 was adequate to detect differences in habitat parameters between LBC and WOC.

The GIS provided rapid visual assessment of hab-

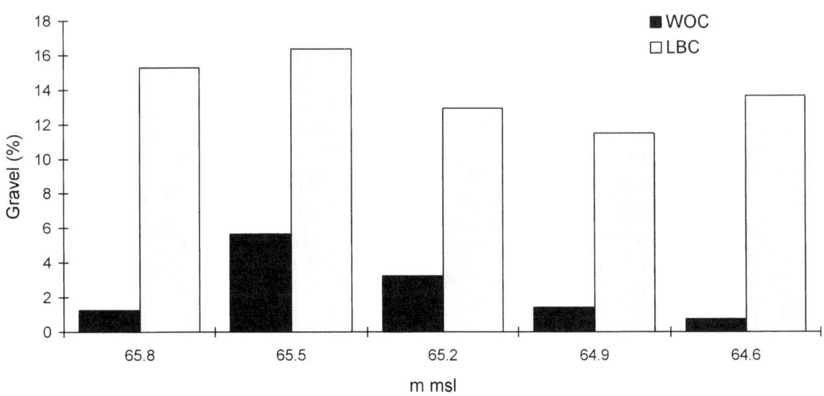

FIGURE 7.—Mean percent gravel in Little Beaver Creek (LBC) and White Oak Creek (WOC) at five water levels (m msl). Grand mean percent gravel was 14.0 and 2.5 for LBC and WOC, respectively.

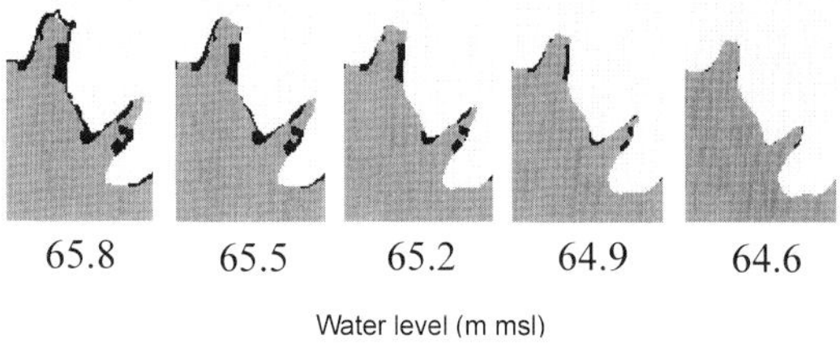

FIGURE 8.—Map (generated with IDRISI software) of gravel substratum (illustrated in solid black) along 770 m of shoreline in Little Beaver Creek showing reduction of habitat and fragmentation of patches as water level declines from 65.8 to 64.6 m msl.

itat differences between LBC and WOC. It was apparent from GIS maps that WOC had a higher percentage of sand substratum than LBC. The on-site measures and GIS produced similar characterization of habitat with the exception of slope. The slopes calculated with the GIS indicated that WOC has shallower sloping shorelines than LBC, whereas no differences in slope were detected using on-site measures.

Use of GIS in conjunction with field data can advance our ability to make predictions and management decisions in aquatic systems. A GIS for habitat database management allows quantification of habitat parameters on large (embayment) scales and requires fewer worker hours for analysis. For example, a team of three workers can sample 25 habitat sites (on-site; total 250 m of shoreline) during one work day, whereas a team of only two workers can record the entire shoreline (up to 12 km) in several hours with video techniques during drawdown. Even when transcription of video data is considered, GIS analysis is more efficient than on-site measures, unless cover data are required.

The GIS provided data regarding habitat changes at decreased water levels. Although only limited control over reservoir operations is possible in multiple use reservoirs, identification of critical water levels (those at which fish habitat becomes limiting) may influence decision-making when recreational fishing is a primary management objective. Knowledge of the impacts of water level manipulation on nursery areas may help avert negative effects on littoral fishes. Because spatially oriented fisheries data such as these are common, we believe that GIS can help to quantify relationships between biotic and abiotic components of landscapes, and will increase the ability to manage complex heterogeneous systems.

Acknowledgments

This research was supported by Federal Aid in Sport Fish Restoration Project F-30-3, administered by the North Carolina Wildlife Resources Commission, Division of Boating and Inland Fisheries. Additional funds were provided through an American Fishing Tackle Manufacturers Association Fellowship awarded to E. Irwin. We express gratitude to Bob Becker and Jean Marie McManus at the North Carolina State University GIS Laboratory for their assistance with Atlas*GIS and IDRISI. Data collections were facilitated by the expertise of James Jackson and field workers Kin Hodges, Andrew Leonard and John Copeland. James Rice, JoAnn Burkholder, James Gregory, Mike Allen, Kevin Rogers, and an anonymous reviewer provided valuable comments on the manuscript.

References

Beauchamp, D. A., E. R. Byron, and W. A. Wurtsbaugh. 1994. Summer habitat use by littoral-zone fishes in Lake Tahoe and the effects of shoreline structures. North American Journal of Fisheries Management 14:385–394.

Block, W. M., K. A. With, and M. L. Morrison. 1987. On measuring bird habitat: influence of observer variability and sample size. Condor 89:241–251.

Bovee, K. D. 1982. A guide to stream habitat analysis using the Instream Flow Incremental Methodology. U. S. Fish and Wildlife Service Instream Flow Information Paper 12, Washington, DC.

Copeland, J. R., and R. L. Noble. 1994. Movements by young-of-year and yearling largemouth bass and their implications for supplemental stocking. North American Journal of Fisheries Management 14:119–124.

Cullinan, V. I., and J. M. Thomas. 1992. A comparison of quantitative methods of examining landscape pattern and scale. Landscape Ecology 7:211–227.

Dibble, E. D. 1993. A patch dynamics approach for the study of habitat use by juvenile centrarchids in an Ozark reservoir: an investigation of factors affecting habitat availability and an experimental test of the predator avoidance hypothesis. Doctoral dissertation. University of Arkansas, Fayetteville.

Gasith, A., and S. Gafny. 1990. Effects of water level fluctuation on the structure and function of the littoral zone. Pages 156–171 *in* M. M. Tilzer and C. Serruya, editors. Large lakes: ecological structure and function. Springer-Verlag, New York.

Giles, R. H., and L. A. Nielsen. 1992. The uses of geographic information systems in fisheries. Pages 81–94 *in* R. H. Stroud, editor. Fisheries management and watershed development. American Fisheries Society Symposium 13.

Irwin, E. R. 1994. Habitat mediation of juvenile largemouth bass population dynamics. Doctoral dissertation. North Carolina State University, Raleigh, North Carolina.

Jackson, J. R., J. A. Rice, R. L. Noble, and S. C. Mozley. 1991. Mechanisms of reservoir fish community dynamics. North Carolina Agricultural Research Service, North Carolina State University, Federal Aid in Sport Fish Restoration, Project F-30-1, Final Report, Raleigh.

Jenkins, R. M. 1967. The influence of some environmental factors on standing crop and harvest of fishes in U.S. reservoirs. Pages 298–321 *in* Reservoir Committee, editors. Reservoir fishery resources symposium. American Fisheries Society, Southern Division, Reservoir Committee, Bethesda, Maryland.

Jenkins, R. M. 1982. The morphoedaphic index and reservoir fish production. Transactions of the American Fisheries Society 111:133–140.

Kohler, C. C., R. J. Sheehan, and J. J. Sweatman. 1993. Largemouth bass hatching success and first-winter survival in two Illinois reservoirs. North American Journal of Fisheries Management 13:125–133.

Lodge, D. M., and seven coauthors. 1988. Spatial heterogeneity and habitat interactions in lake communities. Pages 181–208 *in* S. R. Carpenter, editor. Complex interactions in lake communities. Springer-Verlag, New York.

Meaden, G. J., and J. M. Kapetsky. 1991. Geographical information systems and remote sensing in inland fisheries and aquaculture. FAO (Food and Agriculture Organization of the United Nations) Fisheries Technical Paper 318.

Meals, K. O., and L. E. Miranda. 1991. Variability in abundance of age-0 centrarchids among littoral habitats of flood control reservoirs in Mississippi. North American Journal of Fisheries Management 11:298–304.

NCNRCD (North Carolina Department of Natural Resources and Community Development). 1989. 1988 North Carolina Lakes Monitoring Report. NCNRCD report 89-04, Raleigh.

Noble, R. L., J. R. Jackson, E. R. Irwin, J. M. Phillips, and T. N. Churchill. 1994. Reservoirs as landscapes: implications for stocking programs. Transactions of the North American Wildlife and Natural Resources Conference 59:281–288.

Phillips, J. M. 1993. Spatial and temporal variation in year-class strength of age-0 largemouth bass in a southeastern reservoir. Doctoral dissertation. North Carolina State University, Raleigh.

Ploskey, G. R. 1986. Effects of water-level changes on reservoir ecosystems with implications for fisheries management. Pages 86–97 *in* G. E. Hall and M. J. Van Den Avyle, editors. Reservoir fisheries management: strategies for the 80's. American Fisheries Society, Southern Division, Reservoir Committee, Bethesda, Maryland.

Siler, J. R., W. J. Foris, and M. C. McInerny. 1986. Spatial heterogeneity in fish parameters within a reservoir. Pages 86–97 *in* G. E. Hall and M. J. Van Den Avyle, editors. Reservoir fisheries management: strategies for the 80's. American Fisheries Society, Southern Division, Reservoir Committee, Bethesda, Maryland.

Tisa, M. S., and J. J. Ney. 1991. Compatibility of alewives and gizzard shad as reservoir forage fish. Transactions of the American Fisheries Society 120:157–165.

Van Den Avyle, M. J., R. S. Hayward, R. A. Krause, and A. J. Spells. 1982. Spatial variation in abundance of phytoplankton, zooplankton, and larval fishes in Center Hill Reservoir. Canadian Water Resources Journal 7:189–213.

Zalewski, M., B. Brewinska-Zaras, P. Frankiewicz, and S. Kalinowski. 1990. The potential for biomanipulation using fry communities in a lowland reservoir: concordance between water quality and optimal recruitment. Hydrobiologia 200/201:549–556.

Aquatic Macrophytes as Habitat for Reservoir Fish

American Fisheries Society Symposium 16:335–346, 1996

Habitat Heterogeneity and Fish Community Structure: Inferences from North Temperate Lakes

MELISSA J. WEAVER[1] AND JOHN J. MAGNUSON
University of Wisconsin–Madison, Center for Limnology
680 North Park Street, Madison, Wisconsin 53706, USA

MURRAY K. CLAYTON
University of Wisconsin–Madison, Department of Statistics
1210 Dayton Avenue, Madison, Wisconsin 53706, USA

Abstract.—We review a broad range of studies of north temperate lake fish communities that quantify the role of habitat heterogeneity and biotic interactions in determining community structure. In community analysis, distribution patterns of species are identified, and these patterns are related to environmental and biotic factors using multivariate analysis. The community structure (i.e., species composition, abundance, and distribution) represents the response of the community to past and current socioenvironmental factors, both locally and regionally. Fish community structure and stability (i.e., constancy or resilience) have been associated with the presence, abundance, species composition, growth form, and structural heterogeneity of macrophytes in various studies. The mechanisms that maintain or alter community structure (e.g., predation, anoxia, and invasion), and their relative importance, differ among communities and vary with spatial and temporal scale. However, models can be developed to predict the abundance, species richness, or species distribution of fishes within and among lakes or reservoirs. Through comprehensive analysis of fish communities, fisheries biologists can understand better the structure and dynamics of the systems they study and manage.

Fish biologists have traditionally viewed fisheries at the population level, considering individual species in individual lakes, rather than at the community level. Aquatic ecosystems are inherently complex, and understanding the processes that determine the distribution and abundance of single fish populations is demanding. Further, because biotic interactions can be important, understanding the processes that structure a community requires understanding those processes that affect population interactions as well. An alternative approach is to examine the fish community as a whole, searching for similarities and differences in species distribution patterns, and relating them to environmental and biotic factors. This can be accomplished using multivariate community analysis techniques that consider several variables simultaneously and reduce a complex system to a tractable series of variables (Gauch 1982; Digby and Kempton 1987; Krzanowski 1988).

Fish distribute themselves within and among lakes or reservoirs according to their behavioral and physiological requirements and the availability of preferred habitat (Werner 1986). For a single species, the population is most dense where environmental conditions are optimal for that species. However, because habitats are multidimensional, all aspects of the habitat are unlikely to be optimal in the same area. Social factors may lead to displacement of fishes from their preferred habitat (Magnuson et al. 1979). Intraspecific and interspecific interactions (e.g., density dependence, competition, and predation) can alter distributions. One species may displace a competitor to a suboptimal habitat, resulting in reduced growth or survival; however, preferential predation on the stronger competitor may facilitate coexistence (Werner 1986). Thus, it is reasonable to consider the distribution of fishes as a community response to the environment rather than as separate species responses to abiotic and biotic factors.

The structure of a community (e.g., species composition, abundance, and distribution) can be viewed as the response of the community to past and current socioenvironmental factors. Fishes with similar responses tend to co-occur, forming distinct assemblages. Thus, classification of assemblages may indicate similarities (within assemblages) and differences (among assemblages) in abiotic conditions as perceived by fishes, or in the strength and direction of biotic interactions (Tonn and Magnuson 1982; Rahel 1984). Geophysical barriers or historical events (e.g., glaciation) may also restrict or alter the distribution of fishes and enable the estab-

[1]Present address: The Ecosystems Center, Marine Biological Laboratory, Woods Hole, Massachusetts 02543, USA.

lishment of disjoint regional faunas (Hinch et al. 1991). Processes that structure a community may be discerned by examining the differential responses of assemblages, as well as of individual species, to environmental factors.

To understand the relationship between the distribution of fishes and habitat, it is necessary to characterize and quantify aspects of the environment that are important to fishes. Abiotic factors, such as temperature, turbidity, and acidity can affect fish distributions (Rahel 1984; Brazner and Magnuson 1994). Littoral structure (e.g., substrate, macrophytes, or debris) provides habitat for spawning and forage as well as refuge from predation (Heck and Crowder 1991; Brazner and Magnuson 1994), and the amount and complexity of littoral structure may be important measures of the value of littoral habitat to fish (Benson and Magnuson 1992). The challenge is to choose environmental variables, such as abiotic factors and habitat characteristics, that describe the habitat as perceived by fishes in the community.

By relating similarities and differences among fish assemblages to environmental variables, biologists and managers gain insight into fish habitat use and community interactions. Based on relationships between fish distribution patterns and environmental variables, mechanisms regulating community structure can be inferred and testable hypotheses generated. Models and indices derived from these analyses can be applied to other lakes to predict fish assemblages from environmental variables (Tonn et al. 1983; Carline 1986). Using these models, the potential impacts of proposed management actions, such as species introductions and habitat modifications, can be investigated (Carline 1986).

In this paper we review several studies of fish communities in small (<100 km^2 in surface area) north temperate lakes, most of which are located on the Laurentian Shield (especially in Wisconsin, Ontario, and Quebec). The studies we review range in scope from littoral fish assemblages within one lake to fish assemblages in multiple lakes on two continents. Moreover, these studies use a broad range of environmental variables to elucidate mechanisms (e.g., predation and colonization) important in structuring communities. We emphasize methods for characterizing habitat, especially macrophyte habitat, and for discerning associated processes. We also discuss sampling biases and issues of scale, topics of interest to both fisheries biologists and community ecologists. Finally, we speculate on how generalizations derived from studies of reservoir fish communities may differ from those derived from studies of natural lake communities.

Characterization of Habitat

Abiotic Conditions and Predation

The respective roles of abiotic conditions and predation in structuring fish communities is apparent from several regional studies (Table 1). Only a few environmental variables characterizing severe conditions were necessary to discern two distinct fish assemblages in a study of 18 small lakes (0.02–0.90-km^2 surface area) in northern Wisconsin (Tonn and Magnuson 1982). Centrarchid–*Esox* assemblages occurred in lakes with high winter oxygen levels or with a stream or lake connection affording refuge during periods of anoxia. Low winter oxygen lakes without piscivores contained *Umbra*–cyprinid assemblages. Similarly, in an analysis of 43 Wisconsin bog lakes (0.004–0.247 km^2), Rahel (1984) identified three fish assemblages separated by the level of tolerance to winter anoxia and low pH and by the presence of piscivores. Acidic lakes subject to winter anoxia contained only the *Umbra–Perca* assemblage or were fishless. A cyprinid assemblage dominated in lakes with low winter oxygen and moderate pH; centrarchids were absent due to their higher oxygen requirements. Acidic lakes with moderate winter oxygen contained a centrarchid assemblage; cyprinids were absent due to their intolerance of low pH. While the centrarchid and cyprinid assemblages were distinct, the *Umbra–Perca* assemblage consisted of species common in both of the other assemblages. This is not surprising, because yellow perch *Perca flavescens* and central mudminnow *Umbra limi* tolerate both low pH and low oxygen conditions and are found in a variety of habitats. Lakes with moderate pH and winter oxygen concentrations were dominated by centrarchids, and golden shiner *Notemigonus crysoleucas* was the only cyprinid or soft-rayed fish present. Adults of all other cyprinids are small and vulnerable to predation. Thus, two environmental conditions (acidity and winter oxygen) and one biotic interaction (predation) determined the distribution of fishes among small north temperate lakes.

In a study of coastal Green Bay, Wisconsin (Brazner and Magnuson 1994), regional differences in water temperature and turbidity separated the littoral community into a lower-bay, warm-water, turbidity-tolerant assemblage (e.g., gizzard shad *Dorosoma cepedianum*, white bass *Morone chrysops*,

TABLE 1.—Fish community studies at differing spatial, temporal, and numerical scales, using various abiotic conditions and habitat characterizations to explain distribution patterns and ecosystem processes. Numerical scale: rank abundance = rank of catch abundance, relative abundance = proportion of total catch abundance, abundance = species total catch abundance.

Reference	Spatial and temporal scale	Numerical scale	Processes, mechanisms, and patterns[a]	Abiotic conditions and habitat characteristics[a]
Benson and Magnuson (1992)	Wisconsin lakes, and within lake (littoral)	Community heterogeneity and rarefied richness	Habitat heterogeneity	Depth, macrophyte, and substrate diversity
Brazner and Magnuson (1994)	Green Bay (Lake Michigan) coastal wetlands and beaches	Abundance (juvenile and adult, species and ecotype)	Human disturbance, habitat usage, adaptation	Proportion vegetated, turbidity, macrophyte richness, development
Eadie and Keast (1984)	Northern and southern Ontario lakes	Relative abundance, species richness and diversity	Glaciation, colonization, resource diversity	Zooplankton, benthos, shoreline development, plant complexity
Hinch et al. (1991)	Central Ontario lakes (littoral)	Rank abundance, abundance	Predation, postglacial colonization	Lake morphology, percent littoral, elevation
Hinch and Collins (1993)	Central Ontario lakes (littoral)	Relative abundance	Predation, habitat cover, forage	Macrophyte cover, productivity, lake morphology
Pierce et al. (1994)	Southern Quebec lakes (littoral), early and late summer	Species richness, biomass	Prey resources, habitat usage	Area, chironomid biomass
Rahel (1984)	Wisconsin bog lakes	Presence–absence	Physiological constraints, predatory exclusion, succession	Winter anoxia, pH, proportion hard substrate or vegetated, productivity, area
Tonn and Magnuson (1982)	Northern Wisconsin lakes, winter and summer	Presence–absence	Physiological constraints, predation, refugia, disturbance	Isolation, winter anoxia, depth, substrate, vegetation diversity
Tonn et al. (1990)	Wisconsin and Finnish small lakes	Presence–absence, relative biomass	Contemporary versus historical, local versus regional	Area, pH, isolation, conductivity, maximum depth
Weaver et al. (1993)	Lake Mendota, Wisconsin (littoral)	Presence–absence, rank and absolute abundance	Sampling gear biases, numerical resolution	
Weaver et al. (in press)	Lake Mendota, Wisconsin (littoral)	Abundance (age-0 and adult)	Lake-wide community heterogeneity, fish–macrophyte interactions, vegetation	Macrophyte abundance and biomass, species richness, patchiness, structural variability, milfoil abundance

[a]Only some of the processes and mechanisms or abiotic conditions and habitat characteristics related to fish community structure in these studies are listed here.

and common carp *Cyprinus carpio*) and an upper-bay, cool- and clear-water assemblage (e.g., centrarchids, northern pike *Esox lucius*, and golden shiner). Locally, where there was human development near shore, fish abundance and richness were lower and characteristic species were more disturbance tolerant (e.g., bullheads *Ameiurus* spp. and rainbow smelt *Osmerus mordax*) than where there was no development. The effects of abiotic conditions (turbidity and water temperature) and nearshore development on the distribution of littoral fishes in Green Bay were apparent at both regional and local spatial scales.

Lake Morphology

Habitat and resource (e.g., prey) diversity can increase with lake area and shoreline development, so that larger lakes or lakes with more convoluted shorelines can harbor not only a more abundant but also a richer and more diverse fish community. Fish species richness in 10 southern Quebec lakes (0.3–90.1 km^2) was positively correlated with lake area, but neither total biomass nor biomass of individual species were correlated with lake area (Pierce et al. 1994). Species richness in 70 lakes in northern Ontario (approximately 0.02–40 km^2) was correlated with both lake area and shoreline development (Eadie and Keast 1984).

Lake morphology can also predict whether certain fish species or assemblages are likely to occur. Wisconsin bog lakes with centrarchid assemblages tended to be larger and have more hard substrate and vegetation than did lakes with other assemblages (Rahel 1984). For small lakes in Wisconsin, lake surface area was positively related to summer species richness in both assemblage types (centrarchid–*Esox* and *Umbra*–cyprinid) (Tonn and Magnuson 1982). Among 25 central Ontario lakes (0.29–11.42 km^2), smallmouth bass *Micropterus dolomieu* tended to occur in deep lakes with proportionally less littoral area and brown bullhead *Ameiurus nebulosus* in shallow lakes with more littoral area (Hinch et al. 1991). Additionally, Benson and Magnuson (1992) found a positive correlation between fish species diversity and within-lake variation in the nearshore depth gradient for six lakes (0.37–39.86 km^2). In all of these studies, lake morphology explained some variation in fish species diversity or in the distribution of particular fish species or assemblages among lakes.

Prey Abundance and Diversity

A few studies have successfully related factors considered to be correlates of prey availability (e.g., mean depth, primary productivity, and benthos) to fish production, growth, and biomass. For example, Ryder's morphoedaphic index (total dissolved solids/mean depth) has been correlated with both benthos production (Brinkhurst 1975) and fish yield (Ryder 1965, 1982). However, among several littoral prey resources in southern Quebec lakes, only the biomass of chironomids (major prey items) was related to biomass of the dominant fishes (yellow perch and golden shiner) and total fish biomass (Pierce et al. 1994). There was no correlation between total benthic invertebrate biomass and total fish biomass, even when total benthos was standardized by mean depth for each lake. In northern Ontario lakes, littoral fish species diversity was correlated with family or order level benthos diversity and fish species richness with genus level zooplankton diversity, but not in southern Ontario lakes (Eadie and Keast 1984). Prey biomass and diversity are apparently inadequate measures of prey availability, perhaps due to the patchy distributions of fishes and their prey (Pierce et al. 1994).

Macrophytes as Habitat for Fishes

Macrophytes provide forage and refuge from predation for fishes (Heck and Crowder 1991). Because biotic interactions among fishes can be mediated by macrophytes (Heck and Crowder 1991), the relationship between fish presence and abundance and macrophytes may not be simple (e.g., correlations). Yet, fish community structure and stability (i.e., constancy or resilience) have been associated with the presence, abundance, species composition, growth form, and structural heterogeneity of macrophytes in various studies (Table 1).

Macrophyte Cover

In the study of Wisconsin bog lakes, Rahel (1984) determined the percentage of stations, chosen randomly, that had aquatic vegetation or hard substrate and scored the lakes from 0 to 9 according to these percentages. Lakes with centrarchid assemblages tended to have more areas of vegetation and hard substrate than did other lakes. Hinch and Collins (1993) categorized macrophyte cover in 15 central Ontario lakes as high (25–70%) or low (<10%), and ranked 10 northeastern Ontario lakes, based on subjective estimates of the percentage of littoral area covered by submerged and emergent, nonrosette forms. Pumpkinseeds *Lepomis gibbosus* or bluegills *L. macrochirus* were abundant in lakes with high primary productivity and extensive macrophyte cover. Hinch and Collins (1993) concluded that high productivity and macrophyte cover were important for growth and survival of both adults and juveniles of these species.

Macrophyte and Substrate Diversity

Tonn and Magnuson (1982) categorized depth, substrate (i.e., proportions of leaf litter, muck, sand, and gravel), and vegetation type (i.e., emergent, floating, submergent, and all combinations) at stations along sampling transects and calculated Shannon–Wiener diversity index values for habitat in each lake based on the proportion of all stations in each category. Greater habitat diversity was associated with higher summer fish species richness in both assemblage types (centrarchid–*Esox* and *Umbra*–cyprinid). Benson and Magnuson (1992) compared within-lake fish community heterogeneity and within-lake habitat heterogeneity in six Wisconsin lakes. They used habitat categories similar to those of Tonn and Magnuson (1982), but separated submergent macrophytes into two growth forms (rosette and flexuous) and categorized substrate as muck, sand, gravel, cobble, or some combination.

Both within-lake fish community heterogeneity (calculated as average percent dissimilarity over all pairs of seine sites within a lake) and species richness were correlated with depth gradient diversity (Shannon–Wiener) and, more weakly, with macrophyte diversity (Shannon–Wiener).

Similarly, in the study of Green Bay coastal wetlands and beaches, Brazner and Magnuson (1994) calculated a macrophyte diversity index composed of percent cover (to the nearest 10%), species richness, and growth form diversity. This composite index was greater for wetlands than for beaches. Wetlands also tended to have more vegetation cover, substrate diversity, and sediment organic carbon and smaller waves than did beaches. Additionally, undeveloped sites had more vegetation cover and lower conductivity than did developed sites. Undeveloped wetlands had the highest fish diversity and abundance of all sites and were characterized by fishes associated with macrophytes (e.g., yellow perch, sunfish, and northern pike). In general, fish were more abundant and species rich where macrophyte cover and richness were high.

Structural Complexity

Although difficult to assess, macrophyte structural complexity has been estimated in several ways. Eadie and Keast (1984) sampled plant cover and substrate composition within randomly selected quadrats in northern and southern Ontario lakes. Percent cover of each plant species, height of the plant nearest the center, and percentages of each of eight substrate types were estimated, and substrate and plant species diversity indices were calculated. Horizontal vegetation patchiness was calculated as the coefficient of variation (CV; standard deviation/mean) of plant cover. As a measure of vertical vegetation complexity, the proportion of plant height in each of five categories (0–0.25, 0.25–0.50, 0.50–0.75, 0.75–1.0, and >1.0 m) was estimated and used to calculate foliage height diversity. Fish species diversity in southern Ontario lakes was best predicted by substrate diversity and vertical vegetation complexity and not by plant species diversity or horizontal patchiness (Eadie and Keast 1984). Vertically complex macrophyte beds may afford prey species better cover from visual predators. Among northern Ontario lakes, none of these measures of within-lake habitat heterogeneity was correlated with fish species richness and diversity, although fish species diversity was positively correlated with the diversity of invertebrate prey (benthos and zooplankton).

Weaver et al. (personal observation) related the distribution of the Lake Mendota (40 km^2) littoral fish community to the abundance and structural complexity of macrophytes. Macrophytes were sampled along transects set approximately perpendicular to depth contours at 20 littoral sites (1) using a double-headed rake to estimate species rank abundances (Deppe and Lathrop 1993) and (2) using acoustical equipment (sonar) to estimate water depth, plant height, and vegetation density at 1-m horizontal intervals. From these data several measures of macrophyte abundance, structural complexity, and species diversity were derived. Eurasian milfoil *Myriophyllum spicatum* was more abundant off shore (2–5-m depth) than near shore, and macrophyte species richness was reduced at sites where milfoil was most abundant. Unvegetated sites were dominated by smallmouth bass and rock bass *Ambloplites rupestris*. Age-0 fishes were more abundant and species rich near shore (<2-m depth) than off shore. Furthermore, where vegetation was abundant, species rich, and irregularly structured (i.e., large CV of horizontal patch length), the community of age-0 fishes tended to be more species rich and age-0 fishes and adult yellow perch more abundant. Age-0 bluegills were more abundant in patchy milfoil beds than elsewhere. In summary, the littoral fish community in Lake Mendota responded to a suite of macrophyte characteristics, including relative abundance, species richness, and structural complexity.

Potential Sampling Biases

Analysis of fish community structure and stability requires measures of fish abundance and diversity with sufficient accuracy and resolution to allow identification of spatial and temporal patterns in the community. Because these measures are sensitive to sampling biases and to choice of analysis, preliminary investigation of biases and analytical methods may facilitate more accurate pattern identification and interpretation.

Gear Biases in Sampling

Perceived differences in assemblages among sampled sites may result, in part, from changes in gear biases among habitats and fish, rather than from actual differences in fish abundances. If gear biases are similar among sampled sites, site catch abundance is an appropriate estimate of relative abundance for comparing patterns in community structure. To overcome sampling biases, a composite of catch data from multiple gear types often is used

(Hinch et al. 1991; Weaver et al. 1993). It is assumed that biases inherent in each gear type are countered by those of other types. To estimate species catch abundances, either (1) total catch per effort for each gear type is assumed to be proportional to that gear's share of the catch by all gears, and species catches are summed across gears; or (2) catch per effort is assumed to be equal among gear types, and species catches are scaled so that total catch per gear is the same for all gears before species catches are summed across gears (Weaver et al. 1993). The method for merging gear data, either (1) or (2), producing larger differences among site catches is more likely to discern differences in assemblage structure.

In addition to choosing between methods for merging catch data from multiple gears, Weaver et al. (1993) chose the sampling gear(s) that contributed most to differences in species abundances. Because community level patterns are most apparent when differences in assemblage structures are maximized, it may be prudent to choose the gear providing the greatest differences among site catch abundances. This gear may be effective at discerning structural differences simply because it most readily catches species that differ most among sites, given there is no habitat bias. Catches in small fyke nets differed more in species abundance among littoral sites in Lake Mendota than did catches in beach seines, even though fyke nets were no less precise (i.e., did not have a higher CV) than seines (Weaver et al. 1993). Catches in gill nets did not differ among sites. Also, catches of age-0 bluegills and age-0 black crappies *Pomoxis nigromaculatus* differed among sites more than did those of other fishes, and they were higher in fyke nets than in other gear types. Therefore, fyke nets were considered more effective than other gear types in determining within-lake littoral fish community structure and heterogeneity.

Variability in capture efficiency often can be associated with differences in habitat and fish community composition. For example, daytime seining efficiency was higher for midwater fishes (e.g., cyprinids) than for benthic fishes (darters *Etheostoma* spp.) (Lyons 1986). Pierce et al. (1991) developed a model for improving seine capture efficiency by correcting for habitat and community variability. Regression models indicated, for example, that higher macrophyte biomass was associated with increased capture efficiency. Fish size was a significant predictor of capture efficiency for benthic fish only; smaller benthic species escaped the seine more readily. These relationships yielded corrections to capture efficiency estimates based on aspects of habitat and fish behavior.

Biases in Estimating Species Richness

Estimated species richness may depend on number of individuals caught and sampling regime. Tonn and Magnuson (1982) performed several preliminary analyses to identify possible biases in sampling small Wisconsin lakes. Lakes were sampled with baited minnow traps and small fyke nets. Two lakes were sampled twice and one lake was then resampled with twice the original effort. Species richness was not correlated with number of individuals caught, and no correlation was found between sample richness and sampling sequence. Additionally, similarity in species composition based on presence–absence data among sampling dates, and the addition of only one species when catch effort was doubled or tripled, indicated that the original sampling effort was sufficient to include almost all susceptible species. However, species richness in the sampled lakes was low (1–11); a more species-rich or dynamic community may require greater sampling effort or a more stringent sampling protocol.

To determine the number of seine hauls required to estimate species richness, Eadie and Keast (1984) compared cumulative number of species caught in northern and southern Ontario lakes with the number of seine hauls. The cumulative species curve became asymptotic after two to three seine hauls. Therefore, they conducted a minimum of four seine hauls per site in northern Ontario lakes and usually three or four hauls in southern Ontario lakes. In contrast, in a study of species richness and turnover (i.e., loss or gain of species) in seven north temperate lakes over 11 years, Magnuson et al. (1994) noted that the overall average number of new species caught in each of 18 sequential seine hauls was greater than zero for all 18 hauls, indicating that apparent species turnover may be due to missing rare species in some years. Sampling sufficiency is particularly critical in studies of invasions and extinctions, because these events occur when populations are quite small.

Benson and Magnuson (1992) adjusted species richness for differences in number of individuals caught using the technique of rarefaction. Rarefied species richness is a measure of species diversity, containing aspects of both richness and evenness, and was calculated as the expected number of species if the number of individuals caught had been the same for all lakes. Rarefaction was especially suitable for Benson and Magnuson (1992), because the total number of fish caught per year varied among lakes and years by a factor of about 70.

Components of Community Heterogeneity

Benson and Magnuson (1992) tested whether observed differences in within-lake community heterogeneity among lakes resulted solely from differences in species diversity (as measured by rarefied species richness). In a computer simulation, they randomly reassigned seine samples to sampling sites in 1,000 combinations for each lake and year, derived a heterogeneity value for each combination, and calculated the probability that a simulated value would exceed the observed value. Probabilities of 0.5 or more would indicate that community heterogeneity was due to sampling variability. However, probabilities were well below 0.5 for most lakes and years. Mean simulated heterogeneity was positively correlated with rarefied richness, but the difference between the observed and mean simulated heterogeneity increased with rarefied richness, suggesting that richness was not the only contributor to heterogeneity. The other contributor was habitat variability; both community heterogeneity and rarefied richness were correlated with several measures of habitat diversity. In lakes with more habitat diversity, among-site heterogeneity strongly contributed to community heterogeneity. In lakes with more homogeneous habitat, random variation (e.g., sampling phenomena) was a larger component of community heterogeneity than was among-site heterogeneity.

Issues of Scale

The apparent structure and stability of a fish community, and the apparent complexity of the habitat, may depend on the scale (extent and resolution) of the study (Tonn et al. 1990; Hinch et al. 1991). Fishes that are evenly distributed at a regionwide scale (large extent and coarse resolution) may be aggregated at local scales (small extent and fine resolution). Habitats that appear homogeneous at a lake-wide or region-wide scale may include microhabitats apparent only at a smaller scale. Similarly, patterns in fish assemblages may change over seasons as fish migrate to spawning sites or recruit into fisheries. Care must be taken to choose a scale appropriate to the question of interest (Allen and Starr 1982).

Apparent species distribution may vary with numerical resolution (absolute abundance, relative or rank abundance, presence–absence) (Table 1). Distribution patterns may be evidenced by changes in numerical dominance in the absence of changes in species presence or absence (Rahel 1990). If fluctuations in fish populations are synchronized, or if different fish species respond similarly to habitat cues, then absolute abundance may vary while relative abundance remains constant. Weaver et al. (1993) examined the distribution of littoral fishes in a single lake using different levels of numerical resolution and determined that absolute abundance data were required to discern species level variations among sites. Analyses of presence–absence and rank abundance data did not distinguish among sites, indicating that the within-lake variation in littoral habitat was not sufficient to exclude species from particular sites (presence–absence) nor to elicit a strong differential response among species (rank abundance).

The apparent relative importance of abiotic and biotic factors in structuring fish communities also may depend on study scale. Based on a survey of fisheries papers, Hinch (1991) argued that much of the debate over the relative importance of abiotic and biotic factors in structuring fish assemblages reflects a dichotomy in types of studies done by fisheries biologists. Large-scale studies that incorporate lakes representing a broad range of abiotic conditions, and that determine co-occurrences based on the relative abundance or presence or absence of many fish species, find that abiotic factors are of primary importance. In contrast, small-scale studies that include lakes with similar abiotic conditions, involve intensive sampling, and base comparisons on the abundance or biomass of a few fish species, usually find that biotic interactions are of primary importance. Because most of the studies reviewed here are large in scale, we may inadvertently accentuate the role of environment and habitat over biotic factors in structuring communities.

Taxonomic and Behavioral Groups

Merging species into taxonomic or behavioral groups can clarify distribution patterns. Variation in *Lepomis* abundance among central Ontario lakes was associated with differences in lake primary productivity and nearshore macrophyte cover (Hinch and Collins 1993). Also, in lakes where bluegill was not present, variation in pumpkinseed abundance was correlated with variability in nearshore macrophyte cover and flushing rate. At higher numerical resolution (e.g., rank abundance) or finer taxonomic resolution (e.g., species) these trends would not be apparent.

Separation of fish abundance data by young of year and adult facilitated identification of site catch differences in Green Bay (Brazner and Magnuson 1994) and in Lake Mendota (Weaver et al. 1993), presumably because young and adult fishes differ in their behaviors with respect to habitat. In contrast, by grouping littoral fishes of Green Bay according to life history strategies, Brazner and Magnuson (1994) identified ecotypes characterized by traits with adaptive significance and functional groups (mores) appropriate as indicators of habitat condition (see also Austen et al. 1994). Fishes in undeveloped wetlands were characterized as dark, highly maneuverable or ambush predators, that were nest-guarding, turbidity-intolerant, phytolithophilic spawners, with periodic or equilibrium life history strategies. Beach fishes were light-colored rovers, often psammophilic or pelagophilic spawners, with opportunistic life history strategies. These groupings allowed an ecological interpretation of the littoral fish community in Green Bay that could not be made with any taxonomic classification. Examination of community data at several taxonomic or behavioral, as well as numerical, scales may facilitate understanding of inherently complex systems.

Regional and Historical Studies

Processes that occur at very large spatial and temporal scales (e.g., regional and historical) may also affect relationships between the local, extant fish community and littoral habitat (Table 1). In southern Ontario lakes, fish species diversity was correlated with foliage height diversity, substrate diversity, and water depth, while fish species richness was correlated with plant cover, plant height, and foliage height diversity (Eadie and Keast 1984). In contrast, there were few correlations between littoral fish diversity or richness and measures of habitat complexity in northern Ontario lakes. Rather, fish species diversity was correlated with benthic invertebrate diversity, and fish species richness was correlated with zooplankton diversity. Some of the differences between northern and southern correlations are likely related to the littoral zone habitats themselves. The northern Ontario lake littoral habitat is relatively uniform, and vegetation is sparse, taxonomically and structurally simple, and varies little among sites. Characteristic southern Ontario littoral habitats are structurally and vegetatively complex; consequently, the fish community was expected to be more diverse than in northern Ontario. Northern Ontario lakes were formed more recently, during the Wisconsin glaciation, and are more distant from recolonization sources than are southern Ontario lakes. Because there are fewer littoral fish species in northern than in southern lakes (7 common species compared with 12), and their habitats are relatively new, there may be no clear response of the littoral fish community to habitat structure. Eadie and Keast (1984) concluded that the regional differences in fish ecology between northern and southern Ontario lakes likely relate to glacial history as well as to characteristics of the different habitats.

Similarly, Hinch et al. (1991) concluded that the natural distribution of rock bass and northern pike among Ontario lakes is probably related to colonization during the retreat of the Wisconsin glacier. These fishes are present in the lowland basin formed by glacial Algonquin Lake about 12,000 years ago, but are absent from highland lakes. As the glacier retreated, coldwater fishes of the Mississippi Refugium recolonized the highland via outlets. But after the ice sheet receded, the land rebounded, eliminating these immigration routes. Rock bass is a warmwater species and probably was unable to recolonize the highland area before the outlets closed. Although a coldwater fish, northern pike is the only Ontario species from the Missouri Refugium. Its recolonization route was much longer and it may not have reached the lowlands until after the highlands rebounded and is thus absent from highland lakes. White sucker *Catostomus commersoni*, a coldwater fish, recolonized the highlands, and is more abundant in highland lakes than in lowland lakes, presumably due to the absence of piscivorous pike in highland lakes. Therefore, the extant distribution of ubiquitous fishes among Ontario lakes has been shaped, in part, by the sequence of events during postglacial colonization.

Intercontinental Study

In an intercontinental comparison of the fish assemblages of small northern forest lakes of Wisconsin (0.002–0.640 km^2) and Finland (0.002–0.869 km^2), Tonn et al. (1990) identified differences and similarities in community level patterns at regional and local scales. By studying historically distinct faunas that occur in environmentally similar locales, they assessed the relative role of regional (historical) and local (contemporary) processes. Although the size of the regional species pool present in Finland was half that present in Wisconsin, local species richness was the same, and variability in species composition among assemblages in Finland was lower than that in Wisconsin. Based on presence–absence data, there were three distinct fish assemblages in Wisconsin lakes. Finnish assemblage types were identifiable on the basis of species relative abundance, but not on the basis of presence or absence. Wisconsin assemblages dominated by largemouth bass *Micropterus salmoides*, northern pike, and mudminnows occupied lakes that were environmentally similar to those occupied by Finnish assemblages dominated by Eurasian perch *Perca fluviatilis*, roach *Rutilus rutilus*, and crucian carp *Carassius carassius*, respectively. Community level patterns were interpreted by examining lower-level phenomena: the distributions of individual species, the co-occurrence of species pairs, and the addition of different species among lakes. Habitat specialization and predatory exclusion separated assemblage types among environmentally distinct lakes in Wisconsin, while biotic interactions and severe environmental conditions altered population densities without excluding species in Finnish lakes. Using discriminate functions of lake characteristics from one region, Tonn et al. (1990) were able to identify assemblage types occurring in similar lakes in the other region. This intercontinental community level predictability suggests that lakes in both regions are influenced by a common set of local environmental factors even though they are populated by historically distinct fauna. Tonn et al. (1990) concluded that for ecological interpretation of community level patterns, ecologists must consider community structure and function at a hierarchy of scales.

Lake Succession

Rahel (1984) inferred the temporal succession of the fish community in a single bog lake from the communities associated with habitat gradients across bog lakes in Wisconsin. A decline in fish species richness was associated with physical lake succession to a bog state, which is characterized by declines in lake size and depth, amounts of aquatic vegetation and hard substrate, and pH. As bog characteristics are acquired, specialized zooplanktivores (e.g., black crappie), species requiring hard substrates (e.g., pumpkinseed), and species intolerant of low pH (e.g., golden shiner) are lost from the community. Habitat generalists that are tolerant of low pH remain during intermediate stages of succession. Only yellow perch and central mudminnow remain at the final stages, when pH is low, winter anoxia is frequent, and littoral zones are soft-bottomed and sparsely vegetated. Mudminnow, because of its ability to breathe air, is the last to be extirpated. Thus, the order in which species are lost as a lake senesces sometimes can be predicted from comparative studies of fishes along regional habitat gradients associated with lake succession.

Predictive Models

Fisheries models and indices can be derived to predict fish assemblage type, species richness, and total fish abundance from lake characteristics (Tonn et al. 1983; Carline 1986). Based on habitat characteristics in a suite of north temperate lakes, factors responsible for the maintenance of fish assemblages were identified (Tonn and Magnuson 1982) and used to predict assemblage type in lakes of a similar but distinct suite (Tonn et al. 1983). Discriminant analysis (Gauch 1982; Digby and Kempton 1987; Krzanowski 1988) was used to separate the fish faunas of 18 lakes into three distinct assemblage types (*Umbra*–cyprinid, *Esox*, and centrarchid) based on five environmental variables that were easily measured and available from lake data files (lake area, maximum depth, pH, watershed size, and conductivity). From the discriminant analysis functions, fish-assemblage type was correctly predicted in nearly all of 11 additional lakes from the same lake district. These functions were then refined by repeating the discriminant analyses using all 29 lakes. Using this model, it is possible to predict the assemblage type of other lakes within the same district, so as to determine, for example, whether a lake is likely to support a centrarchid fishery.

Similarly, Brazner and Magnuson (1994) developed a regression model based on environmental variables to predict species richness and total fish abundance at littoral sites in Green Bay, Lake Michigan. Turbidity, a factor that differed by region (upper and lower bay), and either growth form or species richness of plants (local factors) best predicted fish species richness. Total fish abundance was best predicted by turbidity, water temperature (another regional factor), and habitat development (a local factor). Thus, indices of littoral fish species richness and total abundance included both regional and local habitat factors.

Summary and Concluding Remarks

In the studies reviewed here (Table 1), a broad range of processes were suggested to explain the structure and dynamics of fish communities. These include biotic interactions such as predation, distributional limitations due to physiological constraints under severe conditions such as anoxia or to historical events such as glaciation, and the use of habitat for forage or refuge. Scale varied among studies from a single lake or large bay to multiple lakes from one or more regions or continents. Some studies addressed ecosystem processes at only one scale, whereas others investigated the relative role of processes in structuring communities at multiple scales. Abiotic conditions and habitat characteristics were represented by morphometric variables such as mean depth and shoreline development, by limnological variables such as winter oxygen level and primary productivity, and by resource variables such as macrophyte cover and biomass of benthos. The scale of characterizations ranged from within-lake, site-specific estimates of macrophyte abundance and complexity to lake-wide or region-wide estimates of abiotic conditions and other habitat variables. Although the scope of these studies is quite broad, the fundamental approach is the same; namely, to evaluate the structure or stability of the fish community, to characterize the habitat, and to identify relationships between distribution patterns in the community and variations in habitat.

Distribution patterns and the distinctiveness of these patterns differ among communities. In very small north temperate Wisconsin lakes, distinct fish assemblages were readily identified based on presence–absence data, and the distribution pattern was explained by a combination of severe abiotic conditions (winter anoxia and low pH), predatory exclusion, and lake connectivity providing refuge from anoxia or predation. In contrast, the fish community of small Finnish lakes based on presence–absence data was better represented as a hierarchy of species additions with a core of common species than as distinct assemblages. Biomass relative abundance data were required to discern separate assemblages, presumably because predation in Finnish lakes reduced prey abundances without excluding prey species from lakes. In Ontario, fish species diversity increased with littoral habitat heterogeneity in southern lakes but not in northern lakes. Northern Ontario lakes were less diverse in littoral habitat and fish species and were more recently formed and colonized than were southern Ontario lakes. Perhaps the lack of an apparent response of the northern fish community to habitat structure relates to this uniformity of habitat and the relative youth of the community. Similarly, among six Wisconsin lakes, fish communities in lakes with greater habitat heterogeneity were characterized by spatial community heterogeneity and fish communities in lakes with more homogeneous habitat showed more random variation. The Green Bay littoral fish community was separated into upper-bay and lower-bay faunas by differences in temperature and turbidity, into beach and wetland ecotypes by distinct habitat characteristics, and (within regions) into species-rich and disturbance-tolerant assemblages based on the presence or absence of nearshore development. In Lake Mendota, unvegetated sites had higher relative abundances of fishes known to associate with structures such as boulders than did vegetated sites. Among vegetated sites, young-of-year fish community richness, and abundance of some young-of-year fishes, increased with macrophyte heterogeneity, but distinct assemblages were not discernible. In summary, the distinctiveness of fish assemblages and the strength of relationships between community structure and habitat, as identified in these studies, depended on the scale (extent and resolution) of the study, the processes that structured the community, and the variability among habitats within studies.

Grasp of lake dynamics permits an understanding of the more variable and individual characteristics of reservoirs. Hydrodynamics are complex and unique for each reservoir in relation to influent variations, reservoir morphology, characteristics of draw-down and discharge, and other factors. In particular, the distribution of oxygen and dissolved organic carbon in reservoirs is highly variable horizontally, vertically, and seasonally. Further, because a higher proportion of the reservoir volume is exposed to and interacts with the soil and sediments relative to natural lakes, macrophyte development and overall production may be higher. Thus, flow regime, production variability, and habitat spatial complexity may be especially important to reservoir fish communities.

Biotic relationships are more poorly developed in reservoirs than in lakes (Noble 1986). When rivers are impounded, species composition changes but diversity remains relatively high, perhaps because reservoirs incorporate aspects of both rivers and lakes. Artificial fluctuations in water level and stocking of native and nonnative species may interrupt development of homeostasis in the reservoir system. Thus, predicting whether a water body might support particular fisheries may be more difficult for a reservoir than for a lake. By examining the distribution patterns of fish among reservoirs of different ages, perhaps the processes that guide the development of reservoir fish communities can be inferred.

Through comprehensive analyses of fish community structure and dynamics, fisheries biologists can gain insights into the systems they study and manage. The relative roles of environmental factors and biotic interactions in structuring fish communities can be examined. By relating habitat complexity to community structure at various scales, models of complex fish–habitat interactions can be refined. Predictions under changing habitat (e.g., invasion of exotics, macrophyte harvesting) or fishing regimes (e.g., regulation and stocking) can be derived to assist agencies in managing fisheries. Additionally, these analyses can suggest testable hypotheses of mechanisms (e.g., piscivory and habitat modification) to explain observed patterns.

Acknowledgments

We thank B. Benson, R. Carline, S. Gafny, and F. Rahel for thorough and constructive reviews. Financial support for this work was provided by the Federal Aid in Sport Fish Restoration Act, through the Wisconsin Department of Natural Resources (F-95-P) and the Center for Limnology, and by the Anna Grant Birge Scholarship Fund through the Oceanography and Limnology Graduate Program at the University of Wisconsin–Madison.

References

Allen, T. F. H., and T. B. Starr. 1982. Hierarchy: perspectives for ecological complexity. University of Chicago Press, Chicago.

Austen, D. J., P. B. Bayley, and B. W. Menzel. 1994. Importance of the guild concept to fisheries research and management. Fisheries 19(6):12–20.

Benson, B. J., and J. J. Magnuson. 1992. Spatial heterogeneity of littoral fish assemblages in lakes: relation to species diversity and habitat structure. Canadian Journal of Fisheries and Aquatic Sciences 49:1493–1500.

Brazner, J. C., and J. J. Magnuson. 1994. Patterns of fish species richness and abundance in coastal marshes and other nearshore habitats in Green Bay, Lake Michigan. Verhandlungen der Internationale Vereinigung fur Theoretische und Angewandte Limnologie 25:2098–2104.

Brinkhurst, R. O. 1975. The benthos of lakes. St. Martin's Press, New York.

Carline, R. F. 1986. Indices as predictors of fish community traits. Pages 46–56 *in* G. E. Hall and M. J. Van Den Avyle, editors. Reservoir fisheries management: strategies for the 80's. American Fisheries Society, Southern Division, Reservoir Committee, Bethesda, Maryland.

Deppe, E. R., and R. C. Lathrop. 1993. Recent changes in the aquatic macrophyte community of Lake Mendota. Transactions of the Wisconsin Academy of Sciences, Arts and Letters 81:47–58.

Digby, P. G. N., and R. A. Kempton. 1987. Multivariate analysis of ecological communities. Chapman and Hall, New York.

Eadie, J. M., and A. Keast. 1984. Resource heterogeneity and fish species diversity in lakes. Canadian Journal of Zoology 62:1689–1695.

Gauch, H. G. 1982. Multivariate analysis in community ecology. Cambridge University Press, New York.

Heck, K. L., and L. B. Crowder. 1991. Habitat structure and predator–prey interactions in vegetated aquatic systems. Pages 282–299 *in* S. S. Bell, E. D. McCoy, and H. R. Mushinsky, editors. Habitat structure: the physical arrangement of objects in space. Chapman and Hall, New York.

Hinch, S. G. 1991. Small- and large-scale studies in fisheries ecology: the need for cooperation among researchers. Fisheries 16(3):22–27.

Hinch, S. G., and N. C. Collins. 1993. Relationships of littoral fish abundance to water chemistry and macrophyte variables in central Ontario lakes. Canadian Journal of Fisheries and Aquatic Sciences 50:1870–1878.

Hinch, S. G., N. C. Collins, and H. H. Harvey. 1991. Relative abundance of littoral zone fishes: biotic interactions, abiotic factors, and postglacial colonization. Ecology 72:1314–1324.

Krzanowski, W. J. 1988. Principles of multivariate analysis: a user's perspective. Oxford University Press, New York.

Lyons, J. 1986. Capture efficiency of a beach seine for seven freshwater fishes in a north temperate lake. North American Journal of Fisheries Management 6:288–289.

Magnuson, J. J., B. J. Benson, and A. S. McLain. 1994. Insights on species richness and turnover from long-term ecological research: fishes in north temperate lakes. American Zoologist 34:437–451.

Magnuson, J. J., L. B. Crowder, and P. A. Medvich. 1979. Temperature as an ecological resource. American Zoologist 19:331–343.

Noble, R. L. 1986. Management of reservoir fish communities by influencing species interactions. Pages 137–143 *in* G. E. Hall and M. J. Van Den Avyle, editors. Reservoir fisheries management: strategies for the 80's. American Fisheries Society, Southern Division, Reservoir Committee, Bethesda, Maryland.

Pierce, C. L., J. B. Rasmussen, and W. C. Leggett. 1991. Sampling littoral fish with a seine: corrections for variable capture efficiency. Canadian Journal of Fisheries and Aquatic Sciences 47:1004–1010.

Pierce, C. L., J. B. Rasmussen, and W. C. Leggett. 1994. Littoral fish communities in southern Quebec lakes: relationships with limnological and prey resource variables. Canadian Journal of Fisheries and Aquatic Sciences 51:1128–1138.

Rahel, J. F. 1984. Factors structuring fish assemblages along a bog lake successional gradient. Ecology 65: 1276–1289.

Rahel, J. F. 1990. The hierarchical nature of community persistence: a problem of scale. The American Naturalist 136:328–344.

Ryder, R. A. 1965. A method for estimating the potential fish production of north-temperate lakes. Transactions of the American Fisheries Society 94:214–218.

Ryder, R. A. 1982. The morphoedaphic index—use,

abuse, and fundamental concepts. Transactions of the American Fisheries Society 111:154–164.

Tonn, W. M., and J. J. Magnuson. 1982. Patterns in the species composition and richness of fish assemblages in northern Wisconsin lakes. Ecology 63:1149–1166.

Tonn, W. M., J. J. Magnuson, and A. M. Forbes. 1983. Community analysis in fishery management: an application with northern Wisconsin lakes. Transactions of the American Fisheries Society 112:368–377.

Tonn, W. M., J. J. Magnuson, M. Rask, and J. Toivonen. 1990. Intercontinental comparison of small-lake fish assemblages: the balance between local and regional processes. The American Naturalist 136:345–375.

Weaver, M. J., J. J. Magnuson, and M. K. Clayton. In press. The distribution of littoral fishes in structurally complex macrophytes. Canadian Journal of Fisheries and Aquatic Sciences.

Weaver, M. J., J. J. Magnuson, and M. K. Clayton. 1993. Analyses for differentiating littoral fish assemblages with catch data from multiple sampling gears. Transactions of the American Fisheries Society 122:1111–1119.

Werner, E. E. 1986. Species interactions in freshwater fish communities. Pages 344–357 *in* J. Diamond and T. J. Case, editors. Community Ecology. Harper and Row, New York.

American Fisheries Society Symposium 16:347–356, 1996

Establishing Native Submersed Aquatic Plant Communities for Fish Habitat

R. MICHAEL SMART, ROBERT D. DOYLE, AND JOHN D. MADSEN
U.S. Army Corps of Engineers, Waterways Experiment Station
Lewisville Aquatic Ecosystem Research Facility
Rural Route 3, Box 446, Lewisville, Texas 75056, USA

GARY O. DICK
AScI Corporation, Lewisville Aquatic Ecosystem Research Facility
Rural Route 3, Box 446, Lewisville, Texas 75056, USA

Abstract.—Aquatic plants are desirable in reservoirs for fish habitat and for water quality benefits. Unfortunately, unvegetated turbid reservoirs may never develop an aquatic plant community because of insufficient plant propagules and inhospitable environments for establishing seedlings. Weedy exotic species such as hydrilla *Hydrilla verticillata* and Eurasian watermilfoil *Myriophyllum spicatum* may establish in reservoirs, but these are poor long-term ecosystem components that may lead to serious management and ecological problems in reservoirs. Native plants, in contrast, offer excellent habitat and water quality benefits with few management problems. We review basic ecological considerations for selecting appropriate native plant species and planting locations. Revegetation efforts are most successful when a strategy of establishing founder populations within protective exclosures at carefully selected sites is used. Once successfully established, founder populations change environmental conditions within their vicinity, making the area more hospitable to further colonization and providing propagules to other sites within the reservoir.

Many fish species depend on submersed structure for growth and survival. Structure provides substrate for food items and is used for spawning. The role of structure in predator–prey relationships is well documented; structure provides refuge for forage species (Savino and Stein 1982). Rocks, uneven submersed terrain, flooded timber, logs, and aquatic plants are all types of structure.

Submersed aquatic macrophytes arguably provide the most valuable structure to freshwater fisheries because their physical characteristics provide excellent habitat (Engel 1985). However, benefits derived from aquatic plants vary across species. Ultimately, the spatial complexity of a plant community may determine its value to fisheries. Mixed native plant communities often consist of diverse species assemblages with varied growth forms, which yield habitat useful to a wide range of fish species and size classes (French 1988).

Centrarchids in particular use aquatic vegetation as habitat. Mixed size class sunfish *Lepomis* spp. and largemouth bass *Micropterus salmoides* densities were much higher in vegetated reservoir sites than in areas devoid of vegetation (Killgore et al. 1989). Sunfish and largemouth bass often dominate larval and juvenile age-classes in surveys of vegetated sites (Dewey and Jennings 1992), and larger individuals of these species have also been associated with aquatic plants (Bain and Boltz 1992). The role of submersed aquatic plants in fisheries, particularly largemouth bass populations, is demonstrated in the slogan "Grass = Bass" promulgated by bass angler organizations. While this slogan simplifies a complex issue, it conveys the perceived desirability of submersed aquatic vegetation to many anglers.

In this article we will (1) review why plant establishment efforts are needed in constructed reservoirs, (2) identify plant species appropriate for establishment in reservoirs, and (3) identify important considerations for plant establishment efforts, including propagule acquisition and methods for establishment. In some cases it also may be desirable to promote the establishment of emergent macrophytes along the shoreline. While beyond the scope of this paper, information on establishing emergent plants is available from various sources (Hammer 1992; Doyle and Smart 1993).

Establishment of Aquatic Plants

Two factors appear to be primarily responsible for maintaining reservoirs in an unvegetated state. First, reservoirs are constructed systems that do not initially have large numbers of aquatic plant propagules to colonize available habitat. At construction, new reservoirs usually cover large areas of potential aquatic plant habitat, but initially have little or no aquatic vegetation occupying these areas. While plant communities in natural lakes have developed over hundreds of years of ecological interactions, those in even the oldest reservoirs are very young in an ecological sense (Godshalk and Barko 1985). Reservoirs should not be expected to spontaneously develop desirable aquatic plant communities, particularly in the absence of local native plant communities to provide propagules.

In addition to being ecologically young, new reservoirs are not hospitable environments for establishment of most submersed aquatic plants, particularly if the plants are introduced as seed or small shoot fragments. Because many reservoirs are operated for flood control or water supply, their water levels may fluctuate over broad ranges. While many aquatic plant species are adapted to natural hydrologic cycles, water level fluctuations in reservoirs may be asynchronous with seasonal weather cycles. Excessive inorganic turbidity, planktonic algae, and epiphyte growth all can reduce light levels, and small plants may not receive adequate light for photosynthesis and growth. Also, bottom substrates of new reservoirs are flooded terrestrial soils, that may not be initially suitable for rooted aquatic plant growth, particularly if soils are rocky or hard-packed clay. Finally, intense pressure from herbivores in the reservoir may prevent establishment of small populations of relatively fragile, immature submersed plants.

Some reservoirs appear to remain turbid and unvegetated for many years, while others develop extensive macrophyte communities and are much clearer. Ecological theory has long recognized the possibility that ecosystems may have multiple stable equilibria that can differ drastically from one another (May 1977). Aquatic environments with intermediate nutrient inputs appear to have two stable equilibrium states: a state characterized by highly transparent water and abundant aquatic macrophytes, and an alternative state characterized by high turbidity (plankton and resuspended sediments) and a virtual absence of aquatic macrophytes (Scheffer 1990; Scheffer et al. 1993).

Through time, reservoirs may develop a submersed aquatic flora. However, this process can take many years, and the resultant plant community may not be adequate for management needs. Exotic weedy species such as hydrilla *Hydrilla verticillata* or Eurasian watermilfoil *Myriophyllum spicatum* are highly adapted for colonizing disturbed sites and often establish prior to the development of native plant communities. Management problems associated with these species far outweigh short-term benefits they may provide to fisheries.

Because completely unvegetated reservoirs are generally inhospitable to seedling establishment, management intervention may be required in order to break the turbid state and begin to drive the system towards the vegetated, clear state. Plantings of desirable native aquatic plant species may initiate this process.

The concept of stocking desirable organisms to manipulate the species composition of freshwater ecosystems is not new; for decades fisheries managers have stocked desirable fishes into reservoirs. Periodic fish community surveys may lead to additional stockings to achieve and maintain desired fish communities. This practice should be extended to the management of aquatic plant communities, which are an important component of the ecosystem on which fishes and other organisms depend.

Species Selection

In an ecological sense, construction of a reservoir can be considered a major disturbance. Ecological theory predicts that following a disturbance that destroys the plant community at a given site, the plant community will recover in a predictable successional sequence. The first colonizers are pioneer (*r*-selected) species that grow quickly and produce numerous, widely dispersed propagules (Stearns 1977). The prolific rates of reproduction and dispersal allow pioneer species to spread quickly to fill available niches. Pioneer species are eventually displaced by longer-lived, slower-growing, competitive plants (Grime 1979; Sheldon 1986). However, while pioneer species decline in abundance as competitive species proliferate, their seeds remain in the seed bank. This seed bank allows quick recovery of the vegetation if a subsequent disturbance sets the community back to an earlier successional state (van der Valk 1981; Sheldon 1986). Unvegetated reservoirs usually do not have existing seed banks of submersed aquatic species and require another means of establishing pioneer species.

Pioneer plants modify the reservoir environment as they grow and expand, and cause localized changes that promote further expansion. Several mechanisms can be involved, including increased sedimentation resulting in more shallow areas suitable for macrophyte growth (Carpenter 1981), increased water transparency due to reduced turbidity (Carpenter and Lodge 1986), reduced algal populations resulting from decreased water column nutrients (Kufel and Ozimek 1994), and shifts in interactions among fish, zooplankton, and phytoplankton (Schriver et al. 1995). Pioneer species thus promote stability and provide conditions favorable

TABLE 1.—Species of submersed aquatic plants used by the authors for habitat enhancement efforts in reservoirs. Establishment method ratings: (−) not recommended, (?) currently under investigation, (+) recommended, (++) strongly recommended, (±) variable results.

Plant types and species	Common means of propagation in nature	Methods of establishment attempted by authors and ratings
Summer annuals		
Muskgrass *Chara vulgaris*	Propagates by spores or vegetative fragments.	(++) peat-potted transplants (−) dry mudballs or burlap strips (?) sediment laden with spores
Southern naiad *Najas guadalupensis*	Propagates by vegetative expansion, seed, or stem fragments.	(++) peat-potted transplants (?) sediment laden with seeds
Winter annuals		
Small pondweed *Potamogeton pusillus*	Propagates by vegetative expansion, turions, seed, or stem fragments.	(++) peat-potted transplants (?) sediment laden with seeds
Horned pondweed *Zannichellia palustris*	Propagates by seed, vegetative expansion, or stem fragments.	(++) peat-potted transplants (?) sediment laden with seeds
Tuber-forming perennials		
American pondweed *Potamogeton nodosus*	Propagates by vegetative expansion (rhizome), seed, stem fragments, or tuber production.	(++) tubers (++) peat-potted transplants (?) sediment laden with seeds
Sago pondweed *Potamogeton pectinatus*	Propagates by vegetative expansion (rhizome), seed, stem fragments, or tuber production.	(±) tubers (+) weighted bags with tubers
Vallisneria (northern) ecotype *Vallisneria americana*	Propagates by vegetative expansion (stolon), dormant tuber, or seed.	(++) tubers (++) peat-potted transplants
Evergreen perennials		
American elodea *Elodea canadensis*	Propagates by vegetative expansion or stem fragments.	(++) peat-potted transplants (+) stem fragments
Water star grass *Heteranthera dubia*	Propagates by vegetative expansion, stem fragments, or seed (in the South only)	(++) peat-potted transplants (+) stem fragments
Vallisneria (southern ecotype) *Vallisneria americana*	Propagates by vegetative expansion (stolon) or seed.	(++) peat-potted transplants (−) bare-root transplants (?) seed

for succession (Grime 1979; Bertness and Callaway 1994). In addition, the expanding plant community will develop greater tolerance of adverse environmental conditions as plant size increases and energy reserves accumulate (Olesen and Sand-Jensen 1994). This supplies positive feedback in the process of community development.

Reservoirs with a limited distribution of only a few pioneer species will require active management to increase both the area and species diversity of the plant community. Habitat heterogeneity will increase with plant community diversity, increase the number of food items for fish (Schramm and Jirka 1989; Butler et al. 1992), and add stability to the reservoir ecosystem (Scheffer et al. 1993).

Life History Considerations

Ecological benefits and ease of establishment vary greatly among native aquatic plants. Table 1 lists native species used in reservoir research plantings. Plants are categorized according to basic life histories exhibited in most southern reservoirs (summer annual, winter annual, tuber-forming perennial, or evergreen perennial) and most common means of propagation (seed, stem fragments, tubers, turions, or vegetative expansion).

Plants that complete their entire life cycle in 1 year and survive from year to year strictly by seed can be broadly classified as summer or winter annuals depending on the season of maximum biomass. The seeds produced are typically resistant to environmental stress, but contain relatively small energy reserves (Madsen 1991). Because of this limitation, developing seedlings are often susceptible to unfavorable conditions or disturbances (e.g., varying water levels, high turbidity, water currents, herbivory) during the initial establishment period, when seedlings are fragile and the population is sparse (Titus and Hoover 1991). However, because seeds (or spores) are tolerant to stress (such as desiccation), these can be established in areas of a reservoir exposed during drawdowns, if low water levels coincide with the plant's dormant period. In fact, the drawdown zone of Guntersville Reservoir, Alabama, which has a 50-cm winter draw, contains a diverse assemblage of native annual plant species, in sharp contrast to deeper waters which are dominated by Eurasian watermilfoil (Webb 1993).

Perennial plants produce some form of vegetative tissue that survives for more than 1 year. Some perennials survive winter in a dormant state by forming specialized perennating structures such as root crowns, tubers, or modified stem buds (turions). These dormant structures usually contain more energy reserves than seed (Madsen 1991) and can usually survive drawdowns or other unfavorable conditions. Subterraneous tubers, in particular, allow vigorous growth once dormancy is broken (Titus and Adams 1979; Kunii 1982). Other perennials are evergreen and maintain photosynthetically active leaves throughout the year. Depth of establishment is particularly important for these plants, because they lack dormant perennating organs and are generally not well suited for quick recovery from exposure.

Plants to Avoid

The harmful effects of certain exotic species (e.g., hydrilla and Eurasian watermilfoil) on water resources such as water supply, hydropower production, navigation, and many forms of water-based recreation are well known (Langeland 1990). Like native pioneer species, hydrilla has characteristics that allow it to rapidly colonize open sites and has been advocated for use in unvegetated reservoirs. However, unlike native pioneer species, hydrilla is not easily displaced by more desirable natives in a natural successional sequence (Smart and Doyle 1995).

Many growth characteristics of hydrilla make it a very poor choice. Hydrilla grows rapidly and readily from small stem fragments, which gives it a high reproductive capacity and tremendous dispersal capabilities. Hydrilla also tolerates low light levels, has a well-developed ability to elongate, and produces a very dense, highly branched canopy at the water surface (Pieterse 1981). The surface mat can greatly reduce rates of mixing, water exchange, light penetration, and gas exchange, and ultimately leads to dissolved oxygen depletion (Honnell et al. 1993).

When selecting species for establishment, hydrilla and Eurasian watermilfoil should be avoided entirely, even in completely unvegetated reservoirs. Any short-term benefits they provide in unvegetated reservoirs are greatly outweighed by serious, long-term consequences. There are many submersed aquatic plant species native to the United States that provide quality fish and wildlife habitat and do not typically cause significant problems with water resource use.

Native Plants to Consider

Annuals.—Muskgrass *Chara* spp. and southern naiad *Najas guadalupensis*, both summer annuals, are excellent pioneer species. Muskgrass is a macrophytic algae and lacks vascular tissues, including roots. This species is easily established and provides broad benefits, including reduction of water column nutrients (Kufel and Ozimek 1994), increased water clarity, and reduced growth of nuisance species (Denike and Geiger 1976). However, it does not grow well in soft water (Prescott 1978). As pioneer species, muskgrass and southern naiad expand and quickly cover the sediment with a carpet of vegetation. Both plants are replaced by higher successional, longer-lived plants during natural community development.

Two desirable species of winter annuals are horned pondweed *Zannichellia palustris* and small pondweed *Potamogeton pusillus*. These plants grow when many other species are dormant and facilitate the establishment of other species, including both summer annuals and perennials.

Tuber-forming perennials.—American pondweed *Potamogeton nodosus*, sago pondweed *Potamogeton pectinatus*, and the northern ecotype of vallisneria *Vallisneria americana* produce tubers. These tubers are a vigorous propagule that can be refrigerated in dark containers for several months with no apparent decline in viability.

The northern ecotype of vallisneria (typified by plants growing in Iowa, Wisconsin, Minnesota, and New York) is perhaps the most commonly used species in enhancement efforts and there is more information available on this species than any other (e.g., Korschgen and Green 1988). Unfortunately, vallisneria is highly preferred by several herbivores, including red-eared turtles *Trachemys scripta elegans* (Dick et al., in press), and is often subject to intense herbivory. Consequently, protection from herbivores is an essential part of an establishment program for this species (see below). However, with this caveat, plantings of vallisneria tubers by the authors and others (e.g., Carter and Rybicki 1985) often have been successful. Vallisneria tubers are commonly available from northern suppliers.

American pondweed is particularly easy to establish from tubers (e.g., Doyle and Smart 1993) and, because it forms floating leaves, can be used in turbid reservoirs.

Evergreen perennials.—Three species of evergreen perennial have been successfully established in reservoirs: vallisneria (southern ecotype), water star grass *Heteranthera dubia*, and elodea *Elodea canadensis*.

The southern ecotype of vallisneria (typified by plants growing in Texas, Louisiana, Alabama, Georgia, and Florida) is an evergreen perennial that does not produce tubers (Smart and Dorman 1993). Therefore, this ecotype must be propagated by either transplants or seed. This ecotype is typically larger and spreads more rapidly than the northern ecotype when planted in southern waters.

Water star grass is often observed in reservoirs in the southern United States. Because it grows in small, relatively dense clumps it may prove an excellent species for providing structure and edge environments for fisheries. Our field establishments have mostly used mature transplants, although we have also established the species from stem fragments rooted in small peat pots, and even from fresh, unrooted stem cuttings (Smart et al. 1996).

Elodea is a common, turbidity-tolerant component of northern plant communities (Nichols 1992). Its lush growth forms a stabilizing carpet over the sediment, which in combination with the filtering effect of its leafy shoots, improves water clarity. Elodea remains active during winter in the southern United States, and thus provides the benefits of submersed vegetation when many other plant species are dormant. Although growth of elodea may be depressed by high midsummer temperatures in the southern United States, this may be an advantage because the vegetative mat will be less likely to interfere with recreational uses at that time of year. Also, more heat-tolerant species are afforded an opportunity to develop. Elodea can be established from peat-potted transplants or weighted stem sections.

Although elodea superficially resembles hydrilla, it does not cause as many problems with water resources in reservoirs. Elodea exhibits a lower growth profile than hydrilla, usually forming a canopy well below the water surface. Elodea should also not be confused with Brazilian elodea *Egeria densa*, another exotic species that can cause management problems.

Project Implementation

Propagule Acquisition

Propagule acquisition accounts for a large part of the effort for any establishment project. Careful consideration at this point is worthwhile, because various options exist.

Commercial nurseries.—Propagules of some species of submersed aquatic plants are available from commercial nurseries or suppliers. However, when purchasing plants, the genetic source and origin of the plants must be considered. In general, establishment efforts should use plants obtained from a latitude or climate similar to that of the reservoir to be stocked. Particular care must be taken relative to vallisneria, which has ecologically distinct northern and southern ecotypes. Another consideration is that shipments from commercial vendors may be contaminated with viable propagules of unwanted species.

An alternative to commercial suppliers is to contract with local aquatic nurseries to obtain and propagate local ecotypes of species required for specific projects. This approach offers the convenience of purchasing the propagules, while maintaining control over the genetic stock of the plants.

Collection from the field.—Rooted, growing plants or dormant tubers collected from donor sites often have been used in restoration efforts. However, given the potential impact on the donor site and the high transplant mortality that often occurs, we do not recommend this option unless trained personnel are available, transplants are planted immediately after collection, and pilot scale plantings have proven successful. Otherwise, the restoration effort may fail, and the donor site may simultaneously be damaged. In cases where abundant local donor sites and trained personnel are available, logistic obstacles are minimal, and pilot tests have demonstrated a high potential for success, collection from the field can produce good results. The Florida Game and Fresh Water Fish Commission reports that transplants of vallisneria, carefully collected from local donor sites and immediately planted within fenced exclosures in Lake Monroe, suffered little mortality and provided excellent growth (L. Ager, Florida

Game and Fresh Water Fish Commission, personal communication).

Collection of seed from donor sites offers a preferred alternative, given that potential damage to the donor site is minimized. Successful collection of seed requires careful monitoring of the donor population to ensure that seed are harvested after ripening but prior to release by the plant. Viable seed are also likely to be present in the sediments within and adjacent to established populations and seed-laden sediment might be easily collected during low water periods.

Propagation.—For large-scale, long-term efforts where greenhouse or pond facilities are available, it may be worthwhile to mass culture plants for propagation stock. Plants can be propagated to produce seed, tubers, stem fragments, or live transplants.

Ponds are ideal for producing seeds of many species. The research ponds at the Lewisville Aquatic Ecosystem Research Facility, Texas, are former fish hatchery ponds that have been manipulated to enhance production of muskgrass spores and southern naiad and American pondweed seeds (Dick and Smart 1992). When ponds are kept flooded during the fall and winter months, other species, including horned pondweed and small pondweed, often contribute a significant portion of the plant community. After adequate time for seed and spore production, the ponds are drained and allowed to dry. The surface sediment is scraped down to 5 cm depth to collect spores and seeds. This seed-laden sediment can be stored and later surface broadcasted over appropriately protected sites.

Seed can also be harvested from numerous other species such as vallisneria. Fruit pods of vallisneria are produced in summer and can be collected in the fall when the seeds turn black. Air drying ripe seed pods for about 2 weeks appears to increase the germination percentage. To date we have only used field-collected seed of vallisneria to start transplants. Although we have not yet attempted to establish this species directly from seed, we have observed establishment of seedlings in research ponds and in Toledo Bend Reservoir, Texas.

Tuber-forming species may be grown to produce tubers in containers in a greenhouse or pond. After plants senesce, the containers can be removed from water and stored for several months.

Transplanting actively growing plants offers the best odds of short-term success. However, greenhouse culture of most submersed aquatic plant species requires rather exacting conditions (Smart and Barko 1985), and should not be attempted without proper facilities. The primary requirements are fertile sediment (pond sediment, if possible), low phosphorus (<10 μg/L) water to prevent excessive growth of algae, moderate water temperature (20–28°C), and adequate light (35–65% of full sunlight). However, given suitable culture conditions, peat-potted transplants can be grown in outdoor tanks, raceways, or small ponds.

Establishment in the Field

We recommend, given the usual financial constraints and harsh environmental conditions associated with poorly vegetated reservoirs, that restoration efforts focus on establishing small "founder" populations of native plants. This approach involves establishing small protected plots at carefully selected sites known to provide appropriate hydrologic, edaphic, and environmental conditions for the species and propagule type(s) being used. Established populations are then expected to provide propagules to vegetate the remainder of the available habitat in the reservoir. This method also offers the advantage of being easy to monitor and quantify. While macrophyte communities are adapted to environmental changes such as water level fluctuations (Brock 1988), in our experience the difficulty of initial establishment will increase with the amplitude of water level fluctuations during the growing season and with decreasing water clarity.

Protection of plantings.—Protection must be provided for plants during establishment. Because of their small size, these populations are inherently susceptible to destruction by herbivores and other disturbances. The establishment efforts with which we have been involved have benefited dramatically from exclosures or protective wave barriers. In most cases, plantings without protection were completely lost within the first few weeks (Madsen et al. 1993; Doyle and Smart 1995).

Small protective exclosures can be constructed from commercially available plastic fencing and steel rebar (Figure 1). Orange plastic fencing is superior to wire netting because it is highly visible to boat traffic and, because it does not rust, will last two seasons in the water. This exclosure protects the plantings from large herbivores and also acts as a wave break, reducing stress on developing plants.

Planting options.—Ideally, the most cost-effective method for establishing native plants is to broadcast seed on appropriate sites. However, the small energy reserves contained in seeds requires that the developing seedling quickly achieve photosynthetic self-sufficiency (Madsen 1991). Therefore, seeds should be planted in shallow, protected areas. As a

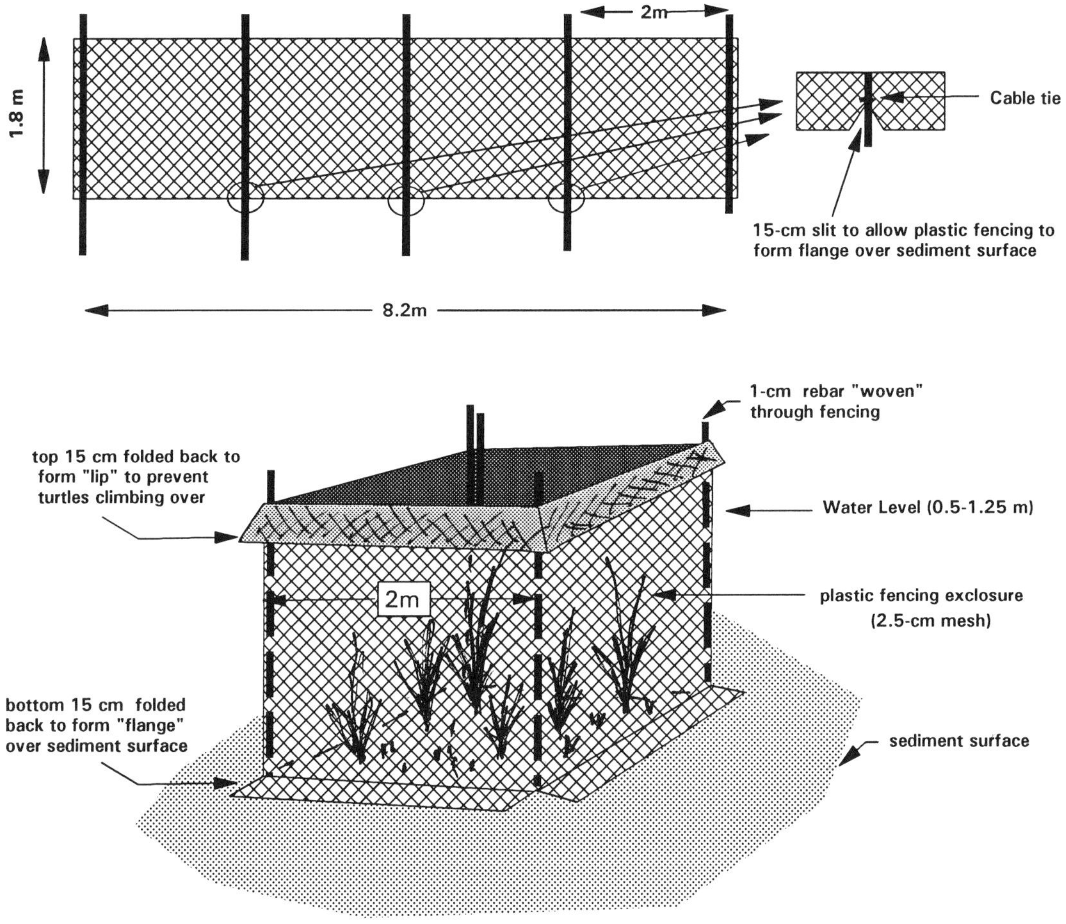

FIGURE 1.—Design for a small exclosure that can be constructed of plastic barrier fencing and rebar. The exclosure can be assembled on land and later deployed in up to 1.25 m of water.

general rule, they should be planted at a depth less than or equal to the average Secchi disk transparency during the growing season.

Many plants produce viable stem fragments that can serve as an effective means of propagation. This trait offers the advantage that a single plant in culture or in the field can be "harvested" to provide numerous propagules. However, because stem cuttings are delicate and subject to desiccation, they must be handled with care to retain viability. Also, stem fragments must sustain high photosynthetic rates during the first weeks as they establish roots, and should therefore be planted in relatively high light (at a depth of less than or equal to the average Secchi disk transparency) and low turbulence environments.

Dormant tubers provide one of the most economical and effective means of establishment. The anchoring ability and energy reserves provided by tubers enable newly established plants to withstand moderate currents and survive lower light levels. Tubers tolerate handling during transplanting to the field and can be planted at water depths up to twice the average Secchi disk transparency. To prevent tubers from being washed away, they can be planted by hand to a depth of 5 cm beneath the sediment surface or scattered in weighted cotton mesh bags.

If culture facilities are available, most species can be effectively established in the field as live transplants. Transplants contain large energy reserves and can tolerate water depths up to twice the average Secchi disk transparency. Commercially available peat pots offer an excellent way to culture submersed plants and allow transport to the field with minimal disturbance to the root system. The advantage of using transplants for establishing native plants is that the high-light and low-turbulence environment needed for initial seedling germination and growth can be provided under controlled conditions. Thus, selection of planting sites is more flexible.

Transplanting actively growing plants to the field offers the most reliable guarantee of short-term survival. However, our research indicates that transplants uprooted before transplanting (bare-root transplants) suffer higher mortality than peat-potted plants transplanted with an undisturbed root system (Doyle and Smart 1993). We have also observed that because the tissues of many plants are buoyant, bare-root transplants were very easily uprooted and floated away.

Monitoring.—Monitoring the results of establishment projects is critical for long-term evaluation of the benefits provided by these efforts. A common problem in wetland creation has been inadequate monitoring (Kentula et al. 1993). Monitoring submersed plantings, because they are less visible and more susceptible to damage, is even more critical. Without information on possible causes of failed efforts, little progress will be made in this area. Several simple monitoring methods are available for the types of plantings discussed.

If small numbers of transplants are used and planted in a pattern, short-term monitoring involves counting surviving clumps within exclosures. As the clumps begin to grow together, visual estimates of the percent cover of the plants within the plot or exclosure can be made.

A method useful for larger-scale and longer-term evaluations is the line transect. This method can be used for either permanent or random transects and offers the advantage of providing an easily interpreted estimate to monitor long-term development of a community (Titus 1993). Monitoring should continue after the initial establishment phase to "fine tune" the species composition of the community. For example, desirable species which are found to be rare or absent can be selectively added to the plant community.

Conclusion

Reservoirs are complex and dynamic ecosystems, and the development of mixed native plant communities can provide new and important dimensions in habitat and ecosystem stability. In addition to their benefit to fisheries, native plant communities can further enhance the aquatic environment by providing valuable habitat for waterfowl, water quality benefits (including increased clarity as well as nutrient uptake and immobilization), resilience to disturbances, and resistance to invasion by nuisance exotics. While research into the reasons for success or failure of specific plantings continues, our experience points to the necessity of diversifying efforts by using a variety of species and locations. In multiple, and seemingly identical efforts, some plantings fail while others flourish. Establishment of aquatic plants under reservoir conditions is still far from an exact science, and as a result remains difficult and expensive. In spite of current limitations, however, documented efforts to establish native plants will ultimately yield rich dividends. Such efforts will serve to establish quality shallow water habitats that are often lacking in reservoir systems.

References

Bain, M. B., and S. E. Boltz. 1992. Effect of aquatic plant control on the microdistribution and population characteristics of largemouth bass. Transactions of the American Fisheries Society 121:94–103.

Bertness, M. D., and R. Callaway. 1994. Positive interactions in communities. Trends in Ecology and Evolution 9:191–193.

Brock, M. A. 1988. Flexibility of life cycle pattern as a mechanism for tolerance of fluctuations of environmental conditions by aquatic plants. Internationale Vereinigung für Theoretische und Angewandte Limnologie 23:1949–1953.

Butler, R. S., E. J. Moyer, M. W. Hulon, and V. P. Williams. 1992. Littoral zone invertebrate communities as affected by a habitat restoration project on Lake Tohopekaliga, Florida. Journal of Freshwater Ecology 7:317–328.

Carpenter, S. R. 1981. Submersed vegetation: an internal factor in lake ecosystem succession. American Naturalist 118:372–383.

Carpenter, S. R., and D. M. Lodge. 1986. Effects of submersed macrophytes on ecosystem processes. Aquatic Botany 26:341–370.

Carter, V., and N. B. Rybicki. 1985. The effects of grazers and light penetration on the survival of *Vallisneria americana* Michx in the tidal Potomac River, Maryland. Aquatic Botany 23:197–213.

Denike, T. J., and R. W. Geiger. 1976. The utilization of Chara in water management. Hyacinth Control Journal 12:18–20.

Dewey, M. R., and C. A. Jennings. 1992. Habitat use by larval fishes in a backwater lake of the upper Mississippi River. Journal of Freshwater Ecology 7:363–372.

Dick, G. O., and R. M. Smart. 1992. Management of endemic aquatic macrophytes in warmwater ponds.

Proceedings, 26th annual meeting, aquatic plant control research program. U.S. Army Corps of Engineers Waterways Experiment Station, Miscellaneous Paper A-92-2:26–29, Vicksburg, Mississippi.

Dick, G. O., R. M. Smart, and E. D. Keiser. In press. Turtles and their potential impact on aquatic plants in Guntersville Reservoir, Alabama. Guntersville Joint Agency Project, Tennessee Valley Authority Technical Report.

Doyle, R. D., and R. M. Smart. 1993. Potential use of native aquatic plants for long-term control of problem aquatic plants in Guntersville Reservoir, Alabama; report 1, establishing native plants. U.S. Army Corps of Engineers Waterways Experiment Station, Technical Report A-93-6, Vicksburg, Mississippi.

Doyle R. D., and R. M. Smart. 1995. Potential use of native aquatic plants for long-term control of problem aquatic plants in Guntersville Reservoir, Alabama; report 2, competitive interactions between beneficial and nuisance species. U.S. Army Corps of Engineers, Waterways Experiment Station Technical Report A-95-6, Vicksburg, Mississippi.

Engel, S. 1985. Aquatic community interactions of submerged macrophytes. Wisconsin Department of Natural Resources Technical Bulletin 156.

French, J. R. P. 1988. Effect of submersed aquatic macrophytes on resource partitioning in yearling rock bass (*Ambloplites rupestris*) and pumpkinseeds (*Lepomis gibbosus*) in Lake St. Clair. Journal of Great Lakes Research 14:291–300.

Godshalk, G. L., and J. W. Barko. 1985. Vegetation succession and decomposition in reservoirs. Pages 59–77 *in* D. Gunnison, editor. Microbial processes in reservoirs. Dr. W. Junk Publishers, Dordrecht.

Grime, J. P. 1979. Plant strategies and vegetation processes. Wiley, New York.

Hammer, D. A. 1992. Creating freshwater wetlands. Lewis Publishers, Chelsea, Michigan.

Honnell, D. R., J. D. Madsen, and R. M. Smart. 1993. Effects of exotic and native aquatic plant communities on water temperature and dissolved oxygen. U.S. Army Corps of Engineers, Waterways Experiment Station, Aquatic Plant Control Research Program, Information Exchange Bulletin A-93-2, Vicksburg, Mississippi.

Kentula, M. E., and five coauthors. 1993. An approach to improving decision making in wetland restoration and creation. Edited by A. J. Hairston. U.S. Environmental Protection Agency, Environmental Research Laboratory, Corvallis, Oregon.

Killgore, K. J., R. P. Morgan, and N. B. Rybicki. 1989. Distribution and abundance of fishes associated with submersed aquatic plants in the Potomac River. North American Journal of Fisheries Management 9:101–111.

Korschgen, C. E., and W. L. Green. 1988. American wild celery (*Vallisneria americana*): ecological considerations for restoration. U.S. Fish and Wildlife Service Technical Report 19.

Kufel, L., and T. Ozimek. 1994. Can chara control phosphorus cycling in Lake Luknojno (Poland)? Hydrobiologia 275/276:277–283.

Kunii, H. 1982. Life cycle and growth of *Potamogeton crispus* L. in a shallow pond, Ojaga-ike. Botanical Magazine Tokyo 95:109–124.

Langeland, K. A. 1990. Hydrilla: a continuing problem in Florida waters. University of Florida Institute of Food and Agricultural Sciences Cooperative Extension Service, Circular 884.

Madsen, J. D. 1991. Resource allocation at the individual plant level. Aquatic Botany 41:67–86.

Madsen, J. D., and five coauthors. 1993. Submersed littoral vegetation distribution: field quantification and experimental analysis of sediment types from Onondaga Lake, New York. U.S. Army Corps of Engineers Waterways Experiment Station, Technical Report A-93-14, Vicksburg, Mississippi.

May, R. M. 1977. Thresholds and breakpoints in ecosystems with a multiplicity of stable states. Nature 269: 471–477.

Nichols, S. A. 1992. Depth, substrate, and turbidity relationship of some Wisconsin lake plants. Transactions of the Wisconsin Academy of Sciences Arts and Letters 80:97–118.

Olesen, B., and K. Sand-Jensen. 1994. Patch dynamics of eelgrass Zostera marina. Marine Ecology Progress Series 106:147–156.

Pieterse, A. H. 1981. Hydrilla verticillata–a review. Abstracts on Tropical Agriculture 7:9–34.

Prescott, G. W. 1978. Freshwater algae. Brown, Dubuque, Iowa.

Savino, J. F., and R. A. Stein. 1982. Predator–prey interaction between largemouth bass and bluegills as influenced by simulated, submersed vegetation. Transactions of the American Fisheries Society 111: 255–266.

Scheffer, M. 1990. Multiplicity of stable states in freshwater systems. Hydrobiologia 200/201:475–486.

Scheffer, M., S. H. Hosper, M.-L. Meijer, B. Moss, and E. Jeppesen. 1993. Alternative equilibria in shallow lakes. Trends in Ecology and Evolution 8:275–279.

Schramm, J. L., Jr., and K. J. Jirka. 1989. Effects of aquatic macrophytes on benthic macroinvertebrates in two Florida lakes. Journal of Freshwater Ecology 5:1–12.

Schriver, P., J. Bogestrand, E. Jeppesen, and M. Sondergaard. 1995. Impact of submerged macrophytes on fish–zooplankton–phytoplankton interactions: large-scale enclosure experiments in a shallow eutrophic lake. Freshwater Biology 33:255–270.

Sheldon, S. P. 1986. The effects of short-term disturbance on a freshwater macrophyte community. Journal of Freshwater Ecology 3:309–317.

Smart, R. M., and J. W. Barko. 1985. Laboratory culture of submersed freshwater macrophytes on natural sediments. Aquatic Botany 21:251–263.

Smart, R. M., and J. D. Dorman. 1993. Latitudinal differences in growth strategy of a submersed aquatic plant: ecotypic differentiation in *Vallisneria americana*? Bulletin of the Ecological Society of America 74(Supplement):439.

Smart, R. M., and R. D. Doyle. 1995. Ecological theory and management of submersed aquatic plant communities. U.S. Army Corps of Engineers Waterways Experiment Station, Aquatic Plant Control Research Program, Information Exchange Bulletin A-95-3, Vicksburg, Mississippi.

Smart, R. M., R. D. Doyle, J. D. Madsen, and G. O. Dick. 1996. Establishing native submersed aquatic plant communities in southern reservoirs. U.S. Army Corps of Engineers, Waterways Experiment Station Technical Report A-96-2, Vicksburg, Mississippi.

Stearns, S. C. 1977. The evolution of life history traits: a critique of the theory and a review of the data. Annual Review of Ecology and Systematics 8:145–171.

Titus, J. E. 1993. Submersed macrophyte vegetation and distribution within lakes: line transect sampling. Lake and Reservoir Management 7:155–164.

Titus, J. E., and M. S. Adams. 1979. Comparative carbohydrate storage and utilization patterns in the submerged macrophytes *Myriophyllum spicatum* and *Vallisneria americana*. American Midland Naturalist 102: 263–272.

Titus, J. E., and D. T. Hoover. 1991. Towards predicting reproductive success in submersed freshwater angiosperms. Aquatic Botany 41:111–136.

van der Valk, A. G. 1981. Succession in wetlands: a Gleasonian approach. Ecology 62:688–696.

Webb, D. H. 1993. Aquatic macrophyte assessment—Joint Agency Guntersville Project. Proceedings, 27th annual meeting, aquatic plant control research program. U.S. Army Corps of Engineers Waterways Experiment Station, Miscellaneous Paper A-93-2:33–38, Vicksburg, Mississippi.

American Fisheries Society Symposium 16:357–372, 1996

Assessment of Fish–Plant Interactions

Eric D. Dibble[1], K. Jack Killgore, and Sherry L. Harrel[1]

U.S. Army Corps of Engineers, Waterways Experiment Station
Vicksburg, Mississippi 39180-6199, USA

Abstract.—We review the published literature to investigate: (1) the functional importance of aquatic plants to fish, (2) how aquatic plant and fish populations are measured in vegetated habitats, (3) the spatial scale at which previous investigators have quantified fish–plant interactions, and (4) how proximate fish behaviors influence population structure at a macroscale. Based on results of comparative studies, the typical conclusion has been that intermediate levels of plants promote high species richness and are optimal for growth and survival of fishes. Predictable responses by fishes to aquatic plants were noted: vegetated habitats supported higher fish densities than unvegetated areas, aquatic plants led to reduced risk of predation, and structurally oriented fish exploited aquatic plant beds. Pelagic species and benthic omnivores often declined in abundance with increased plant cover, and phytophilic fishes showed rapid population increases during plant growing seasons. When plants occupied an entire water body, fish growth became stunted due to depletion of food resources. These interactions have been assessed largely at a macroscale where aquatic plants are generally mapped from aerial photography or surface measurements and fish data are averaged as standing crop, density, catch per unit effort, or percent abundance relative to plant coverage. Because direct observation of fish in dense plant beds is difficult, few attempts have been made to define and quantify structural complexity of plants at a scale perceived by fishes. We provide aquatic plant attributes potentially important to growth and survival of fishes and suggest that microscale assessment of fish behaviors can be linked to macroscale fishery management strategies through analysis of areal distribution of aquatic plants.

Associations between aquatic plants and fish assemblages are demonstrated in scientific literature with a frequently drawn conclusion that "intermediate" plant densities enhance fish diversity, feeding, growth, and reproduction. Comparison of results among studies can be ambiguous and contradictory, however, because investigators have characterized plant distributions and fish responses on different scales.

We consider two scales in this paper: macro and micro. The method and scale of measurement in relation to fish and plants distinguish these two scales. Fish–plant interactions have been assessed largely at a macroscale using indirect measures. Macroscale refers to either an entire water body or a water body divided into zones (e.g., littoral zone, cove) based on the extent to which shoreline and bottom characteristics influence aquatic habitat (Busch and Sly 1992). Aquatic plants are generally mapped from aerial photography or surface measurements and expressed as hectares of plants, percent coverage, or biomass per hectare. Fish may be collected from specific locations, but data are averaged as standing crop, density, catch per unit effort, or percent abundance relative to areal plant coverage.

[1]Present address: Mississippi State University, Department of Wildlife and Fisheries, Mississippi State, Mississippi 39762, USA.

Microscale is a measurement of plant complexity at a scale perceived and exploited by an individual or group of fishes. Microscale assessment focuses on behavioral ecology of fishes: the processes by which fishes interact with the environment, and the consequence of behaviors (Noakes and Baylis 1990). In this paper, the location of a microscale sample is referred to as a patch. Rather than areal coverage, underwater architectural features and surface spatial patterns are used to characterize plant complexity within a patch. Behavioral responses by fish include dispersion, preference, and rates of a particular activity (i.e., foraging).

Microscale assessments are uncommon for several reasons. Direct observation of fish in spatially complex habitats is difficult, yet it is the proximate response of individual fish to variation in habitat complexity that determines success in foraging, reproduction, and predator avoidance. Furthermore, most studies have approached fish–plant interactions from human perspectives (e.g., elimination of nuisance growths, enhancement of recreational fisheries) rather than that of individual fish (e.g., exploitation of specific habitats). Previous literature reviews on relationships between aquatic plants and fishes have been limited in scope, emphasized only a few species of plants and fish (Hinkle 1986; Engel 1995), or covered only a specific geographical region (Janecek 1988).

In this paper, we review the literature to investigate the functional importance of aquatic plants to fishes, how and at what scale previous investigators have quantified aquatic plants and fishes in vegetated habitats, and how proximate behaviors influence population structure at a macroscale. It is at the macroscale that fisheries management decisions are made. We submit that once proximate behaviors defined at a microscale are quantified, relationships to macroscale fisheries management strategies can be drawn through analysis of areal distribution of aquatic plants. We discuss published studies according to topical areas and identify pertinent conclusions, and include most of the primary literature on fish–plant interactions published over the last 40 years.

Aquatic Plants as Fish Habitat

Areas of Concentration

Many juvenile and adult fishes have been reported in habitats containing aquatic vegetation. Janecek (1988) compiled a list of 112 different species representing 19 families that were collected in aquatic plant beds in the upper Mississippi. The families Clupeidae, Cyprinidae, Ictaluridae, Esocidae, Cyprinodontidae, Atherinidae, Percidae, and particularly Centrarchidae are well-represented (Table 1). When compared to unvegetated areas, vegetated sites contain higher fish densities (Borawa et al. 1979). Up to seven times more fish were collected in areas with plants than in areas without them (Killgore et al. 1989). Similarly, Barnett and Schneider (1974) reported fish density in vegetated habitats as high as 2 million fish/ha. Angling, at least in part, may influence population demography of exploitable fish that are concentrated in vegetated areas (Hoyer and Canfield 1996).

Results of previous studies indicate that fish are attracted to aquatic plants. Sunfish *Lepomis* and bass *Micropterus* abundance were positively related to plant abundance (Forester and Lawrence 1978; Durocher et al. 1984). Submersed vegetation was the key factor in the distribution and habitat use of adult northern pike *Esox lucius* (Cook and Bergersen 1988). Age-0 northern pike were 10-times more abundant in vegetated than unvegetated areas (Holland and Huston 1984). Younger and smaller fishes become more abundant as plant density increases (Barnett and Schneider 1974; Borawa et al. 1979; Moxley and Langford 1985). However, pelagic species, such as white bass *Morone chrysops*, gizzard shad *Dorosoma cepedianun*, and inland silverside *Menidia beryllina*, generally decline in abundance as plants increase in areal coverage (Bailey 1978; Maceina and Shireman 1985; Bettoli et al. 1990).

TABLE 1.—A taxonomic list of fish families and their life stages reported in studies related to aquatic plant habitats. Numbers correspond to individual papers in the list of references.

Fish family	Reference
	Adult
Lepisosteidae	8, 14, 18, 172
Amiidae	15, 18, 78, 148, 172
Anguillidae	18, 100, 166, 172
Clupeidae	4, 17, 15, 18, 95, 100, 115, 148, 172
Salmonidae	11
Cyprinidae	8, 15, 18, 78, 89, 95, 97, 98, 99, 100, 133, 148, 159, 166, 172
Catostomidae	8, 15, 18, 78, 148, 172
Ictaluridae	8, 15, 18, 95, 100, 133, 148, 172, 200
Esocidae	8, 15, 18, 38, 78, 80, 100, 148, 159, 172
Umbridae	143, 184
Aphredoderidae	8, 15
Cyprinodontidae	8, 15, 18, 25, 83, 97, 109, 133, 148, 155, 172
Poeciliidae	8, 15, 100, 133, 172
Atherinidae	15, 16, 18, 21, 89, 90, 100, 133, 166, 172
Cottidae	148
Percichthyidae	15, 18, 100
Centrarchidae	1, 4, 5, 8, 14, 15, 18, 24, 29, 31, 32, 39, 40, 41, 44, 58, 60, 64, 68, 72, 78, 80, 85, 89, 93, 95, 97, 98, 99, 100, 103, 115, 126, 130, 148, 155, 157, 159, 166, 172, 194, 198, 199, 200
Percidae	8, 15, 18, 19, 71, 80, 95, 97, 99, 100, 144, 148, 166, 172, 184, 199
Scianenidae	15, 48
	Juvenile
Salmonidae	75
Cyprinidae	187
Ictaluridae	200
Esocidae	88
Cyprinodontidae	37
Poeciliidae	37
Atherinidae	37
Centrarchidae	1, 2, 14, 32, 37, 58, 60, 64, 68, 71, 72, 78, 93, 103, 106, 129, 130, 131, 133, 200
Percidae	37,187
	Larval
Lepisosteidae	140
Clupeidae	17, 30, 37
Cyprinidae	30, 31, 74, 142, 162
Ictaluridae	30
Catostomidae	162
Esocidae	30, 142
Umbridae	30
Cyprinodontidae	37, 162
Poeciliidae	37, 162
Atherinidae	30, 37, 47
Centrarchidae	30, 37, 47, 74, 140, 162
Percidae	30, 37, 74, 140, 162

Natural senescence of aquatic macrophytes is related to decreased fish abundance in the littoral zone, presumably due to reduction of invertebrate density and cover (Whitfield 1984). Even plant disturbance due to boat traffic decreases fauna and

TABLE 2.—Scale and topical emphasis of fish–plant interaction studies. Numbers correspond to individual papers in the list of references.

Topical emphasis	References
	Macroscale
Fish abundance and composition	1, 8, 14, 17, 18, 37, 54, 58, 74, 85, 89, 100, 123, 125, 133, 140, 141, 142, 146, 148, 172, 184, 195
Habitat use and distribution	1, 37, 38, 74, 77, 88, 96, 125, 129, 140, 146, 151, 172, 187, 194, 198
Foraging and diets	5, 36, 96, 138, 181
Fish growth	14, 32, 39, 77, 115, 117, 129, 138
Reproduction and rearing	1, 30, 37, 47, 74, 128, 133, 140, 175
Plant control effects	4, 5, 14, 15, 16, 17, 31, 32, 33, 34, 35, 59, 60, 64, 110, 115, 117, 154, 167, 171, 180, 181, 189, 190, 199, 200
Recreation and sportfishing	32, 33, 35, 133, 135, 151
Habitat restoration	57
Plant senesence or eutrophication effects	91, 196
Fish induced alterations on plants	176
	Microscale
Foraging and diets	51, 57, 67, 68, 80, 107, 130, 155, 196
Foraging efficiency and predator risk	2, 40, 41, 44, 51, 68, 83, 111, 130, 157, 169, 158, 178, 177, 188, 195
Fish growth	67, 160
Habitat use and distribution	27, 31, 44, 45, 57, 71, 72, 78, 80, 95, 103, 107, 111, 126, 155, 162, 193, 195
Fish effects on plants	24, 122
Fish effects on macroinvertebrates	29, 51, 131, 186, 196
Interspecific competition	80, 103, 131
Plant senescence	196
Behavioral response to pH or DO	166

habitat important to fish communities (Murphy and Eaton 1981).

Foraging Efficiency and Refugia

Aquatic plant beds contain food and provide refuge for younger and smaller fishes. Macroinvertebrate abundance and diversity are higher in aquatic plants than in unvegetated areas because leaves and stems provide substrate for attachment and protection from predators (Gilinsky 1984; Keast 1984; Beckett et al. 1992). Morphology of aquatic plants and depths at which they grow influence production of epiphytes (Cattaneo and Kalff 1980; Keast 1984). Epiphytic invertebrates serve as prey for a variety of fishes (e.g., Centrarchidae, Cyprinodae, Percidae, and Cyprinidontidae) (Hall et al. 1970; Keast 1985a, 1985b; Hoover et al. 1988).

Numerous microscale studies have been conducted on foraging efficiency of fishes (Table 2). Structural complexity provided by plants may reduce predation risk by mediating the extent to which fish interact with prey (Glass 1971; Saiki and Tash 1979; Savino and Stein 1982). Visual and swimming barriers created by dense stems and foliage can reduce foraging success of sunfishes and killifish (Heck and Thoman 1981; Savino and Stein 1982; Dionne and Folt 1991). This effect is due to increased search, encounter, and capture times, as well as reduced encounter, attack, and capture rates, and reduced swimming velocities (Anderson 1984; Diehl 1988). Prey capture rates decline with an increase in structural complexity (Crowder and Cooper 1979b); thus, foraging efficiency declines as habitat becomes more spatially complex.

Some species change foraging tactics as aquatic plants become more complex, or as coverage increases. Largemouth bass foraging in spatially complex habitats may switch from actively pursuing prey to ambushing them, which minimizes the energy cost of prey capture (Savino and Stein 1982). Shade in spatially complex habitats is an important attribute. Shaded areas attract fish (Helfman 1979, 1981; Johnson 1993) and may improve vigilance and foraging behavior by increasing visual acuity (Diehl 1988; Lynch and Johnson 1989).

Fish Growth

The size at which age-0 fish enter their first winter is critical to survival and subsequently influences fish recruitment and production (Gutreuter and Anderson 1985; Adams and DeAngelis 1987). Most studies of aquatic plant effects on fish growth were conducted at a macroscale (Table 2). Their conclusions suggest that aquatic plant abundance mediates fish growth and condition, and that both limited and excessive plant growth may decrease fish growth rates, while moderate levels are optimal.

Excessive plant growth reduces growth and condition of largemouth bass *Micropterus salmoides*, bluegill *Lepomis macrochirus*, black crappie *Pomoxis nigromaculatus*, white crappie *P. annularis*, and redear sunfish *L. microlophus* (Colle and Shireman 1980; Wiley et al. 1984; Maceina and Shireman 1985), presumably by reducing foraging efficiency (Wiley et al. 1984). Colle and Shireman (1980) predicted that largemouth bass growth would significantly decrease in a system with 40% or greater total coverage of aquatic plants relative to a system with less than 40% coverage. Wiley et al. (1984) suggested optimal mean standing crop of pondweeds (e.g., *Potamogeton* and *Najas*) at 52 g dry weight/m^3 would improve foraging efficiency in largemouth bass. Total removal of aquatic plants increases growth of largemouth bass, black and white crappies (Maceina et al. 1991), bluegill, and redear sunfish (Bailey 1978), and may alter foraging behaviors of largemouth bass by initiating piscivory sooner in smaller age-classes, resulting in rapid growth (Bettoli et al. 1992).

Young sunfishes and perch *Perca fluviatilis* often showed the opposite trend from larger piscivores; increased vegetation density was positively related to their growth (Gerking 1962; Hall and Werner 1977; Blindow et al. 1993). When plants were sparse, competition increased, resulting in slower growth rates due to reduced caloric intake (Mittelbach 1981; Mittelbach and Chesson 1987; Diehl 1993). However, stunted growth also occurred when plants occupied the entire water body, particularly in shallow systems without any deep, unvegetated areas (Colle and Shireman 1980; Engel 1988).

TABLE 3.—Methods and parameters used to classify aquatic plant habitats. Numbers correspond to individual papers in the list of references.

Method or parameter	Reference
Aerial photography, digital imagery	13, 54, 55, 59, 82, 92, 108, 113, 116, 121, 170, 199
Circular core	81, 116, 120
Divers	4, 24, 42, 53, 59, 85, 95, 108, 111, 118, 120, 132, 146, 169, 196, 201, 202
Fathometer, acoustics	53, 108, 116, 118, 173, 179, 182
Grab, grapnel, dredge, rake	61, 136, 156, 165, 174
Plant removal (by hand)	20, 32, 66, 202
Quadrat	42, 52, 54, 59, 88, 100, 108, 118, 120, 132, 145, 146, 149, 161, 167, 196, 199
Transect	22, 23, 24, 32, 75, 85, 104, 111, 118, 120, 136, 173, 196
Aerial coverage, % composition	4, 14, 23, 32, 38, 61, 75, 85, 104, 135, 148, 195, 199
Biomass measurements	20, 23, 42, 52, 89, 92, 100, 104, 108, 111, 118, 139, 146, 147, 149, 156, 161, 163, 165, 167, 179, 199, 202
Biovolume	182
Canopy, plant density	42, 74, 100, 104, 165
Submerged versus emerged	121, 126
Mat buoyancy	145
Plant morphology	20, 111, 145, 147
Plant weight, wet	81, 145
Presence and absence	4, 14, 32, 31, 98, 106, 140, 162, 172, 183, 198

Spawning and Rearing

Many North American fishes are obligatory plant spawners; these include members of Amiidae (e.g., *Amia*), Esocidae (e.g., *Esox*), Cyprinidae (e.g., *Cyprinus*, *Cyprinella*, *Notemigonus*), Catostomidae (e.g., *Ictiobus*), Cyprinodontidae (e.g., *Fundulus*), Atherinidae (e.g., *Labidesthes*), Umbridae (e.g., *Umbra*), Centrarchidae (e.g., *Elassoma*), and Percidae (e.g., *Perca*, some *Etheostoma*) (Pflieger 1975; Robison and Buchanan 1988). However, most empirical data on spawning success relative to structural complexity and plants come from studies of adult sport fishes that construct nests (i.e., largemouth bass and bluegill).

Adult largemouth bass and bluegill select sites protected from wave action (Tester 1930; Kramer and Smith 1962; Miller and Kramer 1971) and keep their nests cleared of vegetation, sometimes influencing littoral vegetation spatial patterns (Carpenter and McCreary 1985). Although nest spawners successfully spawn in areas devoid of vegetation, they prefer sites with aquatic plants or some other type of structure nearby for refugia (Vogele and Rainwater 1975; Mesing and Wicker 1986; Hoff 1991; Annett et al. 1996, this volume). However, dense vegetation throughout the littoral zone can hinder spawning adults by decreasing the availability of nest sites (Colle and Shireman 1980).

Aquatic vegetation is used as nursery habitat for larvae by at least 12 families (Table 1). Larval stages of sunfish, brook silverside *Labidesthes sicculus*, yellow perch *Perca flavescens*, golden shiner *Notemigonus crysoleucas*, northern pike, and certain species of darters are more abundant in vegetation than in open water (Floyd et al. 1984; Gregory and Powles 1985; Paller 1987; Dewey and Jennings 1992). Some species exhibit ontogenetic shifts in habitat use. For example, prolarvae of yellow perch prefer shallow, dense macrophyte areas, while postlarvae prefer deep, low-density macrophyte zones (Gregory and Powles 1985).

Quantifying Fish–Plant Interactions

Quantifying Plants

Measurements of plant biomass, area coverage, percent species composition, and presence or absence are typically used to quantify growth characteristics of aquatic plants (Table 3). Aerial photography, and more recently digital imaging using geographic information systems (GIS) (Lukens 1967; Harvey et al. 1988; Jennings et al. 1992; Marshall and Lee 1994), are remote sensing techniques to map and estimate acres of aquatic plants at the macrohabitat scale. Ground measurements (e.g., quantifying morphology of individual plants or small groups of plants and use of quadrant and transect data) are common (Table 3), and these data are often extrapolated to the entire system (Forsberg 1959; Edwards and Moore 1975; Cassani and Caton 1985; Smart and Barko 1988). Plant biomass has been quantified using direct hand removal of plants, modified dredges, grabs, and rakes within defined areas (Sabol 1984; Sliger et al. 1990) (Table 3). Where water conditions are favorable, samples are taken directly by divers using scuba or snorkel gear (Kautsky et al. 1981; Pringle 1984; Downing and Anderson 1985; Machena and Kautsky 1988).

Architectural features are microscale assessment of morphology and plant spacing. A fathometer has been used to estimate plant height and map aquatic plant distributions (Maceina and Shireman 1980; Maceina et al. 1984; Schloesser and Manny 1984; Stent and Hanley 1985; Duarte 1987; Pine et al. 1989; Thomas et al. 1990). A more recent approach to quantification of plant architecture is to measure interstitial spaces and leaf and stem morphology (Johnson et al. 1988; Lynch and Johnson 1989; Walters et al. 1991; Lillie and Budd 1992; Wychera et al. 1993; Dibble and Killgore 1994).

Quantifying Fish in Aquatic Plants

Divers have successfully quantified relative abundance of fish species in the littoral zone (Table 4). Under suitable conditions (i.e., high water clarity and moderate plant density), divers can rapidly census fish populations and measure species composition and abundance in habitats that are difficult to sample with traditional methods (Northcote and Wilkie 1963; Keast and Harker 1977; Heggenes et al. 1990; Dibble 1991). However, excessive plant growth may hinder direct observation of fish (Heggenes et al. 1990; Rodgers et al. 1992).

Boat-mounted electroshockers are commonly used to sample fish in aquatic plants (Table 4), but dipping efficiency is reduced in dense plant beds (Killgore et al. 1989). Frame electroshocking equipment used to sample fish in rivers (Bain et al. 1985a) has been modified for use in dense vegetation (Dewey 1991; Vadas and Orth 1993). A time delay between disturbance (i.e., setting up the frame electroshocker) and the sample can decrease the effect of fright response by fish (Bain et al. 1985a).

TABLE 4.—Methods and tools used to quantify fish in or near vegetated areas. Numbers correspond to individual papers in the list of references.

Method	Reference
Angling	203
Belt transect	24, 78, 98, 138, 195
Divers	24, 48, 56, 78, 98, 99, 137, 138, 152, 195, 203
Drop or throw nets	7, 8, 28, 30, 65, 192
Echosounder	11
Electro-shocker and block net	10, 119, 185
Electro-shocker	5, 6, 14, 31, 32, 39, 75, 105, 152, 183, 198
Explosives	3, 9, 10, 63, 105, 127
Fyke nets	184
Gill nets	14, 38, 85, 127, 146, 187, 196
Helicopter	137
Hose pump or net	140
Light trap or minnow trap	47, 74, 85, 95, 107, 184
Modified nets	14, 196
Modified traps	65, 85, 199
Popnets	62, 46, 100, 132, 166, 167, 168
Push net	30, 124, 162
Radio telemetry	5, 31, 38, 126, 183, 198
Rotenone and block net	18, 54, 125, 129, 172
Rotenone	1, 4, 48, 105, 115, 127, 133, 137
Seine	14, 46, 65, 88, 89, 97, 99, 132, 152, 185
Shore observations	137
Stationary nets	14
Strip counts	98, 99
Tow nets	37
Trapnetting	85
Trawl	148
Underwater camera	11, 49, 56

A variety of nets has been used to sample fish in aquatic plants (Table 4). Pop nets and drop nets measure distribution, diversity, and abundance of adult and juvenile fishes in densely vegetated areas where traditional methods (i.e., seining and electrofishing gear) are ineffective (Freeman et al. 1984; Morgan et al. 1988; Serafy et al. 1988; Dewey et al. 1989; Espegren and Bergersen 1990). Underwater observations of pop nets in use in pools and reservoirs demonstrated they were accurate for sampling small fish in complex habitats (Larson et al. 1986), and pop nets may be one of the better gears to collect young fishes in aquatic plants.

Vegetated areas have been blocked off with nets

and sampled with rotenone (Lambout 1959), but collection efficiency decreased as plant density increased (Shireman et al. 1981). Catch depletion techniques, in which a series of samples are collected and differences among samples are plotted on a depletion curve to estimate abundance, eliminated the need to remove fish from the net (Morgan et al. 1988; Maceina et al. 1995). Sampling with rotenone in vegetated areas enclosed with a block net was less expensive and provided a more realistic assessment of largemouth bass than cove rotenone sampling (Maceina et al. 1995).

Seines were commonly used to sample fishes near vegetation (Table 4), but were difficult to use in dense plant beds. Light traps were efficient for determining larval fish abundance and species composition in aquatic plants (Faber 1981; Gregory and Powles 1985). Modified ichthyoplankton nets have been used to sample larval fishes in structurally complex habitats where traditional tow nets could not be easily used (Barnett 1973; Meador and Bulak 1987).

Discussion

Based on our assessment of the literature, there are predictable responses by fish in relation to aquatic plants, albeit mostly derived from macroscale studies of plant control operations (Table 2). Vegetated areas support fish densities from 15,000 to over 2 million fish/ha, higher than unvegetated areas. Structurally oriented fish exploit aquatic plant beds, with juvenile sunfishes being numerically dominant in vegetation in most North American water bodies. In contrast, pelagic species and benthic omnivores (e.g., Catostomidae) often decline in abundance as plants increase in areal coverage. At least 19 families of freshwater fishes have been documented to occupy vegetated habitats during at least one of their life stages.

Aquatic plants, like other sources of structural complexity in habitats, reduce risk of predation by providing refugia for smaller fish and mediating the extent to which fish interact with prey. Both sight and bottom feeders are hampered by interference from plants and stems. Phytophilic fishes increase rapidly during the plant growing season, but if plants occupy an entire water body, growth becomes stunted because food resources are depleted.

Most comparative studies of plant and fish abundance conclude that intermediate vegetation levels, defined as 10–40% coverage of study sites, including areas ranging from individual coves to entire water bodies, promote high species richness and are optimal for growth and survival. Theoretically, because plants provide spatial complexity, intermediate densities may promote community stability by providing habitat heterogeneity (Stenseth 1980), yet mechanisms governing population dynamics as a function of plant coverage remain speculative. In addition, the lack of consistent measures of plant coverage and the problem of defining intermediate density at different scales hamper comparisons among aquatic systems and lead to variable responses by fish populations.

Fish responses are more predictable at the extremes of plant coverage. When aquatic plants cover an entire water body, foraging by piscivores is hampered by stems and leaves, small phytophilic insectivores increase in abundance due to lower predation and higher prey abundance, and spawning by nest builders is confined to limited areas that may increase competition and decrease spawning success. Conversely, water bodies that lack vegetation generally have lower densities of littoral fishes, although standing crop may not differ substantially, and fishes become more aggregated (Aboul and Downing 1994). Comparisons of vegetated and unvegetated areas within the same water body generally show that fish assemblages in unvegetated areas have lower densities and fewer species.

Long-term studies that monitor changes in fish populations as a function of changing plant coverage provide important insight into fish–plant interactions. When plants were completely eliminated in Lake Conroe, Texas, the littoral fish community shifted from sunfish and shad to include sizeable numbers of cyprinids, inland silversides, and channel catfish *Ictalurus punctatus* (Bettoli et al. 1993). Scott (1993) reported shifts in fish assemblages over a 30-year period as Eurasian watermilfoil *Myriophyllum spicatum* increased in Chickamauga Reservoir coves, Tennessee. Midwater insectivores (e.g., golden shiner, sunfishes, brook silverside, yellow perch) and ambush predators (e.g., largemouth bass) increased in abundance while benthic insectivores omnivores (e.g., smallmouth buffalo *Ictiobus bubalus*, spotted sucker *Minytrema melanops*, channel catfish, and freshwater drum *Aplodinotus grunniens*) declined in abundance. Others found that plant reduction had little effect on fish populations, and that factors other than aquatic plants may have greater effect (Bailey 1978). Studies of the effects of plants on specific fish species also were inconclusive when considered together. Wiley et al. (1984) and Noble (1981) showed increases in the number, recruitment, and survival of catfish after plants were removed, whereas, Borawa et al. (1979) reported opposite trends.

Studies of shifts in fish assemblages following changes in plant coverage have produced conflicting results. Most studies are based on indirect measurements of causal mechanisms regulating fish populations (e.g., Kushlan 1974; Freeman et al. 1984; Gregory and Powles 1985), which may lack the precision required to determine important fish–habitat relationships in vegetated areas. Consequently, proximate habitat factors and fish behavioral responses are seldom quantified.

The choice of scales for observing natural interactions must be a primary consideration in study design. Large-scale measurements generally have low resolution but are inherently stable whereas small-scale, site-specific measurements have high resolution and low stability (Busch and Sly 1992). We suggest that vegetated water bodies be viewed and described by first studying their individual parts. Through integration of the parts, biological processes can be defined. This approach is analogous to patch analysis. However, few attempts have been made to define and quantify habitat variables at a scale important to fishes. Microscale measurements have been used to quantify fish habitat in streams (Orth and Maughan 1982; Price 1982; Bain et al. 1985b), and similar approaches are needed to delineate habitat criteria for aquatic plants in reservoirs.

A variety of structural and functional habitat criteria measured at a microscale can be used to better evaluate aquatic plants as fish habitat (Table 5). For example, aquatic plant species differ in morphology and spatial distribution (Lillie and Budd 1992; Wychera et al. 1993; Dibble et al. 1996), and these differences likely influence fish behavior. Young bluegill preferred smaller interstitial spaces (40–150 mm) within structural habitat over larger ones (350 mm), and largemouth bass preferred structure with medium-sized spaces (150 mm) (Johnson et al. 1988). Thus, proximate or microscale studies that quantify fish behavioral responses such as habitat preference, foraging efficiency, predator avoidance, and social attraction in vegetated areas are required to clarify the role of aquatic plants as fish habitats.

Aquatic plant management is usually performed on a macroscale which necessitates the management of fisheries on a compatible scale. For example, there may be a trade off of catching fewer but larger fish at lower plant coverage (Maceina and Reeves 1996). Plants are mapped from remotely sensed data and GIS provide estimates on the areal coverage of submersed, floating, or emergent growth forms. Because boundaries of plant beds can be delineated from remotely sensed data (Marshall and Lee 1994), heterogeneity of different patches of plants can be identified. Thus, remote sensing techniques can be used to spatially extrapolate fish–plant relationships developed at the microscale to a larger scale.

TABLE 5.—Microscale studies on structural and functional role of fish habitat and variables potentially important to evaluate the role of aquatic plants. Numbers correspond to individual papers in the references section.

Variable addressed	Reference
	Structural
Interstitial size and abundance	93, 111, 114, 178
Plant morphology and architecture	20, 26, 51, 111, 153, 197, 202
Plant diversity	194, 197
Plant strata	58, 59, 111, 197
Shade effects	50, 84, 94, 111, 114, 197, 202
Spatial complexity	2, 20, 39, 40, 41, 50, 68, 69, 93, 94, 158, 164, 194
Stem density	2, 20, 40, 41, 51, 69, 71, 72, 83, 158, 160
	Functional
Effects on competition	103, 131
Effects on reproduction and recruitment	69, 76, 86, 101
Influence on foraging	2, 29, 39, 40, 41, 44, 50, 51, 59, 68, 69, 71, 76, 77, 83, 87, 130, 13 1, 155, 157, 158, 165, 177
Prey resources and availability	12, 26, 50, 51, 58, 59, 68, 70, 73, 77, 79, 95, 102, 103, 112, 165, 131, 141, 142, 150, 153, 164, 177, 178, 191
Refugia	39, 40, 41, 43, 45, 68, 72, 77, 83, 87, 131, 134, 155, 157, 158, 164, 188

In conclusion, most investigations of fisheries resources associated with aquatic plants emphasize static responses of only a few fish species at gross spatial levels. Conversely, most behavioral data on fish–plant interactions come from studies of small sunfishes. Few empirical data are available to bridge the theoretical predictions of responses by fish in plants at a microscale to population responses of fishes at a macroscale. Addressing scale, structural variables within aquatic plants, and fish behavior will allow direct effects of aquatic plants to be identified and make it possible to extrapolate results. The role of aquatic plants as fish habitat and their value as a management tool in reservoirs then can be better defined.

Acknowledgments

We thank Mark Bain, Jan Hoover, Gary Mittelbach, and an anonymous reviewer for many helpful comments. Thanks goes to Robin Kimbrough and Catherine Murphy for their assistance on the literature review. This work was funded by the Aquatic Plant Control Research Program of the U.S. Army Corps of Engineers Waterways Experiment Station. Permission was granted by the Chief of Engineers to publish this document.

References[2]

Aboul H. W., and J. A. Downing. 1994. Influence of cover on the spatial distribution of littoral-zone fishes. Canadian Journal of Fisheries and Aquatic Sciences 51:1832–1838.

Adams, S. M., and D. L. DeAngelis. 1987. Indirect effects of early bass–shad interactions on predator population structure and food web dynamics. Pages 103–117 *in* W. C. Kerfoot and A. Sih, editors. Predation: direct and indirect impacts on aquatic communities. University Press of New England, Hanover, New Hampshire.

(1) Aggus, L. R., and G. V. Elliot. 1975. Effects of cover and food on year-class strength of largemouth bass. Pages 317–322 *in* R. H. Stroud and H. Clepper, editors. Black bass biology and management. Sport Fishing Institute, Washington, DC.

(2) Anderson, O. 1984. Optimal foraging by largemouth bass in structured environments. Ecology 65:851–861.

Annett, C., J. Hunt, and E. D. Dibble. 1996. The compleat bass: habitat use patterns of all stages of the life cycle of largemouth bass. Pages 306–314 *in* L. E. Miranda and D. R. DeVries, editors. Multidimensional approaches to reservoir fisheries management. American Fisheries Society Symposium 16.

(3) Averett, R. C., and R. M. Stubbs. 1962. Toward a safe and effective method of using dynamite to sample fish populations and determine species ranges in large rivers. Journal of the Tennessee Academy of Sciences 37:20–22.

(4) Bailey, W. M. 1978. A comparison of fish populations before and after extensive grass carp stocking. Transactions of the American Fisheries Society 107: 181–206.

(5) Bain, M. B., and S. E. Boltz. 1992. Effect of aquatic plant control on the microdistribution and population characteristics of largemouth bass. Transactions of the American Fisheries Society 121:94–103.

(6) Bain, M. B., J. T. Finn, and H. E. Booke. 1985a. A quantitative method for sampling riverine microhabitats by electrofishing. North American Journal of Fisheries Management 5:489–493.

Bain, M. B., J. T. Finn, and H. E. Booke. 1985b. Quantifying stream substrate for habitat analysis studies. North American Journal of Fisheries Management 5:499–506.

(7) Barnett, B. S. 1973. A technique for fish population sampling in dense submersed vegetation. Progressive Fish-Culturist 35:181–182.

[2]Numbers in parentheses correspond to Tables 1–5.

(8) Barnett, B. S., and R. W. Schneider. 1974. Fish populations in dense submersed plant communities. Hyacinth Control Journal 12:12–14.

(9) Bass, Q. G., and V. G. Hitt. 1979. Qualitative sampling of warm-water fish with detonating cord. Proceedings of the Annual Conference Southeastern Association of Fish and Wildlife Agencies 31(1977): 519–521.

(10) Bayley, P. B., and D. J. Austen. 1988. Comparison of detonating cord and rotenone for sampling fish in warmwater impoundments. North American Journal of Fisheries Management 8:310–316.

(11) Beauchamp, D. A., B. C. Allen, R. C. Richards, W. A. Wurtzbaugh, and C. R. Goldman. 1992. Lake trout spawning in Lake Tahoe: egg incubation in deepwater macrophyte beds. North American Journal of Fisheries Management 12:442–449.

(12) Beckett, D. C., T. P. Aartila, and A. C. Miller. 1992. Invertebrate abundance on *Potamogeton nodosus* effects of plant surface area and condition. Canadian Journal Zoology 70:300–306.

(13) Benton, A. R., and R. M. Newnan. 1976. Color aerial photography for aquatic plant monitoring. Journal of Aquatic Plant Management 14:14–16.

(14) Bettoli, P. W., M. J. Maceina, R. L. Noble, and R. K. Betsill. 1992. Piscivory in largemouth bass as a function of aquatic vegetation abundance. North American Journal of Fisheries Management 12:509–516.

(15) Bettoli, P. W., M. J. Maceina, R. L. Noble, and R. K. Betsill. 1993. Response of a reservoir fish community to aquatic vegetation removal. North American Journal of Fisheries Management 13:110–124.

(16) Bettoli, P. W., J. E. Morris, and R. L. Noble. 1991. Changes in the abundance of two atherinid species after aquatic vegetation removal. Transactions of the American Fisheries Society 120:90–97.

(17) Bettoli, P. W., T. Springer, and R. Noble. 1990. A deterministic model of the response of threadfin shad to aquatic macrophyte control. Journal of Freshwater Ecology 5:445–454.

Blindow, I., G. Andersson, A. Hargeby, and S. Johansson. 1993. Long-term pattern of alternative stable states in two shallow eutrophic lakes. Freshwater Biology 30:159–167.

(18) Borawa, J. C., J. H. Kerby, M. T. Huish, and A. W. Mullis. 1979. Currituck Sound fish populations before and after infestation by Eurasian water-milfoil. Proceedings of the Annual Conference Southeastern Association of Fish and Wildlife Agencies 32(1978): 520–528.

(19) Braasch, M. E., and P. W. Smith. 1967. Life history of the slough darter, *Etheostoma gracile* (Pisces, Percidae). Illinois Natural History Survey, Biological Notes 58:1–12.

(20) Budd, J., R. A. Lillie, and P. Rasmussen. 1995. Morphological characteristics of the aquatic macrophyte, *Myriophyllum spicatum* L., in Fish Lake, Wisconsin. Journal of Freshwater Ecology 10:19–29.

Busch, W.-D., and P. G. Sly. 1992. The development of an aquatic habitat classification system for lakes. CRC Press, Boca Raton, Florida.

(21) Cahn, A. R. 1927. An ecological study of southern

Wisconsin fishes. The brook silversides (*Labidesthes sicculus*) and the cisco (*Leucichthys artedi*) in their relations to the region. Illinois Biological Monograph 11:1–151.

(22) Canfield, R. 1941. Application of the line interception method in sampling range vegetation. Journal of Forestry 39:388–394.

(23) Canfield, M. V., Jr., M. V. Hoyer, and C. M. Duarte. 1990. An empirical method for characterizing standing crops of aquatic vegetation. Journal of Aquatic Plant Management 28:64–69.

(24) Carpenter, S. R., and N. J. McCreary. 1985. Effects of fish nests on pattern and zonation of submersed macrophytes in a softwater lake. Aquatic Botany 22: 21–32.

(25) Carranza, J., and H. E. Winn. 1954. Reproductive behavior of the blackstripe topminnow, *Fundulus notatus*. Copeia 1954:273–278.

Cassani, J. R., and W. E. Caton. 1985. Effects of chemical and biological weed control on the ecology of a south Florida pond. Journal of Aquatic Plant Management 23:51–58.

(26) Cattaneo, A., and J. Kalff. 1980. The relative contribution of aquatic macrophytes and their epiphytes to the production of macrophyte beds. Limnology and Oceanography 25:280–289.

(27) Charnov, E. L., G. H. Orians, and K. Hyatt. 1976. Ecological implication of resource depression. American Naturalist 100:247–259.

(28) Chick, J. H., F. Jordan, J. P. Smith, and C. C. Mcivor. 1992. A comparison of four enclosure traps and methods used to sample fishes in aquatic macrophytes. Journal of Freshwater Ecology 7:353–361.

(29) Chilton, E. W., and F. J. Margraf. 1990. Effects of fish predation on invertebrates associated with a macrophyte in Lake Onalaska, Wisconsin. Journal of Freshwater Ecology 5:289–296.

(30) Chubb, S. L., and C. R. Liston. 1986. Density and distribution of larval fishes in Pentwater Marsh, a coastal wetland on Lake Michigan. Journal of Great Lakes Research 12:332–343.

(31) Colle, D. E., R. L. Cailteux, and J. V. Shireman. 1989. Distribution of Florida largemouth bass in a lake after elimination of all submersed aquatic vegetation. North American Journal of Fisheries Management 9:213–218.

(32) Colle, D. E., and J. V. Shireman. 1980. Coefficients of condition for largemouth bass, bluegill, and redear sunfish in hydrilla-infested lakes. Transactions of the American Fisheries Society 109:521–531.

(33) Colle, D. E., J. V. Shireman, D. E. Canfield, W. T. Haller, and J. C. Joyce. 1986. Economic loss results from hydrilla infestation. Aquaphyte 6:1–7.

(34) Colle, D. E., J. V. Shireman, R. D. Gasaway, R. L. Stetler, and W. T. Haller. 1978a. Utilization of selective removal of grass carp (*Ctenopharyngodon idella*) from an 80-hectare Florida lake to obtain a population estimate. Transactions of the American Fisheries Society 107:724–729.

(35) Colle, D. E., J. V. Shireman, W. T. Haller, J. C. Joyce, and D. E. Canfield, Jr. 1987. Influence of hydrilla on harvestable sport-fish populations, angler use, and angler expenditures at Orange Lake, Florida. North American Journal of Fisheries Management 7:410–417.

(36) Colle, D. E., J. V. Shireman, and R. W. Rottmann. 1978b. Food selection by grass carp fingerlings in a vegetated pond. Transactions of the American Fisheries Society 107:149–152.

(37) Conrow, R., A. V. Zale, and R. W. Gregory. 1990. Distributions and abundances of early life stages of fishes in a Florida lake dominated by aquatic macrophytes. Transactions of the American Fisheries Society 119:521–528.

(38) Cook, M. F., and E. P. Bergersen. 1988. Movements, habitat selection, and activity periods of northern pike in Eleven Mile Reservoir, Colorado. Transactions of the American Fisheries Society 117:495–502.

(39) Crowder, L. B., and W. E. Cooper. 1979a. Effects of macrophyte removal on the feeding efficiency and growth of sunfishes: evidence from pond studies. Pages 251–268 *in* J. E. Creck, R. T. Prentki, and O. L. Louks, editors. Proceedings, aquatic plants lake management, and ecosystem consequences of lake harvesting. Institute for Environmental Studies, University of Wisconsin, Madison.

(40) Crowder, L. B., and W. E. Cooper. 1979b. Structural complexity and fish–prey interactions in ponds: a point of view. Pages 2–10 *in* D. L. Johnson and R. A. Stein, editors. Response of fish to habitat structure in standing water. American Fisheries Society, North Central Division, Special Publication 6, Bethesda, Maryland.

(41) Crowder, L. B., and W. E. Cooper. 1982. Habitat structural complexity and the interaction between bluegills and their prey. Ecology 63:1802–1813.

(42) Crowell, W., N. Troelstrup, Jr., L. Queen, and J. Perry. 1994. Effects of harvesting on plant communities dominated by Eurasian watermilfoil in Lake Minnetonka, Minnesota. Journal of Aquatic Plant Management 32:56–60.

(43) Danehy, R. J., N. H. Ringler, and J. E. Gannon. 1991. Influence of nearshore structure on growth and diets of yellow perch (*Perca flavescens*) and white perch (*Morone americana*) in Mexico Bay, Lake Ontario. Journal of Great Lakes Research 17:183–193.

(44) DeVries, D. R., R. A. Stein, and P. L. Chesson. 1989. Sunfish foraging among patches: the patch-departure decision. Animal Behavior 37:455–464.

(45) DeVries, D. R. 1990. Habitat use by bluegill in laboratory pools: where is the refuge when macrophytes are sparse and alternative prey are present? Environmental Biology of Fishes 29:27–34.

Dewey, M. R. 1991. Quantitative sampling of juvenile fishes in vegetation. U.S. Fish and Wildlife Service Research Information Bulletin 91–66.

(46) Dewey, M. R., L. E. Holland-Bartels, and S. J. Zigler. 1989. Comparison of fish catches with buoyant popnets and seines in vegetated and non-vegetated habitats. North American Journal of Fisheries Management 9:249–253.

(47) Dewey, M. R., and C. A. Jennings. 1992. Habitat use by larval fishes in a backwater lake of the Missis-

sippi River. Journal of Freshwater Ecology 7:363–372.

(48) Dibble, E. D. 1991. A comparison of diving and rotenone methods for determining relative abundance of fish. Transactions of the American Fisheries Society 120:663–666.

Dibble, E. D., G. O. Dick, and K. J. Killgore. 1996. Measurement of plant architecture in seven aquatic plants. Journal of Freshwater Ecology 11:311–318.

(49) Dibble, E. D., and K. J. Killgore. 1994. A habitat-based approach for studying fish–plant interactions. U.S. Army Corps of Engineers Waterways Experiment Station, Miscellaneous Paper A-94-2, Vicksburg, Mississippi.

(50) Diehl, S. 1988. Foraging efficiency of 3 fresh-water fishes: effects of structural complexity and light. Oikos 53:207–214.

Diehl, S. 1993. Effects of habitat structure on resource availability, diet and growth of benthivorous perch, *Perca fluviatilis*. Oikos 67:403–414.

(51) Dionne, M., and C. L. Folt. 1991. An experimental analysis of macrophyte growth forms as fish foraging habitat. Canadian Journal of Fisheries and Aquatic Sciences 48:123–131.

(52) Downing, J. A., and M. A. Anderson. 1985. Estimating the standing biomass of aquatic macrophytes. Canadian Journal of Fisheries and Aquatic Sciences 42:1860–1869.

(53) Duarte, C. M. 1987. Use of echo sounder tracings to estimate the above ground biomass of submerged plants in lakes. Canadian Journal of Fisheries and Aquatic Sciences 44:732–735.

(54) Durocher, P. P., W. C. Provine, and J. E. Kraai. 1984. Relationship between abundance of largemouth bass and submerged vegetation in Texas reservoirs. North American Journal of Fisheries Management 4:84–88.

Edwards, D. J., and E. Moore. 1975. Control of water seeds by grass carp in a drainage ditch in New Zealand. New Zealand Journal of Marine and Freshwater Research 9:283–292.

(55) Edwards, R. W., and M. W. Brown. 1960. An aerial photograph method for studying the distribution of aquatic macrophytes in shallow waters. Journal of Ecology 48:161–163.

(56) Ellis, D. V. 1961. Diving and photographic techniques for observing and recording salmon. Journal of the Fisheries Research Board of Canada 18:1159–1166.

(57) Engel, S. 1984. Restructuring littoral zones: a different approach to an old problem. Pages 463–466 *in* Lake and reservoir management. Proceedings of the third annual conference, 1983. North American Lake Management Society, Knoxville, Tennessee. U.S. Environmental Protection Agency 440/5-84-001.

(58) Engel, S. 1985. Aquatic community interaction of submerged macrophytes. Wisconsin Department of Natural Resources Technical Bulletin 156.

(59) Engel, S. 1988. The role and interactions of submersed macrophytes in a shallow Wisconsin lake. Journal of Freshwater Ecology 4:329–342.

(60) Engel, S. 1990. Ecological impacts of harvesting macrophytes in Halverson Lake, Wisconsin. Journal of Aquatic Plant Management 28:41–45.

Engel, S. 1995. Eurasian watermilfoil as a fishery management tool. Fisheries 20(3):20–27.

(61) Engel, S., and S. A. Nichols. 1994. Aquatic macrophyte growth in a turbid windswept lake. Journal of Freshwater Ecology 9:97–109.

(62) Espegren, G. D., and E. P. Bergersen. 1990. Quantitative sampling of fish populations with a mobile rising net. North American Journal of Fisheries Management 10:469–487.

Faber, D. J. 1981. A light trap to sample littoral and limnetic regions of lakes. International Vereinigung fuer Theoretische und Angewandte Limnologie Verhandlungen 21:744–749.

(63) Ferguson, R. G. 1962. The effects of underwater explosives on yellow perch (*Perca flavescens*). Canadian Fish Culturist 29:31–39.

Floyd, K. B., R. D. Hoyt, and S. Timbrook. 1984. Chronology of appearance and habitat partitioning by stream larval fishes. Transactions of the American Fisheries Society 113:217–223.

Forsberg, C. 1959. Quantitative sampling of subaquatic vegetation. Oikos 10:233–240.

(64) Forester, T. S., and J. M. Lawrence. 1978. Effects of grass carp and carp on populations of bluegill and largemouth bass in ponds. Transactions of the American Fisheries Society 107:172–175.

(65) Freeman, B. J., H. S. Greening, and J. D. Oliver. 1984. Comparison of three methods for sampling fishes and macroinvertebrates in a vegetated freshwater wetland. Journal of Freshwater Ecology 2:603–606.

(66) Gerking, S. D. 1957. A method of sampling the littoral macrofauna and its application. Ecology 38:219–225.

(67) Gerking, S. D. 1962. Production and food utilization in a population of bluegill sunfish. Ecological Monographs 32:31–78.

(68) Gilinsky, E. 1984. The role of fish predation and spatial heterogeneity in determining benthic community structure. Ecology 4:455–468.

(69) Glass, N. R. 1971. Computer analysis of predation energetics in the largemouth bass. Pages 325–363 *in* B. C. Patten, editor. Systems analysis and simulation in ecology, volume 1. Academic Press, New York.

(70) Goldsborough, L. G., and G. G. C. Robinson. 1985. Seasonal succession of diatom epiphyton on dense mats of *Lemna minor*. Canadian Journal of Botany 63:2332–2339.

(71) Gotceitas, V. 1990. Plant stem density as a cue in patch choice by foraging juvenile bluegill sunfish. Environmental Biology of Fishes 29:227–232.

(72) Gotceitas, V., and P. Colgan. 1987. Selection between densities of artificial vegetation by young bluegills avoiding predation. Transactions of the American Fisheries Society 116:40–49.

(73) Gregg, W. W., and F. L. Rose. 1985. Influences of aquatic macrophytes on invertebrate community structure, guild structure, and microdistribution in streams. Hydrobiologia 128:45–56.

(74) Gregory, R. S., and P. M. Powles. 1985. Chro-

nology, distribution, and sizes of larval fish sampled by light traps in macrophytic Chemung Lake. Canadian Journal of Zoology 19:2569–2577.

(75) Griffith, J. S., and R. W. Smith. 1995. Failure of submersed macrophytes to provide cover for rainbow trout throughout their first winter in Henrys Fork of the Snake River, Idaho. North American Journal of Fisheries Management 15:42–48.

(76) Gutreuter, S. J., and R. O. Anderson. 1985. Importance of body size to the recruitment process in largemouth bass populations. Transactions of the American Fisheries Society 114:317–327.

(77) Hall, D. J., W. E. Cooper, and E. E. Werner. 1970. An experimental approach to the production dynamics and structure of freshwater animal communities. Limnology and Oceanography 15:839–928.

(78) Hall, D. J., and E. E. Werner. 1977. Seasonal distribution and abundance of fishes in the littoral zone of a Michigan lake. Transactions of the American Fisheries Society 106:545–555.

(79) Hanson, J. M. 1990. Macroinvertebrate size-distribution of two contrasting fresh-water macrophyte communities. Freshwater Biology 24:481–491.

(80) Hanson, J. M., and W. C. Leggett. 1986. Effect of competition between two freshwater fishes on prey consumption and abundance. Canadian Journal of Fisheries and Aquatic Sciences 43:1363–1371.

(81) Hargeby, A. 1990. Macrophyte associated invertebrates and the effect of habitat permanence. Oikos 57:338–346.

(82) Harvey, R. M., G. G. Patterson, and J. R. Pickett. 1988. An automated positioning system for determining aquatic macrophyte distribution. Journal of Aquatic Plant Management 26:38–43.

(83) Heck, K. L., Jr., and T. A. Thoman. 1981. Experiments on predator–prey interactions in vegetated aquatic habitats. Journal of Experimental Marine Biology and Ecology 53:125–134.

Heggenes, J., A. Brabrand, and S. J. Saltveit. 1990. Comparison of three methods for studies of stream habitat use by young brown trout and Atlantic salmon. Transactions of the American Fisheries Society 119:101–111.

Helfman, S. 1979. Fish attraction to floating objects in lakes. Pages 49–57 *in* D. L. Johnson and R. A. Stein, editors. Response of fish to habitat structure in standing water. American Fisheries Society, North Central Division, Special Publication 6, Bethesda, Maryland.

(84) Helfman, S. 1981. The advantage to fishes of hovering in shade. Copeia 1981:392–400.

(85) Hinch, S. G., and N. C. Collins. 1993. Relationships of littoral fish abundance to water chemistry and macrophyte variables in central Ontario lakes. Canadian Journal of Fisheries and Aquatic Sciences 50: 1870–1878.

Hinkle, J. 1986. A preliminary literature review on vegetation and fisheries with emphasis on the largemouth bass, bluegill and hydrilla. Aquatics 8:9–14.

(86) Hoff, M. H. 1991. Effects of increased nesting cover on nesting and reproduction of smallmouth bass in northern Wisconsin lakes. Pages 39–43 *in* D. C. Jackson, editor. The first international smallmouth bass symposium. Mississippi Agricultural and Forestry Experiment Station, Mississippi State University, Mississippi State.

(87) Holbrook, S. J., and R. J. Schmitt. 1988. Effects of predation risk on foraging behavior: mechanisms altering patch choice. Journal of Experimental Marine Biology and Ecology 121:151–163.

(88) Holland, L. E., and M. L. Huston. 1984. Relationship of young-of-the-year northern pike to aquatic vegetation types in backwaters of the upper Mississippi river. North American Journal of Fisheries Management 4:514–522.

(89) Hoover, J. J., K. J. Killgore, and R. P. Morgan, II. 1988. Food habits of fishes associated with hydrilla beds and open water in Lake Seminole, Florida–Georgia. U.S. Army Corps of Engineers, Waterways Experiment Station, Miscellaneous Paper A-89-1, Vicksburg, Mississippi.

Hoyer, M. V., and D. E. Canfield, Jr. 1996. Largemouth bass abundance and aquatic vegetation in Florida lakes: an empirical analysis. Journal of Aquatic Plant Management 34:23–32.

(90) Hubbs, C. L. 1921. An ecological study of the life-history of the fresh-water Atherine fish *Labidesthes sicculus*. Ecology 2:262–276.

(91) Irvine, K., B. Moss, and H. Balls. 1989. The loss of submerged plants with eutrophication II. Relationships between fish and zooplankton in a set of experimental ponds, and conclusions. Freshwater Biology 22:89–107.

Janecek, J. A. 1988. Literature review on fishes interaction with aquatic macrophytes with special reference to the Upper Mississippi River System. Upper Mississippi River Conservation Committee, Fish Section, Rock Island, Illinois.

(92) Jennings C. A., P. A. Vohs, and M. R. Dewey. 1992. Classification of wetland area along the upper Mississippi River with aerial videography. Wetlands 12: 163–170.

(93) Johnson, D. L., R. A. Beaumier, and W. E. Lynch, Jr. 1988. Selection of habitat structure interstice size by bluegills and largemouth bass in ponds. Transactions of the American Fisheries Society 117:171–179.

(94) Johnson, S. L. 1993. Cover choice by bluegills: orientation of underwater structure and light intensity. Transactions of the American Fisheries Society 122: 148–154.

Kautsky, H., B. Widbom, and F. Wulff. 1981. Vegetation, macrofauna and benthic meiofauna. Ophelia 20:53–77.

(95) Keast, A. 1984. The introduced aquatic macrophyte, *Myriophyllum spicatum*, as habitat for fish and their invertebrate prey. Canadian Journal of Zoology 62:1289–1303.

(96) Keast, A. 1985a. Development of dietary specializations in a summer community of juvenile fishes. Environmental Biology of Fishes 13:211–224.

(97) Keast, A. 1985b. Planktivory in a littoral-dwelling lake fish association: prey selection and seasonality. Canadian Journal of Fisheries and Aquatic Sciences 42:1114–1126.

(98) Keast, A., and J. Harker. 1977. Strip counts as

means of determining densities and habitat utilization patterns in lake fishes. Environmental Biology of Fishes 1:181–188.

(99) Keast, A., J. Harker, and D. Turnbull. 1978. Nearshore fish habitat utilization and species associations in Lake Opinicon (Ontario, Canada). Environmental Biology of Fishes 3:173–184.

(100) Killgore, K. J., R. P. Morgan, and N. Rybicki. 1989. Distribution and abundance of fishes associated with submersed aquatic plants in the Potomac River. North American Journal of Fisheries Management 9:101–111.

(101) Kramer, R. H., and L. L. Smith, Jr. 1962. Formation of year classes in largemouth bass. Transactions of the American Fisheries Society 91:29–41.

(102) Krecker, F. H. 1939. A comparative study of the animal population of certain submerged aquatic plants. Ecology 20:553–562.

Kushlan, J. A. 1974. Quantitative sampling of fish populations in shallow freshwater environments. Transactions of the American Fisheries Society 103:348–352.

Lambout, V. W. 1959. Block-off net for taking fish population samples. Progressive Fish-Culturist 21:143–144.

Larson E. W., D. L. Johnson, and W. E. Lynch. 1986. A buoyant popnet for accurately sampling artificial habitat structures. Transactions of the American Fisheries Society 115:351–355.

(103) Laughlin, D. R., and E. E. Werner. 1980. Resource partitioning in coexisting sunfish: pumpkinseed (*Lepomis gibbosus*) and northern longear sunfish (*Lepomis megalotis*). Canadian Journal of Fisheries and Aquatic Sciences 27:1411–1420.

(104) Lauridsen T. L., E. Jeppesen, and M. Sondergaard. 1994. Colonization and succession of submerged macrophytes in shallow Lake Vaeng during the first five years following fish manipulation. Hydrobiologia 275/276:233–242.

(105) Layher, W. G., and O. E. Maughan. 1984. Comparison in efficiencies of three sampling techniques for estimating fish populations in small streams. Progressive Fish-Culturist 46:180–184.

(106) Layzer, J. B., and M. D. Clady. 1987. Phenotypic variation of young-of-year bluegills (*Lepomis macrochirus*) among microhabitats. Copeia 1987:702–707.

(107) Layzer, J. B., and M. D. Clady. 1991. Microhabitat and diet segregation among coexisting young-of-year sunfishes. NOAA (National Oceanic and Atmospheric Administration) NMFS (National Marine Fisheries Service) Technical Report 95:99–108.

(108) Lehmann, A., J. M. Jaquet, and J. B. Lachavanne. 1994. Contribution of GIS to submerged macrophyte biomass estimation and community structure modeling, Lake Geneva, Switzerland. Aquatic Botany 47: 99–117.

(109) Leitholf, E. 1917. *Fundulus chrysotus*. Aquatic Life 2:141–142.

(110) Lembi, C. A., B. G. Ritenour, E. M. Iverson, and E. C. Forss. 1978. The effects of vegetation removal by grass carp on water chemistry and phytoplankton in Indiana ponds. Transactions of the American Fisheries Society 107:161–171.

(111) Lillie, R. A., and J. Budd. 1992. Habitat architecture of *Myriophyllum spicatum* L. as an index to habitat quality for fish and macroinvertebrates. Journal of Freshwater Ecology 7:113–125.

(112) Lodge, D. M. 1985. Macrophyte–gastropod associations: observations and experiments on macrophyte choice by gastropods. Freshwater Biology 15:695–708.

(113) Lukens, J. E. 1968. Color aerial photography for aquatic vegetation surveys. Pages 441–446 *in* Proceedings of the 5th international symposium on remote sensing of environment. Environmental Research Institute of Michigan, Ann Arbor.

(114) Lynch, W. E., Jr., and D. L. Johnson. 1989. Influences of interstice size, shade, and predators on the use of artificial structures by bluegills. North American Journal of Fisheries Management 9:219–255.

(115) Maceina, M. J., P. W. Bettoli, W. G. Klussmann, R. K. Betsill, and R. L. Noble. 1991. Effect of aquatic macrophyte removal on recruitment and growth of black crappies and white crappies in Lake Conroe Texas. North American Journal of Fisheries Management 11:556–563.

Maceina, M. J., and W. C. Reeves. 1996. Relations between submersed macrophyte abundance and largemouth bass tournament success on two Tennessee River impoundments. Journal of Aquatic Plant Management 34:33–38.

(116) Maceina, M. J., and J. V. Shireman. 1980. The use of a recording fathometer for determination of distribution and biomass of hydrilla. Journal of Aquatic Plant Management 18:34–39.

(117) Maceina, M. J., and J. V. Shireman. 1985. Influence of dense hydrilla infestation on black crappie growth. Proceedings of the Annual Conference Southeastern Association of Fish and Wildlife Agencies 36 (1982):394–402.

(118) Maceina, M. J., J. V. Shireman, K. A. Langeland, and D. E. Canfield, Jr. 1984. Prediction of submersed plant biomass by use of a recording fathometer. Journal of Aquatic Plant Management 22:35–38.

(119) Maceina, M. J., W. B. Wrenn, and D. R. Lowery. 1995. Estimating harvestable largemouth bass abundance in a reservoir with an electrofishing catch depletion technique. North American Journal of Fisheries Management 15:103–109.

(120) Machena, C., and N. Kautsky. 1988. A quantitative diving survey of benthic vegetation and fauna in Lake Kariba, a tropical man-made lake. Freshwater Biology 19:1–14.

(121) Marshall, T. R., and P. F. Lee. 1994. Mapping aquatic macrophytes through digital image analysis of aerial photographs: an assessment. Journal of Aquatic Plant Management 32:61–66.

(122) Martin, T. H., L. B. Crowder, C. F. Dumas, and J. M. Burkholder. 1992. Indirect effects of fish on macrophytes in Bays Mountain Lake: evidence for littoral trophic cascade. Oecologia 89:476–481.

(123) McDonough, T. A., and J. P. Buchanan. 1991. Factors affecting abundance of white crappies in Chickamauga Reservoir, Tennessee, 1970–1989. North

American Journal of Fisheries Management 11:513–524.

(124) Meador, M. R., and J. S. Bulak. 1987. Quantifiable ichthyoplankton sampling in congested shallow-water areas. Journal of Freshwater Ecology 4:65–69.

(125) Meals, K. O., and L. E. Miranda. 1991. Variability in abundance of age-0 centrarchids among littoral habitats of flood control reservoirs in Mississippi. North American Journal of Fisheries Management 11:298–304.

(126) Mesing, C. L., and A. M. Wicker. 1986. Home range, spawning migrations, and homing of radio-tagged Florida largemouth bass in two central Florida lakes. Transactions of the American Fisheries Society 115:286–295.

(127) Metzger, R. J., and P. L. Shafland. 1986. Use of detonating cord for sampling fish. North American Journal of Fisheries Management 6:113–118.

(128) Miller, D., and R. H. Kramer. 1971. Spawning and early life history of largemouth bass (*Micropterus salmoides*) in Lake Powell. Pages 78–83 *in* G. E. Hall, editor. Reservoir fisheries and limnology. American Fisheries Society Special Publication 8.

(129) Miranda, L. E., W. L. Shelton, and T. D. Bryce. 1984. Effects of water level manipulation on abundance, mortality and growth of young-of-year largemouth bass in West Point Reservoir, Alabama–Georgia. North American Journal of Fisheries Management 4:314–320.

(130) Mittelbach, G. G. 1981. Foraging efficiency and body size: a study of optimal diet and habitat use by bluegills. Ecology 62:1370–1386.

(131) Mittelbach, G. G. 1988. Competition among refuging sunfishes and effects of fish density on littoral zone invertebrates. Ecology 69:614–623.

Mittelbach, G. G., and P. L. Chesson. 1987. Predation risk: indirect effects on fish populations. Pages 315–330 *in* C. W. Kerfoot and A. Sih, editors. Predation: direct and indirect impacts on aquatic communities. University Press of New England, Hanover, New Hampshire.

(132) Morgan, R. P., K. J. Killgore, and N. H. Douglas. 1988. Modified popnet design for collecting fishes in varying depths of submersed aquatic vegetation. Journal of Freshwater Ecology 4:533–539.

(133) Moxley, D. J., and F. H. Langford. 1985. Beneficial effects of *Hydrilla* on two eutrophic lakes in central Florida. Proceedings of the Annual Conference Southeastern Association of Fish and Wildlife Agencies 36(1982) 280–286.

(134) Murdoch, W. W., and A. Oaten. 1975. Predation and population stability. Advances in Ecological Research 9:1–131.

(135) Murphy, K. J., and J. W. Eaton. 1981. Waterplants, boat traffic and angling in canals. Pages 173–187 *in* British freshwater fisheries second conference proceedings. University of Liverpool, Liverpool, UK.

(136) Nichols, S. J., D. W. Schloesser, and J. W. Geis. 1988. Seasonal growth of the exotic submersed macrophyte *Nitellopsis obtusa* in the Detroit River of the Great Lakes. Canadian Journal of Botany 66:116–118.

Noakes, D. L. G., and J. R. Baylis. 1990. Behavior. Pages 555–584 *in* C. B. Schreck and P. B. Moyle, editors. Methods for fish biology. American Fisheries Society, Bethesda, Maryland.

Noble, R. L. 1981. Management of forage fishes in impoundments of the southern United States. Transactions of the American Fisheries Society 110:738–750.

(137) Northcote, T. G., and D. W. Wilkie. 1963. Underwater census of stream fish populations. Transactions of the American Fisheries Society 92:146–151.

Orth, D. J., and O. E. Maughan. 1982. Evaluation of the incremental methodology for recommending instream flows for fishes. Transactions of the American Fisheries Society 111:413–445.

(138) Osenberg, C. W., E. E. Werner, G. G. Mittelbach, and D. J. Hall. 1987. Growth patterns in bluegill (*Lepomis macrochirus*) and pumpkinseed (*L. gibbosus*) sunfish: environmental variation and the importance of ontogenetic niche shifts. Canadian Journal of Fisheries and Aquatic Sciences 45:17–26.

(139) Owens, M., M. A. Learner, and P. J. Maris. 1967. Determination of the biomass of aquatic plants using optical method. Journal of Ecology 55:671–676.

(140) Paller, M. H. 1987. Distribution of larval fish between macrophyte beds and open channels in a southeastern floodplain swamp. Journal of Freshwater Ecology 4:191–200.

(141) Pardue, G. B. 1973. Production response of the bluegill sunfish, *Lepomis macrochirus* to added attachment surface for fish-food organisms. Transactions of the American Fisheries Society 102:622–626.

(142) Pardue, G. B., and L. A. Nielsen. 1979. Invertebrate biomass and fish production in ponds with added attachment surface. Pages 34–43 *in* D. L. Johnson and R. A. Stein, editors. Response of fish to habitat structure in standing water. American Fisheries Society, North Central Division, Special Publication 6, Bethesda, Maryland.

(143) Peckham, R. S., and C. F. Dineen. 1957. Ecology of the central mudminnow *Umbra limi* (Kirtland). American Midland Naturalist 58:222–231.

(144) Petravicz, J. J. 1936. The breeding habits of the least darter, *Microperca punctulata* Putnam. Copeia 1936:77–82.

(145) Petrell, R. J., L. O. Bagnall, and G. H. Smerage. 1991. Physical description of water hyacinth mats to improve harvester design. Journal of Aquatic Plant Management 29:45–50.

Pflieger, W. L. 1975. The fishes of Missouri. Missouri Department of Conservation, Jefferson City.

(146) Pihl, L., H. Wennhage, and S. Nilsson. 1994. Fish assemblage structure in relation to macrophytes and filamentous epiphytes in shallow non-tidal rocky-and soft-bottom habitats. Environmental Biology of Fishes 39:271–288.

(147) Pine, R. T., L. W. J. Anderson, and S. S. O. Hung. 1989. Non-destructive estimation of aquatic macrophyte biomass. Journal of Aquatic Plant Management 27:47–49.

(148) Poe, T. P., C. O. Hatcher, C. L. Brown, and D. W. Schloesser. 1986. Comparison of species composition and richness of fish assemblages in altered and

unaltered littoral habitats. Journal of Freshwater Ecology 3:525–536.

Price, D. G. 1982. A fishery resource sampling methodology for small streams. Pages 54–91 *in* N. Armantrout, editor. Acquisition and utilization of aquatic habitat inventory information. American Fisheries Society, Western Division, Bethesda, Maryland.

(149) Pringle, J. D. 1984. Efficiency of various quadrate sizes used in benthic sampling. Canadian Journal of Fisheries and Aquatic Sciences 41:1485–1489.

(150) Quade, H. W. 1969. Cladoceran faunas associated with aquatic macrophytes in some lakes in northwestern Minnesota. Ecology 50:170–179.

(151) Reynolds, A. J., and J. W. Eaton. 1983. The role of vegetation structure in a canal fishery. Pages 192–202 *in* British freshwater fisheries third conference proceedings. University of Liverpool, Liverpool, UK.

Robison, H. W., and T. M. Buchanan. 1988. Fishes of Arkansas. The University of Arkansas Press, Fayetteville.

(152) Rodgers, J. D., M. F. Solazzi, S. L. Johnson, and M. A. Buckman. 1992. Comparison of three techniques to estimate juvenile coho salmon populations in small streams. North American Journal of Fisheries Management 12:79–86.

(153) Rosine, W. N. 1955. The distribution of invertebrates on submerged aquatic plant surfaces in Juskee Lake, Colorado. Ecology 36:308–314.

(154) Rottman, R. W., and R. O. Anderson. 1978. Limnological and ecological effects of grass carp in ponds. Proceedings of the Annual Conference Southeastern Association of Fish and Wildlife Agencies 30(1976): 24–39.

(155) Rozas, L. P., and W. E. Odum. 1988. Occupation of submerged aquatic vegetation by fishes: testing the roles of food and refuge. Oecologia 77:101–106.

(156) Sabol, B. M. 1984. Development and use of Waterways Experiment Station's hydraulically operated submersed aquatic plant sampler. Pages 46–57 *in* W. M. Dennis and B. G. Isom, editors. Ecological assessment of macrophyton: collection, use, and meaning of data. American Society for Testing and Materials Special Technical Publication 843, Philadelphia.

(157) Saiki, M. K., and J. C. Tash. 1979. Use of cover and dispersal by crayfish to reduce predation from largemouth bass. Pages 44–48 *in* D. L. Johnson and R. A. Stein, editors. Response of fish to habitat structure in standing water. American Fisheries Society, North Central Division, Special Publication 6, Bethesda, Maryland.

(158) Savino, J. F., and R. A. Stein. 1982. Predator–prey interaction between largemouth bass and bluegills as influenced by simulated, submersed vegetation. Transactions of the American Fisheries Society 11: 255–266.

(159) Savino, J. F., and R. A. Stein. 1989. Behavioural interactions between fish predators and their prey: effects of plant density. Animal Behavior 37:311–321.

(160) Savino, J. F, E. A. Marschall, and R. A. Stein. 1992. Bluegill growth as modified by plant-density: an exploration of underlying mechanisms. Oecologia 89: 153–160.

(161) Scheffer, M., A. A. Achterberg, and B. Beltman. 1984. Distribution of macro-invertebrates in a ditch in relation to the vegetation. Freshwater Biology 14: 367–370.

(162) Scheidegger, K. J., and M. B. Bain. 1995. Larval fish distribution and microhabitat use in free-flowing and regulated rivers. Copeia 1995:125–135.

(163) Schloesser, D. W., and B. A. Manny. 1984. Rapid qualitative method for estimating the biomass of submersed macrophytes in large water bodies. Journal of Aquatic Plant Management 22:102–104.

(164) Schmitt, R. J., and S. J. Holbrook. 1985. Patch selection by juvenile black surfperch (Embiotochidae) under variable risk: interactive influence of food quality and structural complexity. Journal of Experimental Marine Biology and Ecology 85:269–285.

(165) Schramm, H. L., Jr., L. J. Jirka, and M. V. Hoyer. 1987. Epiphytic macroinvertebrates on dominant macrophytes in two central Florida lakes. Journal of Freshwater Ecology 4:151–162.

Scott, E. M. 1993. The effects of aquatic macrophytes on fish populations of Chickamauga reservoir coves, 1970–1990. Tennessee Valley Authority (TVA/ WM—93/24), Norris.

(166) Serafy, J. E., and R. M. Harrell. 1993. Behavioural response of fishes to increasing pH and dissolved oxygen: field and laboratory observations. Freshwater Biology 30:52–61.

(167) Serafy, J. E., R. M. Harrell, and L. M. Hurley. 1994. Mechanical removal of *Hydrilla* in the Potomac River, Maryland: local impacts on vegetation and associated fishes. Journal of Freshwater Ecology 9:135–143.

(168) Serafy, J. E., R. M. Harrell, and J. C. Stevenson. 1988. Quantitative sampling of small fish in dense vegetation: design and field testing of portable "popnets." Journal of Applied Ichthyology 4:149–157.

(169) Sheldon, R. B., and C. W. Boylen. 1978. An underwater survey method for estimating submerged macrophyte population density and biomass. Aquatic Botany 4:65–72.

(170) Shima, L. J., R. R. Anderson, and V. P. Carter. 1976. The use of aerial color infrared photography in mapping the vegetation of a freshwater marsh. Chesapeake Science 17:74–85.

(171) Shireman, J. V., D. E. Canfield, D. E. Colle, D. F. DuRant, and W. T. Haller. 1984. Evaluation of biological, chemical, and mechanical aquatic vegetation control upon fish populations in 0.2 ha. research ponds. USDA/SEA/ARS Final report 7B30-0-177, University of Florida, Center for Aquatic Weeds, Gainesville.

(172) Shireman, J. V., D. E. Colle, and D. F. DuRant. 1981. Efficiency of rotenone sampling with large and small block nets in vegetated and open-water habitats. Transactions of the American Fisheries Society 110:77–80.

(173) Shireman, J. V., and M. J. Maceina. 1979. Techniques utilizing a recording fathometer in determining distribution and biomass of *Hydrilla verticillata*

Royle. Final report of Aquatic Plant Control Research Program to U.S. Army Corps of Engineers Waterways Experiment Station, Vicksburg, Mississippi.

(174) Sliger, W. A., J. W. Henson, and R. C. Shadden. 1990. A quantitative sampler for biomass estimates of aquatic macrophytes. Journal of Aquatic Plant Management 28:100–102.

Smart, R. M., and J. W. Barko. 1988. Effects of water chemistry on aquatic plants: interrelationships among biomass production, plant nutrition, and water chemistry. U.S. Army Corps of Engineers Waterways Experiment Station, Final Report A-88-5, Vicksburg, Mississippi.

(175) Smith, S. L., and J. E. Crumpton. 1979. Interrelationships of vegetative cover and sun fish population density in suppressing spawning in largemouth bass. Proceedings of the Annual Conference Southeastern Association of Fish and Wildlife Agencies. 31(1977): 141–157.

(176) Spencer, C. N., and D. L. King. 1984. Role of fish in regulation of plant and animal communities in eutrophic ponds. Canadian Journal of Fisheries and Aquatic Sciences 41:1851–1855.

(177) Stein, R. A. 1977. Selective predation, optimal foraging, and the predator–prey interaction between fish and crayfish. Ecology 58:1237–1253.

(178) Stein, R. A., and J. J. Magnuson. 1976. Behavioral response of crayfish to a fish predator. Ecology 57: 751–761.

Stenseth, N. C. 1980. Spatial heterogeneity and population stability: some evolutionary consequences. Oikos 35:165–184.

(179) Stent, C. J., and S. Hanley. 1985. A recording echo sounder for assessing submerged aquatic plant populations in shallow lakes. Aquatic Botany 21:377–394.

(180) Strange, R. J., C. R. Berry, and C. B. Schreck. 1975. Aquatic plant control and reservoir fisheries. Pages 513–521 *in* R. H. Stroud and H. Clepper, editors. Black bass biology and management. Sport Fishing Institute, Washington DC.

(181) Terrell, J. W., and T. T. Terrell. 1975. Macrophyte control and food habitats of the grass carp in Georgia ponds. International Vereinigung fuer Theoretische und Angewandte Limnologie Verhandlungen 19:2415–2420.

(182) Thomas, G. L., S. L. Thiesfeld, S. A. Bonar, R. N. Crittenden, and G. B. Pauley. 1990. Estimation of submergent plant bed biovolume using acoustic range information. Canadian Journal of Fisheries and Aquatic Sciences 47:805–812.

(183) Todd, B. L., and C. F. Rabeni. 1989. Movement and habitat use by stream-dwelling smallmouth bass. Transactions of the American Fisheries Society 118: 229–242.

(184) Tonn, W. M., and C. A. Paszkowski. 1986. Size-limited predation, winterkill, and the organization of *Umbra–Perca* fish assemblages. Canadian Journal of Fisheries and Aquatic Sciences 43:194–202.

(185) Vadas, R. L., Jr., and D. J. Orth. 1993. A new technique for estimating the abundance and habitat use of stream fishes. Journal of Freshwater Ecology 8:305–317.

(186) Van Dolah, R. F. 1978. Factors regulating the distribution and population dynamics of the amphipod (*Gammarus palustris*) in an intertidal salt marsh community. Ecological Monographs 48:191–217.

(187) Venugopal, M. N., and I. J. Winfield. 1993. The distribution of juvenile fishes in a hypereutrophic pond: can macrophytes potentially offer a refuge for zooplankton? Journal of Freshwater Ecology 8:389–396.

(188) Vince, S., I. Valiela, N. Backus, and J. M. Teal. 1976. Predation by the salt marsh killifish *Fundulus heteroclitus* L. in relation to prey size and habitat structure: consequences for prey distribution and abundance. Journal of Experimental Marine Biology and Ecology 23:255–266.

Vogele, L. E., and W. C. Rainwater. 1975. Use of brush shelters as cover by spawning black basses (*Micropterus*) in Bull Shoals reservoir. Transactions of the American Fisheries Society 2:264–269.

Walters, D. A., W. E. Lynch, Jr., and D. L. Johnson. 1991. How depth and interstice size of artificial structures influence fish attraction. North American Journal of Fisheries Management 11:319–329.

(189) Ware, F. J., and R. D. Gasaway. 1978. Effects of grass carp on native fish populations in two Florida lakes. Proceedings of the Annual Conference Southeastern Association of Fish and Wildlife Agencies 30(1976):324–335.

(190) Ware, F. J., R. D. Gasaway, R. A. Martz, and T. F. Drda. 1975. Investigation of herbivorous fishes in Florida. Pages 79–84 *in* P. L. Brezonik and J. L. Fox, editors. Proceedings, symposium on water quality management through biological control. U.S. Environmental Protection Agency ENV-07-75-1, University of Florida, Gainesville.

(191) Watkins, C. E., II, J. V. Shireman, and W. T. Haller. 1983. The influence of aquatic vegetation upon zooplankton and benthic macroinvertebrates in Orange Lake, Florida. Journal of Aquatic Plant Management 21:78–83.

(192) Wegener, W., D. Holcomb, and V. Williams. 1974. Sampling shallow water fish populations using the Wegener ring. Proceedings of the Annual Conference Southeastern Association of Game and Fish Commissioners 27(1973):663–673.

(193) Werner, E. E., and five coauthors. 1977. Habitat partitioning in a freshwater fish community. Journal of the Fisheries Research Board of Canada 34:360–370.

(194) Werner, E. E., and D. J. Hall. 1979. Foraging efficiency and habitat switching in competing sunfishes. Ecology 60:256–264.

(195) Werner, E. E., D. J. Hall, and M. D. Werner. 1978. Littoral zone fish communities of two Florida lakes and a comparison with Michigan lakes. Environmental Biology of Fishes 3:163–172.

(196) Whitfield, A. K. 1984. The effects of prolonged aquatic macrophyte senescence on the biology of the dominant fish species in a southern African coastal lake. Estuarine Coastal and Shelf Science 18:315–329.

(197) Wilcox, D. A., and J. E. Meeker. 1992. Implications for faunal habitat related to altered macrophyte structure in regulated lakes in northern Minnesota. Wetlands 12:192–203.

(198) Wildhaber, M. L., and W. H. Neill. 1992. Activity and distribution of northern and Florida largemouth bass in a Texas impoundment. Journal of Freshwater Ecology 7:293–302.

(199) Wile, I. 1978. Environmental effects of mechanical harvesting. Journal of Aquatic Plant Management 16:14–20.

(200) Wiley, M. J., R. W. Gorden, S. W. Waite, and T. Powless. 1984. The relationship between aquatic macrophytes and sport fish production in Illinois ponds: a simple model. North American Journal of Fisheries Management 4:111–119.

(201) Wood, R. D. 1963. Adapting scuba to aquatic plant ecology. Ecology 44:416–419.

(202) Wychera, U., R. Zoufal, P. Christof-Dirry, and G. A. Janauer. 1993. Structure and environmental factors in macrophyte stands. Journal of Aquatic Plant Management 31:118–122.

(203) Zubik, R. J., and J. J. Fraley. 1988. Comparison of snorkel and mark–recapture estimates for trout populations in large streams. North American Journal of Fisheries Management 8:58–62.

American Fisheries Society Symposium 16:373–381, 1996

Management of Nonnative Aquatic Vegetation in Large Impoundments: Balancing Preferences and Economic Values of Angling and Nonangling Groups

JIM E. HENDERSON

U.S. Army Corps of Engineers, Waterways Experiment Station
3909 Halls Ferry Road, Attn: CEWES-EN-R, Vicksburg, Mississippi 39180-0699, USA

Abstract.—Control of nonnative aquatic vegetation in multipurpose reservoirs requires a balance of values and preferences of angling and nonangling user groups. Through on-site visitor interviews, and mailed surveys to lakefront residential users, the economic value and preference for five plant control scenarios were estimated at Lake Guntersville, Alabama. Estimated annual visits totaled 3.1 million and were composed of a 35:65% angling to nonangling user ratio. Plant eradication scenarios ranged from minimal or no control to complete eradication. Economic valuation of willingness to pay for recreation was obtained through a contingent valuation method survey. Annual economic benefits (willingness to pay) were highest at the minimal control scenario for anglers (US$790), and at the scenario that provided 10% lake coverage for nonanglers ($1,050). The aggregate annual economic value for all user groups ranged from $52 million to $122 million. The scenario that provided highest economic value had a mix of angling and nonangling use, and provided ample fishery habitat without hindering nonangling recreation by excessive plant coverage. An increase in plant coverage positively affected boat anglers and shore-based recreation, and negatively affected boaters. Increases in open-water areas benefitted boaters but reduced available fish habitat desired by anglers. Reduced plant coverage around residential shorelines was favored by bank anglers.

Establishment of nonnative plant species in large multipurpose reservoirs has complicated management decisions because angler and nonangler user groups have different preferences for aquatic plant coverage. The edge habitat provided by aquatic macrophytes such as hydrilla *Hydrilla verticillata* and Eurasian watermilfoil *Myriophyllum spicatum* has produced popular and abundant sportfishing areas in reservoirs in the southeastern United States. Decisions about aquatic plant control programs must balance needs and preferences of anglers with those of boaters and other nonangling recreational interests. Plant control decisions should be based on information about user group preferences, perceptions, and values related to plant control as well as control strategies. User groups may hold conflicting preferences for plant control that must be balanced and accommodated in vegetation management. Conflicts arise between anglers' desire for abundant aquatic vegetation (fisheries habitat) and nonanglers' desire for minimal vegetation (Henderson 1995).

During the late 1980s, a series of drought years resulted in near ideal growing conditions for hydrilla and watermilfoil, resulting in rapid expansion of nonnative vegetation in Lake Guntersville, Alabama. Clear water, ample nutrients, and low flows of the drought years allowed rapid plant growth and colonization of areas that were devoid of submersed aquatic plants. In response to increased amounts of vegetation and difficulties in maintaining reservoir operations, the U.S. Congress authorized the Guntersville Joint Agency Project (GJAP) in 1988 to serve as a 5-year demonstration project at Lake Guntersville. Lake Guntersville is a 68,000-acre reservoir in northeastern Alabama, operated by the Tennessee Valley Authority (TVA) to provide navigation, flood control, hydropower, recreation, and fish and wildlife benefits (GJAP 1992). In 1988, nonnative vegetation reached a peak of 24,000 acres and covered 30% of Lake Guntersville's surface area.

The intent of the GJAP was to research and demonstrate state-of-the-art aquatic plant control technologies that could be used in other reservoirs. Major components or study efforts of the GJAP were: (1) application of the U.S. Army Corps of Engineers (COE) master planning process to develop long-range aquatic plant control plans; (2) further documentation of previously developed control technologies; (3) research in on-going control technologies; and (4) investigation of public perceptions and economic valuations of aquatic plant control scenarios. These major components were the joint responsibility of the two agencies conducting the GJAP: TVA and the COE Waterways Experiment Station and Nashville District.

In this paper, I present results of the public perception and economic valuation study by the GJAP. The goal of the study was to evaluate differences in visitation and economic values associated with various levels of aquatic vegetation. Specific objectives were to (1) estimate existing recreational use; (2) assess perceptions and preferences of all users regarding aquatic plant occurrence, abundance, and control; (3) assess the importance and satisfaction of all users with reservoir recreation and resource management conditions; (4) estimate recreational use and economic value for aquatic plant control scenarios; and (5) estimate economic impacts to the local economy resulting from different aquatic plant control scenarios.

Methods

Economic evaluation of plant control scenarios requires obtaining user preferences for plant conditions which may be different from existing conditions. Surveys were used to obtain preference and valuation data from individuals living on or visiting Lake Guntersville. Three separate surveys were conducted to obtain baseline recreational use and preference data, and economic impact information.

Baseline Recreational Use and Preferences

Baseline recreational use and preference, as well as importance and satisfaction data were obtained from anglers and nonanglers through opinion surveys (Anderson et al. 1983; Sheatsley 1983; Miller 1991). Respondents were asked to identify the recreational activities in which they participated, and the amount of time spent at each activity. Information on user perceptions about aquatic plants and plant control were obtained by asking respondents to state their level of agreement with various statements. Opinions on recreation and resource attributes were obtained by measuring respondent perceptions about the importance of, and level of satisfaction with, selected attributes including opportunity to catch numerous fish, availability of open-water areas, and availability of shore-based recreation such as camping, swimming, and picnic facilities (Guadagnalo 1985; Mengak et al. 1986; Hollenhorst et al. 1992). Responses to these statements were collected on a Likert scale.

Baseline recreational use data and perceptions about aquatic plants were collected through face-to-face (on-site) interviews with lake visitors, and through surveys mailed to lake residents. Access to the lake was available at public access points, such as parks or boat ramps, or at private lakefront properties and common access lots. Face-to-face surveys were conducted at public use marinas, boat ramps, and parks. Mailed surveys were sent to a sample of residential users.

On-site survey.—On-site surveys of lake visitors were conducted during a 1-year period at recreation areas around the lake using stratified random sampling (Cochran 1977; Kalton 1983). A comprehensive inventory was made of all access points, including developed (e.g., parks) and undeveloped (areas that provide access but do not have paved roads or are not maintained) areas. A stratified random sample of survey sites was obtained from a list of developed and undeveloped recreation areas. The on-site sample was stratified based on season (fall–winter, spring, and summer), day type (weekend–holiday, weekday), and three geographical zones of the lake corresponding to different recreation environments (open-water, backwater, and riverine areas). All questions asked were based on the recreational use that occurred the day of the survey.

A total of 4,021 surveys were conducted at the recreation areas. The participants were interviewed as they exited the public access points; no on-the-lake or boat surveys were conducted. The interviews were conducted by contract personnel who had experience conducting similar surveys.

Residential survey.—The questions asked in the residential survey were similar to those used in the on-site questionnaire. Residential respondents were instructed to report their use based on their activities during the 4 weeks prior to receiving the survey. A sample of all lake residents was drawn from tax records. For the residential survey, Dillman's total design method for mailed surveys was followed (Dillman 1978). Of 1,050 residential surveys mailed, 699 (66.6%) were returned.

Recreational Use and Economic Value of Plant Control Scenarios

The economic valuation of plant control scenarios was conducted with a contingent valuation method (CVM) survey (Mitchell and Carson 1989; Bergstrom et al. 1990). The CVM survey was administered by mail, using the residential survey sample and names collected as part of the lake visitor survey. For the CVM survey, a modified form of Dillman's total design method for mail surveys was used (Dillman 1978). After the initial CVM survey mailing, reminder cards were sent 1 week later followed by a second survey 2 weeks later. Of 3,611 CVM surveys mailed, 1,587 (43.9%) were returned.

The CVM approach presented alternative plant abundance scenarios that were depicted visually and verbally in enough detail to allow respondents to provide a response to alternative plant abundance conditions. Five plant abundance and control scenarios were presented including a "no management" scenario, minimal control of plants resulting in an estimated 34,000 acres of plant coverage (50% of the lake surface area) and four management scenarios: control to 20,000 acres (scenario A), 14,000 acres (scenario B), 8,000 acres (scenario C), and complete eradication (scenario D). Scenarios A, B, and C represented the vegetation coverage during 1988, 1989, and 1990, respectively. Plant coverage at the time of this survey was near the level of scenario C and was used as a baseline for comparison with other scenarios.

The presentation of scenarios in the CVM survey included artist representations and written descriptions of plant conditions at six recreation environments: public boat launch, public recreation area, undeveloped shoreline, marina, developed shoreline, and open water. Historic aerial photography and records were used to develop the drawings and written descriptions.

The CVM survey obtained valuation responses for the five scenarios by asking whether respondents would continue to use the lake if annual lake use costs increased by a specified bid. After describing the plant conditions for a scenario, the respondent was asked:

> If the total costs to you of using Lake Guntersville increased by $_____ per year (from the total you spent last year), would you still use the lake if management scenario A (20,200 acres of aquatic plants) existed instead of the "minimum control" scenario? (check one)
> ____ Yes - About how many trips would you take each year in this situation? (fill in blank) ____ trips
> ____ No

Bid amounts ranged from US$10 to $4,500 and were assigned randomly. At the time of the survey, there was no annual fee to use the lake; thus, an accepted increase in annual costs was used as the valuation measure. The last part of the CVM survey collected information about household size, age, gender, income, education, and categorization of respondents' residence location (i.e., city, suburban area, small town, or rural area). A detailed explanation of the CVM survey methodology used is given by Mitchell and Carson (1989), and a summary of the economic models is given by Bergstrom et al. (1994).

Respondents were classified as anglers and nonanglers and separate valuation equations were estimated for each. Logistic regression equations were used to estimate recreational visits as well as the probability that respondents would accept the bid amounts. Variables included in the regression equations were the bid amount, management scenario, demographic variables, availability of recreation substitutes, and an index of the amount of plants preferred by the individual. The probability functions were then integrated over the bid variable values.

Economic Impact on the Local Economy

To develop estimates of economic impact, respondents were asked to summarize their recreational expenditures for trips to Lake Guntersville and for annual recreational expenditures (e.g., licenses, boats, motors, fishing tackle). Expenditure surveys were given to on-site respondents (to be completed at home and mailed back) and were mailed to residential respondents. Surveys were given to 1,747 on-site respondents and 684 (39.1%) were returned; 699 were mailed to residential respondents and 370 (52.9%) were returned. The follow-up procedure used with the CVM survey was applied for both on-site and residential surveys.

The expenditure summaries were used in an economic input–output model (IMPLAN) that estimates the economic impact to a local economy (Palmer and Siverts 1985; Probst 1988). The IMPLAN model allocates expenditures to the appropriate economic sector. Purchases of fishing tackle at a store at Lake Guntersville, for example, represent the direct effects of recreational spending. Increases in demand for fishing tackle caused by increased recreation result in increases in orders for tackle. Tackle manufacturers purchase the raw materials and labor necessary to meet the new orders for tackle. The increase in purchase of raw materials, labor, and other items by tackle manufacturers are known as indirect effects. The increase in employment and income resulting from additional tackle production puts more income in the hands of tackle manufacturers, suppliers, and their employees; this income is then spent on additional goods and services. Spending that results from increased income is known as an induced effect. The IMPLAN model accounts for and allocates the direct, indirect, and induced spending effects to the appropriate economic sectors and estimates economic impacts in terms of changes in total gross output or production (e.g., food service, tackle, boats, income, and employment) associated with each of the scenarios. Economic impacts were estimated in terms of total gross output, income, and employment as a result of user group spending associated with the five scenarios. Economic impacts were estimated

TABLE 1.—Perceptions (%) of visiting (N = 1,808) and resident (N = 682) reservoir users of aquatic plant impacts on various recreational activities at Lake Guntersville, Alabama, 1991–1992.

Main activity	User group proportions (%)	Helps	No effect	Bothersome sometimes	Bothersome most of the time
Visiting users					
Bank angling	10	20.1	54.2	15.3	10.4
Boat angling	29	47.1	36.0	9.9	7.1
Boating	10	4.1	74.4	11.9	9.6
Shore-based recreation	51	7.2	86.4	4.0	2.4
All visitors		26.9	58.1	8.7	6.2
Resident users					
Bank angling	15	12.9	12.9	41.9	32.2
Boat angling	22	50.5	20.8	18.6	9.8
Boating	43	2.6	26.5	51.3	19.4
None or other	20	5.5	40.6	34.4	19.3
All residents		16.0	29.4	36.3	18.1

for the two-county area surrounding Lake Guntersville.

Results

Baseline Recreational Use and Preferences

Anglers made an estimated 976,508 visits, of which 645,224 were by boaters and 331,284 by nonboaters. Nonanglers made an estimated 1,868,209 visits, of which 271,807 were by boaters and 1,596,402 by nonboaters. A total of 3,050,000 annual visits composed of 2,844,718 on-site and 208,221 residential visits were estimated at Lake Guntersville.

User groups expressed differences in perceptions about aquatic plants. Most visiting boaters and shore-based recreators indicated that the plants had no effect on their activities (Table 1). Approximately half (47.1%) of the visiting boat anglers viewed the plants as a help to their recreational experience. Bank anglers were more widely split in their evaluation: only a fifth (20.1%) considered plants a help, half (54.2%), said no effect, and a fourth (25.7%) considered plants bothersome at least part of the time. Residential users (Table 1) reported having experienced greater negative impact by the plants. With the exception of resident boat anglers, most groups of residents perceived aquatic plants as bothersome, at least part of the time.

Respondents who indicated plants had an effect on their recreation were asked about their preferences for various levels of plant coverage. An increase in plant coverage was favored by 62% of the

TABLE 2.—Preferences (%) for levels of aquatic plant coverage by visiting reservoir users in Lake Guntersville, Alabama, 1991–1992. Only those visitors who said aquatic plants have some effect on their recreational activity (N = 757) responded to this question.

Main activity	As much as possible	More than currently exists	Same as currently exists	Less than at present	None at all
Bank angling	11.1	17.3	35.8	24.7	11.1
Boat angling	28.9	33.1	20.3	13.6	4.1
Boating (nonangling)	0.0	1.4	29.6	42.3	26.8
Shore-based recreation	20.9	9.1	22.7	36.4	10.9
All visitors	23.9	26.0	22.8	19.8	7.5

boat anglers (Table 2). The bank anglers were about evenly divided: 28.4% preferred more plants or as much as possible, 35.8% were satisfied with existing conditions, and 35.8% wanted less or none at all. Of the nonangling boaters, 69.1% preferred less plant coverage or none at all, whereas 29.6% preferred present conditions. About half (47.3%) of the shore-based recreators preferred less plant coverage than existed at the time of the survey.

Residential respondents were asked to describe the aquatic plant coverage near their homes (Table 3). Residential respondents using public access areas were accounted for in the on-site survey results. Residential bank anglers fishing from piers, boat houses, and residential shoreline property were the group most affected by residential shoreline conditions. They reported heavier plant coverage than the other user groups. A greater percentage of resident bank anglers (51.5%) indicated moderate or heavy to severe conditions than did resident boat anglers (32.6%) or resident nonangling boaters (36.8%). Additionally, 74.1% of the resident bank anglers perceived plants as bothersome at least part of the time (Table 1).

An important part of aquatic plant control programs is maintaining access from open water of a

TABLE 3.—Residents' description (%) of aquatic plant coverage near residential homes in Lake Guntersville, Alabama, 1991–1992; N = 682.

Main activity	Not noticeable	Slightly noticeable	Moderate	Heavy to severe
Bank angling	16.1	32.2	32.2	19.3
Boat angling	42.3	25.0	23.9	8.7
Boating (nonangling)	23.6	39.4	22.8	14.0
No main activity or other	22.9	32.4	29.3	15.2
All permanent residents	27.1	32.7	26.4	13.7

TABLE 4.—Importance of various plant control attributes to residential reservoir users (N = 682), Lake Guntersville, Alabama, 1991–1992. Means are based on a 5-point Likert scale, 1 = strongly agree, 5 = strongly disagree.

Attribute	Bank fishing	Boat fishing	Boating	No main activity	Overall
Like clear shoreline for access	1.57	1.66	1.25	1.55	1.49
Support spraying aquatic plants near my boathouse	1.87	1.92	1.36	1.66	1.66
Aquatic plants decrease value of my property	1.87	2.04	1.50	1.72	1.74
Prefer clear boat lanes leading from my boathouse	1.55	1.65	1.32	1.46	1.47
Aquatic plant control necessary for management of residential access areas	1.71	1.74	1.35	1.57	1.56

reservoir to private homes, boat docks, boat houses, and piers. Overall, residential reservoir users strongly supported control efforts around residential areas (Table 4). Satisfaction levels with boating and angling attributes were higher with nonangling boaters (Table 5).

Recreational Use and Economic Value of Five Aquatic Plant Control Scenarios

Annual recreational use for the five aquatic plant management scenarios was highest for scenario A and scenario B (Table 6). The user group composition of visitation was different, with scenario A having a majority of angler visits, whereas scenario B had a higher proportion of nonangler visits.

Valuation models were developed for angler and nonangler groups. The variables bid amount, respondent income, residence status (whether the respondent lived on the lake), and substitute site availability (for anglers only), affected valuation significantly and were included in the models. Anglers gave the highest valuation to the minimum control scenario, whereas nonanglers valued scenario C highest (Table 6). Scenarios A and B had significant economic values for both user groups. The negative values (Table 6) indicated recreation users would have to be compensated to use the lake.

The aggregate annual economic value of each scenario was calculated based on the number of visits and the mean annual economic values of the two user groups. Although scenario B had the highest annual economic value with an aggregate total of $122 million (all values given herein represent 1990 dollars), followed by scenario C and scenario A (Table 6), scenario B was not the highest valued scenario for either the angler or the nonangler user groups.

TABLE 5.—Satisfaction with various boating and boat-fishing attributes by visiting reservoir users (N = 1,808), Lake Guntersville, Alabama, 1991–1992. Means are based on a 5-point Likert scale, 1 = terrible, 5 = delighted.

Attribute	Bank fishing	Boat fishing	Boating (nonangling)	Shore-based
Open water areas	3.85	3.82	4.13	4.02
Water skiing opportunities	3.56	3.95	4.08	3.92
Opportunity to catch a lot of fish	3.58	3.49	3.80	3.67
Opportunity to catch large fish	3.52	3.50	3.72	3.68

Economic Impact to the Local Economy

Economic impacts to the local economy result from visitor spending on lodging, food, gas, supplies, fishing tackle and other recreational goods and services. Nonangler expenditures were $108 per trip, whereas those of anglers were $88 per trip (Table 7). Scenario A produced the greatest increases in gross production, income, and employment (Table 8). However, scenario B was only about 1% lower than scenario A. The patterns of economic impact across scenarios were similar among all three measures. Considering that scenario A represents 30% aquatic plant coverage of the lake and scenario B 20%, the difference in economic impact resulting from this large change in plant coverage is minimal.

Discussion

Decisions about aquatic plant control programs should take into account the preferences, values, and behavior responses (visits) of angler and nonangler groups. The quality of a recreational experience is affected by many factors directly related to the experience such as fishing quality, and by factors related to the recreational environment such as water level, resource management factors, and demographic characteristics (Ward et al. 1996) that are outside the control of the reservoir manager.

The Lake Guntersville study of recreational preferences and valuation is the most extensive valuation of aquatic plant control undertaken to date (Thunberg 1991). The results represent angler and nonangler preferences and values for Lake Guntersville, which in size, operation purpose (navigation,

TABLE 6.—Estimated annual visits, mean economic value (US$), and aggregate net economic value ($) for five plant control scenarios in Lake Guntersville, Alabama. Vegetation coverages (acres) are given in parentheses. N = 1,569.

User group	Management scenario: Minimal control (34,000)	A (20,000)	B (14,000)	C (8,000)	D (near 0)
Visits					
Angler	1,644,440	1,847,554	1,429,608	976,508	476,536
Nonangler	855,640	1,434,785	1,765,458	1,868,209	1,750,512
All	2,500,080	3,282,339	3,195,066	2,844,718	2,227,048
Economic value by user group					
Angler	790.47	778.46	503.77	9.43	−639.87
Nonangler	−334.80	424.83	918.84	1,050.37	999.94
Aggregate net economic value					
All users	21,591,800	93,979,867	122,795,210	101,799,932	52,721,076

flood control, recreation), and aquatic plant characteristics is similar to other reservoirs in the southeastern United States (GJAP 1992). The major conclusion from this study is that anglers and nonanglers with opposite preferences for aquatic plant abundance can be accommodated at a level of aquatic plant coverage that results in high levels of use, economic value, and recreation quality for both groups. Additionally, the aquatic plant level identified is within historic record, and attainable by available control technologies. Significant economic benefits were attributed to recreation at Lake Guntersville. The greatest positive impact on the local economy from recreation associated with Lake Guntersville—$122 million annually—is achieved at 20% plant coverage of the lake. This amount of plant coverage provides highest economic value when angling and nonangling activity are at high levels, not hindered by excessive plant growth. This level of plant coverage meets recreation needs of angling and nonangling recreators.

The economic value and scenario evaluations examined herein provide insight into understanding angler and nonangler responses to changes in aquatic plant control programs, and may be used to evaluate public acceptance of different levels of aquatic plant coverage. Plant management decisions can be made with information from studies of relative benefits to different user groups, effects on visitation, economic value, and economic impacts.

In evaluating the public response to changes in aquatic plant control programs, there are two questions to be considered. First, what are the angler and nonangler user group perceptions related to the change in amount of vegetation? The second and most important question is, how will recreational behavior likely change in response to altered plant conditions? Specifically, what will be the changes in visitation and economic value resulting from a change in plant coverage?

In evaluating changes in aquatic plants, three situations are likely: (1) reduction in control resulting in an increase in plants; (2) increase in plant control to increase open-water areas; or (3) increase in control to improve shoreline access (Henderson 1995).

Reduction in Control

A reduction in plant control, resulting in increased amounts of aquatic plants, can be implemented through reduced control efforts by the reservoir operations agency or can result from changes in conditions that improve plant growth (e.g., drought, Henderson 1992). Based on the preferences expressed, a change in plant control that in-

TABLE 7.—Mean expenditures per trip (US$) for anglers and nonanglers at Lake Guntersville, Alabama, 1991–1992.

User type	Expenditure category: Lodging	Food	Transportation	Activities	Miscellaneous	Total
Angler	22.88	24.23	32.12	4.17	4.98	88.38
Nonangler	4.28	47.09	40.21	2.82	13.92	108.23

TABLE 8.—Projected economic impact (increase) to the local economy including total annual gross output (US$ millions), total annual income ($millions), and total employment (jobs) due to recreational spending, by plant-control scenario, Lake Guntersville, Alabama, 1991–1992. Vegetation coverage (acres) for each scenario is given in parentheses; N = 673.

User group	Management scenario: Minimal control (34,000)	A (20,000)	B (14,000)	C (8,000)	D (near 0)
Total annual gross output					
Angler	155.19	174.36	134.92	92.16	44.69
Nonangler	86.36	144.81	178.19	188.55	176.68
All	241.55	319.17	313.11	280.71	221.37
Total annual income					
Angler	87.78	98.63	76.31	52.12	25.43
Nonangler	48.70	81.66	100.48	106.33	99.63
All	136.48	180.29	176.79	158.45	125.06
Total employment					
Angler	4,608	5,177	4,066	2,736	1,335
Nonangler	2,681	4,496	5,532	5,854	5,485
All	7,290	9,673	9,538	8,590	6,820

creased the amount of plant coverage would be favored by boat anglers and would negatively impact nonangling boaters and shore-based recreators.

The impact on visitation by a group should be considered in using the preference information to make plant management decisions. Although the degree of preference may be high, the effect on total visitation may be minimal. A majority of the nonangling boaters preferred fewer plants, but that preference represented only 10% of the total number of visits. Visitation affected by reductions in plants can be balanced against benefits to other groups from increased plants. That is, boat anglers favored increased plant coverage, and that group represented over twice as many visits as nonangling boaters. More visitors would be pleased than displeased by increased plant coverage. The above comparison illustrates how the study results can be used to evaluate trade-offs in plant management decisions.

Increase in Open-Water Areas

Many lake managers are under pressure to increase open-water areas for use by pleasure boats and jet skis which require wide expanses of open water to avoid congestion and conflict. Increasing availability of open-water areas positively affects the nonangling boater group by providing more recreational areas, but negatively affects boat anglers by reducing fishery habitat.

Satisfaction levels for characteristics affecting nonangling boaters are higher than satisfaction levels for the boat anglers, indicating that boating conditions are already highly satisfactory. It may be difficult to substantially increase the satisfaction level of boaters. Satisfaction levels for the boat anglers may be further diminished if further reductions in plants occur. If increased open-water areas are created, it may be possible to redistribute boat angling use to other parts of the reservoir through restrictions or construction of new access points in areas of extensive plant coverage.

Change in Plant Control Around Shorelines

In aquatic plant control programs, a major emphasis is placed on maintaining access to the lake by cutting boat lanes to residential boat houses, boat docks, and residential shorelines. Also, public swimming beaches and other areas of high use are kept free of plants.

The minimal plant conditions for the entire lake during the period of the surveys may imply that residential users experienced similar minimal conditions. The responses to perceptions of existing conditions suggest the contrary. Bank anglers were the group most affected by residential shoreline conditions. About half of the residential bank anglers described plant coverage as moderate or heavy to severe. The bank anglers perceived plant coverage as more severe than boat anglers or the nonangling boater group, and most bank anglers perceived the plants as bothersome at least part of the time.

Increased plant control around shorelines would probably be welcomed by bank anglers. Residential groups showed strong support of control efforts, including use of pesticide sprays, around the residential areas.

Aquatic plants may have an impact on residential land values. Driscoll et al. (1994) evaluated the influence of the presence and extent of aquatic plant coverage on residential land values. Their analysis showed that eradication of plants in front of lakefront homes (not backlots or undeveloped lots) at Lake Guntersville increased residential property values by $12 million annually. These findings are in contrast to Milon (1989), however, who found no effect of hydrilla on residential land values in a similar study at Lake Harris, Florida.

The Lake Guntersville study evaluated scenarios depicted for respondents as six different recreational environments. Numerous comments by respondents, resource managers, and recreation specialists noted that the total amount of plant coverage may not be as important as the distribution of the plants. Preliminary work for the CVM study examined distribution in a cursory manner. A plant distribution plan was developed for Lake Guntersville by a panel made up of representatives of various lake management disciplines (aquatic plant management, fisheries, wildlife, recreation, and water quality). Criteria were developed from each discipline for types of lake areas where the plants should be controlled and for areas where plants were acceptable or desired. A consensus plant distribution was then developed by combining criteria from all disciplines. The resulting plant distribution map encompassed approximately 14,000 acres of plants, roughly corresponding to scenario B. Future studies should address the distribution as well as coverage of plants in a more extensive and explicit manner.

Acknowledgments

The work presented here was supported by the GJAP conducted by the TVA and the COE Waterways Experiment Station, Vicksburg, Mississippi. The recreation surveys were organized, implemented and analyzed by the Environmental Resources Assessment Group (ERAG), Athens, Georgia. H. Ken Cordell, U.S. Forest Service, Athens, Georgia, and John C. Bergstrom, University of Georgia were principal investigators for ERAG. George M. Farrell, TVA–Norris, Tennessee, was coinvestigator with J. E. Henderson on the recreation study for the GJAP.

References

Anderson, A. B., A. Basilevsky, and D. P. J. Hum. 1983. Measurement: theory and techniques. Pages 231–287 *in* P. H. Rossi, J. D. Wright, and A. B. Anderson, editors. Handbook of survey research. Academic Press, New York.

Bergstrom, J. C., and six coauthors. 1994. Aquatic plant coverage and outdoor recreation at Lake Guntersville, Alabama: a study of user preferences, economic values, and economic impacts. Tennessee Valley Authority and U.S. Army Corps of Engineers Waterways Experiment Station, Guntersville Joint Agency Project Report TVA/RG/WM-94005, Muscle Shoals, Alabama.

Bergstrom, J. C., J. R. Stoll, J. P. Titre, and V. L. Wright. 1990. Economic value of wetlands-based recreation. Ecological Economics 2:129–147.

Cochran, W. G. 1977. Sampling techniques, 3rd edition. Wiley, New York.

Dillman, D. A. 1978. Mail and telephone surveys, Wiley, New York.

Driscoll, P., and five coauthors. 1994. The effect of aquatic plants on residential shoreline land values at Guntersville reservoir. Tennessee Valley Authority and U.S. Army Corps of Engineers Waterways Experiment Station, Guntersville Joint Agency Project Report TVA/RG/WM-94-007, Muscle Shoals, Alabama.

GJAP (Guntersville Joint Agency Project). 1992. Master plan—Guntersville reservoir aquatic plant management. Tennessee Valley Authority and U.S. Army Corps of Engineers Waterways Experiment Station, Guntersville Joint Agency Project Report TVA/WR-92-21, Muscle Shoals, Alabama.

Guadagnalo, F. 1985. The importance-performance analysis: an evaluation and marketing tool. Journal of Park and Recreation Administration 2:13–22.

Henderson, J. E. 1992. Economics and aquatic plant control. Pages 6–17 *in* Proceedings, twenty-sixth annual meeting, aquatic plant control research program. U.S. Army Corps of Engineers Waterways Experiment Station, Miscellaneous Paper A-92-2, Vicksburg, Mississippi.

Henderson, J. E. 1995. Use of economic information in the evaluation of aquatic plant control programs. Pages 8–18 *in* Proceedings, twenty-ninth annual meeting, aquatic plant control research program. U.S. Army Corps of Engineers Waterways Experiment Station Miscellaneous Paper A-95-3, Vicksburg, Mississippi.

Hollenhorst, S., D. Olson, and R. Fortney. 1992. Use of importance-performance analysis to evaluate state park cabins: the case of the West Virginia state park system. Journal of Park and Recreation Administration 10(1):1–11.

Kalton, G. 1983. Introduction to survey sampling. Sage Publications, Newbury Park, California.

Mengak, K. K., F. D. Dottavio, and J. T. O'Leary. 1986. Use of importance-performance analysis to evaluate a visitor center. Journal of Interpretation 11(2):1–13.

Miller, D. 1991. Handbook of research design and social measurement. Sage Publications, Newbury Park, California.

Milon, J. W. 1989. The effects of hydrilla infestation in Lake Harris on the market price of adjacent residential property. Institute of Food and Agricultural Sciences Economic Information Report 262, Gainesville, Florida.

Mitchell, R. C., and R. T. Carson. 1989. Using surveys to value public goods: the contingent valuation method. Resources for the Future, Washington, DC. (Available from The Johns Hopkins University Press, Baltimore, Maryland.)

Palmer, C., and E. Siverts. 1985. IMPLAN analysis guide. U.S. Forest Service, Ft. Collins, Colorado.

Probst, D. B. 1988. Use of IMPLAN with public area recreation visitor survey pretest data: findings and recommendations, U.S. Army Corps of Engineers Waterways Experiment Station, Miscellaneous Paper R-88-1, Vicksburg, Mississippi.

Sheatsley, P. F. 1983. Questionnaire construction and

item writing. Pages 195–230 *in* P. H. Rossi, J. D. Wright, and A. B. Anderson, editors. Handbook of survey research. Academic Press, New York.

Thunberg, E. M. 1991. Literature review of economic valuation of aquatic plant control. U.S. Army Corps of Engineers Waterways Experiment Station, Miscellaneous Paper A-91-1, Vicksburg, Mississippi.

Ward, F. A., B. A. Roach, J. B. Loomis, R. C. Ready, and J. E. Henderson. 1996. Regional recreation demand models for large reservoirs: database development, model estimation, and management applications. U.S. Army Corps of Engineers Waterways Experiment Station, Technical Report R-96-2, Vicksburg, Mississippi.

American Fisheries Society Symposium 16:382–393, 1996

Relationships between Largemouth Bass and Aquatic Plants in Guntersville Reservoir, Alabama

W. B. Wrenn and D. R. Lowery
Tennessee Valley Authority, Water Management, Muscle Shoals, Alabama 35660, USA

M. J. Maceina
Auburn University, Department of Fisheries and Allied Aquacultures
Alabama Agricultural Experiment Station, Auburn University, Alabama 36849, USA

W. C. Reeves
Alabama Department of Conservation and Natural Resources
64 North Union Street, Montgomery, Alabama 36130, USA

Abstract.—Population dynamics and angling success for largemouth bass *Micropterus salmoides* in the 27,500-ha Guntersville Reservoir, Tennessee River, Alabama, were evaluated in relation to submersed aquatic macrophyte coverage. Surface coverage fluctuated from a peak of 29% of the total reservoir area in 1988 to a low of 7% in 1991, and then increased to 14% by 1994. An optimum level of macrophyte coverage for largemouth bass could not be identified, but coverage of greater than 20% of the surface area contributed to a decline in growth for fish younger than age four. Stronger year-classes were produced when spring and early summer water retention time exceeded 16 d, independent of macrophyte coverage. Biomass estimates of catchable largemouth bass (>250 mm total length) between 1983–1988 and 1993–1994 were similar, exceeding 30 kg/ha. However, size structure of the population during the 1980s was skewed towards smaller fish. This condition was also reflected in data from organized fishing tournaments examined for 1986–1993. Tournament weights per hour were unchanged over time. Tournament anglers caught fewer, but larger fish because aquatic plant coverage was positively correlated to catch rate but negatively correlated to average weight of largemouth bass. During 1990–1994, nontournament angler average catch rate (release and harvest) ranged from 0.43 to 0.99/h, but angling effort declined 46%. High densities (>300/ha) of age-0 largemouth bass occurred in heavily vegetated habitats, but their size and growth rates were usually significantly less than those of age-0 largemouth bass from unvegetated sites, apparently resulting in size-dependent mortality and lower than expected recruitment to age 1.

Aquatic macrophyte management apparently was not a major issue in reservoir fisheries management when the first two Reservoir Symposium proceedings were published. In the first symposium proceedings Boyd (1967) addressed aspects of aquatic plant ecology relative to nutrient requirements and uptake by macrophytes, and reported that from a fish production standpoint, fisheries managers considered macrophytes to be detrimental. However, Boyd also noted that reliable information on the relationship between macrophytes and fish production was scarce. Between the first and second symposia, knowledge or concern about this topic apparently changed little as it was not mentioned in Summerfelt's (1986) summary of the second symposium. This situation changed abruptly, due perhaps to the controversy surrounding the Lake Conroe project in Texas (Klussmann et al. 1988; Bettoli et al. 1993) which was under way as the second reservoir symposium was convened in 1983. Conflicting issues arose in Lake Conroe, a multipurpose water project, where an extensive area of submersed aquatic macrophytes was removed by stocking grass carp *Ctenopharyngodon idella*. Various segments of the angling public considered removal of macrophytes to be a loss of critical fish habitat.

In Guntersville Reservoir, a 27,500-ha mainstream impoundment of the Tennessee River in northeastern Alabama (Figure 1), the extent of submersed macrophyte coverage and the sequence of events that led to control measures were similar to those that occurred at Lake Conroe. The preliminary information from Lake Conroe and related assessments (Noble et al. 1986) prompted us to proceed more cautiously at Guntersville Reservoir. Following several years of only partial success in the management of Eurasian watermilfoil *Myriophyllum spicatum* in Guntersville Reservoir and with the concern for the spread of hydrilla *Hydrilla verticillata*, a test demonstration for using monosex grass carp as a biological control agent was conducted during 1983–1988 in an enclosed segment of the Town Creek embayment (TVA 1989). By 1988, aquatic macrophytes colonized 29% (>8,000 ha) of the reservoir's surface area, including more than 1,000 ha of hydrilla which was considered a long-term threat to multiple uses of Guntersville Reservoir (Webb et al. 1995). At Guntersville Reservoir the 1980s may best be described as a period of low reservoir discharge, high water clarity, and prime angling for largemouth bass *Micropterus salmoides*. Mean water retention time (volume/discharge) for this relatively shallow reservoir (mean depth <5 m) is 13 d, usually less than 10 d in winter to greater than 13 d in summer.

In 1990, the Joint Agency Guntersville Project (JAGP) was established by the U.S. Congress. Sponsored by the Tennessee Valley Authority (TVA) and the U.S. Army Corps of Engineers (COE), this 5-year demonstration (1990–1994) was initiated to assess a broad array of issues pertaining to aquatic plant management, including the effectiveness and impact of various plant control methods. In this paper we examine relationships between the largemouth bass population and fishery, and the surface coverage of submersed aquatic macrophytes in Guntersville Reservoir. In our analyses we include data from monitoring conducted on this reservoir during the 1980s, and data collected during 1990–1994 as part of the JAGP.

Methods

Although testing of various aquatic plant control techniques as part of the JAGP was expected to result in a reduction of submersed macrophytes, the major part of this reduction was expected to be attributed to foraging by grass carp. Between April and July 1990, 100,000 triploid grass carp (20–25 cm total length [TL]) were stocked at 23 locations in the lower two-thirds of the reservoir. Also, during 1987–1989, private groups legally released about 20,000 grass carp that probably included both diploid and triploid fish. The estimated stocking density of grass carp in 1990 was approximately 20 per vegetated hectare, which met the objective of stocking grass carp at a low to intermediate rate to substantially reduce hydrilla and obtain only limited control of other plant taxa, particularly Eurasian watermilfoil (TVA 1990). Other aspects of the evaluation of grass carp in this reservoir are presented by Bain (1993) and Wrenn et al. (1995).

Aquatic macrophyte surface coverage and composition during 1980–1994 were determined by the TVA's aquatic plant management section (Webb et al. 1995). Surface coverage was determined from large-scale (1:7,200) color aerial photography in late summer or early autumn when submersed macrophytes generally reach maximum coverage. Ground surveys were conducted about the same time of the aerial photography to determine species composition of the macrophyte communities and facilitate interpretation of aerial photographs. Colonies of aquatic macrophytes were delineated on transparent mylar overlays on photographic prints and labeled by species or species groups. Coverage was then determined using an electronic planimeter.

To estimate recreational (nontournament) angler use (total effort, harvest, catch rate, and trip expenditures plus willingness to pay), a roving creel survey with nonuniform probability sampling was conducted annually during February through November, 1990–1994. The survey was stratified spatially and temporally (Crow and Malvestuto 1996, this volume). Temporal stratification included 10 monthly blocks (February–November) which were further divided into weekdays (Monday–Thursday) and weekend days (Friday–Sunday). The reservoir was divided into three strata, each approximately 8,000 ha, designated as lower, middle, and upper reservoir (Figure 1). Sampling effort for each of two roving clerks was 20 d/month.

Tournament fishing records for Guntersville Reservoir collected by the Bass Anglers Information Team (BAIT, Reeves and McHugh 1994) were analyzed for 1986–1993. We analyzed records for 78 tournaments (14,646 h fishing), conducted in July–November (peak macrophyte coverage), that had a tournament-imposed 304-mm minimum length limit. Data were summed for each year and the relations among catch-per-hour, mean weight, effort to catch a largemouth bass heavier than 2.27 kg, and macrophyte coverage were described using correlative and regression analyses. Variability of tournament-enforced possession limits ranged from an average of 8.7 fish in 1986 to 5.0 fish in 1993, and was considered in our analysis.

Population Characteristics of Age-2 and Older Largemouth Bass

We examined standing stock, size structure, and age and growth of age-2 and older largemouth bass (age-2 largemouth bass in this reservoir generally equate to fish at least 250 mm TL, defined as harvestable or catchable in TVA's rotenone sampling size classification). Biomass of catchable fish during the 1980s was derived from rotenone samples collected at a single site (11 ha) in September 1983, 1986, and 1988 (TVA 1989). This site was within a 160-ha segment of the Town Creek embayment that

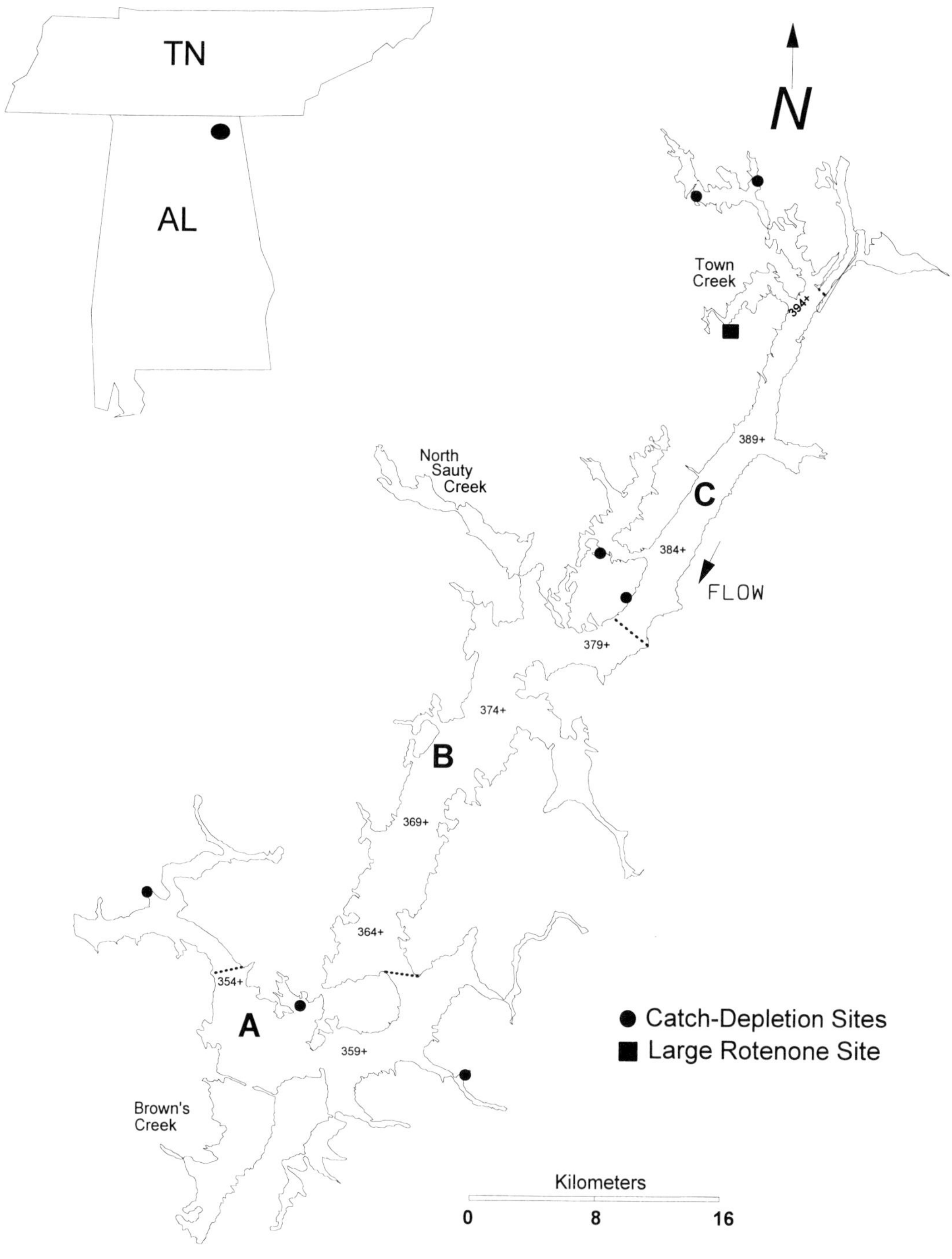

FIGURE 1.—Guntersville Reservoir, a 27,500-ha mainstream impoundment of the Tennessee River in northeastern Alabama. Creel survey strata, each about 8,000 ha, are designated as lower (A), middle (B), and upper (C) reservoir. Large rotenone samples in Town Creek collected 1983–1988 (solid square), and catch depletion samples collected 1993–1994 (solid circles).

was separated by a barrier from the rest of the reservoir. In 1993–1994 an electrofishing catch depletion technique (Maceina et al. 1995a) was used to estimate biomass in seven coves (each 2.8–3.3 ha; Figure 1).

Size-structure designations followed Gabelhouse (1984) and were estimated for largemouth bass collected by catch depletion (1993–1994) and rotenone samples (1983–1988). Three ratios were used, including proportional stock density (PSD = 100[$N \geq$ 304 mm TL]/[$N \geq$ 203 mm TL]), relative stock density–preferred (RSD–P = 100[$N \geq$ 381 mm TL]/[$N \geq$ 203 mm TL]), and relative stock density–memorable (RSD–M = [$N \geq$ 508 mm TL]/[$N \geq$ 203 mm TL]). These indices were also applied to autumn electrofishing samples (collected in 1990–1992 at 10 sites as part of the JAGP), and to spring electrofishing samples collected by the Alabama Game and Fish Division during 1983–1993 as part of a general reservoir monitoring program.

Age and growth was determined using otoliths (Hoyer et al. 1985) from all sizes of largemouth bass collected by electrofishing in 1990. Before 1990, age and growth analyses of largemouth bass were made with the scale method; these estimates were not used in our evaluation because estimates were believed to be inaccurate. In 1993, otoliths from a subsample (N = 190) of fish of all sizes were taken from the full sample (N = 653) to describe age structure, growth, mortality, and year-class strength. Length-at-age was computed and the time to reach 304 mm TL (tournament minimum length until October 1993) and 381 mm TL (new length limit as of October 1993) were determined after fitting a von Bertalanffy (1957) model to the age and length data. In 1994, otoliths were taken from a subsample of largemouth bass less than 375 mm TL to identify and separate age-1 and age-2 fish to describe growth. Data were pooled for the entire reservoir, but population comparisons were also made between vegetated and unvegetated habitats. Length comparisons were made using Student t-tests. Catch-curve analysis was used to estimate mortality (Ricker 1975). Multiple regression was applied to the catch-curve analysis to determine if selected environmental variables (e.g., macrophyte coverage) were related to year-class strength.

Abundance, Growth, and Relative Survival of Age-0 and Age-1 Largemouth Bass

Cohorts of age-0 largemouth bass were collected at 10 sites during July–August and October–November 1991–1994. Age-1 fish were sampled in March, April, and June. Fish age was determined using otoliths. Collection sites spanned that zone of the reservoir covered by the catch depletion sites (Figure 1). Three of the 10 sites were vegetated and seven unvegetated. At least 30 km separated the vegetated and unvegetated habitats; thus we assumed that migration of age-0 and age-1 fish between habitats was nil (Copeland and Noble 1994). Collection methods included small (0.2 ha) shoreline rotenone samples (Timmons et al. 1979), large (1.2–1.3 ha) coves sampled with rotenone; small plot (0.1 ha) electrofishing catch depletion (Maceina et al. 1993), and the larger (2.8–3.3 ha) electrofishing catch depletion.

Vegetation abundance was determined at these sites with a recording fathometer (Maceina and Shireman 1980), and various limnological parameters were measured (Maceina et al. 1994). Zooplankton samples were collected using a 20-cm diameter conical net, 70 cm long. Net tows were vertical, bottom to surface, and volume was calculated by the distance the net was raised times the diameter of the net. Crustacean zooplankton were enumerated and identified, including all cladocerans, copepods, and copepod nauplii.

For the 1991–1994 year classes of largemouth bass, we used otolith ring counts and measurements to estimate daily growth and swim-up date for fish collected in late July to early August from vegetated and unvegetated habitats. Otolith preparation and processing followed Secor et al. (1991), and daily ring counts and measurements followed Maceina et al. (1995b). To assign swim-up dates for all fish, age-frequency distributions were estimated from a subsample of fish that were aged using an age-length key (Ricker 1975). Student t-tests and one-way analysis of variance (ANOVA) were used to test for differences in daily growth rates, density, biomass, and size by habitat.

Results and Discussion

Macrophytes

Surface coverage of submersed aquatic macrophytes in the reservoir peaked at 29% in 1988, declined to 7% in 1991, and then increased to 14% by 1994 (Figure 2). Similar declines and subsequent increases in macrophyte coverage occurred during 1990–1993 on most other mainstream reservoirs (without grass carp) in the TVA system (Webb et al. 1995). Thus, the establishment and much of the change in submersed macrophyte coverage on Guntersville Reservoir during 1980–1994 was likely related primarily to climatic conditions, particularly

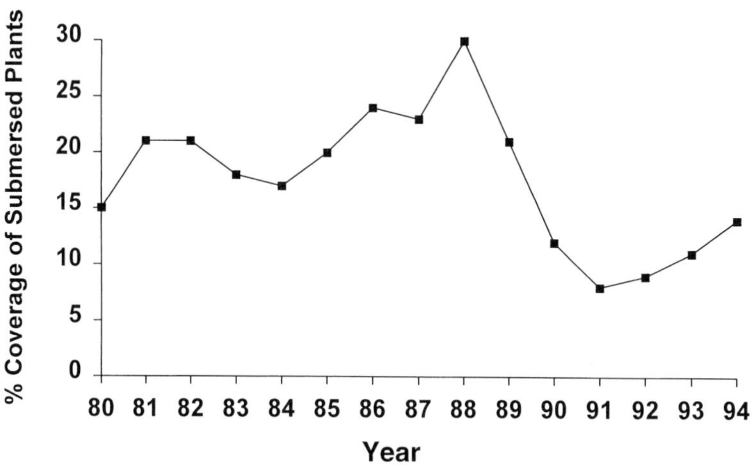

FIGURE 2.—Percentage of surface coverage of aquatic macrophytes in Guntersville Reservoir, total area 27,500 ha, 1980–1994.

the extreme drought during the mid-1980s and the return of normal to high rainfall since 1989. The combination of climatic effects and grass carp nearly eliminated hydrilla by 1992. Eurasian watermilfoil accounted for more than 90% of the submersed macrophytes during 1991–1994.

Surface coverage of submersed macrophytes within the screened 160-ha segment of the Town Creek embayment (large rotenone site) exceeded 70% of the total area in 1983–1986. However, in 1987 coverage declined to less than 20%, but increased to greater than 40% by October 1988.

Sport Fishery

Targeted angling effort for largemouth bass made up at least two-thirds of the total angler effort in this reservoir during 1990–1994. Effort approached 1 million hours in 1990, but subsequently showed a net decline of 46% between 1991 and 1994 (Table 1). Biomass harvested (removed from the reservoir) was near 145,000 kg in 1990, but was lower during 1992–1994, due to the reduction in effort. Total catch rate (release + harvest) ranged from 0.43/h in 1992 to 0.99/h in 1994. Although not statistically significant ($P = 0.2$), a modest correlation ($r = 0.70$) existed between plant coverage and catch rate of largemouth bass during this 5-year period. Before enactment of length limit restrictions, mean weight of largemouth bass harvested was higher in 1993 (0.79 kg) than in 1990 (0.62 kg). Due partly to the large numbers of fish harvested in 1990–1991, a 381-mm minimum length limit was imposed by the Alabama Department of Conservation and Natural Resources in October 1993. Therefore, release of any largemouth bass before October 1993 was strictly voluntary. In 1994, the first creel-year under the 381-mm length limit, the release rate was 0.88/h.

Targeted angling effort for largemouth bass was higher in the middle and lower zones and lowest in the upper zone (Figure 3). However, surface coverage of submersed macrophytes during 1990–1994 in these zones reflected an opposite trend: 24–30% in the upper, 6–9% in the middle, and 0.2–4% in the lower zone. Thus, nontournament anglers did not seem to select for vegetated habitat. Although macrophyte coverage varied among strata, catch rates of largemouth bass in vegetated and in relatively unvegetated areas were similar (Figure 4).

More than 90% of the total angling effort (all species combined) was by Alabama residents. Eco-

TABLE 1.—Targeted angling effort, catch rate, and harvest of largemouth bass from Guntersville Reservoir, 1990–1994.

Variable	Year				
	1990	1991	1992	1993	1994
Effort (h)	960,417	770,705	631,472	406,484	516,242
Total catch (*N*)	633,875	439,302	271,533	178,853	511,079
Total harvest (kg)	144,719	112,430	49,905	23,389	48,712
Total catch rate (*N*/h)	0.66	0.57	0.43	0.44	0.99
Catch and release (*N*/h)	0.48	0.39	0.37	0.38	0.88
Harvest (*N*/h)	0.18	0.14	0.06	0.06	0.11

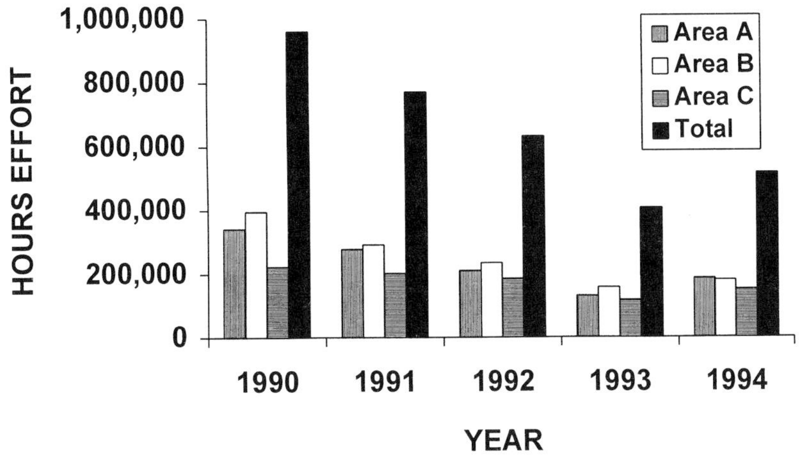

FIGURE 3.—Angler effort for largemouth bass for total reservoir and by lower (A), middle (B), and upper (C) reservoir creel, Guntersville Reservoir, 1990–1994.

nomic value, based on trip expenditures and willingness to pay, ranged from US$4.15 million in 1990 to $1.65 million in 1993. Trip expenditures alone declined $1.4 million during 1991–1993, concurrent with a 63% decline in total angling effort. An expanded economic analysis of recreational fishing and other uses relative to macrophyte coverage on Guntersville Reservoir is given by Henderson (1996, this volume).

As macrophytes declined in Guntersville Reservoir, anglers competing in organized tournaments during 1986–1993 caught and weighed in fewer fish, but the average weight of these fish increased (Figure 5). The weight-per-hour rate was unchanged during this period. However, at high levels of macrophyte coverage, the effort to catch a largemouth bass larger than 2.27 kg increased (Figure 5). Although the possession limit declined from 1986 to 1993, correlative and multiple regression analyses indicated that possession limit was not a significant ($P > 0.25$) factor affecting catch statistics (Maceina and Reeves 1996).

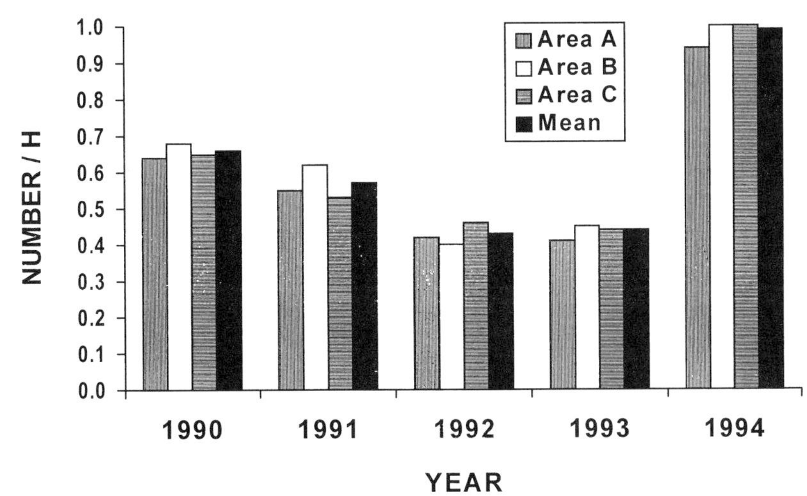

FIGURE 4.—Catch rate (*N*/h, release and harvest) of largemouth bass for lower (A), middle (B), and upper (C) reservoir creel strata and the whole reservoir mean, Guntersville Reservoir, 1990–1994.

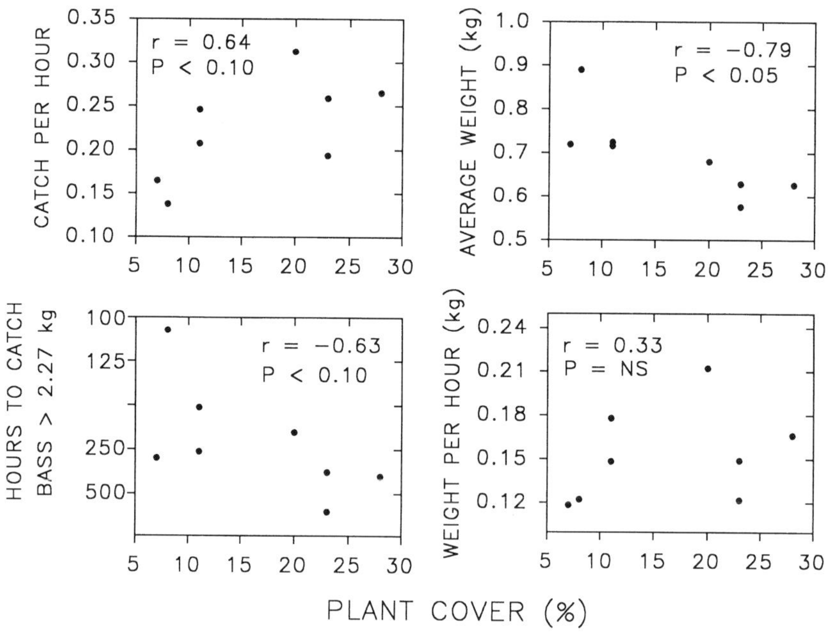

FIGURE 5.—Largemouth bass tournament catch statistics, 1986–1993, in relation to percentage surface coverage of aquatic macrophytes on Guntersville Reservoir.

Largemouth Bass Population Status

Biomass of catchable (>250 mm TL) largemouth bass determined by electrofishing catch depletion in 1993 and 1994 was 33 and 32 kg/ha, respectively (Table 2). In comparison, biomass estimates from the Town Creek large rotenone sample ranged from 36 kg/ha in 1983 to 30 kg/ha in 1988 (Table 2). Given the possible overestimation or underestimation of the standing crop of major species by cove sampling as compared with sampling whole embayments or reservoirs (Aggus et al. 1980), the accuracy of biomass and density estimates both from our large rotenone (11 ha) and catch depletion samples (21 ha, total) is uncertain. However, coves larger than 4.9 ha would be expected to produce a more representative sample of whole embayments than would samples of smaller areas (Aggus et al. 1980). It is unlikely that the true biomass of largemouth bass in Guntersville Reservoir in the 1980s or 1993–1994 was much higher than the estimates reported here. Under similar conditions (i.e., relatively high water clarity, increasing aquatic macrophytes, and long water retention time) in 1986 in Kentucky Reservoir (350 km downstream), the biomass of

TABLE 2.—Biomass of catchable (>250 mm total length, TL) largemouth bass and population size structure in Guntersville Reservoir, 1983–1994.

Year	Method[a]	Biomass (kg/ha)	Stock density[b] PSD (40–70)[c]	RSD–P (10–30)[c]	RSD–M (0–10)[c]
1983	LR	36	40	8	1
	SF		17	5	0
1984	SF		12	5	0
1986	LR	31	30	7	0
	SF		56	8	0
1988	LR	30	16	7	1
	SF		46	12	3
1989	SF		44	10	2
1990	AF		48	21	7
	SF		46	18	6
1991	AF		43	18	4
1992	AF		53	25	8
1993	CD	33	63	26	7
	SF		54	13	10
1994	CD	32	46	17	5

[a]Method: LR–large rotenone; AF–autumn electrofishing, SF–spring electrofishing; CD–catch depletion electrofishing.

[b]Stock density (Gabelhouse 1984), PSD = 100 (quality-length fish / stock-length fish; RSD–P = percentage of bass >381 mm TL; RSD–M = percentage of bass >508 mm TL.

[c]Target range for healthy populations according to Gabelhouse (1984).

TABLE 3.—Mean length-at-age for largemouth bass from Guntersville Reservoir, and Alabama statewide average (M. J. Maceina, unpublished analysis of 32 Alabama reservoirs). Age_{304} and Age_{381} refer to the mean time in years for largemouth bass to reach 304 mm total length (TL) and 381 mm TL, respectively. Data for 1990 were provided by the Alabama Game and Fish Division. Within-column mean lengths-at-age with different letters are significantly different ($P \leq 0.05$).

Site	Year	Age (years) 1	2	3	4	5	Age_{304}	Age_{381}
Guntersville	1990		232 x	311 y	382 z	454 z	2.80	3.88
Guntersville	1993	158 y	261 y	326 z	388 z	416 y	2.65	4.02
Guntersville	1994	190 z	298 z					
Alabama statewide	1989–1994	176	270	339	392	436	2.47	3.79

largemouth bass, including all size groups, was 21 kg/ha (Buynak et al. 1991).

Both the large rotenone and catch depletion electrofishing surveys provided large samples ($N > 500$ specimens) for examination of population size structure under disparate levels of aquatic macrophyte coverage. Based on size structure ratios, the largemouth bass population in this reservoir was satisfactory (i.e., within target range proposed by Gabelhouse 1984) in 1993–1994, whereas the structure of the population in the 1980s reflected an imbalance toward fish smaller than 304 mm long (Table 2). Accordingly, the RSD–P values in the 1980s were more often below Gabelhouse's (1984) recommended target, 10–30% (Table 2). Also, RSD–M values in 1993–1994 were higher than in the 1980s (Table 2).

Decline in PSD and RSD–P ratios in 1994 occurred primarily because of highly successful recruitment of the 1992 year-class to stock size (203–303 mm TL), and lengths of age-1 and age-2 fish increased from 1993 values (Table 3). Age-2 largemouth bass averaged 232 mm long in 1990 contrasted to 298 mm in 1994. Similarly, age-3 fish were larger in 1993 than in 1990. By 1994, mean total length of age-1 and age-2 largemouth bass in this reservoir exceeded the Alabama statewide mean lengths for these age groups (M. J. Maceina, Auburn University, unpublished data for 32 Alabama reservoirs). Growth rates of age-4 and older fish did not appear to vary with macrophyte levels. In 1990 and 1993, it took 3.88 and 4.02 years for largemouth bass in Guntersville Reservoir to reach 381 mm TL (the legal size imposed on 1 October 1993), a slightly longer period than the statewide average of 3.79 years to reach this size. Age analysis was not conducted for age-3 and older fish in Guntersville Reservoir in 1994, but it is likely that recruitment to the 381-mm length limit was occurring at a faster rate. The most likely cause for below average growth of age-3 and younger fish through 1990 was the combination of density-dependent mechanisms and peak macrophyte coverage. This phenomenon has been observed in other waters (Colle and Shireman 1980; Bettoli et al. 1992).

From 1991 to 1994, density of age-0 largemouth bass in August was 3.5–17 times greater in vegetated habitats (>90% surface coverage) than in unvegetated habitats, and biomass of age-0 fish in vegetated habitats was 3.7–10 times greater for 3 of the four year-classes for which this measure differed between habitat types (Table 4). However, abundance of age-1 fish in vegetated habitats was significantly higher ($P < 0.05$) than in unvegetated sites for only one (1991) of the four year-classes (Table 5). In June 1993 abundance of age-1 fish from the 1992 year-class produced in vegetation was significantly lower than that produced in unvegetated sites (Table 5).

Otolith analyses showed that size differences of age-0 largemouth bass between vegetated and unvegetated habitats was due largely to differences in growth rates rather than spawning time. Differences in back-calculated growth rates between habitat types usually were apparent by age 20 d. For the 1991 to 1994 year-classes, daily growth rates were 15–23% higher in unvegetated than in vegetated habitats. In 1991 and 1993, largemouth bass tended to spawn slightly later in vegetated sites, and in

TABLE 4.—Mean density, biomass, and total length of age-0 largemouth bass from the 1991–1994 year-classes collected from vegetated (>90% coverage) and unvegetated habitats in late July and August. Mean values within columns for a specific year-class followed by the same letter were not significantly different ($P > 0.05$).

Year	Habitat	Density (N/ha)	Biomass (kg/ha)	Total length (mm)
1991	Vegetated	377 y	2.645 y	70 z
	Unvegetated	22 z	0.253 z	92 y
1992	Vegetated	883 y	2.535 y	64 z
	Unvegetated	104 z	0.683 z	72 y
1993	Vegetaed	366 y	1.464 y	69 z
	Unvegetated	103 z	1.333 y	97 y
1994	Vegetated	305 y	1.301 y	70 y
	Unvegetated	26 z	0.138 z	74 y

TABLE 5.—Mean density (*N*/ha), relative abundance (*N*/h), and size of age-0 and age-1 largemouth bass collected from vegetated and unvegetated habitats in Guntersville Reservoir for the 1990–1994 year-classes. Mean values within rows for the same variable followed by the same letter are not significantly different ($P > 0.05$).

Year class	Collection time	Age	Vegetated (Density or abundance)	Unvegetated (Density or abundance)	Vegetated (Total length, mm)	Unvegetated (Total length, mm)
1990	Aug 90	0	48/ha y	71/ha z	83 y	98 z
	Nov 91	1	16/h z	20/h z	244 z	236 z
1991	Aug 91	0	387/ha y	22/ha z	70 y	92 z
	Nov 91	0	5/h z	7/h z	123 y	145 z
	Apr 92	1	14/h y	6/h z	150 z	151 z
1992	Aug 92	0	883/ha y	104/ha z	64 y	72 z
	Oct 92	0	332/ha z	267/ha z	74 y	132 z
	Mar 93	1	9/h z	8/h z	144 y	183 z
	Jun 93	1	14/h y	31/h z	185 z	184 z
1993	Aug 93	0	336/ha y	104/ha z	69 z	97 y
	Oct 93	0	42/h z	43/h z	137 z	142 z
	Mar 94	1	28/h z	38/h z	176 y	203 z
	Jun 94	1	39/h z	29/h z	171 y	188 z
1994	Aug 94	0	305/ha y	26/ha z	70 z	74 z

conjunction with slower growth rates, this resulted in the widest between-habitat discrepancy in size of age-0 fish by August observed during the study (Table 5). In 1994, when peak spawning occurred earlier in vegetated sites, size of age-0 fish in August was similar between habitat types (Table 5). However, growth rates were 0.97 mm/d in unvegetated sites contrasted to 0.77 mm/d in vegetated sites.

In early spring–summer 1991–1994, crustacean zooplankton densities were either similar or higher in unvegetated habitats compared to vegetated habitats (Table 6). Although largemouth bass are zooplanktivores in the first few weeks of life (Hirst and DeVries 1994), zooplankton production did not appear to explain differences in age-0 largemouth bass density and growth between habitat types. Likely mechanisms for greater age-0 largemouth bass abundance in vegetation included a reduction in predation and an abundant supply of macroinvertebrate foods (Watkins et al. 1983; Schramm and Jirka 1989). Possible explanations for reduced growth rates of age-0 fish in vegetated habitats include increased largemouth bass density and a possible reduction in feeding efficiency due to greater structural complexity in vegetated sites. Slower growth in vegetated habitats generally conferred smaller age-0 fish by autumn, and size-dependent mortality (Miranda and Hubbard 1994) apparently reduced abundance of age-1 fish produced in vegetation.

Mortality Rates and Year-Class Strength

Catch curve analysis of fish age 2–11 showed alternating year-class strength, with the exception of the 1985 year-class (Figure 6). Age explained 79.9% ($r^2 = 0.799$) of the variation in number caught at each age. The slope coefficient of the regression between $\log_e$ of abundance and age was −0.521, which translates into an average annual survival rate of 59%. Assuming 20% of the mortality of fish age 2 and older was due to natural causes, this would suggest an additional 21% mortality was due to exploitation by humans. Using the mean harvest of 5.2 kg/ha in 1990–1991 from the creel survey and a catchable biomass of 32.5 kg/ha from the 1993–1994 catch depletion survey, the annual exploitation rate was 16%.

Variable year-class strength was not related to aquatic macrophyte coverage ($P = 0.65$) when this term was entered into the catch curve multiple regression analysis. However, year-class strength was associated with water retention time (volume/discharge). Between 1982 and 1991, mean retention time (RET) for Guntersville Reservoir for April–

TABLE 6.—Mean crustacean zooplankton (i.e., copepod nauplii, copepods, and cladocerans) densities (*N*/L) from April through June, Guntersville Reservoir, Alabama, 1991–1994. Mean annual values in rows within the same habitat type followed by the same letter are not significantly different ($P > 0.05$). An asterisk between mean values for each habitat type from the same sampling date indicate values are significantly different ($P \leq 0.05$). No samples were collected from vegetated habitats in April of 1991.

Month	Habitat type	Year 1991	1992	1993	1994
April	Unvegetated	30.8 y	30.8 y	23.3 y	40.8 y
			*		
	Vegetated		10.0 z	35.4 y	26.0 yz
May	Unvegetated	12.0 z	42.7 y	14.8 z	43.2 y
			*		*
	Vegetated	10.3 y	11.8 y	13.2 y	15.7 y
June	Unvegetated	103.4 v	18.6 y	6.5 z	35.2 x
		*			*
	Vegetated	40.8 v	17.5 x	2.6 z	9.5 y

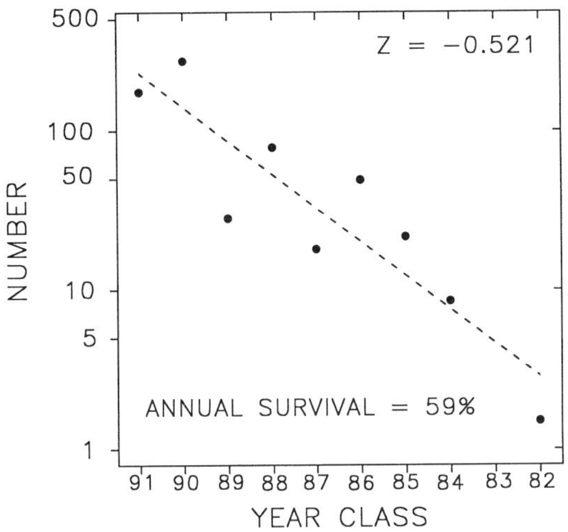

FIGURE 6.—Number of age-2 (1991 year-class) to age-11 (1982 year-class) largemouth bass plotted against year-class, Guntersville Reservoir. The slope coefficient of the regression (dashed line) is represented as Z and is the instantaneous mortality rate. The catch-curve was based on 653 fish representing nine year-classes.

July varied from 10 to 49 d. The multiple regression was:

$$\log_e \text{number} = 5.780 - 0.526(\text{AGE}) + 0.034788(\text{RET}).$$

Besides age (in years) retention time was a significant ($P = 0.05$) variable, and its incorporation explained an additional 8.7% of the variation in $\log_e$ number (total catch-at-age) for the overall model. With the exception of 1984, mean retention times less than 16 d between April and July were associated with weaker year-class abundance. Macrophyte coverage was high from 1986 to 1989, yet strong and weak year-classes were produced during this period. Also, a strong and a weak year-class were produced in 1990 and 1991, respectively, when plant abundance was low.

Management Implications

Based on results examined through 1994, the effect of reductions or changes in surface coverage of aquatic macrophytes on the largemouth bass population in Guntersville Reservoir has been positive. Size structure of the largemouth bass population, which receives two-thirds of the total fishing effort on this reservoir, was balanced, growth rates were high, and biomass was substantial. All of the above are important, particularly in relation to the current management objectives for this reservoir, which include a 381-mm minimum length limit for largemouth bass. Conversely, one point of concern is the stockpiling of fish below the legal length limit. This condition can be accelerated or compounded by depressed growth of largemouth bass younger than age 4 due to fish density and excessive aquatic plant structure. Therefore, frequent monitoring of the size structure of this population is recommended. When other environmental conditions are favorable (e.g., low flushing rate), high levels of aquatic macrophyte coverage (>90% in embayments) are not likely to enhance recruitment of largemouth bass to age 1 in this reservoir, despite the high density (>300/ha) of age-0 fish that commonly occur in heavily vegetated habitats. However, when abundance of age-0 fish is low in unvegetated habitats, the presence of aquatic macrophytes may augment recruitment of age-1 fish to the population. Water retention times longer than 16 d from April to July were associated with higher year-class abundance.

Although the presence of grass carp or the reduction of plant coverage have not affected the largemouth bass population negatively, these factors may have affected the quality of the sport fishery and the level of angler effort in this reservoir. Both of these variables are determined largely by the angling public. The interaction between quality and level of effort cannot be discounted. The 46% decline in angler effort for largemouth bass after 1990 was a concern, due to the impact on revenue to the local economy and possible impact on fishing license sales. We did not analyze or determine the multiple variables that can affect angler effort or the definition of quality. However, the following conditions could be considered. Relative to the high largemouth bass catch rate in 1990 (0.66/h), a decline in the catch rate during 1992–1993 may have represented a decline in quality to anglers. However, catch rate increased to 0.99/h in 1994. Catch rates appeared to be related to plant coverage, but recruitment of strong year-classes to the fishery undoubtedly affected catch. Catch rates from other established largemouth bass fisheries in various lakes or reservoirs seldom exceed 0.3 black bass/h (Farman et. al. 1982; Van Horn and Birchfield 1984; Chapman and Fish 1985; Dolman 1991). Because total angler effort measured at a given reservoir was ultimately determined by where individual anglers decided to fish, the key issue may be how this decision was reached. Was it based on firsthand experience, or was it influenced unduly by the media? As an example of the complexity of angler choice, some anglers at Guntersville Reservoir probably consider the decline in angler effort after 1990 as a benefit.

Frequently, the underlying issue in angler opposition to aquatic plant control or management is the amount of surface coverage of aquatic plants. Presumably there is an optimum level of coverage, but this level probably varies widely depending upon the water body. Also, any systematic determination of optimum plant coverage in mainstream impoundments of large rivers, such as the Tennessee River, usually is confounded by dynamic changes in water discharge and quality that can affect both submersed macrophytes and fish. Accordingly, an optimum level of aquatic macrophytes for the fish community as a whole or for the largemouth bass population alone in Guntersville Reservoir was not determined in the present evaluation. However, when surface coverage exceeded about 20% of the total area of Guntersville Reservoir, growth of largemouth bass declined and size structure was skewed towards smaller fish. In this case, the primary concern would be the continuation of the current 381-mm minimum length limit. The observations by Canfield and Hoyer (1992) from their survey of 60 Florida lakes may be applicable to Guntersville Reservoir. They suggested that when considering the total fish community, a moderate macrophyte coverage of 15% seems to reduce the probability of adverse fisheries problems.

Acknowledgments

We thank Anita and Jim Hausheer, Bob Warden, Rob James, Kevin Chalk, and other personnel from Fish and Wildlife Associates for their assistance in the creel survey and field collections. Steve Rider of Auburn University assisted in field and laboratory efforts to assess recruitment. Larry Neill, Al Brown, Gary Hickman, Gary Jenkins, Frank Taylor, Jack Tuberville, Virginia Brown, Gladys Hassell, and other TVA personnel provided assistance in various phases of the study, particularly equipment preparation and maintenance, electrofishing and rotenone sampling, and creel survey data processing. Steve Malvestuto and Earl Meredith provided creel survey design and analysis. Funding to M. J. Maceina was provided by TVA through the Alabama Cooperative Fish and Wildlife Research Unit (cooperative agreement 14-16-00009-1550, number 25). The Unit is sponsored by the National Biological Service, the Alabama Department of Conservation and Natural Resources, the Wildlife Management Institute, and Auburn University.

References

Aggus, L. R., D. C. Carver, L. L. Olmsted, L. L. Rider, and G. L. Summers. 1980. Evaluation of standing crops of fishes in Crooked Creek Bay, Barkley Lake, Kentucky. Proceedings of the Annual Conference Southeastern Association of Fish and Wildlife Agencies 33(1979):710–722.

Bain, M. B. 1993. Assessing impacts of introduced aquatic species: grass carp in large systems. Environmental Management 17:211–224.

Bettoli, P. W., M. J. Maceina, R. K. Betsill, and R. L. Noble. 1992. Piscivory in largemouth bass as a function of aquatic vegetation abundance. North American Journal of Fisheries Management 12:509–517.

Bettoli, P. W., M. J. Maceina, R. L. Noble, and R. K. Betsill. 1993. Response of a reservoir fish community to aquatic vegetation removal. North American Journal of Fisheries Management 13:110–124.

Boyd, C. E. 1967. Some aspects of aquatic plant ecology. Pages 114–139 *in* Reservoir fishery resources symposium. American Fisheries Society, Southern Division, Reservoir Committee, Bethesda, Maryland.

Buynak, G. L., W. N. McLemore, and B. Mitchell. 1991. Changes in largemouth bass populations at Kentucky and Barkley lakes: environmental or regulatory responses? North American Journal of Fisheries Management 11:285–292.

Canfield, D. E., and M. V. Hoyer. 1992. Aquatic macrophytes and their relation to the limnology of Florida lakes. Final report to Bureau of Aquatic Plant Management, Florida Department of Natural Resources, Tallahassee.

Chapman, P., and W. V. Fish. 1985. Largemouth bass tournament catch results in Florida. Proceedings of the Annual Conference Southeastern Association of Fish and Wildlife Agencies 37(1983):495–505.

Colle, D. E., and J. V. Shireman. 1980. Coefficients of condition for largemouth bass, bluegill, and redear sunfish in hydrilla-infested lakes. Transactions of the American Fisheries Society 109:521–531.

Copeland, J. R., and R. L. Noble. 1994. Movements by young-of-year and yearling largemouth bass and their implications for supplemental stocking. North American Journal of Fisheries Management 14:119–124.

Crow, D. G., and S. P. Malvestuto. 1996. Evaluation of a roving creel survey design for a large reservoir with an emphasis on subsampling within days. Pages 442–449 *in* L. E. Miranda and D. R. DeVries, editors. Multidimensional approaches to reservoir fisheries management. American Fisheries Society Symposium 16.

Dolman, W. B. 1991. Comparison of bass club reports and creel survey data from Texas reservoirs. North American Journal of Fisheries Management 11:177–184.

Farman, R. S., L. A. Nielsen, and M. D. Norman. 1982. Estimating largemouth bass abundance using creel census and tournament data in the fishing success method. North American Journal of Fisheries Management 2:249–256.

Gabelhouse, D. W., Jr. 1984. A length-categorization system to assess fishstocks. North American Journal of Fisheries Management 4:273–285.

Henderson, J. E. 1996. Management of nonnative aquatic vegetation in large impoundments: balancing preferences and economic values of angling and nonangling groups. Pages 373–381 *in* L. E. Miranda and D. R. DeVries, editors. Multidimensional approaches to reservoir fisheries management. American Fisheries Society Symposium 16.

Hirst, S. C., and D. R. DeVries. 1994. Assessing the potential for direct feeding interaction among larval black bass and larval shad in two southern reservoirs. Transactions of the American Fisheries Society 123: 173–181.

Hoyer, M. V., J. V. Shireman, and M. J. Maceina. 1985. Use of otoliths to determine age and growth of largemouth bass in Florida. Transactions of the American Fisheries Society 114:307–309.

Klussmann, W. G., and seven coauthors. 1988. Control of aquatic macrophytes by grass carp in Lake Conroe, Texas, and the effect on the reservoir ecosystem. Texas Agriculture Experiment Station MP-1664, Texas A&M University, College Station, Texas.

Maceina, M. J., and W. C. Reeves. 1996. Relations between submersed macrophyte abundance and largemouth bass tournament success on two Tennessee River impoundments. Journal of Aquatic Plant Management 34:33–38.

Maceina, M. J., S. J. Rider, and D. R. Lowery. 1993. Use of a catch-depletion method to estimate population density of age-0 largemouth bass in submersed vegetation. North American Journal of Fisheries Management 13:847–851.

Maceina, M. J., S. J. Rider, and S. T. Szedlmayer. 1995b. Density, temporal spawning patterns, and growth of age-0 and age-1 largemouth bass *Micropterus salmoides*, in vegetated and unvegetated areas of Lake Guntersville, Alabama. Pages 497–512 *in* D. C. Secor, S. E. Campana, and J. M. Dean, editors. Recent developments in fish otolith research. University of South Carolina Press, Columbia.

Maceina, M. J., S. J. Rider, S. T. Szedlmeyer, and D. R. DeVries. 1994. Assessment of factors affecting growth and abundance of age-0 largemouth bass and crappie in Guntersville Reservoir, Alabama. Final report of Auburn University to the Tennessee Valley Authority, Muscle Shoals, Alabama.

Maceina, M. J., and J. V. Shireman. 1980. Use of recording fathometer for the determination of distribution and biomass of hydrilla. Journal of Aquatic Plant Management 18:34–39.

Maceina, M. J., W. B. Wrenn, and D. R. Lowery. 1995a. Estimating harvestable largemouth bass abundance in a reservoir using an electrofishing catch-depletion technique. North American Journal of Fisheries Management 15:103–109.

Miranda, L. E., and W. D. Hubbard. 1994. Length-dependent winter survival and lipid composition of age-0 largemouth bass in Bay Springs Reservoir, Mississippi. Transactions of the American Fisheries Society 123:80–87.

Noble, R. L., P. W. Bettoli, and R. J. Betsill. 1986. Considerations for the use of grass carp in large, open systems. Pages 46–48 *in* G. Redfield, J. F. Taggart, and L. M. Moore, editors. Lake and reservoir management: volume II. North American Lake Management Society, Washington, DC.

Reeves, W. C., and J. J. McHugh. 1994. B.A.I.T. bass anglers information team 1993 annual report. Alabama Game and Fish Division, Montgomery.

Ricker, W. E. 1975. Computation and interpretation of biological statistics of fish populations. Fisheries Research Board of Canada Bulletin 191.

Schramm, H. L., Jr., and K. J. Jirka. 1989. Effects of aquatic macrophytes on benthic macro-invertebrates in two Florida lakes. Journal of Freshwater Ecology 5:1–12.

Secor, D. H., J. M. Dean, and E. H. Laban. 1991. Manual for otolith removal and preparation for microstructural examination. Electric Power Research Institute and Belle Baruch Institute for Marine Biology and Coastal Research, Columbia, South Carolina.

Summerfelt, R. C. 1986. Summarization of the symposium. Pages 314–327 *in* G. E. Hall and M. J. Van Den Avyle, editors. Reservoir fisheries management: strategies for the 80s. American Fisheries Society, Southern Division, Reservoir Committee, Bethesda, Maryland.

Timmons, T. J., W. L. Shelton, and W. D. Davies. 1979. Sampling reservoir fish populations in littoral areas with rotenone. Proceedings of the Annual Conference Southeastern Association of Fish and Wildlife Agencies 32(1978):474–485.

TVA (Tennessee Valley Authority). 1989. Supplement to white amur project report. Tennessee Valley Authority (TVA/WR/AB-89-1), Muscle Shoals, Alabama.

TVA (Tennessee Valley Authority). 1990. Final environmental assessment: demonstration of use of grass carp in management of aquatic plants in Guntersville Reservoir. Tennessee Valley Authority (TVA/RDG/EQS-90/3), Muscle Shoals, Alabama.

Van Horn, S. L., and L. J. Birchfield. 1984. The North Carolina catch survey. Proceedings of the Annual Conference Southeastern Association of Fish and Wildlife Agencies 35(1981):453–459.

von Bertalanffy, L. 1957. Quantitative laws of metabolism and growth. Quarterly Review of Biology 32: 217–231.

Watkins, C. E., II, J. V. Shireman, and W. T. Haller. 1983. The influence of aquatic vegetation upon zooplankton and benthic macroinvertebrates in Orange Lake, Florida. Journal of Aquatic Plant Management 21: 78–83.

Webb, D. H., L. N. Magnum, A. L. Bates, and H. D. Murphy. 1995. Aquatic vegetation in Guntersville Reservoir following grass carp stocking. Pages 199–209 *in* Proceedings of the grass carp symposium. U.S. Army Corps of Engineers, Waterways Experiment Station, Vicksburg, Mississippi.

Wrenn, W. B., M. J. Maceina, and D. R. Lowery. 1995. Grass carp in Guntersville Reservoir: a review of fisheries impact assessments. Pages 216–225 *in* Proceedings of the grass carp symposium. U.S. Army Corps of Engineers, Waterways Experiment Station, Vicksburg, Mississippi.

Data Collection and Analysis

American Fisheries Society Symposium 16:397–409, 1996

Reservoir Fisheries Sampling and Experimental Design

GENE R. WILDE[1]
Texas Parks and Wildlife Department, 4200 Smith School Road, Austin, Texas 78744, USA

WILLIAM L. FISHER
National Biological Service, Oklahoma Cooperative Fish and Wildlife Research Unit
Oklahoma State University, Department of Zoology, Stillwater, Oklahoma 74078, USA

Abstract.—Fisheries biologists frequently conduct studies to assess abundance and size structure of reservoir fish populations. These investigations include: (1) surveys and monitoring studies designed to inventory fisheries resources and track trends in various fisheries parameters, and (2) experimental and manipulative studies conducted to evaluate fish population responses to management programs or system manipulations. The distinction between these types of studies is often blurred, however, because survey and monitoring data frequently are used to assess experimental manipulations or management recommendations. Reservoir survey and monitoring programs often follow procedures that lack proper design and are infrequently evaluated and updated. Many studies use various forms of nonprobability sampling and subjectively chosen fixed stations, which limit statistical inference to those sites sampled and restricts potential uses of the data. With the increasing cost of data collection and the potential consequences of management strategies based on biased data, alternative sampling methods are needed. Various forms of probability sampling (especially stratified random sampling) provide a means for increasing the utility, reliability, and credibility of fisheries data. Probability sampling procedures also yield data that can be applied more easily to unanticipated questions and issues that may arise in the future. Coupled with geographic information systems, probability samples enhance the ability of reservoir fisheries managers to visualize and interpret spatial patterns in fish populations and to predict fishery responses to management actions or habitat manipulations.

Information on abundance and size structure of reservoir fish populations is obtained from general fisheries surveys or monitoring programs. On occasion, this information is supplemented with data from specially conducted research studies. Although data from such surveys and studies have formed the basis for many management recommendations, their usefulness as reliable indicators of spatial and temporal population trends throughout a reservoir frequently is limited by inadequacies in sampling design and precision. Samples collected from fixed sites, commonly used in fisheries surveys (King et al. 1981), generally can be used to assess trends in population parameters only at those sites actually sampled and cannot be extrapolated to a larger spatial coverage. Important changes in fish abundance associated with, for example, changes in habitat quality or quantity may not be detected; or it may be impossible to quantify such changes if one does not know whether the fixed sampling sites are representative of conditions everywhere in the lake. Failure to detect changes in fish abundance or other population characteristics also may result from an inadequate sample size, a common limitation of fisheries surveys and monitoring programs (Peterman 1990; Cyr et al. 1992).

The sampling design of a reservoir fishery survey or experiment also has important implications for future uses of the data. At present, information collected by fisheries management agencies is almost exclusively used for monitoring fish populations, and proposal and evaluation of protective regulations. In the future, the nature of questions posed by managers may change. For example, greater emphasis may be placed on modeling fish population responses to habitat or water quantity and quality changes to aid in refinement of management programs or to mitigate losses. Accordingly, careful attention should be given to sampling design so that collected data will meet not only immediate objectives, but will be useful in addressing unanticipated questions or issues that may arise in the future.

In this paper we evaluate current reservoir sampling procedures, and discuss advantages of probability-based sampling procedures. We also discuss the use of probability sampling in conjunction with geographic information systems (GIS) technology that promises to improve our ability to manage reservoir fisheries resources. Our focus is toward design of reservoir fisheries sampling surveys; however, much of the discussion is also applicable to

[1]Present address: Texas Tech University, Department of Range, Wildlife, and Fisheries Management, Lubbock, Texas 79409, USA.

experimental studies because they commonly involve sample collection protocols similar to those used in routine surveys. We suggest that fisheries surveys, including both monitoring and experimental studies, should be viewed as points along a continuum for three reasons. First, most reservoirs are the subjects of both surveys and manipulative studies. Fisheries information gathered from surveys may prompt manipulative studies whose outcome is assessed through subsequent monitoring programs or surveys. Second, survey data collected for management purposes often become experimental data used to evaluate management recommendations (e.g., length limits, bag limits) or other system manipulations. Third, and possibly most important, it is often necessary as a cost-saving measure to use all available data, regardless of its original purpose, when evaluating an experimental manipulation. Compatibility of survey and experimental sampling designs can ensure that data collected for one purpose are of the greatest possible use.

Survey Sampling Designs

Survey sampling designs frequently are categorized according to whether probability or nonprobability sampling procedures are used (Deming 1966; Cochran 1977; Levy and Lemeshow 1991). Estimates from probability sampling procedures can be applied to an entire population and estimates of parameter reliability (e.g., standard errors) can be made which allow sampling and selection biases to be eliminated or contained within known limits. In contrast, estimates made from nonprobability sampling methods, such as convenience or judgment sampling, are applicable only to those individuals or locations actually sampled, and the magnitude of sampling biases and errors cannot be determined. Cochran (1977) gives four characteristics of a probability sampling procedure: (1) it must be possible to list or identify all possible distinct samples, $S_1, S_2, S_3, \ldots, S_N$, that could be selected from a population; (2) each possible sample, S_i, is assigned a known probability of selection, π_i; (3) a sample, S_i, is selected by some random process with each S_i receiving its appropriate probability, π_i, of being selected; and (4) the method for computing estimates from the sample, for example the mean $\bar{x}$ or variance s^2, must lead to a unique estimate for any specific sample. Also, it is assumed that samples are independent (uncorrelated). Electrofishing samples are an example of how fisheries sampling methods tend to violate these criteria. Samples from adjacent sites are likely to be correlated; high catch rates at one site would lead one to expect high catches at adjacent sites. Thus samples from adjacent sites fail to provide independent estimates of fish abundance.

For procedures that meet Cochran's (1977) criteria, it is possible to develop a theory of sampling; the goal of such a theory is to produce the most efficient (lowest cost, greatest precision) sampling plan possible for the given purposes and resources. Sampling programs most commonly are designed to maximize precision, a measure of the deviations from the mean obtained by repeated sampling. High precision implies little deviation from the mean and a small sample variance, whereas low precision implies a larger sample variance.

The most basic probability sampling procedure used in fisheries is simple random sampling (Johnson and Nielsen 1983; Gunderson 1993), in which some number of sample sites, n, is selected from among the total number of possible sample sites N such that every potential site has an equal chance $(n \cdot N^{-1})$ of being selected (Figure 1a). Random selection of sample sites prevents the investigator from using judgment in site selection and reduces potential bias in parameter estimates. Simple random sampling is usually outperformed by other probability sampling procedures that offer greater convenience or increased precision in sample estimates.

In stratified sampling (Figure 1b), the total population of sample sites N is divided into subpopulations of $N_1, N_2, N_3, \ldots, N_L$ units. These subpopulations, or strata, are nonoverlapping and together sum to N. Once strata are defined, samples of size $n_1, n_2, n_3, \ldots, n_L$ are independently drawn from each stratum. If sampling within strata is at random, the procedure is described as stratified random sampling. The goal of stratification is to divide a heterogeneous population into a series of relatively homogeneous strata; that is, to maximize the variance among strata while minimizing within-stratum variance. Stratification can result in substantial precision gains when variation among strata is large and an effective stratification plan is used. A stratified random sampling design, blocked on environment, may reduce variance and increase precision without adding biases (see systematic sampling), at the cost of a few degrees of freedom.

Although systematic and cluster sampling (Figures 1c and 1d) at first appear to be quite different—the former involving collection of regularly spaced samples and the latter collection of irregularly spaced clusters of samples—these procedures are closely related. In both, the population is di-

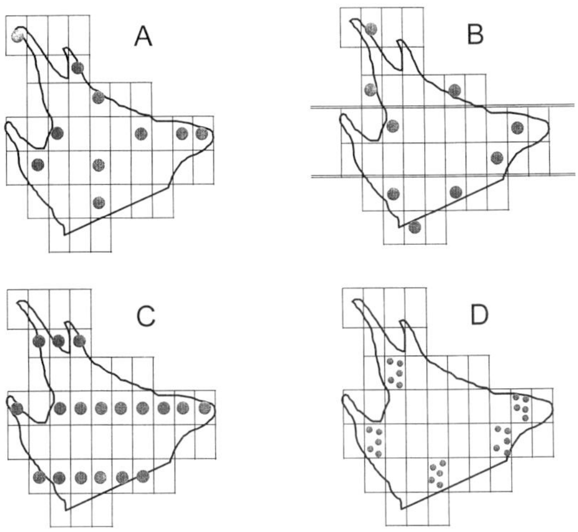

FIGURE 1.—Common forms of probability sampling used in reservoir fisheries studies. In each case, the lake is divided into potential sampling areas by a grid. **A**. Simple random sampling. All potential sites (cells) are identified and *n* sites are sampled at random (circles). **B**. Stratified random sampling. Sites are grouped into strata according to some criterion, such as into upper, middle, and lower reservoir locations, and samples are collected at random from within strata (circles). **C**. Systematic sampling. Primary sampling sites (transects) are identified and secondary sites (circles) along every nth transect are sampled. **D**. Cluster sampling. Primary sampling sites are identified (grid cells) and some number *k* of samples (circles) is collected from within randomly selected primary sampling sites.

vided into a number of primary sampling units (areas, transects, or clusters) and secondary sampling units (sample sites). When a primary unit is selected for sampling, all secondary units within that primary unit are sampled. Variance estimates for systematic and cluster sampling are based on differences among primary units; therefore, sample variance is minimized by making primary units as similar as possible or, equivalently, by maximizing the variance within primary units (Thompson 1992). In systematic sampling, a single primary unit consists of a number of secondary units spaced throughout the population in some regular manner. A starting area or transect is chosen at random and every n^{th} area or transect is sampled. Systematic sampling may outperform random sampling when the variable of interest has a clumped distribution or when an environmental gradient is present; however, this procedure may yield extremely biased estimates if the variable of interest is regular or periodic in occurrence.

In cluster sampling, a primary unit is selected at random and a number *k* of secondary units, usually located in close proximity to one another, are sampled. This procedure is appropriate in extremely heterogeneous habitats where samples from adjacent sites represent independent observations (e.g., results from one site provide no insight into results from other nearby sites). However, cluster sampling should be undertaken with caution. If samples from secondary sites (within clusters) are not independent, they represent replicates rather than samples and variance estimates will be biased.

In many cases probability sampling is impractical (e.g., in small areas such as reservoir tailraces or specialized habitats of rare organisms, or where no baseline information is available with which to design a probability sampling procedure) or is abandoned for purposes of convenience. Under the right conditions, nonprobability sampling procedures can yield useful information. Data from such procedures form the basis for many management actions (e.g., length and bag limits, stocking programs) that have resulted in the current health of our reservoir fisheries resources. In fisheries surveys, perhaps the most frequently used nonprobability sampling procedure is subjectively chosen fixed-site sampling. In this sampling procedure, permanent sample sites are chosen by judgment, either as being representative of the entire lake or reservoir or to maximize catches of targeted species. Data gathered from such fixed sites, or other nonprobability sampling procedures, generally are interpreted as "indices" rather than estimates of actual parameters and are used to detect changes over time in the variable of interest (Johnson and Nielsen 1983). The major disadvantage of fixed-site sampling, and other nonprobability sampling procedures, is that parameter estimates and statistical tests of significance are applicable only to those sites sampled unless whole-system or probability samples are also available (Cochran 1977). Therefore, it is not possible to determine whether sample results reflect local, site-specific, or more general phenomena. For example, electrofishing sample sites purposefully located in preferred largemouth bass *Micropterus salmoides* habitat may fail to detect changes in abundance because, as fish are lost from preferred areas, individuals immigrate from other less desirable areas. Referring to such data as indices rather than estimates neither gains them credibility nor insures that decisions based on these data are even qualitatively correct.

Another justification for use of subjectively chosen fixed sites is that, by avoiding nonpreferred or little-used habitats, sample variation is reduced, allowing greater ability to detect changes in fish populations. However, such habitats, or those in which sampling gear cannot be deployed or efficiency is reduced, can be quantified and treated as unsampled strata in data summary and analysis. Probability sampling procedures can then be applied to remaining habitats.

Designs based upon probability sampling procedures apparently are used infrequently in reservoir fishery surveys. King et al. (1981) surveyed state management agencies and found that of 46 responding agencies, 21 used fixed stations exclusively, 5 used random sites exclusively, and 20 used a combination of the two methods. These results may overrepresent the frequency with which probability sampling procedures are used in fisheries work because many state agencies use a randomized (probability) sampling design for creel surveys, but use fixed-site sampling for fishery-independent (monitoring) surveys.

Although nonprobability sampling procedures, including fixed-site sampling, are widely used in fisheries studies, there has been little critical examination of the assumptions involved in their use. The major assumption behind the use of nonprobability sampling is that trends in abundance and size structure at fixed sites are representative of lakewide trends. Hubbard and Miranda (1988) compared electrofishing catch rates, size-structure indices, and relative weights of largemouth bass collected from both randomly and subjectively chosen sampling sites. Catch rates and size structure indices generally were overestimated at subjectively chosen sites compared with randomly chosen sites (Figure 2), but there was no difference in relative weight between samples collected from randomly and subjectively chosen sites. To assess the potential bias of abundance and size structure information collected from subjectively chosen sites, we used linear regression (with the intercept suppressed) to model data from Hubbard and Miranda (1988). Catch per effort (CPUE) at randomly chosen sites was 77% of CPUE at subjectively chosen sites. Proportional stock density (PSD) and relative stock density (RSD-14) at randomly chosen sites were 87% and 78%, respectively, of those measures at subjectively chosen sites, indicating that bias associated with subjectively chosen sites increases with fish size.

If the sampling bias associated with use of subjectively chosen sites was consistent among reservoirs, our regression results could be used to adjust for this bias; however, this is apparently not the

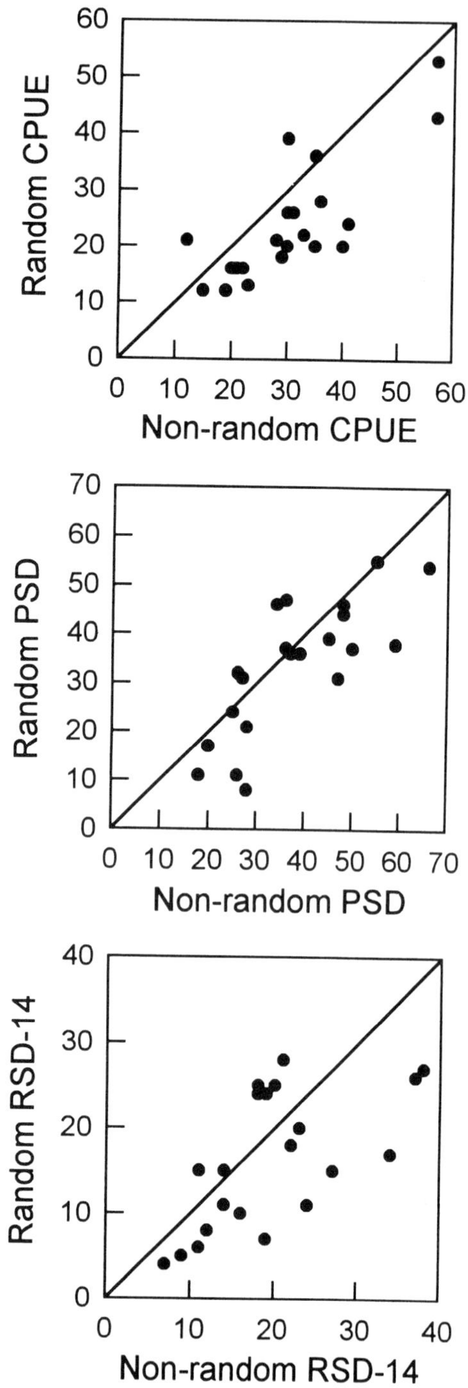

FIGURE 2.—Relationships between CPUE (catch per unit effort), PSD (proportional stock density), and RSD-14 (relative stock density) for largemouth bass collected from randomly and nonrandomly selected stations. Diagonal lines indicate a 1:1 relationship between sample types. Data are from Hubbard and Miranda (1988).

case. Hubbard and Miranda (1988) observed that differences between randomly and subjectively chosen sample sites were related to the quantity of "prime largemouth bass habitat" in their study reservoirs. In impoundments with smaller proportions of prime habitat, there were greater deviations between CPUE, PSD, and RSD-14 estimates from randomly and subjectively chosen sites, which suggests sampling biases depend on local conditions. Osburn (1988) compared bag seine catches of five finfish species from sites selected by three different procedures: random sites chosen from among all possible shoreline locations; fixed sites chosen from a list of sites accessible by vehicle; and sites chosen for convenience. Sampling procedure had no effect on catches of two species; however, for the remaining three species there were statistically significant main effects of sampling procedure and significant interaction effects for sampling procedure by bay and sampling procedure by month. These interactions indicate that the magnitude of differences varies with location and timing of sampling. Abundance and size structure estimates for reservoir fish populations are known to vary considerably within and among seasons (Carline et al. 1984; Van Horn et al. 1993). Thus, interpretation of these fisheries statistics may be further complicated if sampling bias is related to sampling location and season.

A second argument used to support fixed-site sampling is that sampling variability is reduced, relative to randomized sampling, and therefore more precise estimates of fish population characteristics are possible. Gutreuter (1993) tested this assumption by comparing within- and among-site variance components for several freshwater fish species using data collected with electrofishing, fyke nets, hoop nets, and seines in the Long Term Resource Monitoring Program of the upper Mississippi River. If the use of fixed sites reduces sampling variance, Gutreuter reasoned the within-site component of variance should be less than the among-site component. He found no evidence that within- and among-site variance differed for any of the gears and concluded that use of fixed sites did not increase sampling precision. Gutreuter's results cannot be viewed as definitive, however, because his study was not specifically designed to estimate within- and among-site variance components and, consequently, suffers from low statistical power (i.e., a limited ability to reject the null hypothesis that within- and among-site variance components are equal).

In view of the statistical advantages of probability sampling and the convenience of fixed-site sampling, King et al. (1981) and Brown (1991) explored the compromise procedure of random selection of fixed sampling sites. Means and variances of electrofishing catches showed little difference between random sampling and randomly chosen fixed sites in West Point Reservoir, suggesting fixed stations can be used to monitor the whole reservoir (King et al. 1981). In contrast, Brown (1991) observed significantly greater catch rates for certain sizes of largemouth bass at randomly chosen sites. These conflicting results emphasize an important caveat to random sampling procedures. Random sampling will, on average, provide unbiased estimates of population parameters; however, there is no guarantee that any particular set of samples, or sample sites, will be unbiased. To insure data from randomly chosen fixed stations accurately represent lakewide trends, comparisons with randomly chosen sites must be made for several sampling periods and over the entire range of conditions (seasons, water level regimes, vegetation coverages) in which sampling will be conducted.

Estimating Sample Size Requirements

There is perhaps no more neglected aspect of fisheries survey and experimental design than determination of the number of samples required to meet study objectives. Too few samples may result in an investigator being unable to meet survey goals or to decisively reject a hypothesis. The importance of this aspect of survey and study design is underscored by the low precision characteristic of many reservoir sampling gears (Cyr et al. 1992; Hardin and Connor 1992; Wilde 1995). Sample size requirements can be estimated based on several criteria (Cochran 1977). For fisheries surveys Gunderson (1993) recommended designing to achieve some desired level of precision, and Peterman and Bradford (1987) and Peterman (1990) advocated basing sample size on statistical power.

Precision of a sample can be measured as the coefficient of variation of the mean ($CV_{\bar{x}} = SE / \bar{x}$, where SE and $\bar{x}$ are the sample standard error and mean, respectively). The number of samples N required to achieve a desired degree of precision can be determined given an estimate of $\bar{x}$ and s^2 as:

$$N = CV_{\bar{x}}^{-2} \cdot \bar{x}^{-2} \cdot s^2. \tag{1}$$

Using this formula, or variants of it, Baganal (1972), Craig and Fletcher (1982), and Craig et al. (1986) found 6–40 gill net sets were required to estimate catches to within ±50% of the geometric mean for several species. Hardin and Connor (1992) reported

4–110 (median = 19) and 16–329 (median = 68) samples were required to estimate electrofishing catches of largemouth bass and bluegill *Lepomis macrochirus*, respectively, to within ±10% of the mean. Sample size estimates may vary seasonally (Craig and Fletcher 1982) and Hardin and Connor (1992) observed that median sample size requirements for various size classes of largemouth bass increased directly with length.

Fish population data characteristically have a sample variance greater than the mean, $s^2 > \bar{x}$. This implies a relationship between variance and mean, such that the variance increases with the mean, that can be modeled using either nonlinear regression (s^2 and $\bar{x}$ untransformed) or linear regression of $\log s^2$ on $\log \bar{x}$ allowing development of an equation predicting s^2 as a function of $\bar{x}$. Substitution of this predictive relationship into equation 1 allows estimation of sample size requirements requiring only an estimate of the expected mean catch (Cyr et al. 1992):

$$N = \mathrm{CV}_{\bar{x}}^{-2} \cdot x^{-2} \cdot ax^{b} \text{ or,} \quad (2)$$

$$N = \mathrm{CV}_{\bar{x}}^{-2} \cdot a \cdot x^{b-2}, \quad (3)$$

where a and b are the intercept and slope, respectively, of the variance–mean regression (note, if a and b are obtained from regression of $\log s^2$ on $\log \bar{x}$, they must first be back-transformed to a linear scale). Sample size N is inversely related to $\mathrm{CV}_{\bar{x}}$ and $\bar{x}$ as b is generally about 0.5. Holding $\bar{x}$ constant, N increases by a factor of 4× as the desired precision is doubled (as $\mathrm{CV}_{\bar{x}}$ decreases from 0.50 to 0.25, for example). For any given level of precision, N increases as $\bar{x}$ decreases indicating fewer samples are required when catch rates are high than when catches are low. This relationship may explain the seasonal differences in gill net requirements observed by Craig and Fletcher (1982) and is consistent with the length–sample size relationship observed for largemouth bass by Hardin and Connor (1992). Equations for estimating sample size requirements for larval fish trawl samples and for gill net samples are given by Cyr et al. (1992) and Wilde (1995), respectively.

Calculating sample size estimates based on desired precision is relatively simple; however, this method has two important limitations. First, most common methods for estimating sample size requirements tend to underestimate the necessary sample size (Kupper and Hafner 1989). Second, because the population mean and variance are unknown, even though sample estimates are available, there is an uncertainty in sample size estimates that is not reflected by equations 1–3. Methods for estimating sample sizes based on criteria other than precision (estimating the mean to within some value or percentage, among others) explicitly accommodate this uncertainty. The results of Cyr et al. (1992) and Wilde (1995), for example, can be used to estimate average, or statewide, sample size requirements; however, as noted by Wilde (1995), sample size requirements for individual lakes may be underestimated.

There is no universally accepted level of precision for which to design a study. However, Robson and Regier (1964) recommended three alternative goals for precision based on the nature of the study: for preliminary surveys they recommended a confidence interval about the mean of ±50% (equivalent to $\mathrm{CV}_{\bar{x}} = 0.25$); for management studies they recommended estimating the mean to within ±25% ($\mathrm{CV}_{\bar{x}} = 0.125$); and for research studies they recommended a confidence interval about the mean of ±10% ($\mathrm{CV}_{\bar{x}} = 0.05$). Standard fisheries sampling protocols (e.g., Erikson 1978; TPWD 1993) generally call for modest sample requirements (≤15) based more upon logistic and allocation considerations than statistical design; as a result, these protocols often fall short of meeting Robson and Regier's recommendations. For Texas statewide mean catch rates, Wilde (1995) estimated 12–38 gill nets would be required to achieve a precision of $\mathrm{CV}_{\bar{x}} = 0.25$ depending on the species being sampled, 45–150 gill nets for a precision of $\mathrm{CV}_{\bar{x}} = 0.125$, and 279–937 gill nets for a precision of $\mathrm{CV}_{\bar{x}} = 0.05$.

The ability of a statistical test to reject the null hypothesis when, in fact, it is false is termed the power of the test. The power of a statistical test or procedure is measured as $1-\beta$, where β is the probability of making a type II error (i.e., accepting a false null hypothesis), and is a function of sample size, the probability of making a type I error, α (i.e., the probability of rejecting a true null hypothesis), and the magnitude of the difference between the hypothesized and observed values of the variable under investigation (Cohen 1977; Kraemer and Thiemann 1987). This latter difference is termed the effect size. Sample size estimates based on power increase as the desired level of power is increased, as either α or effect size are reduced, or as the sample variance is increased. The advantage of estimating N based on power is that one can have an idea of how likely a certain outcome is. For example, Peterman and Routledge (1983) used power analysis to show a proposed experimental release of 48 million smolts over a 3-year period would have relatively little chance of showing the presence of a smolt–adult relationship whereas release of 88 million smolts would have a high probability (>0.80) of success. Cyr et al. (1992) estimated sample sizes required to detect a difference in larval fish abundance between years using a *t*-test. At low larval fish densities, only a very large difference could be detected with four samples, the median sample number used in studies they reviewed; even at higher densities, only an eightfold difference could be detected with four samples. Sixty-four (if $\bar{x} = 1$) and 11 (if $\bar{x} = 1{,}000$) samples per year would be required to detect a 50% change in mean abundance with a power of 0.95 and $\alpha = 0.05$. Power is used less frequently than precision as a basis for estimating sample size requirements, possibly because of the greater computational demand. Power calculations require specification of a particular model or hypothesis; for example, testing whether abundance in year 1 is greater than in year 2, or abundance increases linearly between years 1 and 3. Because a change in the hypothesis to be tested requires recalculation of sample sizes (Gerrodette 1987; Kraemer and Thiemann 1987), this approach may be better suited for designed experiments than monitoring surveys.

Stratification and Allocation of Samples

Depending on the degree and nature of spatial heterogeneity in a reservoir, sampling procedures such as stratified, systematic, and cluster sampling can reduce sample variability compared with simple random sampling. Systematic sampling designs are commonly used for surveys of pelagic fisheries, especially in marine systems (Gunderson 1993). The distribution and abundance of many freshwater sport and prey fishes generally are related to habitat variables such as depth, substrate type, and presence of aquatic plants or other structure, which can vary spatially in reservoirs in both regular, upstream–downstream, and patchy patterns (Summerfelt 1971; Siler et al. 1986; Bunyak et al. 1989; Smith and Orth 1991). Therefore, stratified sampling procedures, using habitat types or classes as strata, are likely to yield greater precision than are systematic or cluster sampling procedures in reservoir fisheries surveys.

Stratifying sampling effort has several purposes, including: (1) obtaining precise estimates of parameters within all sampled areas or habitat types; (2) facilitating sampling by assigning certain strata (areas, lakes, etc.) to certain sampling crews; (3) allowing customization of sampling intensity or design for specific classes of strata (e.g., systematic sampling might be used in riverine reservoir habitats and random sampling in open water reservoir habitats; and (4) obtaining more precise estimates of population parameters by combining subpopulation estimates from each stratum. The low-precision characteristic of reservoir fisheries sampling programs increases the importance of methodological or other aspects of sampling design that can reduce sampling variation. In reservoir fisheries sampling, stratification generally is based on habitat characteristics thought to be important to the fish populations of interest. One commonly used stratification variable in reservoir sampling is location along the reservoir axis (i.e., up- or downlake). Texas Parks and Wildlife Department (TPWD 1993) stratifies reservoirs into three zones: headwater, midsection, and mainpool areas. A similar plan was used by Siler et al. (1986) with an additional stratification based on proximity to power plant effluents. Other habitat variables such as presence of aquatic vegetation, submerged trees or other cover, substrate type, and littoral versus limnetic zone, can be used. However, to provide any meaningful increase in the precision of whole-population estimates, the stratification variable must be a good surrogate measure of the variable of interest. For example, in designing a population survey of striped bass *Morone saxatilis* in a reservoir, the best possible stratification variable is the true abundance of striped bass; however, this information would not be available prior to the survey. Therefore, stratification might be based upon habitat variables that can reliably predict relative abundance of striped bass, such as depth, temperature, and oxygen, certain combinations of which are known to be either preferred or avoided by striped bass (Coutant 1985; Matthews 1985; Zale et al. 1990).

Sampling effort can be allocated among strata in three ways. With equal allocation, samples are allocated equally among strata, regardless of stratum size. For example, if 15 samples are to be collected from a reservoir stratified into headwater, midsection, and mainpool areas, equal allocation would result in five samples being collected from each stratum regardless of its size. With proportional allocation, samples are collected from each stratum in proportion to the size or area of that stratum. If a stratum occupies 25% of the total sampling area, then 25% of samples are collected from that stratum. With optimal allocation, samples are collected from strata according to a number of cost functions which attempt to optimize sampling to meet some criterion. Depending on the criterion to be optimized, larger samples may be taken from larger strata, strata that are more internally variable, or from strata that are less expensive to sample (and which, therefore, allow a greater number of samples to be collected).

Variances in stratified random sampling are calculated as weighted sums of the variances for individual strata. For example, the variance of the sample mean ($\bar{x}_{st}$), or standard error, is given by:

$$V(\bar{x}_{st}) = \sum W_h^2 \cdot S_h^2 \cdot n_h^{-1}, \tag{4}$$

where W_h is the weight for stratum h, S_h^2 is the variance for stratum h, and n_h is the sample size for stratum h. In the simplest case, equal allocation, $W_h = 1/h$ for all strata, where h is the number of strata sampled. Cochran (1977), Levy and Lemeshow (1991), and Thompson (1992) give methods for determining stratum weights under proportional and optimum allocation. Variances of estimated population parameters, based on the three methods of allocation, can be ordered: $V_{\text{optimum}} \leq V_{\text{proportional}} \leq V_{\text{equal}}$ (Cochran 1977). However, any of these methods of sample allocation will, on average, yield a smaller variance than simple random sampling.

To illustrate the potential gains in precision that can be achieved using stratification, we simulated a series of fish catches (abundance estimates) based on 6, 9, 12, 15, and 30 samples (Table 1). For each simulation we calculated the population standard error (assuming no stratification) and standard errors for samples assigned to three strata containing equal numbers of samples based on both abundance (low, medium, and high catches) and on random assignment. Stratification based on abundance resulted in at least a 100% increase in precision (i.e., sample standard errors were reduced by 50% or more) whereas the random stratification resulted in no increase in precision. These results also show the importance of choosing an effective surrogate of the target variable, in this case abundance, when constructing strata. If the stratification variable is poorly correlated with the target variable, stratification will yield little, if any, increase in precision.

Sampling precision gains from stratification generally increase with the number of strata used. However, increases in precision often are small when more than three or four strata are used (Table 2). Although it is tempting to create strata that represent all unique habitat types, excessive stratification should be avoided unless there is evidence that a substantial increase in precision can be gained. Regardless of the number of strata used, at least two samples must be collected from each stratum in order to calculate within-stratum variances (Cochran 1977; Green 1979). Additionally, estimates of the sample variance generally decrease as sample size increases (Cochran 1977); if a fixed number of samples is to be taken, within-stratum variance estimates will increase as strata number increases. Therefore, proliferation of strata can increase sample requirements while lowering sampling precision.

TABLE 1.—Gains in sampling precision resulting from the use of stratified sampling. A series of fish catches, based on 6, 9, 12, 15, and 30 samples, were simulated. For each series the sample (unstratified) standard error was calculated, as were standard errors based on stratification by catch abundance (low, medium, high) and by random assignment. Sample sizes were equal in all strata.

Number of samples	Unstratified standard error	Standard error for samples stratified on: Abundance	Random assignment
6	2.57	1.28	2.57
9	2.62	1.54	2.80
12	2.35	1.12	2.24
15	1.72	0.85	1.84
30	1.20	0.49	1.24

Geographic Information Systems in Fisheries Management

Reservoirs are heterogeneous environments within which fish, habitat, water quality, and anglers exhibit considerable spatial variability (Siler et al. 1986). Understanding the nature of this heterogeneity is inherently a geographic problem. Geographic Information Systems (GIS) are a rapidly evolving technology that offer enormous potential to fisheries biologists in the spatial decision-making

TABLE 2.—Gains in sampling precision from increasing the number of strata in stratified sampling. A series of fish catches of sample size 30 was simulated. The sample standard error was calculated for all 30 catches and for samples allocated to various numbers of strata based on catch abundance. The percent reduction in the standard error, compared with the whole sample standard error, measures the increase in sampling precision.

Number of strata	Standard error	Reduction in standard error (%) compared to no stratification (strata number = 0)
0	1.42	
2	0.41	70
3	0.29	79
4	0.27	81
5	0.27	81
6	0.22	85
10	0.21	85

process. Geographic Information Systems comprise an integrated set of computer hardware and software used to input, store, manipulate, analyze, and display spatial and associated nonspatial information. The primary use of GIS is for analysis of data for patterns in, and relationships among, geographic variables (Bailey 1994). Recently, emphasis has been placed on linking statistical procedures and mathematical models to GIS (Rogerson and Fotheringham 1994). General and detailed accounts of GIS and applications can be found in Star and Estes (1990), Berry (1993), Fotheringham and Rogerson (1994), and Ripple (1994).

Although GIS applications in natural resource disciplines have become widespread (Ripple 1994), use of this technology in fisheries research and management has been explored only recently (Meaden and Kapetsky 1991; Giles and Nielsen 1992). One of the more common applications is site selection for aquaculture and fish production facilities (Meaden 1987; Kapetsky et al. 1988; Ross et al. 1993). Other reported fisheries applications include mapping riverine fish distributions and habitat (Nielsen 1989; Parsley 1992), assessing reservoir fish habitats (Cross 1991; O'Bara and Hubbs, in press), managing and analyzing aquatic habitats in the Great Lakes (Horvath 1992), and mapping distributions of rare aquatic species for conservation measures (Angermeier and Bailey 1991).

The efficacy of sampling designs and their precision can be enhanced through stratification of the reservoir by physical or chemical attributes, and GIS can facilitate this process. Data layers, or themes, generated from habitat structure, depth, temperature, salinity, and other data collected in the reservoir can be incorporated into GIS. One or a combination of these thematic maps can be layered on the gridded base map to develop spatial strata for the reservoir. Rogala (1992) used GIS to implement a stratified random design in the upper Mississippi River for the Long Term Resource Monitoring Program. The area was divided into channel border and contiguous backwater areas, together these included six classes of aquatic habitat. A rectangular grid of 50 × 50-m cells was placed over the study area, and the center of each cell was labeled automatically with the GIS software (ARCINFO). One hundred cells, 50 within each area, were selected at random for sampling. Rogala noted several advantages of using GIS for developing sampling designs, including: the efficiency of GIS over traditional approaches; the ease with which multiple designs can be generated to sample all aquatic organisms of interest; the ability to omit grid cells that do not meet specific gear requirements (e.g., electrofishing run length); and the capability of incorporating additional data such as bathymetry, vegetation, and water flow to increase the realism of the strata. A noted disadvantage of GIS was the potential to overstratify an area which potentially could reduce statistical power within strata.

Wilkerson (1995) used GIS to develop a stratified random sampling design to sample available habitat and monitor locations of telemetered striped bass in an Arkansas River reservoir in Oklahoma. The reservoir was stratified into four areas based on depth, velocity, and water clarity, and a grid with 2,000 × 2,000-m cells was overlain on the map (Figure 3). Samples were randomly selected from within each stratum during the 18-month study at frequencies of 1–26 samples per cell. These data were subsequently used to develop seasonal maps of water temperature, current velocity, and dissolved oxygen to identify suitable habitat for striped bass. Low precision (CV = 500–1200%) of the interpolated maps was believed by Wilkerson to result from the irregular distribution of sample sites used; he recommended the use of a systematic sampling design to improve precision.

Data stored in GIS are conceptually a mixture of signal (truth) and noise (error). The error comprises two components, measurement error associated with any data collection, and locational error, or uncertainty surrounding the exact geographic coordinates of a datum. Either component can affect the accuracy of maps produced with GIS. Investigators should report these errors and any propagation that results from modeling and spatial analysis (Berry 1994). One method for doing this is to produce maps showing standard errors or variances, in addition to maps showing mean responses, for variables modeled or subjected to analysis.

Experimental Design and Fisheries Management

Fisheries managers conduct a variety of experimental studies to provide information on specific issues such as hooking mortality, gear effectiveness, and effects of artificial structures. Because of the specialized nature of these studies, sampling is often conducted according to study-specific protocols. Fishery responses to habitat or other system manipulations commonly are evaluated with sampling gears and designs similar to those used in routine surveys. Additionally, many actions undertaken by reservoir fisheries managers, including imposition

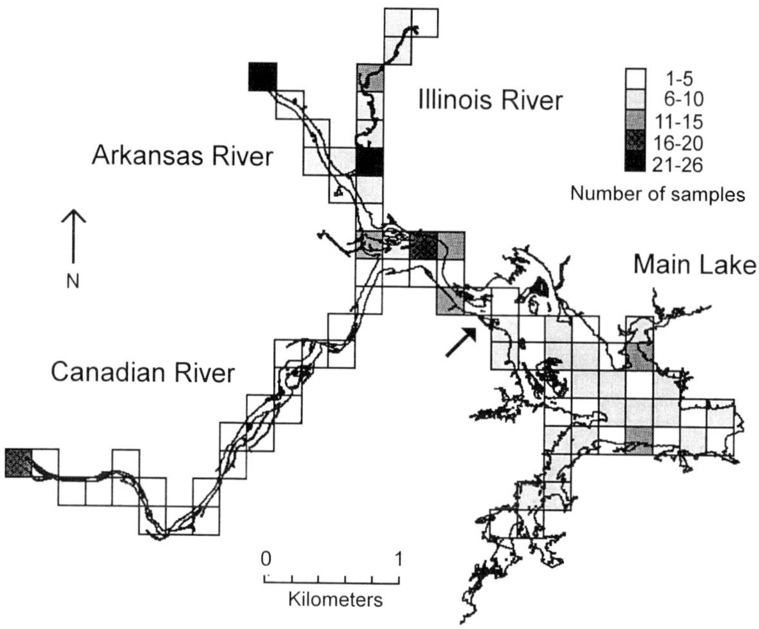

FIGURE 3.—Map of Kerr Reservoir, Oklahoma, generated using geographic information systems technology. The frequency and distribution of samples collected to determine available striped bass habitat are illustrated. The arrow in the middle of the reservoir indicates the break between the upper river and main lake strata.

of length limits and stocking programs, can be considered as experimental manipulations (Walters et al. 1988; Collie et al. 1990; McAllister and Peterman 1992). It is often desirable, if not necessary, to evaluate the efficacy of these management actions, and such evaluations rely heavily on survey data.

McAllister and Peterman (1992) recently reviewed the status of experimentation in fisheries management and observed the same general shortcomings that Romesburg (1981) and Hurlbert (1984) previously observed in experimental studies in wildlife and ecology, namely absence of controls, lack of adequate replication, and failure to randomize. The latter shortcoming is in many cases a sampling issue, which is our interest here. Much of the discussion above relates directly to the sampling design of fisheries experiments. As is the case with general and monitoring surveys, experimental studies commonly are conducted using data collected from subjectively chosen sampling stations. Consequently, study results and statistical tests are limited to the sites actually sampled and cannot be generalized to the entire reservoir. Also, because it is usually unknown whether data collected from nonrandomly chosen sites are representative of lakewide phenomena, the magnitude of fisheries responses to management manipulations can only be assessed qualitatively.

The inherent variability of fish populations, low precision of many sampling gears, and high cost of collecting samples limits detection of population responses to manipulations with two consequences. First, survey and other available data are commonly used to supplement experimental data. The utility of these data is largely determined by whether they represent probability samples. Unless supplemental data are collected from the same sites studied in an experimental setting, at best they allow assessment of treatment effects at an independent set of sites. At worst, they provide little more than background information because the relationship between the two sets of data is unknown. Are survey data more or less variable? Will they exhibit a greater or smaller treatment effect? When both survey and experimental samples are collected by probability sampling procedures, with appropriate weighting, survey data can be directly incorporated into statistical analyses and increase the quantity of data available for assessing interannual or long-term trends and for modeling fishery responses to habitat and system manipulations. Second, the limited ability of many fishery experiments to detect responses requires the use of analytical and design features that increase the sensitivity of these experiments. Stratified sampling, within an experimental setting, can provide an increase in the precision with which treatment effects are measured and in the power of statistical tests of hypotheses. By designing experiments so each treatment (including controls and experimental treatments) occurs within each stratum, the inherent variation among strata can be estimated and modeled, thereby reducing the size of the residual error term used for testing the significance of treatment effects.

Conclusions and Recommendations

We believe a combination of factors will result in changes in the sampling design of fisheries surveys in the future. First, there is an increasing number of regional and national natural resources inventory programs, such as the U.S. Environmental Protection Agency's Environmental Monitoring and Assessment Program, the U.S. Geological Survey's National Water Quality Assessment Program, and various initiatives of the U.S. National Biological Service, including gap analysis and global climate change programs, which have limited budgets for collection of original data and depend on existing databases. Cost-sharing or other agreements among entities that collect and those that use data may require sampling procedure modifications to satisfy the needs of all parties. Second, reviewers of manuscripts and proposals increasingly are questioning sampling designs. Federal aid requirements for demonstrating sampling adequacy might best be satisfied by adopting probability sampling procedures. Third, changes in the nature of the questions asked by fisheries managers will require changes in sampling designs. Most reservoir fisheries are now adequately regulated so overexploitation poses less of a problem than does change or loss of aquatic habitat. With the age of many reservoirs approaching 50 years, fisheries issues resulting from degradation of aquatic habitat and water quality through sedimentation and eutrophication probably will dominate reservoir management in the next century. Finally, the need for credible data to assist in natural resources management, and the increasing consequences of decision-making based on biased data (e.g., Engelmeyer 1993) can best be mitigated by probability sampling designs.

Except possibly in localized habitats or areas of special interest, such as reservoir tailraces and known ranges of endangered species, we see no advantages to nonprobability sampling procedures and recommend the adoption of probability sampling procedures. Fish catches during routine random sampling may decrease slightly; however, this can be offset by increased precision that can be achieved by stratification. For example, the Texas Parks and Wildlife Department currently uses gillnets to sample striped bass with a precision $CV_{\bar{x}} = 0.31$. If adoption of random sampling resulted in a 50% decrease in catches, precision would decrease ($CV_{\bar{x}} = 0.37$); however, by the use of as few as three strata, overall precision could be increased by 32% ($CV_{\bar{x}} = 0.21$).

We also recommend sample size requirements be estimated before undertaking any survey or experimental study. In many cases, it may not be possible to collect a sufficient number of samples to achieve study goals because of a lack of available resources. Whether to proceed with the study then becomes an administrative decision; however, funding organizations, administrators, and fisheries biologists should be fully aware of study limitations before the study is conducted. Additionally, this exercise should stimulate interest in more efficient sampling designs. Specialized designs such as those discussed by Thompson (1992) and Gunderson (1993) may provide fisheries managers with a greater ability to detect changes in fish populations than do the basic designs presented here.

Finally, we see the growing use of GIS as an important impetus for adopting probability sampling. By overlaying distribution and abundance information for sport and prey fishes on data layers containing information on habitat type, physicochemical conditions, and other factors, GIS can be used to identify and delimit sampling strata, and to facilitate examination and modeling of intrareservoir variation in fish community composition and predator–prey relationships. Managers will be able to better identify locations of spawning or other life history events of interest using GIS. Data layers can be created from data or descriptions in the literature on habitats used during key life history stages. By overlaying survey data on these layers, managers can quickly test and refine predictions of habitat use. Similarly, fisheries responses to habitat and other environmental changes can be explored and measured by overlaying pre- and postmanipulation fishery data on the corresponding data layers representing the environmental variable of interest. However, these benefits will accrue only as information from an increasing number of sample locations becomes available. The combination of probability sampling, spatial visualization, and analytical tools such as GIS have the potential to enhance and transform reservoir fisheries management in the future.

References

Angermeier, P. L., and A. Bailey. 1991. Use of a geographic information system as a conservation tool for rivers in Virginia, U.S.A. Pages 151–160 *in* P. Boon, J. Calow, and G. E. Peets, editors. River conservation and management. Wiley, New York.

Baganal, T. B. 1972. The variability of the catch from gill nets set for pike *Esox lucius*. Freshwater Biology 2:77–82.

Bailey, T. C. 1994. A review of statistical spatial analysis in geographic information systems. Pages 13–44 *in* A. Fotheringham and P. Rogerson, editors. Spatial analysis and GIS. Taylor and Francis, Inc., Bristol, Pennsylvania.

Berry, J. K. 1993. Beyond mapping: concepts, algorithms and issues in GIS. GIS World Books, Fort Collins, Colorado.

Berry, J. K. 1994. Beyond mapping: avoid dis-information. GIS World 7:26.

Brown, M. E. 1991. Sampling implications of seasonal and spatial variation in largemouth bass length frequencies and catch rates in mainstream impoundments. Master's thesis. Auburn University, Auburn, Alabama.

Bunyak, G. L., L. E. Kornman, A. Surmont, and B. Mitchell. 1989. Longitudinal differences in electrofishing catch rates and angler catches of black bass in Cave Run Lake, Kentucky. North American Journal of Fisheries Management 9:226–230.

Carline, R. F., B. L. Johnson, and T. J. Hall. 1984. Estimation and interpretation of proportional stock density for fish populations in Ohio impoundments. North American Journal of Fisheries Management 4:139–154.

Cochran, W. G. 1977. Sampling techniques, 3rd edition. Wiley, New York.

Cohen, J. 1977. Statistical power analysis for the behavioral sciences. Academic Press, New York.

Collie, J. S., R. M. Peterman, and C. J. Walters. 1990. Experimental harvest policies for a mixed-stock fishery: Fraser River sockeye salmon, *Oncorhynchus nerka*. Canadian Journal of Fisheries and Aquatic Sciences 47:145–155.

Coutant, C. C. 1985. Striped bass, temperature, and dissolved oxygen: a speculative hypothesis for environmental risk. Transactions of the American Fisheries Society 114:31–61.

Craig, J. F., and J. M. Fletcher. 1982. The variability in the catches of charr, *Salvelinus alpinus* L., and perch, *Perca fluviatilis* L., from multi-mesh gill nets. Journal of Fish Biology 20:517–526.

Craig, J. F., A. Sharma, and K. Smiley. 1986. The variability in the catches from multi-mesh gillnets fished in three Canadian lakes. Journal of Fish Biology 28: 671–678.

Cross, D. 1991. Reservoirs, GIS, and bass habitat. U.S. Forest Service General Technical Report RM-207:121–125.

Cyr, H., J. A. Downing, S. LaLonde, S. B. Baines, and M. L. Pace. 1992. Sampling larval fish populations: choice of sample number and size. Transactions of the American Fisheries Society 121:356–368.

Deming, E. W. 1966. Some theory of sampling. Dover Publications, Inc., New York.

Engelmeyer, P. 1993. Coastal coho survey flawed. Riverkeeper 9(1):1,8.

Erikson, K. E. 1978. Standardized sampling procedures for lake and reservoir management recommendations. Final report, Oklahoma Department of Wildlife Conservation (Project F-38-R, Job 1), Oklahoma City.

Fotheringham, S., and P. Rogerson, editors. 1994. Spatial analysis and GIS. Taylor and Francis, Inc., Bristol, Pennsylvania.

Gerrodette, T. 1987. A power analysis for detecting trends. Ecology 68:1364–1372.

Giles, R. H., and L. A. Nielsen. 1992. The uses of geographic information systems in fisheries. Pages 81–94 *in* R. H. Stroud, editor. Fisheries management and watershed development. American Fisheries Society Symposium 13.

Green, R. H. 1979. Sampling design and statistical methods for environmental biologists. Wiley, New York.

Gunderson, D. R. 1993. Surveys of fisheries resources. Wiley, New York.

Gutreuter, S. 1993. A statistical review of sampling of fishes in the long term resource monitoring program. Environmental Management Technical Center (EMTC 93-T004), Onalaska, Wisconsin.

Hardin, S., and L. L. Connor. 1992. Variability of electrofishing crew efficiency, and sampling requirements for estimating reliable catch rates. North American Journal of Fisheries Management 12:612–617.

Horvath, F. J. 1992. The use of a geographical information system to inventory, classify, and analyze Great Lakes aquatic habitats. Pages 210–215 *in* W. D. N. Busch and P. G. Sly, editors. The development of an aquatic habitat classification system for lakes. CRC Press, Boca Raton, Florida.

Hubbard, W. D., and L. E. Miranda. 1988. Competence of non-random electrofishing sampling in assessment of structural indices. Proceedings of the Annual Conference Southeastern Association of Fish and Wildlife Agencies 40(1986):79–84.

Hurlbert, S. H. 1984. Pseudoreplication and the design of ecological field experiments. Ecological Monographs 54:187–211.

Johnson, D. L., and L. A. Nielsen. 1983. Sampling considerations. Pages 1–21 *in* L. A. Nielsen and D. L. Johnson, editors. Fisheries techniques. American Fisheries Society, Bethesda, Maryland.

Kapetsky, J. M., J. M. Hill, and W. L. Dorsey. 1988. A geographical information system for catfish farming development. Aquaculture 68:311–320.

King, T. A., C. J. Williams, W. D. Davies, and W. J. Shelton. 1981. Fixed versus random sampling of fishes in a large reservoir. Transactions of the American Fisheries Society 110:563–568.

Kraemer, H. C., and S. Thiemann. 1987. How many subjects? Statistical power analysis in research. Sage Publications Ltd., Newbury Park, California.

Kupper, L. L., and K. B. Hafner. 1989. How appropriate

are popular sample size formulas? American Statistician 43:101–105.
Levy, P. S., and S. Lemeshow. 1991. Sampling of populations: methods and applications. Wiley, New York.
Matthews, W. J. 1985. Summer mortality of striped bass in reservoirs of the United States. Transactions of the American Fisheries Society 114:62–66.
McAllister, K. M., and R. M. Peterman. 1992. Experimental design in the management of fisheries: a review. North American Journal of Fisheries Management 12:1–18.
Meaden, G. J. 1987. Where should trout farms be in Britain? Fish Farmer 10:10–20.
Meaden, G. L., and J. M. Kapetsky. 1991. Geographical information systems and remote sensing in inland fisheries and aquaculture. FAO (Food and Agriculture Organization of the United Nations) Fisheries Technical Paper 318.
Nielsen, L. A. 1989. Development of an Ohio River commercial fisheries atlas. Final report, West Virginia Department of Wildlife Resources (Project 3-418-D), Charleston.
O'Bara, C. J., and D. W. Hubbs. In press. The use of geographic information systems technology to predict available spawning habitat for lake trout in Dale Hollow Reservoir, Tennessee and Kentucky. Proceedings of the Annual Conference Southeastern Association of Fish and Wildlife Agencies 48(1994).
Osburn, H. R. 1988. Comparison of random and nonrandom bag seine station selection methods. Texas Parks and Wildlife Department, Management Data Series 117, Austin.
Parsley, M. J. 1992. Use of a raster-structured GIS in fisheries research activities on the Columbia River. Pages 74–81 *in* Proceedings of the third national U.S. Fish and Wildlife Service geographic information systems workshop, Environmental Management Technical Center, Onalaska, Wisconsin.
Peterman, R. M. 1990. Statistical power analysis can improve fisheries research and management. Canadian Journal of Fisheries and Aquatic Sciences 47:2–15.
Peterman, R. M., and M. J. Bradford. 1987. Statistical power of trends in fish abundance. Canadian Journal of Fisheries and Aquatic Sciences 44:1879–1889.
Peterman, R. M., and R. D. Routledge. 1983. Experimental management of Oregon coho salmon (*Oncorhynchus kisutch*): designing for yield of information. Canadian Journal of Fisheries and Aquatic Sciences 40:1212–1223.
Ripple, W. J., editor. 1994. The GIS applications book examples in natural resources: a compendium. American Society for Photogrammetry and Remote Sensing, Bethesda, Maryland.
Robson, D. S., and H. A. Regier. 1964. Sample size in Petersen mark–recapture experiments. Transactions of the American Fisheries Society 93:215–226.
Rogala, J. T. 1992. Use of GIS in stratified random sampling design. Pages 182–185 *in* Proceedings of the third national U.S. Fish and Wildlife Service geographic information systems workshop, Environmental Management Technical Center, Onalaska, Wisconsin.
Rogerson, P. A., and A. S. Fotheringham. 1994. GIS and spatial data analysis: introduction and overview. Pages 1–10 *in* A. Fotheringham and P. Rogerson, editors. Spatial analysis and GIS. Taylor and Francis, Inc., Bristol, Pennsylvania.
Romesburg, H. C. 1981. Wildlife science: gaining reliable knowledge. Journal of Wildlife Management 45:293–313.
Ross, L. G., Q. M. E. A. Mendoza, and M. C. M. Beveridge. 1993. The application of geographical information systems to site selection for coastal aquaculture: an example based on salmonid cage culture. Aquaculture 11:165–178.
Siler, J. R., W. J. Foris, and M. C. McInerny. 1986. Spatial heterogeneity in fish parameters within a reservoir. Pages 122–136 *in* G. E. Hall and M. J. Van Den Avyle, editors. Reservoir fisheries management: strategies for the 80's. American Fisheries Society, Southern Division, Reservoir Committee, Bethesda, Maryland.
Smith, S. M., and D. J. Orth. 1991. Distributions of largemouth bass in relation to submerged aquatic vegetation in Flat Top Lake, West Virginia. Proceedings of the Annual Conference Southeastern Association of Fish and Wildlife Agencies 44(1990):36–44.
Star, J. L., and J. E. Estes. 1990. Geographic information systems: an introduction. Prentice Hall, Englewood Cliffs, New Jersey.
Summerfelt, R. C. 1971. Factors influencing the horizontal distribution of several fishes in an Oklahoma reservoir. Pages 425–439 *in* G. E. Hall, editor. Reservoir fisheries and limnology. American Fisheries Society Special Publication 8.
Thompson, S. K. 1992. Sampling. Wiley, New York.
TPWD (Texas Parks and Wildlife Department). 1993. Fishery assessment procedures. Texas Parks and Wildlife Department, Austin.
Van Horn, S. L., J. W. Kornegay, B. Wynne, and W. R. Chapman. 1993. Intra-season variation in largemouth bass electrofishing catch-per-unit-effort: implications for management decisions. Proceedings of the Annual Conference Southeastern Association of Fish and Wildlife Agencies 45(1991):254–260.
Walters, C. J., J. S. Collie, and T. Webb. 1988. Experimental designs for estimating transient responses to management disturbances. Canadian Journal of Fisheries and Aquatic Sciences 45:530–538.
Wilde, G. R. 1995. Gill net sample size requirements for temperate basses, shads, and catfishes. Proceedings of the Annual Conference Southeastern Association of Fish and Wildlife Agencies 47(1993):588–595.
Wilkerson, M. L. 1995. Movements, distribution and habitat preferences of striped bass in Robert S. Kerr Reservoir, Oklahoma. Master's thesis. Oklahoma State University, Stillwater.
Zale, A. V., J. D. Weichman, R. L. Lochmiller, and J. Burroughs. 1990. Limnological conditions associated with summer mortality of striped bass in Keystone Reservoir, Oklahoma. Transactions of the American Fisheries Society 119:72–76.

American Fisheries Society Symposium 16:410–421, 1996

Adaptive Strategies for the Management of Ecosystems: The Columbia River Experience

Willis E. McConnaha and Peter J. Paquet
Northwest Power Planning Council
851 S.W. 6th Avenue, Suite 1100, Portland, Oregon 97204, USA

Abstract.—We describe the application of adaptive management to development of a system-wide approach for protecting and enhancing populations of salmonids *Oncorhynchus* spp. in the Columbia River basin. This represents one of the few applications of the concept to an ecosystem level recovery program. The use of adaptive management in the Columbia River basin has had mixed results. While a variety of actions in the basin have been promoted on the basis of their potential for adaptive learning, few were implemented as part of an explicit learning plan. As a result, it is difficult to identify significant scientific questions that have been resolved in the basin through the use of adaptive management. Many of the scientific questions that prompted the call for adaptive management in the basin remain. However, the development of a system model and the broad-based participation encouraged by adaptive management have aided development of a vision of the basin as an integrated system, a key feature in managing an area of the size and jurisdictional diversity of the Columbia River basin. It also has fostered an appreciation among decision makers of the impacts of scientific uncertainty and biological variability. Despite the practical difficulties of using an adaptive approach, the learning emphasis of adaptive management will be required if we are to manage ecosystems. An adaptive approach implies, foremost, an overt commitment to learning while moving forward with needed management actions. Because of the broad range of questions likely to face the ecosystem manager, an adaptive approach needs to include a similarly broad range of tools including manipulative management experiments, research, and monitoring. These should be focused through the use of an explicit learning plan that links goals, uncertainties, and actions. Such a plan presents an organized approach to learning and could prevent the diffusion of activities that can occur under the banner of adaptive management.

The decline of many of the world's natural resources has increased the need for the development of effective, scientifically based management solutions. Scientists are being called on to participate in the formulation of these solutions by actively working within the public policy process (Babbitt 1995). A major challenge is to develop public policy that can accommodate the variability and scientific uncertainties that are characteristic of most biological systems. The impact of scientific uncertainties on the development of public policy will become increasingly apparent as reservoirs are viewed as ecosystems, and as reservoir management incorporates the ideas of ecosystem management (Grumbine 1994). The manager's focus will have to expand beyond the set of species deemed valuable for commercial or recreational needs to encompass the biotic and abiotic dimensions of the ecosystem. At the same time, the pace of human impacts on ecosystems will rarely allow the accumulation of needed information prior to taking management action.

Adaptive management (Holling 1978) has been advanced as a means to develop effective natural resource policy in an atmosphere of scientific uncertainty. It calls for learning to be an explicit goal of the management process and permits management to proceed in the absence of a complete scientific foundation for action (McAllister and Peterman 1992). It suggests that management actions be implemented as experiments to investigate critical uncertainties. Although the concept has been discussed in the scientific literature, there are relatively few examples of its actual use in a management situation, particularly for large and complex ecosystems such as the Columbia River basin. Much of the literature on the topic discusses the theoretical application of management experiments pertaining to commercial fisheries (e.g., Walters and Hilborn 1976; Smith and Walters 1981; Peterman and Rutledge 1983).

In this paper we discuss the application of adaptive management to a large-scale, ecosystem level restoration program in the Columbia River basin, in the northwestern United States (Figure 1). Construction of hydroelectric, flood control, and irrigation dams has converted the Columbia River and its tributaries into a series of reservoirs whose habitat conditions differ markedly from those that existed prior to development. The restoration program responds to sharp declines in populations of fish and wildlife, especially Pacific salmon *Oncorhynchus* spp., associated with development of the river's hy-

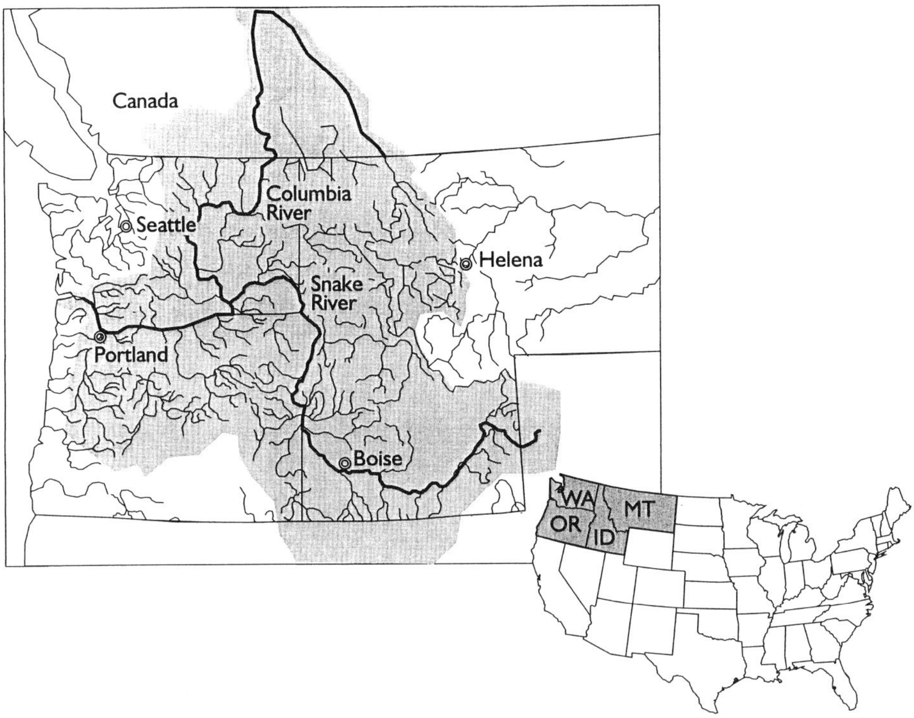

FIGURE 1.—Map of the Columbia River basin.

droelectric potential and other development activities throughout this century. The experience has suggested features that may be of general management application.

Adaptive Management

Throughout most of this century, fisheries and forestry managers have attempted to apply scientific knowledge to the management of natural resources (Callicott 1991). However, the relationship between political and scientific processes often is uneasy, reflecting fundamentally different approaches. The public policy process can be forced to act, with or without the comfort of scientific certainty. The Endangered Species Act (16 U.S.C. §§ 1531 to 1543), for example, requires that federal agencies rule promptly on the status of petitioned species, relying on existing scientific evidence (Kilbourne 1991). In the Columbia River basin, the Pacific Northwest Electric Power Planning and Conservation Act (16 U.S.C. §§ 839 et seq.) requires the Northwest Power Planning Council to develop a plan for the restoration of fish and wildlife resources of the Columbia River basin using the "best available scientific information" without waiting for scientific certainty (Blumm 1986). Decision makers trained to cope with economic and social problems can be dismayed at the scientific uncertainty and the lack of clear solutions characteristic of natural resources management.

Research scientists, on the other hand, are typically cautious, recognizing the inherent variability of ecological systems. Their reaction is often to seek additional study and research prior to offering advice regarding proposed actions. At other times, scientists may confront decision makers with passionately held, but diametrically opposed, views based on similar information. This may leave decision makers with confusion and scant scientific direction regarding decisions which, for legal or other reasons, must be made. The political process will rarely wait for scientific certainty. In addition, the dire condition of many natural resources often requires immediate action. Due to the pervasive nature of human impacts, natural resource managers frequently find themselves fighting a rear guard action. Action must be taken immediately to ward off one or more biological crises. In short, biological considerations may also preclude waiting for scientific certainty.

Adaptive management provides a method that allows managers to proceed in the face of scientific uncertainty. Adaptive management is an explicit commitment to learning while a system is being actively managed. Actions, research, and monitoring are implemented in a structured learning process aimed at eliminating alternative hypotheses affecting the achievement of management goals. In effect, adaptive management calls for the management process to adopt many of the features of the classical experimental method (Platt 1964) to make strong inferences regarding management direction. Lee (1993:53) defined an adaptive approach as one that is "designed from the outset to test clearly formulated hypotheses about the behavior of an ecosystem being changed by human use . . . [T]his experimental approach produces reliable knowledge from experience instead of the slow, random accumulation from unexamined error."

Thus, while adaptive management is not the only way for managers to increase knowledge, it may be the most efficient. Few would suggest that natural resource management does not involve learning, but rather that the learning process is slow and often without direction or purpose (Hilborn and Winton 1993). Adaptive management, however, calls for management to be structured to enhance learning and to seek answers to outstanding problems deliberately. Properly instituted, it represents a significant departure from normal management practices. It requires decision makers to recognize uncertainties and to accept the need to learn. Importantly, it requires the willingness to treat actions as experiments and to structure actions to address important questions.

To promote learning, adaptive management calls for actions to be implemented as part of an experimental design to solve problems critical to the achievement of management goals (Holling 1978). For example, adaptive management frequently has been suggested as a way to address scientific questions associated with fisheries harvest management (Walters and Hilborn 1976; Smith and Walters 1981; Sainsbury 1988; Peterman 1991; McAllister and Peterman 1992). An adaptive approach might call for manipulating exploitation rates to provide information on key parameters for models underlying harvest management (e.g., Walters and Hilborn 1976). Populations could be deliberately over- or underharvested to observe their response and provide insights regarding the underlying production function. Such harvest manipulations would be instituted under a strict experimental design to ensure that population responses could be incorporated into a management model with an acceptable level of confidence.

The emphasis on learning in adaptive management leads to an important social aspect involving the regulatory process as well. A regulatory process that is open to learning and to viewing actions as experiments is also likely to be open to public participation in the decision-making process. Holling (1978), in particular, emphasized the use of a participatory process to develop solutions and experimental approaches to public resource problems. His approach, termed "adaptive environmental assessment and management," used computer models of the system as focal points for public workshops designed to bring together managers, users, and interested parties. The models and workshops served to identify critical uncertainties that could be approached through adaptively implemented actions.

In the Columbia River basin, the use of adaptive management resulted from two circumstances. The first was an environmental crisis brought on, in part, by human development of the basin. The second was a legislative response that permitted the concept of adaptive management to be considered.

The Columbia River Biological Setting

The environmental crisis that spurred the use of adaptive management in the Columbia River basin was the decline of Pacific salmon populations coincident with the development of the river's economic potential. The Columbia River is one of the major river systems in North America, draining an area of 670,810 km^2 in the United States and Canada (Figure 1). Since the arrival of European civilization in the late Nineteenth Century, the basin has been extensively modified for hydroelectric generation, transportation, flood control, irrigation, and recreation. The Columbia River is now the largest integrated hydroelectric system in the United States; it has a generating capacity of about 12,000 MW (NPPC 1991).

The benefits of this development have come with a significant impact on the other natural resources of the river, especially salmon abundance and diversity. Prior to this century, the river was one of the world's major sources of Pacific salmon and had the largest runs of chinook salmon *O. tshawytscha* in the world (Van Hyning 1973; Smith 1979). Based on expansion of cannery records and anecdotal information, the annual return of salmon and steelhead *O. mykiss* to the river prior to 1850 was estimated at 10–16 million fish (NPPC 1986). However, the

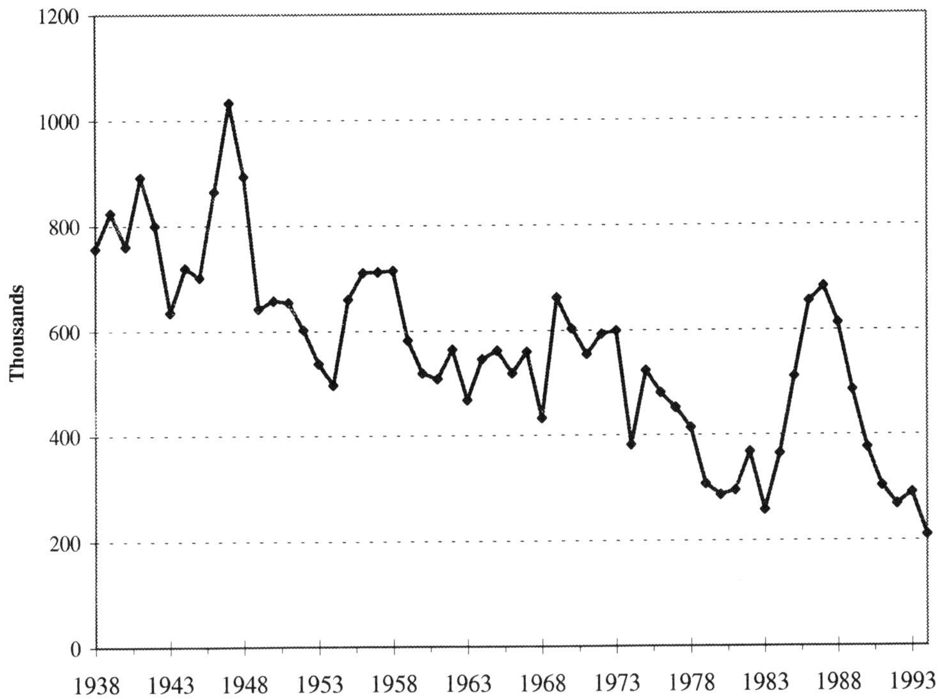

FIGURE 2.—Returns of chinook salmon to the Columbia River above Bonneville Dam, 1938–1993. Bonneville is the lower-most dam on the Columbia River, situated at river kilometer 204 from the ocean. It was completed in 1938 and provided the opportunity for accurate enumeration of returns to upriver areas.

record of returns of salmon to the Columbia River during much of this century, especially returns to the upper basin, shows a pattern of overall decline (Figure 2). Returns of all salmon and steelhead to the basin are now less than 2 million annually, while returns of some upriver populations in 1994 and 1995 were the lowest on record. Most alarming is that approximately 70% of the present return is of hatchery origin; many wild populations hover on the brink of extirpation (Nehlsen et al. 1991). At present, three races of chinook salmon and the only remaining population of sockeye salmon *O. nerka* in the Snake River are classified as endangered under the Endangered Species Act (NMFS 1995). The overall decline in fish numbers during this century has been the result of numerous factors, but is strongly correlated with the development and exploitation of the river basin by people of European traditions. Ebel et al. (1989) and NMFS (1995) discussed the reasons for the decline in some detail, and the causes for the decline can be summarized as follows.

1. Early in this century, logging, mining, and other activities resulted in substantial loss of habitat (Wissmar et al. 1994) and elimination of many salmonid populations (Nehlsen et al. 1991).
2. At the same time, fisheries resulted in significant reduction in the numbers of returning salmon and steelhead (Van Hyning 1973; Smith 1979).
3. Construction of Grand Coulee Dam on the Columbia River and Hells Canyon on the Snake River blocked more than 30% of the habitat available to anadromous salmonids (NPPC 1986), and additional blockages were created on smaller tributaries.
4. Since 1930, development of the Columbia River hydroelectric system has resulted in the construction of more than 92 U.S. federal, U.S. nonfederal, and Canadian hydroelectric dams on the Columbia and Snake rivers and their major tributaries (COE 1985). This has resulted in significant alteration of the Columbia River ecosystem (Ebel et al. 1989; NMFS 1995).
5. Agriculture, logging, and urbanization have degraded water quantity and quality, especially in tributary streams (McIntosh et al. 1994).
6. The impact of these human-caused limitations on salmon production are overlain on natural

variation in ocean survival and freshwater conditions (Beamish and Bouillon 1993). In recent years, unusually persistent ocean warmth (Hill 1995) and drought have likely contributed to record low returns for many chinook salmon populations in the Columbia River and elsewhere.

Many federal and state programs have attempted to address the impacts of development on the basin's fisheries resources. Many of these have focused on construction of hatcheries to mitigate for lost wild fish production. Sixty-four hatcheries and related facilities have been constructed in the basin, releasing approximately 200 million juvenile salmon annually into the Columbia River (USFWS 1990). In the mid-1970s, restoration efforts began including modification of hydroelectric and other facilities to facilitate fish migration along with the restoration and improvement of spawning and rearing habitat.

The Columbia River Jurisdictional Setting

In addition to these biological circumstances, the opportunity to use adaptive management in Columbia River was created by the Pacific Northwest Electric Power Planning and Conservation Act of 1980 (the Act). The Act was passed primarily to facilitate marketing of electric power from the Columbia River hydroelectric system by the Bonneville Power Administration (BPA) and to provide a regional voice in power planning decisions made by BPA (Blumm 1986). To provide this, Congress created the Northwest Power Planning Council (the Council) which is made up of representatives appointed by the governors of Montana, Idaho, Oregon, and Washington. Congress also recognized that hydroelectric development had come with a heavy price in terms of the declining fish and wildlife in the basin. For this reason, the Council was charged to develop a program to "protect, mitigate and enhance" fish and wildlife resources of the Columbia River as affected by hydroelectric development.

In developing its fish and wildlife program, the U.S. Congress directed the Council to treat the Columbia River as a system, which set the stage for viewing the basin as an ecosystem. The Council's authority is limited, and, for its program to be successful, it has had to bring a disparate set of state, federal, and tribal jurisdictions and interests together. Recognizing that fish runs had declined alarmingly, Congress directed the Council to take prompt action; it was to develop its first fish and wildlife plan within 1 year of the Council's creation and regularly update the plan as needed. The limitations on existing knowledge were also recognized. The Council was directed to take significant action to rebuild fish and wildlife populations in the basin and not simply call for new studies. It was to act on the basis of the best available scientific knowledge and not wait for scientific certainty (Lee 1993).

When the Council began work on its first fish and wildlife program in 1981, it quickly became apparent that there was no clear path to restoring the fish and wildlife resource. What was apparent was that the decline in fish and wildlife abundance coincided with development of the basin's economic potential during this century (NPPC 1986). The region was deeply divided about the direction the Council's program should take, with the various sides of the debate marshaling scientific information to support their positions (Blumm 1986). Reflecting its mandate from Congress, the Council's efforts were particularly directed at the impact of hydroelectric development on anadromous salmon runs, a subject that had engendered acrimonious debate within the region for many years. Commensurate with this, the Council's use of adaptive management has focused on efforts to restore anadromous salmonid populations in the basin (Lee and Lawrence 1986).

Predictably, utilities and other nonfisheries users of the river generally favored a cautious and deliberate approach to salmonid restoration and have stressed the limitations of scientific support relative to changing river operations (e.g., McNeil 1995). On the other hand, environmental groups and most natural resource management agencies called for radical changes to prevent the loss of important resource components.[1] While recognizing the limitations of present knowledge, they concluded that enough information existed to justify emergency actions.

The concept of adaptive management was introduced to the region as the Council began debate on the first revision of its fish and wildlife program in 1984 (NPPC 1984; Lee and Lawrence 1986). The rationales for the Council's consideration of adaptive management included the need for a legislative mandate to take significant action despite uncertainties, the existence of a declining resource base, the existence of significant scientific uncertainties that seriously impeded program development, and a need to bring together a wide range of disparate

[1]Strategies for recovery of Snake River salmon: recommendations for Endangered Species Act pre-decisional management plan. State of Idaho, Office of the Governor, Boise.

governmental and nongovernmental bodies to form a regional plan. Lee (1993) and Volkman and McConnaha (1993) provided additional discussion of the reasoning and setting behind the Council's use of adaptive management.

Unique Elements of the Council's Approach to Adaptive Management

Although both the biological and the jurisdictional settings pointed to the applicability of an adaptive approach for the Council's fish and wildlife program, the Columbia River basin differed from the usual adaptive management context. Unique features have influenced the Council's success in implementing the concept.

Adaptive management has been discussed most frequently in connection with harvest management and its utility in developing optimal harvest strategies (e.g., Walters and Hilborn 1976). In this context, adaptive management has been proposed as a way to gain information regarding a rather narrow set of questions affecting the regulation of a single human activity, especially the harvest of salmon in the offshore ocean troll fisheries of the Pacific Northwest. For most areas, the number of regulatory bodies and interest groups was relatively small. For example, determination of annual harvest rates, while often a complicated and contentious exercise, involves a single industry and a limited number of state and federal agencies. Fish populations being harvested off Washington, British Columbia and southeastern Alaska were large, and were perceived to be relatively robust. For these reasons, it might be reasonable to propose an experimental management plan that included deliberate overharvest for several years. Efforts to restore fish and wildlife in the Columbia River basin, however, differed from this scenario in terms of ecosystem scope, jurisdictional complexity, and strength of the fish populations.

Ecosystem scope.—The Council has attempted to apply adaptive management to a program that encompasses most, if not all, of the salmonid ecosystem.[2] It recognizes that this ecosystem encompasses all of the Columbia River basin plus a marine component that stretches into the Bering Sea (NPPC 1994). Thus, scientific questions associated with its program deal with a wide array of subjects including basic salmonid life history, genetics, and artificial production; riverine, estuarine, and marine ecology; and passage problems at hydroelectric projects, in addition to harvest. Each of these presents unique experimental challenges. At the same time, the diversity of potential questions suggests that a variety of experimental approaches is needed.

Jurisdictional complexity.—The scope of the salmonid ecosystem means that the Council's program must span multiple local, state, tribal, and international governments and that it will have significant impacts on a wide array of economic interests. Columbia River fish and wildlife resources are managed at various points in their life cycle by the natural resource agencies of 4 state governments, 2 national governments and 13 tribal governments. The Council's program calls for actions that have substantial social and economic impacts on a number of public and private electric utilities, many industries, local governments, agricultural interests, and a host of public interest and environmental groups. By dealing particularly with environmental conditions in the main-stem Snake and Columbia rivers, the Council's efforts strike at the lifeblood of many Pacific Northwest economies. Manipulations of this system to test experimental hypotheses are guaranteed to be difficult and painful.

Weakened fish populations.—Importantly, the Council's program is treating what appears to many to be a terminally ill patient. When the Power Act was passed in 1980, spring chinook salmon in the Snake River (the largest tributary of the Columbia River, Figure 1) were being considered for listing under the Endangered Species Act. This was forestalled because the Council's program was expected to provide a regional recovery plan. Over the last decade, however, runs have generally continued their decline, and salmon populations have recently dropped to new lows (Figure 2). Nehlsen et al. (1991) listed 36 natural salmonid populations in the Columbia River basin as being at a "high risk" of extinction, an additional 14 at moderate risk, and 26 of special concern. Spring, summer, and fall races of chinook and sockeye salmon in the Snake River have now been listed as endangered under the Endangered Species Act (NMFS 1995). Other populations in the Columbia River are under consideration for listing. Thus, experimental manipulation of critically weakened populations to test hypotheses runs abruptly into legal and ethical concerns (Volkman and McConnaha 1993).

[2]The Council's activities are focused on the impacts on fish and wildlife due to development and operation of the Columbia River hydroelectric system. Its program also addresses many aspects of the freshwater tributary habitat of the Columbia River. To develop a complete restoration plan, however, the Council's program also makes recommendations regarding harvest, but does not actively engage the harvest management process.

Impacts of the Council's Use of Adaptive Management

Since the Council first considered the use of adaptive management, the region has struggled to mold the theoretical aspects of the concept to the pragmatic needs of the Columbia River situation. It is revealing that, while the subject of adaptive management has a rich theoretical literature, there are relatively few discussions of its application to management situations. The record in the Columbia River reveals major difficulties with experimental approaches, particularly as applied to ecosystem level problems. On the other hand, the effort has resulted in significant, if more subtle, progress on other fronts and has indicated directions for development of an experimental approach to the Council's program.

The most obvious reason for using an adaptive management approach is to permit action in the face of scientific uncertainty. In this regard, the Council's experience has been disappointing. It is difficult to point to important scientific questions that have been addressed in the region through processes that could properly be called adaptive management (Volkman and McConnaha 1993). Virtually all of the critical questions facing the Council in 1984 when it adopted the concept remain, and they continue to stymie recovery efforts (NMFS 1995).

Part of the problem has been the lack of a rigorous definition or outline for adaptive management actions connected with the Council's program. Initially, the concept was enthusiastically received because it appeared to offer a way for the region to move ahead with needed action without waiting for the years of research that would have been required to provide a complete scientific foundation for salmon recovery. The term quickly became a part of the standard lexicon of the Council's program.

Unfortunately, the key precepts of adaptive management (i.e., an institutional commitment to learning, actions implemented as part of an experimental design, and so forth) were lost rather quickly. The idea of "learning by doing" became a rationale for virtually any action that might offer the possibility of learning. Over time, management might change in response to this new information. Such "passive adaptive management" (Silvert 1978) simply says that by doing enough things enough times we might eventually learn better ways of achieving management goals. This is a far cry from the rigorous experimental approach called for by adaptive management (Walters 1986).

Halbert (1993) noted a similar tendency in applying adaptive management to the complexities of managing timber harvest and its effect on fish and wildlife in Washington State. In this process, which involved an array of interests and questions similar to that of the Council's program, the experimental aspects were eventually abandoned in favor of "flexible management." This allowed managers the flexibility of changing regulations and rules to meet new circumstances and new information, while experimental actions and research languished.

The Council's implementation of an adaptive approach at the ecosystem level also was hindered by significant experimental problems. As originally proposed, the concept focused on the use of manipulative experiments to probe scientific uncertainties (Walters 1986). However, ecosystem level experimentation presents particularly difficult experimental design problems (Hurlbert 1984; Eberhart and Thomas 1991). The necessary replication is rarely possible, and untreated controls may be difficult to define. For example, a major question in Columbia River management is to what degree the survival of downstream migrating juvenile salmonids has decreased because of the modification of the river's hydrograph by development of upstream storage projects (NPPC 1994; NMFS 1995). It could be argued that an adaptive approach would be to fluctuate river flows experimentally to observe impacts on juvenile fish survival and adult returns. However, because juvenile survival and migrational characteristics are known to vary within years and among populations (Achord et al. 1994), the opportunities for replication and controls for such an experiment are limited.

Experimental problems are overlain on social and political problems that make such broad-scale ecosystem experimentation difficult. To manipulate the Columbia River hydroelectric system to produce the range of flows needed for the experiment discussed above, for example, potentially would have significant impacts on many facets of the Pacific Northwest economy. These would include impacts on power production, transportation, and irrigation, among others. While an experiment with such impacts is not impossible, it is obviously difficult and could be undertaken only with an experimental design that would ensure a clear resolution of the problem.

On a more subtle basis, however, the Council's efforts to use adaptive management have greatly benefited its efforts to devise and implement a regionally based recovery plan. For example, it has facilitated the formation of a regional or basin-wide vision of the Columbia River. For a system of the extent of the Columbia River basin with its diversity of jurisdictions and interests, a basin-wide vision is essential to a coordinated approach to fish and wildlife recovery. One of the first steps the Council took in its adaptive approach was to organize a series of regional workshops for developing the first salmonid life cycle model based on the features of the Columbia River basin (ESSA 1986). Use of such workshops to develop a broad-based concept of natural resource management was a key feature of Holling's (1978) original formulation of adaptive environmental assessment and management. The Council's workshops and the resulting model were instrumental in shifting the provincial view of many management agencies to a more system-wide vision of the Columbia River basin.

Development of the system-wide vision continues as the concept of ecosystem management is incorporated into Columbia River management. While it is inaccurate to call the Council's program an example of ecosystem management in the contemporary sense (Grumbine 1994), it is a program of ecosystem dimensions. The program represents an attempt to address the problems of natural resource components, such as salmon, with an appreciation of ecosystem functions (Apollonio 1994). The attempt to implement adaptive management has helped this by contributing to the vision of the basin as a system and through the use of simulation modeling and other activities that have increased the appreciation for the complexities and synergistic relationships among the many components of the system (McConnaha 1993).

In addition to the development of a system-wide approach to fish and wildlife restoration, the Council's efforts to promote adaptive management have moved the region toward accepting an experimental approach. The idea of viewing actions as experiments runs counter to the usual political thinking (Volkman and McConnaha 1993); it requires a major shift to think experimentally. The Council's most recent fish and wildlife program (NPPC 1994) provides evidence that this shift may be occurring.

The Council's 1994 program contains two key provisions—one programmatic, the other procedural—in this regard. The programmatic provision calls for a major regional experiment to examine key scientific questions that have thwarted the development of a regional recovery plan to date. The amount of benefit that would be afforded downstream-migrating salmonids by augmenting river flow and velocity during spring and summer migration periods is a primary question to be addressed experimentally. To augment flow or velocity is expensive and requires either release of upstream water stored for power or major modifications to hydroelectric projects. Although the effect of flow and velocity on fish movement and survival has been researched for many years, the results have been inconclusive (NMFS 1995). Because of the high cost of providing flow, this is one of the most contentious salmon recovery issues in the region. The Council's experiment also would address the benefits afforded by transporting juvenile salmonids around the hydroelectric system in barges and trucks, often seen as an alternative to providing flows or modifying the hydroelectric projects to improve survival during downstream migration. Transportation has been used for several years in the Columbia River, but runs have continued to decline. Research evaluating the benefits of transportation has been equivocal (Mundy et al. 1994). In devising a program to deal with the impacts of hydroelectric development, the Council has had to make decisions regarding these actions in the face of uncertain science and considerable cost-related controversy.

The Council's experiment is closely linked to a set of actions designed to lessen human-caused mortality on anadromous fish populations. The program outlines an ambitious schedule for major modifications to the hydroelectric system and its operations, but decisions regarding future modifications are to be based on a better understanding of flow and transportation benefits—understanding to which the experiment would be a major contributor. The program also lays out a series of hypotheses regarding the effects of flow, velocity, and transportation on juvenile fish survival and adult returns. For the first time, a plan of action for the Columbia River basin is coupled explicitly with experiments designed to test key hypotheses.

The procedural provision in the Council's program is the process to be used for developing the experiment described above and other associated research. The program does not provide a detailed experimental design but calls on a group of independent scientists to oversee the development of a basin-wide experimental design. The purpose of this is to bring scientific oversight to the design of the experiment that is independent of the parties having an interest in a particular outcome. This design should integrate the flow and transportation projects with other research and monitoring activities, and with the decision-making schedule for major modifications of the hydroelectric system.

An important point regarding this form of experimental management is that it does not call for explicit manipulation of the system to provide a range of experimental conditions, but instead relies on annual variation in natural conditions. While this may result in slower learning than might be possible with a manipulative experiment, it is more compatible with the realities of the Columbia River situation. The measures in the Council's program provide a framework for linking actions, hypotheses, and research. The program thereby retains the structured commitment to learning required of adaptive management, but takes advantage of the learning opportunities provided within the social and biological context of the system.

Although these adaptive management experiments are incorporated into the Council's program, their implementation faces major hurdles. Since their adoption by the Council in December 1994, the political tenor of the region has shifted dramatically in the direction of less ambitious environmental programs and toward greater emphasis on the economic and social impacts of salmon recovery efforts. This has slowed implementation of the Council's experiment. It is not clear that the region retains the will to tackle these questions aggressively. At the same time, the condition of many fish runs has deteriorated to the point that research and experimentation may not be possible. The 1995 upper Columbia River spring chinook salmon returns, for example, are the lowest on record by a substantial amount. Because these populations have been declared endangered under the Endangered Species Act, it remains to be seen if the experimentation needed to address regional problems will be possible.

Adaptive Management of Ecosystems

Over the last decade, fish and wildlife restoration in the Columbia River has served as a laboratory for the use of adaptive management strategies in large ecosystems. The experience has suggested that the range of questions likely to confront an ecosystem manager cannot be addressed solely through the use of manipulative experiments; the manager will need to take advantage of all opportunities for learning and focus those opportunities by devising explicit learning plans. The essential tenet of adaptive management remains the explicit, structured approach to learning even while needed management actions are taken. In this sense, adaptive management is essential to the management of ecosystems when critical information is lacking. Opportunistic experimentation, learning plans, and adaptive learning can form the basis of adaptive management of ecosystems.

Opportunistic Ecosystem Experimentation

Adaptive management often is thought of as the use of manipulative management experiments to address critical uncertainties. The reason for proposing manipulative experiments is to maximize the learning rate. We agree that such experiments can increase the learning rate in some cases, but opportunities to implement them at the ecosystem scale will be limited. Further, manipulative experiments may not be appropriate for all questions pertaining to ecosystem management. Manipulative experiments must accommodate trade-offs between the need for rapid learning and the biological and social costs of the experiment (Sainsbury 1988).

Manipulative experiments may not be possible at the ecosystem level for three reasons. First, socially and legally, such experiments are extremely difficult to implement, and the social difficulties in instituting such experiments are a major factor inhibiting their use (McAllister and Peterman 1992; Halbert 1993). This has been true even for the relatively focused problems of ocean harvest management, with which experimentation has been most frequently associated, and will be the case for the more socially complex problems of a restoration program for an ecosystem such as the Columbia River basin.

A second problem with manipulative experiments is that they require relatively robust populations upon which to conduct experiments. An action known to be detrimental (overharvest, for example) would be questionable for an already depressed population even though it could provide useful information. The experiment could drive the target populations to extinction, rendering the information meaningless for that population.

The third problem with relying on manipulative experiments for adaptive learning is that not all questions are amenable to system manipulation. The myriad questions likely to be asked in an ecosystem level program argue for an equally diverse set of experimental tools. Questions of toxicology, genetics, or life history, for example, may be vital to ecosystem restoration and management, and may be best addressed through laboratory research or directed field research.

Opportunistic learning does not imply the acceptance of less scientific rigor. Management based on bad science can lead to incorrect action and false confidence, whereas management that acknowledges scientific limitations recognizes the need for caution and learning. The social, legal, and logistic constraints imposed on ecological management experiments, however, will likely result in design compromises or less statistical confidence than might be possible under a controlled situation. Campbell (1969) discussed such compromises under the banner of "quasi-experiments," and noted that management experiments must be evaluated in terms of threats to their validity: "Can we reasonably proceed on the basis of this information?" Opportunistic learning implies that if one all-encompassing experiment is not possible, then all learning resources need to be marshaled to address questions from many different directions. As Lee (1993:73) noted: "Insisting on an idealistic approach to science does not entail refusing to do science unless it is invulnerable to criticism." Managers should demand a rigorous approach to science, but they must also recognize the limitations of scientific research and structure their learning and actions accordingly.

Explicit Plans for Learning

Adaptive management calls for an explicit approach to learning as part of a program of actions designed to address the needs of natural resource management. We suggest that this involves an overt and directed plan for learning and management. Such a framework should contain (1) the goals for the management process; (2) the hypotheses, assumptions, and information describing the current beliefs about ecosystem function, which form the basis for current management action; (3) actions designed to address immediate biological information needs, including management experiments: (4) critical uncertainties regarding achievement of management goals based on the existing belief system; and (5) research, monitoring, and management programs (including experiments) designed to address these uncertainties. It is the organized and directed learning provided by this adaptive management framework that distinguishes this approach from the undirected accumulation of knowledge and experience that often typifies natural resource management (Hilborn and Winton 1993; Lee 1993). In addition to providing a conscientious plan for learning, an adaptive management framework should provide a clear plan for adaptive activities. In the absence of a plan, the intuitive appeal of "learning by doing" results in a situation where virtually any activity can be justified because of the possible opportunity for learning.

A key element of the adaptive management framework is the explicit recognition of the existing belief system that drives management actions. Ideally, the belief system summarizes the existing knowledge into a model of the system being managed. This is consistent with Holling's (1978) use of a system model to focus development of policy options and adaptive management strategies. The model provides a vision for the ecosystem by bounding it geographically and biologically within an explicit set of elements and providing a logical basis for taking action. Nodes within the logic structure indicate divergence of alternative hypotheses based on the existing information. These indicate critical uncertainties that should be the focus of experimental actions. The belief system or model will evolve over time as hypotheses are tested.

Development of an adaptive management framework in the Columbia River basin is called for as part of the Northwest Power Planning Council's latest fish and wildlife program (NPPC 1994). The program provides a set of overall goals, hypotheses regarding major elements of main-stem passage by juvenile salmonids, a basin-wide experiment to compare strategies for passing juveniles downstream (described above), and a process for further development of a logical, scientific framework linking hypotheses to research and monitoring activities. While the gulf between measures in the Council's program and the development and implementation of these ideas is often vast, the elements are there for a new approach to adaptive management tempered by the experience of adapting the theoretical concept of adaptive management to the pragmatic needs of an ecosystem level program.

Adaptive Learning

We have identified several difficulties encountered in applying adaptive management to the situation in the Columbia River basin, but management of this system still will require adaptive learning. In light of the depressed condition of this, and many other, ecosystems, the need for a strategy that maximizes the rate of learning while permitting action is inescapable.

Management of single species for the economic benefit of society is the paradigm that has guided natural resource management through much of this century (Callicott 1991). As habitat for these species has been destroyed or populations have been overharvested, we have attempted to engineer solutions. Hatcheries, for example, often are proposed as compensation for habitat destroyed by development (e.g., COE 1975) or for depressed harvests of wild fish (Jefferies 1975). The value of hatcheries in the maintenance of population levels has been questioned (Hilborn 1992). Similarly, in the Columbia River, salmon populations continue to decline despite the use of sophisticated technology to retrofit the hydroelectric system to the biological needs of anadromous salmonids (Ebel et al. 1989; NPPC 1994; NMFS 1995). Shifting the management focus from single, economically important species to the ecosystem is likely to require a different suite of information than we have relied on in the past. If our ability to manage single species with existing knowledge is limited (Ludwig et al. 1993), then surely our ability to manage entire ecosystems is even more limited. Gaining the knowledge to manage ecosystems successfully requires that learning be an integral part of the management process.

Learning about ecosystems as we attempt to manage them leads inevitably to the consideration of an adaptive management strategy. Our experience suggests that adaptive management itself must be adapted to meet the needs of an ecosystem level management program. Its fundamental feature remains the commitment to learning as a part of management. We have suggested that this can be facilitated through an explicit learning plan that links goals, questions, and actions. Adaptive strategies for the management of ecosystems will be developed even as the concept of ecosystem management is developed.

Acknowledgments

We thank Kai N. Lee for his helpful review of a draft of this paper and for his leadership in advocating learning as a management philosophy. Many of the thoughts in this paper arose from discussions with our colleague John Volkman. The review by Gustavo Bisbol is also gratefully acknowledged. Cindy Halbert and an anonymous reviewer provided review and guidance on the preparation of this manuscript. We also thank Carlotta Collette and Jane McConnaha for their editorial reviews.

References

Achord, S., G. M. Matthews, D. M. Marsh, B. P. Sanford, and D. J. Kamikawa. 1994. Monitoring the migrations of wild Snake River spring and summer chinook salmon smolts, 1992. Bonneville Power Administration Report DOE/BP-18800-1, Portland, Oregon.

Apollonio, S. 1994. The use of ecosystem characteristics in fisheries management. Reviews in Fisheries Science 2:157–180.

Babbitt, B. 1995. Science: opening the next chapter of conservation history. Science 267:1954–1955.

Beamish, R. J., and D. R. Bouillon. 1993. Pacific salmon production trends in relation to climate. Canadian Journal of Fisheries and Aquatic Sciences 50:1002–1016.

Blumm, M. 1986. Reexamining the parity promise: more challenges than successes to the implementation of the Columbia Basin Fish and Wildlife Program. Environmental Law 16:461–515.

Callicott, J. B. 1991. Conservation ethics and fisheries management. Fisheries 16(2):22–28.

Campbell, D. T. 1969. Reforms as experiments. Pages 172–204 *in* F. G. Caro, editor. Readings in evaluation research. Russel Sage Foundation, Document 1977, New York.

COE (U.S. Army Corps of Engineers). 1975. A special report: lower Snake River fish and wildlife compensation plan. U.S. Army Corps of Engineers, Walla Walla District, Walla Walla, Washington.

COE (U.S. Army Corps of Engineers). 1985. Project data and operating limits. U.S. Army Corps of Engineers, North Pacific Division, Portland, Oregon.

Ebel, W., C. Becker, J. Mullan, and H. Raymond. 1989. The Columbia River—toward a holistic understanding. Canadian Special Publication of Fisheries and Aquatic Sciences 106:205–219.

Eberhart, L. L., and J. Thomas. 1991. Designing environmental field studies. Ecological Monographs 61:53–73.

ESSA (Environmental and Social Systems Analysts, Ltd.). 1986. Final report on the adaptive management workshops. Report to Northwest Power Planning Council, ESSA, Vancouver, British Columbia.

Grumbine, R. E. 1994. What is ecosystem management? Conservation Biology 8:27–38.

Halbert, C. 1993. How adaptive is adaptive management? Implementing adaptive management in Washington state and British Columbia. Reviews in Fisheries Science 1:261–283.

Hilborn, R. 1992. Hatcheries and future of salmon in the northwest. Fisheries 17(1):5–8.

Hilborn, R. M., and J. Winton. 1993. Learning to enhance salmon production: lessons from the Salmonid Enhancement Program. Canadian Journal of Fisheries and Aquatic Sciences 50:2043–2056.

Hill, D. K. 1995. Pacific warming unsettles ecosystems. Science 267:1911–1912.

Holling, C. S., editor. 1978. Adaptive environmental assessment and management. Wiley, New York.

Hurlbert, S. 1984. Pseudoreplication and the design of ecological field experiments. Ecological Monographs 54:187–211.

Jefferies, E. R. 1975. Present: role and challenge of fish culture in the northwest. Pages 167–174 *in* J. Brophy, editor. 26th Northwest Fish Culture Conference. Oregon Aqua-Foods, Newport.

Kilbourne, J. C. 1991. The Endangered Species Act under the microscope: a closeup look from a litigator's perspective. Environmental Law 21:499–586.

Lee, K. N. 1993. Compass and gyroscope: integrating science and politics for the environment. Island Press, Washington, DC.

Lee, K. N., and J. Lawrence. 1986. Adaptive management: learning from the Columbia River Basin Fish and Wildlife Program. Environmental Law 16:431–460.

Ludwig, D., R. Hilborn, and C. Walters. 1993. Uncertainty, resource exploitation, and conservation: lessons from history. Science 260:17–19.

McAllister, M. K., and R. M. Peterman. 1992. Decision analysis of a large-scale fishing experiment designed to test for a genetic effect of size-selective fishing on British Columbia pink salmon (*Oncorhynchus gorbuscha*). Canadian Journal of Fisheries and Aquatic Sciences 49:1305–1314.

McConnaha, W. E. 1993. Biological modeling in the Columbia River basin: an organized approach to dealing with uncertainty. Pages 142–146 *in* K. Hon, editor. Water management in the 90's: a time for innovation. American Society of Civil Engineers, New York.

McIntosh, B. A., and six coauthors. 1994. Historical changes in fish habitat for select river basins of eastern Oregon and Washington. Northwest Science 68: 36–53.

McNeil, W. J. 1995. Water velocity and migration of juvenile salmon: is faster necessarily better? Hydro Review 14(2):2–6.

Mundy, P. R., and nine coauthors. 1994. Transportation of juvenile salmonids from hydroelectric projects in the Columbia River basin: an independent peer review. Final report to U.S. Fish and Wildlife Service, Portland, Oregon.

Nehlsen, W., J. E. Williams, and J. A. Lichatowich. 1991. Salmon at the crossroads: stocks at risk from California, Oregon, Idaho, and Washington. Fisheries 16(2): 4–21.

NMFS (National Marine Fisheries Service). 1995. Proposed recovery plan for Snake River salmon. NOAA (National Oceanic and Atmospheric Administration) Technical Report NMFS.

NPPC (Northwest Power Planning Council). 1984. 1984 Columbia River basin fish and wildlife program. NPPC, Portland, Oregon.

NPPC (Northwest Power Planning Council). 1986. Compilation of information on salmon and steelhead losses in the Columbia River basin. NPPC, Portland, Oregon.

NPPC (Northwest Power Planning Council). 1991. 1991 Northwest conservation and electric power plan. NPPC, Portland, Oregon.

NPPC (Northwest Power Planning Council). 1994. 1994 Columbia River basin fish and wildlife program. NPPC, Portland, Oregon.

Peterman, R. M. 1991. Density-dependent marine processes in north Pacific salmonids: lessons for experimental design of large-scale manipulations of fish stocks. ICES (International Council for the Exploration of the Sea) Marine Science Symposium 192:69–77.

Peterman, R. M., and R. D. Rutledge. 1983. Experimental management of Oregon coho salmon (*Oncorhynchus kisutch*): designing for yield of information. Canadian Journal of Fisheries and Aquatic Sciences 40:1212–1223.

Platt, J. R. 1964. Strong inference. Science 146:347–352.

Sainsbury, K. J. 1988. Application of an experimental approach to management of a tropical multispecies fishery with highly uncertain dynamics. ICES (International Council for the Exploration of the Sea) Marine Science Symposium 193:301–320.

Silvert, W. 1978. The price of knowledge: fisheries management as a research tool. Journal of the Fisheries Research Board of Canada 35:208–212.

Smith, C. 1979. Salmon fishers of the Columbia. Oregon State University Press, Corvallis.

Smith, A. D. M., and C. J. Walters. 1981. Adaptive management of stock recruitment systems. Canadian Journal of Fisheries and Aquatic Sciences 38:690–703.

USFWS (U.S. Fish and Wildlife Service). 1990. Review of the history, development and management of anadromous fish production facilities in the Columbia River basin. Columbia Basin Fish and Wildlife Authority, Portland, Oregon.

Van Hyning, J. M. 1973. Factors affecting the abundance of fall chinook salmon in the Columbia River. Research Reports of the Fish Commission of Oregon 4(1).

Volkman, J. R., and W. E. McConnaha. 1993. Through a glass, darkly: Columbia River salmon, the Endangered Species Act, and adaptive management. Environmental Law 23:1249–1272.

Walters, C. J. 1986. Adaptive management of renewable resources. Macmillan, New York.

Walters, C. J., and R. Hilborn. 1976. Adaptive control of fishing systems. Journal of the Fisheries Research Board of Canada 33:145–159.

Wissmar, R. C., and five coauthors. 1994. A history of resource use and disturbance in riverine basins in eastern Oregon and Washington (early 1800s–1990s). Northwest Science 68:1–35.

American Fisheries Society Symposium 16:422–441, 1996

Predicting Black Bass Reproductive Success from Bull Shoals Reservoir Hydrology

G. R. Ploskey and J. M. Nestler
U.S. Army Corps of Engineers, Waterways Experiment Station, CEWES-ES-Q
3909 Halls Ferry Road, Vicksburg, Mississippi 39180-6199, USA

W. M. Bivin
Arkansas Game and Fish Commission
220 South Locust Street, Fayetteville, Arkansas 72701, USA

Abstract.—Methods of predicting effects of changes in reservoir operations and water levels on in-pool fish populations are needed for inclusion of reservoir fisheries in multipurpose evaluations of water usage. We describe three methods for relating fish reproductive success in reservoirs to hydrology; responses of young black bass (*Micropterus* spp.) in Bull Shoals Reservoir, Arkansas, are used as an example. Reproductive success was quantified as biomass (kg/ha) in two length-classes of young largemouth bass *Micropterus salmoides*, spotted bass *M. punctulatus*, and smallmouth bass *M. dolomieu* from 34 years of August cove rotenone samples. In the first method, we derived hydrology variables from inflow, release, volume, mean area, change in area, and selected ratios of inflow to release or release to volume from time segments ranging from one season to 2 years before fish sampling in August. This segmented-temporal approach produced 36 hydrology variables, many of which were intercorrelated. Biomass of small and intermediate length-classes of black bass were regressed on all hydrology variables in simple-linear regression and on subsets of independent hydrology variables in multiple-linear regression. Although time consuming, this approach usually provided the best-fitting and most easily interpreted predictive models. In the second method, biomass was regressed on variables derived from principal components analysis of 14 segmented-temporal variables. Four components explained 91% of the variation in hydrology variables, but only the first two components (explaining 82%) could be readily interpreted and were significant in regression models. This method was fast and avoided intercorrelations, but interpretation of third and fourth principal components was difficult, and the method did not produce the best-fit models. In the third method, we fit cosine functions to monthly discharge of the unregulated White River (1920–1949) and to end-of-month area of Bull Shoals Reservoir (1955–1993). We derived eight independent variables related to amplitude, phase, differences in the phase of the reservoir from the preproject river, and error in fit of the harmonic function to reservoir hydrology. This harmonic approach provided a good explanation of variations in surface area ($r^2 = 0.84$; $N = 465$) and in largemouth bass and spotted bass biomass; however, it did not yield predictive models that were significantly better than the principal components method. All three methods are potentially useful for quantifying effects of operations and hydrology on reservoir fish populations for multiple-use assessments of water resources.

Methods of predicting effects of changes in reservoir operations and water levels on in-pool fish populations are needed for inclusion of reservoir fisheries in multipurpose evaluations of water usage. Our objective was to describe three approaches for deriving regression equations for predicting fish reproductive success, as measured by annual catches of young fish. Bull Shoals Reservoir, northwestern Arkansas, was selected to derive models for predicting black bass *Micropterus* spp. reproductive success within a reservoir, not necessarily within reservoirs in general. Equations derived for Bull Shoals probably would have limited utility for other impoundments because the magnitude of hydrological effects appears to vary widely among reservoirs (Ploskey et al. 1984). We selected Bull Shoals Reservoir for study because its fish populations had been sampled with consistent methods for many consecutive years. In addition, black bass populations were intensively studied in and described for some of those years (Bryant and Houser 1971; Aggus and Elliott 1975; Houser and Rainwater 1975; Keith 1975; Rainwater and Houser 1975; Vogele 1975, 1981; Vogele and Rainwater 1975). We wanted to compare predictors of biomass of young black bass developed in earlier assessments with ones developed from a much expanded data set.

We focus here on quantifying empirical relations for black bass species collected in cove rotenone samples; however, the methods are applicable to data for other gears and species. Fisheries data must be acquired with consistent methods for enough years to allow regression on hydrology variables; about 5 years of data should be acquired for every independent variable to be included in a model.

Background

Fisheries biologists often associate strong year-classes of warmwater fishes with years of above-average inflow and water levels in reservoirs (Ploskey 1986). The positive effects on year-class strength typically result from substantial increases in inundated area that occur over several seasons or years; the effects may be accentuated by topography, soil conditions, and vegetation (Wood and Pfitzer 1960). In contrast, daily or weekly hydrological fluctuations may have negative effects on spawning and hatching (Shields 1957; Bennett 1975; Heisey et al. 1980; Bennett et al. 1985; Kohler et al. 1993), although not necessarily on eventual year-class strength (Gasaway 1970; Estes 1971; Kohler et al. 1993). Responses of many species to hydrological increases are positive and more pronounced in hydropower storage reservoirs than they are in mainstream hydropower impoundments (Aggus and Lewis 1978). Fish biomass in storage reservoirs increases in response to increased water exchange rate and surface area (Aggus and Lewis 1978). In years when these occur, both flushing rate and age-0 biomass approach values more typically observed in productive mainstream impoundments. Negative correlations between young-of-year (age-0) fish catch and flushing rate variables in run-of-river reservoirs (Ploskey et al. 1984, 1993) may result from high rates of water exchange that limit the time available for nutrient processing and that flush many young-of-year fish downstream from the reservoir.

Successful reproduction and development of strong year-classes of fish are commonly associated with years of high water that inundates terrestrial vegetation (e.g., Benson 1968; Beckman and Elrod 1971; Nelson and Walburg 1977; Nelson 1978; Ploskey 1986; Kohler et al. 1993). Survey catches of many young fishes are highest in high-water years, despite substantial nutrient dilution by increased water volume. Years of high inflow in a storage reservoir result (1) in increased water surface area that effectively captures solar insolation, (2) in greater nutrient loadings (Westerdahl et al. 1981; Johnson and Ford 1987) derived in part from inundated terrestrial areas, and (3) in greater primary and secondary production (Benson and Cowell 1967; Mitchell 1975; Vollenweider 1975; Ostrofsky and Duthie 1978; McCammon and von Geldern 1979; Grimard and Jones 1982). Flooded vegetation affords fishes optimum spawning and nursery habitats that enhance survival (Martin et al. 1981) of species such as yellow perch *Perca flavescens* (Beckman and Elrod 1971), northern pike *Esox lucius* (Benson 1968; Hassler 1970), buffalo fishes *Ictiobus* spp. (Moen 1974), and common carp *Cyprinus carpio* (Gabel 1974).

Responses of largemouth bass *Micropterus salmoides* and, to a lesser extent, spotted bass *M. punctulatus* often have been studied because of their prominence in warmwater fisheries and their sensitivity to water level changes (Jenkins 1970). Responses of smallmouth bass *M. dolomieu*, which apparently differ from those of largemouth and spotted bass (Aggus and Elliott 1975), have received less attention. Several factors may be related to increased largemouth bass reproductive success in wet years: increased nutrient loading (Wright 1950; Wood 1951; Shirley and Andrews 1977; Aggus 1979), primary production (Benson 1968), and inundation of terrestrial vegetation (Bryant and Houser 1971; von Geldern 1971; Aggus and Elliott 1975; Houser and Rainwater 1975; Keith 1975; Rainwater and Houser 1975; Shirley and Andrews 1977; Strange et al. 1982; Miranda et al. 1984). Inundation of areas with terrestrial vegetation usually leads to increased food availability, followed by increased fish condition (weight:length ratio) and growth (Moffett 1943; Stroud 1948; Jackson 1958; Applegate et al. 1967; Mullan and Applegate 1968; Aggus and Elliott 1975; Allan and Romero 1975; Houser and Rainwater 1975; Rainwater and Houser 1975; Vogele and Rainwater 1975; Summerfelt and Shirley 1978; Shelton et al. 1979; Timmons et al. 1980).

Bull Shoals Reservoir

Impounded in 1950, Bull Shoals is an 18,400-ha, hydropower storage reservoir on the White River in northwest Arkansas (15,100 ha) and southwest Missouri (3,300 ha). It was authorized for hydroelectric power generation, flood control, water supply, and fish and wildlife. It is managed by the U.S. Army Corps of Engineers District, Little Rock, Arkansas. The fluctuation zone below normal (power) pool is inundated for part of every year and is relatively barren of vegetation except for some standing and fallen timber (Keith 1975). In contrast, the flood zone, which was significantly inundated in only 8 of 39 years (21%) between 1955 and 1993, supports extensive deciduous woody and herbaceous vegetation. In a very wet year, surface area can increase to about 28,000 ha or 1.5 times that of normal pool. The reservoir has a drainage area of 15,672 km^2, mean storage ratio of 0.91 years (range, 0.44–2.93 years), and maximum and mean depths of 61.3 and 20.4 m, respectively. Total dissolved solids average 150 mg/L (Leidy and Jenkins 1977), and median concentrations of total phosphorus, total nitrogen, and chlorophyll *a* were 0.015, 0.530, and 3.9 mg/L, respectively, in the mid-1970s (USEPA 1978).

Methods

Hydrological Variables

Reservoir hydrology data summarized as elevation–area–volume tables and daily inflows, releases, and water surface elevations during 1955–1993 were furnished by the Little Rock District, U.S. Army Corps of Engineers. Hydrology variables were derived from area or volume rather than from elevation so that dimensions would be consistent with those for nutrient loading, reservoir productivity, and fish biomass. Volume and area were calculated from reservoir elevation.

$$\text{VOL} = 105{,}747.84 - 1210.46 \times \text{ELEV} + 3.51 \times \text{ELEV}^2;$$

$$\text{AREA} = 326{,}448.88 - 3786.46 \times \text{ELEV} + 11.24 \times \text{ELEV}^2;$$

VOL is volume in millions of cubic meters, ELEV is elevation in meters referenced to mean sea level, and AREA is surface area in hectares. Both equations had coefficients of determination (r^2) greater than 0.99, ($P < 0.0001$, $N = 40$). We defined the annual hydrograph as covering the period from September through August of the next year so that the last month coincided with annual cove rotenone sampling of fish.

In the segmented-temporal method, we derived variables based on inflow, release, volume, area, or selected ratios of inflow to release or inflow to volume from time segments potentially affecting fish reproductive success (Table 1). Our concern during the variable-creating phase was completeness in terms of including all possible variables rather than assuring independence (no intercorrelation). With the principal components method, we generated four independent variables (SAS Institute 1989) for each of 14 segmented-temporal hydrology variables (Table 2). The initial 14 variables were selected because they were the simplest measures of different aspects of reservoir hydrology. They included annual and seasonal changes in volume, inflow, and release and multiyear and annual changes in surface area. Successive principal components are not intercorrelated. In the harmonic method, we fit a cosine function to each annual hydrograph of surface area (Trost 1991) and derived independent variables from coefficients of resulting equations. We describe the derivation of harmonic variables in detail below because the approach is probably unfamiliar to fisheries biologists.

The primary advantage of harmonic analysis is its ability to describe complex patterns in time series data with relatively few, easily interpretable coefficients. Harmonic analysis is useful if underlying hydrological processes approximate harmonic functions (Nestler 1993; Nestler and Long 1994). Harmonic analysis typically generates four coefficients: (1) the mean of individual values in the time series, as a measure of central tendency; (2) the period of time over which the function repeats (assumed to be 1 year); (3) the phase or time displacement of the start of the function; and (4) the amplitude or distance from the mean to the maximum excursion of the function, as an estimate of dispersion in a time series. These coefficients can be used singly or in various combinations to generate summary variables. For example, the range over which reservoir surface area changes during an annual cycle is approximated by twice the amplitude.

For streams, $\log_{10}$-transformed discharge data can be used as a surrogate for elevation because both water depth and water velocity can usually be expressed as simple power functions of discharge at a given cross section. Harmonic analyses of the White River discharge data were based on stream gauging records from the White River near Flippin, Arkansas (U.S. Geological Survey Gauge: 07055000), from 1920 to 1949, before Bull Shoals Reservoir was impounded. A cosine function was fitted to monthly discharges in nonlinear regression:

$$\text{HAR}_i = \text{AMP} \times \text{Cos} \cdot [(\text{MONTH}_i + \text{PHS}) \times 2.0 \times \pi];$$

HAR_i is stage (m) or $\log_{10}$-transformed discharge (m^3s^{-1}) standardized to a mean of 0.0 for month i, AMP is amplitude, MONTH_i is $(\text{month}_{(i)} - 0.5)/12$ (with i from 1 to 12 beginning with September and ending in August), PHS is phase, and π is pi (3.142). Similarly, a cosine function was fitted to end-of-month surface areas of Bull Shoals Reservoir:

$$\text{HECT}_i = \text{AMP} \times \text{Cos} \cdot [(\text{MONTH}_i + \text{PHS}) \times 2.0 \times \pi];$$

TABLE 1.—Abbreviations and definitions of segmented-temporal hydrological variables.

Variable	Definition
	Multiple season or year
CASUSP	Change in area, summer–spring = mean of hectares on 31 Mar, 30 Apr, and 31 May minus mean on 30 Jun, 31 Jul, and 31 Aug of (year − 1) divided by mean on 30 Jun, 31 Jul, and 31 Aug of (year − 1)
CASUSP2	Change in area, summer–spring (2 years) = mean of hectares on 31 Mar, 30 Apr, and 31 May minus mean on 30 Jun, 31 Jul, and 31 Aug of (year − 2) divided by mean on 30 Jun, 31, Jul, and 31 Aug of (year − 2)
CASUSU	Change in area, summer–summer = mean of hectares on 30 Jun, 31 Jul, and 31 Aug minus mean for the same dates in (year − 1) divided by mean on the same dates of (year − 1)
CASUSU2	Change in area, summer–summer (2 years) = mean of hectares on 30 Jun, 31 Jul, and 31 Aug minus the mean for the same dates in (year − 2) divided by mean on the same dates of (year − 2)
HADFLDS	Hectare-days of flooding during spawning (15 April–1 Jun) = $\log_{10}$ (1 + sum of hectare-days $\times 10^3$ when elevation exceeds 654 m above mean sea level). Elevation 654 m in Bull Shoals is the bottom of flood pool.
HADFLDSS	Hectare-days of flooding, spawning–sampling (15 April–15 Aug) = $\log_{10}$ (1 + sum of hectare-days $\times 10^3$ when elevation exceeds 654 m above mean sea level).
HADFLDPS	Hectare-days of flooding, postspawning (1 Jun–15 Aug) = $\log_{10}$ (1 + sum of hectare-days $\times 10^3$ when elevation exceeds 654 m above mean sea level).
XVOL1_8	Mean volume = mean of $\log_{10}$ (end-of-month $m^3 \times 10^6$), Jan–Aug
SINF1_8	Sum of inflow = $\log_{10}$ ($m^3 \times 10^6$), Jan–Aug
SREL1_8	Sum of release = $\log_{10}$ ($m^3 \times 10^6$), Jan–Aug
FR1_8	Flushing rate = sum of release volume/mean volume, Jan–Aug
RIR1_8	Ratio of inflow to release = inflow/release, Jan–Aug
	Fall (year − 1)
XVOL9_11	Mean volume = mean of $\log_{10}$ ($m^3 \times 10^6$) on 30 Sep, 31 Oct, and 30 Nov
SINF9_11	Sum of inflow = $\log_{10}$ (sum of $m^3 \times 10^6$), Sep–Nov
SREL9_11	Sum of release = $\log_{10}$ (sum of $m^3 \times 10^6$), Sep–Nov
FR9_11	Flushing rate = sum of release/mean volume, Sep–Nov
RIR9_11	Ratio of inflow to release = inflow/release, Sep–Nov
XA9_11	Mean area = mean of $\log_{10}$ (hectares) on 30 Sep, 31 Oct, and 30 Nov
PA9_11	Perimeter area = mean of $\log_{10}$ (hectares over depths $\leq$ 6 m) on 30 Sep, 31 Oct, and 30 Nov
CA9_11	Change in area (hectares) = (30 Nov area − 30 Sep area)/Nov area
	Spring
XVOL3_5	Mean volume = mean of $\log_{10}$ ($m^3 \times 10^6$) on 31 Mar, 30 Apr, and 31 May
SINF3_5	Sum of inflow = $\log_{10}$ (sum of $m^3 \times 10^6$), Mar–May
SREL3_5	Sum of release = $\log_{10}$ (sum of $m^3 \times 10^6$), Mar–May
FR3_5	Flushing rate = sum of release/mean volume, Mar–May
RIR3_5	Ratio of inflow to release = inflow/release, Mar–May
XA3_5	Mean area = mean of $\log_{10}$ (hectares) on 31 Mar, 30 Apr, and 31 May
PA3_5	Perimeter area = mean of $\log_{10}$ (hectares over depths $\leq$ 6 m) on 31 Mar, 30 Apr, and 31 May
CA3_5	Change in area (hectares) = $\log_{10}$ (31 Mar area − 31 May area)/30 Mar area)
	Summer
XVOL6_8	Mean volume = mean of $\log_{10}$ ($m^3 \times 10^6$) on 30 Jun, 31 Jul, and 31 Aug
SINF6_8	Sum of inflow = $\log_{10}$ (sum of $m^3 \times 10^6$), Jun–Aug
SREL6_8	Sum of release = $\log_{10}$ (sum of $m^3 \times 10^6$), Jun–Aug
FR6_8	Flushing rate = sum of release/mean volume, Jun–Aug
RIR6_8	Ratio of inflow to release = inflow/release, Jun–Aug
XA6_8	Mean area = mean of $\log_{10}$ (hectares) on 30 Jun, 31 Jul, and 31 Aug
PA6_8	Perimeter area = mean of $\log_{10}$ (hectares over depths $\leq$ 6 m) on 30 Jun, 31 Jul, and 31 Aug
CA6_8	Change in area (hectares) = (30 Jun area − 31 Aug area)/30 Jun

$HECT_i$ is $\log_{10}$ (surface area, in hectares) standardized to a mean of 0.0 for month i, and the independent variables are as just defined. The pattern of 12 end-of-month surface areas can be described by the coefficients of the harmonic function, such as mean, phase, and amplitude. These coefficients were suitable for regression analysis, whereas original data points were too numerous.

We derived three types of variables for use in regression analysis from coefficients of harmonic equations and related information (Table 3). The first type described the phase shift of a cosine function fitted to end-of-month surface areas for the lake from another cosine function fitted to monthly discharge of the unregulated White River. Similar phases were believed to be conducive to successful reproduction because timing should be keyed to natural hydrographs. The second type quantified patterns of change in reservoir area each year based on a redefined water year (September to August). The third type was based on antecedent hydrological conditions determined by harmonic analysis of surface area data in the year before collection of cove rotenone samples (Table 3).

TABLE 2.—Eigenvectors for the first four principal components (prin 1–4) in an analysis of 14 segmented-temporal hydrogical variables for Bull Shoals Reservoir over 34 years. Eigenvalues for principal components, with cumulative percent of variation explained in parentheses, were prin 1, 8.12 (63%); prin 2, 2.52 (82%); prin 3, 0.73 (88%); and prin 4, 0.45 (91%).

Variable	Prin 1	Prin 2	Prin 3	Prin 4
Mean volume (Jan–Aug)	0.3253	−0.0726	−0.0640	0.1831
Sum of inflow (Jan–Aug)	0.3299	0.0239	0.0529	−0.2535
Sum of release (Jan–Aug)	0.3233	0.1024	0.0624	−0.2223
Mean volume (Sep–Nov)	0.1700	0.4344	−0.2512	0.1102
Sum of inflow (Sep–Nov)	0.1769	0.4564	0.1967	0.0102
Sum of release (Sep–Nov)	0.1053	0.5221	0.1790	0.3843
Mean volume (Mar–May)	0.3212	−0.0887	0.0869	0.2075
Sum of inflow (Mar–May)	0.3142	−0.0306	0.2077	−0.2009
Sum of release (Mar–May)	0.2895	0.1072	0.3213	−0.4427
Change in area (summer–spring)	0.1752	−0.4392	0.2924	−0.0652
Two-year change in area (summer–spring)	0.2351	−0.2523	0.2687	0.5238
Mean volume (Jun–Aug)	0.2959	−0.1557	−0.2691	0.2430
Sum of inflow (Jun–Aug)	0.2493	−0.0166	−0.6001	−0.2453
Sum of release (Jun–Aug)	0.2969	−0.1126	−0.3336	0.1250

Biomass Variables

Rotenone samples of fish biomass were collected annually from three coves, one each representing the upper, middle, and lower regions of Bull Shoals Reservoir. Sampling was conducted by the Arkansas Game and Fish Commission before 1963 and after 1982 and by the former National Reservoir Research Program, with help from state biologists, in intervening years. Fish sampled from 1960 through 1971 were not measured but were categorized as age-0, intermediates, and adults based on lengths. We used biomass data for the age-0 category for these years. Standard 0.4- to 1.9-ha (mean, 1.1 ha) coves were sampled with rotenone (Grinstead et al. 1978; Davies and Shelton 1983) in August every year from 1972 to 1993. Cove depths were sufficient to include part of the thermocline. We made no adjustments for nonrecovery of fish by length-class or for differences in distributions between cove and open-water areas because our comparisons were among years for one lake. Cove rotenone data were obtained from standardized sampling procedures data bases (Arkansas Game and Fish Commission), in which length-frequency distributions were standardized on 25-mm length-classes.

Age-0 and age-1 groups were defined by the maximum length-class that each of the three species attained by August, and the biomass in these groups was summed to create dependent variables; however, classes of variables were referred to as small (<125-mm largemouth bass and <100-mm spotted and smallmouth bass) or intermediate (125–225-mm largemouth bass and 100–200-mm spotted and smallmouth bass) to indicate that they were based on length rather than on age. Decisions about max-

TABLE 3.—Abbreviations and definitions of variables used in harmonic analysis of data on White River discharge and Bull Shoals Reservoir surface area. Year was defined to end in August when cove rotenone samples were collected.

Variable	Definition
PHS_SHFT	Phase shift = absolute value of difference in phases of White River discharge and Bull Shoals surface area; it describes difference in timing between peak reservoir surface area and peak discharge in the unregulated White River
A_PHSSFT	Antecedent phase shift = same as PHS_SHFT above but for (year − 1); it evaluates effect of antecedent conditions
AMP	Amplitude = difference between mean and maximum areas as obtained from harmonic analysis; it describes maximum excursion of surface area in a year as part of general annual pattern
A_AMP	Antecedent amplitude = amplitude in (year − 1)
D_HA	Difference in hectares = difference in mean surface area between year of the cove rotenone sample and the previous year after subtraction of lowest recorded surface area; it describes change in surface area from previous year proportioned to the maximum possible change in surface area
ANTCEDNT	Antecedent conditions = a complex variable that describes antecedent mean areal change, amplitude, and phase shift (i.e., standardized estimates of D_HA plus A_AMP minus A_PHSSFT)
COR_RMSE	Corrected root mean square error = the sum of residuals remaining after harmonic analysis of concurrent years surface areas divided by previous year's amplitude; it describes how well the harmonic fit observed area data

imum lengths of the two age groups in mid-August were simple for years when length-at-age data were available (e.g., Bryant and Houser 1971; Aggus and Elliott 1975; Vogele 1975). We also examined 25-mm length-frequency plots for probable maximum lengths for the first two modal groups. We used kilograms per hectare instead of numbers per hectare because weight is more sensitive to fish length than numbers within 25-mm length-classes. All dependent biomass variables were transformed by taking the $\log_{10}([\text{kg/ha}] + 1)$ because before the transformation, variances were positively correlated with means and data were not normally distributed. We examined within-year coefficients of variation ($100 \cdot \text{SD/mean}$) of geometric mean biomass to assess the relative variation of cove rotenone sampling for each species and length-class.

Data Analysis

Correlation and regression analyses were used to find the best single- or multiple-regression models for predicting the biomass of small and intermediate black bass. Independent hydrology variables were matched by year with biomass variables for small black bass. Biomass variables for intermediate black bass often were referred to as next year's intermediates because they were matched with the previous year's hydrology, which should have influenced their production. Multiple-regression models were derived by regressing biomass variables on sets of hydrology variables with minimal intercorrelation ($\alpha < 0.05$, unless specifically noted). Correlation analysis was used to identify intercorrelations and promising combinations of independent variables for multiple-regression analysis. We used a stepwise regression routine with a maximum R^2 selection option to derive multiple-regression models (SAS Institute 1989). We withheld two years of data from the regression analysis, one high-water year (1979) and one average-water year (1992) to determine whether the best models would accurately predict relative year-class strength in those years. Strong year-classes were defined as meeting or exceeding the 75th-percentile biomass of each species and length-class over all years.

Results

Hydrology

Intercorrelations among hydrological variables at α equals 0.05 were common for segmented-temporal variables, nonexistent for principal component variables, and limited for harmonic variables. An example intercorrelation among segmented-temporal variables was the correlation ($N = 35$) of January–August mean volume with 29 other variables, 13 with r^2 greater than 0.64, 8 with r^2 in the range 0.36–0.64, and 8 with r^2 in the range 0.16–0.36. Only six temporal variables were independent of January–August mean volume. They included the ratio of inflow to release during three periods (September–November, January–August, and March–May) and three previous-fall variables (release, flushing rate, and change in area). Only about one-quarter of all possible intercorrelations among harmonic variables defined in Table 3 were significant. Amplitude was positively correlated with the difference between mean areas in the present and the previous year ($r^2 = 0.29$) and with antecedent condition ($r^2 = 0.24$), which summarizes the previous year's mean change in area, amplitude, and phase. Antecedent condition was inversely correlated with antecedent phase shift ($r^2 = 0.52$) and positively correlated with antecedent amplitude ($r^2 = 0.29$). Antecedent amplitude was inversely related to the difference between mean areas in the present and the previous year ($r^2 = 0.19$).

The first four principal components explained 91% of the variation in 14 segmented-temporal variables; each successive component contributed 63, 19, 6, and 3% to total explained variation (Table 2). Eigenvectors for the first principal component had moderately high positive and uniform loadings for current-year hydrological variables, whereas those for the second component were highest for fall variables and lowest for 1- and 2-year change-in-area variables (Table 2). Eigenvectors of the third and fourth principal components were difficult to interpret.

Harmonic functions fitted to annual patterns in reservoir surface area explained 84% of the variation over 39 years (Figure 1). Harmonic predictions were most accurate for surface areas less than or equal to normal pool area (18,400 ha) and less accurate for areas associated with flood pool elevations (Figures 1, 2).

Harmonic analysis of discharge of the unregulated White River and of end-of-month reservoir area indicated a phase shift (i.e., a time lag in occurrence of the harmonic's maxima) of 4 months at low discharge, 2.5 months at mean discharge, and 1.5 months at high discharge; however, catches of young black bass seldom were correlated with phase shift. An exception was the correlation of biomass of small largemouth bass with phase shift ($r = -0.36$; $P = 0.04$).

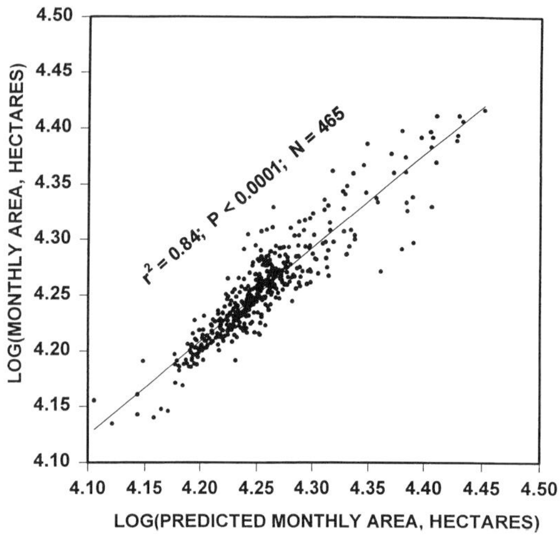

FIGURE 1.—Plot of correlation of observed end-of-month surface areas with predicted surface areas from cosine functions fit to area data of Bull Shoals Reservoir from 1955 through 1993. Logarithms are base 10.

Biomass

Examination of 22 years of coefficients of variation (CV) revealed higher sample variation in biomass estimates for smallmouth bass than in estimates for largemouth bass or spotted bass. For small size-classes, annual CV averaged 72% for smallmouth bass, 49% for largemouth bass, and 39% for spotted bass. For intermediate size-classes, annual CV for smallmouth bass (79%) was 1.8 times higher than that of largemouth bass (44%) and 1.6 times higher than that of spotted bass (50%). Coefficients of variation were less than 50% in 50–59% of the years for largemouth bass, in 68% of the years for spotted bass, but only in 32% of the years for smallmouth bass.

Biomass of intermediate largemouth bass was positively correlated with that of small largemouth bass in the previous year (r^2 = 0.59; P = 0.0001; N = 20). Similar, although less significant, correlations were obtained for small and next-year's intermediate spotted bass (r^2 = 0.30; P = 0.01; N = 20) but not for the two size-classes of smallmouth bass (r^2 = 0.09; P = 0.2; N = 20).

Relations of Biology to Hydrology

The percent contribution of a single species to total biomass of small black bass varied significantly among years depending upon the extent and duration of flooding (Figure 3). Largemouth and spotted basses made up most of the biomass of small black bass under all conditions; however, the percent contribution of an individual species to the total did not always parallel actual biomass trends. For example,

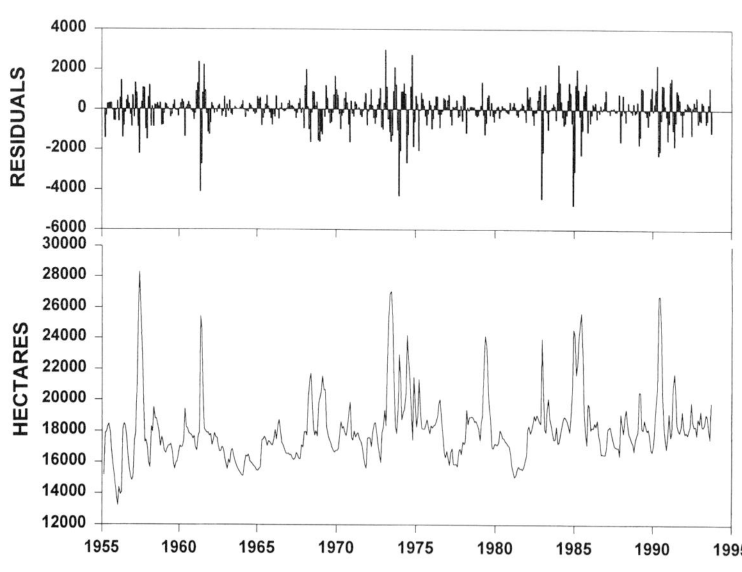

FIGURE 2.—Residuals (predicted − observed) of end-of-month surface areas from harmonic analysis and observed surface areas of Bull Shoals Reservoir from 1955 through 1993.

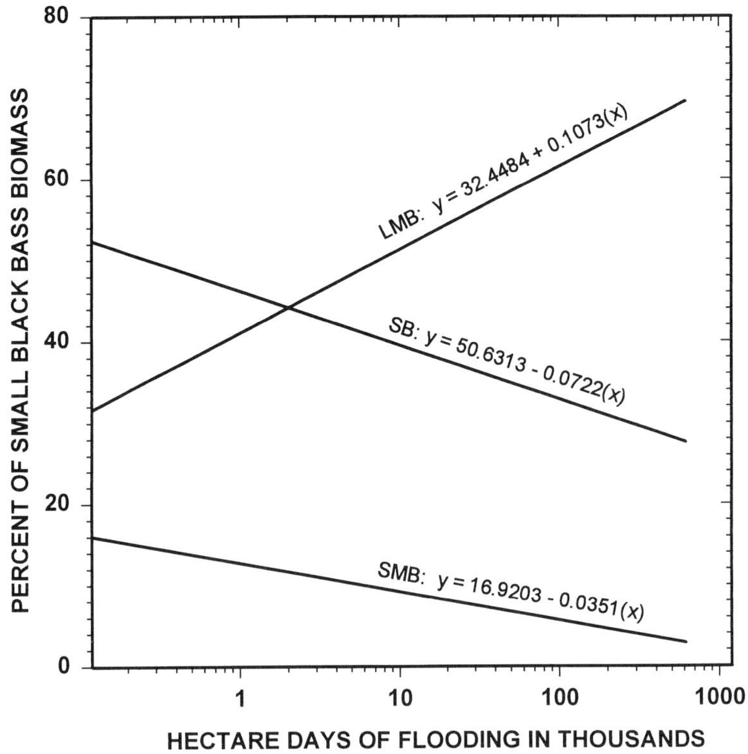

FIGURE 3.—Percent contribution (y) of biomass (kg/ha) of small largemouth bass (LMB; r^2 = 0.38; P = 0.0001), spotted bass (SB; r^2 = 0.35 ; P = 0.0002), and smallmouth bass (SMB; r^2 = 0.16; P = 0.02) to the total biomass of small black bass predicted from hectare-days of flooding (x) from 1 June to 15 August (N = 34) in Bull Shoals Reservoir. Small refers to largemouth bass shorter than 125 mm total length (TL) and to spotted and smallmouth bass shorter than 100 mm TL.

while the biomass of small spotted bass increased in high-water years (Figure 4), its percent contribution to total biomass gradually declined (Figure 3). For small largemouth bass, both the percent contribution to the total for small black bass (Figure 3) and actual biomass (Figures 4 and 5) were higher in wet than in dry or average years. The relative biomass contribution of small smallmouth bass was slightly lower in wet than in dry years (Figure 3). Only one significant correlation was obtained between the biomass of young smallmouth bass and reservoir hydrological variables. The biomass of intermediate smallmouth bass was inversely related to the January–August ratio of inflow to release in the previous year (r = −0.50; P = 0.02).

We found many significant positive correlations between the biomass of both length-classes of largemouth and spotted basses and hydrological variables. Biomasses of small largemouth and spotted basses were significantly correlated with hectare-days of flooding during the spawning, spawning-until-sampling (spawn-summer), and postspawning periods (Figure 4). The fit of regression lines was slightly better (3–6%) for the postspawning period than for the spawning period.

The best single-variable models for predicting the biomass of small and intermediate largemouth bass were derived in the segmented-temporal method. Summer perimeter area was the most significant correlate with the biomass of both length-classes. Regressions based on perimeter area in summer had r^2 statistics that were at least 10% higher than those based on amplitude from the harmonic method or upon the first principal component (Figure 5).

All three methods of deriving hydrological variables produced single-variable regression models of nearly equal quality for young spotted bass (Figure 6). Amplitude from the harmonic method and the first principal component yielded regressions with r^2 statistics that were equal to or only 3–6 percentage points less than those of perimeter area models.

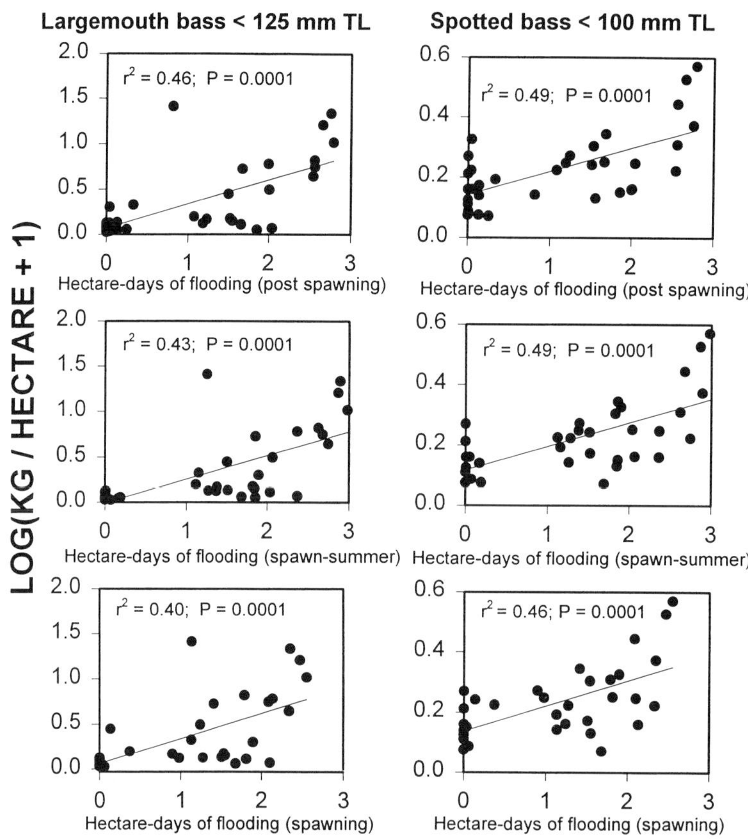

FIGURE 4.—Regression plots of the biomass ($\log_{10}[\{kg/ha\} + 1]$) of small largemouth and spotted bass in Bull Shoals Reservoir in August versus amount of flooding ($\log_{10}$[hectare days]) during three time periods: postspawning, spawning until sampling (spawn-summer), and spawning.

Mean perimeter area was a useful predictor of small spotted bass biomass when calculated for the June–August segment of the hydrograph and of intermediate spotted bass biomass when calculated for the March–May segment.

The biomass of young largemouth bass was predicted equally well by two-variable models derived from all three methods, but the best predictor of next year's intermediate largemouth bass came from the segmented-temporal method (Table 4; Figure 7). Regardless of approach, the first variable entered in the model had a strong positive effect—mean perimeter area for June–August (segmented-temporal method), first principal component, and amplitude (harmonic method). The second variable entered had a less significant negative coefficient in regressions for small largemouth bass (segmented-temporal and harmonic methods) and for intermediate largemouth bass from the principal-component method. The best model for forecasting the biomass of intermediate largemouth bass used the segmented-temporal method and included perimeter area in summer and the ratio of inflow to release in spring (Table 4). This model had an R^2 statistic that was 0.05 and 0.11 higher than those from the best models based upon harmonics and principal components, respectively. It also had no intercorrelation (r^2 = 0.03; P = 0.29), unlike the harmonic model (r^2 = 0.24; P = 0.002).

For spotted bass, R^2 statistics of two-variable models from the segmented-temporal method were only 0.03–0.04 higher than those from the harmonic method, but they were 0.09–0.16 higher than those from the principal-components method (Figure 8; Table 4). The second principal component entered into models for both small and intermediate spotted bass had negative coefficients. Coefficients for variables with second strongest effects for segmented-temporal and harmonic models were both positive and negative. There were no significant intercorre-

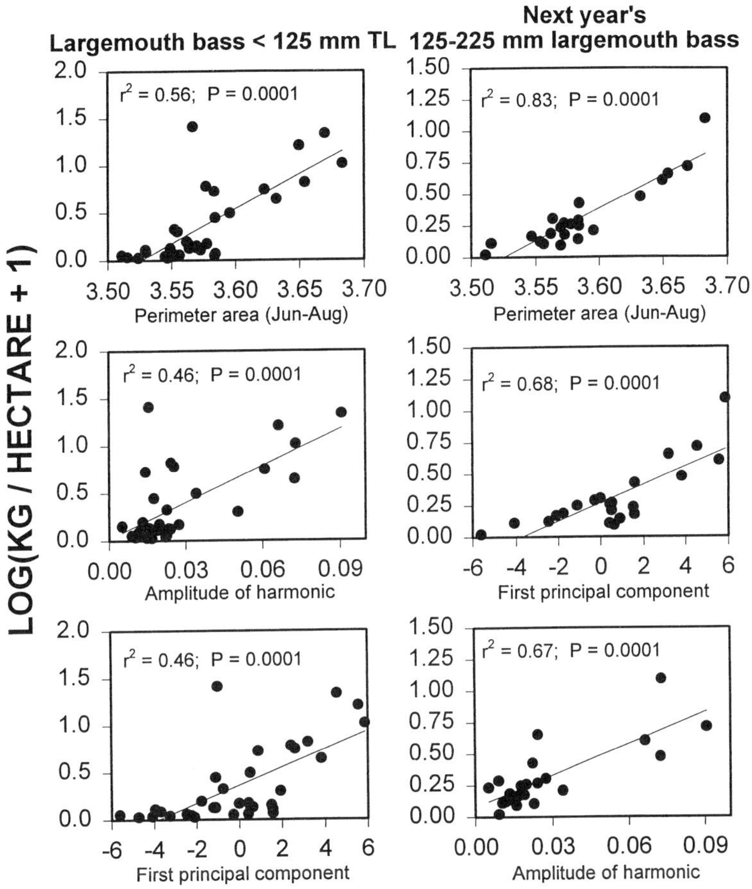

FIGURE 5.—Regression plots of the biomass ($\log_{10}$[{kg/ha}+ 1]) of two size-classes of largemouth bass in August versus June–August mean perimeter area ($\log_{10}$[hectares ≤ 6 m deep], segmented-temporal method), amplitude of a cosine function (harmonic method), or the first principal component describing 14 hydrology variables for Bull Shoals Reservoir.

lations between first and second hydrological variables entered in models derived by the three methods.

Some significant three- and four-variable models were found for predicting the biomass of small and intermediate black basses (Table 4), but rarely was there substantial improvement in fit over two-variable models. Surprisingly, one three-variable model emerged for predicting the biomass of intermediate smallmouth bass, after only one significant one-variable model was obtained. This three-variable model indicated significant negative effects of January–August ratio of inflow to release and of hectare-days of flooding from spawning until sampling. It suggested a positive effect of summer-to-spring change-in-area; however, the model contained a significant intercorrelation between summer-to-spring change-in-area and hectare-days of flooding from spawning until sampling in August (r^2 = 0.27; P = 0.002).

Predicted and observed biomasses of small largemouth bass and spotted bass were greater than or equal to the 75th-percentile biomass in 1979 (a wet year) and less than the 75th-percentile biomass in 1992, a year of average inflow (Table 5). Predicted and observed biomasses of next year's intermediate spotted bass in both 1979 and 1992 were less than the 75th-percentile biomass. Observed biomass of intermediate largemouth bass in 1980 was substantially less than the 75th-percentile biomass and biomass predicted from 1979 hydrology variables.

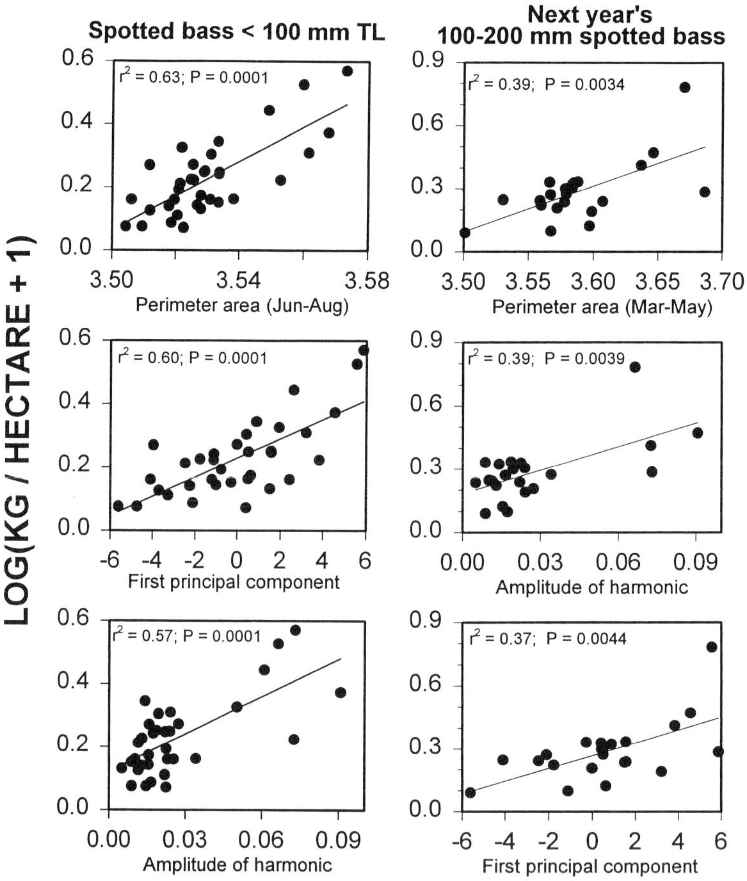

FIGURE 6.—Regression plots of the biomass ($\log_{10}[\{kg/ha\} + 1]$) of two size-classes of spotted bass in August as a function of June–August or March–May mean perimeter area ($\log_{10}$[hectares ≤ 6 m deep], segmented-temporal method), amplitude of a cosine function (harmonic method), or the first principal component describing 14 hydrology variables for Bull Shoals Reservoir.

Discussion

Intercorrelations between segmented-temporal hydrological variables were common because of the way variables were defined or because events in time series were autocorrelated. For example, very high inflow in spring may result in above-average surface area in spring and summer because operators cannot follow the typical rule curve for managing capacity. When independent variables are correlated, regression coefficients and sums of squares are not unique but depend on other variables in a multiple-regression model (Neter and Wasserman 1974).

We accepted some intercorrelation between harmonic variables when we believed that one of a pair of intercorrelated variables provided substantially different information. An example was the retention of change in mean area among years in a two-variable model with amplitude, although they were correlated (r^2 = 0.29; P = 0.0004). Change in mean area measures differences in the baselines about which two consecutive harmonics cycle, whereas amplitude is a function of the deviation in area from the annual mean.

The principal components method was straightforward and foolproof for avoiding intercorrelations, but it did not produce the best regression models. Our first and second principal components, explaining 82% of hydrological variation, could be broadly associated with positive effects of the wetness of the current year (first component) and negative effects of low flow and volume in the previous

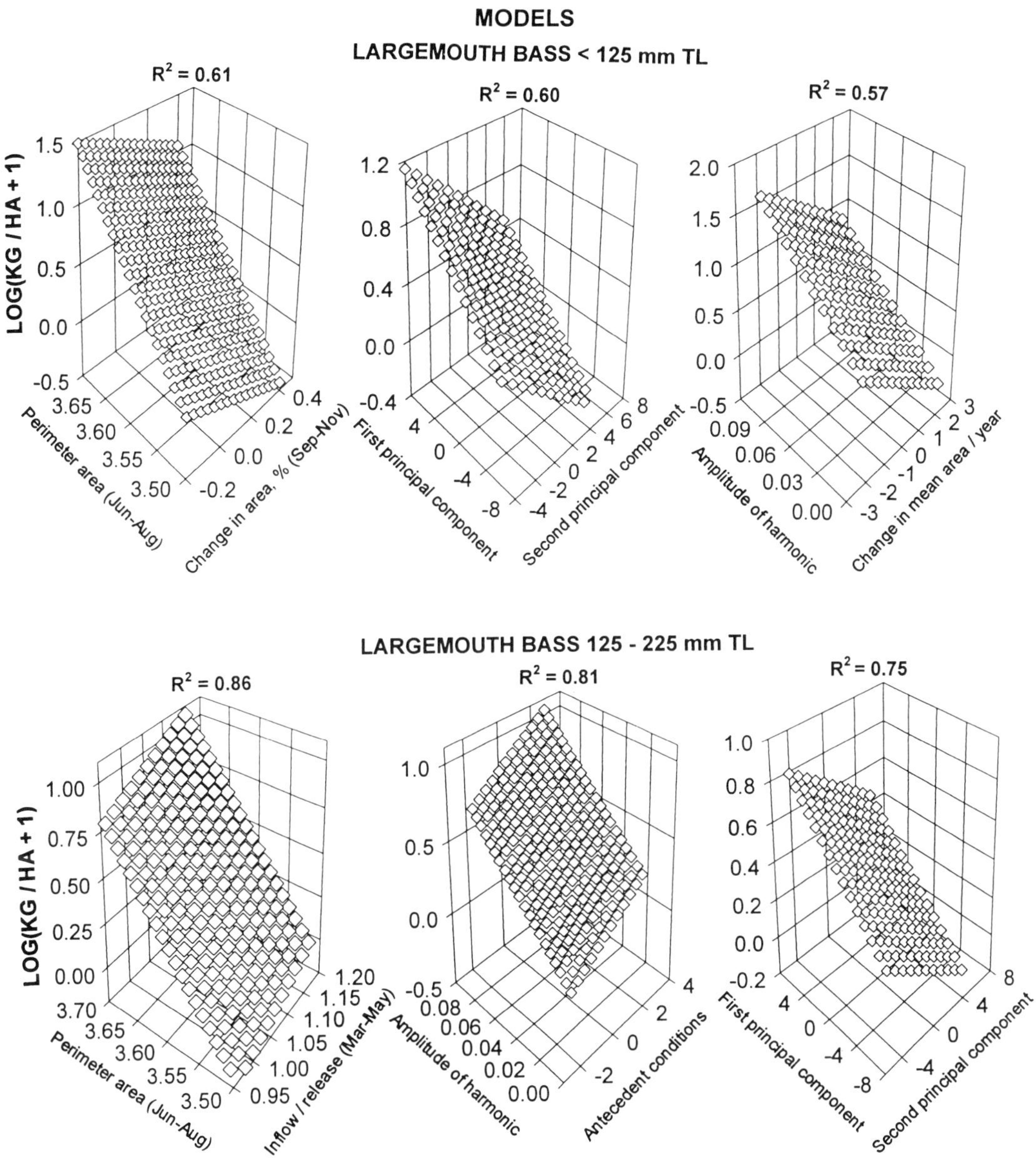

FIGURE 7.—Surface plots of two-variable models predicting the biomass ($\log_{10}[\{kg/ha\} + 1]$) of two size-classes of largemouth bass in August for Bull Shoals Reservoir. Perimeter area has a $\log_{10}$ scale; change in annual area is proportional change from the previous year.

fall (second component). We were unable to interpret the meaning of the third and fourth principal components, and they were not selected in stepwise regression. Principal components analysis may not produce the best regression models because it is not sensitive to specific time segments affecting year-class strength. Also, a few hydrological variables from which principal components were derived may have little or no effect on year-class strength and yet still influence resulting principal components.

The harmonic method provided good explanation of variation in surface area patterns (Figures 1, 2) and of the biomass of young largemouth and spotted basses (Figures 5–8; Table 4) with relatively

TABLE 4.—Multiple-regression models for predicting $\log_{10}$ (kg/ha) of small and next-year's intermediate length-classes of black bass in Bull Shoals Reservoir. Definitions of independent variables associated with each method are defined in Table 1 (segmented-temporal method), Table 2 (principal components method), and Table 3 (harmonic method).

Total length	Independent terms and associated statistics
	Largemouth bass
<125 mm	−26.01 + 7.39 (PA6_8) − 0.07 (CA9_11) R^2 = 0.61; P = 0.0001; N = 32; method: segmented-temporal
	0.37 + 0.06 (PRIN 1) − 0.07 (PRIN 2) R^2 = 0.60; P = 0.0001; N = 32; method: principal components
	0.10 + 10.07 (AMP) + 0.17 (D_HA) R^2 = 0.57; P = 0.0001; N = 32; method: harmonic
125–225 mm	−18.7 + 4.90 (PA6_8) + 1.39 (RIR3_5) R^2 = 0.86; P = 0.0001; N = 20; method: segmented-temporal
	0.12 + 8.60 (AMP) + 0.09 (ANTCEDNT) R^2 = 0.81; P = 0.0001; N = 20; method: harmonic
	0.29 + 0.04 (PRIN 1) − 0.04 (PRIN 2) R^2 = 0.75; P = 0.0001; N = 20; method: principal components
	−3.89 + 5.85 (PA6_8) − 3.18 (XA9_11) + 0.43 (CA9_11) R^2 = 0.90; P = 0.0001; N = 20; method: segmented-temporal
	0.30 + 0.04 (PRIN 1) − 0.04 (PRIN 2) − 0.03 (PRIN 3) R^2 = 0.80; P = 0.0001; N = 20; method: principal components
	0.16 + 8.65 (AMP) − 0.50 (PHS_SHFT) + 0.06 (ANTCEDNT) + 3.44 (COR_RMSE) R^2 = 0.87; P = 0.0001; N = 20; method: harmonic
	Spotted bass
<100 mm	−6.85 + 1.86 (XVOL1_8) + 0.44 (RIR3_5) R^2 = 0.68; P = 0.0001; N = 32; method: segmented-temporal
	0.12 + 3.83 (AMP) − 26.31 (COR_RMSE) R^2 = 0.65; P = 0.0001; N = 32; method: harmonic
	0.24 + 0.02 (PRIN 1) − 0.01 (PRIN 2) R^2 = 0.59; P = 0.0001; N = 32; method: principal components
100–200 mm	−8.01 + 6.40 (CASUSP) + 2.34 (XVOL9_11) R^2 = 0.54; P = 0.0001; N = 20; method: segmented-temporal
	0.17 + 3.28 (AMP) + 0.07 (A_PHSSFT) R^2 = 0.50; P = 0.0001; N = 20; method: harmonic
	0.27 + 0.02 (PRIN 1) − 0.01 (PRIN 2) R^2 = 0.38; P = 0.0001; N = 20; method: principal components
	Smallmouth bass
100–200 mm	3.52 − 3.33 (RIR1_8) + 3.18 (CASUSP) − 0.07 (HADFLDSS) R^2 = 0.56; P = 0.0038; N = 20; method: segmented-temporal

few variables. It is similar to the principal components analysis in that it reduces the number of variables, but unlike higher level principal components, its coefficients were relatively easy to interpret. From a modeling standpoint, the harmonic method has two major advantages over the other methods. First, it can accurately describe complex hydrological patterns with a few coefficients, which facilitates comparisons of operating alternatives. Second, harmonic variables can be easily calculated by using nonlinear regression. In contrast, models based on segmented-temporal or principal components methods require calculation of many variables (Tables 1, 2) and variable reduction by eliminating intercorrelated variables or principal components analysis. Principal components also would have to be verified for simulation of different alternatives. If the harmonic method has a weakness, it may be that resulting coefficients describe an annual pattern and lack specificity for critical segments of the hydrograph, such as the postspawning period.

Phase shift, as identified by harmonic analysis, was unrelated to reproductive success of spotted and smallmouth bass and was only a weak negative correlate with biomass of small largemouth bass (r^2 = 0.13; P = 0.04). This correlation may be coincidental because phase shift was less in wet years with high perimeter area than in dry years, and perimeter area had a strong positive effect (Figure 5). Also, we would expect all species to be

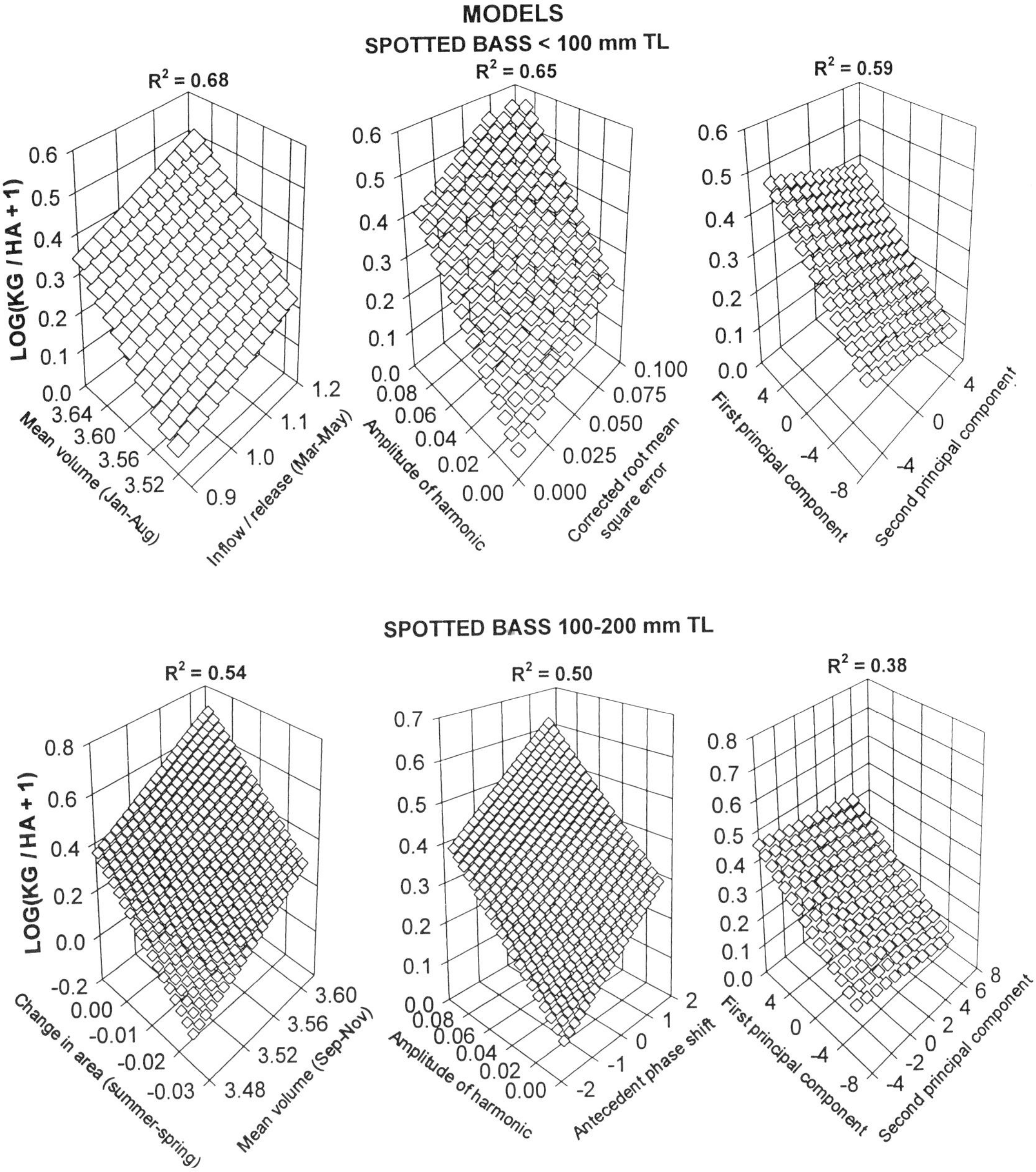

FIGURE 8.—Surface plots of two-variable models predicting the biomass ($\log_{10}[\{kg/ha\} + 1]$) of two size-classes of spotted bass in August for Bull Shoals Reservoir. Mean volume is the mean of $\log_{10}(10^6 m^3)$ for January–August; change in area is the proportional seasonal change in the previous year.

affected similarly by a phase shift that might disrupt spawning or food availability for age-0 black bass at critical times. Effect of phase shift probably is moderated by comparable shifts in temperature cycles, because seasonal temperature changes occur more slowly in the reservoir than in the river.

Perimeter area from June through August, used in the segmented-temporal method, often was the most significant variable in models for largemouth and spotted basses (Figures 5–7). Hectare-days of flooding during the postspawning period also explained slightly more variation in the biomass of small largemouth and spotted bass than did flooding during the spawning season (Figure 4). This

TABLE 5.—Predicted and observed biomasses (kg/ha) of largemouth bass and spotted bass for a high-water year in 1979 and an average-water year in 1992 and the 75th percentile of observed biomass for all years (1955–1993, except 1979 and 1992) in Bull Shoals Reservoir. Ranges in observed biomass among coves are given in parentheses.

Biomass	1979	1992	All other years: 75th percentile
<125-mm largemouth bass			
Predicted			
Segmented-temporal method	6.2	1.1	
Harmonic method	4.5	0.8	
Principal components method	3.4	0.9	
Observed: Cove rotenone	3.4 (1.5–8.6)	0.1 (0.04–0.12)	3.4
Next year's 125- to 225-mm largemouth bass			
Predicted			
Segmented-temporal method	3.3	1.1	
Harmonic method	6.2	0.8	
Principal components method	2.0	0.4	
Observed: Cove rotenone	0.4 (0.1–0.6)	0.7 (0.4–0.9)	1.7
<100-mm spotted bass			
Predicted			
Segmented-temporal method	1.4	0.7	
Harmonic method	1.6	0.4	
Principal components method	1.1	0.8	
Observed: Cove rotenone	1.0 (0.6–1.2)	0.7 (0.6–0.8)	0.9
Next year's 100- to 200-mm spotted bass			
Predicted			
Segmented-temporal method	0.5	1.1	
Harmonic method	0.5	0.6	
Principal components method	0.7	0.8	
Observed: Cove rotenone	0.8 (0.5–1.4)	0.9 (0.5–1.2)	1.3

correlation is consistent with findings of Aggus and Elliott (1975) for these species in Bull Shoals Reservoir, although our correlation based on 32 years of data did not explain as much variation. Aggus and Elliott found coefficients of determination and probabilities of $r^2 = 0.52$ and $P = 0.1$ for the spawning period, $r^2 = 0.69$ and $P = 0.05$ for the spawning-to-sampling period, and $r^2 = 0.83$ and $P = 0.01$ for the postspawning period. Differences in fit may be a function of sample size (32 years in our study versus 7 in theirs) or of the accuracy of identifying age-0 fish. Our use of length to estimate age obviously was less accurate than estimating age from hard body parts.

The importance of postspawning survival of age-0 largemouth bass may be deduced from observations of hatching success versus year-class strength by Kohler et al. (1993). They observed that hatching success was disrupted by rapidly rising or declining water levels in two Illinois reservoirs. Recruitment, as indicated by the fall electrofishing catch of juveniles, was considerably higher in high-water years than in low-water years and apparently was unrelated to hatching success. These results seem intuitive for *r*-selected organisms (Odum 1971) that are highly fecund but succeed in producing strong year-classes only when environmental conditions and carrying capacity permit.

The number of years of data required to identify relations between catches of young fish and reservoir hydrology depends more on the quality of the data than on sample size. Our analysis of 32 years of data produced empirical relations similar to those first obtained with 7–9 years of data by Aggus and Elliott (1975), Rainwater and Houser (1975), and Aggus (1979). Relations between the species composition of small black bass in August and hectare-days of flooding depicted in Figure 4 ($N = 32$) are virtually identical to those ($N = 9$) presented by Aggus (1979). Relationships derived in previous studies were similar to those we obtained despite a fourfold increase in sample size, although coefficients of determination have declined. Reduced explained variation could be a function of sample size. We may now be approaching the true explained variability, or it may be related to how age-0 and age-1 bass were defined, or both. Aggus and Elliott's (1975) use of scales to age fish as age-0 was superior to our assignment of age according to length. We could not legitimately vary maximum length-at-age assignments among years without year-specific age data, most of which were lost after the demise of the former National Reservoir Research Program. Nevertheless, there were obvious differences in length-frequency distributions among years. Species-specific correlations of the biomass of intermediate black bass with that of small black bass the previous year indicate that the biomass of age-0 largemouth and spotted bass in August is a fairly reliable indicator of future year-class strength in Bull Shoals Reservoir. Relationships also suggest that our maximum length-at-age definitions were best for largemouth bass, intermediate for spotted bass, and worst for smallmouth bass. The most likely classification errors are those that would inflate biomass variables by erroneously including weights of bass of the next age-class. These errors should be worst in years of slow growth, which typically coincided with years of below-average water levels. These classification errors should reduce the likelihood of finding significant positive correlations with water level variables by increasing the biomass of small bass in dry years.

Trends in species composition do not necessarily parallel trends in biomass. For example, the biomass of small spotted bass increased in years of high water, but it was so dwarfed by increases in largemouth bass biomass that its percent contribution declined. Apparently, both species respond positively to flooding, but largemouth bass either have a competitive advantage over spotted bass in wet years or simply are more fecund and better able to exploit inundated habitat.

A paucity of age information in historical data sets is common, but it does not preclude a search for empirical relations, although it may hinder detection of weaker effects. A stunted population of older fish in a reservoir could suggest exceptional reproduction every year if examined solely by length-frequency data. A few years of length-at-age data or a single study of age and growth may be sufficient to detect this problem. Chances of finding significant empirical relations for reservoirs where effects are less pronounced than those documented for Bull Shoals Reservoir would be greatly enhanced by accurately estimating age-0 and age-1 catches every year.

Although time consuming, the segmented-temporal approach usually produced the best, most easily interpreted predictive models (Table 4). These variables were the easiest and most intuitive to create and therefore would be particularly amenable for use in multipurpose evaluation models. Model building was the most time-consuming part of this method because we avoided including intercorrelated hydrological variables in multiple-regression models.

Why were empirical relations between hydrology and the production of young bass observed for Bull Shoals Reservoir? Most fisheries biologists who have studied recruitment of largemouth and spotted bass in this reservoir attribute strong year-classes to high inflow and resultant flooding of terrestrial vegetation (Bryant and Houser 1971; Aggus and Elliott 1975; Houser and Rainwater 1975; Keith 1975; Rainwater and Houser 1975; Aggus 1979). Flooding of vegetation apparently alters the trophic base because age-0 largemouth and spotted bass grow faster despite high densities and switch to a more productive diet of fish earlier than they do in drier years (Applegate et al. 1967). An abundance of suitable forage apparently is critical. Accelerated growth and an increased complexity of habitat as refuge from predation were apparently important factors increasing survival, especially during the first summer of life (Aggus and Elliott 1975). Also, predation was considered to be a major factor affecting year-class strength as higher overwinter survival occurred for larger age-0 largemouth bass than for smaller siblings.

If the most important variable affecting production of strong year-classes is postspawning survival, a productive strategy of water level management would consist of assuring high water and acceptable habitat after an acceptably wet spring. We could not determine from correlations the relative importance of high inflow versus inundation of terrestrial vegetation for producing strong year-classes of largemouth and spotted bass. Both factors probably are critical, but more research aimed at quantifying relative contributions is needed. Flooding of terrestrial vegetation in a year of average inflow probably does not provide for increasing largemouth bass growth and recruitment as much as inundating vegetation in a year of naturally high inflow and nutrient loading (Strange et al. 1982; Miranda et al. 1984).

Maintenance of high water in summer after a wet spring may not always be sufficient to assure the development of a strong year-class. Comparisons of observed biomass from rotenone samples with the 75th-percentile biomass indicated that predictions for small largemouth and spotted bass in 1979 and 1992 and of intermediate largemouth and spotted bass in 1992 were reasonably accurate for predicting reproductive success (Table 5); however, all three methods inaccurately forecasted greater than 75th-percentile biomass for next year's 125–225-mm largemouth bass in 1980, a severe drought year that began in fall 1979 with plummeting reservoir area (Figure 2). Mortality resulting in the failure of this promising 1979 year-class must have occurred after August of the first year of life and before August of the second year. A lack of significant correlations of intermediate largemouth bass biomass with current-year hydrological variables (i.e., spring and summer; $N = 20$) suggests that mortality usually occurs in fall or winter rather than in the next spring. Fall and winter are stressful because lower temperatures and shorter days lead to reduced food availability and intake (Adams et al. 1982). At lower latitudes, predation, shelter, and length of young bass are important factors affecting winter survival (Miranda and Hubbard 1994). For management, predictions of above-average biomass of young largemouth and spotted bass can be the first signal of the need to request maintenance of high pools to increase survival throughout the first year of a promising year-class. Fall sampling could be used to verify predictions and reassess the need for continuing with the strategy through winter.

Responses of smallmouth bass to hydrology were different from those of largemouth and spotted bass, indicating that water level management may benefit reproduction of some species at the expense of others. The biomass of intermediate smallmouth bass was inversely related to January–August inflow in the previous year. High variability in estimates of smallmouth bass biomass relative to estimates for other species may account for correlation with a single hydrological variable. Differences in annual coefficients of variation and apparent responses among species may be related to differences in habitat utilization (distributions) and predator avoidance responses. Largemouth bass and especially spotted bass prefer to nest near brush shelters or in flooded terrestrial vegetation, whereas smallmouth bass show no preference for brush areas but prefer gravel or broken rock (Vogele and Rainwater 1975). Smallmouth bass fry schools usually disperse much more rapidly than schools of largemouth and spotted bass (Vogele 1975, 1981). Adult spotted bass have been observed apparently moving schools of fry into areas of heavy cover (Vogele 1975). Based on 18 years of observing age-0 black basses using scuba gear in Bull Shoals Reservoir, L. Vogele (formerly of the National Reservoir Research Program, personal communication) noted that age-0 smallmouth bass prefer rocky areas near the bottom (Vogele 1981) and hide in crevices when threatened. In contrast, age-0 spotted and largemouth bass school in or near heavy cover when available.

All three empirical methods described in this paper provide concordant predictions for spotted and largemouth bass in Bull Shoals Reservoir and would be useful for identifying potential relationships in other reservoirs. The use of principal components analysis is a logical preliminary step in the search for relationships between fish reproductive success and hydrology. It would indicate whether a closer examination with the segmented-temporal or harmonic method was warranted. The latter, more detailed method, would be preferred if regression models were to be incorporated in a large multiple-use model (e.g., Ploskey et al. 1993).

There are several advantages to having empirical models for a specific reservoir. First, many federal agencies will be evaluating multipurpose uses of reservoirs and river systems in the next decade. This evaluation was done in 1992–1994 for the Missouri River system and in 1993–1994 for major river basins in Alabama, Georgia, and Florida. Effects of alternative reservoir operations on important fish populations need to be quantitatively modeled for adequate representation of reservoir fisheries concerns in trade-off analyses. Second, fisheries managers will find resource development agencies more receptive to water level plans that have a quantitative predictive capability.

An incomplete understanding of intricate details of biological responses does not and should not inhibit managers from exploring and using empirical relations to improve reservoir fisheries. Empirical ecological relations have proven very useful for assessing reservoir fisheries potential (Jenkins 1967; Jenkins and Morais 1971; Oglesby 1982). Increasing demands for hydropower, flood control, water supply, irrigation, navigation, instream flow, and recreation (including fisheries), require equitable evaluations that will not wait for comprehensive understanding.

Methods described in this paper should help fisheries managers identify useful empirical predictors for specific reservoirs and stimulate future research. Good predictors can improve water resource decisions to benefit fisheries, but they are not a substitute for careful biological studies of proximate causes and effects. Our methods are empirical in that concurrent biological information that might help explain relationships was not collected every year. We were fortunate to have numerous biological observations published by the former National Reservoir Research Program to help identify probable causes and effects.

Acknowledgments

Special thanks are extended to Randy Sheldon, U.S. Army Corps of Engineers District, Little Rock, Arkansas, for supplying hydrological data for Bull Shoals Reservoir. We also greatly appreciate efforts of many biologists with the Arkansas Game and Fish Commission and the former National Reservoir Research Program who helped collect 34 years of cove rotenone samples. This research was conducted under the Environmental Impact Research Program, Assessing Basin-Wide Impacts of Reservoir Operation on Aquatic Resources work unit of the U.S. Army Corps of Engineers, by the U.S. Army Corps of Engineers Waterways Experiment Station. Permission was granted by the Chief of Engineers to publish this information.

References

Adams, M. S., R. B. Mclean, and J. A. Parrotta. 1982. Energy partitioning in largemouth bass under conditions of seasonally fluctuating prey availability. Transactions of the American Fisheries Society 111:549–558.

Aggus, L. R. 1979. Effects of weather on freshwater fish predator–prey dynamics. Pages 47–56 *in* H. Clepper, editor. Predator–prey systems in fisheries management. Sport Fishing Institute, Washington, DC.

Aggus, L. R., and G. V. Elliott. 1975. Effects of cover and food on year-class strength of largemouth bass. Pages 317–322 *in* Stroud and Clepper (1975).

Aggus, L. R., and S. A. Lewis. 1978. Environmental conditions and standing crops of fishes in predator-stocking evaluation reservoirs. Proceedings of the Annual Conference Southeastern Association of Fish and Wildlife Agencies 30(1976):131–140.

Allan, R. C., and J. Romero. 1975. Underwater observations of largemouth bass spawning and survival in Lake Mead. Pages 104–112 *in* Stroud and Clepper (1975).

Applegate, R. L., J. W. Mullan, and D. I. Morais. 1967. Food and growth of six centrarchids from shoreline areas of Bull Shoals Reservoir. Proceedings of the Annual Conference Southeastern Association of Game and Fish Commissioners 20(1966):469–482.

Beckman, L. G., and J. H. Elrod. 1971. Apparent abundance and distribution of young-of-year fishes in Lake Oahe, 1965–69. Pages 333–347 *in* G. E. Hall, editor. Reservoir fisheries and limnology. American Fisheries Society, Special Publication 8, Bethesda, Maryland.

Bennett, D. H. 1975. Effects of pumped storage project operations on the spawning success of centrarchid fishes in Leesville Lake, Virginia. Doctoral dissertation. Virginia Polytechnic Institute and State University, Blacksburg.

Bennett, D. H., O. E. Maughan, and D. B. Jester, Jr. 1985. Generalized model for predicting spawning success of fish in reservoirs with fluctuating water levels. North American Journal of Fisheries Management 5:12–20.

Benson, N. G. 1968. Review of fishery studies on Missouri River mainstem impoundments. U.S. Fish and Wildlife Service, Bureau of Sport Fisheries and Wildlife Research Report 71, Washington, DC.

Benson, N. G., and B. C. Cowell. 1967. The environment and plankton density in Missouri River reservoirs. Pages 358–373 *in* Reservoir Committee, editors. Reservoir fishery resources symposium. American Fisheries Society, Southern Division, Reservoir Committee, Bethesda, Maryland.

Bryant, H. E., and A. Houser. 1971. Population estimates and growth of largemouth bass in Beaver and Bull Shoals reservoirs. Pages 349–357 *in* G. E. Hall, editor. Reservoir fisheries and limnology. American Fisheries Society, Special Publication 8, Bethesda, Maryland.

Davies, W. D., and W. L. Shelton. 1983. Sampling with toxicants. Pages 199–213 *in* L. A. Nielson and D. L. Johnson, editors. Fisheries techniques. American Fisheries Society, Bethesda, Maryland.

Estes, R. D. 1971. The effects of the Smith Mountain pump storage project on the fishery of the lower reservoir, Leesville, Virginia. Doctoral dissertation. Virginia Polytechnic Institute and State University, Blacksburg.

Gabel, J. A. 1974. Species and age composition of trap net catches in Lake Oahe, South Dakota, 1963–67. U.S. Fish and Wildlife Service, Technical Paper 75.

Gasaway, C. R. 1970. Changes in the fish population in Lake Francis Case in South Dakota in the first 16 years of impoundment. U.S. Fish and Wildlife Service, Technical Paper 56.

Grimard, Y., and H. G. Jones. 1982. Trophic upsurge in new reservoirs: a model for total phosphorus concentrations. Canadian Journal of Fisheries and Aquatic Sciences 39:1473–1483.

Grinstead, B. G., R. M. Gennings, G. R. Hooper, C. A. Schultz, and D. A. Whorton. 1978. Estimation of standing crop of fishes in the PSE reservoirs. Proceedings of the Annual Conference Southeastern Association of Fish and Wildlife Agencies 30(1976): 120–130.

Hassler, T. J. 1970. Environmental influences on early development and year-class strength of northern pike in Lakes Oahe and Sharpe, South Dakota. Transactions of the American Fisheries Society 99:369–380.

Heisey, P. G., D. Mathur, and N. C. Magnusson. 1980. Accelerated growth of smallmouth bass in a pumped storage system. Transactions of the American Fisheries Society 109:371–377.

Houser, A., and W. C. Rainwater. 1975. Production of largemouth bass in Beaver and Bull Shoals lakes. Pages 310–316 *in* Stroud and Clepper (1975).

Jackson, S. W., Jr. 1958. Comparison of the age and growth of four fishes from Lower and Upper Spavinaw lakes, Oklahoma. Proceedings of the Annual Conference Southeastern Association of Game and Fish Commissioners 11(1957):232–249.

Jenkins, R. M. 1967. The influence of some environmental factors on standing crop and harvest of fishes in U.S. reservoirs. Pages 298–321 *in* Reservoir Committee, editors. Reservoir fishery resources symposium. American Fisheries Society, Southern Division, Reservoir Committee, Bethesda, Maryland.

Jenkins, R. M. 1970. The influence of engineering design and operation and other environmental factors on reservoir fishery resources. Water Resources Bulletin 6:110–119.

Jenkins, R. M., and D. I. Morais. 1971. Reservoir sport fishing effort and harvest in relation to environmental variables. Pages 371–384 *in* G. E. Hall, editor. Reservoir fisheries and limnology. American Fisheries Society, Special Publication 8, Bethesda, Maryland.

Johnson, M. C., and D. E. Ford. 1987. Two-dimensional modeling of storm event inflow processes at DeGray Lake, Arkansas. Pages 565–600 *in* R. H. Kennedy and J. Nix, editors. Proceedings of the DeGray Lake symposium. U.S. Army Corps of Engineers Waterways Experiment Station, Technical Report E-87-4, Vicksburg, Mississippi.

Keith, W. E. 1975. Management by water level manipulation. Pages 489–497 *in* Stroud and Clepper (1975).

Kohler, C. C., R. J. Sheenan, and J. J. Sweatman. 1993. Largemouth bass hatching success and first-winter survival in two Illinois reservoirs. North American Journal of Fisheries Management 13:125–133.

Leidy, G. R., and R. M. Jenkins. 1977. The development of fishery compartments and population rate coefficients for use in reservoir ecosystem modeling. U.S. Army Corps of Engineers Waterways Experiment Station, Contract Report Y-77-1, Vicksburg, Mississippi.

Martin, D. B., L. J. Mengel, J. F. Novotny, and C. H. Walburg. 1981. Spring and summer water levels in a Missouri River reservoir: effects on age-0 fish and zooplankton. Transactions of the American Fisheries Society 110:370–381.

McCammon, G. W., and C. von Geldern, Jr. 1979. Predator–prey systems in large reservoirs. Pages 431–442 *in* H. Clepper, editor. Predator–prey systems in fishery management. Sport Fishing Institute, Washington, DC.

Miranda, L. E., and W. D. Hubbard. 1994. Winter survival of age-0 largemouth bass relative to size, predators, and shelter. North American Journal of Fisheries Management 14:790–796.

Miranda, L. E., W. L. Shelton, and T. D. Bryce. 1984. Effects of water level manipulation on abundance, mortality, and growth of young-of-year largemouth bass in West Point Reservoir, Alabama–Georgia. North American Journal of Fisheries Management 4:314–320.

Mitchell, S. F. 1975. Some effects of agricultural development and fluctuations in water level on the phytoplankton productivity and zooplankton of a New Zealand reservoir. Freshwater Biology 5:547–562.

Moen, T. E. 1974. Population trends, growth, and movement of bigmouth buffalo, *Ictiobus cyprinellus*, in Lake Oahe, 1963–73. U.S. Fish and Wildlife Service, Technical Paper 78.

Moffett, J. W. 1943. A preliminary report on the fishery of Lake Mead. Transactions of the North American Wildlife Conference 8:179–186.

Mullan, J. W., and R. L. Applegate. 1968. Centrarchid food habits in a new and old reservoir during and following bass spawning. Proceedings of the Annual Conference Southeastern Association of Game and Fish Commissioners 21(1967):332–342.

Nelson, W. R. 1978. Implications of water management in Lake Oahe for the spawning success of coolwater fishes. Pages 154–158 *in* R. L. Kendall, editor. Selected coolwater fishes of North America. American Fisheries Society, Special Publication 11, Bethesda, Maryland.

Nelson, W. R., and C. H. Walburg. 1977. Population dynamics of yellow perch (*Perca flavescens*), sauger (*Stizostedion canadense*), and walleye (*Stizostedion vitreum vitreum*) in four mainstem Missouri River reservoirs. Journal of the Fisheries Research Board of Canada 34:1748–1763.

Nestler, J. M. 1993. Instream flow incremental methodology: a synopsis with recommendations for use and suggestions for future research. U.S. Army Corps of Engineers Waterways Experiment Station, Technical Report EL-93-3, Vicksburg, Mississippi.

Nestler, J. M., and K. S. Long. 1994. Cumulative impact analysis of wetlands using hydraulic indices. U.S. Army Corps of Engineers Waterways Experiment Station, Wetlands Research Program, Technical Report WRP-SM-3, Vicksburg, Mississippi.

Neter, J., and W. Wasserman. 1974. Applied linear statistical models. Richard D. Irwin, Homewood, Illinois.

Odum, E. P. 1971. Fundamentals of ecology. Saunders, Philadelphia.

Oglesby, R. T. 1982. The MEI symposium—overview and observations. Transactions of the American Fisheries Society 111:171–175.

Ostrofsky, M. L., and H. C. Duthie. 1978. An approach to modelling productivity in reservoirs. Internationale Vereinigung Fuer Theoretische und Angewandte Limnologie Verhandlungen 20:1562–1567.

Ploskey, G. R. 1986. Effects of water-level changes on reservoir ecosystems, with implications for fisheries management. Pages 86–97 *in* G. E. Hall and M. J. Van Den Avyle, editors. Reservoir fisheries management: strategies for the 80s. American Fisheries Society, Southern Division, Reservoir Committee, Bethesda, Maryland.

Ploskey, G. R., L. R. Aggus, and J. M. Nestler. 1984. Effects of water levels and hydrology on fisheries in hydropower storage, hydropower mainstream, and flood control reservoirs. U.S. Army Corps of Engineers Waterways Experiment Station, Technical Report E-84-8, Vicksburg, Mississippi.

Ploskey, G. R., and five coauthors. 1993. Assessing impacts of operations on fish reproduction in Missouri River Reservoirs. U.S. Army Corps of Engineers Waterways Experiment Station, Technical Report EL-93-21, Vicksburg, Mississippi.

Rainwater, W. C., and A. Houser. 1975. Relation of physical and biological variables to black bass crops. Pages 306–309 *in* Stroud and Clepper (1975).

SAS Institute. 1989. SAS/STAT user's guide, version 6, fourth edition, volume 2. SAS Institute, Inc., Cary, North Carolina.

Shelton, W. L., W. D. Davies, T. A. King, and T. J. Timmons. 1979. Variation in the growth of the initial year class of largemouth bass in West Point Reservoir, Alabama and Georgia. Transactions of the American Fisheries Society 108:142–149.

Shields, J. T. 1957. Experimental control of carp reproduction through water drawdowns in Fort Randall Reservoir, South Dakota. Transactions of the American Fisheries Society 87:23–33.

Shirley, K. E., and A. K. Andrews. 1977. Growth, reproduction, and mortality of largemouth bass during the first year of life in Lake Carl, Blackwell, Oklahoma. Transactions of the American Fisheries Society 106: 590–595.

Strange, R. J., W. B. Kittrell, and T. D. Broadbent. 1982. Effects of seeding reservoir fluctuation zones on young-of-the-year black bass and associated species. North American Journal of Fisheries Management 2:307–315.

Stroud, R. H. 1948. Growth of the basses and black crappie in Norris Reservoir, Tennessee. Journal of the Tennessee Academy of Science 23:31–99.

Stroud, R. H., and H. Clepper, editors. 1975. Black bass biology and management. Sport Fishing Institute, Washington, DC.

Summerfelt, R. C., and K. E. Shirley. 1978. Environmental correlates to year-class strength of largemouth bass in Lake Carl Blackwell. Proceedings of the Oklahoma Academy of Science 58:54–63.

Timmons, T. J., J. L. Shelton, and W. D. Davies. 1980. Differential growth of largemouth bass in West Point Reservoir, Alabama and Georgia. Transactions of the American Fisheries Society 109:176–186.

Trost, N. 1991. Approximation of data-rich time series by harmonic analysis. Ecological Modelling 56:259–266.

USEPA (U.S. Environmental Protection Agency). 1978. A compendium of lake and reservoir data collected by the National Eutrophication Survey in the central United States. U.S. EPA, Working Paper 476, Washington, DC.

Vogele, L. E. 1975. Reproduction of spotted bass, *Micropterus punctulatus*, in Bull Shoals Reservoir, Arkansas. U.S. Fish and Wildlife Service, Technical Paper 84.

Vogele, L. E. 1981. Reproduction of smallmouth bass, *Micropterus dolomieui*, in Bull Shoals Lake, Arkansas. U.S. Fish and Wildlife Service, Technical Paper 106.

Vogele, L. E., and W. C. Rainwater. 1975. Use of brush shelters as cover by spawning black basses (*Micropterus*) in Bull Shoals Reservoir. Transactions of the American Fisheries Society 104:264–269.

Vollenweider, R. A. 1975. Input–output models with special reference to the phosphorus loading concept in limnology. Schweizerische Zeitschrift fuer Hydrologie 37:53–84.

von Geldern, C. E., Jr. 1971. Abundance and distribution of fingerling largemouth bass, *Micropterus salmoides*, as determined by electrofishing at Lake Nacimiento. California Fish and Game 57:228–245.

Westerdahl, H. E., W. B. Ford, III, J. Harris, and C. R. Lee. 1981. Evaluation of techniques to estimate annual water quality loadings to reservoirs. U.S. Army Corps of Engineers Waterways Experiment Station, Technical Report E-81-1, Vicksburg, Mississippi.

Wood, R. 1951. The significance of managed water levels in developing the fisheries of large impoundments. Journal of the Tennessee Academy of Science 26: 214–235.

Wood, R., and D. W. Pfitzer. 1960. Some effects of water-level fluctuations on the fisheries of large reservoirs. International Union for Conservation of Nature and Natural Resources, Technical Meeting, Athens, Greece 4:118–138.

Wright, J. C. 1950. The limnology of Atwood Lake, a flood control reservoir. Doctoral dissertation. Ohio State University, Columbus.

American Fisheries Society Symposium 16:442–449, 1996

Evaluation of a Roving Creel Survey Design for a Large Reservoir with an Emphasis on Subsampling within Days

D. GARTH CROW AND STEPHEN P. MALVESTUTO

Fishery Information Management Systems, Inc., Post Office Box 3607
Auburn, Alabama 36831-3607, USA

Abstract.—An estimated US$20 million are spent annually by fish and wildlife agencies to conduct recreational fishery surveys. Sampling designs must be as effective as possible for collecting data that are sufficiently precise to document the effects of management programs. In this paper, we evaluate the roving creel survey design used on Lake Guntersville, Alabama, over a 4-year period (1990–1993) with respect to five response variables: fishing effort (angler-hours), the number of fish harvested per hour, the number of fish caught and released per hour, the total number of fish caught (harvest + release) per hour, and the total weight (lb) of fish harvested per hour. The temporal dimension was stratified into months and day types (weekdays and weekends), and the lake was geographically stratified. Forty sampling days, 20 d for each of two creel clerks, were chosen at random each month to increase daily sampling intensity. Three to four 3.5-h time periods were defined within days (the number depending on daylight length) and two of these time periods were chosen at random for sampling each day. All stratifications together explained 36% of the total variance for fishing effort and 13% for harvest per unit of effort, leaving the large majority of the sampling variance unexplained. In general, stratification across months was most important; geographical strata had very little effect on precision, whereas day type stratification reduced variance for effort estimates but not for fishing success rates. The effect of taking two rather than one subsample within each day was evaluated by simulating single-subsample data (15 runs per year) from the original data provided by the survey. Relative to the distributions provided by the single-subsample simulations, the original data (based on two subsamples), on average, reduced relative standard errors (RSEs) of fishing effort by 14% and RSEs of fishing success measures by 23–32%. The estimates themselves were not affected by increased subsampling. Thus, sampling during more than one time period within a day can substantially improve the precision associated with estimates of fishing effort and fishing success rates.

In the mid-1930s, fishery biologists began to realize the need to evaluate the impact of fishing upon fish populations (Clark 1934; Davis 1935; Eshmeyer 1935; Lord 1935; Smith 1935). Creel surveys have been the most widely used technique for sampling the angling public. Robson (1960, 1961) developed the roving creel census as an unbiased method for assessing recreational fishing effort and total catch. Malvestuto et al. (1978) outlined a stratified, nonuniform probability sampling design for roving creel surveys and Malvestuto (1983) provided an overview of creel survey techniques in common use at that time.

To date, little work has been directed at assessing the effectiveness of various survey designs. Malvestuto and Knight (1991) evaluated the benefits of stratification of days within months into weekdays and weekend days (day types). They found that stratification into day types significantly improved the precision of estimates of fishing effort but not of estimates of harvest rate or total harvest on several reservoirs and a tailwater. Sztramko (1991) found that removing day type stratification increased the variability of estimates of fishing effort and harvest by about 10%; he also found that, for the Long Point Bay fishery in Lake Erie, isolation of the 9-d opening period of the fishing season as a stratum reduced the variability of fishing effort by 27% and that of harvest by 19%.

Malvestuto and Knight (1991) showed that generally most of the variance associated with estimates of harvest per unit of effort (HPUE) resided within sample days rather than between days, suggesting that for estimation of HPUE, more within-day sampling should be conducted, even at the expense of reducing the total number of days sampled. This finding was corroborated by Lester et al. (1991), who found that almost all (90%) of the variance associated with catch per unit of effort (CPUE) resided within days rather than among days, based on historical data from 47 Ontario lakes.

For river fisheries, Meredith and Malvestuto (1991) showed that relatively complex stratification schemes in time and space did little to improve the precision of estimates of fishing effort, HPUE, or harvest. Precision of HPUE estimates improved by about 20% when temporal strata were defined as months rather than as seasons. Dent and Wagner (1991) recommended the use of monthly rather than seasonal strata to improve precision. The smaller the time blocks (e.g., seasonal, monthly, weekly) within the survey period, the more the temporal variability in fishery estimates can be controlled by the survey design (Hayne 1991).

Because it is inherently difficult to sample anglers, estimates of catch-related characteristics are often highly variable; for example, relative standard errors (RSE = 100 · SE/mean) typically exceed 50%. In general, RSEs larger than 20% are not desirable (Malvestuto 1983). Fishing effort appears to be inherently less variable than CPUE (Bayley et al. 1991; Malvestuto 1991; Malvestuto and Knight 1991; Meredith and Malvestuto 1991; Palsson 1991). Relatively precise estimates of fishing effort (RSEs < 15–20%) are attainable for most fisheries with moderate levels of monthly sampling effort (6–12 d) stratified by day type. It can be difficult, however, to obtain the required precision for estimates of CPUE because of within-day variability in catchability among species and because success varies among fishing parties. It has been suggested that within-day stratification (e.g., morning, afternoon, and evening; Bayley et al. 1991) and geographical stratification for large water bodies (Malvestuto 1991) will be necessary to consistently improve estimates of fishing success rates.

There is a need to identify alternative design options that will increase the precision of fishery statistics gathered through creel surveys, particularly for estimates of catch-related parameters. The objectives of this study were to (1) evaluate potential increases in the precision of estimates of fishing effort and catch rates associated with the incorporation of temporal and spatial stratifications into creel survey designs, and (2) assess the benefits associated with increased levels of within-day sampling on the precision of estimates of fishing effort and catch rates. The data used in the analyses were collected annually from Lake Guntersville, Alabama, over a 4-year period.

Methods

Lake Guntersville is a 27,479-ha mainstream impoundment encompassing 121.8 km of the Tennessee River in northeast Alabama (Figure 1). This multipurpose reservoir has an established nationwide fishery reputation for its black bass *Micropterus* spp. Periodically, the reservoir has experienced moderate to severe problems with aquatic macrophytes. In 1988, 30% of the reservoir surface area was covered with aquatic macrophytes, and control measures were undertaken (Bain and Boltz 1992). In February 1990, a roving creel survey was initiated to evaluate changes in the sport fishery resulting from aquatic macrophyte control measures. Four years of survey data (1990–1993) are examined in this study.

Years were stratified into months, and sampling was conducted from February to November each year. Months were stratified into weekdays (Monday–Thursday) and weekend days (Friday–Sunday). Friday was included in the weekend stratum because angling effort was known (from previous survey data) to increase on Fridays and stay relatively high over Saturday and Sunday. Also, it was estimated that about 60% of the fishing effort would occur on weekends and the remaining 40% on weekdays.

The reservoir was divided into three geographical strata, each with five to seven sample sections (Figure 1). From preliminary estimates of fishing effort, it was projected that stratum A (closest to the dam) would receive about 35% of the effort, stratum B (midreservoir) would receive 50% of the effort, and stratum C (upper riverine area) would receive the remaining 15%. Initially, sampling effort was allocated to the strata according to these percentages. After the initial study year, the survey design was evaluated. Based on observed fishing effort, stratum weights were modified and sampling effort was reallocated among strata as follows: strata A and B each received 40% of the sampling effort, and stratum C received the remaining 20%. Other features of the survey design remained unchanged. Sample sections within geographical strata were assigned equal probabilities.

Two creel clerks each sampled 20 d/month (40 sampling days total), 11 weekend days and 9 weekdays per clerk. Sampling schedules for each clerk were developed independently. Creel days were randomly selected within day-type strata with equal probabilities. On days when both clerks worked, a different geographical stratum was sampled by each clerk (Table 1). Sampling days were divided into four 3.5-h sampling periods (0600–0930, 0930–1300, 1300–1630, and 1630–2000 hours) from April through September; during the winter, there were only three sampling periods (0700–1030, 1030–1400, and 1400–1730 hours) because of shorter daylight length. On a given sampling day, for each clerk, two time periods were randomly selected with equal probabilities and then randomly assigned to lake sections within the geographical strata allocated to that day.

The procedure on the water consisted of the clerk making one complete circuit through each sample section during the allocated time periods on a given

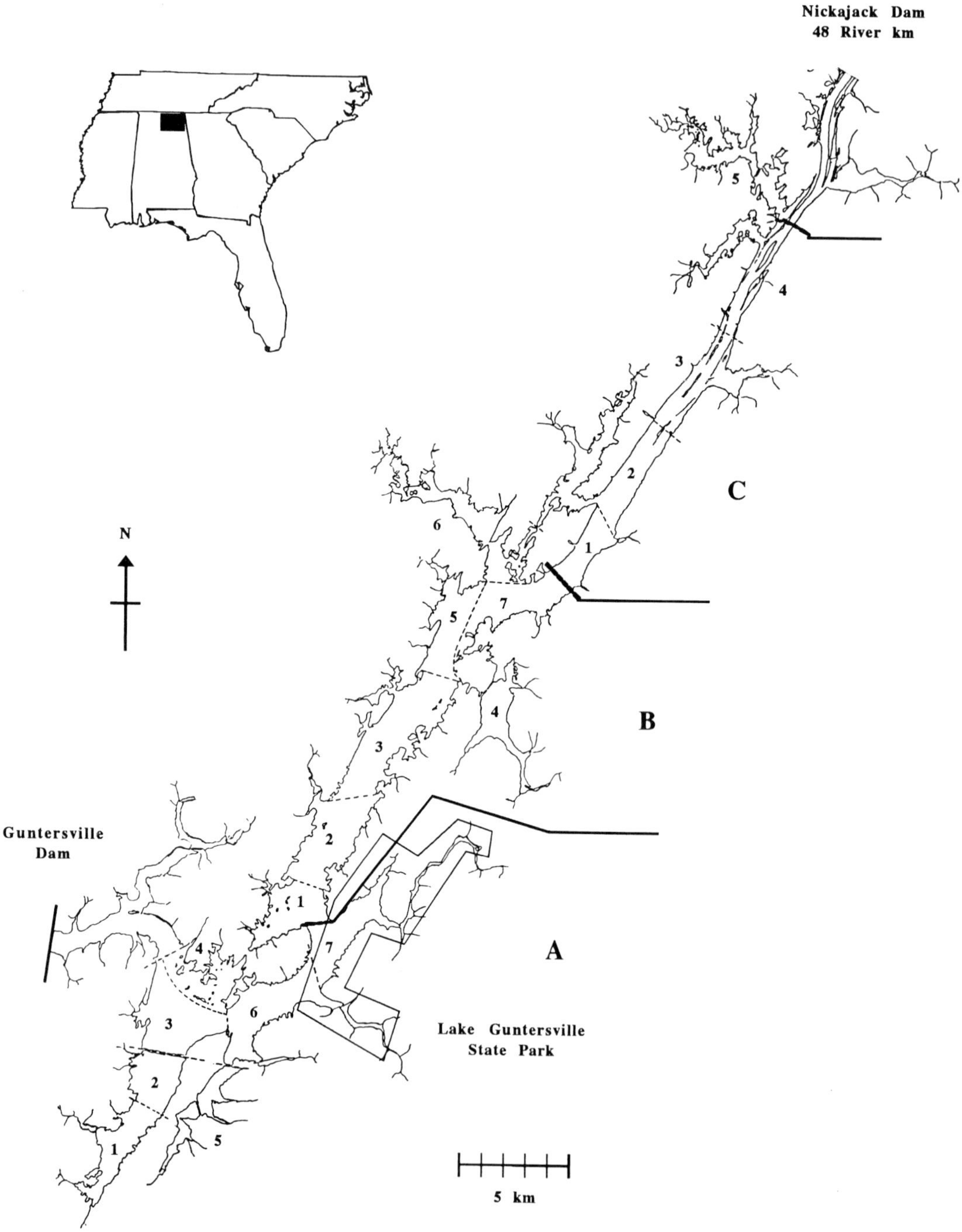

FIGURE 1.—Map of Lake Guntersville, Alabama, showing the three geographical strata (letters) and the sampling sections within strata (numbers) used in this study.

TABLE 1.—Example of the sampling schedule used for the Lake Guntersville creel survey for two creel clerks. On days when both clerks worked (e.g., May 6) they were stationed in different geographical strata.

Day	Stratum	Sections	Time periods
		Clerk 1	
4 May	A	3, 5	2, 3
6 May	C	2, 5	1, 2
7 May	B	3, 4	1, 3
		Clerk 2	
5 May	C	1, 4	1, 3
6 May	A	6, 2	2, 4
9 May	B	2, 7	3, 4

day. Direction of travel for the circuits was randomly determined by coin toss. During the circuit, a count of active anglers was taken and interviews were conducted. Checkpoints were established within each sample section to keep clerks on schedule during the circuit. Harvest was enumerated and then weighed to the nearest 0.1 lb for each species.

Following the procedure established by Malvestuto et al. (1978), estimates of fishing effort, fish harvest per unit of effort (HPUE), released fish per unit of effort (RPUE), catch (harvest + release) per unit of effort (CPUE), and weight (lb) of fish harvested per unit of effort (WPUE) were calculated for each survey day from the two subsamples taken per geographical stratum (Figure 2). For fishing effort, each subsample provided an expanded estimate of effort for the entire day; a mean over the two estimates was taken to provide a "best" single estimate for that day for each geographical stratum sampled. For daily estimates of harvest and catch rates, all interviews were pooled over the two subsamples to provide a single set of interviews from which the daily rate estimates were calculated. Because creel clerks worked different strata on common days, daily estimates for two strata could be generated.

Daily values were expanded to monthly (Figure 2) and then annual estimates for each geographical

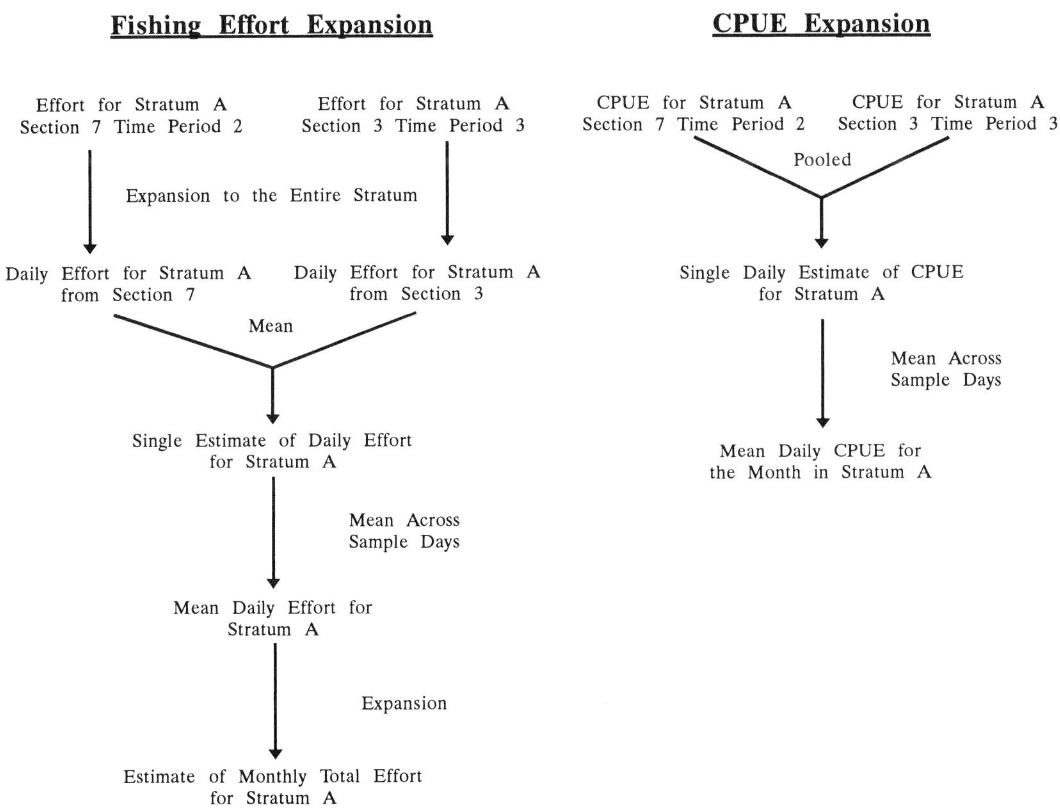

FIGURE 2.—Generalized series of steps for expansion of daily subsample estimates of fishing effort and catch per unit of effort (CPUE) to monthly values. Each series begins with two subsamples from within a hypothetical sample day.

stratum as per Malvestuto et al. (1978). Data were summed across strata to derive whole-lake estimates. Standard errors were calculated at each level of estimation by using daily values as the replicates (between-day variance).

To evaluate benefits associated with temporal (months and day types) and geographical stratification, analysis of variance (ANOVA) was conducted to determine if those sources of variability significantly (α = 0.15) reduced the residual variances of the estimates. We believe that an alpha of 0.15 is an acceptable probability for making a type I error given the objective of determining if a design stratification is meaningful or not. The proportion of the total variability explained by each source in the model also was determined by dividing the sums of squares (SS) associated with each source by the total SS. Only fishing effort and HPUE were subjected to this analysis. Because of potential recall biases associated with release rates, CPUE was not used. The design model treated months, day types, and geographical strata as fixed effects representing the primary stratifications defined for the survey. Randomly chosen days within the temporal and geographical strata provided the between-day variance for testing the significance of the stratification effects. The two-way and three-way interaction terms for the fixed effects were included in the model.

To test the effect of taking multiple subsamples within each sample day, the survey data from each of the 4 years was subjected to a random sampling process that created daily data based on only one of the two possible subsamples per day. Thus, we simulated annual data sets that might arise if daily expansions were based on the common practice of taking only a single subsample each day. This approach kept the number of days in the simulated data constant relative to the original data.

For each annual data set, 15 single-subsample simulations were run that provided an array of estimates and associated RSEs for fishing effort and the four fishing rates defined above. The mean and variance of the distributions (N = 15) generated by the simulation runs provided a basis for statistical comparison with the original annual values. Our approach was to treat the original annual values (from the two-subsample design) as population parameters (with no sampling variance). Thus, for each of the 4 years analyzed, the original survey data provided a parametric mean and RSE for each of the fishery characteristics defined above. Means from the simulated single-subsample distributions were compared to the population parameters with the appropriate version of the t-test (Steel and Torrie 1980:92). A t-test for two independent samples was deemed inappropriate because the two values being compared were not from independent sets of data. The alpha level for these comparisons was set at 0.15.

Results and Discussion

All three stratifications (months, day types, and geographical areas) significantly reduced sampling errors associated with annual estimates of fishing effort; the stratifications together explained 36% of the total variability (SS). Temporal stratification by month and day type accounted for 25% and 8%, respectively, leaving only 3% attributable to geographical stratification. The two-level and three-level interaction terms in the ANOVA model together explained an additional 4% of the variability in fishing effort, and only the two-level interactions involving day type were statistically significant.

The benefit from stratification was substantially less for HPUE. Only 13% of the total variability was accounted for by month (12%), day type (<1%) and geographical area (1%), and month was the only source that was statistically significant. The small contribution of day type stratification to reduction in variability in HPUE was shown by Meredith and Malvestuto (1991) and by Malvestuto and Knight (1991) for an array of lakes, rivers, and tailwaters. The combined interaction terms accounted for an additional 5% of the variability in HPUE, and only the interaction between day types and geographical strata was statistically significant. The full ANOVA model explained 25% of the variability in HPUE and 50% in fishing effort, which is similar to results presented by Meredith and Malvestuto (1991) for large rivers.

The means from the simulated single-subsample-per-day distributions provided good estimates of the original two-subsample values across all characteristics and years; in most cases, the original value was well within the range of simulated estimates based on a single daily subsample (Figure 3, top). There was no pattern to the differences—even when they were significant—between the simulation means and the original values; that is, the original values were not consistently higher or lower than the means across years or characteristics.

The differences in the RSEs, however, between the two-subsample and one-subsample schemes were substantial across all characteristics and years, except in 1992, when RSEs were nearly equal for harvest rate and catch rate (Figure 3, bottom). Even

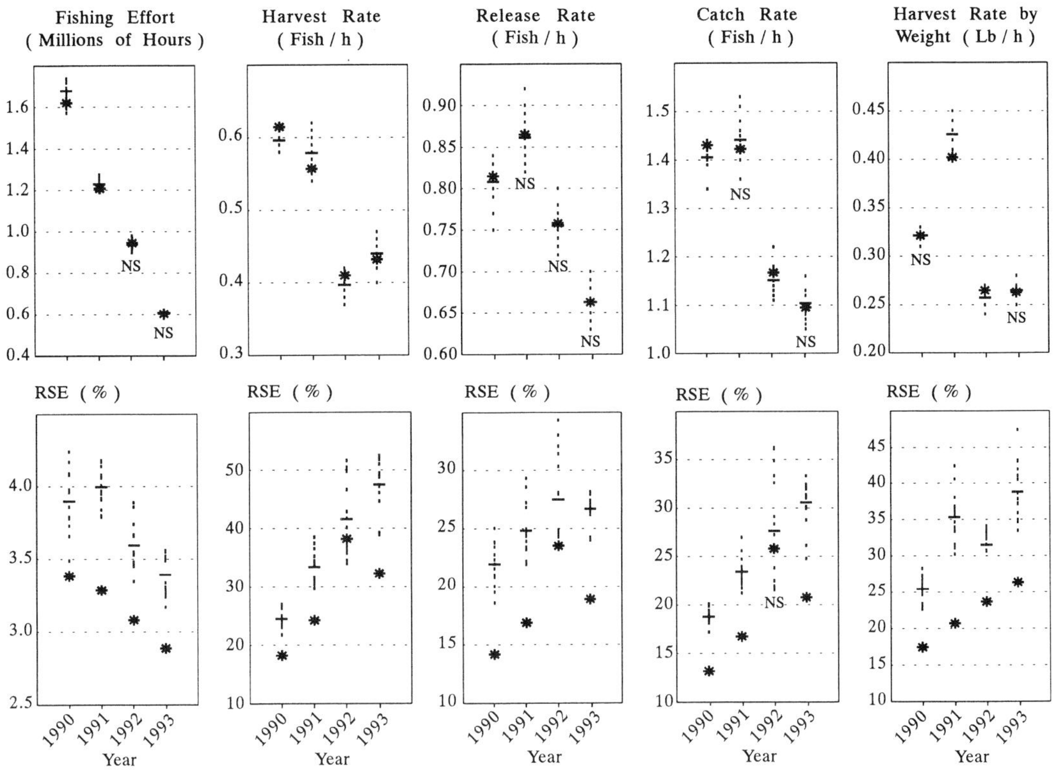

FIGURE 3.—Graphs showing relationships between single-subsample and two-subsample estimates of fishery characteristics and their relative standard errors (RSE) for the roving creel survey at Lake Guntersville, 1990–1993. Horizontal bars represent mean values from the single-subsample-per-day simulations (vertically arrayed points). Asterisks represent estimates from the original two-subsample-per-day survey data; NS denotes nonsignificant differences between single- and two-subsample estimates (α = 0.15).

though all the characteristics did not behave in the same way when summarized as a group over all 4 years, the mean RSE associated with taking two subsamples per day (RSE_2) was 18.2% (Table 2). The mean RSE associated with taking one subsample per day (RSE_1) was 24.7%. The decrease in mean RSE of 6.5 percentage points when subsampling increased represented a relative increase in precision of 26% (relative change = $100[RSE_1 - RSE_2]/RSE_1$). Relative standard errors for the effort estimates were 3.7 and 3.2% for the one- and two-subsample schemes, respectively, both very precise estimates. For the catch-related statistics viewed together, the two-subsample scheme gave a 27% relative decrease in variability. The maximum reduction in the RSE attributable to increased subsampling occurred for harvest weights, which showed an average annual absolute decrease of 10.6 percentage points, or a relative improvement in precision of 32%.

In 1992, differences in precision were smaller than in other years for measures of fishing success based on number of fish per angler-hour (Figure 3). The original two-subsample scheme gave RSEs that were less (by 3% on average) than the mean RSEs from the single-subsample simulations. There was no significant difference between the original and simulation means for catch rate. Gains in precision with increased daily subsampling were more obvious for fishing effort and harvest weight in 1992. The reasons for these contrasts are unknown.

Because the single-subsample simulation data sets contained the same number of sample days as the original survey data, the reduced RSEs associated with the original estimates relative to the single-subsample results are not an artifact of reduced sample size. From a quantitative standpoint, there was no difference in the statistical expansion formulas used for the two subsampling schemes once the daily estimates were formulated (see methods). The

TABLE 2.—Improvement in relative precision (relative standard error, RSE) between simulated single-subsample data (RSE_1) and original two-subsample data (RSE_2) from the roving creel survey on Lake Guntersville, 1990–1993. Improvement is measured as the absolute decrease in mean annual RSE ($RSE_1 - RSE_2$), as well as the relative decrease expressed as a percentage ($100[RSE_1 - RSE_2]/RSE_1$).

Estimate	One subsample per day	Two subsamples per day	Absolute decrease	Relative decrease
Effort (hours)	3.7	3.2	0.5	14%
Release rate (fish/h)	25.2	18.4	6.8	27%
Harvest rate (fish/h)	36.8	28.3	8.5	23%
Harvest rate (lb/h)	32.8	22.2	10.6	32%
Catch rate (fish/h)[a]	25.1	19.1	6.0	24%
Mean across all statistics	24.7	18.2	6.5	26%
Mean across all catch statistics[b]	30.0	22.0	8.0	27%

[a]Released plus harvested fish.
[b]Release, harvest, and total catch rates.

advantage of the increased information gathered within each day appears to be that day-to-day variability was stabilized (reduced) significantly. Given that the two time periods for the subsamples were chosen randomly each day, there was more opportunity for different parts of the same days to be sampled. Thus, it was common for clerks to sample the first 3.5-h time period and the last 3.5-h time period on a given day. (Note that the clerks lived on the lake, thereby making this type of scheduling convenient). We surmise that simply creating longer sampling periods (e.g., 6 h) to increase time in the field, but still choosing only one period per day (either morning or afternoon), probably will not have an equal positive effect because the daily data would still be clustered into a limited number of contiguous hours relative to having a more random mix of times within each day.

Summary

The sampling design for the Lake Guntersville creel survey provided relatively good estimates of fishing effort and fishing success measures. Relative standard errors for fishing effort were about 3% across the four survey years and most of the RSEs for the fishing success measures were between 18 and 28%, around or below the desired 20% level. As indicated above, stratification by month, day type (weekday and weekend), and lake region did not contribute strongly to reducing variability, particularly of catch-related measures. The overall sampling intensity (40 clerk-days per month) and the use of subsampling within days seem to have played the major roles with respect to reducing sampling error.

Although increased within-day sampling had little effect on the mean values of the fishery characteristics evaluated, there was significant improvement in the precision of the estimates of most fishery characteristics. Our results suggest that relative improvements in precision of at least 15% would be expected for fishing effort and of at least 20–25% for catch-related measures if subsampling were increased from one to two subsamples per day, as defined in this study. Given the difficulty of controlling variability in creel survey estimates of fishing rates, we recommend that subsampling within days be conducted whenever possible. Until more effective stratifications are developed for estimation of fishing success, monthly sample sizes will have to remain high and cover the majority of the available fishing days within the survey period to obtain acceptable precision.

Acknowledgments

The Lake Guntersville creel survey was funded by the Tennessee Valley Authority (TVA). We appreciate the contributions of Bill Wrenn and Donny Lowery of the Muscle Shoals Office in the planning and supervision of the survey. We thank the TVA creel clerks, Jim Hausheer and Anitea Owen, for ongoing and rigorous collection of field data.

References

Bain, M. B., and S. E. Boltz. 1992. Effect of aquatic plant control on the microdistribution and population characteristics of largemouth bass. Transactions of the American Fisheries Society 121:94–103.

Bayley, P. B., S. T. Sobaski, M. H. Halter, and D. J. Austen. 1991. Comparisons of Illinois creel surveys and the precision of their estimates. Pages 206–211 *in* Guthrie et al. (1991).

Clark, G. H. 1934. The need for a measure of the anglers' catch. Transactions of the American Fisheries Society 64:49–53.

Davis, H. S. 1935. Stream management in the national forest. Transactions of the American Fisheries Society 65:234–239.

Dent R. J., and B. Wagner. 1991. Changes in sampling design to reduce variability in selected estimates from a roving creel survey conducted on Pomme de Terre Lake. Pages 88–96 *in* Guthrie et al. (1991).

Eshmeyer, R. W. 1935. Analysis of the game-fish catch in a Michigan lake. Transactions of the American Fisheries Society 65:207–233.

Guthrie, D., and seven coeditors. 1991. Creel and angler surveys in fisheries management. American Fisheries Society Symposium 12.

Hayne, D. W. 1991. The access point creel survey: procedures and comparison with the roving-clerk creel survey. Pages 123–138 *in* Guthrie et al. (1991).

Lester, N. P., M. M. Petzold, and W. I. Dunlop. 1991. Sample size determination in roving creel surveys. Pages 25–39 *in* Guthrie et al. (1991).

Lord, R. F. 1935. The 1935 trout harvest from Furnace Brook Vermont's "test stream." Transactions of the American Fisheries Society 65:224–233.

Malvestuto, S. P. 1983. Sampling the recreational fishery. Pages 400–423 *in* L. Nielsen and D. Johnson, editors. Fisheries techniques. American Fisheries Society, Bethesda, Maryland.

Malvestuto, S. P. 1991. The customization of recreational fishery surveys for management purposes in the United States. Pages 201–213 *in* I. G. Cowx, editor. Catch effort sampling strategies: their application in freshwater fisheries management. Fishing News Books, Oxford, UK.

Malvestuto, S. P., W. D. Davies, and W. L. Shelton. 1978. An evaluation of the roving creel survey with nonuniform probability sampling. Transactions of the American Fisheries Society 107:255–262.

Malvestuto, S. P., and S. S. Knight. 1991. Evaluation of components of variance for a stratified, two-stage roving creel survey with implications for sample size allocation. Pages 108–115 *in* Guthrie et al. (1991).

Meredith E. K., and S. P. Malvestuto. 1991. An evaluation of survey designs for the assessment of effort, catch rate and catch for two contrasting river fisheries. Pages 223–232 *in* I. G. Cowx, editor. Catch effort sampling strategies: their application in freshwater fisheries management. Fishing News Books, Oxford, UK.

Palsson, W. A. 1991. Using creel surveys to evaluate angler success in discrete fisheries. Pages 139–154 *in* Guthrie et al. (1991).

Robson, D. S. 1960. An unbiased sampling and estimation procedure for creel census of fishermen. Biometrics 16:261–277.

Robson, D. S. 1961. On the statistical theory of a roving creel census of fishermen. Biometrics 17:415–437.

Smith, O. R. 1935. A creel census from Dry River, New Hampshire. Transactions of the American Fisheries Society 66:313–315.

Steel, R. G. D., and J. H. Torrie. 1980. Principles and procedures of statistics: a biometrical approach. McGraw-Hill, New York.

Sztramko, L. K. 1991. Improving precision of roving-creel-survey estimates: implications for fisheries with a closed season. Pages 116–121 *in* Guthrie et al. (1991).

Summary

American Fisheries Society Symposium 16:453–463, 1996

Multidimensional Approaches to Reservoir Fisheries Management: A Summary

JOHN J. NEY

Virginia Polytechnic Institute and State University, Department of Fisheries and Wildlife Sciences
Blacksburg, Virginia 24061-0321, USA

Abstract.—This synopsis reviews the contributions to these proceedings and offers a brief critical appraisal of the merits and limitations of each toward more effective management of reservoir fisheries. The emphasis in these proceedings on people–biota–habitat interactions is a timely departure from previous reservoir fisheries symposia, which focused largely on fish and their environment. The challenges of twenty-first century reservoir management will be to meet intense demands by diverse user groups for a static resource. These challenges and the emerging tools to address them are discussed relative to the popular mandate to do more and do it better in an era of fiscal conservatism.

Impounded streams continue to be a mainstay of recreational fishing, but maintaining angler satisfaction has proven to be an increasingly complex challenge to reservoir fisheries managers. Between 1980 and 1990, fishing effort on U.S. lakes and reservoirs increased by 42%, representing 51% of the total annual U.S. freshwater angler-days (USFWS 1993), and new technology (e.g., fish locators, high-speed boats) has increased angler efficiency. Few new reservoirs have been developed to accommodate the greater and more proficient demand, however. Construction of large (>500-ha) impoundments peaked in the 1960s, and has declined to only a few per year in the 1990s, as both costs and public resistance to the damming of free-flowing rivers have escalated (Kennedy 1995). Most future impoundments are likely to be smaller systems built to meet local water supply demands, rather than sprawling waterbodies on major rivers operated for flood control, power generation, irrigation, and navigation. Existing large impoundments face increasing threats to fisheries and water quality from development within their expansive watersheds. Nonfishing recreational uses of existing reservoirs are also intensifying, raising the potential for conflicts between anglers and other users.

As the relationship between user demand and reservoir supply has changed, anglers have become both more vocal and more diverse. Reservoir fishery management plans must now consider and sometimes incorporate political concerns, including those arising among anglers themselves. Tournament fishing, catch-and-release practices, and specialization on target species are some of the more prominent phenomena that may promote divisiveness among reservoir anglers.

Fisheries management frequently has been depicted as a Venn diagram of three overlapping circles representing biota, habitat, and people (Nielsen 1993). Throughout its twentieth century development, fisheries management has focused primarily on understanding and manipulating biota and habitat under the implicit assumption that these efforts would satisfy the clientele. That assumption is demonstrably no longer valid; clearly it is time to involve the public actively.

These proceedings represent the fourth symposium on North American reservoir fisheries management, and the third sponsored by the Reservoir Committee of the Southern Division of the American Fisheries Society. The first (1967) and second (1971) symposia focused almost exclusively on biological and habitat components. The third symposium (1983) began to consider some socioeconomic aspects of management. The fourth symposium was designed to offer a timely departure in content from its predecessors. Its theme, "Multidimensional Approaches to Reservoir Fisheries Management," succinctly describes its goal: coverage of diverse but essential topics outside the realm of the reservoir manager's normal operation. Although papers with a biological or habitat focus are included, the emphasis has been shifted purposely to socioeconomic issues and people–biota–habitat interactions. As a result, these proceedings offer a nearly panoramic snapshot of the broad array of challenges that will confront reservoir fisheries managers in the twenty-first century as well as some emerging tools that will help them to meet those challenges.

Organization of this Summary

These proceedings consist of an introductory section followed by eight technical sections organized around particular themes. I will attempt to summarize briefly the content of each section in the order they are presented. My review will be neither comprehensive nor completely objective. Rather, I seek to highlight major (and, I hope, enduring) points and to offer a critical appraisal of the merits and limitations of concepts and procedures. I approach this task as a reservoir scientist whose main concern has been the biota, secondarily the habitat, and only recently (but unavoidably) the people. I conclude with an overview of where I perceive we are and need to go, and offer some suggestions for getting there. Following the lead of Summerfelt (1986) in summarizing the last reservoir fisheries management symposium, papers cited from these proceedings are not listed in the references, and only common names of fishes are used (Robins et al. 1991).

Introductory Presentations

The initial presentations offer a seemingly eclectic mix of history, politics, and biology. Considered collectively, the papers set the stage for the technical sessions by providing an historical perspective of reservoir fisheries management as well as offering a promising peek at its future. Miranda describes the evolution of reservoir fisheries management as a series of paradigms. The Biological Deserts paradigm of the 1930s was replaced by the Biomanipulation paradigm (1940s–present) as biologists came first to appreciate the fishery potential of these new, artificial systems and then to strive to maximize use of their productivity. Angler concerns were largely muted when reservoirs were exploited only lightly, but anglers have become more vocal as reservoir use has increased, and they can no longer be ignored. The paradigm is being reformulated of necessity to encompass socioeconomic research and applications, although the form it eventually will take remains murky: will the need to understand reservoir biology be supplanted by the need to know (and manage) the human dimension? Stein et al. provide a convincing answer that much biology can still be learned and applied profitably to enhance reservoir fish communities. In turbid Ohio impoundments, larval gizzard shad decimate zooplankton, to the detriment of sunfish and eventually largemouth bass recruitment; predator stocking can suppress shad and may restore sport fisheries. Food web management is the core of successful biomanipulation, and we are still too low on the learning curve to abandon good ecological research. A merger of biomanipulation and socioeconomic paradigms is indicated, rather than a shift from the former to the latter.

On a larger scale, Ungate analyzes the unique success of the Tennessee Valley Authority's (TVA) Lake Improvement Plan for upgrading downstream water quality and stabilizing summer lake levels in its many reservoirs. Conflicts over reservoir operation are on the increase throughout the United States as enfranchised uses clash with emerging "quality of life" demands; protracted litigation among jurisdictions and user groups is the order of the day. The TVA success was facilitated because a single decision-making authority was empowered to operate under a broad, multipurpose mandate and continually sought popular as well as professional support. In other large watersheds, reservoir authority and operational objectives are fragmented; creative political and social approaches will be required if the ascendant goal of ecosystem management is to be realized.

Administrating and Marketing Reservoir Fisheries

Many, if not most, reservoir fisheries managers are poorly prepared to meet the rising public demand for participation in the planning and decision-making processes. Management actions that once were announced only in terse press releases are now subject to microscopic dissection and require elaborate, well-publicized justification. Still, the end result is too often confrontation and alienation of segments of the public (known in the current lexicon as stakeholders). What went wrong and how can the same mistakes be avoided next time? Several papers offer some answers garnered from experience.

Brown provides a primer on communication, perhaps the most glaring deficiency of resource agencies. The bottom line message is, know your audience and talk to them. Knowledge of, and appreciation for, the values and goals of stakeholder groups is critical and can be achieved only through repeated personal contact. Group presentations and discussions must be dialogues; managers create unintended barriers when they distance themselves with data and jargon, or assume that the audience shares management's value system.

Other lessons have been learned the hard way. Based on Wisconsin experiences, Addis and Les enjoin administrators to include *all* stakeholders in the dialogue. Letting everybody into the tent may be imperative when a tiny faction can stall a project through political or legal maneuvering. But inclusion also raises the potential for conflict over consensus. Hard decisions may require that stakeholders be valuated according to their commitment to the resource. Securing commitment, in money or effort, also broadens the opportunity for two-way communication and the synthesis of common goals. At Norris Reservoir, Tennessee, Wilson et al. nurtured a partnership among anglers, agency biologists, and university scientists to meet the demands for better fishing. The Norris Lake Task Force successfully fostered public involvement.

The role of the public in formulating reservoir management decisions will be situation specific. In many instances, the traditional public hearing format will be inadequate to an audience clamoring for effective participation. McMullin adapts the Vroom–Yelton leadership model from the social sciences to determine the desired nature of public involvement. Leadership models range from autocratic to shared decision making, and the choice is made on the basis of seven attributes, including need for knowledge and public acceptance. Applications of the model to natural resources problems probably will produce refinements.

Interjurisdictional disputes are an additional plague to the ecosystem level management of the reservoirs in large watersheds. Keenlyne describes an example of collaboration among state and federal agencies to effect more responsive management of the Missouri River reservoir system. The Missouri River Natural Resources Committee has had some success, but navigation needs continue to drive the system and the U.S. Army Corps of Engineers (COE) still controls the spigot. Additional players are the Native American tribes who control upstream water rights. One must wonder if ecosystem management of large multijurisdictional watersheds is only a pipe dream.

Environmental Assessment

Reservoirs function as traps for dissolved and suspended materials, and their water quality thus depends on human activities throughout their watersheds, which average three times larger per hectare of waterbody surface than those of natural lakes (Soballe et al. 1992). Consequently, it is not surprising that reservoirs are especially vulnerable to toxic contamination, which can pose a risk to both the aquatic community and human consumers. Papers on environmental assessment are split evenly between examinations of risks to human health through fish consumption and protocols to assess the health of reservoir communities.

Eating fish from contaminated reservoirs can have consequences ranging from bad taste (Perschbacher et al.) to neurological damage from mercury poisoning (Rodgers) and elevated lifetime risk of cancer (Reinert et al.). Musty and earthy flavors in fish flesh are transitory byproducts of blue-green algae blooms, yet another consequence of eutrophication. Analytic procedures to measure off-flavors are being developed, but control of algae in large reservoirs remains problematic (see Ney in the Habitat Management Section of these proceedings). Mercury contamination is responsible for almost 60% of fish consumption advisories in the United States and Canada. Reservoirs appear to be particularly susceptible to mercury contamination, most often through soil leaching. Rodgers uses bioenergetic modeling to develop biomanipulation tactics to limit mercury levels in edible fish. Tactics include maintaining short food webs, stocking at larger sizes, and regulating to promote harvest of young fish or those low on the food chain. Reinert et al. provide an extremely lucid overview of the complex and controversial subject of fish consumption advisories, emphasizing on how consumption limits are set and how they should be interpreted. Anglers appear to be polarized by these advisories, either ignoring the warnings or dropping fish from their diets. Most times, the appropriate response is somewhere between these extremes, and reservoir fisheries managers have a role to play as risk communicators. Mortality risks are relative, and anglers should be able to make informed decisions if they are informed, for example, that the lifetime death threat from eating one meal of catfish per week from their favorite reservoir may be 0.2% of the threat from automobile travel.

Most responses of fish and other aquatic biota to anthropogenic impairment of environmental quality are sublethal and difficult to detect, prompting the development of assessment indices. Coughlan et al. adapt the fish health assessment index (FHAI) of Adams et al. (1993) and apply it to largemouth bass in the Catawba River. The largemouth bass may be a good representative indicator of ecosystem health because it is a top predator, easily collected, and widely distributed. The FHAI consists of metrics based on gross external and histologic examination as well as blood parameters. It is easy to use, and appears to identify degraded habitats well, although more refinements are suggested. Hickman and McDonough test a more comprehensive measure, the reservoir fish assemblage index (RFAI) based on the index of biotic integrity (IBI) developed by Karr (1981) for lotic communities. A combination of electrofishing and gillnetting provides representative samples of fish community composition in Tennessee Valley impoundments. Samples are scored by metrics of species richness and abundance, tolerance, and trophic level as well as reproductive guild composition. Because reservoirs are themselves modified waterbodies, reference conditions and metrics must be developed somewhat subjectively. The ability of the RFAI to identify particular human impacts remains to be determined.

Bevelheimer and Adams describe a multifaceted approach to assessing the impacts and distribution of chronic contamination of Watts Bar Reservoir by Oak Ridge National Laboratory. Analysis of fish and sediments identified localized hot spots near discharge sites, but ecological impacts were not clearly evident. Reactions of anglers to fish consumption advisories were mixed, possibly indicative of confusion among the general public in interpreting the warnings. Watts Bar Reservoir is a U.S. Environmental Protection Agency Superfund site, and assessment efforts of this magnitude are unlikely to be applied at many other impoundments.

Competitive Fishing and Angler Perspective

Resource users—their diversity, motivation, trends, and fishery impacts—are the focus of several papers. Angler interviews, agency surveys, biological assessment, and opinion born of experience combine to paint a blurry but fascinating portrait of reservoir fisheries. Much more human dimension research is needed to produce a clear image.

Angler diversity is described by Ditton based on a statewide Texas mail survey and by Allen and Miranda from one-on-one interviews of crappie anglers on a Mississippi reservoir. Texas anglers show anticipated differences based on avidity, gender, residence, and other demographic features. Surprisingly, only 18% of Texas black bass anglers compete in tournaments despite the high profile of competitive fishing. Both studies found that anglers could be categorized by their degree of effort and expertise in pursuit of target species (i.e., strangers, tourists, regulars, and insiders). However, categorizing anglers into market segments is the easy part of the challenge; providing a product mix to match their particular desires may be considerably more difficult. The next round of questions should focus on angler expectations. Perhaps fishing satisfactions are more homogeneous than the anglers themselves.

Catch-and-release recycling has been the salvation of many intensive fisheries in recent years, particularly for black basses, trouts, and pikes. Quinn chronicles the rise of voluntary catch-and-release practices and their companions, releases mandated through size and bag limits. Catch-and-release is in particular favor for piscivorous sport species that might attain trophy size, but it is less widely accepted for sunfish, catfish, and other species caught primarily for food. Catch-and-release is not a universal panacea, particularly if it clashes with social or biological objectives; slot limits fail when anglers refuse to harvest small fish. Expect also that catch and release will face growing opposition from animal rights groups. In short, a fishery should not be considered infinitely self-managing just because catch-and-release is the order of the day.

Competitive anglers attempt to release their catch alive, but usually only after they have transported fish to a fixed weigh-in site. As Stang et al. report, black basses released at a tournament weigh-in station are considerably more likely to remain in that vicinity than to swim back to where they were caught. Effects of displacement on growth and reproduction were not determined, although black bass concentrated at weigh-in sites were highly vulnerable to recapture. Displacement coupled with hooking and handling mortality as high as 25% argue for length rather than weight as the competition criterion and for immediate release at the capture site, perhaps after a confirming photograph.

The comments of Dean make provocative reading. He advocates, among other things, a participatory democracy in fish management planning and regulation. Giving every license holder a vote on contemplated actions would maximize public involvement but forfeit management's responsibility for scientifically sound decision making. The average angler is, like me, a middle-aged male reactionary; innovation will not stand a chance unless accompanied by a massive and costly education program. Conversely, Dean's call for better two-way communication between resource managers and media personnel cannot be criticized. Most confrontations between resource agencies and the media result from poor communication.

Facilities Management

Papers on facilities management encompass topics ranging from amenities for enhancing angler satisfaction to hydropower operation under the mandate of the Federal Energy Regulatory Commission (FERC) relicensing process. The opening paper is the only contribution to the proceedings that directly addresses what may be the overriding issue to face reservoir managers in the twenty-first century: how to prevent a lake from being loved to death. Jones, citing the example of Lake Monroe, Indiana (about 300 visitors per hectare per year), recognizes that all recreational uses are not inherently compatible (in particular, power boating is incompatible with everything else). He suggests restrictions on boat motor use and introduces the concept of spatial and temporal segregation of lake user groups (see Klessig 1994 for a wry example of the latter). Such zoning may be inevitable on the most popular reservoirs if quality recreation is to be maintained. Intense resistance can be expected from the advocates of personal freedom whose voices now clamor for the abolition of highway speed limits and bans on assault weapons.

Three papers consider the implications of the FERC relicensing process as they apply to reservoir fisheries management. The Electric Consumers Protection Act (ECPA) of 1986 requires the agency to give consideration to the protection of environmental quality, recreational opportunities, and fish and wildlife resources equal to that formerly accorded chiefly to power production. The ECPA thus offers the potential to redress old wrongs (e.g., downstream flow, fish passage, and reservoir water level fluctuation) in the design and operation of hydroelectric dams. Hill presents a cogent overview of the components involved in the detailed FERC relicensing process, including issues, participants, and opportunities. A possible weakness in this multiyear process is FERC's permission for utilities to prepare their own environmental assessments. Harrell provides a utility's perspective on relicensing; constructive interaction with other stakeholders and willingness to negotiate are requisites of success. Navarro et al. contribute a case history of relicensing for a series of low-head dams in northern Michigan. Cooperation between resource agency and utility biologists led to resolution of flow and passage problems. Collectively, these three papers demonstrate the powerful leverage tool that the ECPA represents. Reservoir fisheries managers must become engaged early and proactively in relicensing issues to realize full advantage of this one-time opportunity. Only one-fourth of the dams on large U.S. impoundments are subject to FERC licensing, however, and no equivalent legislation affects the operation of the others.

Ecosystem management has been embraced, in principle, by the COE, the nation's largest dam operator, as part of its 1994 Natural Resources Mission Statement. Wahus provides a wealth of examples of the COE's work to improve water-based recreation, including fishing. Of special interest are the ongoing studies to determine reservoir carrying capacity for watercraft. Implementation of limits on the density of boaters will be controversial. Despite the earnest efforts of its recreation branch, the COE's operational philosophy too often continues to be dictated by institutionalized objectives, such as flood control and navigation: the love–hate relationship of reservoir fisheries managers with the COE continues (see Keenlyne).

The remaining papers deal with site amenities. Hamilton and Jackson cite the need for reservoir fisheries managers to collaborate with outdoor recreation planners in the design and location of facilities ranging from boat ramps to restrooms. The Wallop–Breaux amendment to the Federal Aid in Sport Fish Restoration Act has accelerated the construction of fishing piers and boat launches under the supervision of fisheries personnel. But many reservoirs are comanaged by federal agencies who need input in directing their recreation budgets. Ott and Webb describe the benefits of a particular amenity, on-site weight certification scales, to both anglers and biologists. The cost of alternative data acquisition, particularly for trophy-sized specimens, can reduce the true expense of such perceived "luxury" items. Fish-cleaning facilities also can serve the dual role of promoting angler satisfaction and data collection.

Socioeconomics and Optimum Yield

The value of a fishery can be measured in monetary terms or by less tangible metrics, such as user satisfaction. Regardless, calculation of fishery value requires questioning the angler, and several papers address the methods, results, and applications of angler surveys. Hunt and Ditton describe the format and procedures of the annual Texas statewide survey. The most notable feature of this mail survey is total design management, a front-end effort to personalize contacts and then follow them up, resulting in an impressive 70% response rate. Public dissemination of results and feedback also reinforce the survey's effectiveness. Hudgins and Malvestuto present an ambitious conceptual framework for determining a fishery's optimum yield, then outline a creel survey questionnaire to obtain the necessary inputs. The framework consists of four interactive accounts: ecosystem health, sociocultural values, economic values, and human health. Primary metrics for each are, in order: catch per effort, user satisfaction, willingness to pay, and daily fish consumption (rated positive or negative depending upon contamination). Integration of the resultant data to evaluate optimality of the fishery was beyond the scope of the paper. Indeed, determination of optimal yield remains elusive even in marine commercial fisheries (Smith 1994) for which harvest and economic return are the only variables. Optimal yield perhaps is considered best as a fisheries management philosophy that recognizes that multiple elements contribute to the value of the resource, rather than as a measurable goal. More practically, one must wonder about the tolerance limits of anglers for responding thoughtfully to progressively more elaborate questionnaires. Ebert presents a well-reasoned argument for adding questions concerning fish consumption, and Meredith and Malvestuto tested methods for determining consumption. Determining who eats how much of what can help risk assessors set more realistic guidelines, but should these important questions be appended to an already lengthy questionnaire or be posed separately on a different occasion? The problem of recall fatigue merits scrutiny.

Costello and Betsill focus their paper on what constitutes fishing success. Analysis of a 10-year database for 15 Texas reservoirs showed that the single best predictor of perceived success is the number of fish caught. This result may seem intuitive, but it should be viewed relative to the alternatives. Anglers preferred to catch more rather than larger fish and generally did not relate success to the time expended to catch them. The most effective management objective for these and similar reservoirs may be maximizing the density (and perhaps catchability) of the fishable population.

Reichers and Fedler present a primer on economic estimates of a fishery's worth. Economic impact can be measured as direct expenditures magnified by a "ripple effect" multiplier; calculations are straightforward and results are as reliable as the multiplier is accurate. Estimating economic value, the total societal worth of the resource, is considerably more tricky. Most commonly, direct expenditures are combined with consumer surplus (willingness-to-pay) numbers generated from questionnaire responses. As Reichers and Fedler demonstrate, willingness-to-pay questions must be designed with extreme care to ensure reasonable responses. Our society continues to rely heavily on economic cost-benefit analysis in deciding issues of natural resource exploitation. Credibility demands that we get the numbers right.

Cole and Ward analyze cost–benefit alternatives directly in a comprehensive management model to manage multiple reservoirs simultaneously with minimum redundancy and maximum angler use. The model, RIOFISH, integrates demographic, hydrologic, fisheries, and economic data to allocate management actions (e.g., stocking, facilities improvements) among lakes. This application will work best where all waterbodies are under a single jurisdiction and relatively homogenous in their aesthetic and recreational attributes. However, the comprehensive management analysis model seems to have many applications beyond site selection, wherever alternative management actions will be evaluated in terms of costs versus user response. As with any good model, the output will be only as useful as the input is accurate. Many of those data come from questionnaires.

Habitat Management

Papers on habitat management and on aquatic macrophytes (next section) examine reservoir habitat from new and emerging perspectives. Several habitat papers focus on the habitat–fish interface; others include the human element. Two of the five papers in the habitat management group are in the latter category. Ney uses case histories and regression analysis to argue that nutrient loading reductions (oligotrophication) can have deleterious effects on reservoir fisheries. As primary production declines, less energy may be available to produce sport fishes. Oligotrophication becomes a people issue when nonfishing reservoir users demand nutrient abatement to eliminate algae blooms. Maceina et al. examine water clarity versus fishery productivity directly with a data set of 32 Alabama reservoirs. Acceptable water clarity and good fisheries for crappies and black basses both occur in oligotrophic and mesotrophic Alabama reservoirs, but growth rates and the catch of trophy-sized specimens are higher in the eutrophic impoundments. It appears that choice of optimum nutrient load may require compromise among reservoir user groups. However, the public should be educated that the popular tenet, clear lakes mean good fishing, is a myth for intensively fished waterbodies.

The other three papers in this group describe fish distribution as it relates to available habitat. Annett et al. used underwater surveys of the littoral zone of a clearwater Arkansas impoundment to determine habitat associations of all life stages of largemouth bass. Not surprisingly, largemouth bass seem to prefer physical structure (brush, logs, vegetation), but the type and complexity is life stage specific. Results of this study have applications where poor recruitment indicates habitat limitation. However, it should be noted that habitat *associations* (where the fish are) are not necessarily equivalent to habitat *requirements* (where the fish need to be to survive and grow). Rogers and Bergersen map the distributions of adult largemouth bass and northern pike in small Colorado impoundments, and Irwin and Noble relate age-0 largemouth bass distribution to habitat in a North Carolina reservoir. Both studies demonstrate the potential of geographic information systems (GIS) technology to become the ultimate habitat assessment tool, because GIS links geographic descriptors to a variety of databases. The resulting product is a composite of map overlays. Applications range from the microscale (e.g., daily movements of a single fish) to the landscape level (e.g., maps of species distributions). Computer-assisted facilitations of database analysis such as GIS do not relieve the burden of field data acquisition, but rather heighten the demands for it. As Rogers and Bergersen illustrate, GIS can pinpoint where a fish is, but a satisfying explanation of why it is there can require multiple descriptors or even experimental manipulations.

Aquatic Macrophytes as Habitat for Reservoir Fish

The management philosophy on rooted aquatic vegetation in lakes and reservoirs continues to evolve. Until recently, macrophyte management was largely synonymous with macrophyte control—preventing or eradicating dense stands of nuisance "weeds," usually exotics, that can interfere with boating and swimming (Pieterse and Murphy 1990). With a single exception (Boyd 1971), the previous three reservoir symposia have ignored the role of aquatic macrophytes in fisheries management. However, fisheries scientists have long appreciated the positive benefits of rooted littoral vegetation as spawning and nursery areas; indeed a major objective of reservoir water level management for fisheries enhancement is the flooding of *terrestrial* vegetation to promote reproduction (Groen and Schroeder 1978; Willis 1986). Judging from the relevant literature, reservoir fisheries managers apparently have dealt with aquatic macrophyte "problems" either by substituting alternative physical structure when macrophytes were absent or by controlling macrophytes when they were overabundant; little effort has been directed toward promoting desirable macrophyte growth. Two recent developments have altered this strategy radically. Widespread introductions of grass carp, with its potential to eliminate *all* rooted aquatic vegetation, has prompted fisheries biologists to ask how much vegetation the fish community requires (Durocher et al. 1984; Bettoli et al. 1993). Concurrently, a people issue has arisen: largemouth bass anglers have embraced aquatic vegetation as important to their target species, which sometimes puts these anglers at odds with other lake users.

The papers on this topic directly address the changing perceptions about aquatic vegetation with respect to both its significance to fish communities and its management to optimize abundance and distribution for reservoir user groups. Weaver and Magnuson review correlative and experimental studies on north-temperate lakes to confirm that fish species diversity benefits from littoral zone complexity, including macrophyte beds. Dibble et al. perform an exhaustive literature review on fish–plant associations. They conclude that data pertaining to why fish need plants and in what abundance often have been contradictory because the scales of examination were inappropriate. A microscale approach is advocated in which fish are directly observed and plant architecture is measured precisely.

Smart et al. describe their pioneering efforts to establish native macrophytes in large reservoirs. Lessons are presented on site selection and protection, and on choice of candidate species. Establishment of thriving stands of wild celery, pondweed, and other native species can pay rich dividends, including control of nuisance exotic plants. But the task is daunting, particularly for turbid systems with fluctuating water levels. If macrophytes cannot become established spontaneously, can we provide enough help to make the difference?

The last two papers on macrophytes are case studies of Lake Guntersville, a 27,000-ha main-stem impoundment of the Tennessee River. A drought in the mid-1980s stimulated expansion of submerged aquatic vegetation to 30% coverage. As the hydrologic cycle reverted to the long-term norm in the 1990s, plant coverage shrank to 10%. Wrenn et al. examine the effects of this drastic macrophyte reduction on the largemouth bass population and angler effort. Henderson conducted a socioeconomic analysis of lake users to identify the optimal percent plant coverage to maximize lake visits and economic benefits over all major user groups. The Lake Guntersville experience offers the following three particularly germane lessons regarding the growing bass–weeds–people issue.

1. *Bass do not require aquatic vegetation.* After vegetation was reduced, bass numbers declined, but growth rate increased and biomass (kg/ha) remained stable. Size structure of the population also improved sharply. Results of spatial or tem-

poral comparisons of vegetation coverage versus bass populations substantiate this response in Florida (Canfield and Hoyer 1992), Texas (Bettoli et al. 1993), and Iowa (Mitzner 1994). Aquatic vegetation apparently promotes both age-0 survival and intraspecific trophic competition.

2. *Bass anglers want lots of vegetation*. Angler effort on Lake Guntersville declined by 63% after macrophyte coverage shrank. Tournament anglers caught fewer but larger bass; weigh-in totals did not decline. Why did bass anglers leave? Perhaps the Texas finding of Costello and Betsill is generally true: the best measure of fishing success is the number, not the size of the catch.
3. *User group compromise over vegetation is possible*. When boaters, anglers, and shore-based recreationists were presented with options for the ideal percentage vegetation coverage of Lake Guntersville, a wide divergence of opinion (12–50%) was apparent. Most anglers and other recreationists would continue to visit the lake at a coverage of 20%; this option maximized economic return. In the future, socioeconomic analyses such as Henderson's model likely will be widely applied in the search for the best management options to prevent and resolve user group conflicts.

Data Collection and Analysis

The four papers in the concluding section offer a potpourri of advice and examples to improve data acquisition and interpretation. Sampling design for reservoir fisheries assessment (Wilde and Fisher) and creel surveys (Crow and Malvestuto) both can benefit from anticipating sources of variability. Reservoir fisheries investigations often follow standard procedures that specify fixed stations and sampling effort. Probability sampling (e.g., stratified random design) offers increased precision over fixed stations for the same effort, and produces data expandable to the entire system. Limitations on sampling effort may be less tractable; precision or power always should be estimated in the earlier stages of sampling design to determine if sufficient effort can be expended to produce acceptable risk of error. Temporal or spatial stratification also improves the precision of creel survey data. Subsampling within days provided a greater benefit than stratification by day, month, or geographic area in the Lake Guntersville study of Crow and Malvestuto. However, most variance remained unexplained, meaning that good creel surveys will remain labor-intensive.

Ploskey et al. analyzed a 34-year data set for Bull Shoals Reservoir, Arkansas, to identify the best hydrologic predictors of year-class strength for black bass populations. Three methodologies (segmented-temporal correlation, principal components analysis, and harmonic analysis) were compared, and all identified strong predictors of year-class success. Sophisticated analyses of hydrologic influences should become increasingly useful as the operating regimes of large reservoirs are subject to scrutiny and possible revision.

McConnaha and Paquet contribute the experience of an ecosystem level attempt at adaptive management. The term "adaptive management" as propounded by Walters (1986) features the idea that large-scale management manipulations can be designed for use as experiments to understand ecosystem structure and function. As applied to the 765,000-km^2 Columbia River ecosystem, this definition of adaptive management has been stifled by political reality: people do not want to take the risk that imperiled fisheries will be damaged further to satisfy scientific curiosity. McConnaha and Paquet offer a broader and more conventional redefinition that stresses taking the most positive actions permitted by the level of uncertainty, then applying the scientific method to interpret results and refine the next round of actions. This new definition of adaptive management harnesses the rigors of scientific inquiry to the all-too-traditional trial-and-error approach, while dropping the nonviable demand for perturbing experimentation. As such, it should be embraced as the guiding framework for reservoir fisheries investigations. As redefined, "adaptive management" should be popular with the public.

Conclusions

These proceedings readily meet the symposium goal of presenting an array of topics that are rapidly becoming essential for effective fisheries management, yet that remain outside the realm of the expertise and experience of many managers. The symposium was not intended to be comprehensive, and all new issues or technological advances are not included. Concerns associated with exotic species, endangered species, biodiversity, and genetic integrity are certainly germane to reservoir fisheries management. New tools for biological assessment—the use of hydroacoustics to estimate prey fish abundance and bioenergetics and the use of individual-based modeling to quantify predator demand—are not featured in these proceedings, although their applications to large reservoirs are obvious. Instead, the emphasis was directed to socioeconomic matters and to situations in which the human component of fisheries management must be considered on a par with biotic and habitat components. As demand for water-based recreation escalates in this time of a static supply, this orientation was most appropriate; user satisfaction will be the overriding but frustratingly difficult goal of twenty-first century reservoir management.

It is tempting to use this forum to recommend a spectrum of intensive and expensive measures to meet both chronic and new challenges, but most of these would require an infusion of tax dollars that will not be forthcoming in this era of fiscal conservatism. We are now in a period of corporate and government downsizing. The mandate of the 1990s is starkly simple: *do more, better, with less*. I contend that meeting any two of these demands would present an onerous burden; meeting all three will be possible only rarely. However, resource managers may be required to demonstrate their responsiveness to each of these commands. This symposium has provided some insights as to how those responses may best be structured.

Doing More

Papers in these proceedings exhort reservoir fisheries managers to expand their traditional biologist roles to become, among other things, advocates, educators, facilitators, interviewers, health assessors, and risk communicators. How to achieve competence in these added human dimensions responsibilities is a major question. Few resource agencies can afford to recruit a parallel staff of social scientists to work with their biologists. Instead, practicing biologists will require continuing education to develop proficiency for everyday public interactions. The percentage of new fisheries professionals who have received some human dimensions training as part of their college curriculum is rapidly increasing. However, both veteran and novice biologists will need to rely on staff human dimension specialists (most state fish and wildlife agencies now employ at least one) for expertise in coping with particular problems.

Monitoring the ecological health of impoundments also will require a greater effort. The conventional approach, coupling water chemistry measurements with tissue analyses for selected contaminants, is clearly inadequate to assess aquatic community response to the suite of potentially interactive stressors that can be produced in large, highly developed watersheds. Biotic indices must be used in conjunction with chemical analyses to make meaningful assessments of reservoir well-being. Some degree of retraining will be necessary in this area as well if the reservoir fisheries manager is to produce biotic indices of ecosystem health.

An even more serious concern is how to find the time to meet these new demands. Assessment and manipulation of reservoir fish and habitat are far from routine tasks, and data acquisition and interpretation remain inevitably time intensive.

Doing Better

In the future, reservoir fisheries management will require catering to diversity among both human users and the biota. Traditionally, management has been directed to maximizing fisheries for a few target species, usually top carnivores. Managing for diversity in angling opportunities and the biota will demand multispecies, multi-trophic-level objectives, as well as the sacrifice of the maximum yield goal to some currently undefined goal of optimum yield. These shifts may temporarily exacerbate user conflicts as the search for optimality proceeds on an empiric basis.

Reworking goals and objectives also may be necessary if the call for ecosystem management endures. Ecosystem management seems to offer a logical framework for the treatment of reservoirs. The watershed provides a natural scale to delimit the ecosystem, and reservoirs are particularly responsive to disturbances in their watersheds. But impoundments are themselves ecosystem perturbations, and they are subject to multiple-use demands. When reservoirs occur in series, each with different operational objectives, discharge regimes, and owners, the vision of ecosystem management becomes nebulous. The present conflicts over control of the hydrology of the Missouri and Columbia rivers provide excellent test cases for the future of holistic ecosystem management (Keenlyne; McConnaha and Paquet).

In the near-term, doing better will raise the potential for group conflicts. Avoiding stalemate conflicts and bad political decisions will require expert efforts in education, advocacy, and negotiation. Reservoir fisheries biologists can and should contribute to these efforts, but success will require a strong support system and talented team work.

Doing With Less?

Reservoir fisheries management programs have been chronically underfunded, as the builders and operators of large impoundments have conceded responsibility cheerfully to state fisheries agencies that lacked the funding to develop the enormous knowledge base needed to manipulate the resource effectively. Direct U.S. federal involvement in reservoir fisheries management has declined since the abolition of the National Reservoir Research Program by the U.S. Fish and Wildlife Service in 1983; calls for reinvestment have been unheeded (Summerfelt 1986). Lack of financial accountability has plagued reservoir fisheries management, but there are some bright spots indicating that fiscal austerity and resulting "black box" management may not be a life sentence. The TVA Lake Improvement Plan (Ungate) is one example, as is the multijurisdictional effort to restore Pacific salmon populations in the Columbia River watershed (McConnaha and Paquet). More generally, the FERC relicensing process (described by Hill) offers a golden opportunity to provide funds for reservoir research and remedial management. However, the enabling legislation should be expanded to make all large reservoir owners accountable for ecologically sound operation. At present, only hydroelectric utilities must consider fisheries values under federal law; only lawsuits can force other operators to behave with equivalent responsibility.

User fees offer another means to provide the funding that will certainly be needed to do more, better. Although fishery license fees frequently are adjusted to market value, other waterbody users (e.g., boaters, swimmers, hikers, lakeshore residents) often are not made to pay. These users absorb fishery agency funds and sometimes promote conflicts with anglers. It is difficult to justify their legitimacy as stakeholders when they have made no tangible contribution to the resource. The user fee is perhaps the most politically palatable form of tax, and resource agencies need to exploit this funding source on the user-pays principle.

Even if additional funding can be secured, reservoir fisheries managers will have to institute some economies to do more and do it better. These proceedings show some promising new tools and procedures to increase management efficiency. Among the tools, GIS may become a standard technology for analyzing spatial relationships among people, fish, and habitat. The comprehensive management analysis methodology of Cole and Ward is another computer-based system that promises to facilitate decision making by incorporating socioeconomic, biological, and habitat information into the working data base.

Human dimension tools to provide angler feedback (e.g., interviews, questionnaire surveys) and to manage user group involvement (e.g., the Vroom–Yelton model) are being tested and refined rapidly for the everyday users for which they increasingly will be required. Similarly, new biotic indices are under development that are based on the health of individual fish species (Coughlan et al.) or of the entire reservoir fish assemblage (Hickman and McDonough). Conceptually, adaptive management will provide the framework for design and conduct of reservoir biomanipulations in a responsible and productive manner, systematically reducing scientific uncertainty while promoting fisheries enhancement.

So the mandate for the future of reservoir fisheries management will be met in some parts and in some places. We will need to do more, and some things we can do better, but we probably will not have the resources to meet these dictates without sacrifice. I end my summary with a plea; do not sacrifice essential biological assessment and research to meet new demands. The predictive knowledge of the structure and function of reservoir fisheries is simply too weak to leave it at the current state of understanding. Even in the twenty-first century, the fisheries themselves will be the foundation on which fisheries management decisions will be made.

References

Adams, S. M., A. M. Brown, and R. W. Goede. 1993. A quantitative health assessment index for rapid evaluation of fish condition in the field. Transactions of the American Fisheries Society 122:63–73.

Bettoli, P. W., M. J. Maceina, R. L. Noble, and R. K. Betsill. 1993. Response of a reservoir fish community to aquatic vegetation removal. North American Journal of Fisheries Management 13:110–124.

Boyd, C. E. 1971. The limnological role of aquatic macrophytes and their relationship to reservoir management. Pages 153–166 *in* G. E. Hall, editor. Reservoir fisheries and limnology. American Fisheries Society Special Publication 8.

Canfield, D. E., Jr., and M. V. Hoyer. 1992. Aquatic macrophytes and their relation to the limnology of Florida lakes. Report of the Center for Aquatic Plants, University of Florida, to Florida Department of Natural Resources, Tallahassee.

Durocher, P. P., W. C. Provine, and J. E. Kraai. 1984. Relationship between abundance of largemouth bass and submerged vegetation in Texas reservoirs. North American Journal of Fisheries Management 4:84–88.

Groen, C. L., and T. A. Schroeder. 1978. Effects of water level management on walleye and other coolwater fishes in Kansas reservoirs. Pages 278–283 *in* R. L. Kendall, editor. Selected coolwater fishes of North America. American Fisheries Society Special Publication 11.

Karr, J. R. 1981. Assessment of biotic integrity using fish communities. Fisheries 6(6):21–27.

Kennedy, R. H. 1995. Reservoir water quality management. Lake Line 15(1):24–26; 40.

Klessig, L. 1994. Load limits for lakes. Lake and Reservoir Management 10:69–73.

Mitzner, L. 1994. Management of aquatic vegetation with grass carp in Iowa, 1973–1993. Lake and Reservoir Management 9:99–100.

Nielsen, L. A. 1993. History of inland fisheries management in North America. Pages 3–31 *in* C. C. Kohler and W. A. Hubert, editors. Inland fisheries management in North America. American Fisheries Society, Bethesda, Maryland.

Pieterse, A. H., and K. J. Murphy, editors. 1990. The ecology and management of nuisance aquatic vegetation. Oxford University Press, London.

Robins, C. R., and six coauthors. 1991. Common and scientific names of fishes from the United States and Canada, 5th edition. American Fisheries Society Special Publication 20.

Smith, T. D. 1994. Scaling fisheries: the science of measuring the effects of fishing, 1855–1955. Cambridge University Press, Cambridge, UK.

Soballe, D. M., B. L. Kimmel, R. H. Kennedy, and R. F. Gaugush. 1992. Reservoirs. Pages 421–474 *in* C. T. Hackney, S. M. Adams, and W. H. Martin, editors. Biodiversity of the southeastern United States: aquatic communities. Wiley, New York.

Summerfelt, R. C. 1986. Summarization of the symposium. Pages 314–327 *in* G. E. Hall and M. J. Van Den Ayvle, editors. Reservoir fisheries management: strategies for the 80's. American Fisheries Society, Southern Division, Reservoir Committee, Bethesda, Maryland.

USFWS (U.S. Fish and Wildlife Service). 1993. 1991 National survey of fishing, hunting and wildlife-associated recreation. U.S. Government Printing Office, Washington, DC.

Walters, C. J. 1986. Adaptive management of renewable resources. Macmillan, New York.

Willis, D. W. 1986. Review of water level management on Kansas reservoirs. Pages 110–114 *in* G. E. Hall and M. J. Van Den Ayvle, editors. Reservoir fisheries management: strategies for the 80's. American Fisheries Society, Southern Division, Reservoir Committee, Bethesda, Maryland.